Audio Engineering Explained

Audio Engineering Explained

Edited by Douglas Self

AMSTERDAM • BOSTON • HEIDELBERG • LONDON • NEW YORK • OXFORD
PARIS • SAN DIEGO • SAN FRANCISCO • SINGAPORE SYDNEY • TOKYO
Focal Press is an imprint of Elsevier

Focal Press is an imprint of Elsevier
Linacre House, Jordan Hill, Oxford OX2 8DP, UK
30 Corporate Drive, Suite 400, Burlington, MA 01803, USA

First edition 2010

Notices

Knowledge and best practice in this field are constantly changing. As new research and experience broaden our understanding, changes in research methods, professional practices, or medical treatment may become necessary.

Practitioners and researchers must always rely on their own experience and knowledge in evaluating and using any information, methods, compounds, or experiments described herein. In using such information or methods they should be mindful of their own safety and the safety of others, including parties for whom they have a professional responsibility.

To the fullest extent of the law, neither the Publisher nor the authors, contributors, or editors, assume any liability for any injury and/or damage to persons or property as a matter of products liability, negligence or otherwise, or from any use or operation of any methods, products, instructions, or ideas contained in the material herein.

British Library Cataloguing in Publication Data
Audio engineering explained : professional audio recording.
1. Sound—Recording and reproducing. 2. Acoustical engineering.
I. Self, Douglas.
621.3′828-dc22

Library of Congress Control Number: 2009933707

ISBN: 978-0-240-81273-1

For information on all Focal Press publications visit our website at focalpress.com

Printed and bound in the United States

09 10 11 12 11 10 9 8 7 6 5 4 3 2 1

CONTENTS

PART VII • Acoustics and Sound Reinforcement

PART VIII • Recording Studios

PREFACE

This is a collection of chapters from contemporary books on audio engineering. It covers a wide range across a large and ever-increasing subject. Obviously a book of practical size cannot give a complete account of even one area of audio technology, and that is not the intention here.

The aim is to give a good picture of the field by focusing on a number of audio topics without either getting too specialized or making do with vague generalities. I have tried hard to make this a collection which overall gives considerable insight into the whole subject of audio engineering, but which also acts as a jumping-off point for further study. Most of the chapters here include comprehensive references and extremely useful suggestions for further reading.

This book is divided into eight sections, which are as follows:

The first section, "Basics," deals with the fundamental principles of audio, acoustics, and human hearing. It comprises one chapter on the "Fundamentals of Audio and Acoustics," by Pat Brown, which covers the physical side of sound, such as decibels, frequency and wavelength, and a complementary chapter by David Howard and Jamie Angus which gives a good introduction to the processes of human hearing. This second chapter includes a wealth of material on the anatomy of the ear, the crucial function of the basilar membrane, and critical band theory. This last topic may sound academically dry, but it is in fact the foundation of audio compression systems such as MP3 which are so widely used today. There is also useful information on the perception of loudness, which does not always work as you might expect: if you want to double the loudness of a string section, you will have to hire *ten times* the number of violinists.

The second section of this book, "Microphones," consists of two chapters taken from *The Microphone Book* by John Eargle. The first chapter deals with the basic ways in which various types of sound field are propagated and how they interact with a microphone. The second chapter focuses on the electrical aspects of microphone operation, such as phantom powering, microphone transformers, mixing console microphone preamplifiers, microphone splitters, cable losses, radio microphones, and the latest so-called "digital" microphones.

The third section "Preamplifiers, Mixers, and Interconnects" gets to grips with the electronics in sound systems. It deals with what might be called the small-signal part of a system—the controlling electronics, usually in the form of a mixing console, and the interconnections between the console and the other components of the system. In Chapter 5, Francis Rumsey and Tim McCormick give a through and comprehensive description of how mixing consoles work. I spent some years engaged in mixing console design, as Chief Engineer at Soundcraft Electronics, and I can tell you that I think this chapter is the best concise description of mixer operation that I have come across, and it was an easy decision to include it in this collection. The fact that one of the illustrations is of a console I had a hand in designing (The Soundcraft Sapphyre) is, I hasten to add, merely a happy coincidence. Apart from mixing console design as such, the chapter also deals with approaches to recording and mixing, and console automation.

Compared with the complexities of mixing console design, Chapter 6 on interconnects, taken from *Handbook for Sound Engineers*, and written by Glen Ballou, may sound uninspiring. It's just a bit of cable, isn't it? No, it isn't. It is no use constructing near-perfect boxes of electronics if the cable leading to them fatally compromises the quality of the signal before it even arrives there. Chapter 6 covers loudspeaker cables, digital audio cables, and video cables, as well as ordinary audio connections.

The final chapter in this section, Chapter 7, is by Bill Whitlock, who is an acknowledged expert in the fields of equipment grounding and interconnection. The technology of audio interconnection is not only a good deal more complex than it appears, but can also present some fascinating (and in the wrong circumstances thoroughly irritating) problems. The greatest danger to a signal in a cable is that unwanted currents flowing down the ground connection, in the classic ground loop, cause hum and noise to be added to the signal. There are several ways of preventing this, of which the best known is the balanced interconnection, which uses two wires to carry the signal instead of one. It is now possible for the receiving equipment to respond to the difference between the two signal conductors and (to a first approximation) ignore the unwanted voltage drop down the ground conductor. Like so many technical matters, the principle is beautifully simple but the practicalities can get rather complicated; this technology is thoroughly dealt with in Chapter 7 by an author who is a master of his subject.

The section on "Power Amplifiers" makes up the fourth part of this book, and I have contributed this myself, having spent a good deal of time working on the apparently simple but, once again, deeply complex and fascinating matter of turning small voltages and currents into large ones without distorting them in the process. The three chapters are taken from my *Audio Power Amplifier Design Handbook*, which is now available in a greatly extended fifth edition.

The first chapter taken from the *Handbook* explains the basic architecture of power amplifiers, and shows how they are put into different classes such as Class-A, Class-B, and so on, depending on their mode of operation, up to Class-S. Some less well-known types such as error-correcting and non-switching amplifiers are covered. The intelligent use of negative feedback is absolutely fundamental to power amplifier design, and it is explained here.

The second chapter goes into more detail on power amplifier functioning, and shows how the apparently hopelessly entangled sources of amplifier distortion can in fact be reduced to a few straightforward mechanisms, most of which can be dealt with quite easily once they are properly identified. Topics such as measuring open-loop gain are also covered.

In the third chapter, I explain the operation and design of Class-G amplifiers, which significantly economize on power consumption and heat dissipation by running from low-voltage rails when reproducing small amplitudes, and switching to high-voltage rails to accommodate the relatively infrequent signal peaks. This technology is of increasing importance, as it has found its niche in subwoofers where relatively large amplifier power outputs are required. When properly implemented—as described here—it can give much better quality than the Class-D amplifiers which are sometimes used in this application.

The fifth section of this book "Loudspeakers" deals with the tricky business of turning electricity into sound. It consists of four chapters, all taken from the book *Loudspeakers* by Philip Newell and Keith Holland.

The first chapter sets out the most basic fundamentals of loudspeaker operation, covering acoustic radiation and the load a loudspeaker presents to a power amplifier.

In the second chapter, diversity in loudspeaker design is explored, and the construction of both moving-coil and electrostatic loudspeakers is gone into in more detail. The third chapter describes loudspeaker cabinets, which are not just enclosures to keep the dust off, but an integral part of the functioning of the loudspeaker.

Finally, the fourth chapter in this section deals with the crossover units that direct the appropriate frequencies to the high-frequency and low-frequency units of a multi-way loudspeaker system. This alone is a large and complex field of study.

The next section of this book, the sixth, addresses the vitally important field of "Digital Audio." I am particularly conscious that this is a huge and ever-growing field of knowledge, and one section of one book can only give a taste of its extent. Digital audio entered most people's lives with the introduction of the CD in 1982, and since then digital methods have extended from relatively simple storage media to highly complex manipulations known as Digital Signal Processing or DSP. Very sophisticated specialized processors have been developed to efficiently handle the enormous amounts of calculation required.

Chapter 15, by Craig Richardson, gives a good account of the essential principles of digital audio, covering sampling theory, Z-transforms (which are not as scary as they sound), and the basics of choosing a DSP chip.

Another aspect of digital technology is the use of the MIDI standard for controlling electronic musical instruments. Back in the days when I had my own eight-track recording studio (a setup which dated back to when eight tracks referred to tracks on magnetic tape, and which unfortunately succumbed to sheer lack of time to make proper use of it) I made very extensive use of MIDI, and I have always been impressed by the robustness of its technical specifications and its remarkable flexibility. In Chapter 16 David Huber gives a very good overview of MIDI technology, including MIDI sequencing software and MIDI time-code.

The seventh section is "Acoustics and Sound Reinforcement". The first two chapters are from *Sound System Engineering* by Don and Carolyn Davis. The first chapter covers the complex subject of large-room acoustics, dealing with reverberation, flutter-echoes, and the reverberant and direct sound fields.

The second chapter deals with the problem of equalizing the sound system and its associated acoustics to get the desired frequency response, describing the filters and equalizers that are used in this endeavor, and how its success is evaluated by specialized measuring equipment.

The final chapter in this section, from *Sound Reproduction* by Floyd Toole, covers the propagation of sound in spaces, and how that interacts with human perception. The Precedence Effect, or Haas Effect, in which delayed sounds are not heard as separate events but reinforce the initial sound, is a fundamental part of sound reproduction in a space, and is dealt with comprehensively here.

The final section "Recording Studios" is extracted from Philip Newell's book *Recording Studio Design*. The first chapter deals with the vital issue of sound isolation. The most sophisticated electronics in the world is helpless against the sound of traffic, road-drills and worse infiltrating the recording studio because the precautions taken against it were inadequate. Upgrading and correcting a poorly-designed sound isolation scheme is almost certainly going to be expensive and disruptive to the business of recording, and it is only sensible to carefully consider the initial plans before putting trowel to mortar. This chapter provides an excellent basis for this sort of planning, dealing with the reflection, transmission and absorption of sound. The mass law, damping and decoupling, and floor, wall and ceiling isolation are all dealt with in detail.

Having spent more time in recording studios than was probably good for me, I can vouch for the fact that the studio environment is important if you want to get the best artistic results and a healthy number of repeat bookings. The second chapter by Philip Newell discusses the environmental issues of studio lighting, ventilation, decoration, and the provision of headphone foldback and clean mains supplies.

This brief introduction can only give a taste of the varied and extensive information in this book. Each chapter has its own preface which gives more detail on its contents.

PART I

Basics

PREFACE

This chapter begins our journey through the world of audio engineering by explaining some fundamental principles. Our hearing, like our other senses, works in a logarithmic fashion, which allows us to cope with the enormous variation in sound levels that occur in the real world. Explanation starts with the decibel, a unit whose logarithmic basis matches the response of our ears, and makes the plotting of graphs and the doing of calculations much simpler and more intuitive than the alternative of grappling with scientific notation and numbers wreathed in zeros.

From considering the mathematics of decibels (which is really very simple), Pat Brown moves on to consider how they are used to measure absolute rather than just relative sound levels. Our ears respond to sound pressure levels. Taking the quietest sound that can be detected, and calling it the Threshold of Hearing, we can assign 0 decibels to this sound pressure level. This is invariably abbreviated to 0 dB. The decibel scale can now be used to establish the loudness of common situations, such as a room with quiet air-conditioning (about 30 dB), normal conversation (about 50 dB), and the traditional jack-hammer at 1 meter (about 100 dB). The threshold of pain is usually defined as 130 dB, a level that hopefully few of us will ever encounter; it is a sobering thought that hearing damage due to long-term exposure can begin as low as 78 dB.

Our perception of frequency, i.e., pitch, is also logarithmic, which is why music has as its foundation the octave, which stands for a doubling of frequency. When drawing graphs of quantities against frequency, it is usual to use decades, or ratios of 10, instead of octaves, as this simplifies the maths. Pat then relates frequency to wavelength and the speed of sound.

It is one of the most important features of human hearing that the frequency response of our ears is not a flat line when plotted against sound pressure level. Worse than that, the deviation from flat increases as the level falls. This disconcerting situation is summed up by the famous Fletcher and Munson equal-loudness curves. This naturally complicates the business of measuring noise levels, and so standard weighting curves have been defined that allow for this, the best-known being the A-weighting curve, which is often used in the measurement of audio electronics. This curve has a pronounced roll-off at the low-frequency end which takes into account the fact that our hearing is markedly less sensitive at, say, 50 Hz, than it is at the reference frequency of 1 kHz. There is then a discussion of how the ear integrates sound, one example being the Precedence effect, in which late-arriving reflections within 40 milliseconds of the original sound reinforce its subjective volume rather than being heard as separate events.

Pat Brown concludes this chapter with a look at some of the simplest cases of sound propagation: the point source and the line source. These are of course mathematical abstractions rather than real sound sources, but they are the basic foundation of more realistic and therefore more complex models, such as the treatment of a loudspeaker cone as a rigid circular piston.

CHAPTER 1

Fundamentals of Audio and Acoustics

Handbook for Sound Engineers by Pat Brown

INTRODUCTION

Many people get involved in the audio trade prior to experiencing technical training. Those serious about practicing audio dig in to the books later to learn the physical principles underlying their craft. This chapter is devoted to establishing a baseline of information that will prove invaluable to anyone working in the audio field.

Numerous tools exist for those who work on sound systems. The most important are the mathematical tools. Their application is independent of the type of system or its use, plus, they are timeless and not subject to obsolescence like audio products. Of course, one must always balance the mathematical approach with real-world experience to gain an understanding of the shortcomings and limitations of the formulas. Once the basics have been mastered, sound system work becomes largely intuitive.

Audio practitioners must have a general understanding of many subjects. The information in this chapter has been carefully selected to give the reader the big picture of what is important in sound systems. In this initial treatment of each subject, the language of mathematics has been kept to a minimum, opting instead for word explanations of the theories and concepts. This provides a solid foundation for further study of any of the subjects. Considering the almost endless number of topics that could be included here, I selected the following based on my own experience as a sound practitioner and instructor. They are:

1. The Decibel and Levels.
2. Frequency and Wavelength.
3. The Principle of Superposition.
4. Ohm's Law and the Power Equation.
5. Impedance, Resistance, and Reactance.
6. Introduction to Human Hearing.
7. Monitoring Audio Program Material.
8. Sound Radiation Principles.
9. Wave Interference.

A basic understanding in these areas will provide the foundation for further study in areas that are of particular interest to the reader. Most of the ideas and principles in this chapter have existed for many years. While I haven't quoted any of the references verbatim, they get full credit for the bulk of the information presented here.

THE DECIBEL

Perhaps the most useful tool ever created for audio practitioners is the decibel (dB). It allows changes in system parameters such as power, voltage, or distance to be related to level changes heard by a listener. In short, the decibel is a way to express "how much" in a way that is relevant to the human perception of loudness. We will not track its long evolution or specific origins here. Like most audio tools, it has been modified many times to stay current with the technological practices of the day. Excellent resources are available for that information. What follows is a short study on how to use the decibel for general audio work.

Most of us tend to consider physical variables in linear terms. For instance, twice as much of a quantity produces twice the end result. Twice as much sand produces twice as much concrete. Twice as much flour produces twice as much bread. This linear relationship does not hold true for the human sense of hearing. Using that logic, twice the amplifier power should sound twice as loud. Unfortunately, this is not true.

Perceived changes in the loudness and frequency of sound are based on the *percentage* change from some initial condition. This means that audio people are concerned with ratios. A given ratio always produces the same result. Subjective testing has shown that the power applied to a loudspeaker must be increased by about 26% to produce an audible change. Thus a ratio of 1.26:1 produces the minimum audible change, regardless of the initial power quantity. If the initial amount of power is 1 watt, then an increase to 1.26 watts (W) will produce a "just audible" increase. If the initial quantity is 100 W, then 126 W will be required to produce a just audible increase. A number scale can be linear with values like 1, 2, 3, 4, 5, etc. A number scale can be proportional with values like 1, 10, 100, 1000, etc. A scale that is calibrated proportionally is called a *logarithmic* scale. In fact, *logarithm* means "proportional numbers." For simplicity, base 10 logarithms are used for audio work. Using amplifier power as an example, changes in level are determined by finding the ratio of change in the parameter of interest (e.g., wattage) and taking the base 10 logarithm. The resultant number is the level change between the two wattages expressed in Bels. The base 10 logarithm is determined using a look-up table or scientific calculator. The log conversion accomplishes two things:

1. It puts the ratio on a proportional number scale that better correlates with human hearing.
2. It allows very large numbers to be expressed in a more compact form, Figure 1.1.

The final step in the decibel conversion is to scale the Bel quantity by a factor of 10. This step converts Bels to decibels and completes the conversion process, Figure 1.2. The decibel scale is more resolute than the Bel scale.

The decibel is always a power-related ratio. Electrical and acoustical power changes can be converted exactly in the manner described. Quantities that are not powers must be made proportional to power—a relationship established by the power equation:

$$W = \frac{E^2}{R} \tag{1.1}$$

where,

W is power in watts,
E is voltage in volts,
R is resistance in ohms.

This requires voltage, distance, and pressure to be squared prior to taking the ratio. Some practitioners prefer to omit the squaring of the initial quantities and simply change the log multiplier from 10 to 20. This produces the same end result.

Figure 1.3 provides a list of some dB changes along with the ratio of voltage, pressure, distance, and power required to produce the indicated dB change. It is a worthwhile endeavor to memorize the changes indicated in bold type and be able to recognize them by listening.

A decibel conversion requires two quantities that are in the same unit, i.e., watts, volts, meters, feet. The unit cancels during the initial division process, leaving the ratio between the two quantities. For this

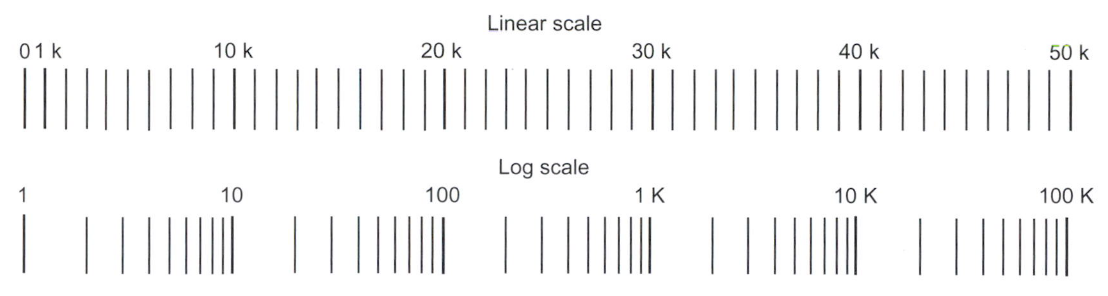

FIGURE 1.1
A logarithmic scale has its increments marked by a fixed ratio, in this case 10 to 1, forming a more compact representation than a linear scale. (Courtesy Syn-Aud-Con.)

1. Compare

Quantity "A" Quantity "B"

$\frac{\text{Watts}}{\text{Watts}}$ $\frac{\text{Volts}^2}{\text{Volts}^2}$ $\frac{\text{Pressure}^2}{\text{Pressure}^2}$ $\frac{\text{Distance}^2}{\text{Distance}^2}$

Results in a ratio between the two quantities

2. Compress

$1=10^0=0$
$10=10^1=1$
$100=10^2=2$
$1{,}000=10^3=3$
$10{,}000=10^4=4$
$100{,}000=10^5=5$
$1{,}000{,}000=10^6=6$

Results in a ratio between the two quantities expressed in Bels (compressed)

3. Scale

Power (x10)
0 dB
10 dB
20 dB
30 dB
40 dB
50 dB
60 dB

Scales the value in Bels to a value in decibels

$$dB = 10\log\frac{W_1}{W_2} \qquad dB = 20\log\frac{E_1}{E_2}$$

FIGURE 1.2
The steps to performing a decibel conversion are outlined. (Courtesy Syn-Aud-Con.)

reason, the decibel is without dimension and is therefore technically not a unit in the classical sense. If two arbitrary quantities of the same unit are compared, the result is a relative level change. If a standard reference quantity is used in the denominator of the ratio, the result is an absolute level and the unit is dB relative to the original unit. Relative levels are useful for live work. Absolute levels are useful for equipment specifications and calibration. Figure 1.4 lists some references used for determining absolute levels.

The decibel was originally used in impedance-matched interfaces and always with a power reference. Power requires knowledge of the resistance that a voltage is developed across. If the resistance value is fixed, changes in applied voltage can be expressed in dB, since the power developed will be directly proportional to the applied voltage. In modern sound systems, few device interfaces are impedance matched. They are actually mismatched to optimize the voltage transfer between components. While the same impedance does not exist at each device interface, the same impedance condition may. If a minimum 1:10 ratio exists between the output impedance and input impedance, then the voltage transfer is essentially independent of the actual output or input impedance values. Such an interface is termed *constant voltage*, and the signal source is said to be operating *open circuit* or *un-terminated*. In constant voltage interfaces, open circuit conditions are assumed when using the decibel. This means that the level change at the output of the system is caused by changing the voltage somewhere in the processing chain and is dependent on the voltage change only, not the resistance that it is developed across or the power transfer. Since open-circuit conditions exist almost universally in modern analog systems, the practice of using the decibel with a voltage reference is widespread and well accepted.

One of the major utilities of the decibel is that it provides a common denominator when considering level changes that occur due to voltage changes at various points in the signal chain. By using the decibel, changes in sound level at a listener position can be determined from changes in the output voltage

Subjective Change	Voltage, Distance, or Pressure Ratio	% of orig.	Power Ratio	dB Change
Barely perceptible	1.12 to 1	89	1.26 to1	1 dB
	1.26 to 1	79	1.58 to 1	2 dB
Noticeable to most	1.41 to 1	71	2 to 1	3 dB
	1.58 to 1	63	2.51 to 1	4 dB
	1.78 to 1	56	3.16 to 1	5 dB
Goal for system changes	2 to 1	50	4 to 1	6 dB
	2.24 to 1	45	5 to 1	7 dB
	2.51 to 1	40	6.3 to 1	8 dB
	2.8 to 1	36	8 to 1	9 dB
Twice as loud or soft	3.16 to 1	32	10 to 1	10 dB
	10 to 1	10	100 to 1	20 dB
	31.6 to 1	3	1000 to 1	30 dB
Limits of audibility	100 to 1	1	10000 to 1	40 dB
	316 to 1	.3	100000 to 1	50 dB
	1000 to 1	.1	1000000 to 1	60 dB
	20 log		10 log	

FIGURE 1.3
Some important decibel changes and the ratios of power, voltage, pressure, and distance that produce them. (Courtesy Syn-Aud-Con.)

Electrical Power	
dBW	1 Watt
dBm	0.001 Watt
Acoustical Power	
dB-PWL or L_w	10^{-12} Watt
Electrical Voltage	
dBV	1 Volt
dBu	0.775 Volts
Acoustical Pressure	
dB SPL or L_p	0.00002 Pascals

FIGURE 1.4
Some common decibel references used by the audio industry.

of any device ahead of the loudspeaker. For instance, a doubling of the microphone output voltage produces a 6 dB increase in output level from the microphone, mixer, signal processor, power amplifier, and ultimately the sound level at the listener. This relationship assumes linear operating conditions in each device. The 6 dB increase in level from the microphone could be caused by the talker speaking 6 dB louder or by simply reducing the miking distance by one-half (a 2:1 distance ratio). The level controls on audio devices are normally calibrated in relative dB. Moving a fader by 6 dB causes the output voltage of the device (and system) to increase by a factor of 2 and the output power from the device (and system) to be increased by a factor of four.

Absolute levels are useful for rating audio equipment. A power amplifier that can produce 100 watts of continuous power is rated at

$$\begin{aligned} L_{out} &= 10 \log W \\ &= 10 \log 100 \\ &= 20 \text{ dBW} \end{aligned} \tag{1.2}$$

This means that the amplifier can be 20 dB louder than a 1 watt amplifier. A mixer that can output 10 volts prior to clipping can be rated at

$$\begin{aligned} L_{out} &= 20 \log E \\ &= 20 \log 10 \\ &= 20 \text{ dBV} \end{aligned} \tag{1.3}$$

If the same mixer outputs 1 volt rms at meter zero, then the mixer has 20 dB of peak room above meter zero.

If a loudspeaker can produce a sound level at 1 meter of 90 dB ref. 20 μPa (micro-Pascals), then at 10 meters its level will be

$$\begin{aligned} L_p &= 90 + 20 \log \frac{1}{10} \\ &= 90 + (-20) \\ &= 70 \text{ dB} \end{aligned} \tag{1.4}$$

In short, the decibel says, "The level difference caused by changing a quantity will depend upon the initial value of the quantity and the *percentage* that it is changed."

The applications of the decibel are endless, and the utility of the decibel is self-evident. It forms a bridge between the amount of change of a physical parameter and the loudness change that is perceived by the human listener. The decibel is the language of audio, Figure 1.5.

LOUDNESS AND LEVEL

The perceived loudness of a sound event is related to its acoustical level, which is in turn related to the electrical level driving the loudspeaker. Levels are electrical or acoustical pressures or powers expressed in decibels. In its linear range of operation, the human hearing system will perceive an increase in level as an increase in loudness. Since the eardrum is a pressure-sensitive mechanism, there exists a threshold below which the signal is distinguishable from the noise floor. This threshold is about 20 μPa of pressure deviation from ambient at midrange frequencies. Using this number as a reference and converting to decibels yields

$$\begin{aligned} L_p &= 20 \log \frac{0.00002}{0.00002} \\ &= 0 \text{ dB (or 0 dB SPL)} \end{aligned} \tag{1.5}$$

This is widely accepted as the threshold of hearing for humans at mid-frequencies. Acoustic pressure levels are always stated in dB ref. 0.00002 Pa. Acoustic power levels are always stated in dB ref. 1 pW (picowatt or 10^{-12} W). Since it is usually the pressure level that is of interest, we must square the Pascals term in the decibel conversion to make it proportional to power. Sound pressure levels are measured using sound level meters with appropriate ballistics and weighting to emulate human hearing. Figure 1.6 shows some typical sound pressure levels that are of interest to audio practitioners.

FREQUENCY

Audio practitioners are in the wave business. A wave is produced when a medium is disturbed. The medium can be air, water, steel, the earth, etc. The disturbance produces a fluctuation in the ambient condition of the medium that propagates as a wave that radiates outward from the source of the disturbance. If one second is used as a reference time span, the number of fluctuations above and

Relative Level Changes

$dB = 10 \log(W_2/W_1)$ where W is power (electric or acoustic)

$dB = 20 \log(P_2/P_1)$ where P is pressure(voltage for electrical circuits)

$dB = 20 \log(D_2/D_1)$ where D is distance in feet or meters

Electrical Levels

$dBV = 20 \log(E/1)$ where E is electromotive force in Volts

$dBu = 20 \log(E/0.775)$ where E is electromotive force in Volts

$dBW = 10 \log(W/1)$ where W is electrical power in Watts

$dBm = 10 \log(W/.001)$ where W is electrical power in Watts

Acoustic Levels

L_P or $SPL = 20 \log(P/0.00002)$ where P is sound pressure

$L_W = 10 \log(W/10^{-12})$ where W is acoustic power

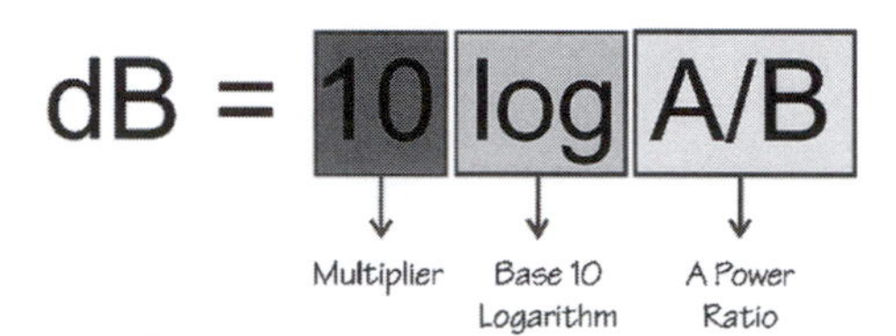

FIGURE 1.5
Summary of decibel formulas for general audio work. (Courtesy Syn-Aud-Con.)

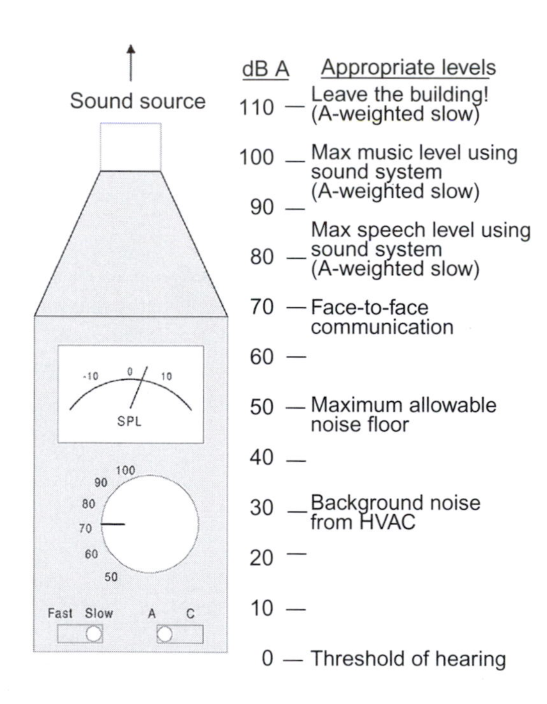

FIGURE 1.6
Sound levels of interest to system designers and operators. (Courtesy Syn-Aud-Con.)

below the ambient condition per second is the frequency of the event, and is expressed in cycles per second, or Hertz. Humans can hear frequencies as low as 20 Hz and as high as 20,000 Hz (20 kHz). In an audio circuit the quantity of interest is usually the electrical voltage. In an acoustical circuit it is the air pressure deviation from ambient atmospheric pressure. When the air pressure fluctuations have a frequency between 20 Hz and 20 kHz they are audible to humans.

As stated in the decibel section, humans are sensitive to proportional changes in power, voltage, pressure, and distance. This is also true for frequency. If we start at the lowest audble frequency of 20 Hz

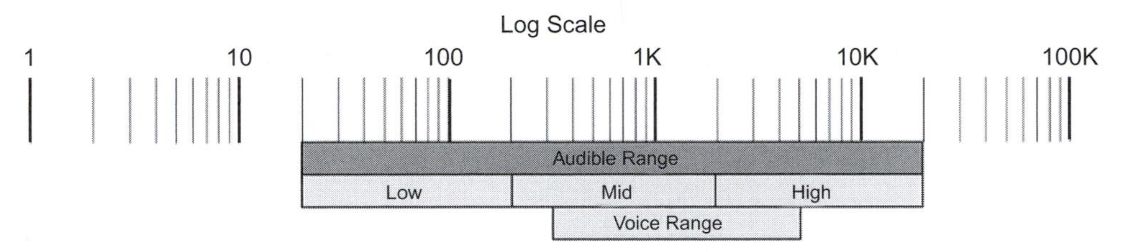

FIGURE 1.7
The audible spectrum divided into decades (a 10 to 1 frequency ratio). (Courtesy Syn-Aud-Con.)

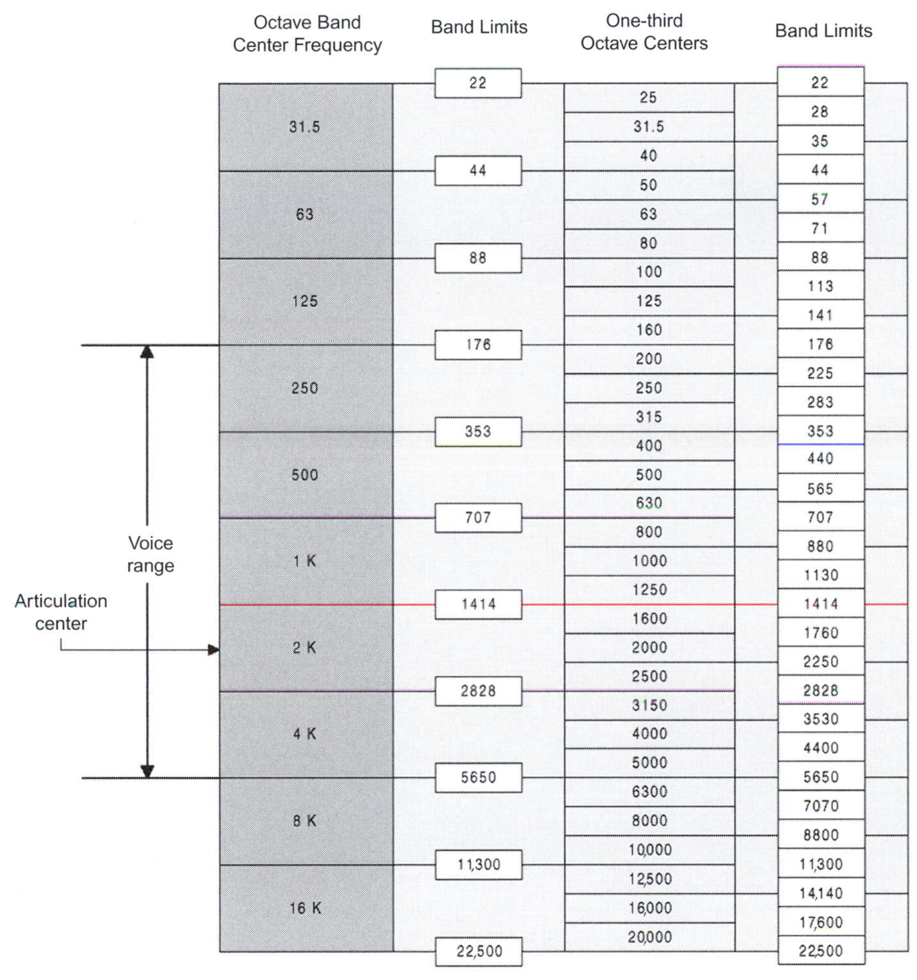

FIGURE 1.8
The audible spectrum divided into octaves (a 2 to 1 ratio) and one-third octaves. (Courtesy Syn-Aud-Con.)

and increase it by a 2:1 ratio, the result is 40 Hz, an interval of one octave. Doubling 40 Hz yields 80 Hz. This is also a one-octave span, yet it contains twice the frequencies of the previous octave. Each successive frequency doubling yields another octave increase and each higher octave will have twice the spectral content of the one below it. This makes the logarithmic scale suitable for displaying frequency. Figures 1.7 and 1.8 show a one-logarithmic frequency scale and some useful divisions. The perceived midpoint of the spectrum for a human listener is about 1 kHz. Some key frequency ratios exist:

- 10:1 ratio—decade.
- 2:1 ratio—octave.

The spectral or frequency response of a system describes the frequencies that can pass through that system. It must always be stated with an appropriate tolerance, such as ±3 dB. This range of frequencies is the bandwidth of the system. All system components have a finite bandwidth. Sound systems are usually bandwidth limited for reasons of stability and loudspeaker protection. A spectrum analyzer can be used to observe the spectral response of a system or system component.

WAVELENGTH

If the frequency f of a vibration is known, the time period T for one cycle of vibration can be found by the simple relationship:

$$T = \frac{1}{f} \tag{1.6}$$

The time period T is the inverse of the frequency of vibration. The period of a waveform is the time length of one complete cycle, Figure 1.9. Since most waves propagate or travel, if the period of the wave is known, its physical size can be determined with the following equation if the speed of propagation is known:

$$\lambda = Tc \tag{1.7}$$

$$\lambda = \frac{c}{f} \tag{1.8}$$

Waves propagate at a speed that is dependent on the nature of the wave and the medium that it is passing through. The speed of the wave determines the physical size of the wave, called its *wavelength*. The speed of light in a vacuum is approximately 300,000,000 meters per second (m/s). The speed of an

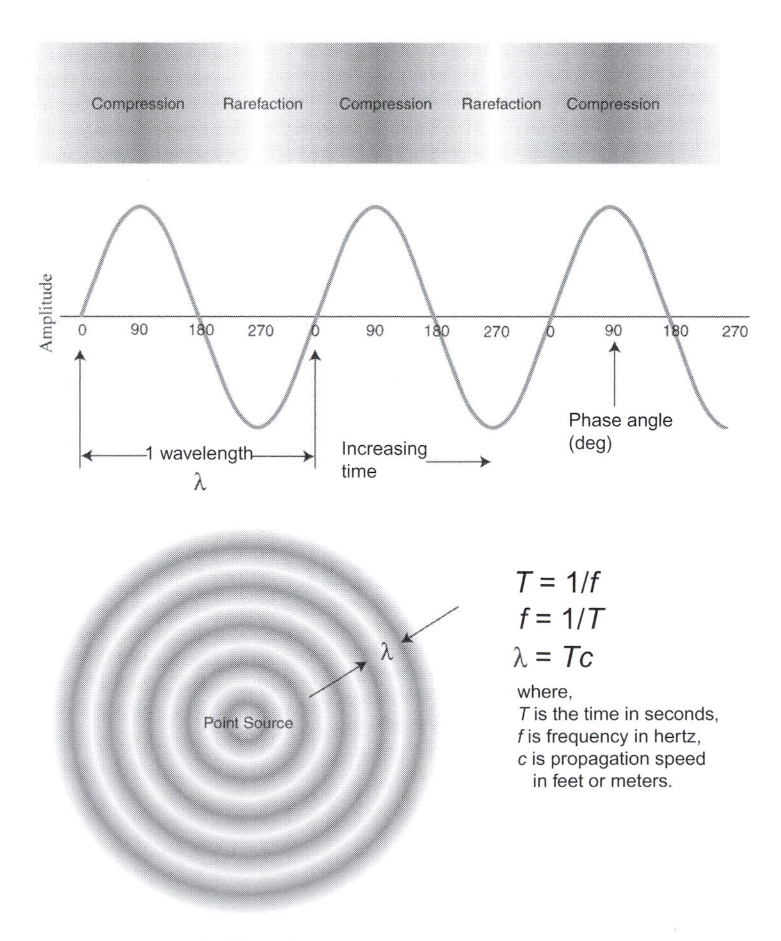

FIGURE 1.9
The wavelength of an event determines how it interacts with the medium that it is passing through. (Courtesy Syn-Aud-Con.)

electromagnetic wave in copper wire is somewhat less, usually 90% to 95% of the speed of light. The fast propagation speed of electromagnetic waves makes their wavelengths extremely long at audio frequencies, Figure 1.10.

At the higher radio frequencies (VHF and UHF), the wavelengths become very short—1 meter or less. Antennas to receive such waves must be of comparable physical size, usually one-quarter to one-half wavelength. When waves become too short for practical antennae, concave dishes can be used to collect the waves. It should be pointed out that the highest frequency that humans can hear (about 20 kHz) is a very low frequency when considering the entire electromagnetic spectrum.

An acoustic wave is one that is propagating by means of vibrating a medium such as steel, water, or air. The propagation speeds through these media are relatively slow, resulting in waves that are long in length compared with an electromagnetic wave of the same frequency. The wavelengths of audio frequencies in air range from about 17 m (20 Hz) to 17 mm (20 kHz). The wavelength of 1 kHz in air is about 0.334 m (about 1.13 ft).

When physically short acoustic waves are radiated into large rooms, there can be adverse effects from reflections. Acoustic reflections occur when a wave encounters a change in acoustic impedance, usually from a rigid surface, the edge of a surface or some other obstruction. The reflection angle equals the incidence angle in the ideal case. Architectural acoustics is the study of the behavior of sound waves in enclosed spaces. Acousticians specialize in creating spaces with reflected sound fields that enhance rather than detract from the listening experience.

When sound encounters a room surface, a complex interaction takes place. If the surface is much larger than the wavelength, a reflection occurs and an acoustic shadow is formed behind the boundary.

If the obstruction is smaller than the wavelength of the wave striking it, the wave diffracts around the obstruction and continues to propagate. Both effects are complex and frequency (wavelength) dependent, making them difficult to calculate, Figure 1.11

The reflected wave will be strong if the surface is large and has low absorption. As absorption is increased, the level of the reflection is reduced. If the surface is random, the wave can be scattered

	Sound in Air		Copper Wire	
Frequency in Hertz	U.S. English (Feet)	SI (Meters)	U.S. English (Miles)	SI (KM)
31.5	36	11	5609	9047
63	18	5.5	2952	4523
125	9	2.7	1476	2261
250	4.5	1.4	738	1130
500	2.3	0.7	369	565
1K	1.13	0.344	184	282
2K	0.56	0.172	92	141
4K	0.28	0.086	46	70
8K	0.14	0.043	23	35
16K	0.07	0.021	11	17.6

FIGURE 1.10
Acoustic wavelengths are relatively short and interact dramatically with their environment. Audio wavelengths are extremely long, and phase interaction on audio cables is not usually of concern. (Courtesy Syn-Aud-Con.)

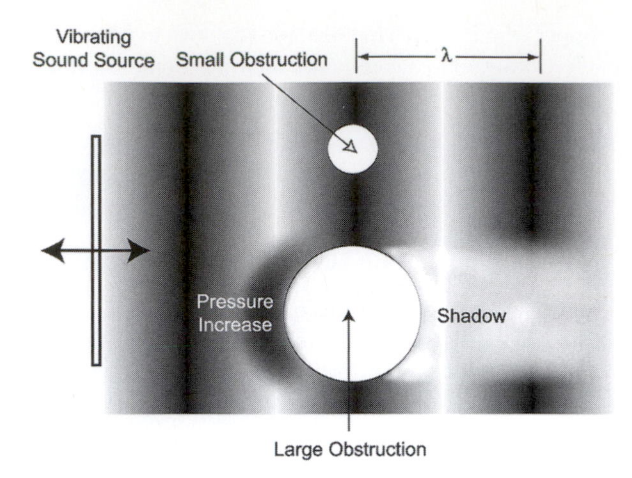

FIGURE 1.11
Sound diffracts around objects that are small relative to the length of the sound wave. (Courtesy Syn-Aud-Con.)

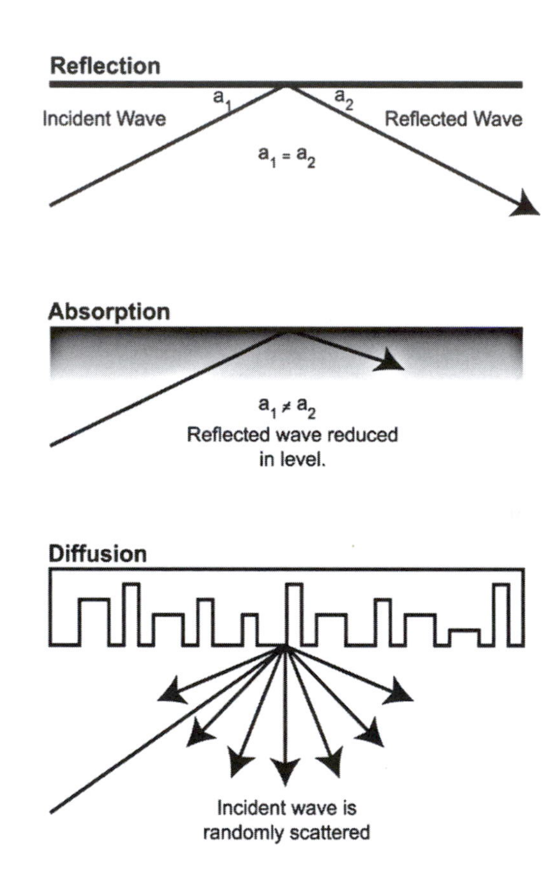

FIGURE 1.12
Sound waves will interact with a large boundary in a complex way. (Courtesy Syn-Aud-Con.)

depending on the size relationship between the wave and the surface relief. Commercially available diffusors can be used to achieve uniform scattering in critical listening spaces, Figure 1.12.

SURFACE SHAPES

The geometry of a boundary can have a profound affect on the behavior of the sound that strikes it. From a sound reinforcement perspective, it is usually better to scatter sound than to focus it. A concave room boundary should be avoided for this reason, Figure 1.13. Many auditoriums have concave rear walls and balcony faces that require extensive acoustical treatment for reflection control. A convex

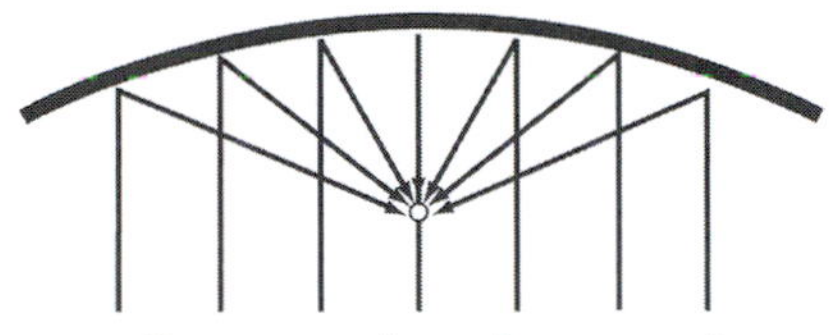

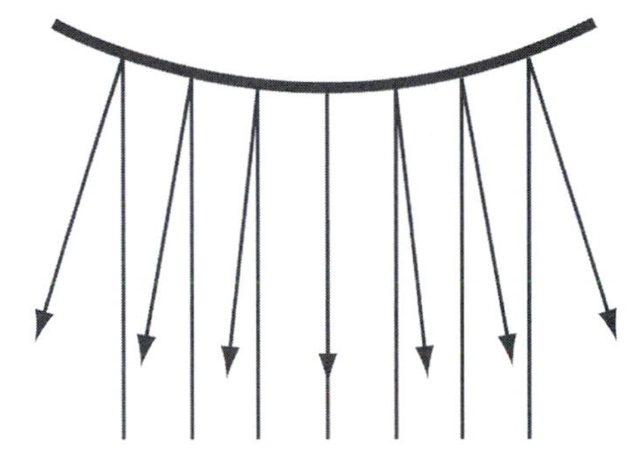

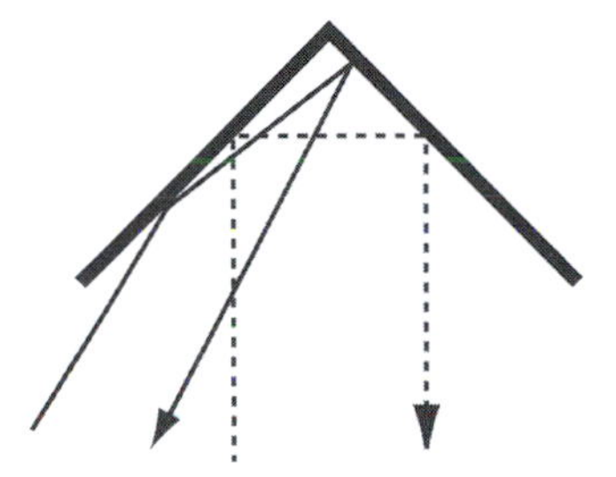

FIGURE 1.13
Some surfaces produce focused reflections. (Courtesy Syn-Aud-Con.)

surface is more desirable, since it scatters sound waves whose wavelengths are small relative to the radius of curvature. Room corners can provide useful directivity control at low frequencies, but at high frequencies can produce problematic reflections.

Electrical reflections can occur when an electromagnetic wave encounters a change in impedance. For such waves traveling down a wire, the reflection is back towards the source of the wave. Such reflections are not usually a problem for analog waves unless there is a phase offset between the outgoing and reflected waves. Note that an audio cable would need to be very long for its length to cause a significant time offset between the incident and reflected wave (many thousands of meters). At radio frequencies, reflected waves pose a huge problem, and cables are normally terminated (operated into a matched impedance) to absorb the incident wave at the receiving device and reduce the level of the reflection. The same is true for digital signals due to their very high frequency content.

SUPERPOSITION

Sine waves and cosine waves are periodic and singular in frequency. These simple waveforms are the building blocks of the complex waveforms that we listen to every day. The amplitude of a sine wave can be displayed as a function of time or as a function of phase rotation, Figure 1.14. The sine wave will serve as an example for the following discussion about superposition. Once the size (wavelength) of a wave is known, it is useful to subdivide it into smaller increments for the purpose of tracking its progression through a cycle or comparing its progression with that of another wave. Since the sine wave describes a cyclic (circular) event, one full cycle is represented by 360°, at which point the wave repeats.

When multiple sound pressure waves pass by a point of observation, their responses sum to form a composite wave. The composite wave is the complex combination of two or more individual waves. The amplitude of the summation is determined by the relative phase of the individual waves. Let's

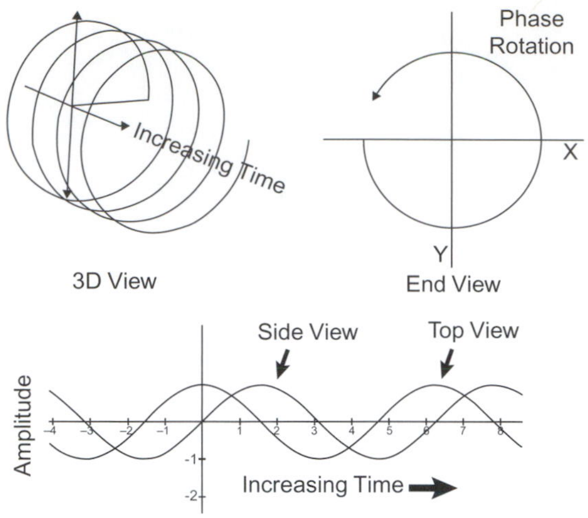

FIGURE 1.14
Simple harmonic motion can be represented with a sine or cosine wave. Both are viewpoints of the same event from different angles. (Courtesy Syn-Aud-Con.)

consider how two waves might combine at a point of observation. This point might be a listener seat or microphone position. Two extremes exist. If there is no phase offset between two waves of the same amplitude and frequency, the result is a coherent summation that is twice the amplitude of either individual wave (+6 dB). The other extreme is a 180° phase offset between the waves. This results in the complete cancellation of the pressure response at the point of observation. An infinite number of intermediate conditions occur between these two extremes. The phase interaction of waves is not a severe problem for analog audio signals in the electromagnetic domain for sound systems, where the wavelengths at audio frequencies are typically much longer than the interconnect cables. Waves reflected from receiver to source are in phase and no cancellation occurs. This is not the case for video, radio frequency, and digital signals. The shorter wavelengths of these signals can be dramatically affected by wave superposition on interconnect cables. As such, great attention must be given to the length and terminating impedance of the interconnect cables to assure efficient signal transfer between source and receiver. The practice of impedance matching between source, cable, and load is usually employed.

In sound reinforcement systems, phase interactions are typically more problematic for acoustical waves than electromagnetic waves. Phase summations and cancellations are the source of many acoustical problems experienced in auditoriums. Acoustic wavelengths are often short relative to the room size (at least at high frequency), so the waves tend to bounce around the room before decaying to inaudibility. At a listener position, the reflected waves "superpose" to form a complex waveform that is heard by the listener. The sound radiated from multiple loudspeakers will interact in the same manner, producing severe modifications in the radiated sound pattern and frequency response. Antenna designers have historically paid more attention to these interactions than loudspeaker designers, since there are laws that govern the control of radio frequency emissions. Unlike antennas, loudspeakers are usually broadband devices that cover one decade or more of the audible spectrum. For this reason, phase interactions between multiple loudspeakers never result in the complete cancellation of sound pressure, but rather cancellation at some frequencies and coherent summation at others. The subjective result is *tonal coloration* and *image shift* of the sound source heard by the listener. The significance of this phenomenon is application-dependent. People having dinner in a restaurant would not be concerned with the effects of such interactions since they came for the food and not the music. Concert-goers or church attendees would be more concerned, because their seat might be in a dead spot, and the interactions disrupt their listening experience, possibly to the point of reducing the information conveyed via the sound system. A venue owner may make a significant investment in good quality loudspeakers, only to have their response impaired by such interactions with an adjacent loudspeaker or room surface, Figure 1.15.

Phase interactions are most disruptive in critical listening environments, such as recording studio control rooms or high quality home entertainment systems. Users of these types of systems often make a large investment to maintain sonic accuracy by purchasing phase coherent loudspeakers and appropriate

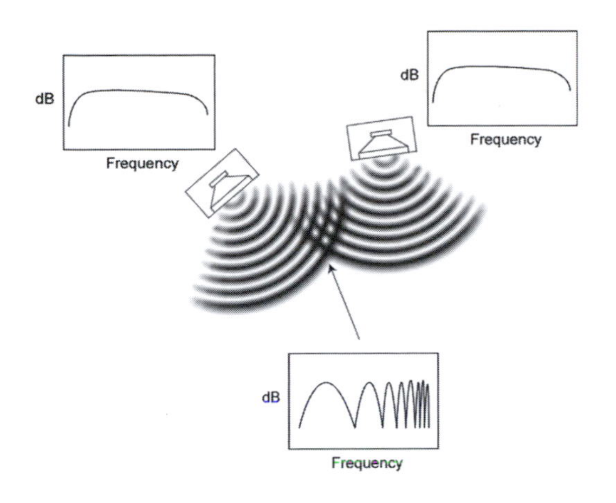

FIGURE 1.15
Phase interference occurs when waves from multiple sources arrive at different times. (Courtesy Syn-Aud-Con.)

acoustical treatments for the listening space. The tonal coloration caused by wave interference may be unacceptable for a recording studio control room but may be artistically pleasing in a home audio system.

Loudspeaker designers can use wave interaction to their advantage by choosing loudspeaker spacings that form useful radiation patterns. Almost all pattern control in the low frequency decade is achieved in this manner. Uninformed system designers create undesirable radiation patterns by accident in the way that they place and stack loudspeakers. The results are poor coverage and reduced acoustic gain.

The proper way to view the loudspeaker and room are as filters that the sound energy must pass through en route to the listener. Some aspects of these filters can be compensated with electronic filters—a process known as equalization. Other aspects cannot, and electronic *equalization* merely aggravates or masks the problem.

OHM'S LAW

In acoustics, the sound that we hear is nature restoring an equilibrium condition after an atmospheric disturbance. The disturbance produces waves that cause the atmospheric pressure to oscillate above and below ambient pressure as they propagate past a point of observation. The air always settles to its ambient state upon cessation of the disturbance.

In an electrical circuit, a potential difference in electrical pressure between two points causes current to flow. Electrical current results from electrons flowing to a point of lower potential. The electrical potential difference is called an *electromotive force* (EMF) and the unit is the volt (V). The rate of electron flow is called *current* and the unit is the *ampere* (A). The ratio between voltage and current is called the *resistance* and the unit is the *ohm* (Ω). The product of voltage and current is the *apparent power*, W, that is produced by the source and consumed by the load. Power is the rate of doing work and power ratings must always include a reference to time. A power source can produce a rated voltage at a rated flow of current into a specified load for a specified period of time. The ratio of voltage to current can be manipulated to optimize a source for a specific task. For instance, current flow can be sacrificed to maximize voltage transfer. When a device is called upon to deliver appreciable current, it is said to be operating under load. The load on an automobile increases when it must maintain speed on an uphill grade, and greater power transfer between the engine and drive train is required. Care must be taken when loading audio components to prevent distortion or even damage. Ohm's Law describes the ratios that exist between voltage, current, and resistance in an electrical circuit.

$$R = \frac{E}{I} \tag{1.9}$$

$$E = IR \tag{1.10}$$

$$I = \frac{E}{R} \tag{1.11}$$

where,

E is in volts,
I is in amperes,
R is in ohms.

Direct current (DC) flows in one direction only. In AC (alternating current) the direction of current flow is alternating at the frequency of the waveform. Voltage and current are not always in sync so the phase relationship between them must be considered. Power flow is reduced when they are not in relative phase (synchronization). Voltage and current are in phase in resistive circuits. Phase *shifts* between voltage and current are produced by reactive elements in a circuit. Reactance reduces the power transferred to the load by storing energy and reflecting it back to the source. Loudspeakers and transformers are examples of sound system components that can have significant reactive characteristics. The combined opposition to current flow caused by resistance and reactance is termed the *impedance* (Z) of the circuit. The unit for impedance is also the ohm (Ω). An impedance can be purely resistive, purely reactive, or most often some combination of the two. This is referred to as a *complex* impedance. Impedance is a function of frequency, and impedance measurements must state the frequency at which the measurement was made. Sound system technicians should be able to measure impedance to verify proper component loading, such as at the amplifier/loudspeaker interface.

$$Z = \sqrt{R^2 + (X_T)^2} \tag{1.12}$$

where,

Z is the impedance in ohms,
R is the resistance in ohms,
X_T is the total reactance in ohms.

Reactance comes in two forms. Capacitive reactance causes the voltage to lag the current in phase. Inductive reactance causes the current to lag the voltage in phase. The total reactance is the sum of the inductive and capacitive reactance. Since they are different in sign one can cancel the other, and the resultant phase angle between voltage and current will be determined by the dominant reactance.

In mechanics, a spring is a good analogy for capacitive reactance. It stores energy when it is compressed and returns it to the source. In an electrical circuit, a capacitor opposes changes in the applied voltage. Capacitors are often used as filters for passing or rejecting certain frequencies or smoothing ripples in power supply voltages. Parasitic capacitances can occur when conductors are placed in close proximity.

$$X_C = \frac{1}{2\pi fC} \tag{1.13}$$

where,

f is frequency in hertz,
C is capacitance in farads,
X_C is the capacitive reactance in ohms.

In mechanics, a moving mass is analogous to an inductive reactance in an electrical circuit. The mass tends to keep moving when the driving force is removed. It has therefore stored some of the applied energy. In electrical circuits, an inductor opposes a change in the current flowing through it. As with capacitors, this property can be used to create useful filters in audio systems. Parasitic inductances can occur due to the ways that wires are constructed and routed.

$$X_L = 2\pi fL \tag{1.14}$$

where,

X_L is the inductive reactance in ohms.

Inductive and capacitive reactance produce the opposite effect, so one can be used to compensate for the other. The total reactance X_T is the sum of the inductive and capacitive reactance.

$$X_T = X_L - X_C \tag{1.15}$$

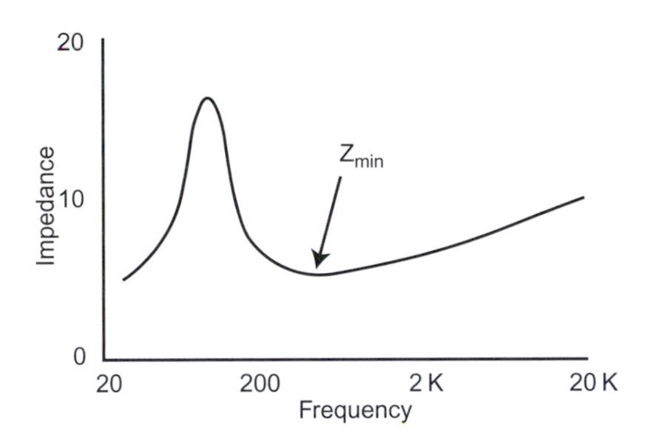

FIGURE 1.16
An impedance magnitude plot displays impedance as a function of the applied frequency. (Courtesy Syn-Aud-Con.)

Note that the equations for capacitive and inductive reactance both include a frequency term. Impedance is therefore frequency dependent, meaning that it changes with frequency. Loudspeaker manufacturers often publish impedance plots of their loudspeakers. The impedance of interest from this plot is usually the *nominal* or *rated* impedance. Several standards exist for determining the rated impedance from the impedance plot, Figure 1.16.

An impedance phase plot often accompanies an impedance magnitude plot to show whether the loudspeaker load is resistive, capacitive, or inductive at a given frequency. A resistive load will convert the applied power into heat. A reactive load will store and reflect the applied power. Complex loads, such as loudspeakers, do both. When considering the power delivered to the loudspeaker, the impedance Z is used in the power equation. When considering the power dissipated by the load, the resistive portion of the impedance must be used in the power equation. The power factor describes the reduction in power transfer caused by the phase angle between voltage and current in a reactive load. Some definitions are useful.

$$Apparent\ Power\ (Total\ Power) = \frac{E^2}{Z} \tag{1.16}$$

$$Active\ Power\ (Absorbed\ Power) = \frac{E^2}{R} \tag{1.17}$$

$$Reactive\ Power\ (Reflected\ Power) = \frac{E^2}{Z\cos\theta} \tag{1.18}$$

where,

θ is the phase angle between the voltage and current.

Ohm's Law and the power equation in its various forms are foundation stones of the audio field. One can use these important tools for a lifetime and not exhaust their application to the electrical and acoustical aspects of the sound reinforcement system.

HUMAN HEARING

It is beneficial for sound practitioners to have a basic understanding of the way that people hear and perceive sound. The human auditory system is an amazing device, and it is quite complex. Its job is to transduce fluctuations in the ambient atmospheric pressure into electrical signals that will be processed by the brain and perceived as sound by the listener. We will look at a few characteristics of the human auditory system that are of significance to audio practitioners.

The dynamic range of a system describes the difference between the highest level that can pass through the system and its noise floor. The threshold of human hearing is about 0.00002 Pascals (Pa) at mid

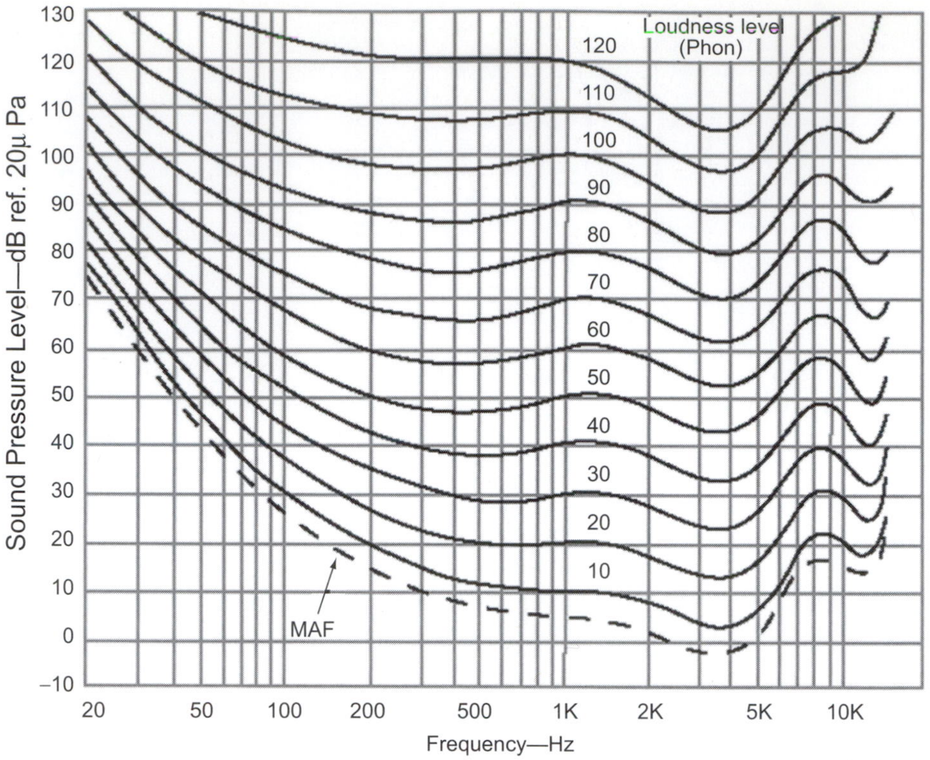

FIGURE 1.17
The equal-loudness contours. (Illustration courtesy Syn-Aud-Con.)

frequencies. The human auditory system can withstand peaks of up to 200 Pa at these same frequencies. This makes the dynamic range of the human auditory system approximately

$$\begin{aligned} DR &= 20\log\frac{200}{0.00002} \\ &= 140\text{ dB} \end{aligned} \tag{1.19}$$

The hearing system cannot take much exposure at this level before damage occurs. Speech systems are often designed for 80 dB ref. 20 μPa and music systems about 90 dB ref. 20 μPa for the mid-range part of the spectrum.

Audio practitioners give much attention to achieving a flat spectral response. The human auditory system is not flat and its response varies with level. At low levels, its sensitivity to low frequencies is much less than its sensitivity to mid-frequencies. As level increases, the difference between low- and mid-frequency sensitivity is less, producing a more uniform spectral response. The classic equal loudness contours, Figure 1.17, describe this phenomenon and have given us the weighting curves, Figure 1.18, used to measure sound levels.

Modern sound systems are capable of producing very high sound pressure levels over large distances. Great care must be taken to avoid damaging the hearing of the audience.

The time response of the hearing system is slow compared with the number of audible events that can occur in a given time span. As such, our hearing system *integrates* closely spaced sound arrivals (within about 35 ms) with regard to level. This is what makes sound indoors appear louder than sound outdoors. While reflected sound increases the perceived level of a sound source, it also adds colorations. This is the heart of how we perceive acoustic instruments and auditoriums. A good recording studio or concert hall produces a musically pleasing reflected sound field to a listener position. In general, secondary energy arrivals pose problems if they arrive earlier than 10 ms (severe tonal coloration) after the first arrival or later than 50 ms (potential echo), Figure 1.19.

The integration properties of the hearing system make it less sensitive to impulsive sound events with regard to level. Peaks in audio program material are often 20 dB or more higher in level than the

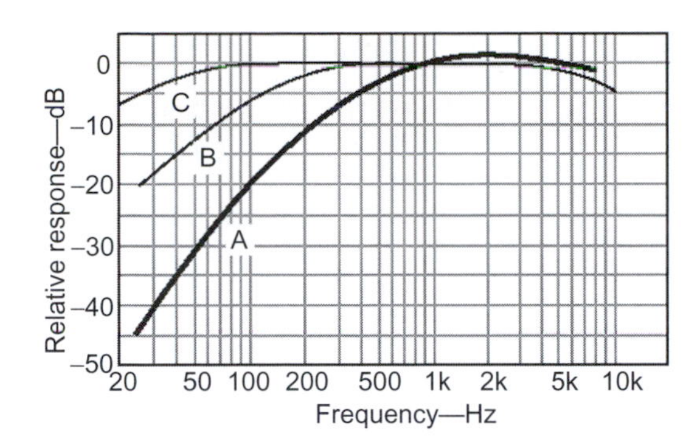

FIGURE 1.18
Weighting scales for measuring sound levels. (Illustration courtesy Syn-Aud-Con.)

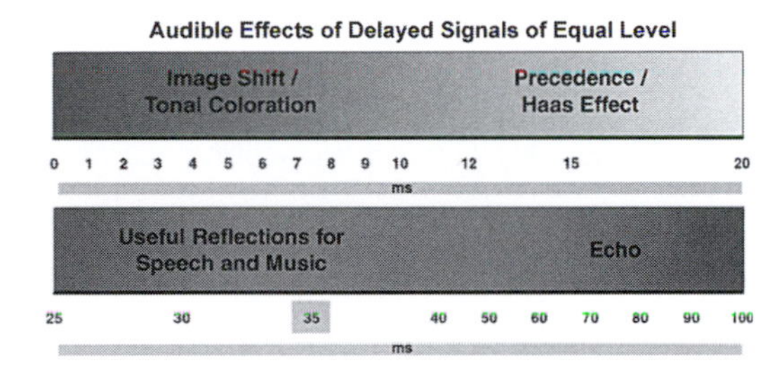

FIGURE 1.19
The time offset between sound arrivals will determine if the secondary arrival is useful or harmful in conveying information to the listener. The direction of arrival is also important and is considered by acousticians when designing auditoriums. (Courtesy Syn-Aud-Con.)

perceived loudness of the signal. Program material that measures 90 dBA (slow response) may contain short-term events at 110 dBA or more, so care must be taken when exposing musicians and audiences to high-powered sound systems.

The eardrum is a pressure-sensitive diaphragm that responds to fluctuations in the ambient atmospheric pressure. Like a loudspeaker and microphone, it has an overload point at which it distorts and can be damaged. The Occupational Safety and Health Administration (OSHA) is responsible for assuring that public spaces remain in compliance regarding sound exposure. Sound systems are a major source of high level sounds and should work within OSHA guidelines. Tinnitus, or ringing in the ears, is one symptom of excessive sound exposure.

MONITORING AUDIO PROGRAM MATERIAL

The complex nature of the audio waveform necessitates specialized instrumentation for visual monitoring. Typical voltmeters are not suitable for anything but the simplest waveforms, such as sine waves. There are two aspects of the audio signal that are of interest to the system operator. The peaks of the program material must not exceed the peak output capability of any component in the system. Ironically the peaks have little to do with the perceived loudness of the signal or the electrical or acoustic power generated by it. Both of these parameters are more closely tied to the rms (root-mean-square) value of the signal. Measurement of the true rms value of a waveform requires specialized equipment that integrates energy over a time span, much like the hearing system does. This integrated data will better correlate with the perceived loudness of the sound event. So audio practitioners need to monitor at least two aspects of the audio signal—its relative loudness (related to the rms level) and peak levels. Due to the complexity of true rms monitoring, most meters display an average value that is an approximation of the rms value of the program material.

Many audio processors have instrumentation to monitor either peak or average levels, but few can track both simultaneously. Most mixers have a VI (volume indicator) meter that reads in VU (volume units). Such meters are designed with ballistic properties that emulate the human hearing system and are useful for tracking the perceived loudness of the signal. Meters of this type all but ignore the peaks

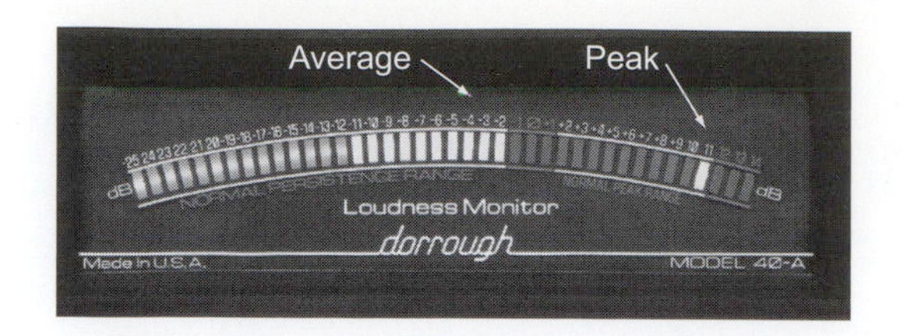

FIGURE 1.20
A meter that can display both average and peak levels simultaneously. (Courtesy Dorrough Electronics.)

in the program material, making them unable to display the available headroom in the system or clipping in a component. Signal processors usually have a peak LED that responds fast enough to indicate peaks that are at or near the component's clipping point. Many recording systems have PPM (peak program meters) that track the peaks but reveal little about the relative loudness of the waveform.

Figure 1.20 shows an instrument that monitors both peak and relative loudness of the audio program material. Both values are displayed in relative dB, and the difference between them is the approximate *crest factor* of the program material. Meters of this type yield a more complete picture of the audio event, allowing both loudness and available headroom to be observed simultaneously.

SOUND PROPAGATION

Sound waves are emitted from acoustic sources—devices that move to modulate the ambient atmospheric pressure. Loudspeakers become intentional acoustic sources when they are driven with waveforms that cause them to vibrate at frequencies within the bandwidth of the human listener. A point source is a device that radiates sound from one point in space. A true point source is an abstract idea and is not physically realizable, as it would be of infinitesimal size. This does not prevent the use of the concept to describe the characteristics of devices that are physically realizable.

Let us consider the properties of some idealized acoustic sources—not ideal in that they would be desirable for sound reinforcement use, but ideal in respect to their behavior as predictable radiators of acoustic energy.

The point source

A point source with 100% efficiency would produce one watt of acoustical power from one watt of applied electrical power. No heat would result, since all of the electrical power is converted. The energy radiated from the source would travel equally in all directions from the source. Directional energy radiation is accomplished by interfering with the emerging wave. Since interference would require a finite size, a true infinitesimal point source would be omnidirectional. We will introduce the effects of interference later.

Using one pW (picowatt) as a power reference, the sound power level produced by one acoustic watt will be

$$\begin{aligned} L_W &= 10\log\frac{1\,\text{W}}{10^{-12}\,\text{W}} \\ &= 120\,\text{dB} \end{aligned} \tag{1.20}$$

Note that the sound power is not dependent on the distance from the source. A sound power level of $L_W = 120\,\text{dB}$ would represent the highest continuous sound power level that could result from one watt of continuous electrical power. All real-world devices will fall short of this ideal, requiring that they be rated for efficiency and power dissipation.

Let us now select an observation point at a distance 0.282 m from the sound source. As the sound energy propagates, it forms a spherical wave front. At 0.282 m this wave front will have a surface area of one square meter. As such, the one watt of radiated sound power is passing through a surface area of $1\,\text{m}^2$.

$$\begin{aligned} L_I &= 10\log\frac{1\,\text{W/m}^2}{10^{-12}\,\text{W/m}^2} \\ &= 120\,\text{dB} \end{aligned} \tag{1.21}$$

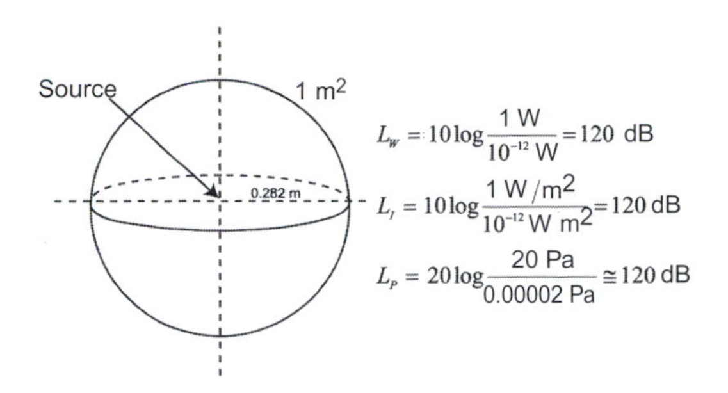

FIGURE 1.21
This condition forms the basis of the standard terminology and relationships used to describe sound radiation from loudspeakers. (Courtesy Syn-Aud-Con.)

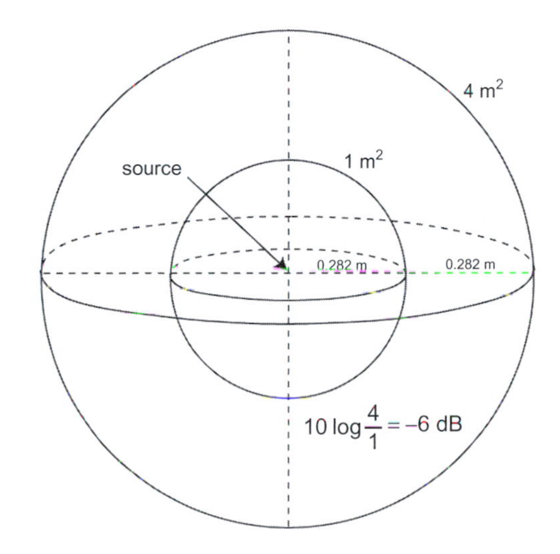

FIGURE 1.22
When the distance to the source is doubled, the radiated sound energy will be spread over twice the area. Both L_I and L_P will drop by 6 dB. (Courtesy Syn-Aud-Con.)

This is the sound intensity level L_I of the source and represents the amount of power flowing through the surface of a sphere of 1 square meter. Again, this is the highest intensity level that could be achieved by an omnidirectional device of 100% efficiency. L_I can be manipulated by confining the radiated energy to a smaller area. The level benefit gained at a point of observation by doing such is called the *directivity index* (DI) and is expressed in decibels. All loudspeakers suitable for sound reinforcement should exploit the benefits of directivity control.

For the ideal device described, the sound pressure level L_P (or commonly *SPL*) at the surface of the sphere will be numerically the same as the L_W and L_I ($L_P = 120\,dB$) since the sound pressure produced by one watt will be 20 Pa. This L_P is only for one point on the sphere, but, since the source is omnidirectional, all points on the sphere will be the same. To summarize, at a distance of 0.282 m from a point source, the sound power level, sound intensity level, and sound pressure level will be numerically the same. This important relationship is useful for converting between these quantities, Figure 1.21.

Let us now consider a point of observation that is twice as far from the source. As the wave continues to spread, its total area at a radius of 0.564 m will be four times the area at 0.282 m. When the sound travels twice as far, it spreads to cover four times the area. In decibels, the sound level change from point one to point two is

$$\Delta L_p = 20\log\frac{0.564}{0.282}$$
$$= 6\ dB$$

This behavior is known as the inverse-square law (ISL), Figure 1.22. The ISL describes the level attenuation versus distance for a point source radiator due to the spherical spreading of the emerging waves.

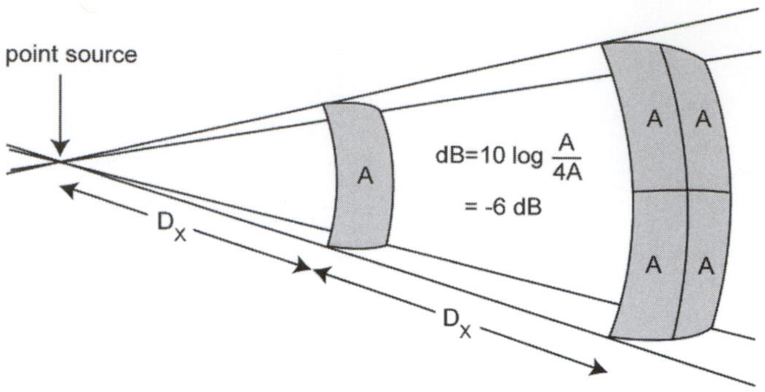

FIGURE 1.23
The ISL is also true for directional devices in their far field (remote locations from the device). (Courtesy Syn-Aud-Con.)

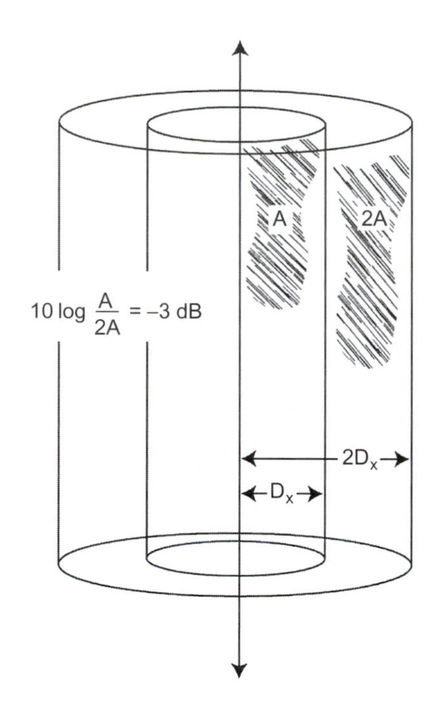

FIGURE 1.24
Line sources radiate a cylindrical wave (ideal case). The level drop versus distance is less than for a point source. Courtesy Syn-Aud-Con.

Frequency-dependent losses will be incurred from atmospheric absorption, but those will not be considered here. Most loudspeakers will roughly follow the inverse square law level change with distance at points remote from the source, Figure 1.23.

The line source

Successful sound radiators have been constructed that radiate sound from a line rather than a point. The infinite line source emits a wave that is approximately cylindrical in shape. Since the diverging wave is not expanding in two dimensions, the level change with increasing distance is half that of the point source radiator. The sound level from an ideal line source will decrease at 3 dB per distance doubling rather than 6 dB, Figure 1.24. It should be pointed out that these relationships are both frequency and line length dependent, and what is being described here is the ideal case. Few commercially available line arrays exhibit this cylindrical behavior over their full bandwidth. Even so, it is useful to allow a mental image of the characteristics of such a device to be formed.

If the line source is finite in length (as all real-world sources will be), then there will be a phase differential between the sound radiated from different points on the source to a specific point in space. All

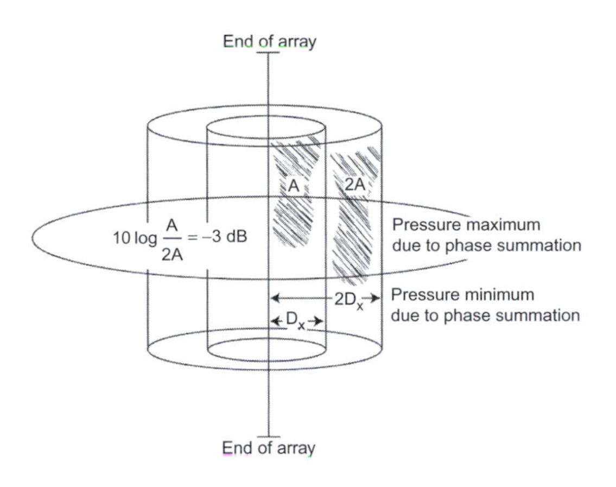

FIGURE 1.25
The finite line array has gained wide acceptance among system designers, allowing wide audience coverage with minimal energy radiation to room surfaces. Courtesy Syn-Aud-Con.

of the points will be the most in phase on a plane perpendicular from the array and equidistant from the end points of the array. As the point of observation moves away from the midpoint, phase interaction will produce lobes in the radiated energy pattern. The lobes can be suppressed by clever design, allowing the wave front to be confined to a very narrow vertical angle, yet with wide horizontal coverage. Such a radiation pattern is ideal for some applications, such as a broad, flat audience plane that must be covered from ear height. Digital signal processing has produced well-behaved line arrays that can project sound to great distances. Some incorporate an adjustable delay for each element to allow steering of the radiation lobe. Useful designs for auditoriums are at least 2 meters in vertical length.

While it is possible to construct a continuous line source using ribbon drivers, etc., most commercially available designs are made up of closely spaced discrete loudspeakers or loudspeaker systems and are more properly referred to as line arrays, Figure 1.25.

CONCLUSION

The material in this chapter was carefully selected to expose the reader to a broad spectrum of principles regarding sound reinforcement systems. As a colleague once put it, "Sound theory is like an onion. Every time you peel off a layer another lies beneath it!" Each of these topics can be taken to higher levels. The reader is encouraged to use this information as a springboard into a life-long study of audio and acoustics. We are called upon to spend much of our time learning about new technologies. It must be remembered that new methods come from the mature body of principles and practices that have been handed down by those who came before us. Looking backward can have some huge rewards.

If I can see farther than those who came before me, it is because I am standing on their shoulders.

Sir Isaac Newton

BIBLIOGRAPHY

Alton Everest, F., 1998. Master Handbook of Acoustics, third ed., TAB Books, .

Ballou, G., 1991. Handbook for Sound Engineers, second ed., Focal Press, Boston.

Baranek, L., 1954. Acoustics, McGraw-Hill New York.

Davis, D. and Davis, C., 1997. Sound System Engineering, Focal Press, Boston.

Olson, H., Acoustical Engineering, Professional Audio Journals, Inc. Philadelphia, PA.

PREFACE

The very basic principles of sound and hearing having been established in the first chapter of this book, it is time to take a closer look at the process of human hearing, which since it must be regarded as the last stage of the audio engineering process, is clearly of fundamental importance. It is a complex but fascinating business, and it shows you just what evolution can do.

In this chapter David Howard and Jamie Angus first demonstrate how the human ear can be divided into three parts: the outer ear, the middle ear, and the inner ear, each with their own vital functions.

The outer ear is not simply a gristly ear-trumpet to raise the sound pressure levels being received, though that is part of its function. The convoluted grooves and ridges that make up its distinctive shape modify the frequency response of sounds reaching us, allowing us to perceive the height of an acoustic source, despite only having two ears arranged in a horizontal plane.

The middle ear comprises the three tiny bones or ossicles that transmit vibration from the ear-drum to a much smaller diaphragm that interfaces with the inner ear; their lever action, and the relative areas of the two diaphragms, act as an acoustic transformer that steps up the pressure by a remarkable 34 times. As an added feature, the acoustic reflex (also called the stapedius reflex) provides protection against loud noises by contracting tiny muscles that restrain the ossicles and reduce sound transmission.

The inner ear is a spiral snail-shaped thing that contains the basilar membrane, which acts rather like a large bank of overlapping band-pass filters, and here the various frequencies are turned into nerve impulses that go to the brain. From there Howard and Angus move on to critical band theory, which gives insight in to how the basilar membrane does its job.

The next section deals with the frequency and pressure sensitivity ranges of human hearing, and includes some depressing data on how our high-frequency response falls with advancing age.

From there we move on to the perception of loudness, which as in most things appertaining to the human senses, is rather more complex than it seems at first. The relationship between sound pressure and subjective loudness is not a simple one, and it varies markedly with frequency (Fletcher and Munson again). Subjective loudness is measured in Phons or Sones, not decibels, and it takes a 10 decibel increase (*not* a 6 decibel increase) in sound pressure level to give a perceived doubling of loudness. That is why, if you want to double the loudness of that string section, you have to hire *ten times* the number of violinists. Expensive.

There is a fascinating account of how hearing is damaged by excessive noise exposure, with the loss usually appearing first at 4 kHz, which is the main resonance in the canal from outer ear to the ear-drum.

This chapter concludes with a detailed look at the perception of sound source direction, and the various mechanisms that make it possible, including the ear ridge reflections and head movement effects.

Introduction to Hearing

Acoustics and Psychoacoustics by David Howard and Jamie Angus

Psychoacoustics is the study of how humans perceive sound. To begin our exploration of psychoacoustics it is first necessary to become familiar with the basic anatomy of the human hearing system to facilitate understanding of:

- the effect the normal hearing system has on sounds entering the ear,
- the origin of fundamental psychoacoustic findings relevant to the perception of music,
- how listening to very loud sounds can cause hearing damage, and
- some of the listening problems faced by the hearing impaired.

This chapter introduces the main anatomical structures of the human hearing system which form the path along which incoming music signals travel up to the point where the signal is carried by nerve fibers from the ear(s) to the brain. It also introduces the concept of "critical bands", which is the single most important psychoacoustic principle for an understanding of the perception of music and other sounds in terms of pitch, loudness and timbre.

It should be noted that many of the psychoacoustic effects have been observed experimentally, mainly as a result of playing sounds that are carefully controlled in terms of their acoustic nature to panels of listeners from whom responses are monitored. These responses are often the result of comparing two sounds and indicating, for example, which is louder or higher in pitch or "brighter". Many of the results from such experiments cannot as yet be described in terms of either where anatomically or by what physical means they occur. Psychoacoustics is a developing field of research. However, the results from such experiments give a firm foundation for understanding the nature of human perception of musical sounds, and knowledge of minimum changes that are perceived provide useful guideline bounds for those exploring the subtleties of sound synthesis.

THE ANATOMY OF THE HEARING SYSTEM

The anatomy of the human hearing system is illustrated in Figure 2.1. It consists of three sections:

- the outer ear,
- the middle ear, and
- the inner ear.

The anatomical structure of each of these is discussed below, along with the effect that each has on the incoming acoustic signal.

Outer ear function

The outer ear (see Figure 2.1) consists of the external flap of tissue known as the "pinna" with its many grooves, ridges and depressions. The depression at the entrance to the auditory canal is known as the "concha". The auditory canal is approximately 25–35 mm long from the concha to the "tympanic membrane," more commonly known as the "eardrum". The outer ear has an acoustic effect on sounds

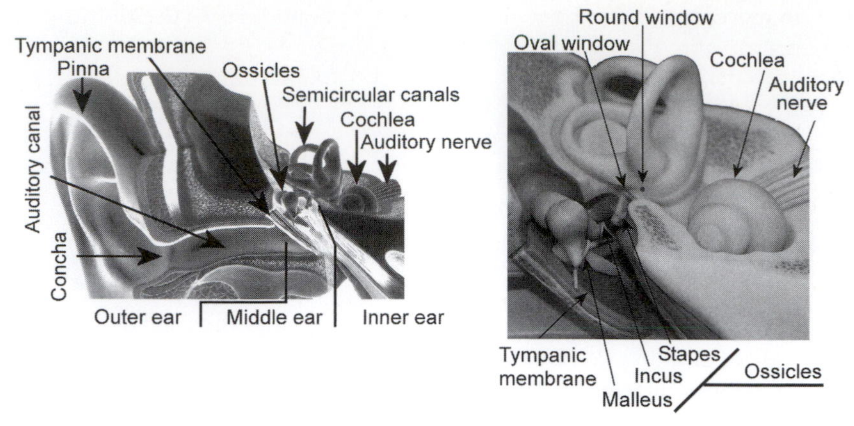

FIGURE 2.1
The main structures of the human ear showing an overall view of the outer, middle and inner ears (left) and a detailed view of the middle and inner ear (right).

entering the ear in that it helps us both to locate sound sources and it enhances some frequencies with respect to others.

Sound localization is helped mainly by the acoustic effect of the pinna and the concha. The concha acts as an acoustic resonant cavity. The combined acoustic effects of the pinna and concha are particularly useful for determining whether a sound source is in front or behind, and to a lesser extent whether it is above or below.

The acoustic effect of the outer ear as a whole serves to modify the frequency response of incoming sounds due to resonance effects, primarily of the auditory canal whose main resonance frequency is in the region around 4 kHz.

The tympanic membrane is a light, thin, highly elastic structure which forms the boundary between the outer and middle ears. It consists of three layers: the outside layer which is a continuation of the skin lining of the auditory canal, the inside layer which is continuous with the mucous lining of the middle ear, and the layer in between these which is a fibrous structure which gives the tympanic membrane its strength and elasticity. The tympanic membrane converts acoustic pressure variations from the outside world into mechanical vibrations in the middle ear.

Middle ear function

The mechanical movements of the tympanic membrane are transmitted through three small bones known as "ossicles", comprising the "malleus", "incus" and "stapes"—more commonly known as the "hammer", "anvil" and "stirrup"—to the oval window of the cochlea (see Figure 2.1). The oval window forms the boundary between the middle and inner ears.

The malleus is fixed to the middle fibrous layer of the tympanic membrane in such a way that when the membrane is at rest, it is pulled inwards. Thus the tympanic membrane when viewed down the auditory canal from outside appears concave and conical in shape. One end of the stapes, the stapes footplate, is attached to the oval window of the cochlea. The malleus and incus are joined quite firmly such that at normal intensity levels they act as a single unit, rotating together as the tympanic membrane vibrates to move the stapes via a ball and socket joint in a piston-like manner. Thus acoustic vibrations are transmitted via the tympanic membrane and ossicles as mechanical movements to the cochlea of the inner ear.

The function of the middle ear is twofold: (1) to transmit the movements of the tympanic membrane to the fluid which fills the cochlea without significant loss of energy, and (2) to protect the hearing system to some extent from the effects of loud sounds, whether from external sources or the individual concerned.

In order to achieve efficient transfer of energy from the tympanic membrane to the oval window, the effective pressure acting on the oval window is arranged by mechanical means to be greater than that acting on the tympanic membrane. This is to overcome the higher resistance to movement of the cochlear fluid compared with that of air at the input to the ear. Resistance to movement can be thought of as "impedance" to movement and the impedance of fluid to movement is high compared with that of air. The ossicles act as a mechanical "impedance converter" or "impedance transformer" and this is achieved essentially by two means:

- the lever effect of the malleus and incus, and
- the area difference between the tympanic membrane and the stapes footplate.

The lever effect of the malleus and incus arises as a direct result of the difference in their lengths. Figure 2.2 shows this effect. The force at the stapes footplate relates to the force at the tympanic membrane by the ratio of the lengths of the malleus and incus as follows:

$$F1 \times L1 = F2 \times L2$$

where

F1 = force at tympanic membrane

F2 = force at stapes footplate

L1 = length of malleus

and L2 = length of incus

Therefore:

$$F2 = F1 \times \frac{L1}{L2} \tag{2.1}$$

FIGURE 2.2
The function of the ossicles of the middle ear.

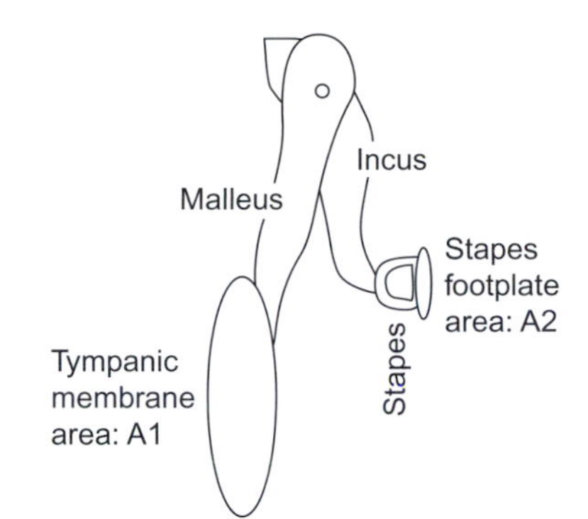

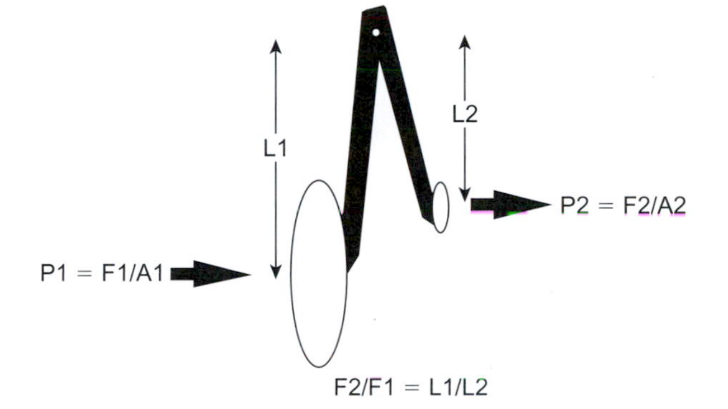

The area difference has a direct effect on the pressure applied at the stapes footplate compared with the incoming pressure at the tympanic membrane since pressure is expressed as force per unit area as follows:

$$\text{Pressure} = \frac{\text{Force}}{\text{Area}} \tag{2.2}$$

The areas of the tympanic membrane and the stapes footplate in humans are represented in Figure 2.2 as A1 and A2 respectively. The pressure at the tympanic membrane (P1) and the pressure at the stapes footplate (P2) can therefore be expressed as follows:

$$P1 = \frac{F1}{A1}$$

$$P2 = \frac{F2}{A2}$$

The forces can therefore be expressed in terms of pressures:

$$F1 = (P1 \times A1) \tag{2.3}$$

$$F2 = (P2 \times A2) \tag{2.4}$$

Substituting Equations 2.3 and 2.4 into Equation 2.1 gives:

$$(P2 \times A2) = (P1 \times A1) \times \frac{L1}{L2}$$

Therefore:

$$\frac{P2}{P1} = \frac{A1 \times L1}{A2 \times L2} \tag{2.5}$$

Pickles (1982) describes a third aspect of the middle ear which appears relevant to the impedance conversion process. This relates to a buckling motion of the tympanic membrane itself as it moves, resulting in a twofold increase in the force applied to the malleus.

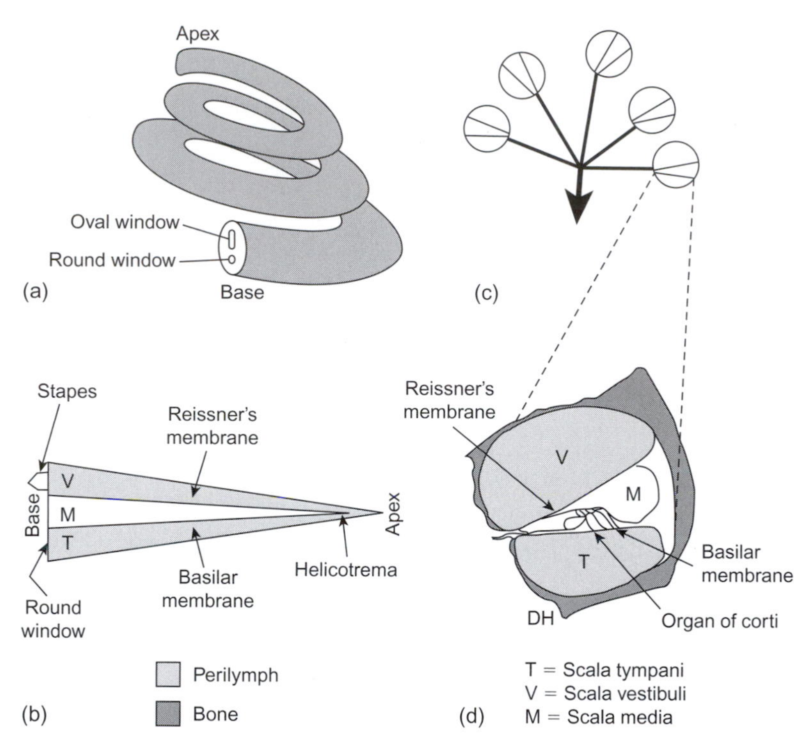

FIGURE 2.3
(a) The spiral nature of the cochlea. (b) The cochlea "unrolled". (c) Vertical cross-section through the cochlea. (d) Detailed view of the cochlear tube.

In humans, the area of the tympanic membrane (A1) is approximately 13 times larger than the area of the stapes footplate (A2), and the malleus is approximately 1.3 times the length of the incus. The buckling effect

EXAMPLE 2.1

Express the pressure ratio between the stapes footplate and the tympanic membrane in decibels.

The pressure ratio is 33.8:1. Equation 1.20 is used to convert from pressure ratio to decibels:

$$\text{dB(SPL)} = 20\log_{10}\frac{P2}{P1}$$

Substituting 33.8 as the pressure ratio gives:

$$20\log_{10}[33.8] = 30.6\,\text{dB}$$

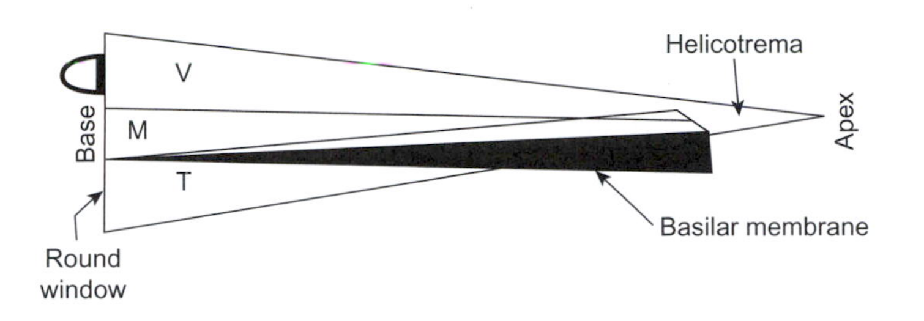

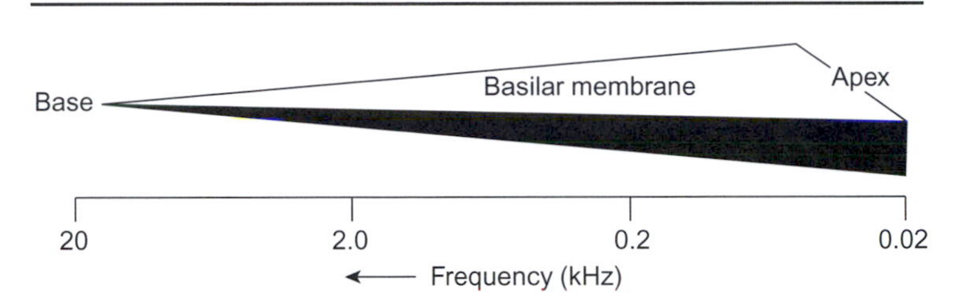

FIGURE 2.4
Idealized shape of basilar membrane as it lies in the unrolled cochlea (upper), and the basilar membrane response with frequency (lower).

of the tympanic membrane provides a force increase by a factor of 2. Thus the pressure at the stapes footplate (P2) is about ($13 \times 1.3 \times 2 = 33.8$) times larger than the pressure at the tympanic membrane (P1).

The second function of the middle ear is to provide some protection for the hearing system from the effects of loud sounds, whether from external sources or the individual concerned. This occurs as a result of the action of two muscles in the middle ear: the tensor tympani and the stapedius muscle. These muscles contract automatically in response to sounds with levels greater than approximately 75 dB(SPL) and they have the effect of increasing the impedance of the middle ear by stiffening the ossicular chain. This reduces the efficiency with which vibrations are transmitted from the tympanic membrane to the inner ear and thus protects the inner ear to some extent from loud sounds. Approximately 12–14 dB of attenuation is provided by this protection mechanism, but this is for frequencies below 1 kHz only. The names of these muscles derive from where they connect with the ossicular chain: the tensor tympani is attached to the "handle" of the malleus, near the tympanic membranes, and the stapedius muscle attached to the stapes.

This effect is known as the "acoustic reflex". It takes some 60–120 ms for the muscles to contract in response to a loud sound. In the case of a loud impulsive sound such as the firing of a large gun, it has been suggested that the acoustic reflex is too slow to protect the hearing system. In gunnery situations, a sound loud enough to trigger the acoustic reflex, but not so loud as to damage the hearing systems, is often played at least 120 ms before the gun is fired.

Inner ear function

The inner ear consists of the snail-like structure known as the "cochlea". The function of the cochlea is to convert mechanical vibrations into nerve firings to be processed eventually by the brain. Mechanical vibrations reach the cochlea at the oval window via the stapes footplate of the middle ear.

The cochlea consists of a tube coiled into a spiral with approximately 2.75 turns (see Figure 2.3(a)). The end with the oval and round windows is the "base" and the other end is the "apex" (see Figure 2.3(b)). Figure 2.3(c) illustrates the effect of slicing through the spiral vertically, and it can be seen in part (d) that the tube is divided into three sections by Reissner's membrane and the basilar membrane. The outer channels, the scala vestibuli (V) and scala tympani (T), are filled with an incompressible fluid known as "perilymph", and the inner channel is the scala media (M). The scala vestibuli terminates at the oval window and the scala tympani at the round window. An idealized unrolled cochlea is shown in Figure 2.3(b). There is a small hole at the apex known as the "helicotrema" through which the perilymph fluid can flow.

Input acoustic vibrations result in a piston-like movement of the stapes footplate at the oval window which moves the perilymph fluid within the cochlea. The membrane covering the round window moves to compensate for oval window movements since the perilymph fluid is essentially incompressible. Inward movements of the stapes footplate at the oval window cause the round window to move outwards and outward movements of the stapes footplate cause the round window to move inwards. These movements cause traveling waves to be set up in the scala vestibuli which displace both Reissner's membrane and the basilar membrane.

The basilar membrane is responsible for carrying out a frequency analysis of input sounds. In shape, the basilar membrane is both narrow and thin at the base end of the cochlea, becoming both wider and thicker along its length to the apex, as illustrated in Figure 2.4. The upper part of Figure 2.4 shows the idealized shape of the basilar membrane where it sits along the unrolled cochlea (compare with Figure 2.3(b)), which illustrates that the width and depth of the basilar membrane are narrowest at the base and they increase towards the apex. The basilar membrane vibrates in response to stimulation by signals in the audio frequency range.

Small structures respond better to higher frequencies than do large structures (compare, for example, the sizes of a violin and a double bass or the strings at the treble and bass ends of a piano). The basilar membrane therefore responds best to high frequencies where it is narrow and thin (at the base) and to low frequencies where it is wide and thick (at the apex). Since its thickness and width change gradually along its length, input pure tones at different frequency will produce a maximum basilar membrane movement at different positions or "places" along its length.

This is illustrated in Figure 2.5 for a section of the length of the membrane. This is the basis of the "place" analysis of sound by the hearing system. The extent, or "envelope," of basilar membrane movement is plotted against frequency in an idealized manner for five input pure tones of different frequencies. If the input sound were a complex tone consisting of many components, the overall basilar membrane response is effectively the sum of the responses for each individual component. The basilar membrane is stimulated from the base end (see Figure 2.3) which responds best to high frequencies, and it is important to note that its envelope of movement for a pure tone (or individual component of a complex sound) is not symmetrical, but that it tails off less rapidly towards high frequencies than towards low frequencies.

The movement of the basilar membrane for input sine waves at different frequencies has been observed by a number of researchers following the pioneering work of von Békésy (1960). They have confirmed that the point of maximum displacement along the basilar membrane changes as the frequency of the input is altered. It has also been shown that the linear distance measured from the apex to the point of maximum basilar membrane displacement is directly proportional to the logarithm of the input frequency. The frequency axis in Figure 2.5 is therefore logarithmic. It is illustrated in the figure as being "back-to-front" (i.e., with increasing frequency changing from right to left, low frequency at the apex and high at the base) to maintain the left to right sense of flow of the input acoustic signal and to reinforce understanding of the anatomical nature of the inner ear. The section of the inner ear which is responsible for the analysis of low-frequency sounds is the end farthest away from the oval window, coiled into the center of the cochlear spiral.

In order that the movements of the basilar membrane can be transmitted to the brain for further processing, they have to be converted into nerve firings. This is the function of the organ of corti which consists of a number of hair cells that trigger nerve firings when they are bent. These hair cells are distributed along the basilar membrane and they are bent when it is displaced by input sounds. The nerves from the hair cells form a spiral bundle known as the "auditory nerve". The auditory nerve leaves the cochlea as indicated in Figure 2.1.

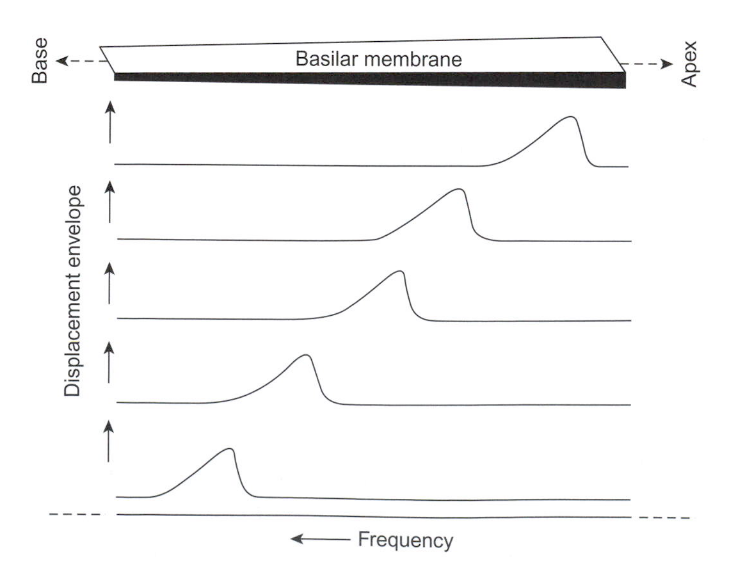

FIGURE 2.5
Idealized envelope of basilar membrane movement to sounds at five different frequencies.

CRITICAL BANDS

Page 25–30 describes how the inner ear carries out a frequency analysis of sound due to the mechanical properties of the basilar membrane and how this provides the basis behind the "place" theory of hearing. The next important aspect of the place theory to consider is how well the hearing system can discriminate between individual frequency components of an input sound. This will provide the basis for understanding the resolution of the hearing system and it will underpin discussions relating to the psychoacoustics of how we hear music, speech and other sounds.

Each component of an input sound will give rise to a displacement of the basilar membrane at a particular place, as illustrated in Figure 2.5. The displacement due to each individual component is spread to some extent on either side of the peak. Whether or not two components that are of similar amplitude and close together in frequency can be discriminated depends on the extent to which the basilar membrane displacements, due to each of the two components, are clearly separated or not.

Suppose two pure tones, or sine waves, with amplitudes A_1 and A_2 and frequencies F_1 and F_2 respectively are sounded together. If F_1 is fixed and F_2 is changed slowly from being equal to or in unison with F_1 either upwards or downwards in frequency, the following is generally heard (see Figure 2.6). When F_1 is equal to F_2, a single note is heard. As soon as F_2 is moved higher or lower than F_1 a sound with clearly undulating amplitude variations known as "beats" is heard. The frequency of the beats is equal to $(F_2 - F_1)$, or $(F_1 - F_2)$ if F_1 is greater than F_2, and the amplitude varies between $(A_1 + A_2)$ and $(A_1 - A_2)$, or $(A_1 + A_2)$ and $(A_2 - A_1)$ if A_2 is greater than A_1. Note that when the amplitudes are equal $(A_1 = A_2)$ the amplitude of the beats varies between $(2 \times A_1)$ and 0.

For the majority of listeners beats are usually heard when the frequency difference between the tones is less than about 12.5 Hz, and the sensation of beats generally gives way to one of a "fused" tone which sounds "rough" when the frequency difference is increased above 15 Hz. As the frequency difference is increased further there is a point where the fused tone gives way to two separate tones but still with the sensation of roughness, and a further increase in frequency difference is needed for the rough sensation to become smooth. The smooth separate sensation persists while the two tones remain within the frequency range of the listener's hearing.

The changes from fused to separate and from beats to rough to smooth are shown hashed in Figure 2.6 to indicate that there is no exact frequency difference at which these changes in perception occur for every listener. However, the approximate frequencies and order in which they occur is common to all listeners, and, in common with most psychoacoustic effects, average values are quoted which are based on measurements made for a large number of listeners.

The point where the two tones are heard as being separate as opposed to fused when the frequency difference is increased can be thought of as the point where two peak displacements on the basilar membrane begin to emerge from a single maximum displacement on the membrane. However, at this point the underlying motion of the membrane, which gives rise to the two peaks, causes them to interfere with each other giving the rough sensation, and it is only when the rough sensation becomes smooth that the separation of the places on the membrane is sufficient to fully resolve the two tones. The frequency difference between the pure tones at the point where a listener's perception changes from rough and separate to smooth and separate is known as the critical bandwidth, and it is therefore

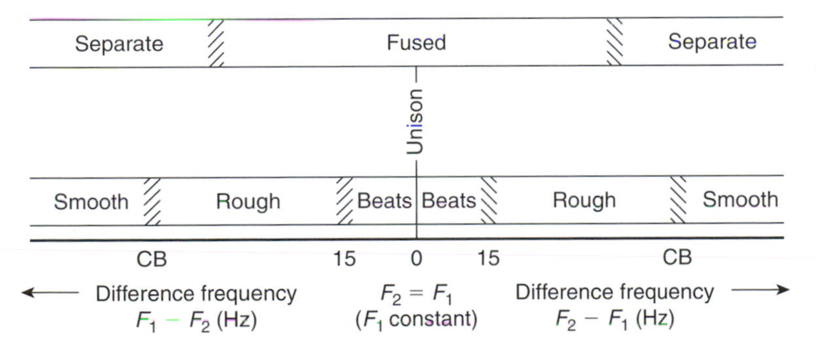

FIGURE 2.6
An illustration of the perceptual changes which occur when a pure tone fixed at frequency F_1 is heard combined with a pure tone of variable frequency F_2.

marked CB in the figure. A more formal definition is given by Scharf (1970), "The critical bandwidth is that bandwidth at which subjective responses rather abruptly change."

In order to make use of the notion of critical bandwidth practically, an equation relating the effective critical bandwidth to the filter center frequency has been proposed by Glasberg and Moore (1990). They define a filter with an ideal rectangular frequency response curve which passes the same power as the auditory filter in question, which is known as the equivalent rectangular bandwidth or ERB. The ERB is a direct measurement of the critical bandwidth, and the Glasberg and Moore equation which allows the calculation of the ERB for any filter center frequency is as follows:

$$\text{ERB} = \{24.7 \times [(4.37 \times f_c) + 1]\}\text{Hz} \tag{2.6}$$

where

f_c = is the filter center frequency in kHz

and ERB = the equivalent rectangular bandwidth in Hz Equation valid for ($100\,\text{Hz} < f_c < 10\,000\,\text{Hz}$)

This relationship is plotted in Figure 2.7 and lines representing where the bandwidth is equivalent to 1, 2, 3, 4, and 5 semitones (or a semitone, whole tone, minor third, major third and perfect fourth respectively) are also plotted for comparison purposes. A third octave filter is often used in the studio as an approximation to the critical bandwidth; this is shown in the figure as the 4 semitone line (there are 12 semitones per octave, so a third of an octave is 4 semitones). A keyboard is shown on the filter center frequency axis for convenience, with middle C marked with a spot.

EXAMPLE 2.2

Calculate the critical bandwidth at 200 Hz and 2000 Hz to three significant figures.

Using Equation 2.6 and substituting 200 Hz and 2000 Hz for f_c (noting that f_c should be expressed in kHz in this equation as 0.2 kHz and 2 kHz respectively) gives the critical bandwidth (ERB) as:

$$\text{ERB at 200Hz} = \{24.7 \times [(4.37 \times 0.2) + 1]\} = 46.3\,\text{Hz}$$

$$\text{ERB at 2000Hz} = \{24.7 \times [(4.37 \times 2) + 1]\} = 241\,\text{Hz}$$

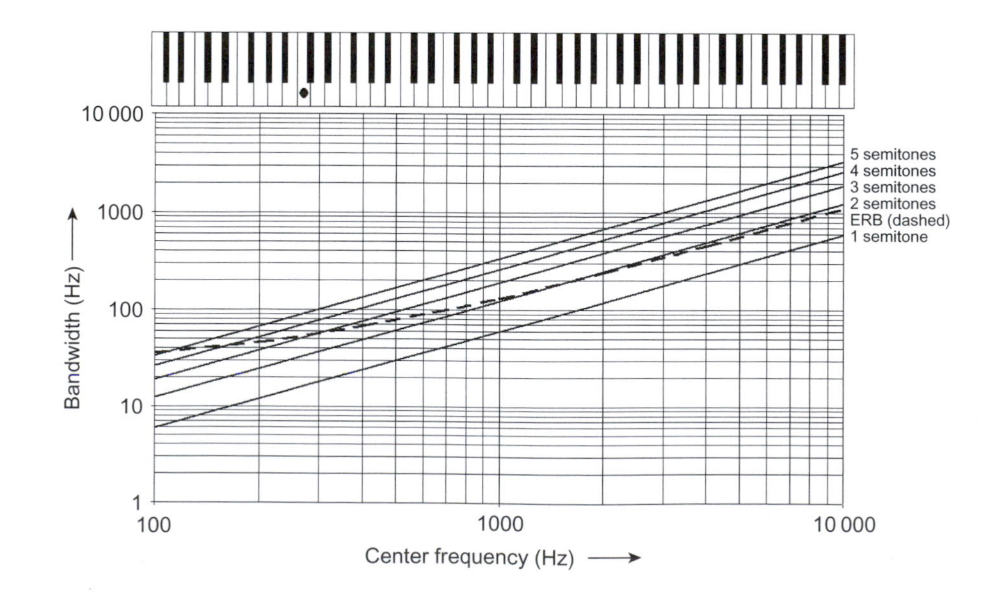

FIGURE 2.7
The variation of equivalent rectangular bandwidth (ERB) with filter center frequency, and lines indicating where the bandwidth would be equivalent to 1, 2, 3, 4 and 5 semitones. (Middle C is marked with a spot on the keyboard.)

The change in critical bandwidth with frequency can be demonstrated if the fixed frequency F_1 in Figure 2.6 is altered to a new value and the new position of CB is found. In practice, critical bandwidth is usually measured by an effect known as masking, in which the "rather abrupt change" is more clearly perceived by listeners.

The response characteristic of an individual filter is illustrated in the bottom curve in Figure 2.8, the vertical axis of which is marked "filter response" (notice that increasing frequency is plotted from right to left in this figure in keeping with Figure 2.5 relating to basilar membrane displacement). The other curves in the figure are idealized envelopes of basilar membrane displacement for pure tone inputs spaced by f Hz, where f is the distance between each vertical line as marked. The filter center frequency F_c Hz is indicated with an unbroken vertical line, which also represents the place on the basilar membrane corresponding to a frequency F_c Hz. The filter response curve is plotted by observing the basilar membrane displacement at the place corresponding to F_c Hz for each input pure tone and plotting this as the filter response at the frequency of the pure tone. This results in the response curve shape illustrated as follows.

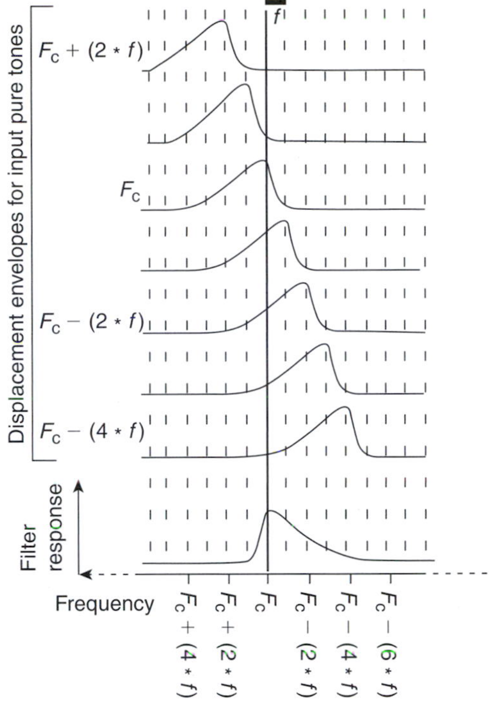

FIGURE 2.8
Derivation of response of an auditory filter with center frequency F_c Hz based on idealized envelope of basilar membrane movement to pure tones with frequencies local to the center frequency of the filter.

As the input pure tone is raised to F_c Hz, the membrane displacement gradually increases with the less steep side of the displacement curve. As the frequency is increased above F_c Hz, the membrane displacement falls rapidly with the steeper side of the displacement curve. This results in the filter response curve as shown, which is an exact mirror image about F_c Hz of the basilar membrane displacement curve. Figure 2.9(a) shows the filter response curve plotted with increasing frequency and plotted more conventionally from left to right.

The action of the basilar membrane can be thought of as being equivalent to a large number of overlapping band-pass filters, or a "bank" of band-pass filters, each responding to a particular band of frequencies. Based on the idealized filter response curve shape in Figure 2.9(a), an illustration of the nature of this bank of filters is given in Figure 2.9(b). Each filter has an asymmetric shape to its response with a steeper roll-off on the high-frequency side than on the low-frequency side; and the bandwidth of a particular filter is given by the critical bandwidth (see Figure 2.7) for any particular center frequency. It is not possible to be particularly exact with regard to the extent to which the filters overlap. A common practical compromise, for example, in studio third octave graphic equalizer filter banks, is to overlap adjacent filters at the −3 dB points on their response curves.

In terms of the perception of two pure tones illustrated in Figure 2.6, the "critical bandwidth" can be thought of as the bandwidth of the band-pass filter in the bank of filters, the center frequencies of which are exactly halfway between the frequencies of the two tones. This ignores the asymmetry of the basilar membrane response (see Figure 2.5) and the consequent asymmetry in the individual filter response curve (see Figure 2.9(a)), but it provides a good working approximation for calculations. Such a filter (and others close to it in center frequency) would capture both tones while they are perceived as "beats," "rough fused" or "rough separate," and at the point where rough changes to smooth, the two tones are too far apart to be both captured by this *or any other* filter. At this point there is no single filter which captures both tones, but there are filters which capture each of the tones individually and they are therefore resolved and the two tones are perceived as being "separate and smooth."

A musical sound can be described by the frequency components which make it up, and an understanding of the application of the critical band mechanism in human hearing in terms of the analysis of the components of musical sounds gives the basis for the study of psychoacoustics. The resolution with which the hearing system can analyze the individual components or sine waves in a sound is important for understanding psychoacoustic discussions relating to, for example, how we perceive:

- melody
- harmony
- chords
- tuning
- intonation

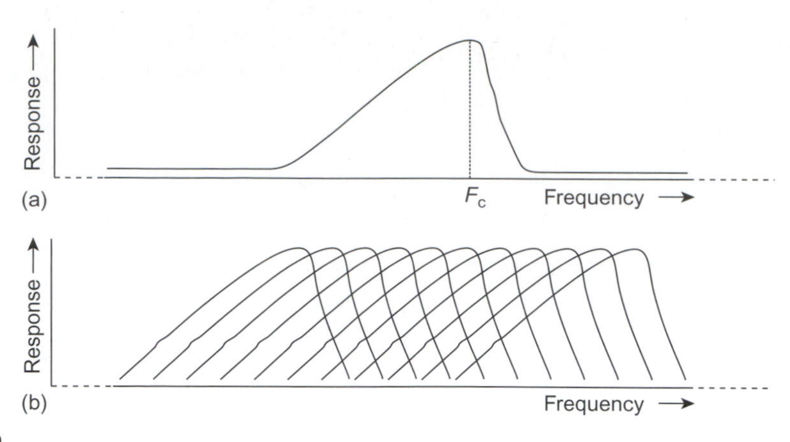

FIGURE 2.9

(a) Idealized response of an auditory filter with center frequency F_c Hz with increasing frequency plotted in the conventional direction (left to right). (b) Idealized bank of band-pass filters model the frequency analysis capability of the basilar membrane.

- musical dynamics
- the sounds of different instruments
- blend
- ensemble
- interactions between sounds produced simultaneously by different instruments.

FREQUENCY AND PRESSURE SENSITIVITY RANGES

The human hearing system is usually quoted as having an average frequency range of 20–20 000 Hz, but there can, however, be quite marked differences between individuals. This frequency range changes as part of the human aging process, particularly in terms of the upper limit which tends to reduce. Healthy young children may have a full hearing frequency range up to 20 000 Hz, but, by the age of 20, the upper limit may have dropped to 16 000 Hz. From the age of 20, it continues to reduce gradually. This is usually known as presbyacusis, or less commonly as presbycusis, and is a function of the normal aging process.

This reduction in the upper frequency limit of the hearing range is accompanied by a decline in hearing sensitivity at all frequencies with age, the decline being less for low frequencies than for high as shown in Figure 2.10. The figure also shows that this natural loss of hearing sensitivity and loss of upper frequencies is more marked for men than for women. Hearing losses can also be induced by other factors such as prolonged exposure to loud sounds, particularly with some of the high sound levels now readily available from electronic amplification systems, whether reproduced via loudspeakers or particularly via headphones.

The ear's sensitivity to sounds of different frequencies varies over a vast sound pressure level range. On average, the minimum sound pressure variation which can be detected by the human hearing system around 4 kHz is approximately 10 micropascals (10 μPa), or 10^{-5} Pa. The maximum average sound pressure level which is heard rather than perceived as being painful is 64 Pa. The ratio between the loudest and softest is therefore:

$$\frac{\text{Threshold of pain}}{\text{Threshold of hearing}} = \frac{64}{10^{-5}} = 6400000 = 6.4 \times 10^{6}$$

This is a very wide range variation in terms of the numbers involved, and it is not a convenient one to work with. Therefore sound pressure level (SPL) is represented in decibels relative to 20 μPa, as dB(SPL) as follows:

$$\text{dB(SPL)} = 20\log\left(\frac{p_{\text{actual}}}{p_{\text{ref}}}\right)$$

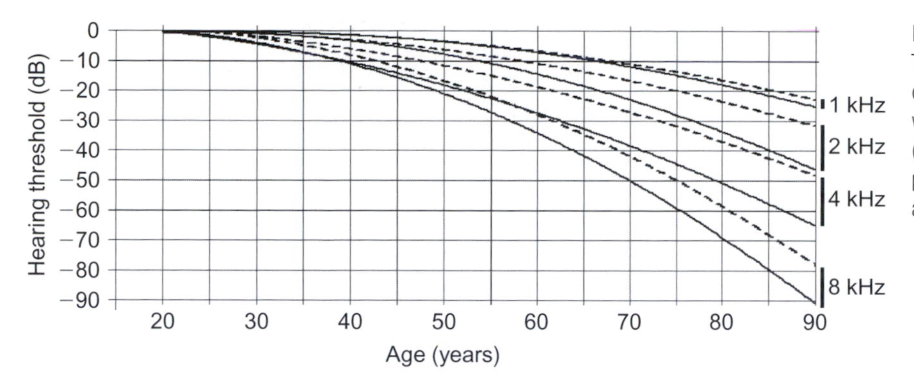

FIGURE 2.10
The average effect of aging on the hearing sensitivity of women (dotted lines) and men (solid lines) usually known as presbyacusis, or less commonly as presbycusis.

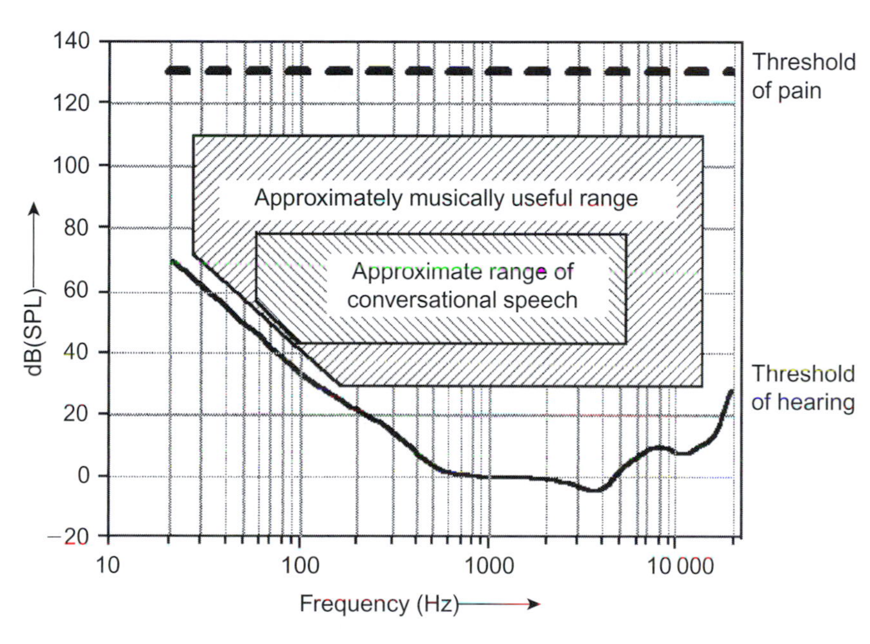

FIGURE 2.11
The general shape of the average human threshold of hearing and the threshold pain with approximate conversational speech and musically useful ranges.

where

p_{actual} = the actual pressure level (in Pa)

and p_{ref} = the reference pressure level (20 μPa)

EXAMPLE 2.3

Calculate the threshold of hearing and threshold of pain in dB(SPL).

The threshold of hearing at 1 kHz is, in fact, p_{ref} which in dB(SPL) equals:

$$20\log\left(\frac{p_{ref}}{p_{ref}}\right) = 20\log\left(\frac{2\times10^{-5}}{2\times10^{-5}}\right) = 20\log(1)$$
$$= 20\times0 = 0\,\text{dB(SPL)}$$

and the threshold of pain is 64 Pa which in dB(SPL) equals:

$$20\log\left(\frac{p_{actual}}{p_{ref}}\right) = 20\log\left(\frac{64}{2\times10^{-5}}\right) = 20\log(6.4\times10^{6})$$
$$= 20\times6.5 = 130\,\text{dB(SPL)}$$

Table 2.1 Typical sound levels in the environment

Example sound/situation	dB(SPL)	Description
Long range gunfire at gunner's ear	140	
Threshold of pain	130	Ouch!
Jet take-off at approximately 100 m	120	
Peak levels on a nightclub dance floor	110	
Loud shout at 1 m	100	Very noisy
Heavy truck at about 10 m	90	
Heavy car traffic at about 10 m	80	
Car interior	70	Noisy
Normal conversation at 1 m	60	
Office noise level	50	
Living room in a quiet area	40	Quiet
Bedroom at night time	30	
Empty concert hall	20	
Gentle breeze through leaves	10	Just audible
Threshold of hearing for a child	0	

Use of the dB(SPL) scale results in a more convenient range of values (0 to 130) to consider, since values in the range of about 0 to 100 are common in everyday dealings. Also, it is a more appropriate basis for expressing acoustic amplitude values, changes in which are primarily perceived as variations in loudness, since loudness perception turns out to be essentially logarithmic in nature.

The threshold of hearing varies with frequency. The ear is far more sensitive in the middle of its frequency range than at the high and low extremes. The lower curve in Figure 2.11 is the general shape of the average threshold of hearing curve for sinusoidal stimuli between 20 Hz and 20 kHz. The upper curve in the figure is the general shape of the threshold of pain, which also varies with frequency but not to such a great extent. It can be seen from the figure that the full 130 dB(SPL) range, or "dynamic range," between the threshold of hearing and the threshold of pain exists at approximately 4 kHz, but that the dynamic range available at lower and higher frequencies is considerably less. For reference, the sound level and frequency range for both average normal conversational speech and music are shown in Figure 2.11, and Table 2.1 shows approximate sound levels of everyday sounds for reference.

LOUDNESS PERCEPTION

Although the perceived loudness of an acoustic sound is related to its amplitude, there is not a simple one-to-one functional relationship. As a psychoacoustic effect it is affected by both the context and nature of the sound. It is also difficult to measure because it is dependent on the interpretation by listeners of what they hear. It is neither ethically appropriate nor technologically possible to put a probe in the brain to ascertain the loudness of a sound.

The concepts of the sound pressure level and sound intensity level have been shown to be approximately equivalent in the case of free space propagation in which no interference effects were present. The ear is a pressure sensitive organ that divides the audio spectrum into a set of overlapping frequency bands whose bandwidth increases with frequency. These are both objective descriptions of the amplitude and the function of the ear. However, they tell us nothing about the perception of loudness in relation to the objective measures of sound amplitude level. Consideration of such issues will allow us to understand some of the effects that occur when one listens to musical sound sources.

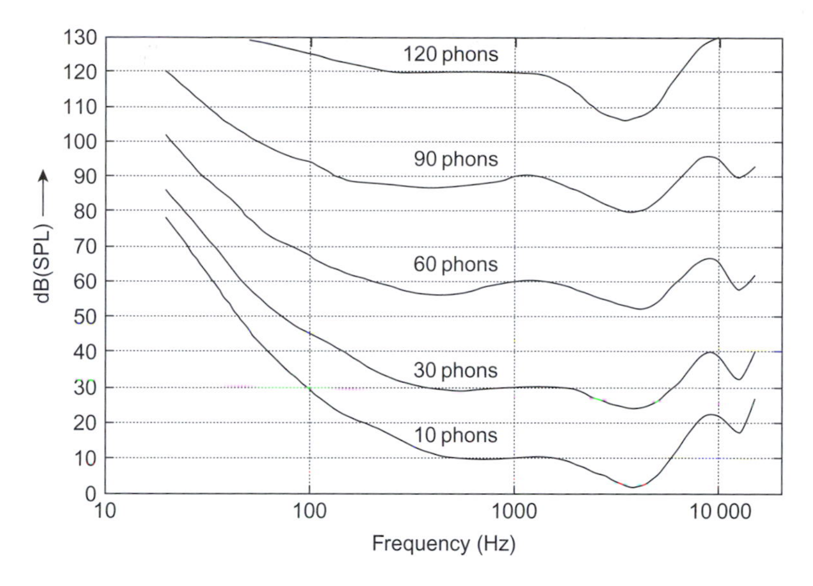

FIGURE 2.12
Equal loudness contours for the human ear.

The pressure amplitude of a sound wave does not directly relate to its perceived loudness. In fact it is possible for a sound wave with a larger pressure amplitude to sound quieter than a sound wave with a lower pressure amplitude. How can this be so? The answer is that the sounds are at different frequencies and the sensitivity of our hearing varies as the frequency varies. Figure 2.12 shows the equal loudness contours for the human ear. These contours, originally measured by Fletcher and Munson (1933) and by others since, represent the relationship between the measured sound pressure level and the perceived loudness of the sound. The curves show how loud a sound must be in terms of the measured sound pressure level to be perceived as being of the same loudness as a 1 kHz tone of a given level. There are two major features to take note of, which are discussed below.

The first is that there are some humps and bumps in the contours above 1 kHz. These are due to the resonances of the outer ear. Within the outer ear there is a tube about 25 mm long with one open and one closed end. This will have a first resonance at about 3.4 kHz and, due to its non-uniform shape, a second resonance at approximately 13 kHz, as shown in the figure. The effect of these resonances is to enhance the sensitivity of the ear around the resonant frequencies. Note that because this enhancement is due to an acoustic effect in the outer ear it is independent of signal level.

The second effect is an amplitude dependence of sensitivity which is due to the way the ear transduces and interprets the sound and, as a result, the frequency response is a function of amplitude. This effect is particularly noticeable at low frequencies but there is also an effect at higher frequencies. The net result of these effects is that the sensitivity of the ear is a function of both frequency and amplitude. In other words the frequency response of the ear is not flat and is also dependent on sound level. Therefore two tones of equal sound pressure level will rarely sound equally loud. For example, a sound at a level which is just audible at 20 Hz would sound much louder if it was at 4 kHz. Tones of different frequencies therefore have to be at different sound pressure levels to sound equally loud and their relative loudness will be also a function of their absolute sound pressure levels.

The loudness of sine wave signals, as a function of frequency and sound pressure levels, is given by the "phon" scale. The phon scale is a subjective scale of loudness based on the judgments of listeners to match the loudness of tones to reference tones at 1 kHz. The curve for *N* phons intersects 1 kHz at *N* dB(SPL) by definition, and it can be seen that the relative shape of the phon curves flattens out at higher sound levels, as shown in Figure 2.12. The relative loudness of different frequencies is not preserved, and therefore the perceived frequency balance of sound varies as the listening level is altered. This is an effect that we have all heard when the volume of a recording is turned down and the bass and treble components appear suppressed relative to the midrange frequencies and the sound becomes "duller" and "thinner." Ideally we should listen to reproduced sound at the level at which it was originally recorded. However, in most cases this would be antisocial, especially as much rock material is mixed at levels in excess of 100 dB(SPL)!

In the early 1970s hi-fi manufacturers provided a "loudness" button which put in bass and treble boost in order to flatten the Fletcher–Munson curves, and so provide a simple compensation for the reduction in hearing sensitivity at low levels. The action of this control was wrong in two important respects:

- Firstly, it directly used the equal loudness contours to perform the compensation, rather than the difference between the curves at two different absolute sound pressure levels, which would be more accurate. The latter approach has been used in professional products to allow nightclubs to achieve the equivalent effect of a louder replay level.
- Secondly, the curves are a measure of the equal loudness for *sine waves* at a similar level. Real music on the other hand consists of many different frequencies at many different amplitudes and does not directly follow these curves as its level changes. We shall see later how we can analyze the loudness of complex sounds. In fact because the response of the ear is dependent on both absolute sound pressure level and frequency, it cannot be compensated for simply by using treble and bass boost.

Measuring loudness

These effects make it difficult to design a meter which will give a reading which truly relates to the perceived loudness of a sound, and an instrument which gives an approximate result is usually used. This is achieved by using the sound pressure level but frequency weighting it to compensate for the variation of sensitivity of the ear as a function of frequency. Clearly the optimum compensation will depend on the absolute value of the sound pressure level being measured and so some form of compromise is necessary.

Figure 2.13 shows two frequency weightings which are commonly used to perform this compensation: termed "A" and "C" weightings. The "A" weighting is most appropriate for low amplitude sounds as it broadly compensates for the low-level sensitivity versus frequency curve of the ear. The "C" weighting on the other hand is more suited to sound at higher absolute sound pressure levels and because of this is more sensitive to low-frequency components than the "A" weighting. The sound levels measured using the "A" weighting are often given the unit dBA and levels using the "C" weighting dBC. Despite the fact that it is most appropriate for low sound levels, and is a reasonably good approximation there, the "A" weighting is now recommended for any sound level, in order to provide a measure of consistency between measurements.

The frequency weighting is not the only factor which must be considered when using a sound level meter. In order to obtain an estimate of the sound pressure level it is necessary to average over at least one cycle, and preferably more, of the sound waveform. Thus most sound level meters have slow and fast time response settings. The slow time response gives an estimate of the average sound level whereas the fast response tracks more rapid variations in the sound pressure level.

Sometimes it is important to be able to calculate an estimate of the equivalent sound level experienced over a period of time. This is especially important when measuring people's noise exposure in order to see if they might suffer noise-induced hearing loss. This cannot be done using the standard fast or slow

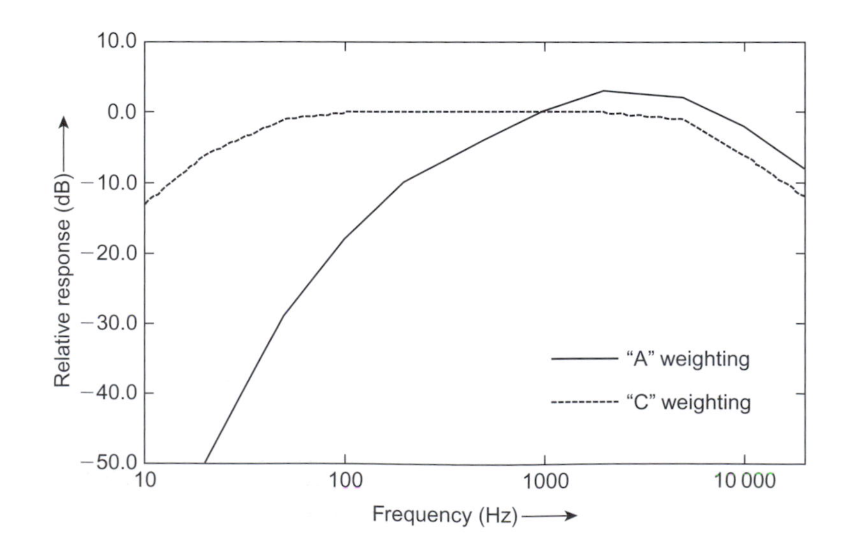

FIGURE 2.13
The frequency response of "A" and "C" weightings.

time responses on a sound level meter; instead a special form of measurement known as the L_{eq} (pronounced L E Q) measurement is used.

This measure integrates the instantaneous squared pressure over some time interval, such as 15 mins or 8 hrs, and then takes the square root of the result. This provides an estimate of the root mean square level of the signal over the time period of the measurement and so gives the equivalent sound level for the time period. That is, the output of the L_{eq} measurement is the constant sound pressure level which is equivalent to the varying sound level over the measurement period. The L_{eq} measurement also provides a means of estimating the total energy in the signal by squaring its output. A series of L_{eq} measurements over short times can also be easily combined to provide a longer time L_{eq} measurement by simply squaring the individual results, adding them together, and then taking the square root of the result, shown in Equation 2.7.

$$L_{eq(total)} = \sqrt{L_{eq1}^2 + L_{eq2}^2 + \ldots + L_{eqn}^2} \tag{2.7}$$

where $L_{eq(1-n)}$ = the individual short time L_{eq} measurements

This extendibility makes the L_{eq} measurement a powerful method of noise monitoring. As with a conventional instrument, the "A" or "C" weightings can be applied.

Loudness of simple sounds

In Figure 2.10 the two limits of loudness are illustrated: the threshold of hearing and the threshold of pain. As we have already seen, the sensitivity of the ear varies with frequency and therefore so does the threshold of hearing, as shown in Figure 2.14. The peak sensitivities shown in this figure are equivalent to a sound pressure amplitude in the sound wave of 10 μPa or about −6 dB(SPL). Note that this is for monaural listening to a sound presented at the front of the listener. For sounds presented on the listening side of the head there is a rise in peak sensitivity of about 6 dB due to the increase in pressure caused by reflection from the head. There is also some evidence that the effect of hearing with two ears is to increase the sensitivity by between 3 and 6 dB.

At 4 kHz, which is about the frequency of the sensitivity peak, the pressure amplitude variations caused by the Brownian motion of air molecules, at room temperature and over a critical bandwidth, correspond to a sound pressure level of about −23 dB. Thus the human hearing system is close to the theoretical physical limits of sensitivity. In other words there would be little point in being much more sensitive to sound, as all we would hear would be a "hiss" due to the thermal agitation of the air! Many studio and concert hall designers now try to design the building such that environmental noises are less than the threshold of hearing, and so are inaudible.

The second limit is the just noticeable change in amplitude. This is strongly dependent on the nature of the signal, its frequency, and its amplitude. For broad-band noise the just noticeable difference in

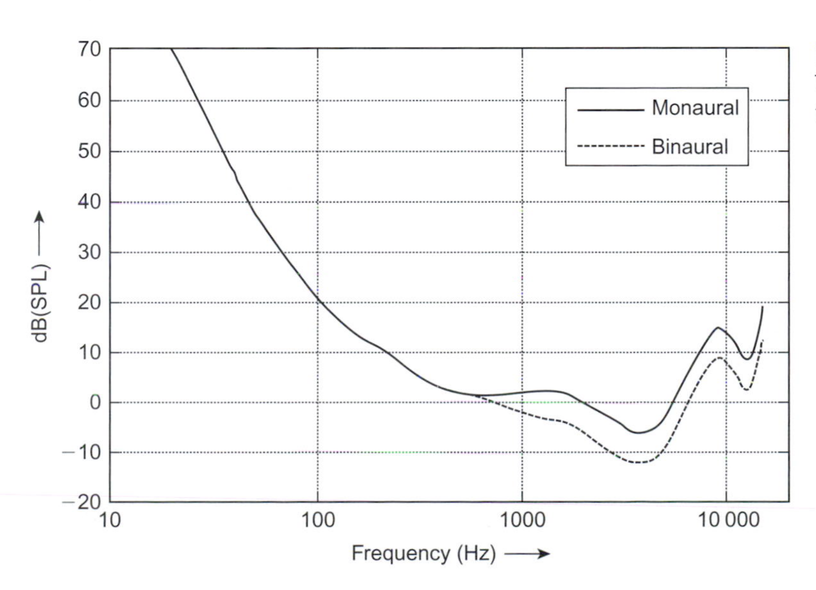

FIGURE 2.14
The threshold of hearing as a function of frequency.

amplitude is 0.5 to 1 dB when the sound level lies between 20 and 100 dB(SPL) relative to a threshold of 0 dB(SPL). Below 20 dB(SPL) the ear is less sensitive to changes in sound level. For pure sine waves, however, the sensitivity to change is markedly different and is a strong function of both amplitude and frequency. For example, at 1 kHz the just noticeable amplitude change varies from 3 dB at 10 dB(SPL) to 0.3 dB at 80 dB(SPL).

This variation occurs at other frequencies as well but in general the just noticeable difference at other frequencies is greater than the values for 1–4 kHz. These different effects make it difficult to judge exactly what difference in amplitude would be noticeable as it is clearly dependent on the precise nature of the sound being listened to. There is some evidence that once more than a few harmonics are present the just noticeable difference is closer to the broad-band case, of 0.5–1 dB, rather than the pure tone case. As a general rule of thumb the just noticeable difference in sound level is about 1 dB.

The mapping of sound pressure change to loudness variation for larger changes is also dependent on the nature of the sound signal. However, for broad-band noise, or sounds with several harmonics, it is generally accepted that a change of about 10 dB in SPL corresponds to a doubling or halving of perceived loudness. However, this scaling factor is dependent on the nature of the sound, and there is some dispute over both its value and its validity.

EXAMPLE 2.4

Calculate the increase in the number of violinists required to double the loudness of a string section, assuming all the violinists play at the same sound level.

The total level from combining several uncorrelated sources is given by:

$$P_{N\,\text{uncorrelated}} = P\sqrt{N}$$

This can be expressed in terms of the SPL as:

$$SPL_{N\,\text{uncorrelated}} = SPL_{\text{single source}} + 10\log_{10}(N)$$

In order to double the loudness we need an increase in SPL of 10 dB. Since 10 log(10) = 10, 10 times the number of sources will raise the SPL by 10 dB.

Therefore we must increase the number of violinists in the string section by a factor of 10 in order to double their volume.

As well as frequency and amplitude, duration also has an effect on the perception of loudness, as shown in Figure 2.15 for a pure tone. Here we can see that once the sound lasts more than about 200 milliseconds then its perceived level does not change. However, when the tone is shorter than this the perceived amplitude reduces. The perceived amplitude is inversely proportional to the length of the tone burst. This means that when we listen to sounds which vary in amplitude the loudness level is not perceived significantly by short amplitude peaks, but more by the sound level averaged over 200 milliseconds.

Loudness of complex sounds

Unlike tones, real sounds occupy more than one frequency. We have already seen that the ear separates sound into frequency bands based on critical bands. The brain seems to treat sounds within a critical band differently to those outside its frequency range and there are consequential effects on the perception of loudness.

The first effect is that the ear seems to lump all the energy within a critical band together and treat it as one item of sound. So when all the sound energy is concentrated within a critical band the loudness is proportional to the total intensity of the sound within the critical band. That is:

$$\text{Loudness} \propto P_1^2 + P_2^2 + \cdots + P_n^2 \tag{2.8}$$

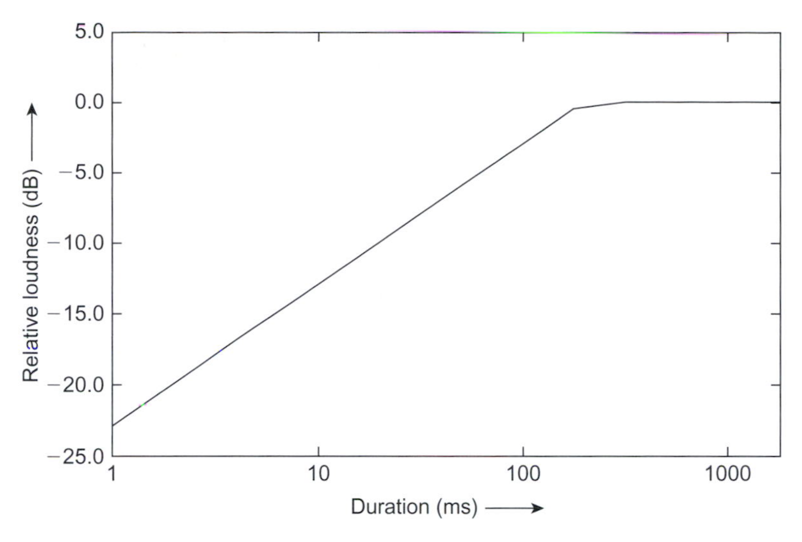

FIGURE 2.15
The effect of tone duration on loudness.

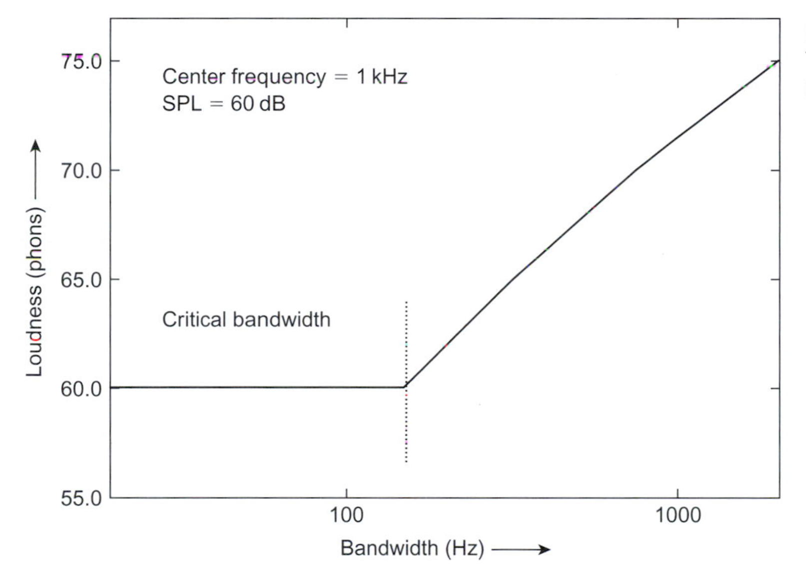

FIGURE 2.16
The effect of tone bandwidth on loudness.

where P_{1-n} = the pressures of the n individual frequency components

As the ear is sensitive to sound pressures, the sound intensity is proportional to the square of the sound pressures. Because the acoustic intensity of the sound is also proportional to the sum of the squared pressure, the loudness of a sound within a critical band is independent of the number of frequency components so long as their total acoustic intensity is constant. When the frequency components of the sound extend beyond a critical band, an additional effect occurs due to the presence of components in other critical bands. In this case more than one critical band is contributing to the perception of loudness and the brain appears to add the individual critical band responses together. The effect is to increase the perceived loudness of the sound even though the total acoustic intensity is unchanged.

Figure 2.16 plots the subjective loudness perception of a sound at a constant intensity level as a function of the sound's bandwidth which shows this effect. In the cochlea the critical bands are determined by the place at which the peak of the standing wave occurs, therefore all energy within a critical band will be integrated as one overall effect at that point on the basilar membrane and transduced into nerve impulses as a unit. On the other hand energy which extends beyond a critical band will cause other nerves to fire and it is these extra nerve firings which give rise to an increase in loudness.

The interpretation of complex sounds which cover the whole frequency range is further complicated by psychological effects, in that a listener will attend to particular parts of the sound, such as the soloist, or conversation, and ignore or be less aware of other sounds, and will tend to base their perception of loudness on what they have attended to.

Duration also has an effect on the perception of the loudness of complex tones in a similar fashion to that of pure tones. As is the case for pure tones, complex tones have an amplitude which is independent of duration once the sound is longer than about 200 milliseconds and is inversely proportionate to duration when the duration is less than this.

NOISE-INDUCED HEARING LOSS

The ear is a sensitive and accurate organ of sound transduction and analysis. However, the ear can be damaged by exposure to excessive levels of sound or noise. This damage can manifest itself in two major forms:

- *A loss of hearing sensitivity*: The effect of noise exposure causes the efficiency of the transduction of sound into nerve impulses to reduce. This is due to damage to the hair cells in each of the organs of corti. Note this is different from the threshold shift due to the acoustic reflex which occurs over a much shorter time period and is a form of built-in hearing protection. This loss of sensitivity manifests itself as a shift in the threshold of hearing of someone who has been exposed to excessive noise, as shown in Figure 2.17. This shift in the threshold can be temporary, for short times of exposures, but ultimately it becomes permanent as the hair cells are permanently flattened as a result of the damage, due to long-term exposure, which does not allow them time to recover.
- *A loss of hearing acuity*: This is a more subtle effect but in many ways is more severe than the first effect. We have seen that a crucial part of our ability to hear and analyze sounds is our ability to separate out the sounds into distinct frequency bands, called critical bands. These bands are very narrow. Their narrowness is due to an active mechanism of positive feedback in the cochlea which enhances the standing wave effects mentioned earlier. This enhancement mechanism is very easily damaged; it appears to be more sensitive to excessive noise than the main transduction system. The effect of the damage though is not just to reduce the threshold but also to increase the bandwidth of our acoustic filters, as shown in idealized form in Figure 2.18. This has two main effects: firstly, our ability to separate out the different components of the sound is impaired, and this will reduce our ability to understand speech or separate out desired sound from competing noise. Interestingly it may well make musical sounds which were consonant more dissonant because of the presence of more than one frequency harmonic in a critical band. The second effect is a reduction in the hearing sensitivity, also shown in Figure 2.18, because the enhancement mechanism also increases the amplitude sensitivity of the ear. This effect is more

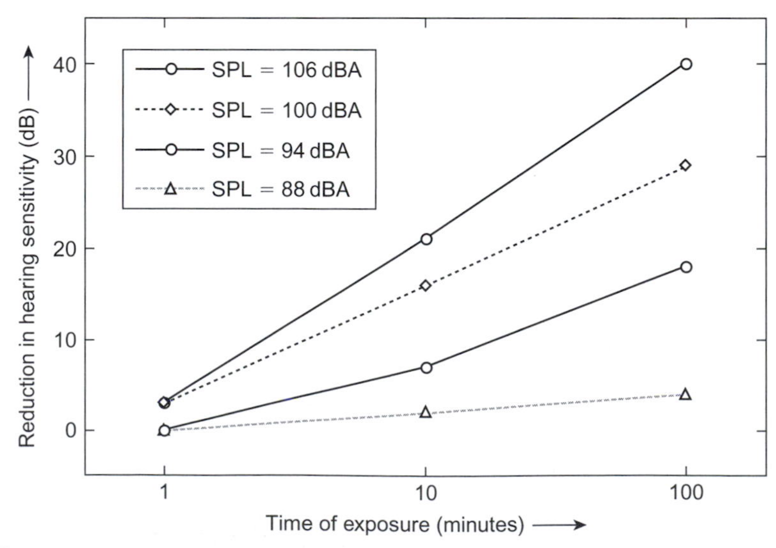

FIGURE 2.17
The effect of noise exposure on hearing sensitivity (data from Tempest, 1985).

insidious because the effect is less easy to measure and perceive; it manifests itself as a difficulty in interpreting sounds rather than a mere reduction in their perceived level.

Another related effect due to damage to the hair cells is noise-induced tinnitus. Tinnitus is the name given to a condition in which the cochlea spontaneously generates noise, which can be tonal, random noises, or a mixture of the two. In noise-induced tinnitus exposure to loud noise triggers this, and as well as being disturbing, there is some evidence that people who suffer from this complaint may be more sensitive to noise-induced hearing damage.

Because the damage is caused by excessive noise exposure it is more likely at the frequencies at which the acoustic level at the ear is enhanced. The ear is most sensitive at the first resonance of the ear canal, or about 4 kHz, and this is the frequency at which most hearing damage first shows up. Hearing damage in this region is usually referred to as an "audiometric notch" because of its distinctive shape on a hearing test, or audiogram (once the results have been plotted following a hearing test), see Figure 2.19. This distinctive pattern is evidence that the hearing loss measured is due to noise exposure rather than some other condition, such as the inevitable high-frequency loss due to aging.

How much noise exposure is acceptable? There is some evidence that levels of noise generated by our normal noisy Western society has some long-term effects because measurements on the hearing of other cultures have a much lower threshold of hearing at a given age compared with that of Westerners. However, this may be due to other factors as well, for example the level of pollution. But strong evidence exists demonstrating that exposure to noises with amplitudes of greater than 90 dBA can cause permanent

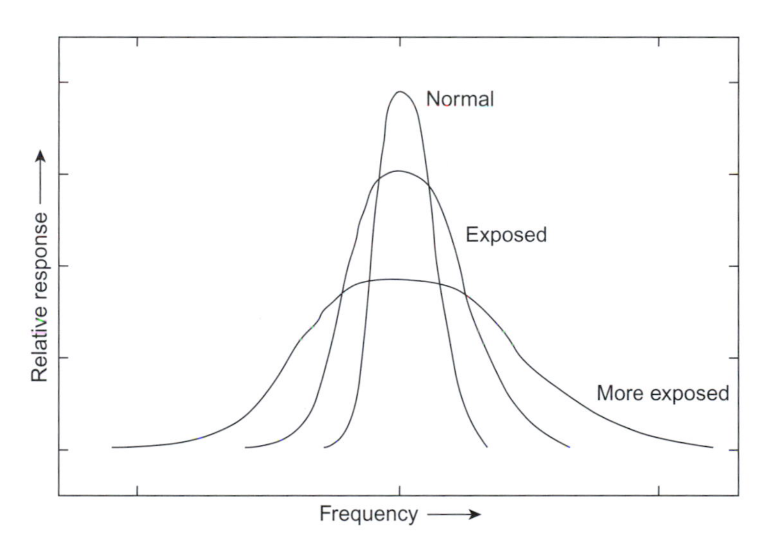

FIGURE 2.18
Idealized form of the effect of noise exposure on hearing bandwidth.

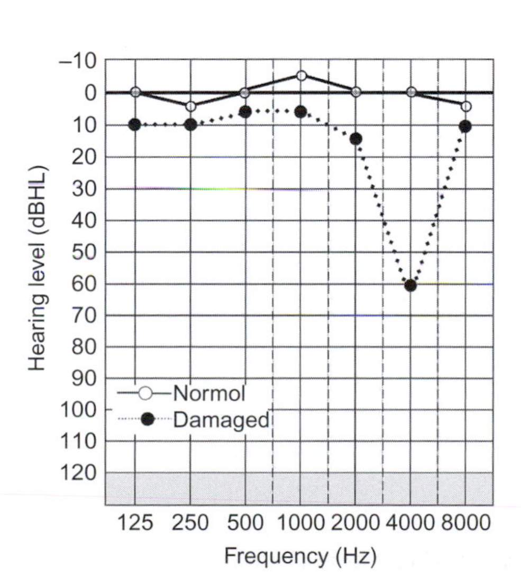

FIGURE 2.19
Example audiograms of normal and damaged (notched) hearing.

hearing damage. This fact is recognized, for example, by UK legislation, which requires that the noise exposure of workers be less than this limit, known as the second action limit. This level has been reduced from April 2006 under European legislation to 85 dBA. If the work environment has a noise level greater than the second action limit, then employers are obliged to provide hearing protection for employees of a sufficient standard to bring the noise level at the ear below this figure. There is also a "first action level," which is 5 dB below the second action level, and if employees are subjected to this level (now 80 dBA in Europe), then employees can request hearing protection, which must be made available.

Please be aware that the regulations vary by country. Readers should check their local regulations for noise exposure in practice.

Integrated noise dose

However, in many musical situations the noise level is greater than 90 dBA for short periods. For example, the audience at a concert may well experience brief peaks above this, especially at particular instants in works such as Elgar's *Dreams of Gerontius*, or Orff's *Carmina Burana*. Also, in many practical industrial and social situations the noise level may be louder than the second action level of 85 dBA, in Europe for only part of the time. How can we relate intermittent periods of noise exposure to continuous noise exposure? For example, how damaging is a short exposure to a sound of 96 dBA? The answer is to use a similar technique to that used in assessing the effect of radiation exposure, that of "integrated dose."

The integrated noise dose is defined as the equivalent level of the sound over a fixed period of time, which is currently 8 hours. In other words the noise exposure can be greater than the second action level providing that it is for an appropriately shorter time, which results in a noise dose that is less than that which would result from being exposed to noise at the second action level for 8 hours. The measure used is the L_{eq} mentioned earlier and the maximum dose is 90 dBL_{eq} over 8 hours. This means that one can be exposed to 88 dBA for 4 hours, 91 dBA for 2 hours, and so on.

Figure 2.20 shows how the time of exposure varies with the sound level on linear and logarithmic time scales for the second action level in Europe. It can be seen that exposure to extreme sound levels, greater than 100 dBA, can only be tolerated for a very short period of time, less than half an hour. There is also a limit to how far this concept can be taken because very loud sounds can rupture the eardrum causing instant, and sometimes permanent, loss of hearing.

This approach to measuring the noise dose takes no account of the spectrum of the sound which is causing the noise exposure, because to do so would be difficult in practice. However, it is obvious that the effect of a pure tone at 85 dBA on the ear is going to be different to the same level spread over the full frequency range. In the former situation there will be a large amount of energy concentrated at a particular point on the basilar membrane and this is likely to be more damaging than the second case in which the energy will be spread out over the full length of the membrane. Note that the specification for noise dose uses "A" weighting for the measurement which, although it is more appropriate for low rather than high sound levels, weights the sensitive 4 kHz region more strongly.

Protecting your hearing

Hearing loss is insidious and permanent, and by the time it is measurable it is too late. Therefore in order to protect hearing sensitivity and acuity one must be proactive. The first strategy is to avoid exposure to excess noises. Although 85 dB(SPL) is taken as a damage threshold if the noise exposure causes ringing in the ears, especially if the ringing lasts longer than the length of exposure, it may be that damage may be occurring even if the sound level is less than 85 dB(SPL).

There are a few situations where potential damage is more likely:

- The first is when listening to recorded music over headphones, as even small ones are capable of producing damaging sound levels.
- The second is when one is playing music, with either acoustic or electric instruments, as these are also capable of producing damaging sound levels, especially in small rooms with a "live" acoustic.

In both cases the levels are under your control and so can be reduced. However, there is an effect called the acoustic reflex, which reduces the sensitivity of your hearing when loud sounds occur. This effect, combined with the effects of temporary threshold shifts, can result in a sound level increase spiral,

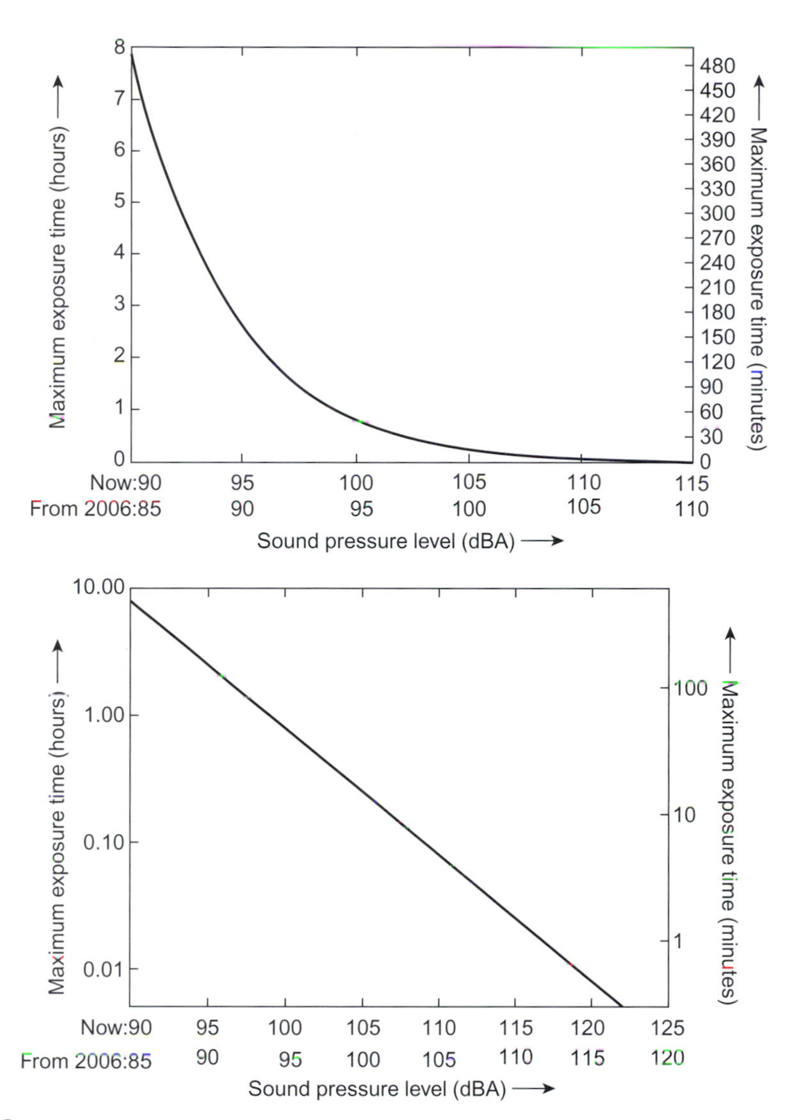

FIGURE 2.20
Maximum exposure time as a function of sound level plotted on a linear scale (upper) and a logarithmic scale (lower).

where there is a tendency to increase the sound level "to hear it better" that results in further dulling, etc. The only real solution is to avoid the loud sounds in the first place. However, if this situation does occur then a rest away from the excessive noise will allow some sensitivity to return.

There are sound sources over which one has no control, such as bands, discos, nightclubs, and power tools. In these situations it is a good idea either to limit the noise dose or, better still, use some hearing protection. For example, one can keep a reasonable distance away from the speakers at a concert or disco. It takes a few days, or even weeks in the case of hearing acuity, to recover from a large noise dose so one should avoid going to a loud concert, or nightclub, every day of the week!

The authors regularly use small "in-ear" hearing protectors when they know they are going to be exposed to high sound levels, and many professional sound engineers also do the same. These have the advantage of being unobtrusive and reduce the sound level by a modest, but useful, amount (15–20 dB) while still allowing conversation to take place at the speech levels required to compete with the noise! These devices are also available with a "flat" attenuation characteristic with frequency and so do not alter the sound balance too much, and cost less than a CD recording. For very loud sounds, such as power tools, then a more extreme form of hearing protection may be required, such as headphone-style ear defenders.

Your hearing is essential, and irreplaceable, for the enjoyment of music, for communicating, and for socializing with other people. Now and in the future, it is worth taking care of.

PERCEPTION OF SOUND SOURCE DIRECTION

How do we perceive the direction that a sound arrives from? The answer is that we make use of our two ears, but how? Because our two ears are separated by our head, this has an acoustic effect which is a function of the direction of the sound. There are two effects of the separation of our ears on the sound wave: firstly, the sounds arrive at different times and secondly, they have different intensities. These two effects are quite different so let us consider them in turn.

Interaural time difference (ITD)

Consider the model of the head, shown in Figure 2.21, which shows the ears relative to different sound directions in the horizontal plane. Because the ears are separated by about 18 cm there will be a time difference between the sound arriving at the ear nearest the source and the one further away. So when the sound is off to the left the left ear will receive the sound first, and when it is off to the right the right ear will hear it first. If the sound is directly in front, or behind, or anywhere on the median plane, the sound will arrive at both ears simultaneously. The time difference between the two ears will depend on the difference in the lengths that the two sounds have to travel. A simplistic view might just allow for the fact that the ears are separated by a distance *d* and calculate the effect of angle on the relative time difference by considering the extra length introduced purely due to the angle of incidence, as shown in Figure 2.22. This assumption will give the following equation for the time difference due to sound angle:

$$\Delta t = \frac{d\sin(\theta)}{c}$$

where

Δt = the time difference between the ears (in s)

d = the distance between the ears (in m)

θ = the angle of arrival of the sound from the median (in radians)

and c = the speed of sound (in ms^{-1})

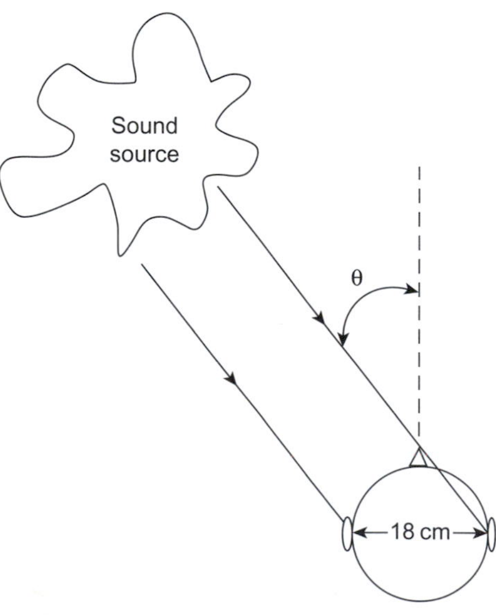

FIGURE 2.21
The effect of the direction of a sound source with respect to the head.

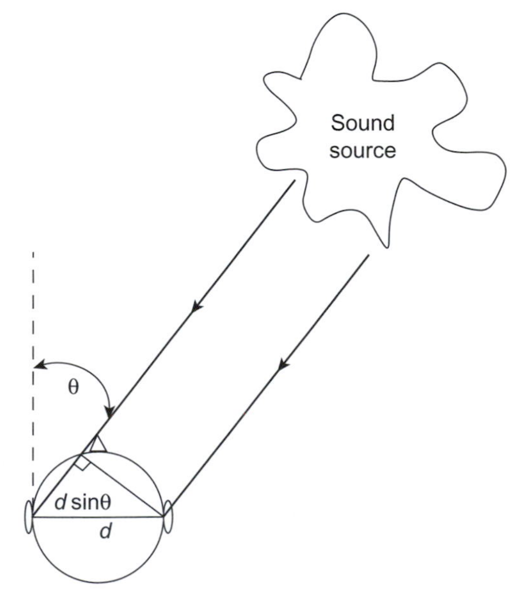

FIGURE 2.22
A simple model for the interaural time difference.

Unfortunately this equation is wrong. It underestimates the delay between the ears because it ignores the fact that the sound must travel around the head in order to get to them. This adds an additional delay to the sound, as shown in Figure 2.23. This additional delay can be calculated, providing one assumes that the head is spherical, by recognizing that the distance traveled around the head for a given angle of incidence is given by:

$$\Delta d = r\theta$$

where

Δd = the extra path round the head at a given angle of incidence (in m)

and r = half the distance between the ears (in m)

This equation can be used in conjunction with the extra path length due to the angle of incidence, which is now a function of r, as shown in Figure 2.24, to give a more accurate equation for the ITD as:

$$ITD = \frac{r(\theta + \sin(\theta))}{c} \tag{2.9}$$

Using this equation we can find that the maximum ITD, which occurs at 90° or ($\pi/2$ radians), is:

$$ITD_{max} = \frac{0.09\,\text{m} \times (\pi/2 + \sin(\pi/2))}{344\,\text{ms}^{-1}} = 6.73 \times 10^{-4}\,\text{s}(673\,\mu\text{s})$$

This is a very small delay but a variation from this to zero determines the direction of sounds at low frequencies. Figure 2.25 shows how this delay varies as a function of angle, where positive delay corresponds to a source at the right of the median plane and negative delay corresponds to a source on the left. Note that there is no difference in the delay between front and back positions at the same angle. This means that we must use different mechanisms and strategies to differentiate between front and back sounds. There is also a frequency limit to the way in which sound direction can be resolved by the

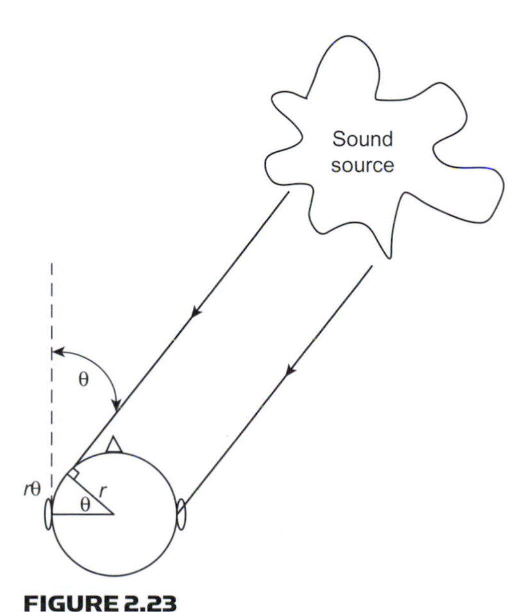

FIGURE 2.23
The effect of the path length around the head on the interaural time difference.

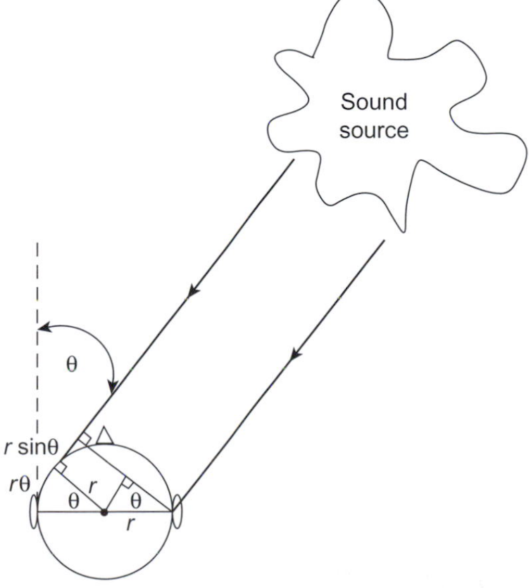

FIGURE 2.24
A better model for the interaural time difference.

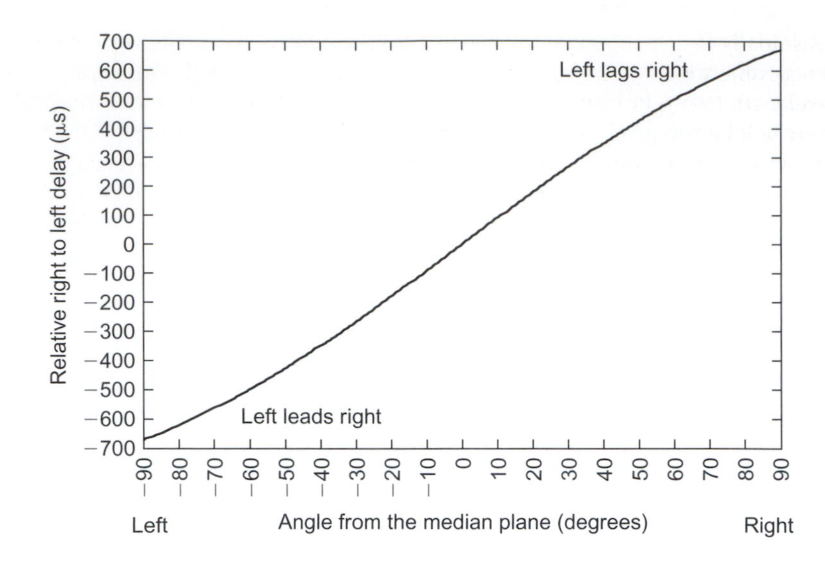

FIGURE 2.25
The interaural time difference (ITD) as a function of angle.

ear in this way. This is due to the fact that the ear appears to use the phase shift in the wave caused by the interaural time difference to resolve the direction; that is, the ear measures the phase shift given by:

$$\Phi_{\text{ITD}} = 2\pi fr(\theta + \sin(\theta))$$

where

Φ_{ITD} = the phase difference between the ears (in radians)

and f = the frequency (in Hz)

When this phase shift is greater than π radians (180°) there will be an unresolvable ambiguity in the direction because there are two possible angles—one to the left and one to the right—that could cause such a phase shift. This sets a maximum frequency, at a particular angle, for this method of sound localization, which is given by:

$$f_{\max}(\theta) = \frac{1}{2 \times 0.09\text{m} \times (\theta + \sin(\theta))}$$

which for an angle of 90° is:

$$f_{\max}(\theta = \pi/2) = \frac{1}{2 \times 0.09\text{m} \times (\pi/2 + \sin(\pi/2))} = 743\,\text{Hz}$$

Thus for sounds at 90° the maximum frequency that can have its direction determined by phase is 743 Hz. However, the ambiguous frequency limit would be higher at smaller angles.

Interaural intensity difference (IID)

The other cue that is used to detect the direction of the sound is the differing levels of intensity that result at each ear due to the shading effect of the head. This effect is shown in Figure 2.26 which shows that the levels at each ear are equal when the sound source is on the median plane but that the level at one ear progressively reduces, and increases at the other, as the source moves away from the median plane. The level reduces in the ear that is furthest away from the source. The effect of the shading of the head is harder to calculate but experiments seem to indicate that the intensity ratio between the two ears varies sinusoidally from 0 dB up to 20 dB, depending on the sound direction, as shown in Figure 2.27. However, an object is not significant as a scatterer or shader of sound until its size is about two thirds of a wavelength (½λ), although it will be starting to scatter an octave below that frequency. This means that there will be a minimum frequency below which the effect of intensity is less useful for localization which will correspond to when the head is about one third of a wavelength in size ($\frac{1}{3}\lambda$). For a head the diameter of which is 18 cm, this corresponds to a minimum frequency of:

$$f_{\min(\theta=\pi/2)} = \frac{1}{3}\left(\frac{c}{d}\right) = \frac{1}{3}\times\left(\frac{344\,\text{ms}^{-1}}{0.18\,\text{m}}\right) = 637\,\text{Hz}$$

Thus the interaural intensity difference is a cue for direction at high frequencies whereas the interaural time difference is a cue for direction at low frequencies. Note that the cross-over between the two techniques starts at about 700 Hz and would be complete at about four times this frequency at 2.8 kHz.

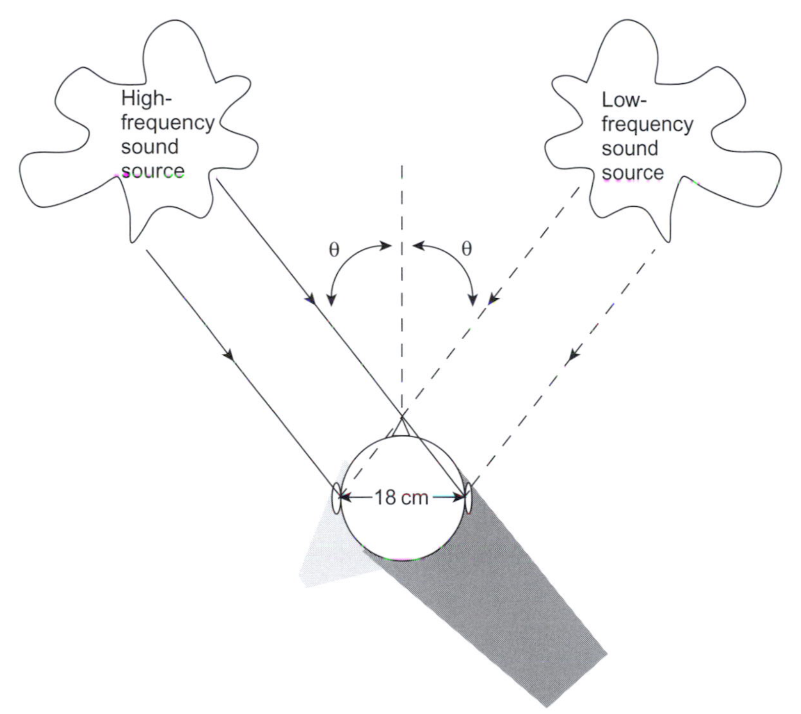

FIGURE 2.26
The effect of the head on the interaural intensity difference.

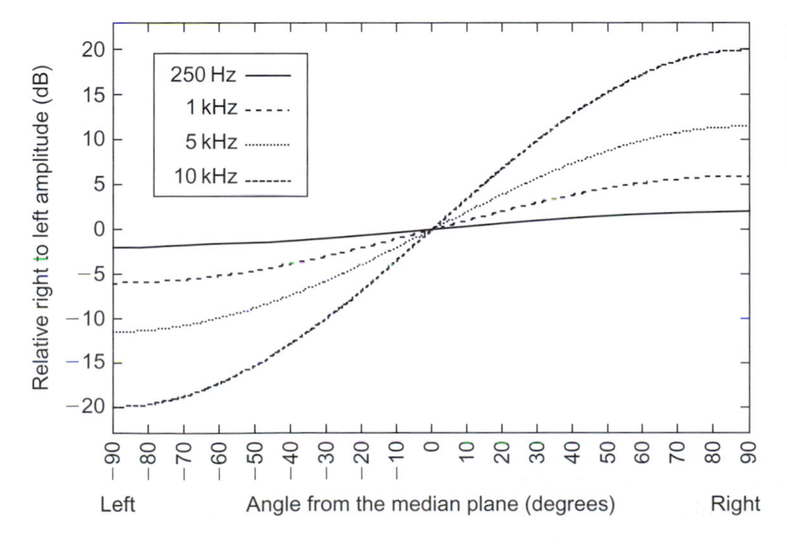

FIGURE 2.27
The interaural intensity difference (IID) as a function of angle and frequency (data from Gulick, 1971).

In between these two frequencies the ability of our ears to resolve direction is not as good as at other frequencies.

Pinnae and head movement effects

The above models of directional hearing do not explain how we can resolve front to back ambiguities or the elevation of the source. There are in fact two ways which are used by the human being to perform these tasks.

The first is to use the effect of our ears on the sounds we receive to resolve the angle and direction of the sound. This is due to the fact that sounds striking the pinnae are reflected into the ear canal by the complex set of ridges that exist on the ear. These pinnae reflections will be delayed, by a very small but significant amount, and so will form comb filter interference effects on the sound the ear receives. The delay that a sound wave experiences will be a function of its direction of arrival, in all three dimensions, and we can use these cues to help resolve the ambiguities in direction that are not resolved by the main directional hearing mechanism. The delays are very small and so these effects occur at high audio frequencies, typically above 5 kHz. The effect is also person specific, as we all have differently shaped ears and learn these cues as we grow up. Thus we get confused for a while when we change our acoustic head shape radically, for example, by cutting very long hair short. We also find that if we hear sound recorded through other people's ears we may have a different ability to localize the sound, because the interference patterns are not the same as those for our ears. In fact, sometimes this localization capability is worse than when using our own ears and sometimes it is better.

The second, and powerful, means of resolving directional ambiguities is to move our heads. When we hear a sound that we wish to attend to, or whose direction we wish to resolve, we turn our head towards the sound and may even attempt to place it in front of us in the normal direction, where all the delays and intensities will be the same. The act of moving our head will change the direction of the sound arrival, and this change of direction will depend on the sound source position relative to us. Thus a sound from the rear will move in a different direction compared with a sound in front of or above the listener. This movement cue is one of the reasons that we perceive the sound from headphones as being "in the head." Because the sound source tracks our head movement it cannot be outside and hence must be in the head. There is also an effect due to the fact that the headphones also do not model the effect of the head. Experiments with headphone listening which correctly model the head and keep the source direction constant as the head moves give a much more convincing illusion.

ITD and IID trading

Because both intensity and delay cues are used for the perception of sound source direction, one might expect the mechanisms to be in similar areas of the brain and linked together. If this were the case one might also reasonably expect that there was some overlap in the way the cues were interpreted such that intensity might be confused with delay and vice versa in the brain. This allows for the possibility that the effect of one cue, for example delay, could be canceled out by the other, for example intensity. This effect does in fact happen and is known as "interaural time difference versus interaural intensity difference trading." In effect, within limits, an interaural time delay can be compensated for by an appropriate interaural intensity difference, as shown in Figure 2.28, which has several interesting features.

Firstly, as expected, time delay versus intensity trading is only effective over the range of delay times which correspond to the maximum interaural time delay of 673 μs. Beyond this amount of delay small intensity differences will not alter the perceived direction of the image. Instead the sound will appear to come from the source which arrives first. This effect occurs between 673 μs and 30 ms. However, if the delayed sound's amplitude is more than 12 dB greater than the first arrival then we will perceive the direction of the sound to be towards the delayed sound. After 30 ms the delayed signal is perceived as an echo and so the listener will be able to differentiate between the delayed and undelayed sound. The implications of these results are twofold: firstly, it should be possible to provide directional information purely through either only delay cues or only intensity cues. Secondly, once a sound is delayed by greater than about 700 μs the ear attends to the sound that arrives first almost irrespective of their relative levels, although clearly if the earlier arriving sound is significantly lower in amplitude, compared with the delayed sound, then the effect will disappear.

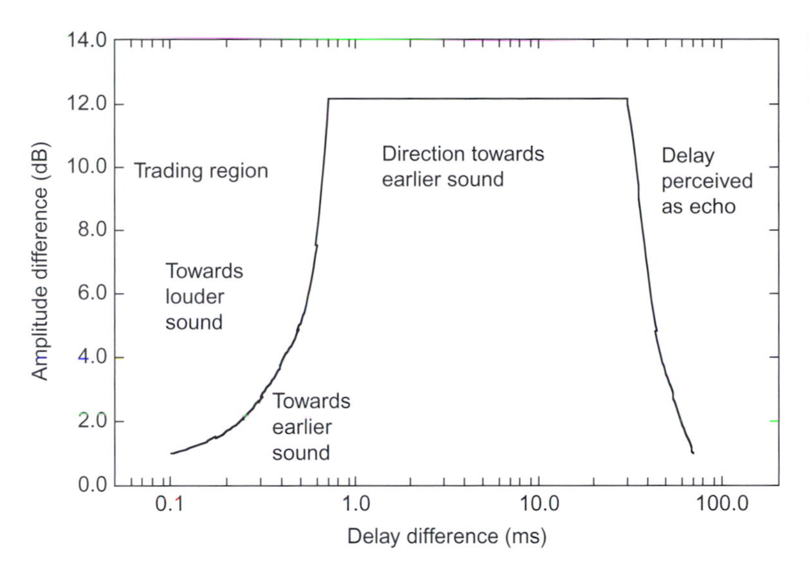

FIGURE 2.28
Delay versus intensity trading (data from Madsen, 1990)

The Haas effect

The second of the ITD and IID trading effects is also known as the "Haas", or "precedence", effect, named after the experimenter who quantified this behavior of our ears. The effect can be summarized as follows:

- The ear will attend to the direction of the sound that arrives first and will not attend to the reflections providing they arrive within 30 ms of the first sound.
- The reflections arriving before 30 ms are fused into the perception of the first arrival. However, if they arrive after 30 ms they will be perceived as echoes.

These results have important implications for studios, concert halls and sound reinforcement systems. In essence it is important to ensure that the first reflections arrive at the audience earlier than 30 ms to avoid them being perceived as echoes. In fact it seems that our preference is for a delay gap of less than 20 ms if the sound of the hall is to be classed as "intimate." In sound reinforcement systems the output of the speakers will often be delayed with respect to their acoustic sound but, because of this effect, we perceive the sound as coming from the acoustic source, unless the level of sound from the speakers is very high.

Stereophonic listening

Because of the way we perceive directional sound it is possible to fool the ear into perceiving a directional sound through just two loudspeakers or a pair of headphones in stereo listening. This can be achieved in basically three ways: two using loudspeakers and one using headphones. The first two ways are based on the concept of providing only one of the two major directional cues in the hearing system; that is using either intensity or delay cues and relying on the effect of the ear's time–intensity trading mechanisms to fill in the gaps. The two systems are as follows:

- *Delay stereo*: This system is shown in Figure 2.29 and consists of two omni-directional microphones spaced a reasonable distance apart and away from the performers. Because of the distance of the microphones a change in performer position does not alter the sound intensity much, but does alter the delay. So the two channels when presented over loudspeakers contain predominantly directional cues based on delay to the listener.
- *Intensity stereo*: This system is shown in Figure 2.30 and consists of two directional microphones placed together and pointing at the left and right extent of the performers' positions. Because the microphones are closely spaced, a change in performer position does not alter the delay between the two sounds. However, because the microphones are directional the intensity received by the two microphones does vary. So the two channels when presented over loudspeakers contain predominantly directional cues based on intensity to the listener. Intensity stereo is the method that is mostly used in pop music production as the pan-pots on a mixing desk, which determine the

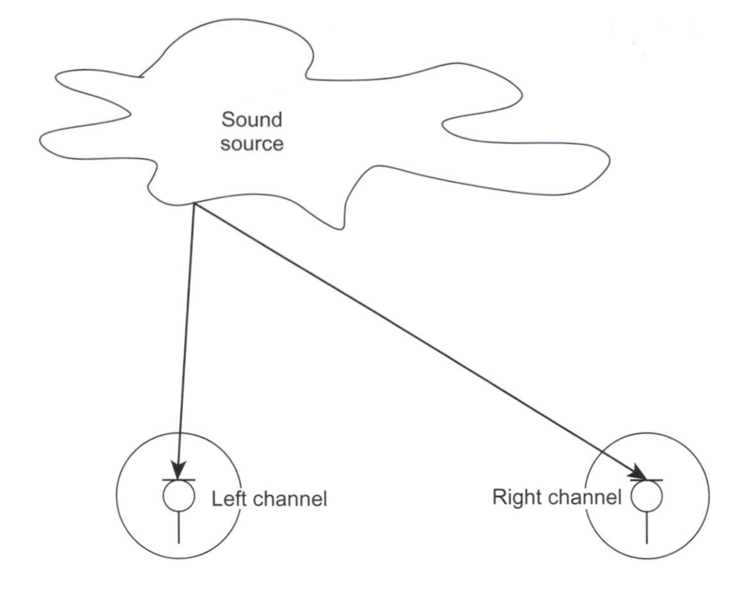

FIGURE 2.29
Delay stereo recording.

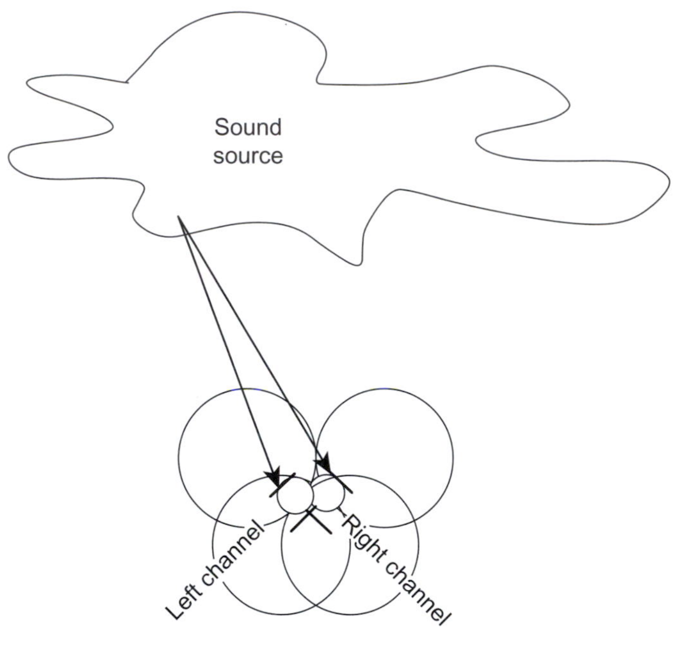

FIGURE 2.30
Intensity stereo recording.

position of a track in the stereo image, vary the relative intensities of the two channels, as shown in Figure 2.31.

These two methods differ primarily in the method used to record the original performance and are independent of the listening arrangement, so which method is used is determined by the producer or engineer on the recording. It is also possible to mix the two cues by using different types of microphone arrangement—for example, slightly spaced directional microphones—and these can give stereo based on both cues. Unfortunately they also provide spurious cues, which confuse the ear, and getting the balance between the spurious and wanted cues, and so providing a good directional illusion, is difficult.

- *Binaural stereo*: The third major way of providing a directional illusion is to use binaural stereo techniques. This system is shown in Figure 2.32 and consists of two omni-directional microphones placed on a head—real or more usually artificial—and presenting the result over headphones. The distance of the microphones is identical to the ear spacing, and they are placed on an object which shades the sound in the same way as a human head, and possibly torso. This means that a change in performer position provides both intensity and delay cues to the listener; the results can be very effective. However, they must be presented over headphones because any

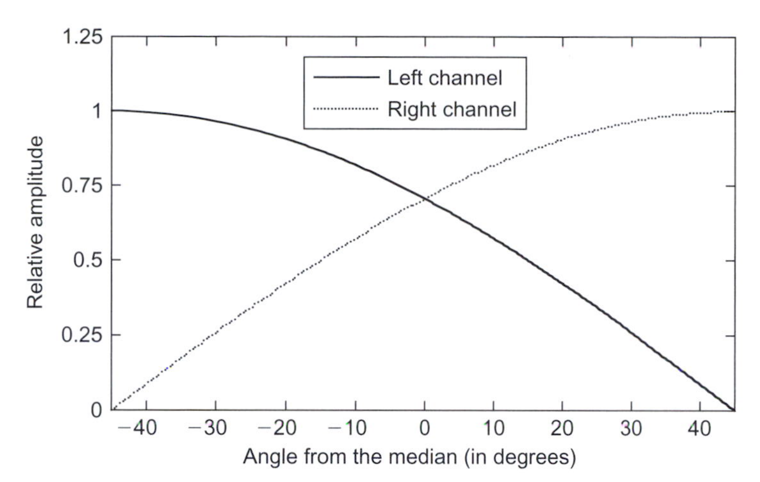

FIGURE 2.31
The effect of the "pan-pots" in a mixing desk on intensity of the two channels.

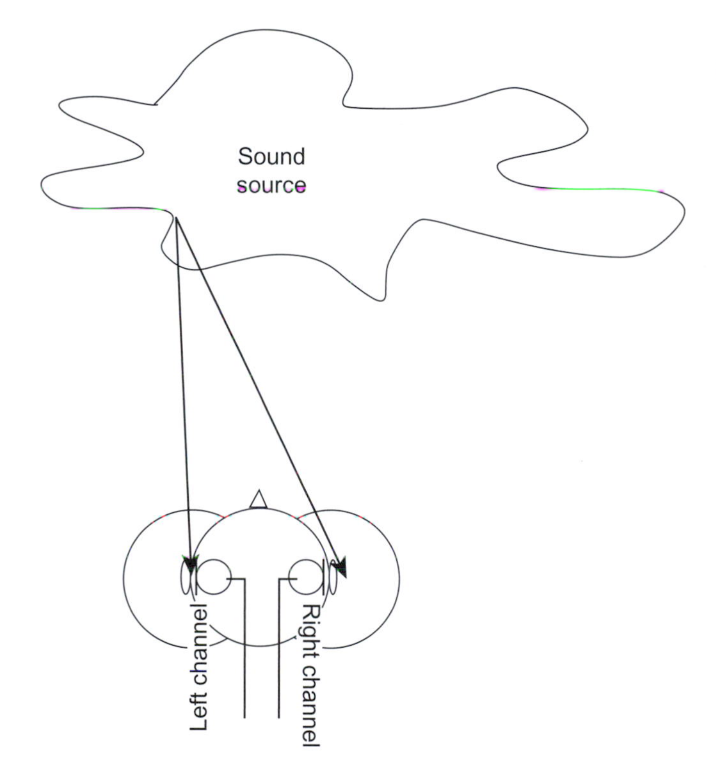

FIGURE 2.32
Binaural stereo recording.

cross-ear coupling of the two channels, as would happen with loudspeaker reproduction, would cause spurious cues and so destroy the illusion. Note that this effect happens in reverse when listening to loudspeaker stereo over headphones, because the cross coupling that normally exists in a loudspeaker presentation no longer exists. This is another reason that the sound is always "in the head" when listening via conventional headphones.

The main compromise in stereo sound reproduction is the presence of spurious direction cues in the listening environment because the loudspeakers and environment will all contribute cues about their position in the room, which have nothing to do with the original recording. More information about directional hearing and stereophonic listening can be found in Blauert (1997) and Rumsey (2001).

REFERENCES

Blauert, J., 1997. Spatial hearing: The psychophysics of Human Sound Location, Revised Edition, Cambridge, MIT press.

Fletcher, H. Munson, W., 1933. Loudness, its measurement and calculation. Journal of the Acoustical Society of America, 5, 82–108.

Glasberg, B.R., Moore, B.C.J., 1990. Derivation of auditory filter shapes from notched-noise data. Hearing Research 47, 103–138.

Gulick, L.W., 1971. Hearing: Physiology and Psychophysics, Oxford, Oxford University Press, pp. 188–189.

Madsen, E.R., 1990. In: Rossing, T.D., (Ed.) The Science of Sound, Second Edition, Addison Wesley, p. 500.

Pickles, J.O., 1982. An Introduction to the Physiology of Hearing, London, Academic Press.

Rumsey, F., 2001. Spatial Audio, Oxford, Focal Press.

Scharf, B., 1970. Critical bands, In: Tobias, J.V., (Ed.) Foundations of Modern Auditory Theory, Volume 1, London, Academic Press, pp. 159–202.

Tempest, W., 1985. The Noise Handbook, Oxford: Academic Press.

von Békésy, G., 1960. Experiments in Hearing, New York: McGraw-Hill.

PART II

Microphones

PREFACE

The first chapter in the microphone section deals with the basic ways in which various types of sound field are propagated and how they interact with a microphone.

The very basis of the acoustical part of audio engineering is the behavior of air. This is the medium in which we live and breathe, and whose pressure fluctuations we are so good at detecting with ears honed by long ages of natural selection.

John Eargle begins with the basics of frequency, wavelength, and the speed of sound (which we note varies somewhat with temperature). He moves quickly on to acoustical power, sound pressure and decibels, and shows how the inverse square law determines the way that intensity falls off with distance in a free sound field, and the relationship between air particle velocity and amplitude.

A free sound field (i.e., one with no reflections) is approximated by taking your microphone out into the great outdoors, but in most circumstances a microphone is instead working in what is called a reverberant field, due to multiple reflections in an enclosed space, and picks up a mixture of the direct sound and the reverberation.

John then considers how sound moves in plane and spherical wave fields, using the relationship between sound pressure and air particle velocity. The link between pressure and pressure gradient is also examined; this is vital to an understanding of how pressure gradient microphones work, and why directional microphones exhibit bass boost when used at close operating distances.

There is next a most interesting section on the effects of atmospheric humidity on sound transmission. Lower than normal humidity can cause quite serious high-frequency losses over distances of, say, 12 meters. This may sound like a long way off to put your microphone from your subject, when the average working distance for rock music vocals is about a millimeter, but it is not uncommon when recording cathedral organs and suchlike.

Moving on, we encounter the ways in which microphones interact with the sound field they are sampling. Microphones are physically small compared with most sound waves, but at short wavelengths diffraction effects start to make themselves felt, and the microphone begins to show a non-uniform directional response. This leads to the concept of a *directivity index,* which is a measure, expressed in decibels, of the ratio of on-axis sound pickup relative to the total sound pickup integrated over all directions.

The response of a microphone is also affected by the shape of its body, and the usual method of mounting the diaphragm at the end of a cylinder, for ease of handling, is shown to introduce off-axis frequency response disturbances of 10 dB or more, which is food for thought. On- and off-axis frequency responses for microphones mounted on the end of a cylinder and on a sphere are given.

Another factor affecting the microphone response is the angle at which the incident sound waves hit the diaphragm. John Eargle shows how this too affects the off-axis response.

CHAPTER 3

Basic Sound Transmission and Operational Forces on Microphones

The Microphone Book by John Eargle

INTRODUCTION

All modern microphones benefit from electrical amplification and thus are designed primarily to sample a sound field rather than take power from it. In order to understand how microphones work from the physical and engineering points of view, we must understand the basics of sound transmission in air. We base our discussion on sinusoidal wave generation, since sine waves can be considered the building blocks of most audible sound phenomena. Sound transmission in both plane and spherical waves will be discussed, both in free and enclosed spaces. Power relationships and the concept of the decibel are developed. Finally, the effects of microphone dimensions on the behavior of sound pickup are discussed.

BASIC WAVE GENERATION AND TRANSMISSION

Figure 3.1 illustrates the generation of a sine wave. The vertical component of a rotating vector is plotted along the time axis, as shown at A. At each 360° of rotation, the wave structure, or waveform, begins anew. The *amplitude* of the sine wave reaches a crest, or maximum value, above the zero reference baseline, and the *period* is the time required for the execution of one cycle. The term *frequency* represents the number of cycles executed in a given period of time. Normally we speak of frequency in *hertz* (Hz), representing cycles per second.

For sine waves radiating outward in a physical medium such as air, the baseline in Figure 3.1 represents the static atmospheric pressure, and the sound waves are represented by the alternating plus and minus values of pressure about the static pressure. The period then corresponds to *wavelength*, the distance between successive iterations of the basic waveform.

The speed of sound transmission in air is approximately equal to 344 meters per second (m/s), and the relations among speed (m/s), wavelength (m), and frequency (1/s) are:

$$c\,(\text{speed}) = f(\text{frequency}) \times \lambda(\text{wavelength})$$

$$f = c/\lambda$$

$$\lambda = c/f \qquad (3.1)$$

For example, at a frequency of 1000 Hz, the wavelength of sound in air will be 344/1000 = 0.344 m (about 13 inches).

Audio Engineering Explained

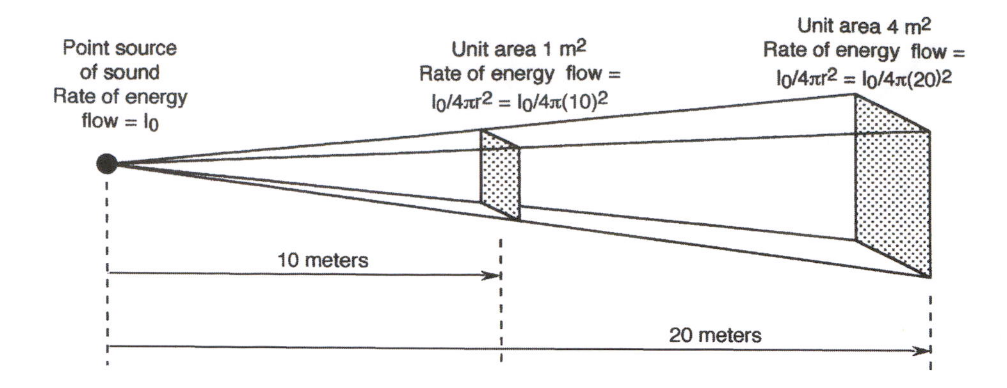

FIGURE 3.3
Sound intensity variation with distance over a fixed solid angle.

The intensity at any distance r from the source is given by:

$$I = W/4\pi r^2 \tag{3.3}$$

The effective sound pressure in pascals at that distance will be:

$$p = \sqrt{I\rho_0 c} \tag{3.4}$$

where $\rho_0 c$ is the specific acoustical impedance of air (405 SI rayls).

For example, consider a point source of sound radiating a power of one watt uniformly. At a distance of 1 meter the intensity will be:

$$\mathrm{I} = 1/4\pi(1)^2 = 1/4\pi = 0.08\ \mathrm{W/m^2}$$

The effective sound pressure at that distance will be:

$$p = \sqrt{(0.08)405} = 5.69\ \mathrm{Pa}$$

RELATIONSHIP BETWEEN AIR PARTICLE VELOCITY AND AMPLITUDE

The relation between air particle velocity (u) and particle displacement (x) is given by:

$$u(t) = j\omega \times (t) \tag{3.5}$$

where $\omega = 2\pi f$ and $x(t)$ is the maximum particle displacement value. The complex operator j produces a positive phase shift of 90°.

Some microphones, notably those operating on the capacitive or piezoelectric principle, will produce constant output when placed in a constant amplitude sound field. In this case $u(t)$ will vary proportional to frequency.

Other microphones, notably those operating on the magnetic induct ion principle, will produce a constant output when placed in a constant velocity sound field. In this case, $x(t)$ will vary inversely proportional to frequency.

THE DECIBEL

We do not normally measure acoustical intensity; rather, we measure sound pressure level. One cycle of a varying sinusoidal pressure might look like that shown in Figure 3.4(a). The peak value of this

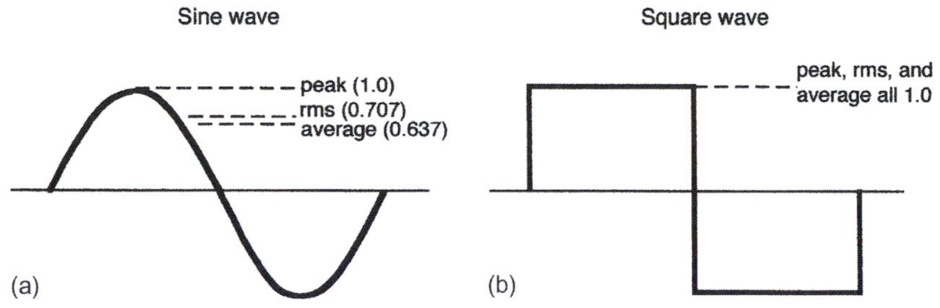

FIGURE 3.4
Sine (a) and square (b) waves: definitions of peak, rms and average values.

signal is shown as unity; the *root-mean-square* value (rms) is shown as 0.707, and the average value of the waveform is shown as 0.637. A square wave of unity value, shown at (b), has *peak*, *rms*, and *average* values that all are unity. The *rms*, or effective, value of a pressure waveform corresponds directly to the power that is delivered or expended in a given acoustical system.

The unit of pressure is the pascal (Pa) and is equal to one newton/m^2. (The newton (N) is a unit of force that one very rarely comes across in daily life and is equal to about 9.8 pounds of force.) Pressures encountered in acoustics normally vary from a low value of 20 μPa (micropascals) up to normal maximum values in the range of 100 Pa. There is a great inconvenience in dealing directly with such a large range of numbers, and years ago the decibel (dB) scale was devised to simplify things. The dB was originally intended to provide a convenient scale for looking at a wide range of power values. As such, it is defined as:

$$\text{Level(dB)} = 10\log(W/W_0) \qquad (3.6)$$

where W_0 represents a reference power, say, 1 watt, and the logarithm is taken to the base 10. (The term *level* is universally applied to values expressed in decibels.) With 1 watt as a reference, we can say that 20 watts represents a level of 13 dB:

$$\text{Level (dB)} = 10\log(20/1) = 13\,\text{dB}$$

Likewise, the level in dB of a 1 milliwatt signal, relative to 1 watt, is:

$$\text{Level (dB)} = 10\log(0.001/1) = -30\,\text{dB}$$

From basic electrical relationships, we know that power is proportional to the *square* of voltage. As an analog to this, we can infer that acoustical power is proportional to the square of acoustical pressure. We can therefore rewrite the definition of the decibel in acoustics as:

$$\text{Level (dB)} = 10\log(p/p_0)^2 = 20\log(p/p_0) \qquad (3.7)$$

In sound pressure level calculations, the reference value, or p_0, is established as 0.00002 Pa, or 20 micropascals (20 μPa).

Consider a sound pressure of one Pa. Its level in dB is:

$$\text{dB} = 20\log(1/0.00002) = 94\,\text{dB}$$

This is an important relationship. Throughout this book, the value of 94 dB L_P will appear time and again as a standard reference level in microphone design and specification. (L_P is the standard terminology for sound pressure level.)

Figure 3.5 presents a comparison of a number of acoustical sources and the respective levels at reference distances.

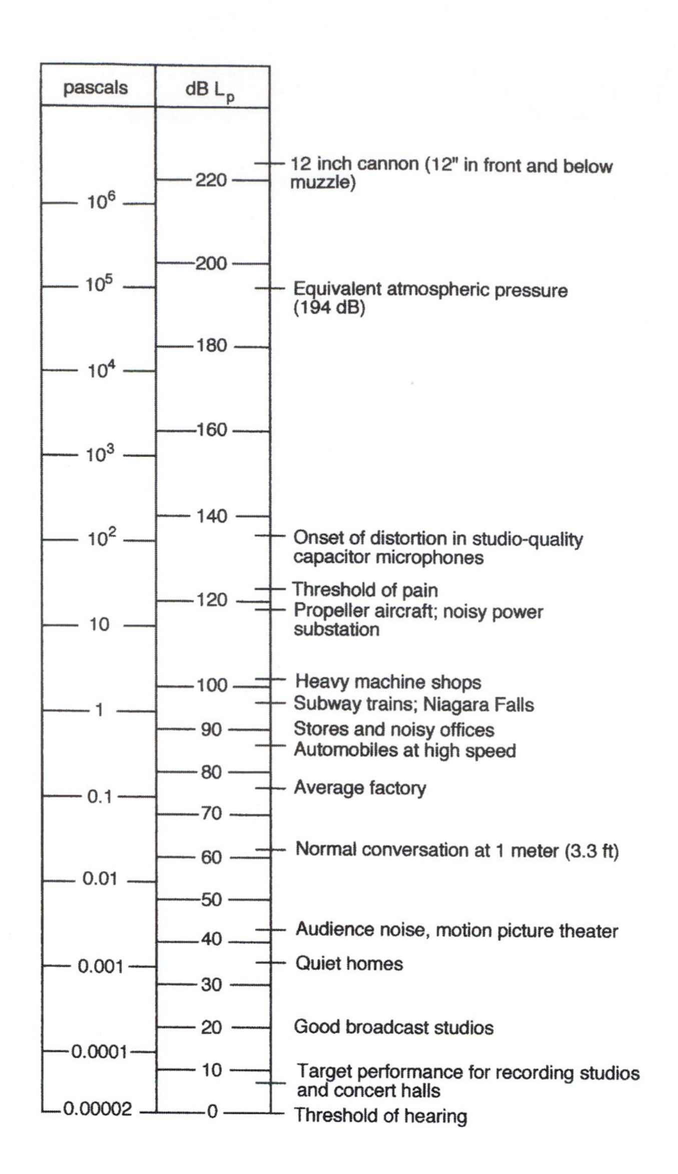

FIGURE 3.5
Sound pressure levels of various sound sources.

The graph in Figure 3.6 shows the relationship between pressure in Pa and L_P. The nomograph shown in Figure 3.7 shows the loss in dB between any two reference distances from a point source in the free field.

Referring once again to equation (3.4), we will now calculate the sound pressure level of 1 acoustical watt measured at a distance of 1 m from a spherically radiating source:

$$L_P = 20 \log (5.69/0.00002) = 109 \text{ dB}$$

It can be appreciated that 1 acoustical watt produces a considerable sound pressure level. From the nomograph of Figure 3.7, we can see that 1 acoustical watt, radiated uniformly and measured at a distance of 10 m (33 feet), will produce $L_P = 89$ dB. How "loud" is a signal of 89 dB L_P? It is approximately the level of someone shouting in your face!

THE REVERBERANT FIELD

A free field exists only under specific test conditions. Outdoor conditions may approximate it. Indoors, we normally observe the interaction of a direct field and a reverberant field as we move away from a sound source. This is shown pictorially in Figure 3.8(a). The reverberant field consists of the ensemble

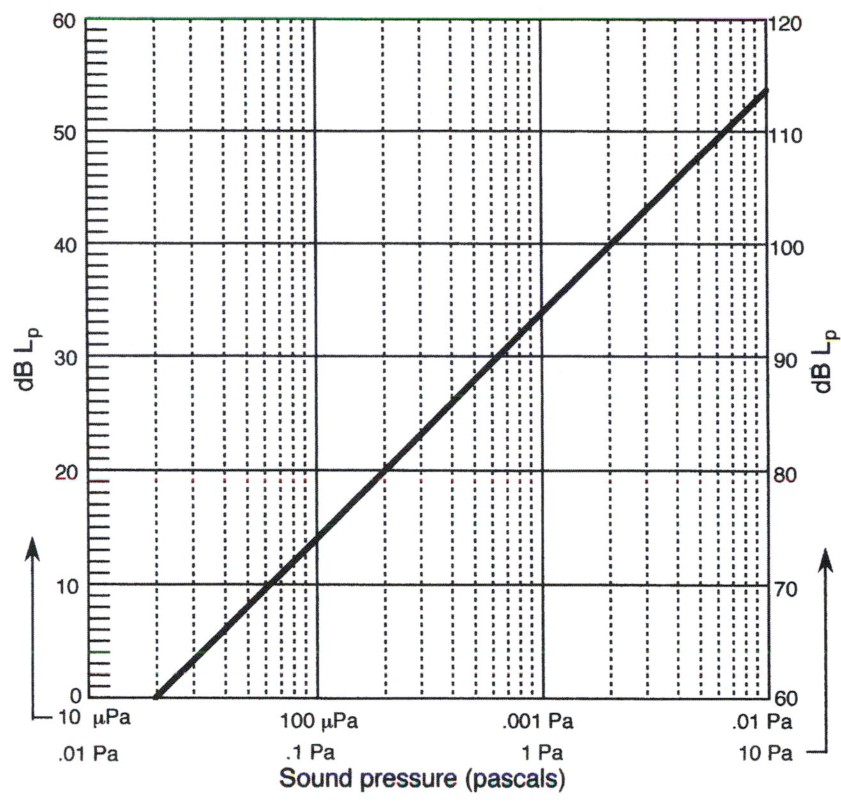

FIGURE 3.6
Relationship between sound pressure and sound pressure level.

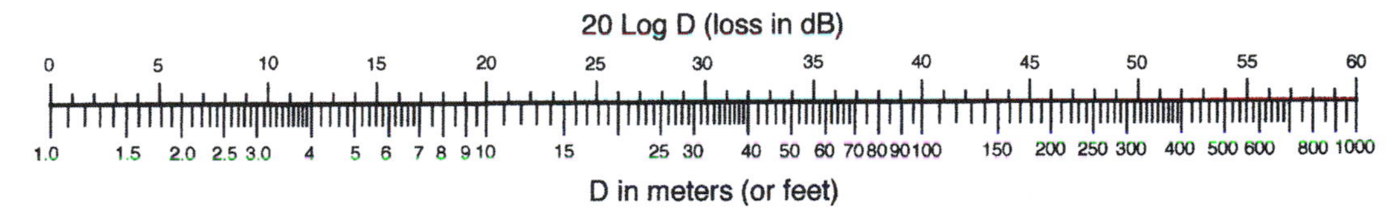

FIGURE 3.7
Inverse square sound pressure level relationships as a function of distance from the source; to determine the level difference between sound pressures at two distances, locate the two distances and then read the dB difference between them; for example, determine the level difference between distances 50 m and 125 m from a sound source; above 50 read a level of 34 dB; above 125 read a level of 42 dB; taking the difference gives 8 dB.

of reflections in the enclosed space, and reverberation time is considered to be that time required for the reverberant field to diminish 60 dB after the direct sound source has stopped.

There are a number of ways of defining this, but the simplest is given by the following equation:

$$\text{Reverbertion time(s)} = \frac{0.16V}{S\overline{\alpha}} \tag{3.8}$$

where V is the room volume in m^3, S is the interior surface area in m^2, and $\overline{\alpha}$ is the average absorption coefficient of the boundary surfaces.

The distance from a sound source to a point in the space where both direct and reverberant fields are equal is called *critical distance* (D_C). In live spaces critical distance is given by the following equation:

$$D_C = 0.14\sqrt{QS\overline{\alpha}} \tag{3.9}$$

where Q is the *directivity factor* of the source.

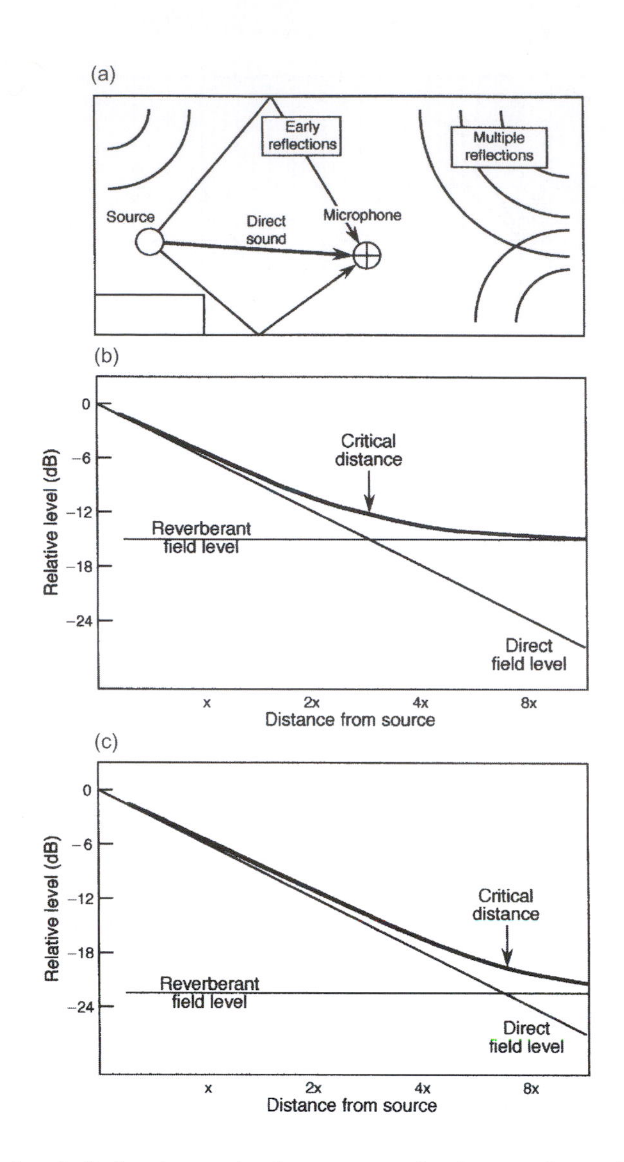

FIGURE 3.8

The reverberant field. Illustration of reflections in an enclosed space compared to direct sound at a variable distance from the sound source (a); interaction of direct and reverberant fields in a live space (b); interaction of direct and reverberant fields in a damped space (c).

In a live acoustical space, $\bar{\alpha}$ may be in the range of 0.2, indicating that, on average, only 20% of the incident sound power striking the boundaries of the room will be absorbed; the remaining 80% will reflect from those surfaces, strike other surfaces, and be reflected again. The process will continue until the sound is effectively damped out. Figures 3.8(b) and (c) show, respectively, the observed effect on sound pressure level caused by the interaction of direct, reflected, and reverberant fields in live and damped spaces.

Normally, microphones are used in the direct field or in the transition region between direct and reverberant fields. In some classical recording operations, a pair of microphones may be located well within the reverberant field and subtly added to the main microphone array for increased ambience.

SOUND IN A PLANE WAVE FIELD

For wave motion in a free plane wave field, time varying values of sound pressure will be in phase with the air particle velocity, as shown in Figure 3.9. This satisfies the conditions described in Table 3.1, in which

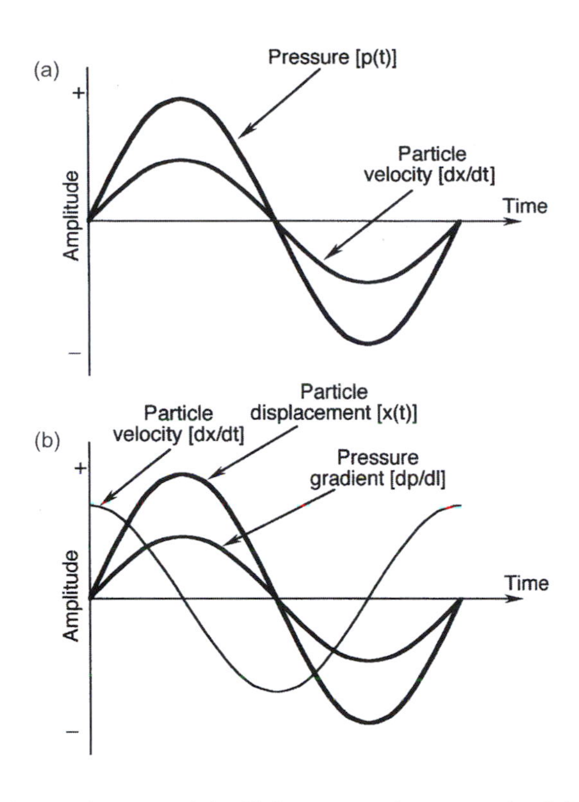

FIGURE 3.9
Wave considerations in microphone performance: relationship between sound pressure and particle velocity (a); relationship among air particle velocity, air particle displacement, and pressure gradient (b); relationship between pressure and pressure gradient (c) (Data presentation after Robertson, 1963).

the product of pressure and air volume velocity define acoustical power. (Volume velocity may be defined here as the product of particle velocity and the area over which that particle velocity is observed.)

If a microphone is designed to respond to sound pressure, the conditions shown in Figure 3.9(a) are sufficient to ensure accurate reading of the acoustical sound field.

Most directional microphones are designed to be sensitive to the air pressure difference, or *gradient*, existing between two points along a given pickup axis separated by some distance *l*. It is in fact this sensitivity that enables these microphones to produce their directional pickup characteristics. Figure 3.9(b) shows the phase relationships at work here. The pressure gradient [dp/dl] is in phase with the particle displacement [x(t)]. However, the particle displacement and particle velocity [dx/dt] are at a 90° phase relationship.

SOUND IN A SPHERICAL WAVE FIELD

Relatively close to a radiating sound source, the waves will be more or less spherical. This is especially true at low frequencies, where the difference in wavefront curvature for successive wave crests will be quite pronounced. As our observation point approaches the source, the phase angle between pressure and particle velocity will gradually shift from zero (in the far field) to 90°, as shown in Figure 3.10(a). This will cause an increase in particle velocity with increasing phase shift, as shown at (b).

As we will see in a later detailed discussion of pressure gradient microphones, this phenomenon is responsible for what is called *proximity effect*: the tendency of directional microphones to increase their LF (low frequency) output at close operating distances.

EFFECTS OF HUMIDITY ON SOUND TRANSMISSION

Figure 3.11 shows the effects of both inverse square losses and HF losses due to air absorption. Values of relative humidity (RH) of 20% and 80% are shown here. Typical losses for 50% RH would be roughly halfway between the plotted values shown.

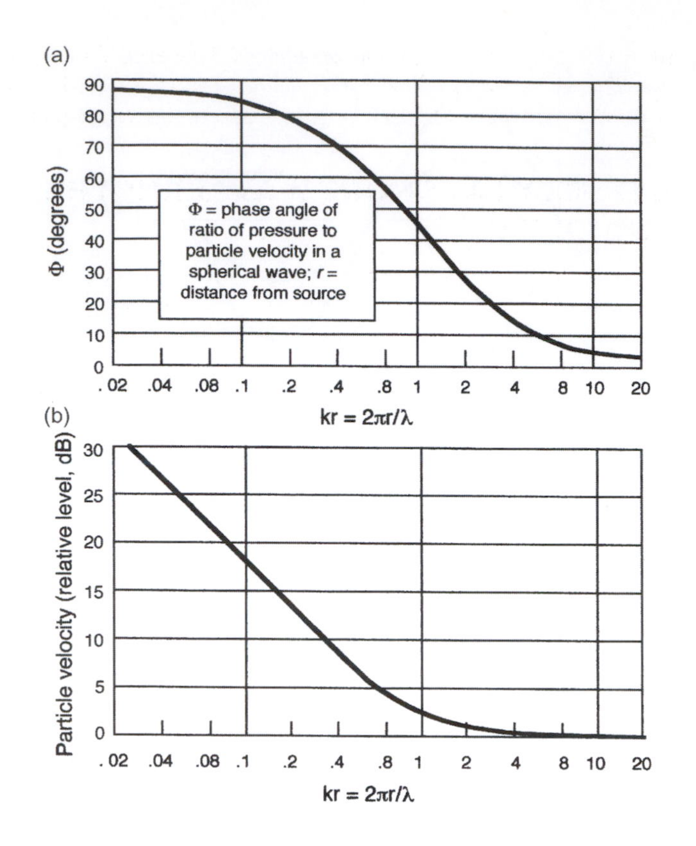

FIGURE 3.10
Spherical sound waves: phase angle between pressure and particle velocity in a spherical wave at low frequencies; r is the observation distance and λ is the wavelength of the signal (a); increase in pressure gradient in a spherical wave at low frequencies (b).

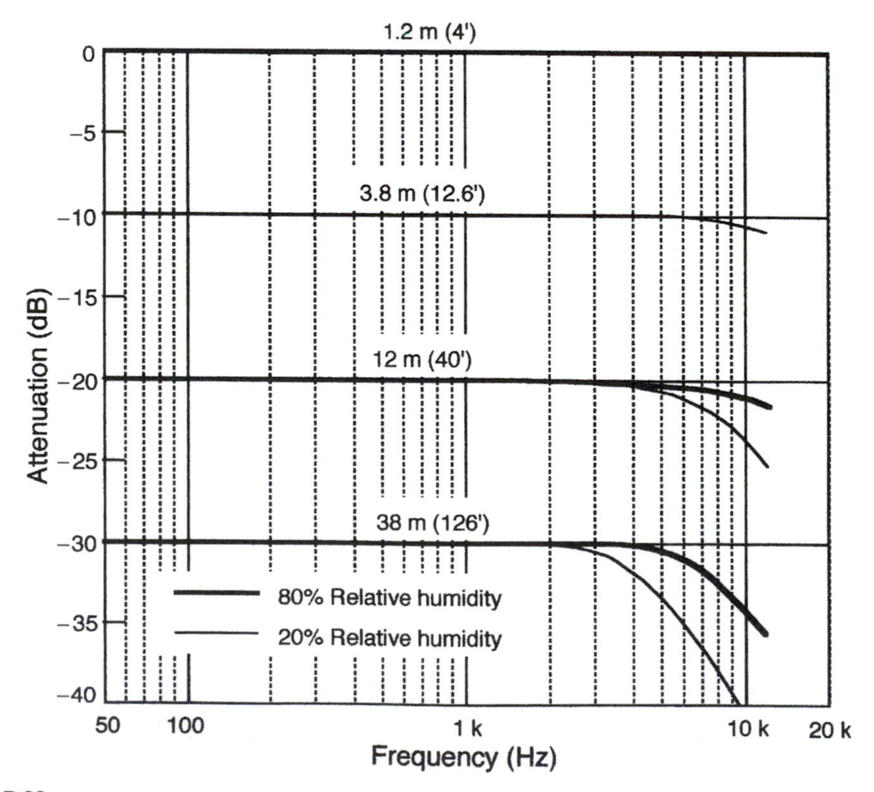

FIGURE 3.11
Effects of both inverse square relationships and HF air losses (20% and 80% RH).

For most studio recording operations HF losses may be ignored. However, if an organ recording were to be made at a distance of 12 m in a large space and under very dry atmospheric conditions, the HF losses could be significant, requiring an additional HF boost during the recording process.

DIFFRACTION EFFECTS AT SHORT WAVELENGTHS; DIRECTIVITY INDEX (DI)

Microphones are normally fairly small so that they will have minimal effect on the sound field they are sampling. There is a limit, however, and it is difficult to manufacture studio quality microphones smaller than about 12 mm (0.5 in) in diameter. As microphones operate at higher frequencies, there are bound to be certain aberrations in directional response as the dimensions of the microphone case become a significant portion of the sound wavelength. *Diffraction* refers to the bending of sound waves as they encounter objects whose dimensions are a significant portion of a wavelength.

Many measurements of off-axis microphone response have been made over the years, and even more theoretical graphs have been developed. We will now present some of these.

Figure 3.12 shows polar response diagrams for a circular diaphragm at the end of a long tube, a condition that describes many microphones. In the diagrams, $ka = 2\pi a/\lambda$, where a is the radius of the diaphragm. Thus, ka represents the diaphragm circumference divided by wavelength. *DI* stands for *directivity index*; it is a value, expressed in decibels, indicating the ratio of on-axis pickup relative to the total pickup integrated over all directions. Figure 3.13 shows the same set of measurements for a

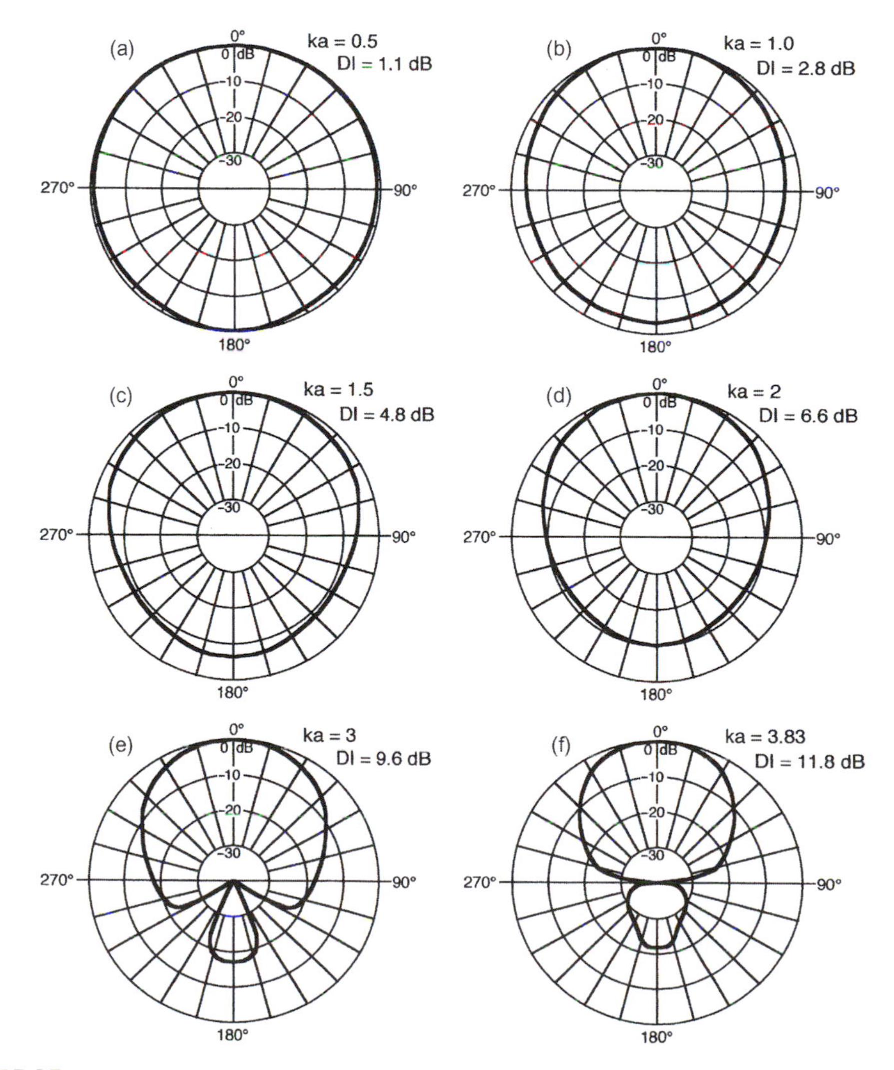

FIGURE 3.12
Theoretical polar response for a microphone mounted at the end of a tube. (Data presentation after Beranek, 1954.)

The second chapter on microphone technology by John Eargle focuses on the electrical aspects of microphone operation, including phantom powering, microphone transformers, mixing console microphone preamplifiers, microphone splitters, cable losses, ground loops, RF microphones, and the latest so-called "digital" microphones.

The powering of microphones is a big subject in itself. Dynamic microphones, in which a coil attached to the diaphragm moves in the field of a permanent magnet, do not require external power, but they are not generally considered to be in the first line of quality. Capacitor microphones do need power, and the standard method is "phantom powering" which applies +48V to the hot and cold signal connection through two 6k8 resistors. This puts a strict limit on the amount of power that a microphone can draw, restricting the flexibility of design for the electronics. To address this problem, work on setting up a "super phantom" standard with 2k2 resistors is proceeding. John also describes the unloved and obsolescent technology known as T-feed or T-powering.

Virtually all high-performance microphones are capacitor types that require a substantial polarizing voltage on the capsule diaphragm. The capsule polarizing voltage is often in the range 60–65 Volts, and since this is greater than the phantom voltage, some sort of DC-DC conversion is required. Low-voltage power is also required for buffer stages to drive the loading of a connecting cable and a mixing console input.

A capacitor capsule has an extremely high output impedance, equivalent to a very small capacitor—in fact the highest impedance I have ever had to deal with. A while ago I designed the electronics for an capacitor microphone, and to avoid bass loss the biasing resistor had to have the astronomical value of 10 gigaOhms. This exotic component came in a glass encapsulation that had to be manipulated with tweezers—one touch of a finger and the insulation properties of the glass were fatally compromised.

In general, microphone output levels are low. Back in the day (say pre-1978) step-up microphone transformers with ratios of up to 1:10 were widely used to get the signal up to a level suitable for low-noise amplification. While this technology had its advantages, the cost and weight of the transformers, and their non-ideal linearity, made a transformerless input amplifier a very desirable thing to aim at. The difficulties of getting the noise low enough and the linearity good enough were at first formidable, but after a good deal of work (some of which I did) they have since become almost universal.

Microphone pads, or attenuators, are used when the output is too high for the mixing console input to cope with; this typically happens when you put a microphone inside a kick-drum. If switchable pads are built into the console, they are at the wrong end of the microphone cable and degrade both noise performance and common-mode rejection. I might modestly mention that I invented the "padless" microphone input amplifier in 1987 to eliminate these problems, and got a patent for it. Looking at the mixer market today, the idea seems to have caught on.

John concludes this chapter by examining "digital" microphones. These are of course only digital after a certain point. The sound pressure impinging on the diaphragm and the movement of the diaphragm itself remain defiantly analog. The digital-to-analog conversion takes place later and the data is sent down an AES-3 cable.

Electrical Considerations and Electronic Interface

The Microphone Book by John Eargle

INTRODUCTION

In this chapter we examine details of electronic performance of microphones and their interface with the following preamplifiers and consoles. Subjects to be covered include: remote powering, microphone output/preamp input circuitry, the stand-alone microphone preamp, microphone cable characteristics and interference, and overall system considerations. A final section deals with capacitor microphones operating on the radio frequency (RF) transmission principle.

POWERING

Most modern capacitor microphones operate on 48V *phantom powering* (also known as simplex powering) provided by the preamplifier or console input module. The basic circuitry for phantom powering is shown in Figure 4.1. Here, a master 48V DC supply provides positive voltage to pins 2 and 3 through a pair of 6800 ohm resistors, while a ground return path is provided through pin 1. The signal is carried on pins 2 and 3 and is unaffected by the presence of identical DC voltages on pins 2 and 3. Today, pin 2 is universally designated as the "hot" lead; that is, a positive-going acoustical signal at the microphone will produce a positive-going voltage at pin 2. Pin 1 provides both the DC ground return path for phantom powering as well as the shield for the signal pair.

The circuit normally used with a transformerless microphone for receiving power from the source is shown in Figure 4.2(a). Here, a split resistor combination is tapped for positive DC voltage which, along with the ground return at pin 1, is delivered to the microphone's electronics and to the capsule for polarizing purposes.

If the microphone has an integral output transformer, then a center-tapped secondary winding is used for receiving the positive voltage. This method is shown in Figure 4.2(b). Phantom powering can also be used when there is neither a microphone output transformer nor a console input transformer, as shown in Figure 4.2(c).

Phantom powering at 48V is normally designated as P48. There are also standards (IEC publication 268-15) for operation at nominal values of 24V (P24) and 12V (P12). Voltage tolerances and current limits for the three standards are shown in Table 4.1. The resistor values are generally held to tolerances of 1%.

T-powering (also known as A-B powering) is rarely encountered today. Circuit details are shown in Figure 4.3. It is normally designated as T12. T-powering is a holdover from earlier years and still may be encountered in motion picture work, where it is built into the many Nagra tape recorders used in that field. Here, the audio signal leads are at different DC voltages, and any residual hum or noise in the DC supply will be reflected through the microphone's output as noise.

Audio Engineering Explained

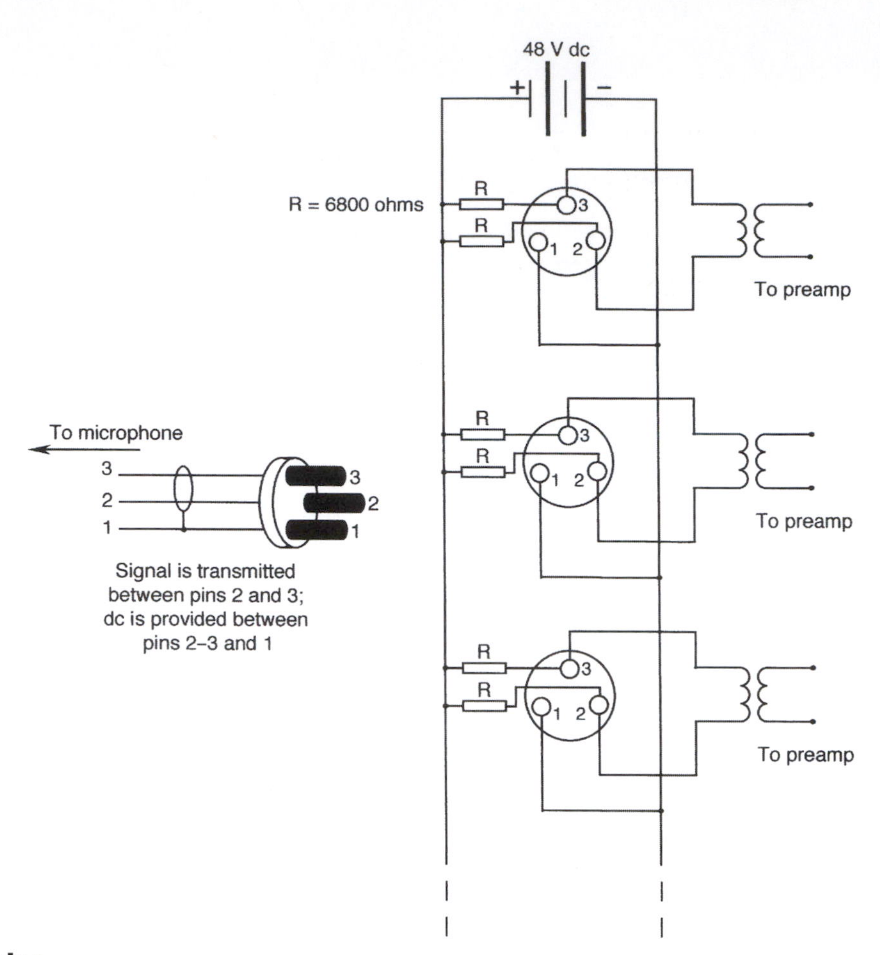

FIGURE 4.1
Basic circuit for phantom (simplex) powering.

Using phantom powering

Most microphone preamplifiers and console input sections have provision for individual switching of phantom power on or off. When using dynamic microphones it is good engineering practice to *turn off* the phantom powering, even though no current will flow through the microphone's voice coil should the phantom power be left on. However, if T12 power is inadvertently applied to a dynamic microphone, the 12V DC will appear across the microphone's voice coil with noticeable deterioration of response and possible damage.

Another important rule is not to turn phantom power on or off when a microphone is bussed on and assigned to the monitor channels. The ensuing loud "pop" could easily burn out an HF transducer in the monitor loudspeaker systems. At the end of a session, it is normal practice to reduce both the master fader and the monitor level control to zero before shutting down all phantom power from the console.

While on the subject of phantom power, *never* attempt, when phantom power is on, to remove or replace the screw-on capsule that many capacitor microphones have. This has been known to burn out the FET in the input circuitry of the impedance converter.

Capacitor microphones vary in their susceptibility to shifts in nominal voltage in phantom powering. Generally, the variation is on the low side, such as may be encountered in very long microphone cable runs. Symptoms may be reduced signal output, increase in noise, as well as distortion. When these conditions occur with normal cable runs, the round-trip cable resistance should be measured, and the power supply itself checked for possible problems.

DC-to-DC conversion

Some capacitor microphones are designed to operate over multiple ranges of phantom powering, for example, from 20 to 52 volts DC to cover requirements of both P24 and P48 powering. What is required here is a circuit that converts the applied voltage to the value required for proper biasing of the capsule

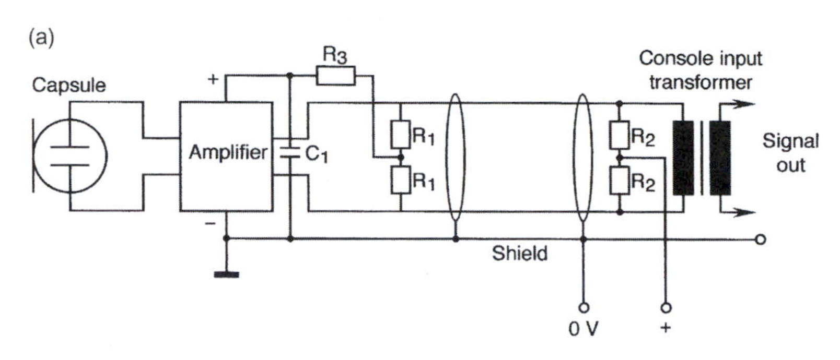

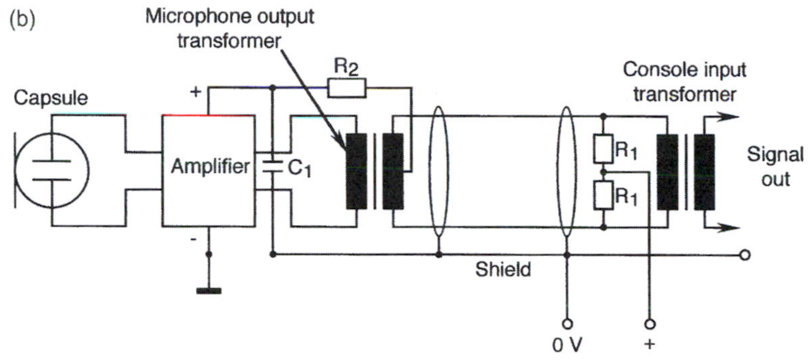

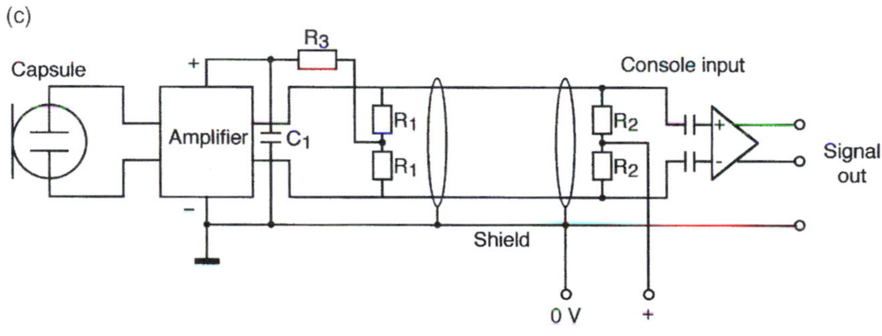

FIGURE 4.2
Microphone power input circuitry using a resistive voltage divider (a); using a center-tapped transformer secondary (b); using no transformers in the powering path (c).

Table 4.1 Voltage tolerances and current limits

Supply voltage	12 ± 1V	24 ± 4V	48 ± 4V
Supply current	max. 15 mA	max. 10 mA	max. 10 mA
Feed resistors	680 Ω	1200 Ω	6800 Ω

and operation of the impedance converting preamp. The circuit for the Neumann TLM107 microphone is shown in Figure 4.4. A major design challenge in such a circuit is the suppression of noise that could result from the high switching rate of the input voltage during the DC-to-DC conversion process.

This circuit provides capsule biasing voltages for its selectable patterns, reducing them accordingly when the –10dB pad is engaged. The 10V DC output is for powering the microphone's electronics.

Recent developments in phantom powering

The present standard for phantom powering ensures that 48V DC in a short circuit loading condition through two parallel 6800 ohm resistors will produce a current of 14mA DC, thus limiting the current

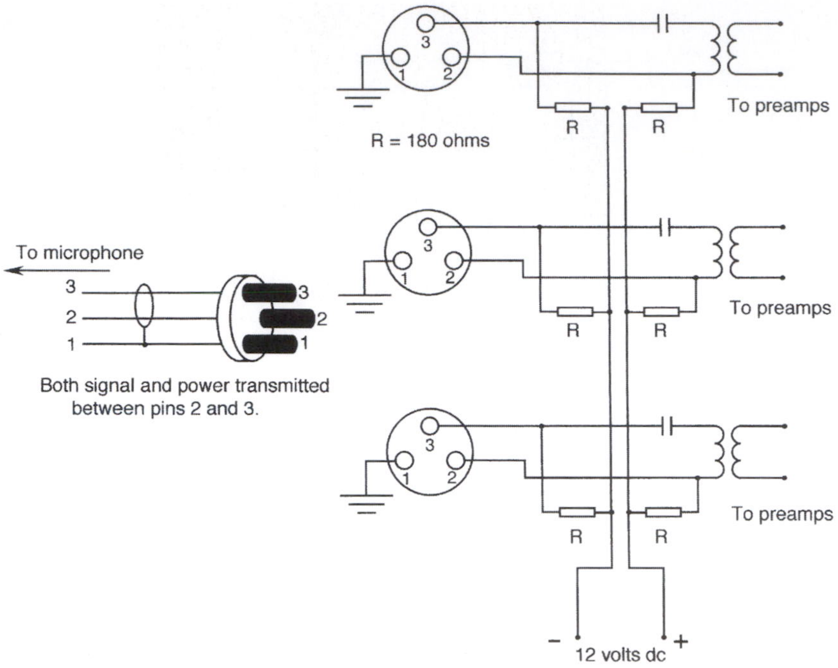

R = 180 ohms
R
To preamps
To microphone
Both signal and power transmitted between pins 2 and 3.
12 volts dc

FIGURE 4.3
Basic circuit for T-powering at 12 V.

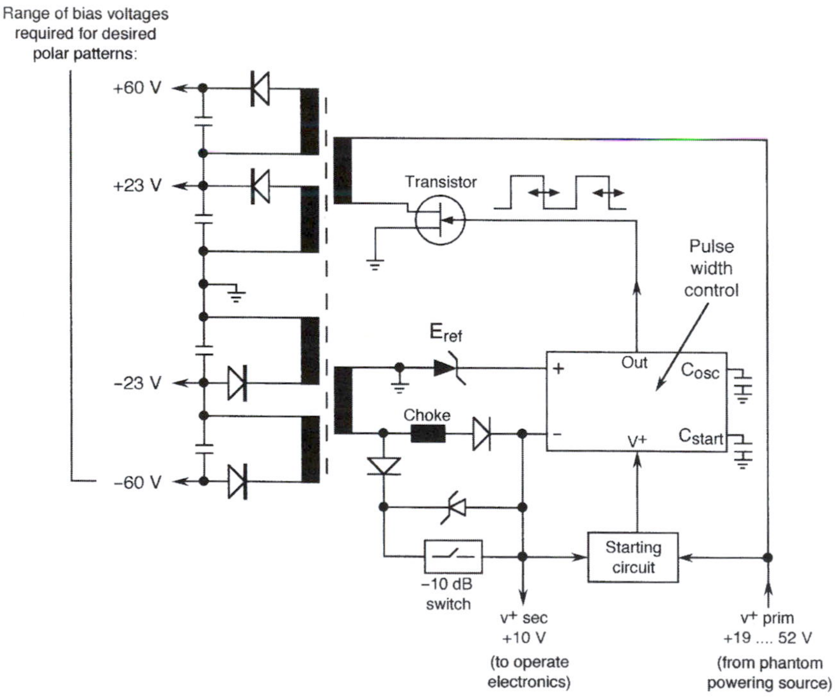

FIGURE 4.4
Details of DC-to-DC conversion for microphone operation at P24 and P48 standards. (Data after Neumann/USA.)

availability for a given microphone model. Some manufacturers have designed microphones that can accommodate greater current for handling higher sound pressure levels in the studio, and such microphones require lower resistance values in the phantom supply in order to receive the higher current. Generally, this has been carried out in proprietary stand-alone power supplies that a manufacturer may provide for a specific new microphone model.

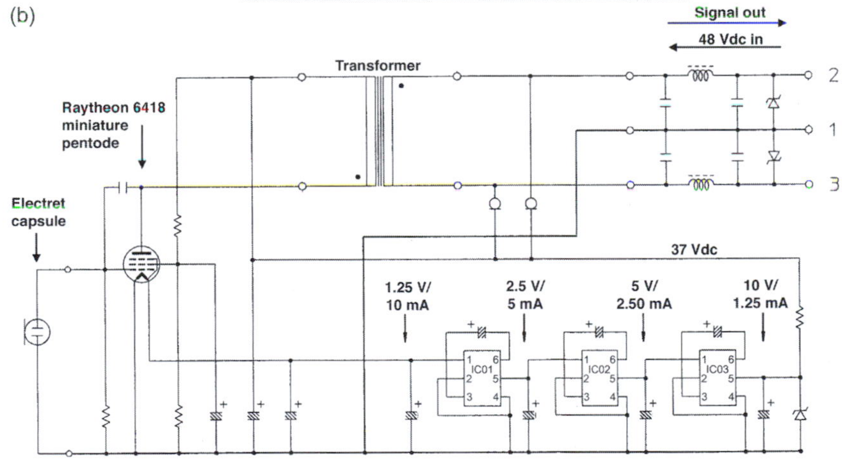

FIGURE 4.5
The Audio-Technica model AT3090 microphone; photo of microphone (a); circuit details (b). (Data courtesy of Audio-Technica USA.)

Two-way compatibility is maintained. A new high-current microphone will work on a standard phantom supply, but it will not be able to attain its highest degree of performance. A standard microphone will work on the new supply, drawing only the current it needs.

Typical here is the "Super Phantom" powering that Josephson Engineering has specified for certain microphone models, in which current fed through a pair of 2200 ohm resistors in each leg of the power supply is directed to the microphone. A short circuit loading condition here would result in a current draw slightly in excess of 43 mA DC. International standardization activities are presently under way in this area.

In an unusual design approach, Audio-Technica has introduced the model AT3060 tube-type microphone which is powered at P48. A photo of the microphone is shown in Figure 4.5(a), and basic circuit details are shown at (b). Using a vacuum tube that will operate at a fairly low plate potential of 37V DC, the tube filament requirement of 1.25V DC at a current draw of 10 mA is attained via an unusual integrated circuit (IC) arrangement shown in the bottom-right portion of Figure 4.5(b). The voltage at the input to the first IC is 10V at a current of 1.25mA. The three ICs in tandem progressively *halve* the voltage, while *doubling* the current, attaining a value of 1.25V at a current draw of 10mA at the output of the final IC. Various diodes are used for maintaining separation between signal and DC paths.

FIGURE 4.6
The Royer Labs Model R-122 powered ribbon microphone: photo (a); polar response (b); frequency response at 1 meter (c). (Data courtesy of Royer Labs.)

The microphone element itself is an electret. The AT3060 has a nominal sensitivity of 25.1 mV/Pa and can handle levels in the studio of 134 dB. The self-noise floor of the microphone is 17dB(A).

Royer Labs has introduced integral electronics for phantom powering in two of their ribbon microphone models. Ribbons are particularly susceptible to variations in downstream loading, and, with their relatively low sensitivity, they may be subject to electrical interference in long runs. Figure 4.6(a) shows a photo of the Royer Labs R-122, with performance curves shown at (b) and (c). The circuit diagram is shown in Figure 4.7.

The most unusual aspect of the circuit is the compound transformer, which has four parallel sets of windings resulting in an optimum impedance match to the ribbon. The secondaries of these four sections are connected in series to attain a voltage gain of 16 dB, looking into the very high input impedance of the buffer stage. The system has a maximum level capability of 135 dB L_p and a self-noise no greater than 20dB(A). The sensitivity is 11 mV/Pa.

Power supplies for older tube microphones

Classic tube-type capacitor microphones are as popular today in studio recording as they have ever been. In a typical power supply (one for each microphone), DC voltages are produced for heating the vacuum tube's filament, biasing the capsule, and providing plate voltage for the vacuum tube amplifier. In many dual diaphragm designs, remote pattern switching is also included in the supply. Such

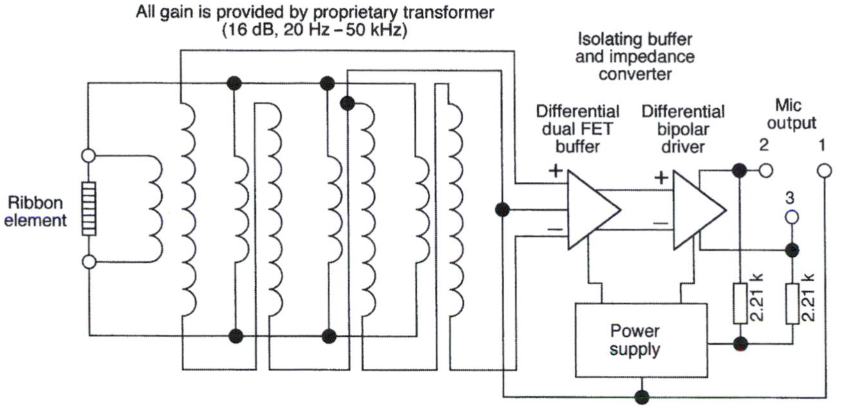

FIGURE 4.7
Circuit diagram for the Royer Labs Model R-122 microphone. (Figure courtesy of Royer Labs.)

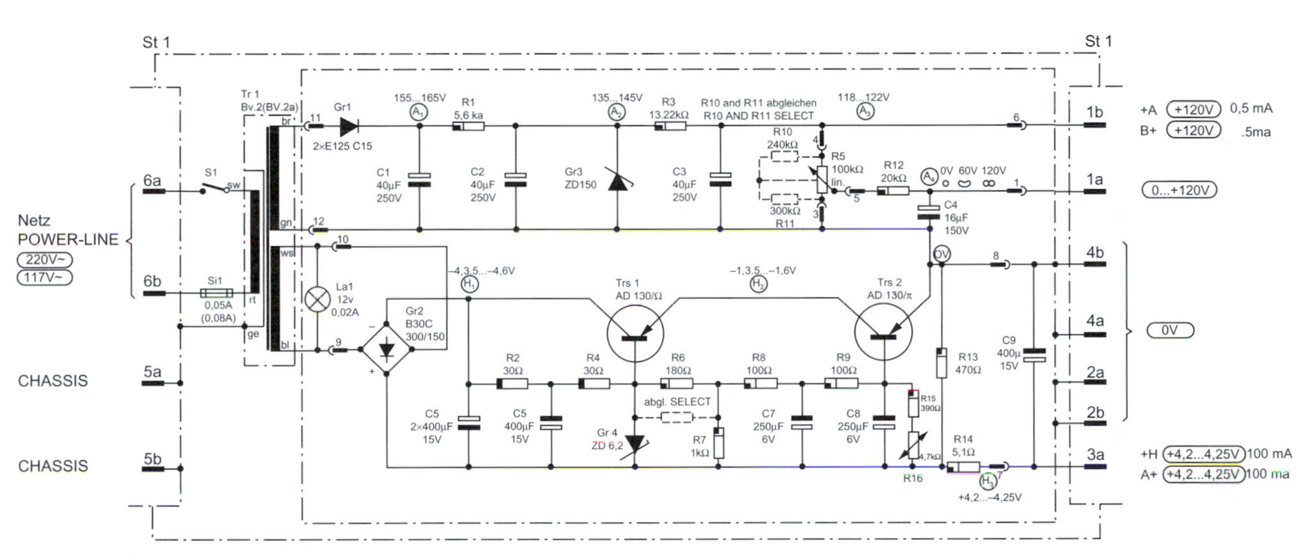

FIGURE 4.8
Power supply schematic for a tube capacitor microphone with variable pattern switching at the power supply; pin 1 A provides the pattern control voltage. (Figure courtesy of Neumann/USA).

a design is shown in Figure 4.8. Note that the cable connecting the power supply to the microphone contains seven conductors.

MICROPHONE OUTPUT/CONSOLE PREAMPINPUT CIRCUITRY

The microphone output pad

Most capacitor microphones have an integral output pad. The effect of the pad is basically to shift the microphone's entire operating range from noise floor to overload point downward by some fixed amount, normally 10 to 12 dB. The effect of this is shown in Figure 4.9. Note that the total dynamic range of the microphone remains fixed, with or without the pad.

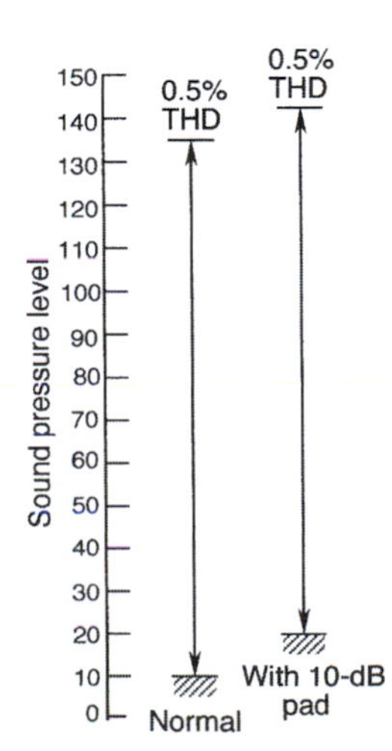

FIGURE 4.9
Typical effect of −10 dB microphone output pad on maximum level capability and self-noise of a capacitor microphone.

The microphone output transformer

A number of capacitor microphone models have an integral output transformer built into the microphone body. Most early tube-type capacitor microphones had output transformers that were usually housed in the power supply unit. The transformer often had split secondary windings and could be strapped for either parallel or series operation, as shown in Figure 4.10. Typically, when the windings are strapped for parallel operation (a), the output impedance is 50 ohms; for series strapping (b) the output impedance is 200 ohms. In either case the output power capability remains the same:

$$\text{Output power} = E^2/R = V^2/50 = (2V)^2/200.$$

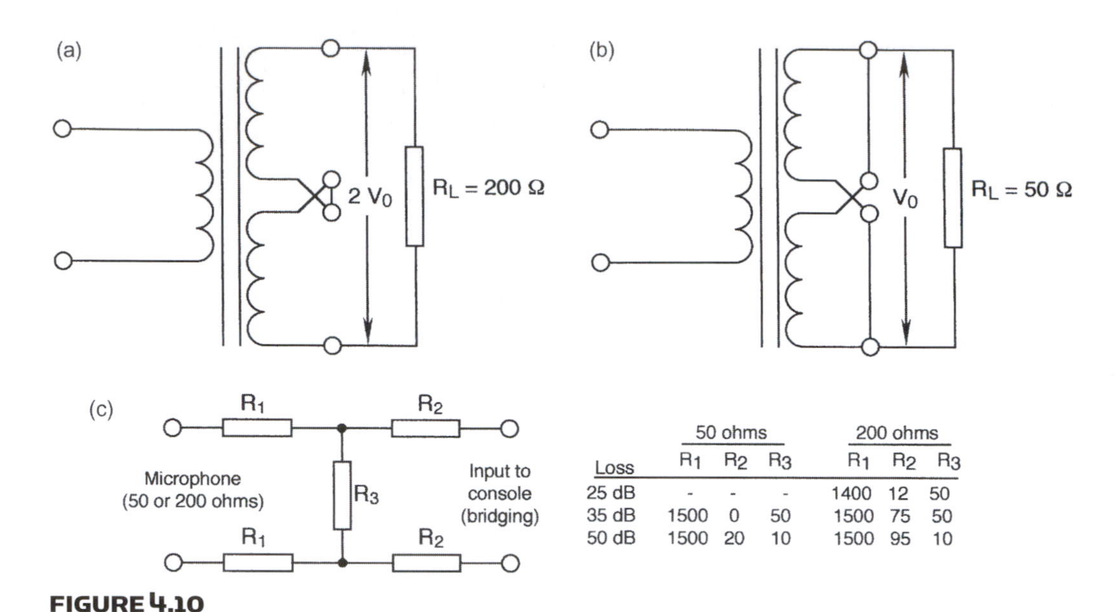

Loss	50 ohms R1	50 ohms R2	50 ohms R3	200 ohms R1	200 ohms R2	200 ohms R3
25 dB	-	-	-	1400	12	50
35 dB	1500	0	50	1500	75	50
50 dB	1500	20	10	1500	95	10

FIGURE 4.10
Transformer strapping for 200 ohms (a) and 50 ohms output impedance (b); H-pad values for balanced attenuation for microphone impedances of 50 and 200 ohms (c).

In many cases it is necessary to reduce the output signal from the microphone even lower than the parallel transformer strapping. H-pads (shown in Figure 4.10(c)) may be used both to maintain the impedance relationship and produce the desired amount of attenuation. Today, most transistorized capacitor microphones are transformerless and operate at a fixed balanced output impedance.

Console input section

In the very early days of broadcasting and recording, the dynamic microphones of the period had relatively low outputs and normally operated into matching input impedances. Typically, a microphone with a 600 ohm source impedance looked into a 600 ohm load, following the matched impedance concept. Impedance matching was a holdover from early telephone practice and found its way into broadcasting, and from there into recording. Today, the *bridging* concept is well established. In a bridging system, all output impedances are relatively low and all input impedances are relatively high throughout the audio chain. Here, the ratio of high to low impedance is normally in the range of 10-to-1 or greater.

Figure 4.11 shows a simplified transformerless console input section showing switchable line/microphone operation. At one time transformers were felt to be indispensable in the design of microphone input circuitry. Their chief advantages are a high degree of electrical balance and consequent high common-mode signal rejection. (A common-mode signal is one that is identical at both inputs; typically, induced noise signals are common mode.) In the era of vacuum tubes the input transformer was of course essential. Today's best solid state balanced input circuitry does not mandate the use of transformers, and there are considerable economic advantages to pass on to the user. Only under conditions of high electrical interference might their use be required.

Today, most microphones have an output impedance in the range of 50 to 200 ohms. Most consoles have an input bridging impedance in the range of 1500 to 3000 ohms.

Improper microphone loading

A typical 250 ohm ribbon microphone may have an impedance modulus as shown in Figure 4.12(a). As long as the microphone looks into a high impedance load, its response will be fairly flat. When used with a modern console having 1500 or 3000 ohm input impedances, the frequency response will

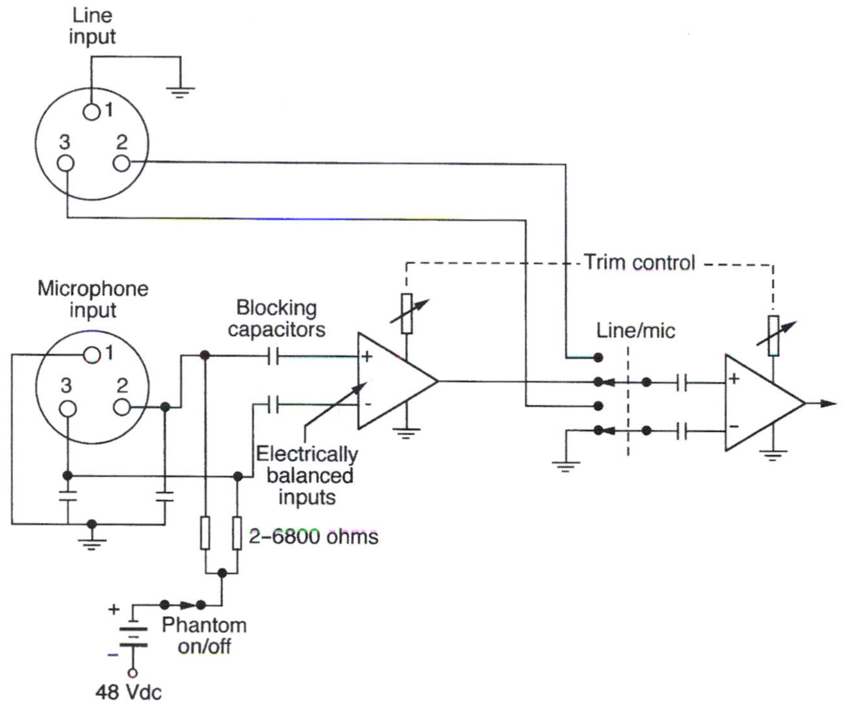

FIGURE 4.11
Simplified circuit for a transformerless console microphone/line input section.

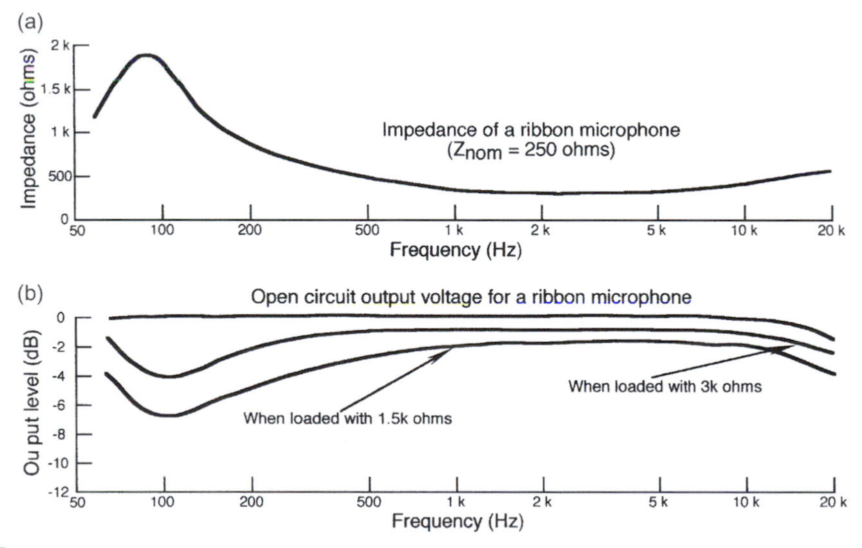

FIGURE 4.12
Effect of loading on ribbon microphone response.

be altered as shown at (b). This is a response problem routinely encountered today and frequently goes unchallenged.

Another problem with improper loading is shown in Figure 4.13. Here, a capacitor microphone is loaded by a console input transformer that has a non-optimal termination on its secondary side. This can produce response variations similar to those shown in the figure as a function of the microphone's output impedance. Note that as the input impedance is reduced, the response develops a rise in the 10 kHz range. This comes as a result of an undamped resonance involving the stray capacitance between the primary and secondary windings of the transformer (Perkins, 1994).

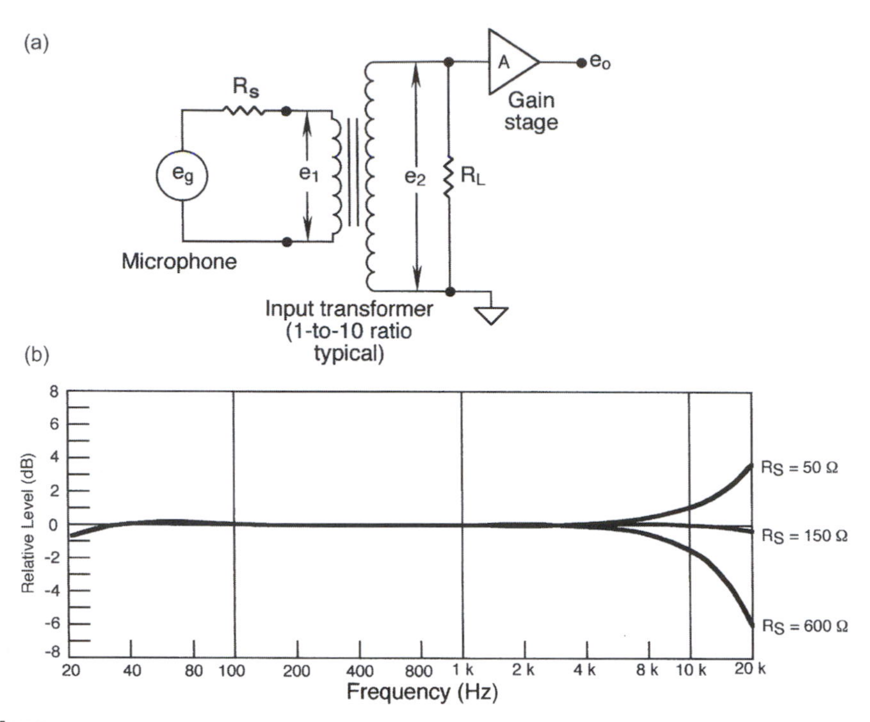

FIGURE 4.13
Response of transformer input circuit to varying values of microphone source impedance.

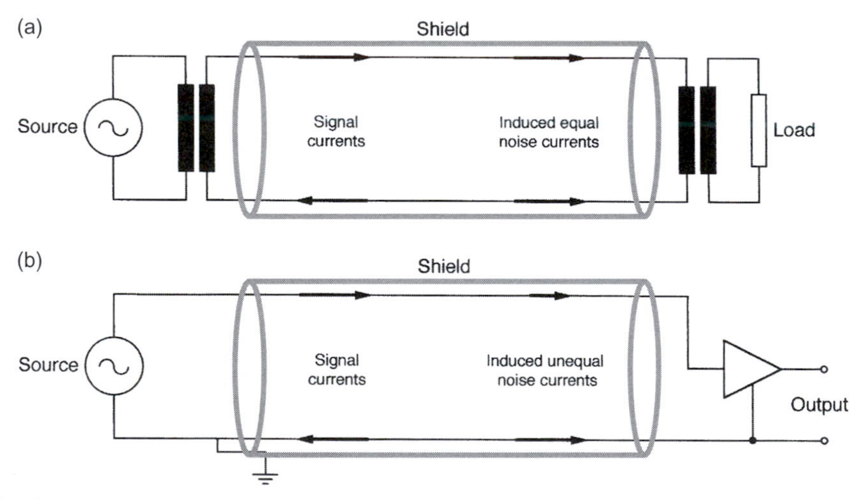

FIGURE 4.14
Balanced (a) versus unbalanced (b) microphone operation.

Unbalanced microphone inputs

Only in the lowest cost paging systems is one likely to come across an unbalanced microphone input. For fairly short cable runs from microphone to amplifier, there may be no problems. For longer runs, where there is greater likelihood for interference, the difference between balanced and unbalanced operation is as shown in Figure 4.14. For balanced operation, shown at (a), induced noise will be equal and in-phase in both signal leads; it will be effectively canceled by the high common-mode rejection of the input circuitry. For unbalanced operation, shown at (b), the induced signal currents will be different between shield and conductor, and the noise will be significant.

Microphone splitters

For many field operations, stage microphones must be fed to both recording and sound reinforcement activities. Microphone splitters are used to provide both an electrically direct feed to one operation,

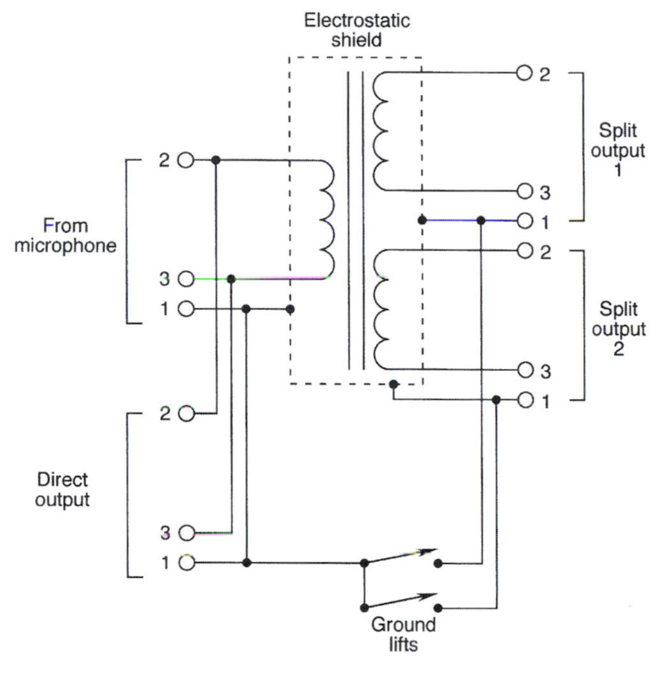

FIGURE 4.15
Details of a passive microphone splitter.

normally the recording activity, and a one-to-one transformer feed to the other activity. Circuit details for a typical passive splitter are shown in Figure 4.15. Here, there are two secondary windings, providing an additional output for broadcast activities. Ground lifts are often used to avoid hum due to ground loops. When using passive splitters, it is essential that all loads fed by the splitter be bridging. Active splitters are often used in order to avoid expensive transformers and to permit feeding the signal to low input impedance loads.

THE STAND-ALONE MICROPHONE PREAMP

Many leading recording engineers prefer to bypass console microphone inputs altogether and use multiple individual stand-alone microphone preamps instead. These microphone preamps are normally highly refined versions of what may be found in the typical console, some offering significant improvements in areas of noise floor, common-mode rejection, increased output level capability, source impedance matching, calibrated step-type gain trim, and rugged construction. Other features in some models include equalization options and metering. These performance attributes do not come cheaply, and a set of 16 individual preamps may cost many times more than an excellent mass-produced 24-input console.

For many recording activities it may be difficult to justify the expense of stand-alone preamps, considering that their outputs will be fed into a host console for normal signal routing and bus assignment. However, there are specific applications where their use is appropriate:

- Direct-to-stereo via two microphones, where no console routing functions are needed.
- Pop multichannel recording, with each microphone assigned to one recording channel. Here, the local host console is relegated to control room and studio monitoring only, with all studio microphones and direct feeds going to their own preamps and from there to their individual recorder inputs. In this regard, the set of external stand-alone preamps has taken the place of the "channel path" in a modern in-line console.

Many producers and engineers prefer to work in this manner when called upon to go into an unknown studio to work with an artist who may be on tour and wishes to lay down tracks, or perhaps do additional work on an album in progress. One great benefit for the producer/engineer is the comfort of working with a consistent set of tools that can be taken from one field location to another, reserving for a later time all post-production activities that will take place in a known, home environment.

Figure 4.16 shows front and back panel views of a high-quality stand-alone microphone preamplifier.

FIGURE 4.16
Front (a) and back (b) views of a stand-alone microphone preamplifier. (Figure courtesy of FM Acoustics.)

LINE LOSSES AND ELECTRICAL INTERFERENCE

To a very great extent, the recording engineer working in a well maintained studio does not have to worry about microphone cables, except to make sure that they are in good repair. Things may not be so simple for the engineer working in a specific remote location for the first time and finding that he may actually run short of cable! There are two concerns with long cable runs: phantom power operation and HF losses due to cable capacitance.

Typical electrical values for professional quality microphone cable are:

- Gauge: #24 AWG stranded copper wire
- Resistance/meter: 0.08 ohms
- Capacitance/meter: 100 pF

P48 phantom powering has a fairly generous design margin for long cable runs. For example, a 100 m run of cable will have a resistance, per leg, of 8 ohms, and the total resistance in the cable will be 16 ohms. Considering the relatively high resistance of the phantom feed network of 6800 ohms per leg, this added value is negligible.

More likely, HF roll-off will be encountered in long cable runs, as shown in Figure 4.17. Here, data is shown for cable runs of 10 and 60 meters, with source impedances of 200 and 600 ohms.

Another problem with long cable runs is their increased susceptibility to RF (radio frequency) and other forms of electromagnetic interference. Local interference may arise from lighting control systems, which can generate sharp "spikes" in the power distribution system; these can be radiated and induced into the microphone cables. Likewise, nearby radio transmitters, including mobile units, can induce signals into microphone cables.

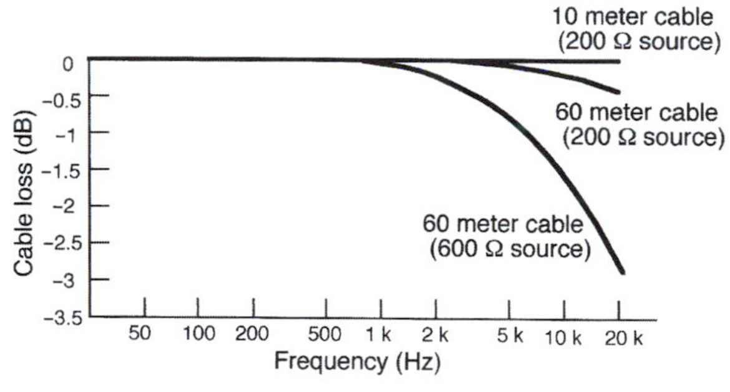

FIGURE 4.17
Effects of cable length and microphone impedance on HF response.

Cables of the so-called "starquad" configuration (developed early in telephony) can reduce interference by up to 10 dB relative to the normal two-conductor configuration. The starquad configuration is shown in Figure 4.18. Here, four conductors within the braided shield are twisted throughout the run of the cable. Diagonally opposite pairs are coupled and connected at each end of the cable to pins 2 and 3. The twisting of the pairs ensures that induced noise components are equal in each leg of the balanced signal pair, resulting in cancellation of noise components at the receiving end.

Physical characteristics of cables

No engineer should ever stint on cable quality. The best cables available are of the starquad configuration, supple, and are easily coiled. In normal use, cables may be stepped on, crimped by doors and wheels on roll-about equipment, and otherwise subjected to daily abuse. In general, braided shield is preferable to wound foil shield; however, in permanent installations this may not be important.

Interconnecting hardware should be chosen for good fit, with an awareness that not all brands of connectors easily work together, even though they nominally meet the standards of XLR male and female receptacles.

Microphone "snakes" are made up of a number of individual microphone cables enclosed in one outer sheath, normally numbering 12 to 16 pairs. In an effort to keep the size down, foil inner shields are normally used on each pair. At the sending end it is preferable to have the snake terminate in a metal box with XLR female receptacles, rather than a fan-out of female XLR receptacles. This recommendation is based on ease of reassignment of microphones by cable number, should that be necessary. At the receiving end a generous fan-out should be provided for easy access to the console's inputs, with each cable number clearly and permanently indicated.

Capacitive coupling between pairs is generally quite low, and for signals at microphone level we can usually ignore it. However, for some applications in sound reinforcement, there may be microphone signals sent in one direction along the snake with concurrent line level monitoring signals sent in the opposite direction. This is not recommended in general recording practice.

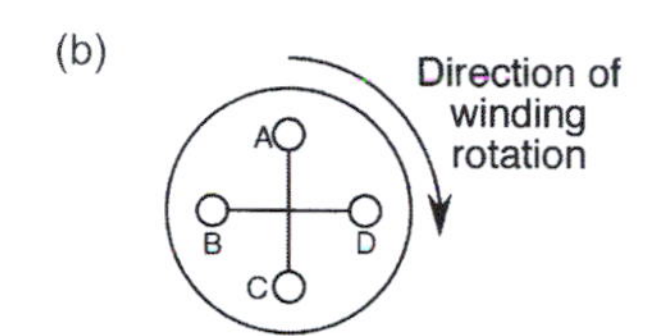

FIGURE 4.18
Details of starquad microphone cable construction.

Cable testing

A microphone emulator circuit is shown in Figure 4.19. The circuit shown is single-ended (unbalanced) and should not necessarily be used to assess details of interference. However, it is excellent for determining HF cable losses over long runs. The 40 dB input pad will reduce an applied signal of, say, 0.2 volts to 20 mV, a typical capacitor microphone sensitivity rating at 1 pascal. The 20 mV signal appears across a 200 ohm resistor and thus simulates the output of an actual microphone.

The circuit shown in Figure 4.20(a) is useful for simple cable continuity and leakage testing. Measurement across like pin numbers should result in a virtual short-circuit. Measurements across different pin numbers should be made with the ohmmeter set to a high resistance range in order to identify any stray leakage between wire pairs. The more sophisticated cable tester shown in Figure 4.20(b) is typical of items offered by a number of companies. As an active device it can be used to test the following:

1. Cable continuity
2. Cable intermittent failure
3. Phantom power integrity
4. In-circuit continuity (via test tones)
5. Cable grounding integrity.

Three types of cables can be tested using this system.

Polarity checking of a microphone/cable combination can be made using a system such as is shown in Figure 4.21. Here, a LF positive-going acoustical pulse is fed to the microphone and analyzed at the other end of the cable run using an analyzer that detects the presence of a positive- or negative-going pulse. If negative polarity is indicated, then there is near certainty that a cable has been miswired or that an item of in-line equipment is inverting the signal. Virtually all modern amplifiers and signal processing devices are non-inverting. Thus, most electrical polarity problems encountered today are the result of local wiring mistakes.

Stage preamplifiers and fiber optic lines

While normal phantom powering can easily handle cable runs up to the 100 meter range, there may be some environments where RF interference is a chronic problem. One option is to use stage microphone-to-line preamplifiers, which will raise the level of the transmitted signals by 40 or 50 dB with very low output impedance, thus providing considerable immunity to induced noise and losses due to long cable runs. Stage preamplifiers are available with remote controlled gain adjustment.

Another option is to provide on-stage digital conversion for each microphone and transmit the individual digital signals via fiber optic cables to a digital console in the control room. In this form, the signals may be sent over extremely long runs with no loss.

SYSTEM CONSIDERATIONS

Ground loops

One of the most common audio transmission problems is the *ground loop*. Figure 4.22 shows how a ground loop is created. Electronic devices are cascaded as shown. Note that there is a ground path, not only in the cables that connect the devices but also in the metal rack that houses them. Any AC power

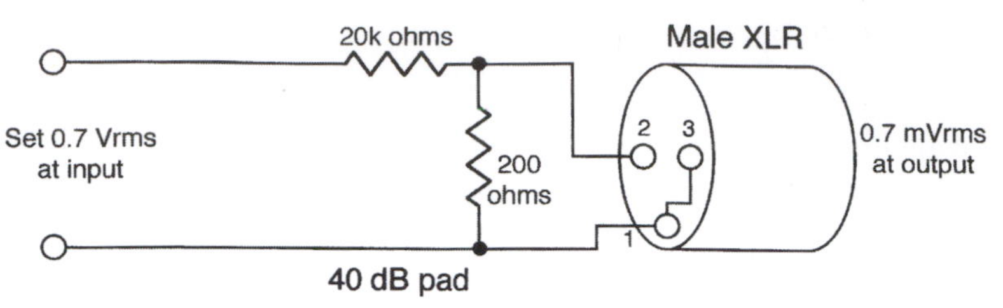

FIGURE 4.19
An unbalanced microphone emulator circuit providing a loss of 40 dB.

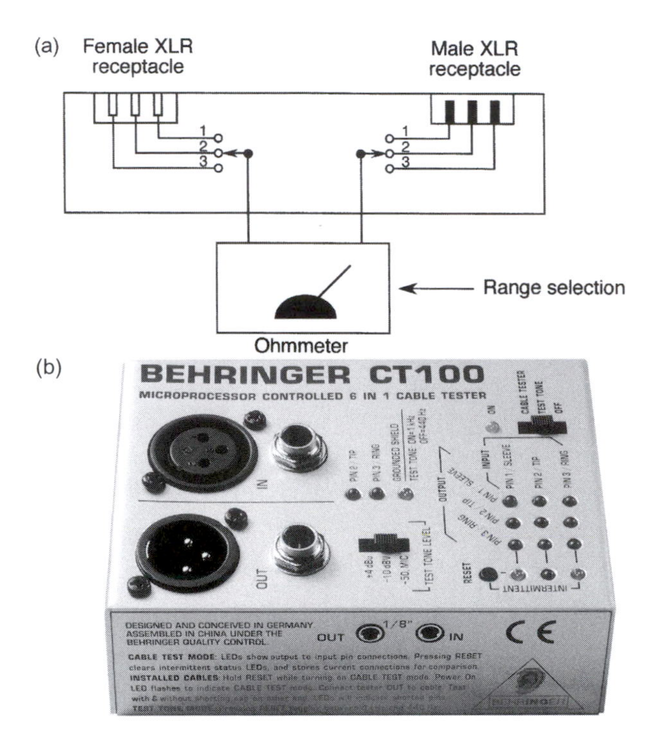

FIGURE 4.20
Details of a simple continuity and leakage detector for microphone cables (a); a more complex cable testing apparatus (b). (Photo courtesy of Behringer Audio Technology.)

flowing in the rack will generate an external magnetic field, and that magnetic flux which flows through the loop will induce a small current through it.

It is necessary to break the continuity in the loop, and this is normally done as shown in Figure 4.23. Microphone cables are grounded at the input to the mixer in order to isolate the microphone from electrostatic (RF) interference and also to provide a continuity path for phantom powering. From the mixer onward, it is customary to connect the wiring shield at the output end only as shown. In this manner the continuity of the ground loop conductive path is broken, thus minimizing the severity of the ground loop.

Gain structure

Many audio transmission problems begin with the microphone and the initial input gain setting looking into the console. Modern consoles are well engineered, and essentially all the operator has to do is to adjust the input and output faders to their nominal zero calibration positions and then set normal operating levels on each input channel using the input trim control. This may sound simple but it requires a little practice.

If the operator sets the microphone trim too low, then there is the risk that input noise in the console will become audible, perhaps even swamping out the noise of the microphone; if the trim is set too

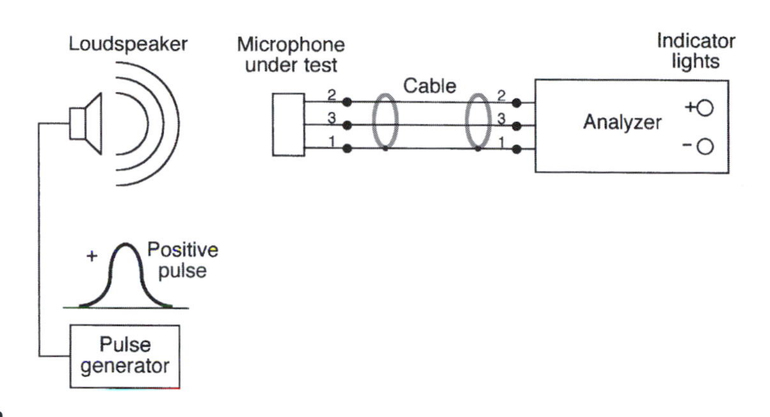

FIGURE 4.21
Details of a system for checking signal polarity of microphones.

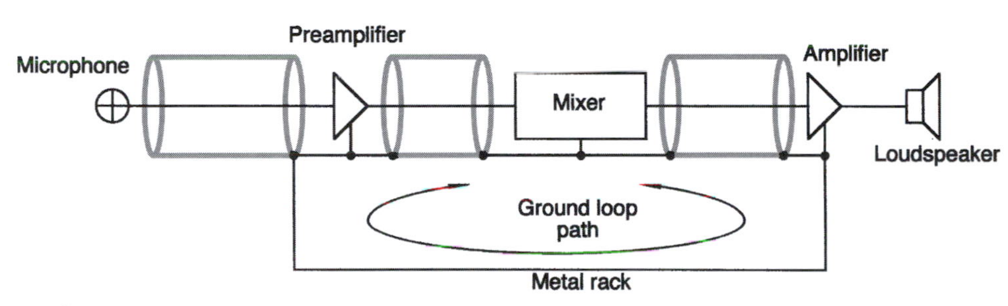

FIGURE 4.22
Generation of a ground loop.

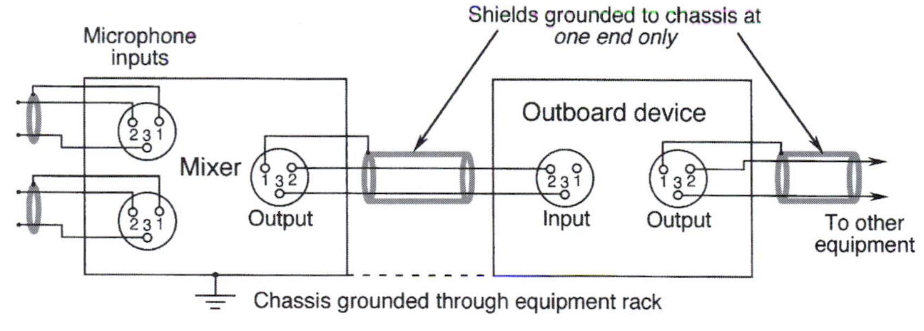

FIGURE 4.23
Breaking the ground loop.

high, then there is the risk that the console's input stage will be overloaded on loud input signals. Since the console has a wider overall operating dynamic range than a microphone, the requirement is only to adjust the trim so that the program delivered from the studio via the microphone will fit comfortably within the total dynamic range of the console.

Figure 4.24 shows a level diagram for a single input channel through a typical console. Let us assume that we have a microphone whose sensitivity is 21 mV for an acoustical signal of 94dB L_P and whose self-noise floor is 10dB(A). Further assume that the microphone's maximum output level (0.5% THD) is 135adB L_P.

The microphone's output is equivalent to –31dBu, where dBu is defined as:

$$\text{dBu} = 20\log(\text{signal voltage}/0.775) \qquad (8.1)$$

A level of 94dB L_P is typical of many instruments in the pop/rock studio when picked up at fairly close quarters and at normal playing levels, and thus the engineer will set the input trim control so that this electrical level will produce console output meter deflections in the normal operating range.

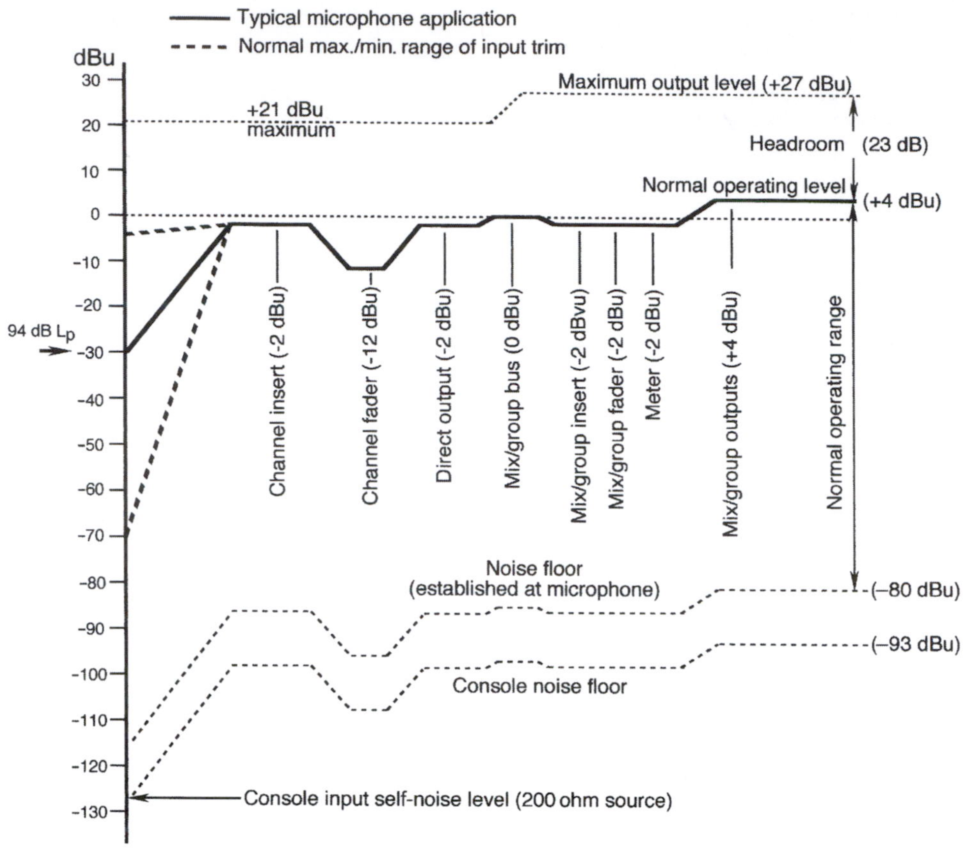

FIGURE 4.24
Typical console gain structure or level diagram.

Note also that with this input setting the microphone's noise floor of 10 dB(A) will correspond to an electrical level 84 dB *lower* than –31 dBu, or –115 dBu. This level is about 13 dB below the noise floor of the console, which is in the range of –128 dBu. Thus, it is clear that the audio channel's noise floor will be essentially that of the microphone, with little contribution from the console's input circuitry. As the signal progresses through the remainder of the console, the noise floor does not change relative to normal operating level.

The microphone input channel has the capability, for undistorted console output, of handling an input signal that is 23 dB higher than 94 dB, or 117 dB L_p. If the microphone's input signal exceeds this amount on a continuing basis, then the engineer must adjust the input trim to accommodate it. If the signal exceeds this amount only occasionally then the engineer may make necessary adjustments at the input fader.

The microphone's undistorted output level extends up to 135 dB L_p, which is 18 dB higher than the console's maximum output capability. As such, this microphone may be used in a wide variety of acoustical environments, and the only adjustments that need to be made are the microphone's output pad (for very loud signal conditions) or resetting the input trim as required.

Summing multiple inputs into a single output channel

During remix operations, a number of microphones are often fed to a given output bus. For each new microphone input added to a given bus it is apparent that the overall input levels must be adjusted so that the signal fed downstream is uniform in level. Figure 4.25 shows the way that individual inputs should be adjusted so that their sum will be consistent, whether one or more microphone inputs are used.

Recording engineers engaged in direct-to-stereo recording, or in making stereo mixes from multitrack tapes, must be aware of the process discussed here.

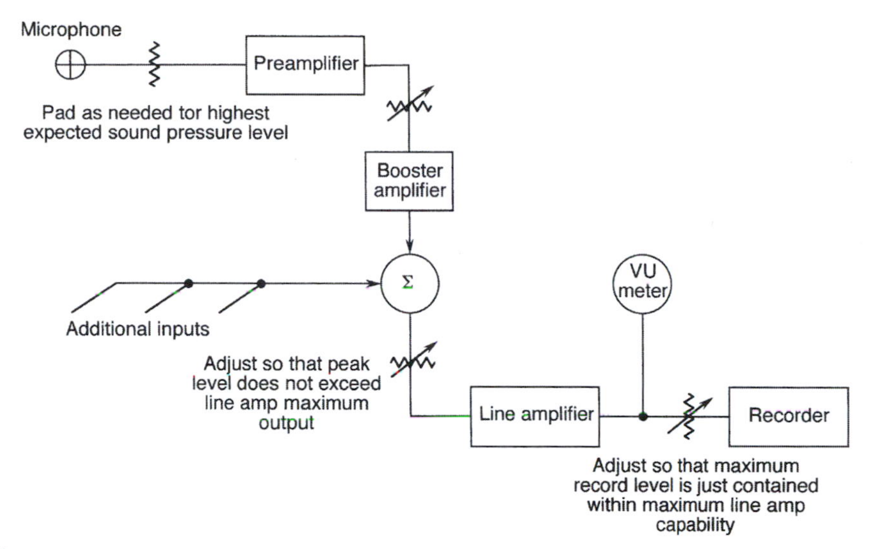

FIGURE 4.25
Combining multiple inputs.

MICROPHONES USING THE RF (RADIO FREQUENCY) TRANSMISSION PRINCIPLE

The capacitor microphone operates on the principle of variable capacitance with fixed charge producing a variable signal output voltage. There is another method of deriving a signal from variable capacitance that has been successfully used by only a few microphone manufacturers, and that is the RF transmission principle. Today, only the Sennheiser company of Germany manufactures such microphones on a large scale.

The RF capacitor microphone has exactly the same acoustical characteristics as a DC polarized microphone of the same physical geometry. The only difference is the method of converting the capacitance variations into a signal output.

Hibbing (1994) gives a complete account of the history and analysis of RF microphones. He describes two general methods, and simplified circuits for each are shown in Figure 4.26. The circuit shown at (a) works on the phase modulation (PM) principle and as such resembles a small FM transmitter-receiver combination in a single microphone package. The variable capacitance alters the tuning of a resonant circuit; adjacent to the tuned circuit are a discriminator section and a fixed oscillator operating in the range of 8 MHz. All three sections are mutually coupled through RF transformers. The alternations of tuning resulting from audio pressure variations affect the balance of upper and lower sections of the discriminator, and an audio signal output is present at the output.

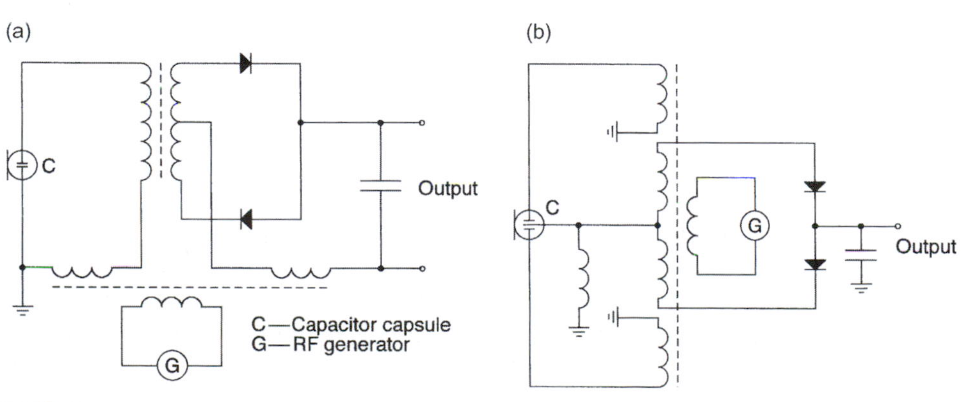

FIGURE 4.26
RF microphones: simplified circuit for phase modulation system (a); simplified circuit for balanced bridge operation (b). (Data after Sennheiser.)

Circuit details of the AM bridge design are shown in Figure 4.26(b). Here, a dual-backplate push-pull diaphragm is used to advantage as a push-pull voltage divider, distributing the RF signal equally to both sides of the bridge circuit when in its rest position. Under audio excitation, the bridge output will be proportional to diaphragm excursion, providing a high degree of linearity. It is this modulation principle that is used in the Sennheiser MKH-20 series of studio microphones.

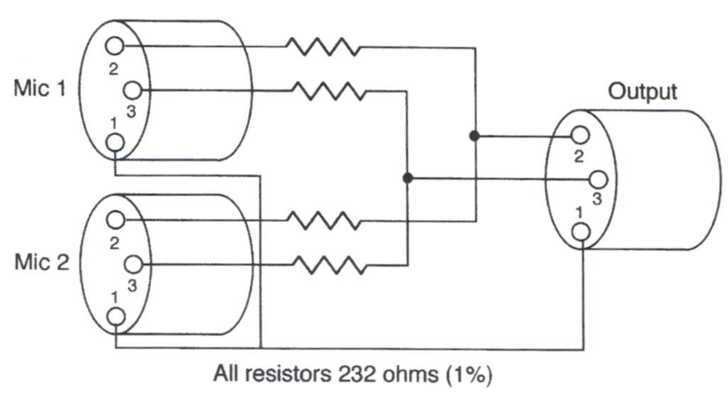

FIGURE 4.27
A circuit for paralleling the output of two microphones. (Data after Shure Inc.)

Early RF microphones were prone to internal instability and on occasion interference from local radio transmissions. Over the years the art has literally "fine-tuned" itself, and the modern Sennheiser MKH-series microphones are superb performers in all respects.

PARALLEL OPERATION OF MICROPHONES

While not generally recommended, it is possible to operate two microphones at a single console input if phantom power is not required. The price paid for this is a reduction in level of about 6 dB for each microphone and of course the inability to adjust levels individually. The circuit shown in Figure 4.27 shows how this can be done.

DIGITAL MICROPHONES

So-called "digital microphones" have entered the market place during the last six or so years. Strictly speaking, these models are not actually "digital" in the specific sense of directly generating a digital output code from the diaphragm. Rather, they make use of traditional DC bias and preamplification of the analog signal at the diaphragm. It is only after this stage that analog-to-digital conversion takes place.

The advantage of these microphones is that certain problems in digital processing can be dealt with earlier, rather than later, in the audio chain. For example, the useful signal-to-noise ratio of a well-designed 25 mm (1 in) condenser diaphragm can be in the range of about 125 to 135 dB. An ideal 20-bit system is capable of a signal-to-noise range of 120 dB, and in a traditional recording system this will require truncation of the available dynamic range of the microphone by about 10 dB. In and of itself, this may or may not be a problem, depending on other electrical and acoustical considerations in the actual studio environment.

In the beyerdynamic model MCD100 series, the capsule looks into a 22-bit conversion system directly when the acoustical level is high (greater than 124 dB L_P). For normal studio levels (less than about 100 dB L_P), −10 or −20 dB padding can be inserted ahead of the digital conversion stage in order to optimize the bit depth. Sophisticated level control prevents the system from

(a)

(b)

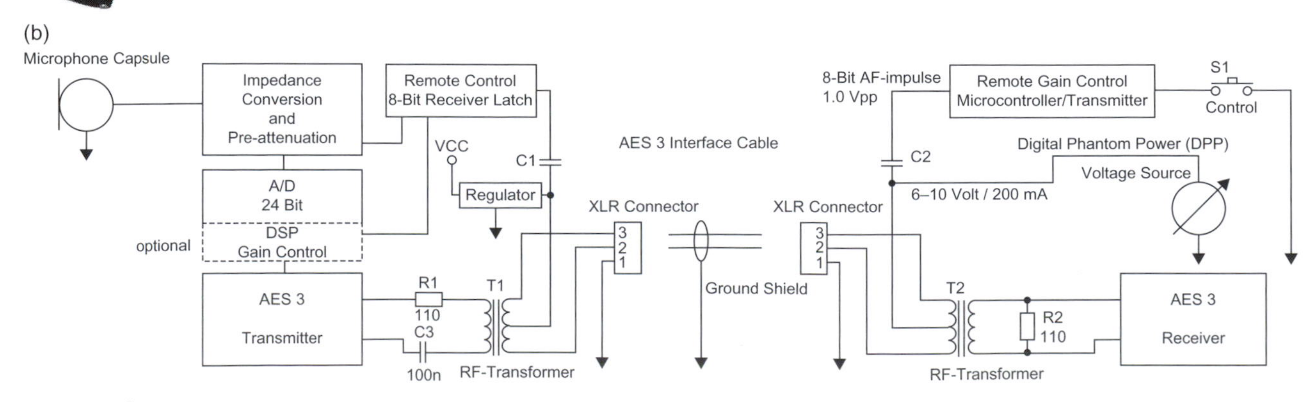

FIGURE 4.28
Details of beyerdynamic digital microphone system: view of microphone (a); signal flow diagram (b). (Data courtesy of beyerdynamic.)

going into digital clipping. The microphone and associated signal flow diagram is shown in Figure 4.28(a) and (b).

The Neumann Solution-D uses two 24-bit A-to-D converters operating in parallel and offset by 24 dB. These two digital signals are seamlessly recombined in the digital domain to produce a single digital output signal with a net resolution of 28 bits (Monforte, 2001). Figure 4.29(a) shows a view of the Solution-D microphone, and a signal flow diagram is shown in Figure 4.29(b).

Both of these microphone systems have additional digital features, including variable sampling rates, various interface formats, some degree of built-in digital signal processing, and the ability to respond to certain user commands via the digital bus. The Audio Engineering Society (AES) is actively pursuing interface standards for this new class of products.

FIGURE 4.29
Details of Neumann Solution-D digital microphone system: view of microphone (a); signal flow diagram (b). (Data courtesy of Neumann/ USA.)

REFERENCES

Hibbing, M., 1994. High Quality RF Condenser Microphones, Chapter 4 in *Microphone Engineering Handbook*, ed. M. Gayford. London: Focal Press.

Monforte, J., 2001. Neumann Solution-D Microphone, *Mix Magazine*.

Perkins, C., 1994. Microphone Preamplifiers – A Primer, Sound & Video Contractor 12, no.2.

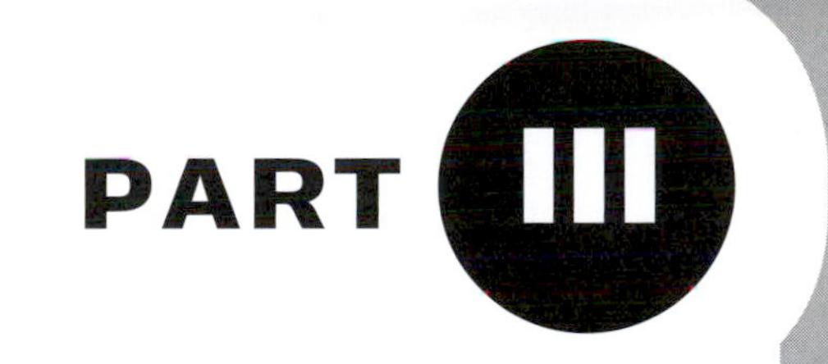

Preamplifiers, Mixers, and Interconnects

PREFACE

The third section of this book "Preamplifiers, Mixers, & Interconnects" gets to grips with the part that small-signal electronics plays in sound systems.

In the first chapter, Francis Rumsey and Tim McCormick give a thorough and comprehensive account of mixing console principles and recording methods, diving in at once with a description of a simple six-channel mixer that has on each channel microphone and line inputs, an input gain control, bass and treble EQ controls, a pan control, a PFL button and a slide fader with +10 dB at the top. Its master section has output faders, two output level meters, and a monitor volume control. This is representative of millions of little mixers, and they can be seen in the window of most music shops.

Each feature of this mixer is carefully examined, with special sections on the faders, the pan-pots with their special control laws, and the PFL (Pre-Fade Listen) system.

From here we move on to the needs of multitrack recording, for which a more sophisticated mixing console is required, because of the need to play back existing tracks at the same time as recording new ones, so the new material can be laid down in synchrony. Twice the number of signal paths are therefore needed; one path from microphone to recorder, and one return path from recorder to the monitor system.

These two paths can be arranged in two different ways, which brings us to the next section of this chapter, the distinction between split and in-line console configurations. The split console is the more obvious of the two; it puts the input channels on one side (normally the left), a master control section in the middle, and the replay monitor mixer on the right. In many ways there are two consoles in one chassis. Since there must be as many replay monitor channels as there are recorder tracks, this section takes up a lot of room, but it has the great advantage that it is conceptually simple.

The in-line configuration is quite different. Now the replay monitor sections are integrated with the input channels, so they can share resources such as EQ and auxiliary sends. This tends to give a console which is a little deeper, due to the increased channel length, but much less wide. The only real drawback is that it is less intuitive and takes a bit of learning.

The next section deals with channel grouping, and is illustrated by a picture of the Soundcraft Sapphyre (yep, one of mine). Both sorts of grouping—mixing several channels to a group and controlling that, and VCA group control—are covered.

Rumsey and McCormick then move on to examine more closely the features of a sophisticated recording console, such as EQ, dynamics sections, monitor controls and effects returns. The chapter concludes by looking at basic operational techniques, technical specifications, metering, and console automation.

Some readers may be aware that I spent a good number of years engaged in mixing console design, as Chief Engineer at Soundcraft Electronics, and I think this chapter is the best concise description of mixer operation that I have come across.

CHAPTER 5

Mixers

Sound and Recording by Francis Rumsey and Tim McCormick

This chapter describes the principles and basic operation of audio mixers. It begins with a description of a simple system and moves on to consider the facilities of large-scale multitrack systems. Because many design and layout concepts of analog mixers find a place in more recent digital mixers, these aspects are covered here in a fairly generic way. Those features found more commonly only in digital systems are described towards the end of the chapter.

In its simplest form an audio mixer combines several incoming signals into a single output signal. This cannot be achieved simply by connecting all the incoming signals in parallel and then feeding them into a single input because they may influence each other. The signals need to be isolated from each other. Individual control of at least the level of each signal is also required.

In practice, mixers do more than simply mix. They can provide phantom power for capacitor microphones; pan control (whereby each signal can be placed in any desired position in a stereo image); filtering and equalization; routing facilities; and monitoring facilities, whereby one of a number of sources can be routed to loudspeakers for listening, often without affecting the mixer's main output.

A SIMPLE SIX-CHANNEL ANALOG MIXER

Overview

By way of example, a simple six-channel analog mixer will be considered, having six inputs and two outputs (for stereo). Figure 5.1 illustrates such a notional six-into-two mixer with basic facilities. It also illustrates the back panel. The inputs illustrated are via XLR-type three-pin latching connectors, and are of a balanced configuration. Separate inputs are provided for microphone and line level signals, although it is possible to encounter systems which simply use one socket switchable to be either mic or line. Many cheap mixers have unbalanced inputs via quarter-inch jack sockets, or even "phono" sockets such as are found on hi-fi amplifiers. Some mixers employ balanced XLR inputs for microphones, but unbalanced jack or phono inputs for line level signals, since the higher-level line signal is less susceptible to noise and interference, and will probably have traveled a shorter distance.

On some larger mixers a relatively small number of multipin connectors are provided, and multicore cables link these to a large jackfield which consists of rows of jack sockets mounted in a rack, each being individually labeled. All inputs and outputs will appear on this jackfield, and patch cords of a meter or so in length with GPO-type jack plugs at each end enable the inputs and outputs to be interfaced with other equipment and tie-lines in any appropriate combination. (The jackfield is more fully described in "Patchfield or jackfield," below.)

The outputs are also on three-pin XLR-type connectors. The convention for these audio connections is that inputs have sockets or holes, outputs have pins. This means that the pins of the connectors "point" in the direction of the signal, and therefore one should never be confused as to which connectors

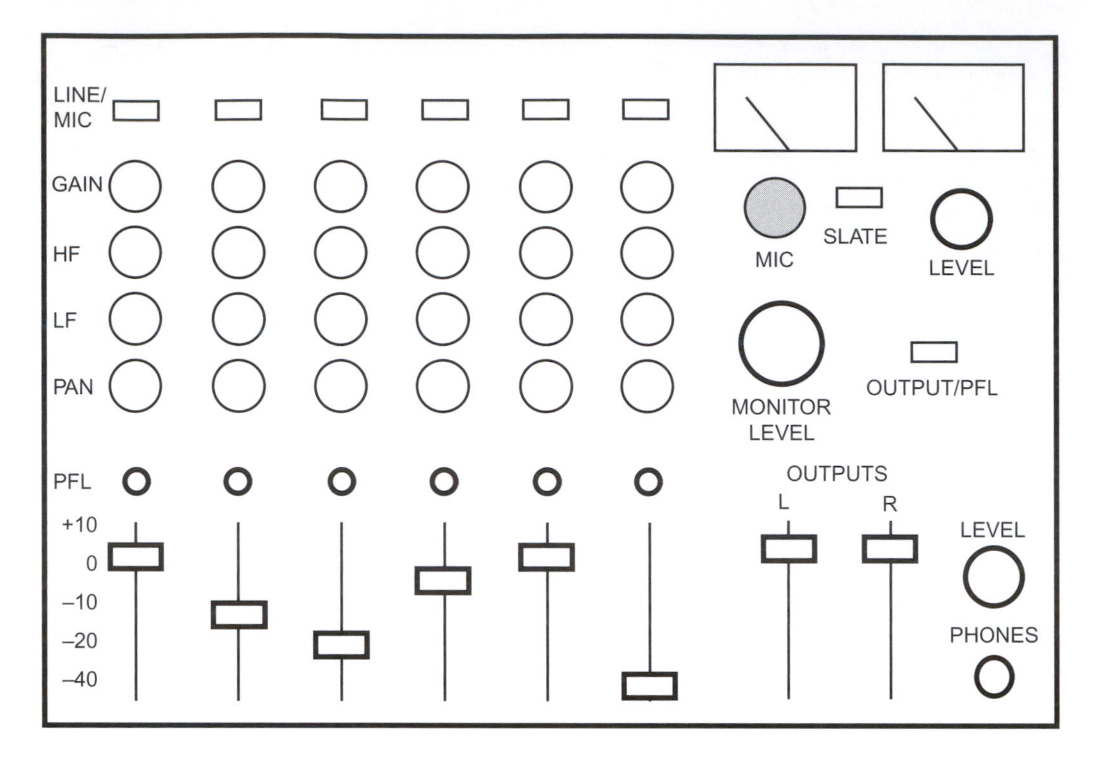

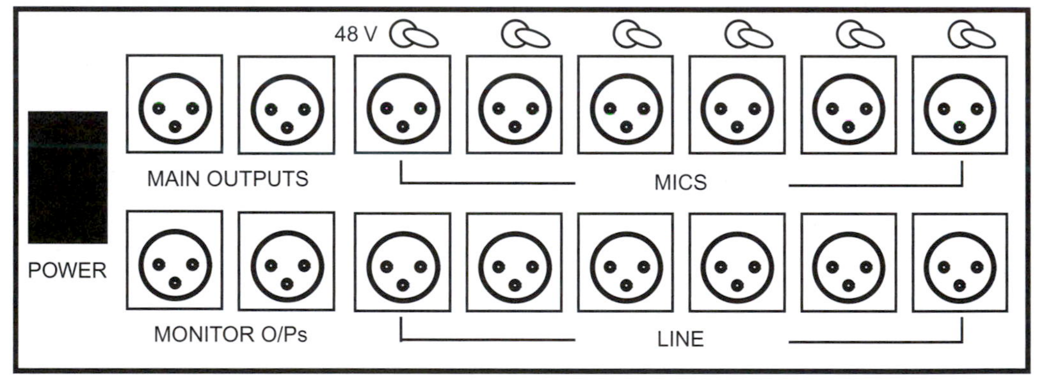

FIGURE 5.1
Front panel and rear connectors of a typical simple six-channel mixer.

are inputs and which are outputs. The microphone inputs also have a switch each for supplying 48V phantom power to the microphones if required. Sometimes this is found on the input module itself, or sometimes on the power supply, switching 48V for all the inputs at once.

The term "bus" is frequently used to describe a signal path within the mixer to which a number of signals can be attached and thus combined. For instance, routing some input channels to the "stereo bus" conveys those channels to the stereo output in the manner of a bus journey in the conventional everyday sense. A bus is therefore a mixing path to which signals can be attached.

Input channels

All the input channels in this example are identical, and so only one will be described. The first control in the signal chain is input gain or sensitivity. This control adjusts the degree of amplification provided by the input amplifier, and is often labeled in decibels, either in detented steps or continuously variable. Inputs are normally switchable between mic and line. In "mic" position, depending on the output level of the microphone connected to the channel, the input gain is adjusted to raise the signal to a suitable line level, and up to 80dB or so of gain is usually available here (see "Miscellaneous features,"

FACT FILE 5.1 FADER FACTS

Fader law

Channel and output faders, and also rotary level controls, can have one of two laws: linear or logarithmic (the latter sometimes also termed "audio taper"). A linear law means that a control will alter the level of a signal (or the degree of cut and boost in a tone control circuit) in a linear fashion: that is, a control setting midway between maximum and minimum will attenuate a signal by half its voltage, i.e., −6 dB. But this is not a very good law for an audio level control because a 6 dB drop in level does not produce a subjective halving of loudness. Additionally, the rest of the scaling (−10 dB, −20 dB, −30 dB, and so on) has to be accommodated within the lower half of the control"s travel, so the top half gives control over a mere 6 dB, the bottom half all the rest.

For level control, therefore, the logarithmic or "log" law is used whereby a non-linear voltage relationship is employed in order to produce an approximately even spacing when the control is calibrated in decibels, since the decibel scale is logarithmic. A log fader will therefore attenuate a signal by 10 dB at a point approximately a quarter of the way down from the top of its travel. Equal dB increments will then be fairly evenly spaced below this point. A rotary log pot ("pot" is short for potentiometer) will have its maximum level usually set at the 5 o'clock position and, the −10 dB point will be around the 2 o'clock position. An even subjective attenuation of volume level is therefore produced by the log law as the control is gradually turned down. A linear law causes very little to happen subjectively until one reaches the lowest quarter of the range, at which point most of the effect takes place.

The linear law is, however, used where a symmetrical effect is required about the central position; for example, the cut and boost control of a tone control section will have a central zero position about which the signal is cut and boosted to an equal extent either side of this.

Electrical quality

There are two types of electrical track in use in analog faders, along which a conductive "wiper" runs as the fader is moved to vary its resistance. One type of track consists of a carbon element, and is cheap to manufacture. The quality of such carbon tracks is, however, not very consistent and the "feel" of the fader is often scrapy or grainy, and as it is moved the sound tends to jump from one level to another in a series of tiny stages rather than in a continuous manner. The carbon track wears out rather quickly, and can become unreliable.

The second type employs a conductive plastic track. Here, an electrically conductive material is diffused into a strip of plastic in a controlled manner to give the desired resistance value and law (linear or log). Much more expensive than the carbon track, the conductive plastic track gives smooth, continuous operation and maintains this standard over a long period of time. It is the only serious choice for professional-quality equipment.

below). In "line" position little amplification is used and the gain control normally provides adjustment either side of unity gain (0 dB), perhaps ±20 dB either way, allowing the connection of high-level signals from such devices as CD players, tape machines and musical keyboards.

The equalization or EQ section which follows (see "Equalizer section," below) has only two bands in this example – treble and bass – and these provide boost and cut of around ±12 dB over broad low- and high-frequency bands (e.g., centered on 100 Hz and 10 kHz). This section can be used like the tone controls on a hi-fi amplifier to adjust the spectral balance of the signal. The fader controls the overall level of the channel, usually offering a small amount of gain (up to 12 dB) and infinite attenuation. The law of the fader is specially designed for audio purposes (see Fact File 5.1). The pan control divides the mono input signal between left and right mixer outputs, in order to position the signal in a virtual stereo sound stage (see Fact File 5.2).

FACT FILE 5.2 PAN CONTROL

The pan control on a mixer is used for positioning a signal somewhere between left and right in the stereo mix image. It does this by splitting a single signal from the output of a fader into two signals (left and right), setting the position in the image by varying the level difference between left and right channels. It is thus not the same as the balance control on a stereo amplifier, which takes in a stereo signal and simply varies the relative levels between the two channels. A typical pan-pot law would look similar to that shown in the diagram, and ensures a roughly constant perceived level of sound as the source is panned from left to right in stereo. The output of the pan-pot usually feeds the left and right channels of the stereo mix bus (the two main summation lines which combine the outputs of all channels on the mixer), although on mixers with more than two mix buses the pan-pot's output may be switched to pan between any pair of buses, or perhaps simply between odd and even groups (see Fact File 5.4).

On some older consoles, four-way routing is provided to a quadraphonic mix bus, with a left–right pot and a front–back pot. These are rare now. Many stereo pan-pots use a dual-gang variable resistor which follows a law giving a 4.5 dB level drop to each channel when panned centrally, compared with the level sent to either channel at the extremes. The 4.5 dB figure is a compromise between the –3 dB and –6 dB laws. Pan-pots which only drop the level by 3 dB in the center cause a rise in level of any centrally panned signal if a mono sum is derived from the left and right outputs of that channel, since two identical signals summed together will give a rise in level of 6 dB. A pot which gives a 6 dB drop in the center results in no level rise for centrally panned signals in the mono sum. Unfortunately, the 3 dB drop works best for stereo reproduction, resulting in no perceived level rise for centrally panned signals.

Only about 18 dB of level difference is actually required between left and right channels to give the impression that a source is either fully left or fully right in a loudspeaker stereo image, but most pan-pots are designed to provide full attenuation of one channel when rotated fully towards the other. This allows for the two buses between which signals are panned to be treated independently, such as when a pan control is used to route a signal either to odd or even channels of a multitrack bus (see "Routing section," below).

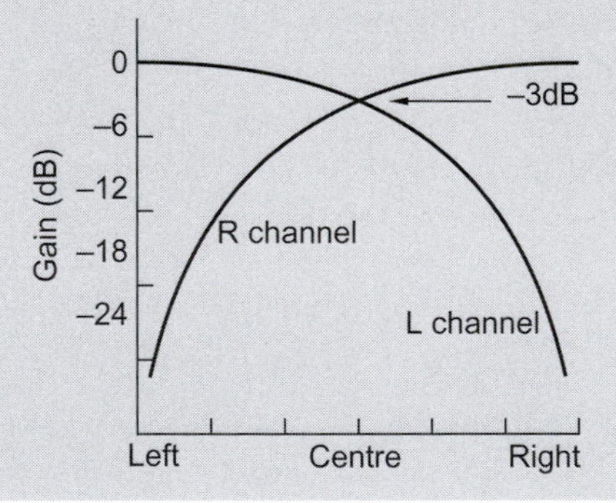

Output section

The two main output faders (left and right) control the overall level of all the channel signals which have been summed on the left and right mix buses, as shown in the block diagram (Figure 5.2). The outputs of these faders (often called the group outputs) feed the main output connectors on the rear panel, and an internal feed is taken from the main outputs to the monitor selector. The monitor selector on this simple example can be switched to route either the main outputs or the PFL bus (see Fact File 5.3) to the loudspeakers. The monitor gain control adjusts the loudspeaker output level without affecting the main line output level, but of course any changes made to the main fader gain will affect the monitor output.

The slate facility on this example allows for a small microphone mounted in the mixer to be routed to the main outputs, so that comments from the engineer (such as take numbers) can be recorded on a tape machine connected to the main outputs. A rotary control adjusts the slate level.

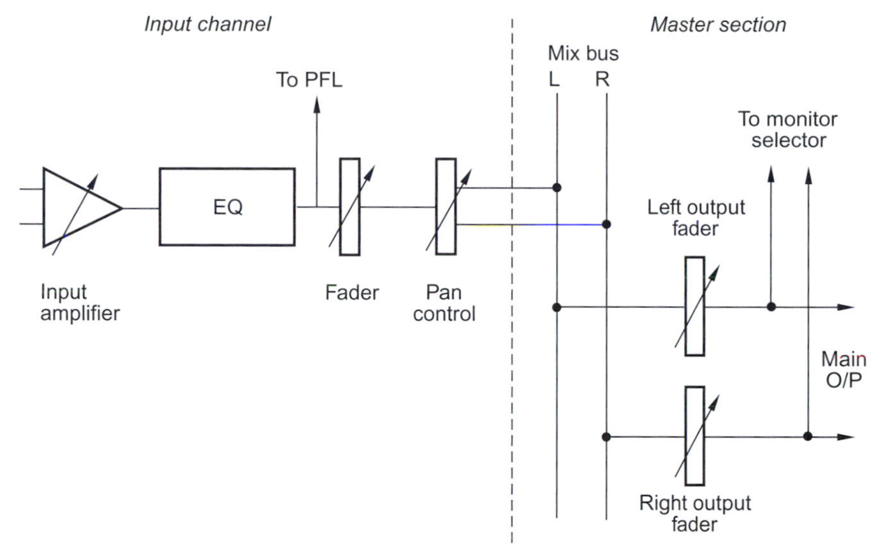

FIGURE 5.2
Block diagram of a typical signal path from channel input to main output on a simple mixer.

FACT FILE 5.3 PRE-FADE LISTEN (PFL)

Pre-fade listen, or PFL, is a facility which enables a signal to be monitored without routing it to the main outputs of the mixer. It also provides a means for listening to a signal in isolation in order to adjust its level or EQ.

Normally, a separate mono mixing bus runs the length of the console picking up PFL outputs from each channel. A PFL switch on each channel routes the signal from before the fader of that channel to the PFL bus (see diagram), sometimes at the same time as activating internal logic which switches the mixer's monitor outputs to monitor the PFL bus. If no such logic exists, the mixer's monitor selector will allow for the selection of PFL, in which position the monitors will reproduce any channel currently with its PFL button pressed. On some broadcast and live consoles a separate small PFL loudspeaker is provided on the mixer itself, or perhaps on a separate output, in order that selected sources can be checked without affecting the main monitors.

Sometimes PFL is selected by "overpressing" the channel fader concerned at the bottom of its travel (i.e., pushing it further down). This activates a microswitch which performs the same functions as above. PFL has great advantages in live work and broadcasting, since it allows the engineer to listen to sources before they are faded up (and thus routed to the main outputs which would be carrying the live program). It can also be used in studio recording to isolate sources from all the others without cutting all the other channels, in order to adjust equalization and other processing with greater ease.

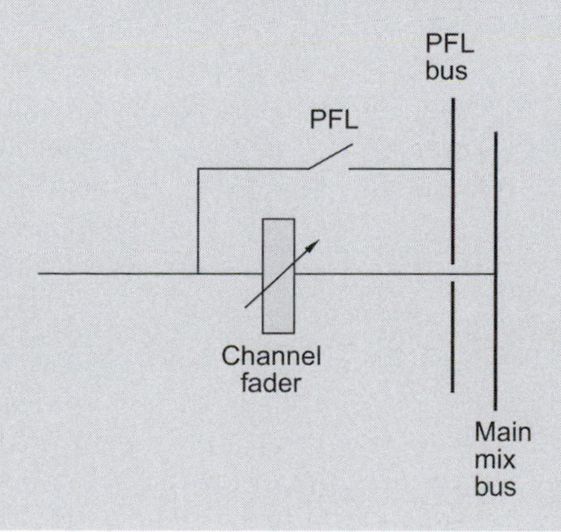

Miscellaneous features

Professional-quality microphones have an output impedance of around 200 ohms, and the balanced microphone inputs will have an input impedance of between 1000 and 2000 ohms ("2 kΩ", k = thousand). The outputs should have an impedance of around 200 ohms or lower. The headphone output impedance will typically be 100 ohms or so. Small mixers usually have a separate power supply which plugs into the mains. This typically contains a mains transformer, rectifiers and regulating circuitry, and it supplies the mixer with relatively low DC voltages. The main advantage of a separate power supply is that the mains transformer can be sited well away from the mixer, since the alternating 50 Hz mains field around the former can be induced into the audio circuits. This manifests itself as "mains hum" which is only really effectively dealt with by increasing the distance between the mixer and the transformer. Large mixers usually have separate rack-mounting power supplies.

FIGURE 5.3
A compact stereo mixer: the Seemix "Seeport." (Courtesy of Seemix Sound AS)

The above-described mixer is very simple, offering few facilities, but it provides a good basis for the understanding of more complex models. A typical commercial example of a compact mixer is shown in Figure 5.3.

A MULTITRACK MIXER

Overview

The stereo mixer outlined in the previous section forms only half the story in a multitrack recording environment. Conventionally, popular music recording involves at least two distinct stages: the "track-laying" phase and the "mixdown" phase. In the former, musical tracks are layed down on a multitrack recorder in stages, with backing tracks and rhythm tracks being recorded first, followed by lead tracks and vocals. In the mixdown phase, all the previously recorded tracks are played back through the mixer and combined into a stereo or surround mix to form the finished product which goes to be made into a commercial release. Since the widespread adoption of electronic instruments and MIDI-controlled equipment, MIDI-sequenced sound sources are often played directly into the mix in the second stage.

For these reasons, as well as requiring mixdown signal paths from many inputs to a stereo bus the mixer also requires signal paths for routing many input signals to a multitrack recorder. Often it will be necessary to perform both of these functions simultaneously – that is, recording microphone signals to multitrack whilst also mixing the return from multitrack into stereo, so that the engineer and producer can hear what the finished result will sound like, and so that any musicians who may be overdubbing additional tracks can be given a mixed feed of any previously recorded tracks in headphones. The latter is known as the monitor mix and this often forms the basis for the stereo mixdown when the tracklaying job is finished.

So there are two signal paths in this case: one from the microphone or line source to the multitrack recorder, and one from the multitrack recorder back to the stereo mix, as shown in Figure 5.4. The path from the microphone input which usually feeds the multitrack machine will be termed the channel path, whilst the path from the line input or tape return which usually feeds the stereo mix will be termed the monitor path.

It is likely that some basic signal processing such as equalization will be required in the feed to the multitrack recorder (see below), but the more comprehensive signal processing features are usually applied in the mixdown path. The situation used to be somewhat different in the American market where there was a greater tendency to record on multitrack "wet," that is with all effects and EQ, rather than applying the effects on mixdown.

In-line and split configurations

As can be seen from Figure 5.4, there are two complete signal paths, two faders, two sets of EQ, and so on. This takes up space, and there are two ways of arranging this physically, one known as the split-monitoring, or European-style console, the other as the in-line console. The split console is the more obvious of the two, and its physical layout is shown in Figure 5.5. It contains the input channels on one side (usually the left), a master control section in the middle, and the monitor mixer on the other

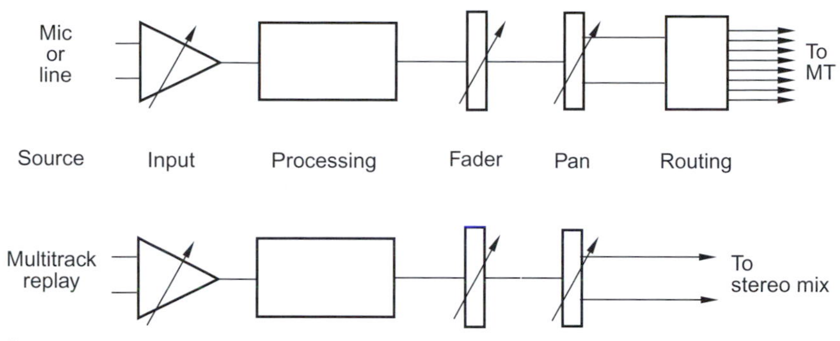

FIGURE 5.4
In multitrack recording two signal paths are needed – one from mic or line input to the multitrack recorder, and one returning from the recorder to contribute to a "monitor" mix.

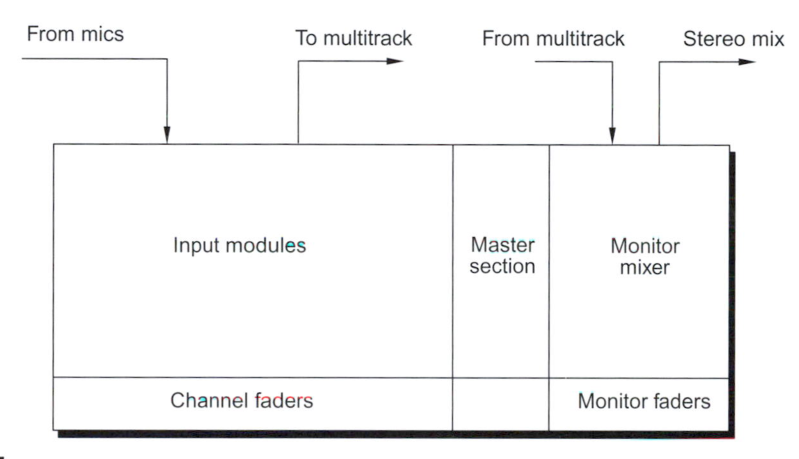

FIGURE 5.5
A typical "split" or "European-style" multitrack mixer has input modules on one side and monitor modules on the other: two separate mixers in effect.

side. So it really is two consoles in one frame. It is necessary to have as many monitor channels as there are tracks on the tape, and these channels are likely to need some signal processing. The monitor mixer is used during track laying for mixing a stereo version of the material that is being recorded, so that everyone can hear a rough mix of what the end result will sound like. On mixdown every input to the console can be routed to the stereo mix bus so as to increase the number of inputs for outboard effects, etc. and so that the comprehensive facilities provided perhaps only on the left side of the console are available for the multitrack returns.

This layout has advantages in that it is easily assimilated in operation, and it makes the channel module less cluttered than the in-line design (described below), but it can make the console very large when a lot of tracks are involved. It can also increase the build cost of the console because of the near doubling in facilities and metalwork required, and it lacks flexibility, especially when switching over from track laying to remixing.

The in-line layout involves the incorporation of the monitor paths from the right-hand side of the split console (the monitor section) into the left side, rather as if the console were sawn in half and the right side merged with the left, as shown in Figure 5.6. In this process a complete monitor signal path is fitted into the module of the same-numbered channel path, making it no more than a matter of a few switches to enable facilities to be shared between the two paths. In such a design each module will contain two faders (one for each signal path), but usually only one EQ section, one set of auxiliary sends (see below), one dynamics control section, and so on, with switches to swap facilities between paths. (A simple example showing only the switching needed to swap one block of processing is shown in Figure 5.7.) Usually this means that it is not possible to have EQ in both the multitrack recording path and the stereo mix path, but some more recent designs have made it possible to split the equalizer so that some frequency-band controls are in the channel path whilst others are in the monitor path. The band ranges are then made to overlap considerably which makes the arrangement quite flexible.

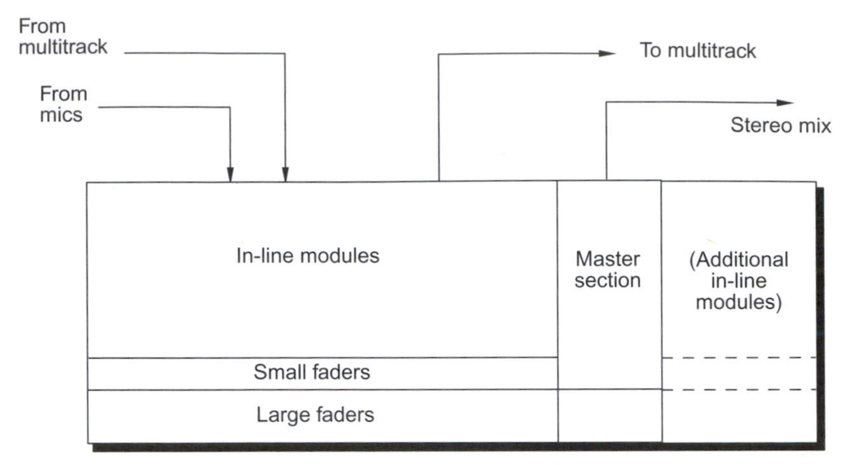

FIGURE 5.6
A typical "in-line" mixer incorporates two signal paths in one module, providing two faders per module (one per path). This has the effect of reducing the size of the mixer for a given number of channels, when compared with a split design.

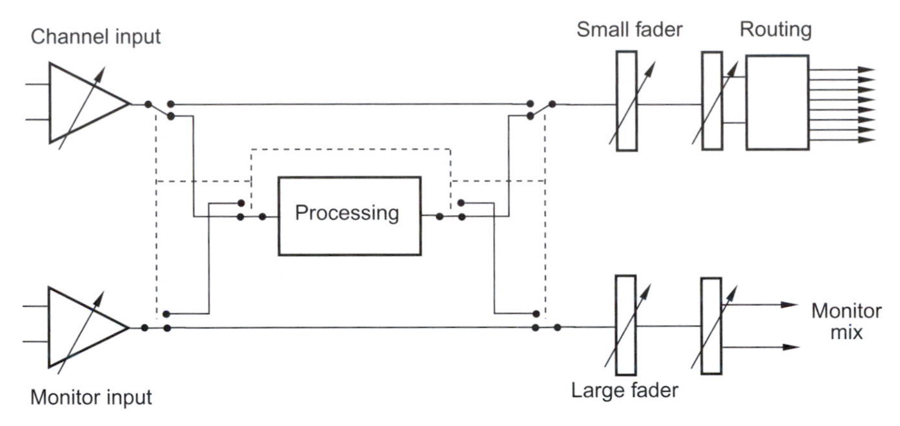

FIGURE 5.7
The in-line design allows for sound processing facilities such as EQ and dynamics to be shared or switched between the signal paths.

Further aspects of the in-line design

It has already been stated that there will be two main faders associated with each channel module in an in-line console: one to control the gain of each signal path. Sometimes the small fader is not a linear slider but a rotary knob. It is not uniformly agreed as to whether the large fader at the bottom of the channel module should normally control the monitor level of the like-numbered tape track or whether it should control the channel output level to multitrack tape. Convention originally had it that American consoles made the large fader the monitor fader in normal operation, while British consoles tended to make it the channel fader. Normally their functions can be swapped over, depending on whether one is mixing down or track laying, either globally (for the whole console), in which case the fader swap will probably happen automatically when switching the console from "recording" to "remix" mode, or on individual channels, in which case the operation is usually performed using a control labeled something like "fader flip," "fader reverse" or "changeover". The process of fader swapping is mostly used for convenience, since more precise control can be exercised over a large fader near the operator than over a small fader which is further away, so the large fader is assigned to the function that is being used most in the current operation. This is coupled with the fact that in an automated console it is almost invariably the large fader that is automated, and the automation is required most in the mixdown process.

Confusion can arise when operating in-line mixers, such as when a microphone signal is fed into, say, mic input 1 and is routed to track 13 on the tape. In such a case the operator will control the monitor level of that track (and therefore the level of that microphone's signal in the stereo mix) on monitor fader 13, whilst the channel fader on module 1 will control the multitrack record level for that mic signal.

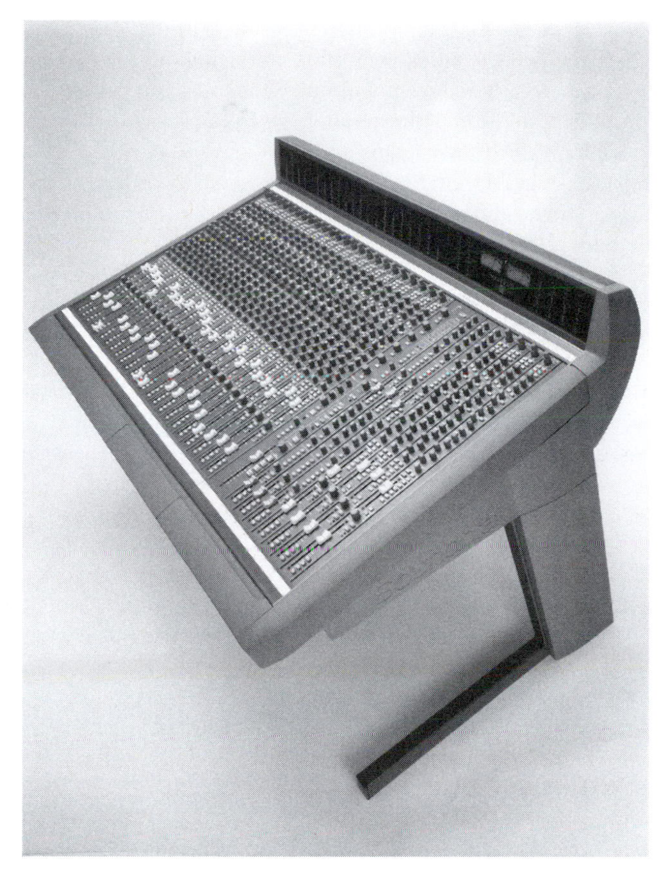

FIGURE 5.8
A typical in-line mixer: the Soundcraft "Sapphyre." (Courtesy of Soundcraft Electronics Ltd)

If a 24 track recorder is in use with the mixer, then monitor faders higher than number 24 will not normally carry a tape return, but will be free for other sources. More than one microphone signal can be routed to each track on the tape (as each multitrack output on the mixer has its own mix bus), so there will be a number of level controls that affect each source's level in the monitor mix, each of which has a different purpose:

- MIC LEVEL TRIM – adjusts the gain of the microphone pre-amplifier at the channel input. Usually located at the top of the module.
- CHANNEL FADER – comes next in the chain and controls the individual level of the mic (or line) signal connected to that module's input before it goes to tape. Located on the same-numbered module as the input. (May be switched to be either the large or small fader, depending on configuration.)
- BUS TRIM or TRACK SUBGROUP – will affect the overall level of all signals routed to a particular track. Usually located with the track routing buttons at the top of the module. Sometimes a channel fader can be made to act as a subgroup master.
- MONITOR FADER –located in the return path from the multitrack recorder to the stereo mix. Does not affect the recorded level on the multitrack tape, but affects the level of this track in the mix. (May be switched to be either the large or small fader, depending on configuration.)

A typical in-line multitrack mixer is shown in Figure 5.8.

CHANNEL GROUPING

Grouping is a term that refers to the simultaneous control of more than one signal at a time. It usually means that one fader controls the levels of a number of slave channels. Two types of channel grouping are currently common: audio grouping and control grouping. The latter is often called VCA grouping, but there are other means of control grouping that are not quite the same as the direct VCA control method. The two approaches have very different results, although initially they may appear to be similar because one fader appears to control a number of signal levels. The primary reason for using group faders of any kind is in order to reduce the number of faders that the engineer has to handle at a time. This can be done in a situation where a number of channels are carrying audio signals that can be faded up and down together. These signals do not all have to be at the same initial level, and indeed one is still free to adjust levels individually within a group. A collection of channels carrying drum sounds, or carrying an orchestral string section, would be examples of suitable groups. The two approaches are described in Fact Files 5.4 and 5.5.

FACT FILE 5.4 AUDIO GROUPS

Audio groups are so called because they create a single audio output which is the sum of a number of channels. A single fader controls the level of the summed signal, and there will be a group output from the console which is effectively a mix of the audio signals in that group, as shown in the diagram. The audio signals from each input to the group are fed via equal-value resistors to the input of a summing or virtual-earth amplifier.

The stereo mix outputs from an in-line console are effectively audio groups, one for the left, one for the right, as they constitute a sum of all the signals routed to the stereo output and include overall level control. In the same way, the multitrack routing buses on an in-line console are also audio groups, as they are sums of all the channels routed to their respective tracks. More obviously, some smaller or older consoles will have routing buttons on each channel module for, say, four audio group destinations, these being really the only way of routing channels to the main outputs.

The master faders for audio groups will often be in the form of four or eight faders in the central section of the console. They may be arranged such that one may pan a channel between odd and even groups, and it would be common for two of these groups (an odd and an even one) to be used as the stereo output in mixdown. It is also common for perhaps eight audio group faders to be used as "subgroups," themselves having routing to the stereo mix, so that channel signals can be made more easily manageable by routing them to a subgroup (or panning between two subgroups) and thence to the main mix via a single level control (the subgroup fader), as shown in the diagram. (Only four subgroups are shown in the diagram, without pan controls. Subgroups 1 and 3 feed the left mix bus, and 2 and 4 feed the right mix bus. Sometimes subgroup outputs can be panned between left and right main outputs.)

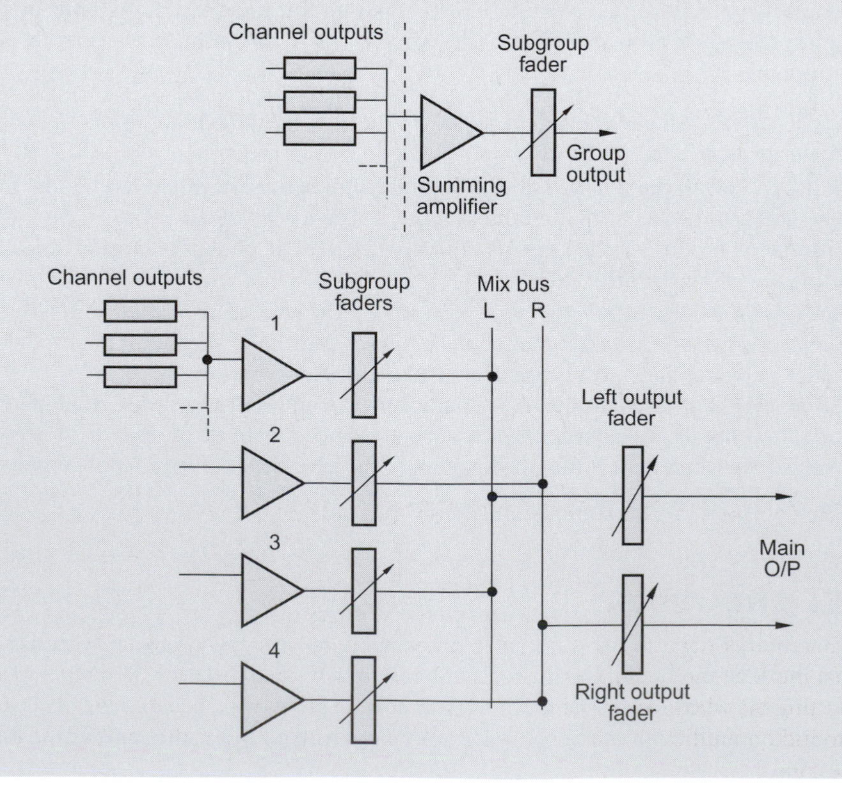

FACT FILE 5.5 CONTROL GROUPS

Control grouping differs from audio grouping primarily because it does not give rise to a single summed audio output for the group: the levels of the faders in the group are controlled from one fader, but their outputs remain separate. Such grouping can be imagined as similar in its effect to a large hand moving many faders at the same time, each fader maintaining its level in relation to the others.

The most common way of achieving control grouping is to use VCAs (voltage-controlled amplifiers), whose gain can be controlled by a DC voltage applied to a control pin. In the VCA fader, audio is not passed through the fader itself but is routed through a VCA, whose gain is controlled by a DC voltage derived from the fader position, as shown in the diagram. So the fader now carries DC instead of audio, and the audio level is controlled indirectly. A more recent alternative to the VCA is the DCA, or digitally controlled attenuator, whose gain is controlled by a binary value instead of a DC voltage. This can be easier to implement in digitally controlled mixers.

Indirect gain control opens up all sorts of new possibilities. The gain of the channel could be controlled externally from a variety of sources, either by combining the voltage from an external controller in an appropriate way with the fader's voltage so that it would still be possible to set the relative level of the channel, or by breaking the direct connection between the DC fader and the VCA so that an automation system could intervene, as discussed in "Automation," below. It becomes possible to see that group faders could be DC controls which could be connected to a number of channel VCAs such that their gains would go up and down together. Further to this, a channel VCA could be assigned to any of the available groups simply by selecting the appropriate DC path: this is often achieved by means of thumbwheel switches on each fader, as shown in the diagram.

Normally, there are dedicated VCA group master faders in a non-automated system. They usually reside in the central section of a mixer and will control the overall levels of any channel faders assigned to them by the thumbwheels by the faders. In such a system, the channel audio outputs would normally be routed to the main mix directly, the grouping affecting the levels of the individual channels in this mix.

In an automated system grouping may be achieved via the automation processor which will allow any fader to be designated as the group master for a particular group. This is possible because the automation processor reads the levels of all the faders, and can use the position of the designated master to modify the data sent back to the other faders in the group (see "Automation," below).

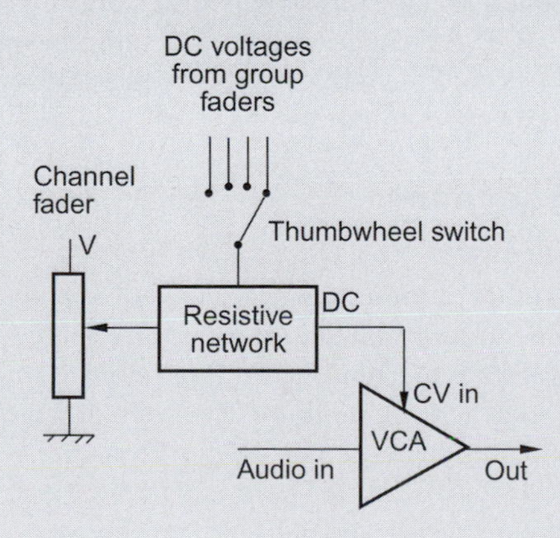

AN OVERVIEW OF TYPICAL MIXER FACILITIES

Most mixing consoles provide a degree of sound signal processing on board, as well as routing to external processing devices. The very least of these facilities is some form of equalization (a means of controlling the gain at various frequencies), and there are few consoles which do not include this. As well as signal processing, there will be a number of switches that make changes to the signal path or operational mode of the console. These may operate on individual channels, or they may function globally (affecting the whole console at once). The following section is a guide to the facilities commonly found on multitrack consoles. Figure 5.9 shows the typical location of these sections on an in-line console module.

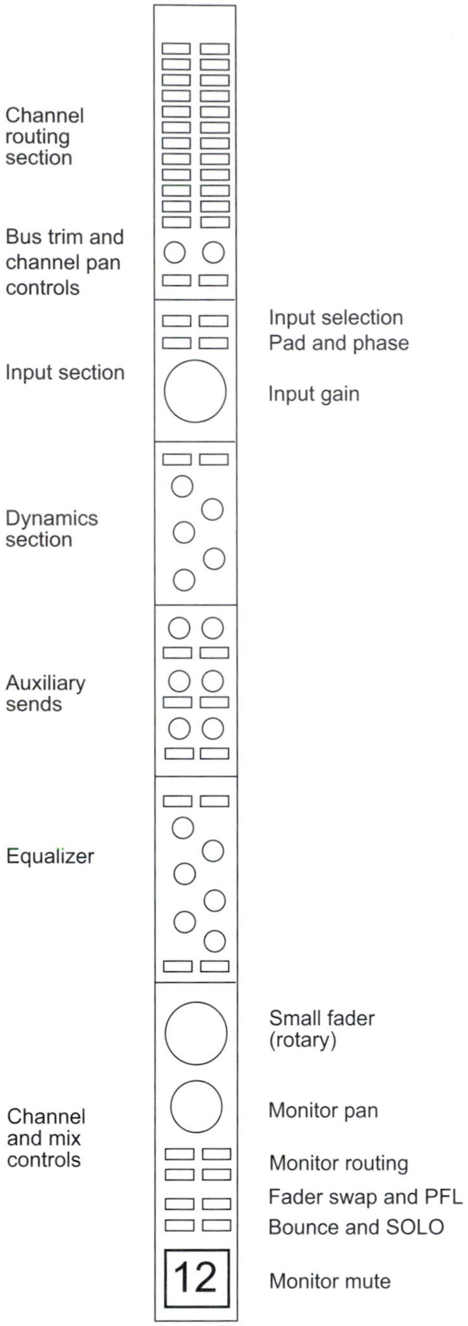

FIGURE 5.9
Typical layout of controls on an in-line mixer module (for description see text).

Input section

- *Input gain control*
  Sets the microphone or line input amplifier gain to match the level of the incoming signal. This control is often a coarse control in 10 dB steps, sometimes accompanied by a fine trim. Opinion varies as to whether this control should be in detented steps or continuous. Detented steps of 5 or 10 dB make for easy reset of the control to an exact gain setting, and precise gain matching of channels.
- *Phantom power*
  Many professional mics require 48 volts phantom powering. There is sometimes a switch on the module to turn it on or off, although most balanced mics which do not use phantom power will not be damaged if it is accidentally left on. Occasionally this switch is on the rear of the console, by the mic input socket, or it may be in a central assignable switch panel. Other methods exist; for example, one console requires that the mic gain control is pulled out to turn on the phantom power.
- *MIC/LINE switch*
  Switches between the channel's mic input and line input. The line input could be the playback output from a tape machine, or another line level signal such as a synth or effects device.
- *PAD*
  Usually used for attenuating the mic input signal by something like 20 dB, for situations when the mic is in a field of high sound pressure. If the mic is in front of a kick drum, for example, its output may be so high as to cause the mic input to clip. Also, capacitor mics tend to produce a higher output level than dynamic mics, requiring that the pad be used on some occasions.
- *Phase reverse or "ϕ"*
  Sometimes located after the mic input for reversing the phase of the signal, to compensate for a reversed directional mic, a mis-wired lead, or to create an effect. This is often left until later in the signal path.
- *HPF/LPF*
  Filters can sometimes be switched in at the input stage, which will usually just be basic high- and low-pass filters which are either in or out, with no frequency adjustment. These can be used to filter out unwanted rumble or perhaps hiss from noisy signals. Filtering rumble at this stage can be an advantage because it saves clipping later in the chain.

Routing section

- *Track routing switches*
  - The number of routing switches depends on the console: some will have 24, some 32 and some 48. The switches route the channel path signal to the multitrack machine, and it is possible to route a signal to more than one track. The track assignment is often arranged as pairs of tracks, so that odd and even tracks can be assigned together, with a pan-pot used to pan between them as a stereo pair, e.g., tracks 3 and 4 could be a stereo pair for background vocals, and each background vocal mic could be routed to 3 and 4, panned to the relevant place in the image. In an assignable console these controls may be removed to a central assignable routing section.
  - It is common for there to be fewer routing switches than there are tracks, so as to save space, resulting in a number of means of assigning tracks. Examples are rotary knobs to select the track, one button per pair of tracks with "odd/even/both" switch, and a "shift" function to select tracks higher than a certain number. The multitrack routing may be used to route signals to effects devices during mixdown, when the track outputs are not being used for recording. In this case one would patch into the track output on the patchfield (see below) and take the relevant signal to an effects input

somewhere else on the patchfield. In order to route monitor path signals to the track routing buses it may be necessary to use a switch which links the output of the monitor fader to the track assignment matrix.

- In theater sound mixers it is common for output routing to be changed very frequently, and thus routing switches may be located close to the channel fader, rather than at the top of the module as in a music mixer. On some recent mixers, track routing is carried out on a matrix which resides in the central section above the main faders. This removes unnecessary clutter from the channel modules and reduces the total number of switches required. It may also allow the storing of routing configurations in memory for later recall.

- *Mix routing switches*
Sometimes there is a facility for routing the channel path output signal to the main monitor mix, or to one of perhaps four output groups, and these switches will often be located along with the track routing.
- *Channel pan*
Used for panning channel signals between odd and even tracks of the multitrack, in conjunction with the routing switches.
- *Bus trim*
Used for trimming the overall level of the send to multitrack for a particular bus. It will normally trim the level sent to the track which corresponds to the number of the module.
- *Odd/Even/Both*
Occasionally found when fewer routing buttons are used than there are tracks. When one routing button is for a pair of tracks, this switch will determine whether the signal is sent to the odd channel only, the even channel only, or to both (in which case the pan control is operative).
- *DIRECT*
Used for routing the channel output directly to the corresponding track on the multitrack machine without going via the summing buses. This can reduce the noise level from the console since the summing procedure used for combining a number of channel outputs to a track bus can add noise. If a channel is routed directly to a track, no other signals can be routed to that track.

Dynamics section

Some advanced consoles incorporate dynamics control on every module, so that each signal can be treated without resorting to external devices. The functions available on the best designs rival the best external devices, incorporating compressor and expander sections which can act as limiters and gates respectively if required. One system allows the EQ to be placed in the side-chain of the dynamics unit, providing frequency-sensitive limiting, among other things, and it is usually possible to link the action of one channel's dynamics to the next in order to "gang" stereo channels so that the image does not shift when one channel has a sudden change in level while the other does not.

When dynamics are used on stereo signals it is important that left and right channels have the same settings, otherwise the image may be affected. If dynamics control is not available on every module, it is sometimes offered on the central section with inputs and outputs on the patchbay. Dynamics control will not be covered further here, but is discussed in more detail in "The compressor/limiter," below.

Equalizer section

The EQ section is usually split into three or four sections, each operating on a different frequency band. As each band tends to have similar functions these will be described in general. The principles of EQ are described in greater detail in "EQ explained," below.

- *HF, MID 1, MID 2, LF*
A high-frequency band, two mid-frequency bands, and a low-frequency band are often provided. If the EQ is parametric these bands will allow continuous variation of frequency (over a certain range), "Q," and boost/cut. If it is not parametric, then there may be a few switched frequencies for the mid band, and perhaps a fixed frequency for the LF and HF bands.

- *Peaking/shelving or BELL*
  Often provided on the upper and lower bands for determining whether the filter will provide boost/cut over a fixed band (whose width will be determined by the Q), or whether it will act as a shelf, with the response rising or rolling off above or below a certain frequency (see "EQ explained," below).
- *Q*
  The Q of a filter is defined as its center frequency divided by its bandwidth (the distance between frequencies where the output of the filter is 3 dB lower than the peak output).
  In practice this affects the "sharpness" of the filter peak or notch, high Q giving the sharpest response, and low Q giving a very broad response. Low Q would be used when boost or cut over a relatively wide range of frequencies is required, while high Q is used to boost or cut one specific region (see Fact File 5.6).
- *Frequency control*
  Sets the center frequency of a peaking filter, or the turnover frequency of a shelf.
- *Boost/cut*
  Determines the amount of boost or cut applied to the selected band, usually up to a maximum of around ±15 dB.
- *HPF/LPF*
  Sometimes the high- and low-pass filters are located here instead of at the input, or perhaps in addition. They normally have a fixed frequency turnover point and a fixed roll-off of either 12 or 18 dB per octave. Often these will operate even if the EQ is switched out.
- *CHANNEL*
  The American convention is for the main equalizer to reside normally in the monitor path, but it can be switched so that it is in the channel path. Normally the whole EQ block is switched at once, but on some recent models a section of the EQ can be switched separately. This would be used to equalize the signal which is being recorded on multitrack tape. If the EQ is in the

FACT FILE 5.6 VARIABLE Q

Some EQ sections provide an additional control whereby the Q of the filter can be adjusted. This type of EQ section is termed a parametric EQ since all parameters, cut/boost, frequency, and Q can be adjusted. The diagram below illustrates the effect of varying the Q of an EQ section. High Q settings affect very narrow bands of frequencies, low Q settings affect wider bands. The low Q settings sound "warmer" because they have gentle slopes and therefore have a more gradual and natural effect on the sound. High Q slopes are good for a rather more overt emphasis of a particular narrow band, which of course can be just as useful in the appropriate situation. Some EQ sections are labeled parametric even though the Q is not variable. This is a misuse of the term, and it is wise to check whether or not an EQ section is truly parametric even though it may be labeled as such.

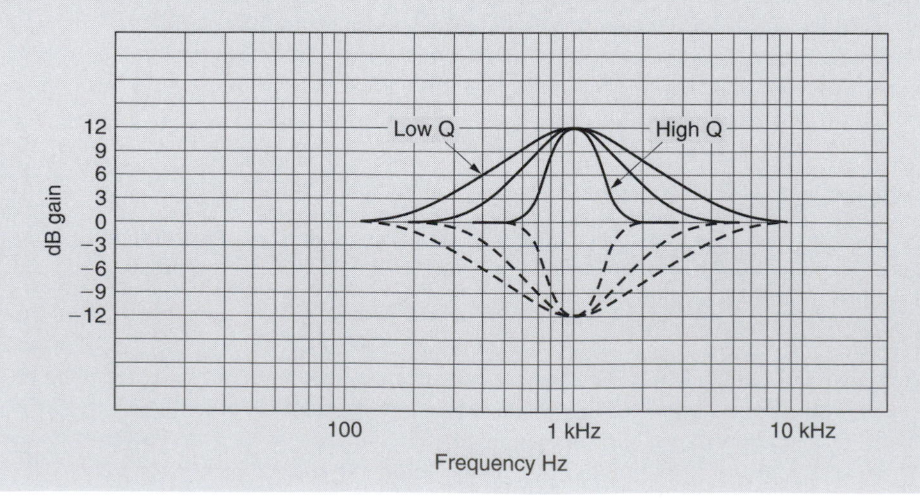

monitor path then it will only affect the replayed signal. The traditional European convention is for EQ to reside normally in the channel path, so as to allow recording with EQ.

- *IN/OUT*
  Switches the EQ in or out of circuit. Equalization circuits can introduce noise and phase distortion, so they are best switched out when not required.

Channel and mix controls

- *Pan*
  This control is a continuous rotary knob, and is used to place the signal of that channel in any desired position in the stereo picture. See Fact File 5.2.
- *Fader reverse*
  Swaps the faders between mix and channel paths, so that the large fader can be made to control either the mix level or the channel level. Some systems defeat any fader automation when the large fader is put in the channel path. Fader reverse can often be switched globally, and may occur when the console mode is changed from recording to mixdown.
- *Line/Tape or Bus/Tape*
  Switches the source of the input to the monitor path between the line output of the same-numbered channel and the return from multitrack tape. Again it is possible that this may be switched globally. In "line" or "bus" mode the monitor paths are effectively "listening to" the line output of the console's track assignment buses, while in "tape" mode the monitor paths are listening to the off-tape signal (unless the tape machine's monitoring is switched to monitor the line input of the tape machine, in which case "line" and "tape" will effectively be the same thing!). If a problem is suspected with the tape machine, switching to monitor "line" will bypass the tape machine entirely and allow the operator to check if the console is actually sending anything.
- *Broadcast, or "mic to mix," or "simulcast"*
  Used for routing the mic signal to both the channel and monitor paths simultaneously, so that a multitrack recording can be made while a stereo mix is being recorded or broadcasted. The configuration means that any alterations made to the channel path will not affect the stereo mix, which is important when the mix output is live (see Figure 5.10).
- *BUS or "monitor-to-bus"*
  Routes the output of the monitor fader to the input of the channel path (or the channel fader) so that the channel path can be used as a post-fader effects send to any one of the multitrack buses (used in this case as aux sends), as shown in Figure 5.11. If a BUS TRIM control is provided on each multitrack output this can be used as the master effects-send level control.

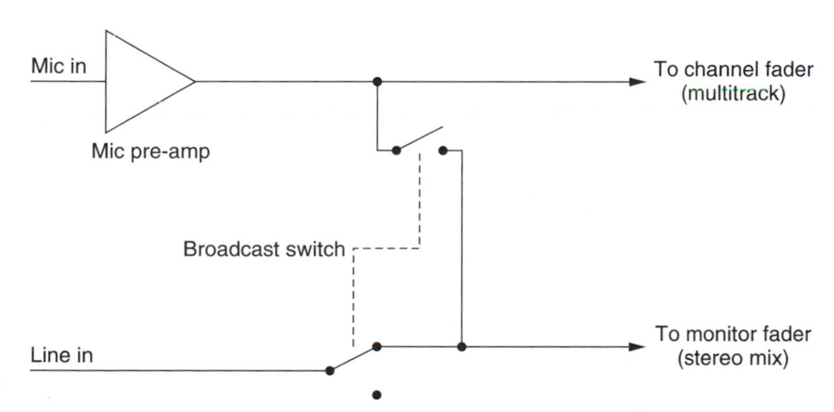

FIGURE 5.10
A "broadcast mode" switch in an in-line console allows the microphone input to be routed to both signal paths, such that a live stereo mix may be made independent of any changes to multitrack recording levels.

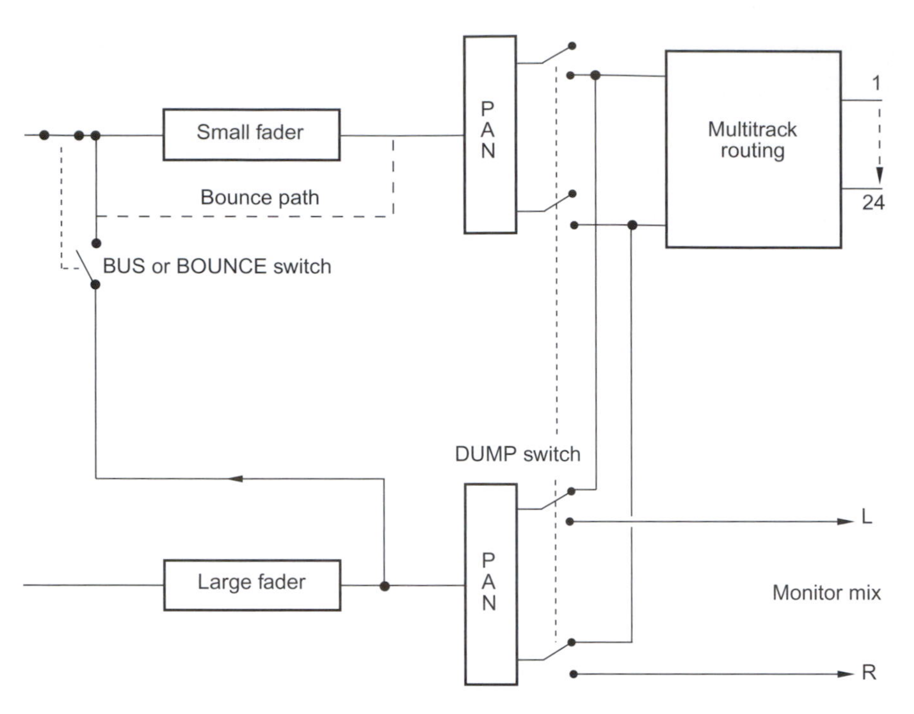

FIGURE 5.11
Signal routings for "bounce," "bus" and "dump" modes (see text).

- *DUMP*
  Incorporated (rarely) on some consoles to route the stereo panned mix output of a track (i.e. after the monitor path pan-pot) to the multitrack assignment switches. In this way, the mixed version of a group of tracks can be "bounced down" to two tracks on the multitrack, panned and level-set as in the monitor mix (see Figure 5.11).
- *BOUNCE*
  A facility for routing the output of the monitor fader to the multitrack assignment matrix, before the pan control, in order that tracks can be "bounced down" so as to free tracks for more recording by mixing a group of tracks on to a lower number of tracks. BOUNCE is like a mono version of DUMP (see Figure 5.11).
- *MUTE or CUT*
  Cuts the selected track from the mix. There may be two of these switches, one for cutting the channel signal from the multitrack send, the other for cutting the mix signal from the mix.
- *PFL*
  See Fact File 5.3.
- *AFL*
  After fade listen is similar to PFL, except that it is taken from after the fader. This is sometimes referred to as SOLO, which routes a panned version of the track to the main monitors, cutting everything else. These functions are useful for isolating signals when setting up and spotting faults. On many consoles the AFL bus will be stereo. Solo functions are useful when applying effects and EQ, in order that one may hear the isolated sound and treat it individually without hearing the rest of the mix. Often a light is provided to show that a solo mode is selected, because there are times when nothing can be heard from the loudspeakers due to a solo button being down with no signal on that track. A solo safe control may be provided centrally, which prevents this feature from being activated.
- *In-place solo*
  On some consoles, solo functions as an "in-place" solo, which means that it actually changes the mix output, muting all tracks which are not solo'ed and picking out all the solo'ed tracks. This may be preferable to AFL as it reproduces the exact contribution of each channel to the mix, at the presently set master mix level. Automation systems often allow the solo functions to be automated in groups, so that a whole section can be isolated in the mix. In certain designs, the function of the automated mute button on the monitor fader may be reversed so that it becomes solo.

Auxiliary sends

The number of aux(iliary) sends depends on the console, but there can be up to ten on an ordinary console, and sometimes more on assignable models. Aux sends are "take-off points" for signals from either the channel or mix paths, and they appear as outputs from the console which can be used for foldback to musicians, effects sends, cues, and so on. Each module will be able to send to auxiliaries, and each numbered auxiliary output is made up of all the signals routed to that aux send. So they are really additional mix buses. Each aux will have a master gain control, usually in the center of the console, for adjusting the overall gain of the signal sent from the console, and may have basic EQ. Aux sends are often a combination of mono and stereo buses. Mono sends are usually used as routes to effects, while stereo sends may have one level control and a pan control per channel for mixing a foldback source.

- *Aux sends 1–n*
  Controls for the level of each individual channel in the numbered aux mix.
- *Pre/post*
  Determines whether the send is taken off before or after the fader. If it is before then the send will still be live even when the fader is down. Generally, "cue" feeds will be pre-fade, so that a mix can be sent to foldback which is independent of the monitor mix. Effects sends will normally be taken post-fade, in order that the effect follows a track's mix level.
- *Mix/channel*
  Determines whether the send is taken from the mix or channel paths. It will often be sensible to take the send from the channel path when effects are to be recorded on to multitrack rather than on to the mix. This function has been labeled "WET" on some designs.
- *MUTE*
  Cuts the numbered send from the aux mix.

Master control section

The master control section usually resides in the middle of the console, or near the right-hand end. It will contain some or all of the following facilities:

- *Monitor selection*
  A set of switches for selecting the source to be monitored. These will include recording machines, aux sends, the main stereo mix, and perhaps some miscellaneous external sources like CD players, cassette machines, etc. They only select the signal going to the loudspeakers, not the mix outputs. This may be duplicated to some extent for a set of additional studio loudspeakers, which will have a separate gain control.
- *DIM*
  Reduces the level sent to the monitor loudspeakers by a considerable amount (usually around 40 dB), for quick silencing of the room.
- *MONO*
  Sums the left and right outputs to the monitors into mono so that mono compatibility can be checked.
- *Monitor phase reverse*
  Phase reverses one channel of the monitoring so that a quick check on suspected phase reversals can be made.
- *TAPE/LINE*
  Usually a global facility for switching the inputs to the mix path between the tape returns and the console track outputs. Can be reversed individually on modules.
- *FADER REVERSE*
  Global swapping of small and large faders between mix and channel paths.
- *Record/Overdub/Mixdown*
  Usually globally configures mic/line input switching, large and small faders and auxiliary sends depending on mode of operation. (Can be overridden on individual channels.)
- *Auxiliary level controls*
  Master controls for setting the overall level of each aux send output.

- *Foldback and Talkback*
  There is often a facility for selecting which signals are routed to the stereo foldback which the musicians hear on their headphones. Sometimes this is as comprehensive as a cue mixer which allows mixing of aux sends in various amounts to various stereo cues, while often it is more a matter of selecting whether foldback consists of the stereo mix, or one of the aux sends. Foldback level is controllable, and it is sometimes possible to route left and right foldback signals from different sources. Talkback is usually achieved using a small microphone built into the console, which can be routed to a number of destinations. These destinations will often be aux sends, multitrack buses, mix bus, studio loudspeakers and foldback.
- *Oscillator*
  Built-in sine-wave oscillators vary in quality and sophistication, some providing only one or two fixed frequencies, while others allow the generation of a whole range. If the built-in oscillator is good it can be used for lining up the tape machine, as it normally can be routed to the mix bus or the multitrack outputs. The absolute minimum requirement is for accurate 1 kHz and 10 kHz tones, the 10 kHz being particularly important for setting the bias of an analog tape machine. The oscillator will have an output level control.
- *Slate*
  Provides a feed from the console talkback mic to the stereo output, often superimposing a low-frequency tone (around 50 Hz) so that the slate points can be heard when winding a tape at high speed. Slate would be used for recording take information on to tape.
- *Master faders*
  There may be either one stereo fader or left and right faders to control the overall mix output level. Often the group master faders will reside in this section.

Effects returns

Effects returns are used as extra inputs to the mixer, supplied specifically for inputs from external devices such as reverberation units. These are often located in the central section of the console and may be laid out like reduced-facility input channels. Returns sometimes have EQ, perhaps more basic than on channels, and they may have aux sends. Normally they will feed the mix, although sometimes facilities are provided to feed one or more returns to the multitrack via assignment switches. A small fader or rotary level control is provided, as well as a pan-pot for a mono return. Occasionally, automated faders may be assigned to the return channels so as to allow automated control of their levels in the mix.

Patchbay or jackfield

Most large consoles employ a built-in jackfield or patchbay for routing signals in ways which the console switching does not allow, and for sending signals to and from external devices. Just about every input and output on every module in the console comes up on the patchbay, allowing signals to be cross-connected in virtually any configuration. The jackfield is usually arranged in horizontal rows, each row having an equal number of jacks. Vertically, it tries to follow the signal path of the console as closely as possible, so the mic inputs are at the top and the multitrack outputs are nearer the bottom. In between these there are often insert points which allow the engineer to "break into" the signal path, often before or after the EQ, to insert an effects device, compressor, or other external signal processor. Insert points usually consist of two rows, one which physically breaks the signal chain when a jack is inserted, and one which does not. Normally it is the lower row which breaks the chain, and should be used as inputs. The upper row is used as an output or send. Normaling is usually applied at insert points, which means that unless a jack is inserted the signal will flow directly from the upper row to the lower.

At the bottom of the patchfield will be all the master inputs and outputs, playback returns, perhaps some parallel jacks, and sometimes some spare rows for connection of one's own devices. Some consoles bring the microphone signals up to the patchbay, but there are some manufacturers who would rather not do this unless absolutely necessary as it is more likely to introduce noise, and phantom power may be present on the jackfield.

EQ EXPLAINED

The tone control or EQ (=equalization) section provides mid-frequency controls in addition to bass and treble. A typical comprehensive EQ section may have first an HF (high-frequency) control similar to a treble control but operating only at the highest frequencies. Next would come a hi-mid control, affecting frequencies from around 1 kHz to 10 kHz, the center frequency being adjusted by a separate control. Lo-mid controls would come next, similar to the hi-mid but operating over a range of, say, 200 Hz to 2 kHz. Then would come an LF (low-frequency) control. Additionally, high- and low-frequency filters can be provided. The complete EQ section looks something like that shown in Figure 5.12. An EQ section takes up quite a bit of space, and so it is quite common for dual concentric or even assignable controls (see below) to be used. For instance, the cut/boost controls of the hi- and lo-mid sections can be surrounded by annular skirts which select the frequency.

Principal EQ bands

The HF section affects the highest frequencies and provides up to 12 dB of boost or cut. This type of curve is called a shelving curve because it gently boosts or cuts the frequency range towards a shelf where the level remains relatively constant (see Figure 5.13(a)). Next comes the hi-mid section. Two controls are provided here, one to give cut or boost, the other to select the desired center frequency. The latter is commonly referred to as a "swept mid" because one can sweep the setting across the frequency range.

Figure 5.13(b) shows the result produced when the frequency setting is at the 1 kHz position, termed the center frequency. Maximum boost and cut affects this frequency the most, and the slopes of the

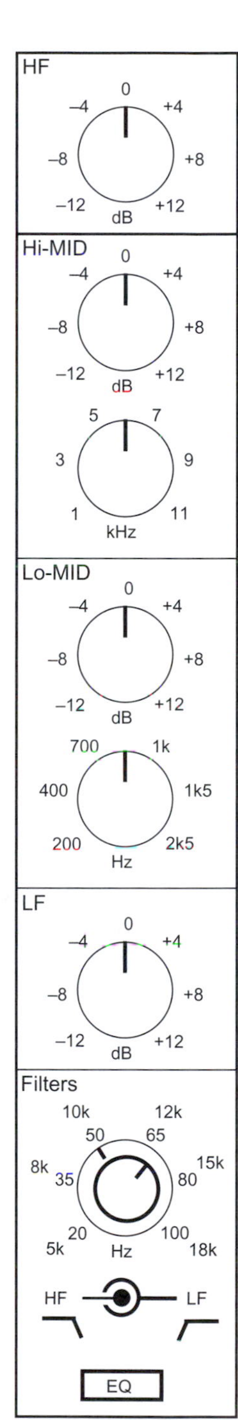

FIGURE 5.12
Typical layout of an EQ section.

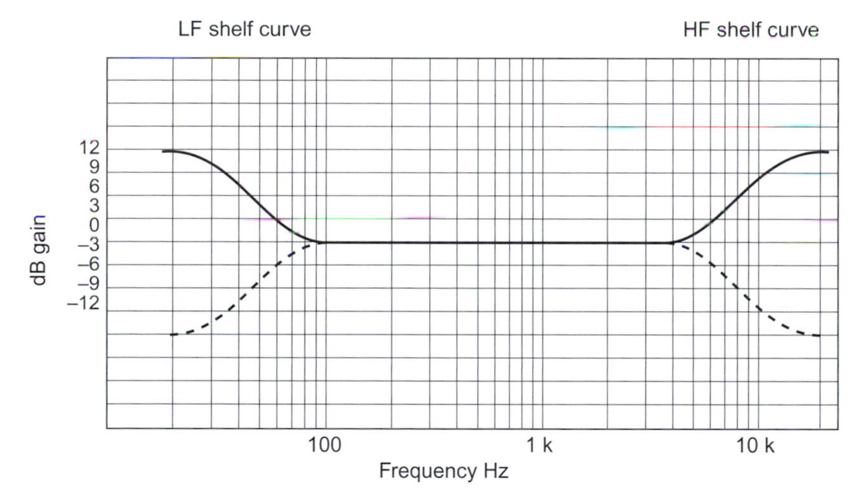

FIGURE 5.13 (a)
Typical HF and LF shelf EQ characteristics shown at maximum settings.

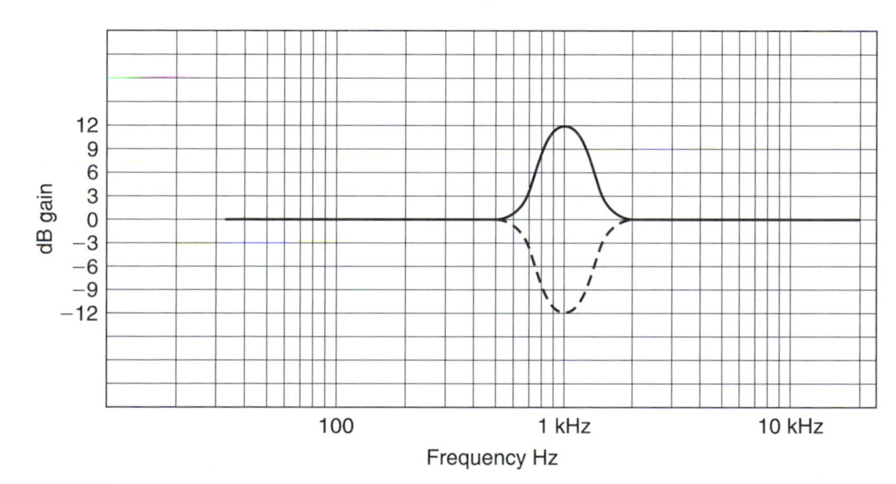

FIGURE 5.13 (b)
Typical MF peaking filter characteristic.

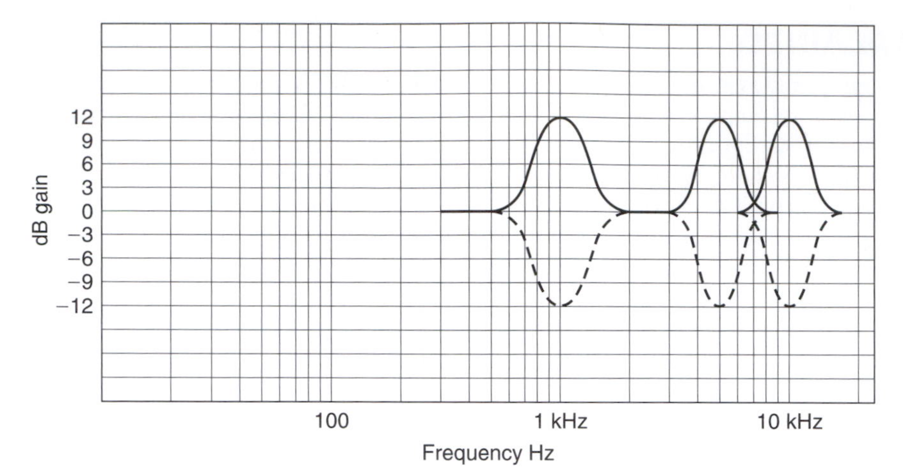

FIGURE 5.13 (c)
MF peaking filter characteristics at 1, 5 and 10 kHz.

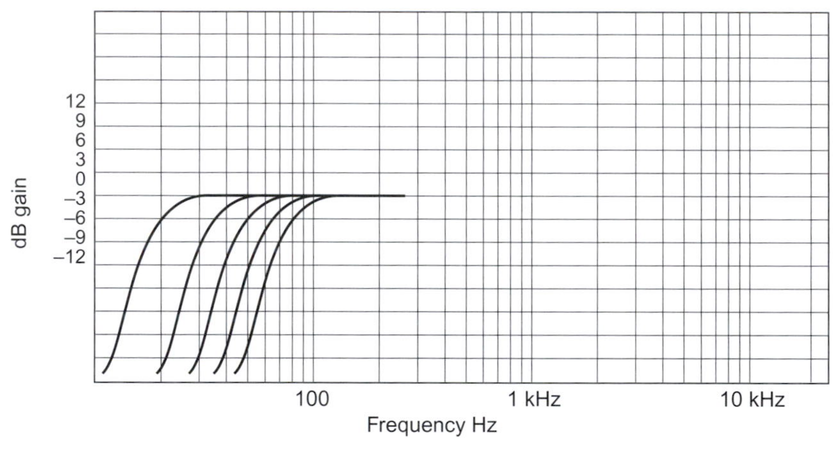

FIGURE 5.13 (d)
High-pass filters with various turnover frequencies.

curve are considerably steeper than those of the previous shelving curves. This is often referred to as a "bell" curve due to the upper portion's resemblance to the shape of a bell. It has a fairly high "Q," that is its sides are steep. Q is defined as:

$$Q = \text{center frequency} \div \text{bandwidth}$$

where the bandwidth is the spacing in hertz between the two points at which the response of the filter is 3 dB lower than that at the center frequency. In the example shown the center frequency is 1 kHz and the bandwidth is 400 Hz, giving $Q = 2.5$.

MF EQ controls are often used to hunt for trouble-spots; if a particular instrument (or microphone) has an emphasis in its spectrum somewhere which does not sound very nice, some mid cut can be introduced, and the frequency control can be used to search for the precise area in the frequency spectrum where the trouble lies. Similarly, a dull sound can be given a lift in an appropriate part of the spectrum which will bring it to life in the overall mix. Figure 5.13(c) shows the maximum cut and boost curves obtained with the frequency selector at either of the three settings of 1, 5 and 10 kHz. The high Q of the filters enables relatively narrow bands to be affected. Q may be varied in some cases, as described in Fact File 5.6.

The lo-mid section is the same as the hi-mid section except that it covers a lower band of frequencies. Note though that the highest frequency setting overlaps the lowest setting of the hi-mid section. This is quite common, and ensures that no "gaps" in the frequency spectrum are left uncovered.

Filters

High- and low-cut filters provide fixed attenuation slopes at various frequencies. Figure 5.13(d) shows the responses at LF settings of 80, 65, 50, 35 and 20 Hz. The slopes are somewhat steeper than is the case with the HF and LF shelving curves, and slope rates of 18 or 24 dB per octave are typical. This enables just the lowest, or highest, frequencies to be rapidly attenuated with minimal effect on the mid band. Very low traffic rumble could be removed by selecting the 20 or 35 Hz setting. More serious low-frequency noise may require the use of one of the higher turnover frequencies. High-frequency hiss from, say, a noisy guitar amplifier or air escaping from a pipe organ bellows can be dealt with by selecting the turnover frequency of the HF section which attenuates just sufficient HF noise without unduly curtailing the HF content of the wanted sound.

STEREO LINE INPUT MODULES

In broadcast situations it is common to require a number of inputs to be dedicated to stereo line level sources, such as CD players, electronic musical instruments, etc. Such modules are sometimes offered as an option for multitrack consoles, acting as replacements for conventional I/O modules and allowing two signals to be faded up and down together with one fader. Often the EQ on such modules is more limited, but the module may provide for the selection of more than one stereo source, and routing to the main mix as well as the multitrack. It is common to require that stereo modules always reside in special slots on the console, as they may require special wiring. Such modules also used to provide facilities for handling LP turntable outputs, offering RIAA equalization.

Stereo microphone inputs can also be provided, with the option for MS (middle and side) format signals as well as AB (conventional left and right) format. A means of control over stereo width can be offered on such modules.

DEDICATED MONITOR MIXER

A dedicated monitor mixer is often used in live sound reinforcement work to provide a separate monitor mix for each musician, in order that each artist may specify his or her precise monitoring requirements. A comprehensive design will have, say, 24 inputs containing similar facilities to a conventional mixer, except that below the EQ section there will be a row of rotary or short-throw faders which individually send the signal from that channel to the group outputs, in any combination of relative levels. Each group output will then provide a separate monitor mix to be fed to headphones or amplifier racks.

TECHNICAL SPECIFICATIONS

This section contains some guidance concerning the meanings and commonly encountered values of technical specifications for mixers.

Input noise

The output from a microphone is in the millivolt range, and so needs considerable amplification to bring it up to line level. Amplification of the signal also brings with it amplification of the microphone's own noise output, which one can do nothing about, as well as amplification of the mixer's own input noise. The latter must therefore be as low as possible so as not to compromise the noise performance unduly. A 200 ohm source resistance on its own generates 0.26 μV of noise (20 kHz bandwidth). Referred to the standard line level of 775 mV (0 dBu) this is −129.6 dBu. A microphone amplifier will add its own noise to this, and so manufacturers quote an "equivalent input noise" (EIN) value which should be measured with a 200 ohm source resistance across the input.

An amplifier with a noise contribution equal to that of the 200 ohm resistor will degrade the theoretically "perfect" noise level by 3 dB, and so the quoted equivalent input noise will be $-129.6 + 3 = -126.6$ dBm. (Because noise contributions from various sources sum according to their power content, not their voltage levels, dBm is traditionally used to express input noise level.) This value is quite respectable, and good-quality mixers should not be noisier than this. Values of

FACT FILE 5.7 COMMON MODE REJECTION

Common mode rejection is the ability of a balanced input to reject interference which can be induced into the signal lines. A microphone input should have a CMRR (common mode rejection ratio) of 70 dB or more; i.e., it should attenuate the interference by 70 dB. But look at how this measurement is made. It is relatively easy to achieve 70 dB at, say, 500 Hz, but rejection is needed most at high frequencies – between 5 and 20 kHz – and so a quoted CMRR of "70 dB at 15 kHz" or "70 dB between 100 Hz and 10 kHz" should be sought. Line level CMRR can be allowed to be rather lower since the signal voltage level is a lot higher than in microphone cabling. CMRRs of as low as 30 dB at 10 kHz are deemed to be adequate.

Common mode rejection is a property of a balanced input, and so it is not applicable to a balanced output. However, output balance is sometimes quoted which gives an indication of how closely the two legs of a balanced output are matched. If the two legs were to be combined in antiphase, total cancelation would ideally be achieved. In practice, around 70 dB of attenuation should be looked for.

around −128 dBm are sometimes encountered, which are excellent, indicating that the input resistance is generating more noise than the amplifier. Make sure that the EIN is quoted with a 200 ohm source, and a bandwidth up to 20 kHz, unweighted. A 150 ohm source, sometimes specified, will give an apparently better EIN simply because this resistor is itself quieter than a 200 ohm one, resistor noise being proportional to ohmic value. Also, weighting gives a flattering result, so one always has to check the measuring conditions. Make sure that EIN is quoted in dBm or dBu. Some manufacturers quote EIN in dBV (i.e. ref. 1 volt) which gives a result that looks 2.2 dB better. An input should have high common mode rejection as well as low noise, as discussed in Fact File 5.7.

Output noise

The output residual noise of a mixer, with all faders at minimum, should be at most −90 dBu. There is no point in having a very quiet microphone amplifier if a noisy output stage ruins it. With all channels routed to the output, and all faders at the "zero" position, output noise (or "mixing" noise) should be at least −80 dBu with the channel inputs switched to "line" and set for unity gain. Switching these to "mic" inevitably increases noise levels because this increases the gain of the input amplifier. It underlines the reason why all unused channels should be unrouted from the mix buses, or their faders brought down to a minimum. Digital mixers with "scene" memories tend to be programmed by copying a particular scene to another vacant scene, then modifying it for the new requirements. When doing this, one needs to ensure that all unwanted inputs and routing from the copied scene are removed so as to maintain the cleanest possible signal. Make sure that the aux outputs have a similarly good output noise level.

Impedance

A microphone input should have a minimum impedance of 1 kΩ. A lower value than this degrades the performance of many microphones. A line level input should have a minimum impedance of 10 kΩ. Whether it is balanced or unbalanced should be clearly stated, and consideration of the type of line level equipment that the mixer will be partnered with will determine the importance of balanced line inputs. All outputs should have a low impedance, below 200 ohms, balanced. Check that the aux outputs are also of very low impedance, as sometimes they are not. If insert points are provided on the input channels and/or outputs, these also should have very low output and high input impedances.

Frequency response

A frequency response that is within 0.2 dB between 20 Hz and 20 kHz for all combinations of input and output is desirable. The performance of audio transformers varies slightly with different source and load impedances, and a specification should state the range of loads between which a "flat" frequency response will be obtained. Above 20 kHz, and probably below 15 Hz or so, the frequency response should fall away so that unwanted out-of-band frequencies are not amplified, for example radio-frequency breakthrough or subsonic interference.

FACT FILE 5.8 CLIPPING

A good mixer will be designed to provide a maximum electrical output level of at least 20 dBu. Many will provide 24 dBu. Above this electrical level clipping will occur, where the top and bottom of the audio waveform are chopped off, producing sudden and excessive distortion (see diagram). Since the nominal reference level of 0 dBu usually corresponds to a meter indication of PPM 4 or −4 VU, it is very difficult to clip the output stages of a mixer. The maximum meter indication on a PPM would correspond in this case to an electrical output of around 12 dBu, and thus one would have to be severely bending the meter needles to cause clipping.

Clipping, though, may occur at other points in the signal chain, especially when large amounts of EQ boost have been added. If, say, 12 dB of boost has been applied on a channel, and the fader is set well above the 0 dB mark, clipping on the mix bus may occur, depending on overload margins here. Large amounts of EQ boost should not normally be used without a corresponding overall gain reduction of the channel for this reason.

An input pad or attenuator is often provided to prevent the clipping of mic inputs in the presence of high-level signals (see "Input section," above).

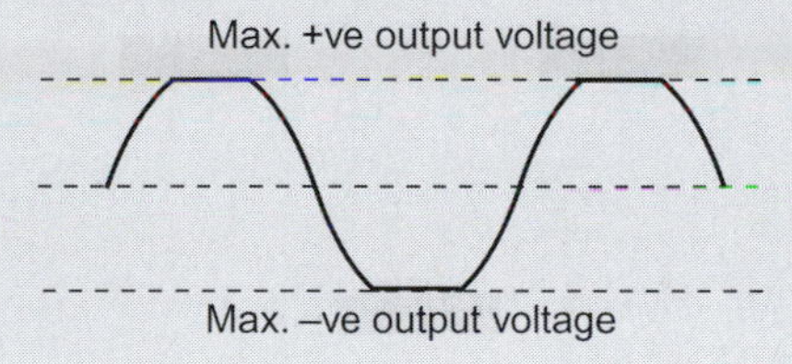

Distortion

With an analog mixer, distortion should be quoted at maximum gain through the mixer and a healthy output level of, say, 10 dBu or more. This will produce a typical worst case, and should normally be less than 0.1% THD (total harmonic distortion). The distortion of the low-gain line level inputs to outputs can be expected to be lower: around 0.01%. The outputs should be loaded with a fairly low impedance which will require more current from the output stages than a high impedance will, this helping to reveal any shortcomings. A typical value is 600 ohms.

Clipping and overload margins are discussed in Fact File 5.8.

Crosstalk

In analog mixers, a signal flowing along one path may induce a small signal in another, and this is termed "crosstalk". Crosstalk from adjacent channels should be well below the level of the legitimate output signal, and a figure of −80 dB or more should be looked for at 1 kHz. Crosstalk performance tends to deteriorate at high frequencies due to capacitive coupling in wiring harnesses, for instance, but a crosstalk of at least −60 dB at 15 kHz should still be sought. Similarly, very low-frequency crosstalk often deteriorates due to the power supply source impedance rising here, and a figure of −50 dB at 20 Hz is reasonable.

Ensure that crosstalk between all combinations of input and output is of a similarly good level. Sometimes crosstalk between channel auxiliaries is rather poorer than that between the main outputs.

METERING SYSTEMS

Metering systems are provided on audio mixers to indicate the levels of audio signals entering and leaving the mixer. Careful use of metering is vital for optimizing noise and distortion, and to the recording of the correct audio level on tape. In this section the merits of different metering systems are examined.

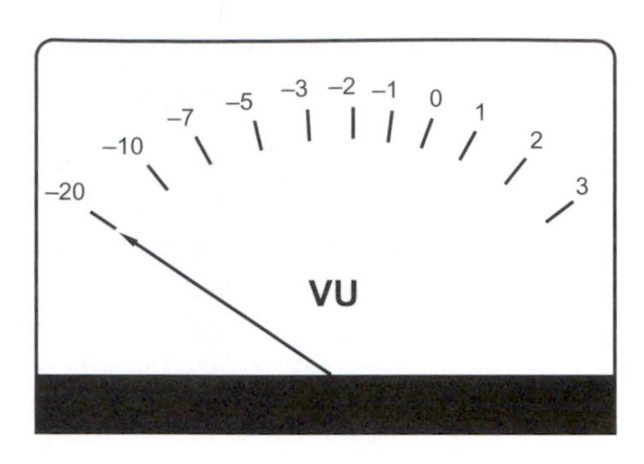

FIGURE 5.14
Typical VU meter scale.

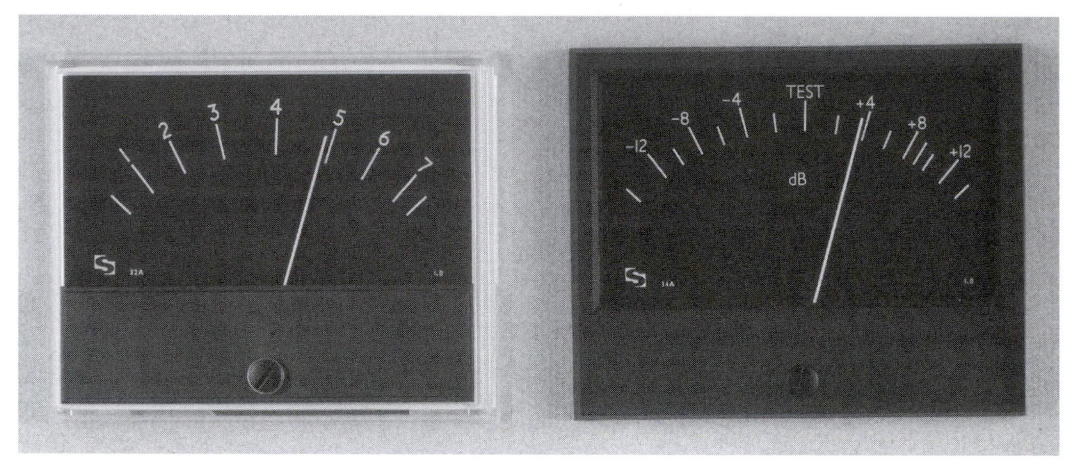

FIGURE 5.15
(Left) BBC-type peak programme meter (PPM). (Right) European-type PPM.

Mechanical metering

Two primary types of mechanical meters have been used: the VU (volume unit) meter (Figure 5.14) and the PPM (peak program meter), as shown in Figure 5.15. These are very different to each other, the only real similarity being that they both have swinging needles. The British, or BBC-type, PPM is distinctive in styling in that it is black with numbers ranging from 1 to 7 equally spaced across its scale, there being a 4 dB level difference between each gradation, except between 1 and 2 where there is usually a 6 dB change in level. The EBU PPM (also shown in Figure 5.15) has a scale calibrated in decibels. The VU, on the other hand, is usually white or cream, with a scale running from −20 dB up to 3 dB, ranged around a zero point which is usually the studio's electrical reference level. Originally the VU meter was associated with a variable attenuator which could vary the electrical alignment level for 0 VU up to +24 dBu, although it is common for this to be fixed these days at 0 VU = +4 dBu.

It is important to know how meter readings relate to the line-up standard in use in a particular environment, and to understand that these standards may vary between establishments and areas of work. Fact File 5.9 discusses the relationship between meter indication and signal levels, considering practical issues such as the onset of distortion.

Problems with mechanical meters

PPMs respond well to signal peaks, that is they have a fast rise-time, whereas VUs are quite the opposite: they have a very slow rise-time. This means that VUs do not give a true representation of the peak level going on to tape, especially in cases when a signal with a high transient content, such as a harpsichord, is being recorded, often showing as much as 10–15 dB lower than a peak-reading meter. This can result in overmodulation of the recording, especially with digital recorders where the system is very sensitive to peak overload. Nonetheless, many people are used to working with VUs, and have

FACT FILE 5.9 METERING, SIGNAL LEVELS AND DISTORTION

Within a studio there is usually a "reference level" and a "peak recording level". In the broadcast domain these are usually referred to as "alignment level" and "permitted maximum level (PML)" as shown in Figure 5.17. The reference or alignment level usually relates to the level at which a 1 kHz line-up tone should play back on the console's meters. In analog mixers this may correspond to PPM 4 on a BBC-type PPM or "Test" on a European PPM. Electrically PPM 4 usually corresponds to a level of 0 dBu, although the German and Nordic metering standards traditionally had it as -3 dBu. In the digital domain, line-up level usually corresponds to either −20 dBFS (SMPTE) or −18 dBFS (EBU), depending on the area of the world and standard concerned. A relationship is therefore established between meter reading and signal level in each domain.

In digital audio systems, where compatibility with other systems is not an issue, it is possible to peak close to 0 dBFS without incurring increases in distortion, and many recording engineers use all this "headroom" in order to maximize dynamic range. Digital systems clip hard at 0 dBFS whereas analog tape tends to give rise to gradually increasing distortion and level compression as levels rise. Because of the typical relationship between analog and digital levels, peak digital level can correspond to electrical levels as high as +18 to +22 dBu, which can be inconvenient in mixed format systems sharing the same meters. In broadcasting it is normal to peak no more than −89 dB above line-up level (PPM 6 in the UK) as higher levels than this can have serious effects on analog transmitter distortion. Limiters are normally used in broadcasting systems, which start to take effect rapidly above this level. Because of the standardized relationship between analog and digital levels in broadcasting, the PML is lower than digital peak level, leading to a degree of undermodulation of the digital system that is considered acceptable in the interests of compatibility.

learned to interpret them. They are good for measuring continuous signals such as tones, but their value for monitoring program material is dubious in the age of digital recording.

VUs have no control over the fall-time of the needle, which is much the same as the rise-time, whereas PPMs are engineered to have a fast rise-time and a longer fall-time, which tends to be more subjectively useful. The PPM was designed to indicate peaks that would cause audible distortion, but does not measure the absolute peak level of a signal. Mechanical meters take up a lot of space on a console, and it can be impossible to find space for one meter per channel in the case of a multitrack console. In this case there are often only meters on the main outputs, and perhaps some auxiliary outputs, these being complemented on more expensive consoles by electronic bargraph meters, usually consisting of LED or liquid crystal displays, or some form of "plasma" display.

Electronic bargraph metering

Unlike mechanical meters, electronic bargraphs have no mechanical inertia to overcome, so they can effectively have an infinitely fast rise-time although this may not be the ideal in practice. Cheaper bargraphs are made out of a row of LEDs (light emitting diodes), and the resolution accuracy depends on the number of LEDs used. This type of display is sometimes adequate, but unless there are a lot of gradations it is difficult to use them for line-up purposes. Plasma and liquid crystal displays look almost continuous from top to bottom, and do not tend to have the glare of LEDs, being more comfortable to work with for any length of time. Such displays often cover a dynamic range far greater than any mechanical meter, perhaps from −50 dB up to +12 dB, and so can be very useful in showing the presence of signals which would not show up on a mechanical PPM. Such a meter is illustrated in Figure 5.16.

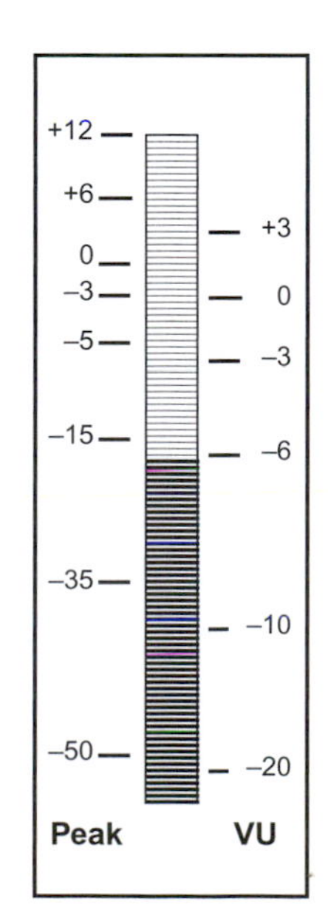

FIGURE 5.16
Typical peak-reading bargraph meter with optional VU scale.

There may be a facility provided to switch the peak response of these meters from PEAK to VU mode, where they will imitate the scale and ballistic response of a VU meter. On more up-market designs it may be possible to use the multitrack bargraphs as a spectrum analyzer display, indicating perhaps a one-third octave frequency-band analysis of the signal fed to it. Occasionally, bargraph displays incorporate a peak-hold facility. A major advantage of these vertical bargraphs is that they take up very little horizontal space on a meter bridge and can thus be used for providing one meter for every channel of the console: useful for monitoring the record levels on a multitrack tape machine. In this case, the feed to the meter is usually taken off at the input to the monitor path of an in-line module.

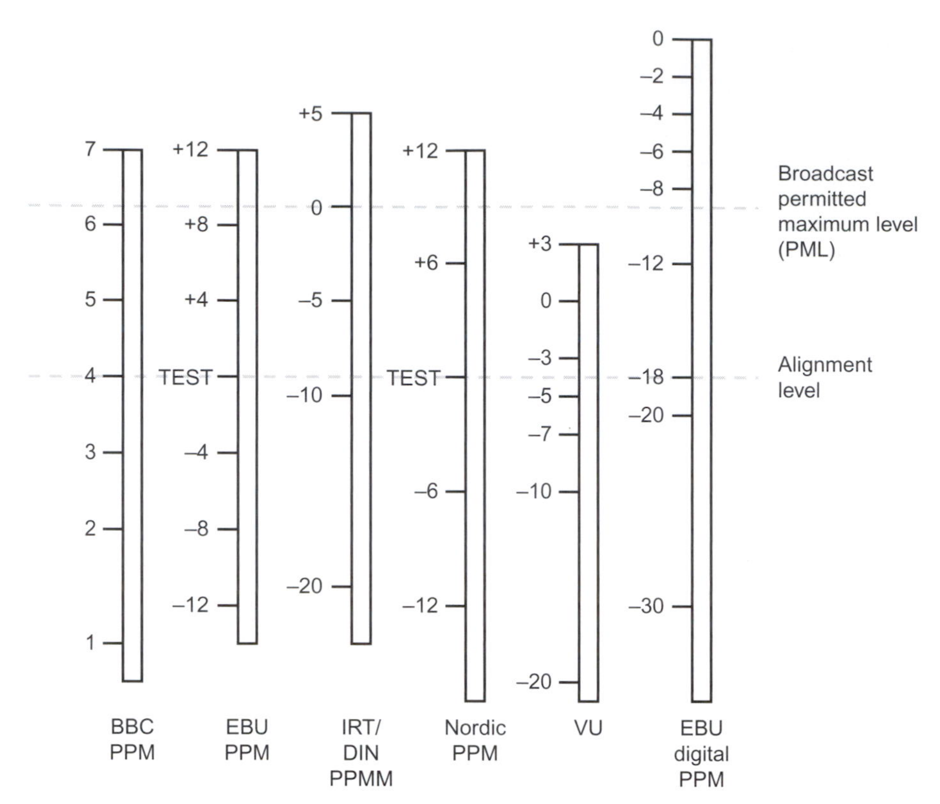

FIGURE 5.17
Graphical comparison of commonly encountered meter scalings and electrical levels in dBu.

Miscellaneous meters may also be provided on the aux send outputs for giving some indication of the level being sent to auxiliary devices such as effects. These are commonly smaller than the main meters, or may consist of LED bargraphs with lower resolution. A phase meter or correlation meter is another option, this usually being connected between the left and right main monitor outputs to indicate the degree of phase correlation between these signals. This can be either mechanical or electronic. In broadcast environments, sum and difference (or M and S) meters may be provided to show the level of the mono-compatible and stereo difference signals in stereo broadcasting. These often reside alongside a stereo meter for left and right output levels.

Relationship between different metering standards

Figure 5.17 shows a number of common meter scales and the relationship between them. The relationship between meter indication and electrical level varies depending on the type of meter and the part of the world concerned. As introduced in Fact File 5.9, there is a further relationship to be concerned with, this being that between the electrical output level of the mixer and the recording or transmitted level, and between the analog and digital domains.

Meter take-off point

Output level meter-driving circuits should normally be connected directly across the outputs so that they register the real output levels of the mixer. This may seem self-evident but there are certain models in which this is not the case, the meter circuit taking its drive from a place in the circuit just before the output amplifiers. In such configurations, if a faulty lead or piece of equipment connected to the mixer places, say, a short-circuit across the output the meter will nevertheless read normal levels, and lack of signal reaching a destination will be attributed to other causes. The schematic circuit diagrams of the mixer can be consulted to ascertain whether such an arrangement has been employed. If it is not clear, a steady test tone can be sent to the mixer's output, giving a high meter reading. Then a short-circuit can be deliberately applied across the output (the output amplifier will not normally be harmed by several seconds of short-circuit) and the meter watched. If the indicated level drastically reduces then the meter is correctly registering the real output. If it stays high then the meter is taking its feed from elsewhere.

AUTOMATION

Background

The original and still most common form of mixer automation is a means of storing fader positions dynamically against time for reiteration at a later point in time, synchronous with recorded material. The aim of automation has been to assist an engineer in mixdown when the number of faders that need to be handled at once becomes too great for one person. Fader automation has resulted in engineers being able to concentrate on sub-areas of a mix at each pass, gradually building up the finished product and refining it.

MCI first introduced VCA (voltage controlled amplifier) automation for their JH500 series of mixing consoles in the mid-1970s, and this was soon followed by imitations with various changes from other manufacturers. Moving fader automation systems, such as Neve's NECAM, were introduced slightly later and tended to be more expensive than VCA systems. During the mid-1980s, largely because of the falling cost of microprocessor hardware, console automation enjoyed further advances resulting in developments such as snapshot storage, total dynamic automation, retrofit automation packages, and MIDI-based automation. The rise of digital mixers and digitally controlled analog mixers with integral automation has continued the trend towards total automation of most mixer controls as a standard feature of many new products.

In the following sections a number of different approaches to console automation will be presented and discussed.

Fader automation

There are two common means of memorizing and controlling the gain of a channel: one which stores the positions of the fader and uses this data to control the gain of a VCA or digitally controlled attenuator (DCA), the other which also stores fader movements but uses this information actually to drive the fader's position using a motor. The former is cheaper to implement than the latter, but is not so ergonomically satisfactory because the fader's physical position may not always correspond to the gain of the channel.

It is possible to combine elements of the two approaches in order that gain control can be performed by a VCA but with the fader being moved mechanically to display the gain. This allows for rapid changes in level which might be impossible using physical fader movements, and also allows for dynamic gain offsets of a stored mix whilst retaining the previous gain profile (see below). In the following discussion the term "VCA faders" may be taken to refer to any approach where indirect gain control of the channel is employed, and many of the concepts apply also to DCA implementations.

With VCA faders it is possible to break the connection between a fader and the corresponding means of level control, as was described in Fact File 5.5. It is across this breakpoint that an automation system will normally be connected. The automation processor then reads a digital value corresponding to the position of the fader and can return a value to control the gain of the channel (see Figure 5.18).

The information sent back to the VCA would depend on the operational mode of the system at the time, and might or might not correspond directly to the fader position. Common operational modes are:

- WRITE: channel level corresponds directly to the fader position.
- READ: channel level controlled by data derived from a previously stored mix.
- UPDATE: channel level controlled by a combination of previously stored mix data and current fader position.
- GROUP: channel level controlled by a combination of the channel fader's position and that of a group master.

In a VCA implementation the fader position is measured by an analog-to-digital convertor, which turns the DC value from the fader into a binary number (usually eight or ten bits) which the microprocessor can read. An eight-bit value suggests that the fader's position can be represented by one of 256 discrete values, which is usually enough to give the impression of continuous movements, although professional systems tend to use ten bit representation for more precise control (1024 steps). The automation computer "scans" the faders many times a second and reads their values. Each fader has a unique

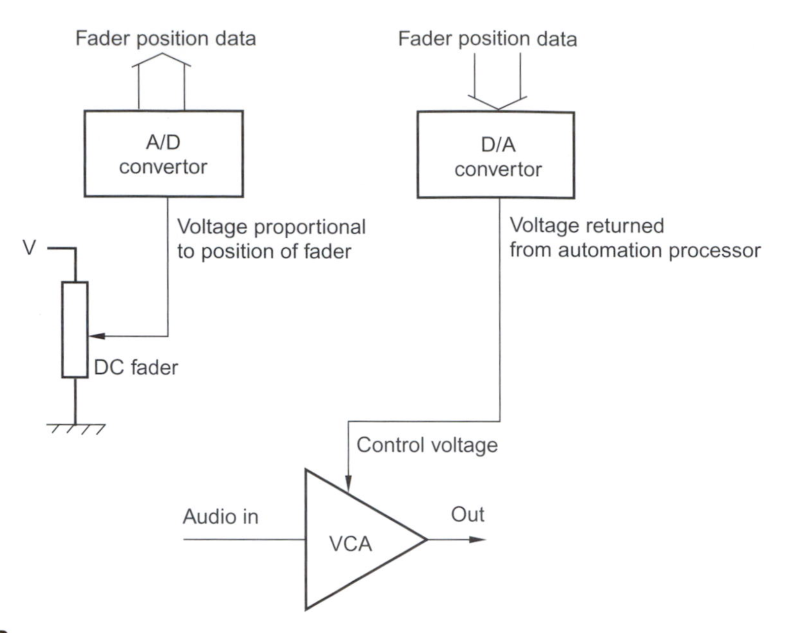

FIGURE 5.18
Fader position is encoded so that it can be read by an automation computer. Data returned from the computer is used to control a VCA through which the audio signal flows.

address and the information obtained from each address is stored in a different temporary memory location by the computer. A generalized block diagram of a typical system is shown in Figure 5.19.

The disadvantage of such a system is that it is not easy to see what the level of the channel is. During a read or update pass the automation computer is in control of the channel gain, rather than the fader. The fader can be halfway to the bottom of its travel whilst the gain of the VCA is near the top. Sometimes a mixer's bargraph meters can be used to display the value of the DC control voltage which is being fed from the automation to the VCA, and a switch is sometimes provided to change their function to this mode. Alternatively a separate display is provided for the automation computer, indicating fader position with one marker and channel gain with another.

Such faders are commonly provided with "null" LEDs: little lights on the fader package which point in the direction that the fader must be moved to make its position correspond to the stored level. When the lights go out (or when they are both on), the fader position is correct. This can sometimes be necessary when modifying a section of the mix by writing over the original data. If the data fed from the automation is different to the position of the fader, then when the mode is switched from read to write there will be a jump in level as the fader position takes over from the stored data. The null lights allow the user to move the fader towards the position dictated by the stored data, and most systems only switch from read to write when the null point is crossed, to ensure a smooth transition. The same procedure is followed when coming out of rewrite mode, although it can be bypassed in favor of a sudden jump in level.

Update mode involves using the relative position of the fader to modify the stored data. In this mode, the fader's absolute position is not important because the system assumes that its starting position is a point of unity gain, thereafter adding the changes in the fader's position to the stored data. So if a channel was placed in update mode and the fader moved up by 3 dB, the overall level of the updated passage would be increased by 3 dB (see Figure 5.20). For fine changes in gain the fader can be preset near the top of its range before entering update mode, whereas larger changes can be introduced nearer the bottom (because of the gain law of typical faders).

Some systems make these modes relatively invisible, anticipating which mode is most appropriate in certain situations. For example, WRITE mode is required for the first pass of a new mix, where the absolute fader positions are stored, whereas subsequent passes might require all the faders to be in UPDATE.

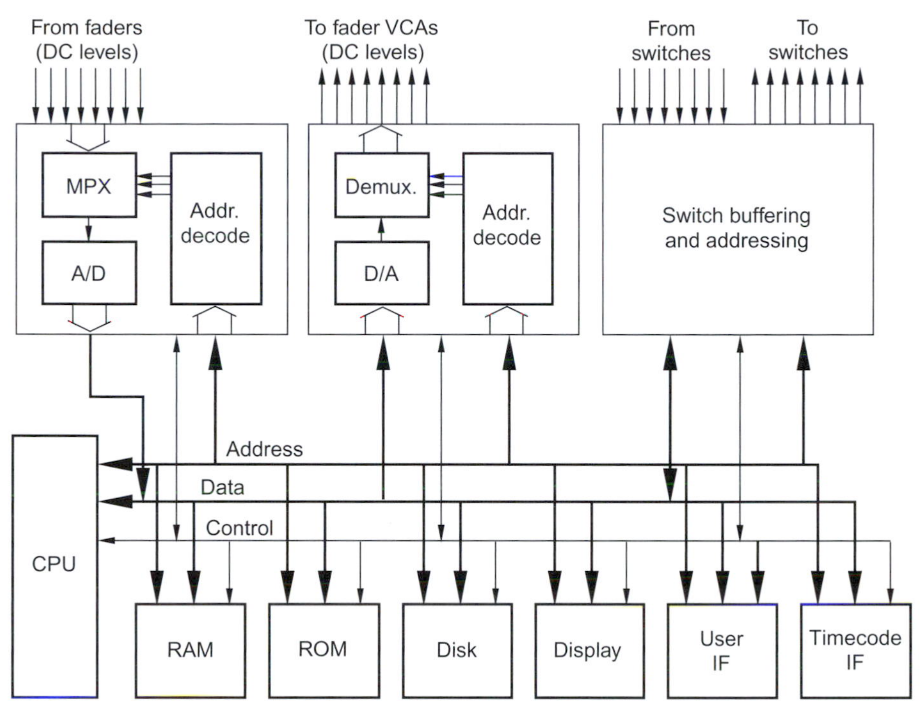

FIGURE 5.19
Generalized block diagram of a mixer automation system handling switches and fader positions. The fader interfaces incorporate a multiplexer (MPX) and demultiplexer (Demux) to allow one convertor to be shared between a number of faders. RAM is used for temporary mix data storage; ROM may hold the operating software program. The CPU is the controlling microprocessor.

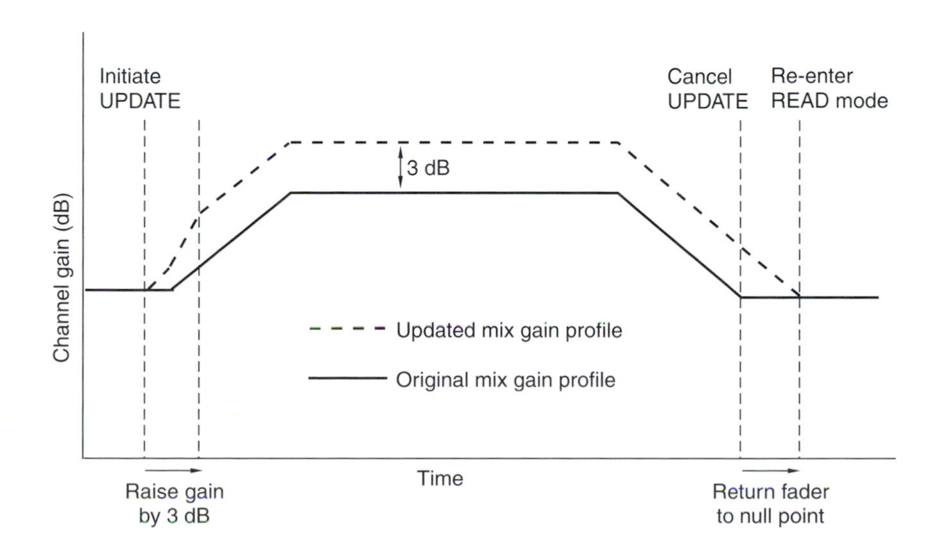

FIGURE 5.20
Graphical illustration of stages involved in entering and leaving an UPDATE or RELATIVE mode on an automated VCA fader.

A moving fader system works in a similar fashion, except that the data which is returned to the fader is used to set the position of a drive mechanism which physically moves the fader to the position in which it was when the mix was written. This has the advantage that the fader is its own means of visual feedback from the automation system and will always represent the gain of the channel.

If the fader were to be permanently driven, there would be a problem when both the engineer and the automation system wanted to control the gain. Clutches or other forms of control are employed to remove the danger of a fight between fader and engineer in such a situation, and the fader is usually made touch-sensitive to detect the presence of a hand on it.

Such faders are, in effect, permanently in update mode, as they can at any time be touched and the channel gain modified, but there is usually some form of relative mode which can be used for offsetting a complete section by a certain amount. The problem with relative offsets and moving faders is that if there is a sudden change in the stored mix data while the engineer is holding the fader, it will not be executed. The engineer must let go for the system to take control again. This is where a combination of moving fader and VCA-type control comes into its own.

Grouping automated faders

Conventional control grouping (Fact File 5.5) is normally achieved by using dedicated master faders. In an automated console it may be possible to do things differently. The automation computer has access to data representing the positions of all the main faders on the console, so it may allow any fader to be designated a group master for a group of faders assigned to it. It can do this by allowing the user to set up a fader as a group master (either by pressing a button on the fader panel, or from a central control panel). It will then use the level from this fader to modify the data sent back to all the other faders in that group, taking into account their individual positions as well. This idea means that a master fader can reside physically within the group of faders to which it applies, although this may not always be the most desirable way of working.

Sometimes the computer will store automation data relating to groups in terms of the motions of the individual channels in the group, without storing the fact that a certain fader was the master, whereas other systems will store the data from the master fader, remembering the fact that it was a master originally.

Mute automation

Mutes are easier to automate than faders because they only have two states. Mute switches associated with each fader are also scanned by the automation computer, although only a single bit of data is required to represent the state of each switch. A simple electronic switch can be used to effect the mute, and in analog designs this often takes the form of a FET (field effect transistor) in the signal path, which has very high attenuation in its "closed" position (see Figure 5.21). Alternatively, some more basic systems effect mutes by a sudden change in channel gain, pulling the fader down to maximum attenuation.

Storing the automation data

Early systems converted the data representing the fader positions and mute switches into a modulated serial data stream which could be recorded alongside the audio to which it related on a multitrack tape. In order to allow updates of the data, at least two tracks were required: one to play back the old data, and one to record the updated data, these usually being the two outside tracks of the tape (1 and 24 in the case of a 24 track machine). This was limiting, in that only the two most recent mixes were

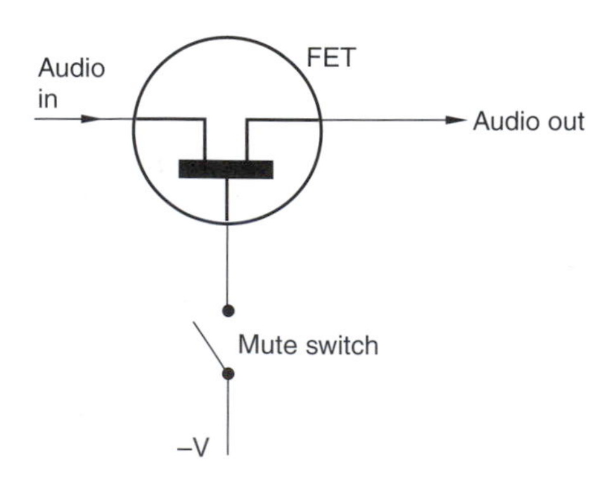

FIGURE 5.21
Typical implementation of an FET mute switch.

ever available for comparison (unless more tracks were set aside for automation), whole tracks had to be mixed at a time (because otherwise the updated track would be incomplete), and at least two audio tracks were lost on the tape. Yet it meant that the mix data was always available alongside the music, eliminating the possibility of losing a disk with the mix data stored separately.

More recent systems use computer hardware to store mix data, in RAM and on disks. Data is synchronized to the audio by recording time-code alongside audio which uniquely identifies any point in time, this being read by the automation system and used to relate recording position to stored data. This method gives almost limitless flexibility in the modification of a mix, allowing one to store many versions, of which sections can be joined together "off-line" (that is, without the recorder running) or on-line, to form the finished product. The finished mix can be dumped to a disk for more permanent storage, and this disk could contain a number of versions of the mix.

It is becoming quite common for some automation systems to use MIDI or MIDI over Ethernet (ipMIDI) for the transmission of automation data. A basic automation computer associated with the mixer converts control positions into MIDI information and transmits/receives it using a device known as a UART which generates and decodes serial data at the appropriate rate for the MIDI standard, as shown in Figure 5.22. MIDI data can then be stored on a conventional sequencer or using dedicated software.

Integrating machine control

Control of recording machines is a common feature of modern mixers. It may only involve transport remotes being mounted in the center panel somewhere, or it may involve a totally integrated autolocator/synchronizer associated with the rest of the automation system. On top-flight desks, controls are provided on the channel modules for putting the relevant tape track into record-ready mode, coupled with the record function of the transport remotes. This requires careful interfacing between the console and the recording machine, but means that it is not necessary to work with a separate recording machine remote unit by the console.

It is very useful to be able to address the automation in terms of the mix in progress; in other words, "go back to the second chorus," should mean something to the system, even if abbreviated. The alternative is to have to address the system in terms of timecode locations. Often, keys are provided which allow the engineer to return to various points in the mix, both from a mix data point-of-view and from the recording machines" point-of-view, so that the automation system locates the recorder to the position described in the command, ready to play.

Retrofitting automation

Automation can sometimes be retrofitted into existing analog consoles that do not have any automation. These systems usually control only the faders and the mutes, as anything else requires considerable modification of the console's electronics, but the relatively low price of some systems makes them attractive, even on a modest budget. Fitting normally involves a modification or replacement of the

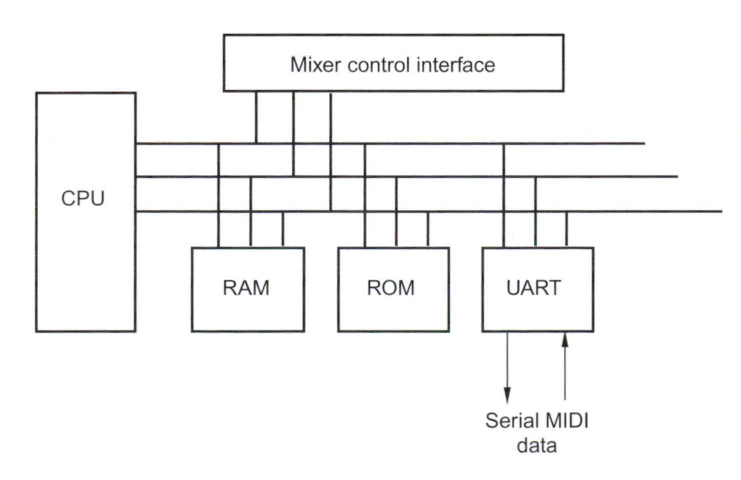

FIGURE 5.22
A UART is used to route MIDI data to and from the automation computer.

fader package, to incorporate VCAs in consoles which don't have them, or to break into the control path between fader and VCA in systems which do. This job can normally be achieved in a day. It is also possible to retrofit moving fader automation.

A separate control panel may be provided, with buttons to control the modes of operation, as well as some form of display to show things like VCA gains, editing modes, and set-up data. The faders will be interfaced to a processor rack which would reside either in a remote bay, or under the console, and this will normally contain a disk drive to store the final mixes. Alternatively a standard desktop computer will be used as the control interface.

Total automation systems

SSL originally coined the term "Total Recall" for its system, which was a means of telling the operator where the controls should be and leaving him or her to reset them him/herself. This saved an enormous amount of time in the resetting of the console in between sessions, because it saved having to write down the positions of every knob and button.

True Total Reset is quite a different proposition and requires an interface between the automation system and every control on the console, with some means of measuring the position of the control, some means of resetting it, and some means of displaying what is going on. A number of options exist, for example one could:

- motorize all the rotary pots
- make all the pots continuously rotating and provide a display
- make the pots into up/down-type incrementers with display
- provide assignable controls with larger displays.

Of these, the first is impractical in most cases due to the space that motorized pots would take up, the reliability problem and the cost, although it does solve the problem of display. The second would work, but again there is the problem that a continuously rotating pot would not have a pointer because it would merely be a means of incrementing the level from wherever it was at the time, so extra display would be required and this takes up space. Nonetheless some ingenious solutions have been developed, including incorporating the display in the head of rotary controls (see Figure 5.23). The third is not ergonomically very desirable, as the human prefers analog interfaces rather than digital ones, and there is no room on a conventional console for all the controls to be of this type with their associated displays. Most of the designs which have implemented total automation have adopted a version of the fourth option: that is, to use fewer controls than there are functions, and to provide larger displays.

The concept of total automation is inherent in the principles of an assignable mixing console, as described below in the section on digital mixers. In such a console, few of the controls carry audio directly as they are only interfaces to the control system, so one knob may control the HF EQ for any channel to which it is assigned, for example. Because of this indirect control, usually via a microprocessor, it is relatively easy to implement a means of storing the switch closures and settings in memory for reiteration at a later date.

Dynamic and static systems

Many analog assignable consoles use the modern equivalent of a VCA: the digitally controlled attenuator, also to control the levels of various functions such as EQ, aux sends, and so on. Full dynamic automation requires regular scanning of all controls so as to ensure smooth operation, and a considerable amount of data is generated this way. Static systems exist which do not aim to store the continuous changes of all the functions, but they will store "snapshots" of the positions of controls which can be recalled either manually or with respect to timecode. This can often be performed quite regularly (many times a second) and in these cases we approach the dynamic situation, but in others the reset may take a second or two which precludes the use of it during mixing. Changes must be silent to be useful during mixing.

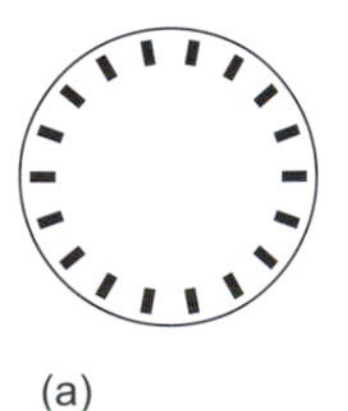

FIGURE 5.23
Two possible options for positional display with continuously rotating knobs in automated systems. (a) Lights around the rim of the knob itself; (b) lights around the knob's base.

Other snapshot systems merely store the settings of switch positions, without storing the variable controls, and this uses much less processing time and memory. Automated routing is of particular use in theater work where sound effects may need to be routed to a complex combination of destinations. A static memory of the required information is employed so that a single command from the operator will reset all the routing ready for the next set of sound cues.

DIGITAL MIXERS

Much of what has been said in this chapter applies equally to both analog and digital mixers, at least in conceptual and operational terms. To complete the picture some features specific to digital mixers will now be described. Digital mixing has now reached the point where it can be implemented cost effectively, and there are a number of reasonably priced digital mixers with full automation. At the other end of the scale companies are manufacturing large-scale studio mixers with an emphasis on ultra-high sound quality and an ergonomically appropriate control interface. At the low cost end of the scale, digital mixers are implemented within computer-based workstations and represented graphically on the computer display. Faders and other controls are moved using a mouse.

Audio handling

In a digital mixer incoming analog signals are converted to the digital domain as early as possible so that all the functions are performed entirely in the digital domain, with as much as 32-bit internal processing resolution to cope with extremes of signal level, EQ settings and other effects. The advantage of this is that once the signal is in the digital domain it is inherently more robust than its analog counterpart: it is virtually immune from crosstalk, and is unaffected by lead capacitance, electromagnetic fields from mains wiring, additional circuit distortion and noise, and other forms of interference. Digital inputs and outputs can be provided to connect recording devices and other digital equipment without conversion to analog. Inputs can be a mixture of analog and digital (the latter configurable via plug-in modules for Tascam, ADAT, Yamaha and AES/EBU formats, for example) with digital and analog main and monitoring outputs. Functions such as gain, EQ, delay, phase, routing, and effects such as echo, reverb, compression and limiting, can all be carried out in the digital domain precisely and repeatably using digital signal processing.

Inputs and outputs, digital and analog, are often provided by a series of outboard rack-mounting units which incorporate the D/A and A/D convertors, microphone amplifiers and phantom power, and these can be positioned where needed, the gain still being adjusted from the mixer. In a recording studio one rack could be by the control surface itself, one or more in the recording studio area, and one by the recording machines. In a theater, units would be placed next to power amplifiers, and in off-stage areas where musicians play. These units are connected in a daisy-chain loop to the main control surface via coaxial BNC cabling, MADI interface, or proprietary fiber optic links, the latter being preferable for longer distances.

Assignable control surfaces

Operationally a digital mixer can remain similar to its analog counterpart, although commercial examples have tended at least partially to use assignable control surface designs as they can be easier to implement in digital mixers. With assignable designs many of the controls of the traditional console such as pan, EQ, aux send and group assign are present only as single assignable sections or multi-function controls, many facilities can be packed into a unit of modest dimensions and quite modest cost.

A fully assignable digital mixing console is ergonomically quite different from its analog counterpart. Typically, the control surface consists of many input channel faders and/or rotary knobs, each channel having "active" and "select" buttons. Much smaller areas of the control surface are given over to single sections of EQ, routing (aux and group) and processing: these sections are automatically assigned to one particular channel when its "select" button is active before adjustments can take place. Thus many processes which affect the signals are not continuously on view or at the fingertips of the operator as is the case with the traditional analog desk. The assignable design is therefore more suitable to recording work (particularly post-session mixdowns) where desk states can be built up gradually and saved to scene memories, rather than to live performance and primary recording work where continuous visual indication of and access to controls remains desirable.

Facilities such as channel delay, effects processing, moving fader automation and fader ganging, scene memories offering total recall of all settings, MIDI (including memory load and dump via separate MIDI data filers), and timecode interface are typically offered, and a display screen shows the status

FIGURE 5.24
Digico D5T. (Courtesy of the RSC)

of all controls: either in simple global formats for the whole console for parameters such as routing, channel delay, scene memory details and the like, or in much greater detail for each individual channel. Metering can also be shown. Cursors facilitate both navigation around the screen displays and adjustments of the various parameters.

Digital mixers – a case study

The difficulty and expense of implementing true "total recall" of an analog mixer, that is automated resetting of all surface controls, has already been discussed. Digital mixers can incorporate such a feature routinely, and the Digico D5T illustrated in Figure 5.24 is a typical example of a console in which all set-up parameters including such things as input gain and phantom power switching are recallable in seconds when a particular project or show is loaded into the mixer or recalled from its memory store. Such mixers are essentially versions of computer mixing systems but with a hardware control surface to provide a more traditional mode of hands-on operation, still an essential feature for live mixing work and many types of recording and broadcast session. Ergonomically, the mixer combines traditional "analog" facilities of channel faders, aux send and EQ knobs, "VCA" and group faders, with a considerable degree of assignability using large touch-sensitive screens and several selectable "layers" across which banks of inputs can be displayed and accessed (Figure 5.25). A master output screen can display a variety of things such as group outputs, automation parameters, scene memory information, matrix settings and the like. A console such as this can offer 96 input channels, 20 aux sends, 24 group sends, and in a theater version a 32-output matrix section. It is not difficult to appreciate the huge size and cost of an equivalent analog console. A QWERTY key pad facilitates the labeling of all sections.

Typically, adjacent to each bank of channel faders will be a row of buttons enabling access to different control layers. Layer 1 could be input channels 1 to 8, the accompanying screen display showing such things as input gain, phantom power, routing, aux send levels, and EQ. Touching the appropriate area of the display expands that area for ease of viewing and adjustment, e.g. touching the EQ section of a channel displays the settings in much more detail and assigns the EQ controls adjacent to the screen to that channel. Layer 2 could display channels 25 to 30 (channels 9 to 24 being provided by adjacent

FIGURE 5.25
Detail of input channels display. (Digico D5T, courtesy of the RSC)

banks of faders). Layer 3 of all the fader banks could give fader control of all the matrix outputs, or all the group outputs, or all the aux master outputs, or a combination. All of these things are chosen by the operator and set up to his or her requirements. The top layer would normally be assigned to inputs which need to be continuously on view, e.g. musicians" microphones and vocal mics, radio mics, and DI inputs. The lower layers would be assigned to things such as CD players, sampler and other replay machine outputs, and probably to some of the effects returns. These inputs do not normally need to be accessed quickly. Other features such as digital delay and EQ on inputs and outputs (the latter particularly useful in live sound work), compressors and limiters, and internal effects processors, are routinely available. This reduces the number of outboard effects processors needed. The settings for these are all programmable and recordable along with the rest of the console settings.

Two main observations can be made regarding the operation of such consoles compared with their analog counterparts. First, a good deal of initial setting up, assigning and labeling needs to be carried out before a session can begin. Input and output channels need to be assigned to appropriate sockets on the outboard units around the building; the various layers have to be assigned to inputs/outputs/auxs/VCAs as appropriate, and labeled; and a series of scene memories has to be created in anticipation of what will be required for the show or recording session. Second, the operation of the console often requires a two-stage thinking process. Although channel faders and some other facilities for a particular layer will be instantly available for adjustment, many other facilities will need to be accessed either on a different layer or by touching an area of a screen before adjustments can be made. Additionally, adjustments need to be stored in a scene memory. Normally, storing changes such as input gain, EQ, and aux send levels in a particular scene will automatically store those changes to the other scene memories. Channel fader adjustments will be stored only to that particular scene. Just what adjustments are stored to the present scene, and which ones are automatically stored to a bank of scenes, can be chosen by the operator. The complete project then needs to be stored to the mixer's hard disk drive, and preferably also to an external backup. This all needs an operator who is familiar with that particular console and its software quirks. Digital consoles necessarily have many common features, but

FIGURE 5.26
The Midas Heritage 3000. (Courtesy of Klark Teknik)

manufacturers have their own proprietary ways of doing things. The typical analog console, in contrast, will be fairly familiar to a user after ten or 15 minutes.

Digitally controlled analog mixers

A digitally controlled analog console will be looked at briefly next. The Midas Heritage 3000 shown in Figure 5.26 is a good example of such a mixer. Its control surface is analog, and the signals remain in the analog domain throughout. Digital control gives such things as mute and mute group automation, VCA assign, and virtual fader automation (a row of LEDs adjacent to each fader displays the audio level of the fader regardless of its physical position; moving the fader to the top lit LED gives the operator manual control). Scene memories can thus be programmed into the desk giving the appropriate fader positions and channel mutes, these being of great value in the live mixing situations for which such consoles are designed. Other consoles also provide automation of EQ in/out, insert, aux send enable, group assign, and moving fader automation, albeit at a somewhat higher cost, and such consoles undoubtedly prove their worth in the live sound market where visiting and freelance sound engineers need to become quickly familiar with a console which does not have too much automation. Such consoles are likely to be in use until well into the second decade of the present century.

MIXERS WITH INTEGRATED CONTROL OF DIGITAL WORKSTATIONS

Integrated control of digital audio workstations is now a growing feature of either analog or digital mixing consoles. In the case of some designs the mixing console has become little more than a sophisticated control surface, enabling the functions of a workstation to be adjusted using more conventional controls. This is offered as an alternative to using a computer display and mouse, which can be inconvenient when trying to handle complex mixes. In such cases most of the audio processing is handled by the workstation hardware, either using desktop computer processing power or dedicated signal

FIGURE 5.27
A typical workstation-integrated mixer control surface from the Digidesign ICON series (D-Control ES). Courtesy of Digidesign.

FIGURE 5.28
Analog mixer with integrated workstation control: SSL Matrix. (Courtesy of Solid-State Logic)

processing cards. Some audio handling can be included, such as monitoring and studio communication control. The mixer control surface is connected to the workstation using a dedicated interface such as MIDI or Ethernet. Such control surfaces often include remote control facilities for the workstation transport and editing functions. An example from the Digidesign ICON series is shown in Figure 5.27.

An alternative is to employ an external analog mixer that has dedicated workstation control facilities, such as the SSL Matrix, pictured in Figure 5.28. In this case the mixer handles more of the audio processing itself and can be used as an adjunct or alternative to the onboard digital processing of the workstation, with comprehensive routing to and from conventional analog outboard equipment. It enables analog mixing of audio channels outside the workstation, which has become a popular way of working for some. The control interface to the workstation is by means of MIDI over Ethernet (ipMIDI), and USB to carry the equivalent of computer keyboard commands. The control protocol can be made to conform either to Mackie's HUI (Human User Interface) or MCU (Mackie Control Universal), or to a special configuration of MIDI controllers, and the mixer's digitally controlled attenuators can be remote controlled using MIDI commands. The latter facility enables the mixer to be automated using MIDI tracks on the workstation. A typical configuration of this system is shown in Figure 5.29.

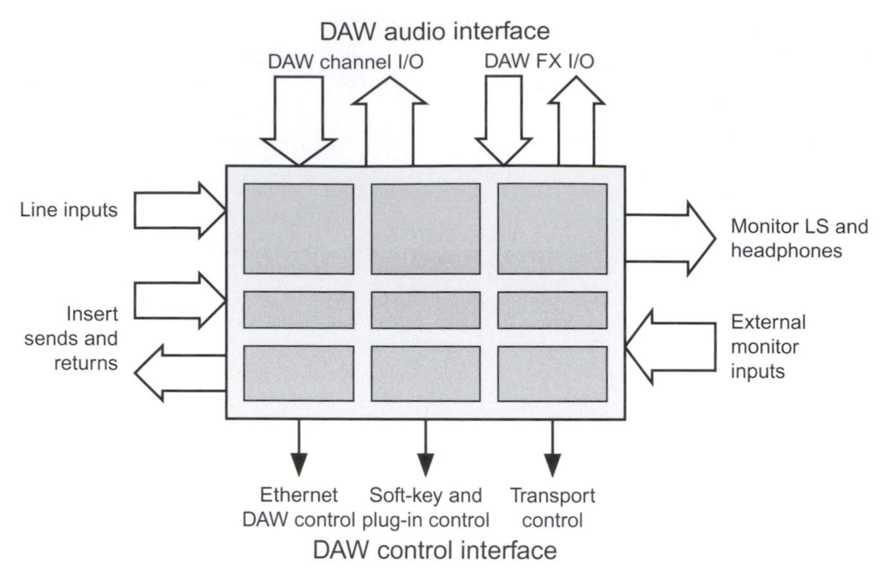

FIGURE 5.29
A typical system configuration of the SSL Matrix console showing interconnections to studio equipment and workstation.

INTRODUCTION TO MIXING APPROACHES

This section provides an introduction to basic mixer operation and level setting.

Acoustic sources will be picked up by microphones and fed into the mic inputs of a mixer (which incorporates amplifiers to raise the low-voltage output from microphones), whilst other sources usually produce so-called "line level" outputs, which can be connected to the mixer without extra amplification. In the mixer, sources are combined in proportions controlled by the engineer and recorded. In "straight-to-stereo" (or surround) techniques, such as a classical music recording, microphone sources are often mixed "live" without recording to a multitrack medium, creating a session master which is the collection of original recordings, often consisting of a number of takes of the musical material. The balance between the sources must be correct at this stage, and often only a small number of carefully positioned microphones are used. The session master recordings will then proceed to the editing stage where takes are assembled in an artistically satisfactory manner, under the control of the producer, to create a final master which will be transmitted or made into a commercial release. This final master could be made into a number of production masters which will be used to make different release formats. In the case of "straight-to-stereo" mixing the console used may be a simpler affair than that used for multitrack recording, since the mixer's job is to take multiple inputs and combine them to a single stereo output, perhaps including processing such as equalization. This method of production is clearly cheaper and less time consuming than multitrack recording, but requires skill to achieve a usable balance quickly. It also limits flexibility in post-production. Occasionally, classical music is recorded in a multitrack form, especially in the case of complex operas or large-force orchestral music with a choir and soloists, where to get a correct balance at the time of the session could be costly and time consuming. In such a case, the production process becomes more similar to the pop recording situation described below.

"Pop" music is rarely recorded live, except at live events such as concerts, but is created in the recording studio. Acoustic and electrical sources are fed into a mixer and recorded on to a multitrack medium, often a few tracks at a time, gradually building up a montage of sounds. The resulting recording then contains a collection of individual sources on multiple tracks which must subsequently be mixed into the final release format. Individual songs or titles are recorded in separate places on the tape, to be compiled later. It is not so common these days to record multitrack pop titles in "takes" for later editing, as with classical music, since mixer automation allows the engineer to work on a song in sections for automatic execution in sequence by a computer. In any case, multitrack machines have comprehensive "drop-in" facilities for recording short inserted sections on individual tracks without introducing clicks, and a pop-music master is usually built up by laying down backing tracks for a complete song (drums, keyboards, rhythm guitars, etc.) after which lead lines are overdubbed using drop-in facilities. Occasionally multitrack recordings are edited or compiled ("comped") early on during a recording session to assemble an acceptable backing track from a number of takes, after which further layers are added. Considerable use may be made of computer-sequenced electronic instruments, under MIDI control, often in conjunction with multitrack disk recording. The computer controlling the electronic instruments is synchronized to the recording machine using time code and the outputs of the instruments are fed to the mixer to be combined with the non-sequenced sources.

Once the session is completed, the multitrack recording is mixed down. This is often done somewhere different from the original session, and involves feeding the outputs of each track into individual inputs of the mixer, treating each track as if it were an original source. The balance between the tracks, and the positioning of the tracks in the stereo image, can then be carried out at leisure (within the budget constraints of the project!), often without all the musicians present, under control of the producer. During the mixdown, further post-production takes place such as the addition of effects from outboard equipment to enhance the mix. An automation system is often used to memorize fader and mute movements on the console, since the large number of channels involved in modern recording makes it difficult if not impossible for the engineer to mix a whole song correctly in one go. Following mixdown, the master that results will be edited very basically, in order to compile titles in the correct order for the production master. The compiled tape will then be mastered for the various distribution media.

BASIC OPERATIONAL TECHNIQUES

Level setting

If one is using a microphone to record speech or classical music then normally a fairly high input gain setting will be required. If the microphone is placed up against a guitar amplifier then the mic's output will be high and a much lower input gain setting can be used. There are essentially three ways of setting the gain control to the optimum position. First, using PFL or prefade listen (see Fact File 5.3).

PFL is pressed, or the fader overpressed (i.e. pressed beyond the bottom of its travel against a sprung microswitch), on the input module concerned and the level read on either a separate PFL meter or with the main meters switched to monitor the PFL bus. The channel input gain should be adjusted to give a meter reading of, say, PPM 5, or 0 VU on older analog desks, and a meter reading of perhaps 6–10 dB below maximum on a digital desk. This gain-setting procedure must be carried out at a realistic input level from the source. It is frequently the case during rehearsals that vocalists and guitarists will produce a level that is rather lower than that which they will use when they actually begin to play.

The pan control should be set next (see Fact File 5.2) to place the source in the stereo image. The main output faders will normally be set to 0 dB on their calibration, which is usually at the top. The channel faders can then be set to give both the desired subjective sound balance and appropriate output meter readings.

The second way of setting the gain is a good way in its own right, and it has to be used if PFL facilities are not provided. First of all both the channel fader and the output faders need to be positioned to the 0 dB point. This will be either at the top of the faders" travels or at a position about a quarter of the way down from the top of their travel. If no 0 dB position is indicated then the latter position should be set. After the pan control and faders have been positioned, the input gain may then be adjusted to give the desired reading on the output level meters. When several incoming signals need to be balanced the gain controls should all be positioned to give both the desired sound balance between them and the appropriate meter readings – normally PPM 6 or just over 0 VU during the loudest passages.

These two gain-setting methods differ in that with the former method the channel fader positions will show a correspondence to the subjective contribution each channel is making towards the overall mix, whereas the latter method places all the channel faders at roughly the same level.

The third way is similar to the second way, but one channel at a time is set up, placing channel and output faders at 0 dB and adjusting the gain for a peak meter reading. That channel fader is then turned completely down and the next channel is set up in a similar way. When all the channels which are to be used have been set up, the channel faders can then be advanced to give both the desired subjective balance and peak meter readings.

Use of the EQ controls often necessitates the resetting of the channel's gain. For example, if a particular instrument requires a bit of bass boost, applying this will also increase the level of signal and so the gain will often need to be reduced a little to compensate. Applying bass or treble cut will sometimes require a small gain increase.

Using auxiliary sends

Aux facilities were described in "Auxiliary sends," above. The auxiliaries are configured either "pre-fade" or "post-fade". Pre-fade aux sends are useful for providing a monitor mix for musicians, since this balance will be unaffected by movements of the faders which control the main mix. The engineer then retains the freedom to experiment in the control room without disturbing the continuity of feed to the musicians.

Post-fade sends are affected by the channel fader position. These are used to send signals to effects devices and other destinations where it is desirable to have the aux level under the overall control of the channel fader. For example, the engineer may wish to add a little echo to a voice. Aux 2, set to post-fade, is used to send the signal to an echo device, probably positioning the aux 2 control around the number 6 position and the aux 2 master at maximum. The output of the echo device is returned to another input channel or an echo return channel, and this fader can be adjusted to set the amount of echo. The level of echo will then rise and fall with the fader setting for the voice.

The post-fade aux could also be used simply as an additional output to drive separate amplifiers and speakers in another part of a hall, for example.

Using audio groups

The group outputs (see "Channel grouping," above) or multitrack routing buses (see "Routing section," above) can be used for overall control of various separate groups of instruments, depending on whether mixing down or track laying. For example, a drum kit may have eight microphones on it. These eight input channels can be routed to groups 1 and 2 with appropriate stereo pan settings. Groups 1 and 2 would then be routed to stereo outputs left and right respectively. Overall control of the drum kit level is now achieved simply by moving group faders 1 and 2.

PREFACE

Whatever sort of audio system you are dealing with, it's a pretty safe bet that cables will be involved somewhere. This chapter, taken from the *Handbook for Sound Engineers*, and written by Glen Ballou, covers loudspeaker cables, digital audio cables, and video cables, as well as audio signal connections. It is no use constructing near-perfect boxes of electronics if the cable leading to them fatally compromises the quality of the signal before it even arrives there.

Glen begins by examining the National Electrical Code which covers the installation of wiring in commercial and residential areas in the USA, the emphasis being on safety. This deals with power wiring rather than signal cables, and focuses on such matters as the specifications for outlet sockets, conduits, circuit-breakers, and so on. In book form the the NEC runs to approximately 1000 pages.

From there we move on to consider some of the most basic properties of wire. Firstly, its size and how that is specified, its resistance and how that is derived from the dimensions and the properties of the conductor metal, and wire current ratings. There is a table of values for the resistivity of metals which will come in handy if you're planning to make cables out of arsenic.

The next topic is the vital one of shielding a signal cable. There are various methods—braided shielding, foil shielding, foil tape shielding, and so on, each with their own advantages.

In the next section, Glen considers the many kinds of insulation that can be applied to cables. Once again there are many types, each with their own benefits and drawbacks for given applications.

The next topic is the characteristic impedance of a cable. This only applies if the length of the cable is a significant fraction of the wavelength of the frequency being carried, and it therefore only applies to analog audio if the cable is very long indeed—the classic example being telephone lines, where the air-spacing of the two conductors of a go and return circuit by about 25 cm gives a characteristic impedance of 600 Ω. This figure persists in audio because it is generally accepted as the heaviest load that a line output stage ought to be able to drive, though a line input normally has an input impedance of at least 10 kΩ.

Glen moves on to consider the requirements for AES/EBU digital audio cables. Digital audio requires a much greater bandwidth than analog audio, and the amount required increases proportionally with the sampling rate. A sample rate of 44.1 kHz needs a bandwidth of 5.6 MHz, whereas a sample rate of 192 kHz demands 24.6 MHz. Coaxial cable is preferred to twisted pair for this application.

Next up are loudspeaker cables, both those for direct connection and those intended for 70V or 100V lines, where the audio is transmitted at high voltage and low current to minimize the resistive losses.

Coaxial cables and CCTV cable requirements are then looked at, and the chapter concludes with data on AC power cables and the international standards for their plugs and outlets.

CHAPTER 6

Transmission Techniques: Wire and Cable

Handbook for Sound Engineers by Glen Ballou

INTRODUCTION

It was not long ago that wire was the only method to inexpensively and reliably transmit sound or pictures from one place to another. Today we not only have wire, but we also have fiber optics, and wireless radio frequency (RF) transmission from Blu-tooth to wireless routers, cell phones, and microwave and satellite delivery. This chapter will discuss the various forms of wire and cable used in audio and video.

Wire is a single conductive element. Wire can be insulated or uninsulated. Cable, on the other hand, is two or more conductive elements. While they theoretically could be uninsulated, the chance of them touching each other and creating a short circuit requires that they are usually both insulated. A cable can be multiple insulated wires, called a *multiconductor cable*, or wires that are twisted together, called a *twisted pair cable*, or cables with one wire in the center, surrounded by insulation and then a covering of metal used as another signal path, called *coaxial cable*.

CONDUCTORS

Wire and cable are used to connect one circuit or component to another. They can be internal, connecting one circuit to another inside a box, or externally connecting one box to another.

Resistance and wire size

Wire is made of metal, or other conductive compounds. All wire has resistance which dissipates power through heat. While this is not apparent on cables with small signals, such as audio or video signals, it is very apparent where high power or high current travels down a cable, such as a power cord. Resistance is related to the size of the wire. The smaller the wire, the greater the resistance.

Calculating wire resistance

The resistance for a given length of wire is determined by:

$$R = \frac{KL}{d^2} \qquad (6.1)$$

where,

R is the resistance of the length of wire in ohms,

K is the resistance of the material in ohms per circular mil foot,

L is the length of the wire in feet,

d is the diameter of the wire in mils.

The resistance, in ohms per circular mil foot (Ω/cir mil ft), of many of the materials used for conductors is given in Table 6.1. The resistance shown is at 20°C (68°F), commonly called *room temperature.*

Table 6.1 Resistance of Metals and Alloys

Material	Symbol	Resistance (Ω/cir mil ft)
Silver	Ag	9.71
Copper	Cu	10.37
Gold	Au	14.55
Chromium	Cr	15.87
Aluminum	Al	16.06
Tungsten	W	33.22
Molybdenum	Mo	34.27
High-brass	Cu-Zn	50.00
Phosphor-bronze	Sn-P-Cu	57.38
Nickel, pure	Ni	60.00
Iron	Fe	60.14
Platinum	Pt	63.80
Palladium	Pd	65.90
Tin	Sn	69.50
Tantalum	Ta	79.90
Manganese-nickel	Ni-Mn	85.00
Steel	C-Fe	103.00
Lead	Pb	134.00
Nickel-silver	Cu-Zn-Ni	171.00
Alumel	Ni-Al-Mn-Si	203.00
Arsenic	As	214.00
Monel	Ni-Cu-Fe-Mn	256.00
Manganin	Cu-Mn-Ni	268.00
Constantan	Cu-Ni	270.00
Titanium	Ti	292.00
Chromel	Ni-Cr	427.00
Steel, manganese	Mn-C-Fe	427.00
Steel, stainless	C-Cr-Ni-Fe	549.00
Chromax	Cr-Ni-Fe	610.00
Nichrome V	Ni-Cr	650.00
Tophet A	Ni-Cr	659.00
Nichrome	Ni-Fe-Cr	675.00
Kovar A	Ni-Co-Mn-Fe	1732.00

When determining the resistance of a twisted pair, remember that the length of wire in a pair is twice the length of a single wire. Resistance in other constructions, such as coaxial cables, can be difficult to determine from just knowing the constituent parts. The center conductor might be easy to determine but a braid or braid + foil shield can be difficult. In those cases, consult the manufacturer.

Table 6.1 show the resistance in ohms (Ω) per foot per circular mil area for various metals, and combinations of metals (alloys). Of the common metals, silver is the lowest resistance. But silver is expensive and hard to work with. The next material, copper, is significantly less expensive, readily available, and lends itself to being annealed. Copper is therefore the most common material used in the manufacture of wire and cable. However, where price is paramount and performance not as critical, aluminum is often used. The use of aluminum as the conducting element in a cable should be an indication to the user that this cable is intended to be lower cost and possibly lower performance.

One exception to this rule might be the use of aluminum foil which is often used in the foil shielding of even expensive high-performance cables. Another exception is emerging for automobile design, where the weight of the cable is a major factor. Aluminum is significantly less weight than copper, and the short distances required in cars means that resistance is less of a factor.

Table 6.1 may surprise many who believe, in error, that gold is the best conductor. The advantage of gold is its inability to oxidize. This makes it an ideal covering for articles that are exposed to the atmosphere, pollution, or moisture such as the pins in connectors or the connection points on insertable circuit boards. As a conductor, gold does not require annealing, and is often used in integrated circuits since it can be made into very fine wire. But, in normal applications, gold would make a poor conductive material, closer to aluminum in performance than copper.

One other material on the list commonly found in cable is steel. As can be seen, this material is almost 10 times the resistance of copper, so many are puzzled by its use. In fact, in the cables that use steel wires, they are coated with a layer of copper, called *copper-clad steel* and signal passes only on the copper layer, an effect called *skin effect* that will be discussed later. Therefore, the steel wire is used for strength and is not intended to carry signals.

Copper-clad steel is also found in cables where cable pulling strength (pulling tension) is paramount. Then a stranded conductor can be made up of many copper-clad steel strands to maximize strength. Such a cable would compromise basic resistive performance. As is often the case, one can trade a specific attribute for another. In this case, better strength at the cost of higher resistance.

Resistance and gage size

In the United States, wire is sized by the American Wire Gage (AWG) method. AWG was based on the previous Brown and Sharpe (B & S) system of wire sizes which dates from 1856. AWG numbers are most common in the United States, and will be referred to throughout this chapter. The wire most often used in audio ranges from approximately 10 AWG to 30 AWG, although larger and smaller gage sizes exist. Wire with a small AWG number, such as 4 AWG, is very heavy, physically strong but cumbersome, and has very low resistance, while wire of larger numbers, such as 30 AWG can be very lightweight and fragile, and has high resistance. Resistance is an important factor in determining the appropriate wire size in any circuit. For instance, if an 8 Ω loudspeaker is being connected to an amplifier 500 ft away through a #19 wire, 50% of the power would be dropped in the wire in the form of heat. This is discussed later regarding loudpeaker cable.

Each time the wire size changes three numbers, such as from 16 AWG to 19 AWG, the resistance doubles. The reverse is also true. With a wire changed from 16 AWG to 13 AWG, the resistance halves. This also means that combining two identical wires of any given gage decreases the total gage of the combined wires by three units, and reduces the resistance. Two 24 AWG wires combined (twisted together) would be 21 AWG, for instance. If wires are combined of different gages, the resulting gage can be easily calculated by adding the circular mil area (CMA) shown in Tables 6.2 and 6.3. For instance, if three wires were combined, one 16 AWG (2583 CMA), one 20 AWG (1022 CMA) and one 24 AWG (404 CMA), the total CMA would be 2583 + 1022 + 404 = 4009 CMA. Looking at Table 6.1, this numbers falls just under 14 AWG. While even number gages are the most common, odd number gages (e.g., 23 AWG) can sometimes be found. There are many Category 6 (Cat 6) premise/data cables that are 23 AWG, for instance. When required, manufacturers can even produce partial gages. There are coaxial cables with 28.5 AWG center conductors. Such specialized gage sizes might require equally special connectors.

There are two basic forms of wire: solid and stranded. A solid conductor is one continuous piece of metal. A stranded conductor is made of multiple smaller wires combined to make a single conductor. Solid wire has slightly lower resistance, with less flexibility and less flex-life (flexes to failure) than stranded wire.

Drawing and annealing

Copper conductors start life as copper ore in the ground. This ore is mined, refined, and made into bars or rod. Five sixteenth inch copper rod is the most common form used for the making of wire and cable. Copper can be purchased at various purities. These commonly follow the ASTM (American Society for Testing and Materials) standards. Most of the high-purity copper is known as ETP, electrolytic tough pitch. For example, many cable products are manufactured with ASTM B115 ETP. This copper is 99.95% pure. Copper of higher purity can be purchased should the requirement arise. Many consumer audiophiles consider these to be oxygen free, when this term is really a discussion of copper purity and is determined by the number of nines of purity. The cost of the copper rises dramatically with each "9" that is added.

To turn 5/16 inch rod into usable wire, the copper rod is drawn through a series of dies. Each time it makes the rod slightly smaller. Eventually you can work the rod down to a very long length of very small wire. To take 5/16 inch rod down to a 12 AWG wire requires drawing the conductor through 11 different dies. Down to 20 AWG requires 15 dies. To take that wire down to 36 AWG requires 28 dies.

The act of drawing the copper work hardens the material, making it brittle. The wire is run through an in-line annealing oven, at speeds up to 7000 feet per minute, and a temperature of 900 to 1000°F (482 to 537°C). This temperature is not enough to melt the wire, but it is enough to let the copper lose its brittleness and become flexible again, to reverse the work hardening. Annealing is commonly done at the end of the drawing process. However, if the next step requires more flexibility, it can be annealed partway through the drawing process. Some manufacturers draw down the wire and then put the entire roll in an annealing oven. In order to reduce oxygen content, some annealing ovens have inert atmospheres, such as nitrogen. This increases the purity of the copper by reducing the oxygen content. But in-line annealing is more consistent than a whole roll in an oven.

Lack of annealing, or insufficient annealing time or temperature, can produce a conductor which is stiff, brittle, and prone to failure. With batch annealing, the inner windings in a roll may not be heated as effectively as the outer windings. Cables made in other countries may not have sufficient purity for high-performance applications. Poor-quality copper, or poor annealing, is very hard to tell from initial visual inspection but often shows up during or after installation.

Plating and tinning

Much of the wire manufactured is plated with a layer of tin. This can also be done in-line with the drawing and annealing by electroplating a layer on the wire. Tinning makes the wire especially resistant to pollutants, chemicals, and salt (as in marine applications). But such a plated conductor is not appropriate for high-frequency applications where the signal travels on the skin of the conductor, called *skin effect*. In that case, bare copper conductors are used. The surface of a conductor used for high frequencies is a major factor in good performance and should have a mirror finish on that surface. Wires are occasionally plated with silver. While silver is slightly more conductive, its real advantage is that silver oxide is the same resistance as bare silver. This is not true with copper, where copper oxide is a semiconductor. Therefore, where reactions with a copper wire are predicted, silver plating may help preserve performance. So silver plating is sometimes used for marine cables, or cables used in similar outdoor environments.

Some plastics, when extruded (melted) onto wires, can chemically affect the copper. This is common, for instance, with an insulation of extruded TFE (tetrafluoroethylene), a form of Teflon™. Wires used inside these cables are often silver-plated or silver-clad. Any oxidizing caused by the extrusion process therefore has no effect on performance. Of course, just the cost of silver alone makes any silver-plated conductor significantly more expensive than bare copper.

Conductor parameters

Table 6.2 shows various parameters for solid wire from 4 AWG to 40 AWG. Table 6.3 shows the same parameters for stranded wire. Note that the resistance of a specific gage of solid wire is lower than stranded wire of the same gage. This is because the stranded wire is not completely conductive; there are spaces (interstices) between the strands. It takes a larger stranded wire to equal the resistance of a solid wire.

Table 6.2 Parameters for Solid Wire from 4 AWG to 40 AWG

AWG	Nominal Diameter	CMA (×1000)	Bare lbs/ft	Ω/1000 ft	Current A	MM^2 Equivalent
4	0.2043	41.7	0.12636	0.25	59.57	21.1
5	0.1819	33.1	0.10020	0.31	47.29	16.8
6	0.162	26.3	0.07949	0.4	37.57	13.3
7	0.1443	20.8	0.06301	0.5	29.71	10.6
8	0.1285	16.5	0.04998	0.63	23.57	8.37
9	0.1144	13.1	0.03964	0.8	18.71	6.63
10	0.1019	10.4	0.03143	1	14.86	5.26
11	0.0907	8.23	0.02493	1.26	11.76	4.17
12	0.0808	6.53	0.01977	1.6	9.33	3.31
13	0.075	5.18	0.01567	2.01	7.40	2.62
14	0.0641	4.11	0.01243	2.54	5.87	2.08
15	0.0571	3.26	0.00986	3.2	4.66	1.65
16	0.0508	2.58	0.00782	4.03	3.69	1.31
17	0.0453	2.05	0.00620	5.1	2.93	1.04
18	0.0403	1.62	0.00492	6.4	2.31	0.823
19	0.0359	1.29	0.00390	8.1	1.84	0.653
20	0.032	1.02	0.00309	10.1	1.46	0.519
21	0.0285	0.81	0.00245	12.8	1.16	0.412
22	0.0254	0.642	0.00195	16.2	0.92	0.324
23	0.0226	0.51	0.00154	20.3	0.73	0.259
24	0.0201	0.404	0.00122	25.7	0.58	0.205
25	0.0179	0.32	0.00097	32.4	0.46	0.162
26	0.0159	0.253	0.00077	41	0.36	0.128
27	0.0142	0.202	0.00061	51.4	0.29	0.102
28	0.0126	0.159	0.00048	65.3	0.23	0.08
29	0.0113	0.127	0.00038	81.2	0.18	0.0643
30	0.01	0.1	0.00030	104	0.14	0.0507
31	0.0089	0.0797	0.00024	131	0.11	0.0401
32	0.008	0.064	0.00019	162	0.09	0.0324
33	0.0071	0.0504	0.00015	206	0.07	0.0255
34	0.0063	0.0398	0.00012	261	0.06	0.0201
35	0.0056	0.0315	0.00010	331	0.05	0.0159
36	0.005	0.025	0.00008	415	0.04	0.0127
37	0.0045	0.0203	0.00006	512	0.03	0.0103
38	0.004	0.016	0.00005	648	0.02	0.0081
39	0.0035	0.0123	0.00004	847	0.02	0.0062
40	0.003	0.0096	0.00003	1080	0.01	0.0049

Table 6.3 Parameters for ASTM Class B Stranded Wires from 4 AWG to 40 AWG

AWG	Nominal Diameter	CMA (×1000)	Bare lbs/ft	Ω/1000 ft	Current A*	MM2 Equivalent
4	0.232	53.824	0.12936	0.253	59.63	27.273
5	0.206	42.436	0.10320	0.323	47.27	21.503
6	0.184	33.856	0.08249	0.408	37.49	17.155
7	0.164	26.896	0.06601	0.514	29.75	13.628
8	0.146	21.316	0.05298	0.648	23.59	10.801
9	0.13	16.9	0.04264	0.816	18.70	8.563
10	0.116	13.456	0.03316	1.03	14.83	6.818
11	0.103	10.609	0.02867	1.297	11.75	5.376
12	0.0915	8.372	0.02085	1.635	9.33	4.242
13	0.0816	6.659	0.01808	2.063	8.04	3.374
14	0.0727	5.285	0.01313	2.73	5.87	2.678
15	0.0647	4.186	0.01139	3.29	4.66	2.121
16	0.0576	3.318	0.00824	4.35	3.69	1.681
17	0.0513	2.632	0.00713	5.25	2.93	1.334
18	0.0456	2.079	0.00518	6.92	2.32	1.053
19	0.0407	1.656	0.00484	8.25	1.84	0.839
20	0.0362	1.31	0.00326	10.9	1.46	0.664
21	0.0323	1.043	0.00284	13.19	1.16	0.528
22	0.0287	0.824	0.00204	17.5	0.92	0.418
23	0.0256	0.655	0.00176	20.99	0.73	0.332
24	0.0228	0.52	0.00129	27.7	0.58	0.263
25	0.0203	0.412	0.01125	33.01	0.46	0.209
26	0.018	0.324	0.00081	44.4	0.36	0.164
27	0.0161	0.259	0.00064	55.6	0.29	0.131
28	0.0143	0.204	0.00051	70.7	0.23	0.103
29	0.0128	0.164	0.00045	83.99	0.18	0.083
30	0.0113	0.128	0.00032	112	0.14	0.0649
31	0.011	0.121	0.00020	136.1	0.11	0.0613
32	0.009	0.081	0.00020	164.1	0.09	0.041
33	0.00825	0.068	0.00017	219.17	0.07	0.0345
34	0.0075	0.056	0.00013	260.9	0.06	0.0284
35	0.00675	0.046	0.00011	335.96	0.04	0.0233
36	0.006	0.036	0.00008	414.8	0.04	0.0182
37	0.00525	0.028	0.00006	578.7	0.03	0.0142
38	0.0045	0.02	0.00005	658.5	0.02	0.0101
39	0.00375	0.014	0.00004	876.7	0.02	0.0071
40	0.003	0.009	0.00003	1028.8	0.01	0.0046

*For both solid and stranded wire, amperage is calculated at 1 A for each 700 CMA.

STRANDED CABLES

Stranded cables are more flexible, and have greater flex-life (flexes to failure) than solid wire. Table 6.4 shows some suggested construction values. The two numbers (65 × 34, for example) show the number of strands (65) and the gage size of each strand (34) for each variation in flexing.

Table 6.4 Suggested Conductor Strandings for Various Degrees of Flexing Severity

Typical Applications	AWG	mm	AWG	mm
	12 AWG		**14 AWG**	
Fixed Service (Hook-Up Wire Cable in Raceway)			Solid	
	19 × 25	19 × 0.455	19 × 27	19 × 0.361
Moderate Flexing (Frequency Disturbed for Maintenance)	65 × 30	65 × 0.254	19 × 27	19 × 0.361
			41 × 30	41 × 0.254
Severe Flexing (Microphones and Test Prods)	165 × 34	165 × 0.160	104 × 34	104 × 0.160
	16 AWG		**18 AWG**	
Fixed Service (Hook-Up Wire Cable in Raceway)	Solid		Solid	
	19 × 29	19 × 0.287	7 × 26	7 × 0.404
			16 × 30	16 × 0.254
Moderate Flexing (Frequently Disturbed for Maintenance)	19 × 29	19 × 0.287	16 × 30	16 × 0.254
	26 × 30	26 × 0.254	41 × 34	41 × 0.160
Severe Flexing (Microphones, Test Prods)	65 × 34	65 × 0.160	41 × 34	41 × 0.160
	104 × 36	104 × 0.127	65 × 36	65 × 0.127
	20 AWG		**22 AWG**	
Fixed Service (Hook-Up Wire Cable in Raceway)	Solid		Solid	
	7 × 28	7 × 0.320	7 × 30	7 × 0.254
	10 × 30	10 × 0.254		
Moderate Flexing (Frequency Distributed for Maintenance)	7 × 28	7 × 0.320	7 × 30	7 × 0.254
	10 × 30	10 × 0.254		
	19 × 32	19 × 0.203	19 × 34	19 × 0.160
	26 × 34	26 × 0.160		
Severe Flexing (Microphones, Test Prods)	26 × 34	26 × 0.160	19 × 34	19 × 0.160
	42 × 36	42 × 0.127	26 × 36	26 × 0.127
	24 AWG		**26 AWG**	
Fixed Service (Hook-Up Wire Cable in Raceway)	Solid		Solid	
	7 × 32	7 × 0.203	7 × 34	7 × 0.160
Moderate Flexing (Frequently Disturbed for Maintenance)	7 × 32	7 × 0.203	7 × 34	7 × 0.160
	10 × 34	10 × 0.160		
Severe Flexing (Microphones, Test Prods)	19 × 36	19 × 0.127	7 × 34	7 × 0.160
	45 × 40	45 × 0.079	10 × 36	10 × 0.127

(Courtesy Belden.)

Table 6.5 Pulling Tension for Annealed Copper Conductors

24 AWG	5.0 lbs
22 AWG	7.5 lbs
20 AWG	12.0 lbs
18 AWG	19.5 lbs
16 AWG	31.0 lbs
14 AWG	49.0 lbs
12 AWG	79.0 lbs

Pulling tension

Pulling tension must be adhered to so the cable will not be permanently elongated. The pulling tension for annealed copper conductors is shown in Table 6.5.

Multiconductor cable pulling tension can be determined by multiplying the total number of conductors by the appropriate value. For twisted pair cables, there are two wires per pair. For shielded twisted pair cables, with foil shields, there is a drain wire that must be included in the calculations. Be cautious: the drain wire can sometimes be smaller gage than the conductors in the pair. The pulling tension of coaxial cables or other cables that are not multiple conductors is much harder to calculate. Consult the manufacturer for the required pulling tension.

Skin effect

As the frequency of the signal increases on a wire, the signal travels closer to the surface of the conductor. Since very little of the area of the center conductor is used at high frequencies, some cable is made with a copper-clad steel-core center conductor. These are known as copper-clad, copper-covered, or Copperweld™ and are usually used by CATV/broadband service providers.

Copper-clad steel is stronger than copper cable so it can more easily withstand pulling during installation, or wind, ice, and other outside elements after installation. For instance, a copper-clad #18 AWG coaxial cable has a pull strength of 102 lbs while a solid copper #18 AWG coax would have a pull strength of 69 lbs. The main disadvantage is that steel is not a good conductor below 50 MHz, between four and seven times the resistance of copper, depending on the thickness of the copper layer.

This is a problem where signals are below 50 MHz such as DOCSIS data delivery, or VOD (video-on-demand) signals which are coming from the home to the provider. When installing cable in a system, it is better to use solid copper cable so it can be used at low frequencies as well as high frequencies.

This is also why copper-clad conductors are not appropriate for any application below 50 MHz, such as baseband video, CCTV, analog, or digital audio. Copper-clad is also not appropriate for applications such as SDI or HD-SDI video, and similar signals where a significant portion of the data is below 50 MHz.

The skin depth for copper conductors can be calculated with the equation:

$$D = \frac{2.61}{\sqrt{f}} \tag{6.2}$$

where,

D is the skin depth in inches,

f is the frequency in hertz.

Table 6.6 Skin Depths at Various Frequencies

Frequency	Skin Depth in Inches	% Used of #18 AWG Conductor
1 kHz	0.082500	100
10 kHz	0.026100	100
100 kHz	0.008280	41
1 MHz	0.002610	13
10 MHz	0.000825	4.1
100 MHz	0.000261	1.3
1 GHz	0.0000825	0.41
10 GHz	0.0000261	0.13

Table 6.6 compares the actual skin depth and percent of the center conductor actually used in an RG-6 cable. The skin depth always remains the same no matter what the thickness of the wire is. The only thing that changes is the percent of the conductor utilized. Determining the percent of the conductor utilized requires using two times the skin depth because we are comparing the diameter of the conductor with its depth.

As can be seen, by the time the frequencies are high, the depth of the signal on the skin can easily be microinches. For signals in that range, such as high-definition video signals, for example, this means that the surface of the wire is as critical as the wire itself. Therefore, conductors intended to carry high frequencies should have a mirror finish.

Since the resistance of the wire at these high frequencies is of no consequence, it is sometimes asked why larger conductors go farther. The reason is that the *surface area*, the *skin*, on a wire is greater as the wire gets larger in size.

Further, some conductors have a tin layer to help prevent corrosion. These cables are obviously not intended for use at frequencies above just a few megahertz, or a significant portion of the signal would be traveling in the tin layer. Tin is not an especially good conductor as can be seen in Table 6.1.

Current capacity

For conductors that will carry large amounts of electrical flow, large amperage or current from point to point, a general chart has been made to simplify the current carrying capacity of each conductor. To use the current capacity chart in Figure 6.1, first determine conductor gage, insulation and jacket temperature rating, and number of conductors from the applicable product description for the cable of interest. These can usually be obtained from a manufacturer's Web site or catalog.

Next, find the current value on the chart for the proper temperature rating and conductor size. To calculate the maximum current rating/conductor multiply the chart value by the appropriate conductor factor. The chart assumes the cable is surrounded by still air at an ambient temperature of 25 °C (77 °F). Current values are in amperes (rms) and are valid for copper conductors only. The maximum continuous current rating for an electronic cable is limited by conductor size, number of conductors contained within the cable, maximum temperature rating of the insulation on the conductors, and environment conditions such as ambient temperature and air flow. The four lines marked with temperatures apply to different insulation plastics and their melting point. Consult the manufacturer's Web site or catalog for the maximum insulation or jacket temperature.

The current ratings of Figure 6.1 are intended as general guidelines for low-power electronic communications and control applications. Current ratings for high-power applications generally are set by regulatory

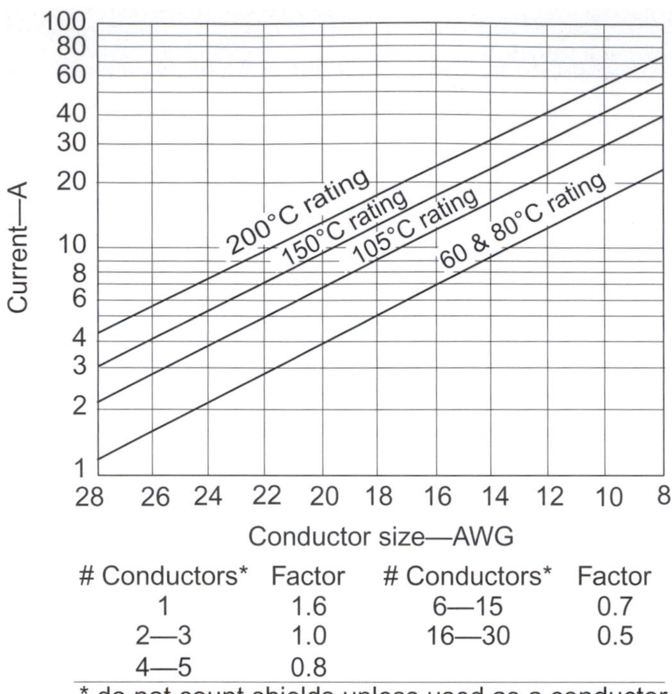

# Conductors*	Factor	# Conductors*	Factor
1	1.6	6—15	0.7
2—3	1.0	16—30	0.5
4—5	0.8		

* do not count shields unless used as a conductor

FIGURE 6.1
Current ratings for electronic cable. (Courtesy Belden.)

agencies such as Underwriters Laboratories (UL), Canadian Standards Association (CSA), National Electrical Code (NEC), and others and should be used before final installation.

Table 310-15(b)(2)(a) of the NEC contains amperage adjustment factors for whenever more than three current carrying conductors are in a conduit or raceway.

Section 240-3 of the NEC provides requirements for overload protection for conductors other than flexible cords and fixture wires. Section 240-3(d), Small Conductors, states that #14 to #10 conductors require a maximum protective overcurrent device with a rating no higher than the current rating listed in the 60°C column. These currents are 15 A for #14 copper wire, 20 A for #12 copper wire, and 30 A for #10 copper wire. These values are familiar as the breaker ratings for commercial installations.

When connecting wire to a terminal strip or another wire, etc., the temperature rise in the connections must also be taken into account. Often the circuit is not limited by the current carrying capacity of the wire but of the termination point.

WIRE CURRENT RATINGS

Current carrying capacity of wire is controlled by the NEC, particularly in Table 310-16, Table 310-15(b)(2)(a), and Section 240-3.

Table 310-16 of the NEC shows the maximum current carrying capacity for insulated conductors rated from 0 to 2000 V, including copper and aluminum conductors. Each conductor amperage is given for three temperatures: 60°C, 75°C, and 90°C. Copper doesn't melt until almost 2000°, so the current limit on a copper wire is not the melting point of the wire but the melting point of the insulation. This number is listed by most manufacturers in their catalog or on their Web site. For instance, PVC (polyvinyl chloride) can be formulated to withstand temperatures from 60°C to as high as 105°C. The materials won't melt right at the specified temperature, but may begin to fail certain tests, such as cracking when bent.

Table 6.7 Comparative Properties of Rubber Insulation (Courtesy Belden)

Properties	Rubber	Neoprene	Hypalon (Chlorosulfonated Polyethylene)	EPDM (Ethylene Propylene Diene Monomer)	Silicone
Oxidation Resistance	F	G	E	G	E
Heat Resistance	F	G	E	E	O
Oil Resistance	P	G	G	F	F-G
Low Temperature Flexibility	G	F-G	F	G-E	O
Weather, Sun Resistance	F	G	E	E	O
Ozone Resistance	P	G	E	E	O
Abrasion Resistance	E	G-E	G	G	P
Electrical Properties	E	P	G	E	O
Flame Resistance	P	G	G	P	F-G
Nuclear Radiation Resistance	F	F-G	G	G	E
Water Resistance	G	E	G-E	G-E	G-E
Acid Resistance	F-G	G	E	G-E	F-G
Alkali Resistance	F-G	G	E	G-E	F-G
Gasoline, Kerosene, etc. (Aliphatic Hydrocarbons) Resistance	P	G	F	P	P-F
Benzol, Toluol, etc. (Aromatic Hydrocarbons) Resistance	P	P-F	F	F	P
Degreaser Solvents (Halogenated Hydrocarbons) Resistance	P	P	P-F	P	P-G
Alcohol Resistance	G	F	G	P	G

P = poor, F = fair, G = good, E = excellent, O = outstanding.

INSULATION

Wire can be bare, often called *bus bar* or *bus wire*, but is most often insulated. It is covered with a nonconducting material. Early insulations included cotton or silk woven around the conductor, or even paper. Cotton-covered house wiring can still be found in perfect operating condition in old houses. Today, most insulation materials are either some kind of rubber or some kind of plastic. The material chosen should be listed in the manufacturer's catalog with each cable type. Table 6.7 lists some of the rubber-based materials with their properties. Table 6.8 lists the properties of various plastics. The ratings in both tables are based on average performance of general-purpose compounds. Any given property can usually be improved by the use of selective compounding.

Plastics and dielectric constant

Table 6.9 is a list of various insulation materials with details on performance, requirements, and special advantages. Insulation, when used on a cable intended to carry a signal, is often referred to as a *dielectric*. The performance of any material, i.e., its ability to insulate with minimal effect to the signal running on the cable, is called the *dielectric constant* and can be measured in a laboratory. Table 6.9 shows some standard numbers as a point of reference.

Table 6.8 Comparative Properties of Plastic Insulation (Courtesy Belden)

Properties	PVC	Low-Density Polyethylene	Cellular Polyethylene	High-Density Polyethylene	Polyethylene	Polyurethane	Nylon	Teflon®
Oxidation Resistance	E	E	E	E	E	E	E	O
Heat Resistance	G-E	G	G	E	E	G	E	O
Oil Resistance	F	G	G	G-E	F	E	E	O
Low Temperature Flexibility	P-G	G-E	E	E	P	G	G	O
Weather, Sun Resistance	G-E	E	E	E	E	G	E	O
Ozone Resistance	E	E	E	E	E	E	E	E
Abrasion Resistance	F-G	F-G	F	E	F-G	O	E	E
Electrical Properties	F-G	E	E	E	E	P	P	E
Flame Resistance	E	P	P	P	P	P	P	O
Nuclear Radiation Resistance	G	G	G	G	F	G	F-G	P
Water Resistance	E	E	E	E	E	P-G	P-F	E
Acid Resistance	G-E	G-E	G-E	G-E	E	F	P-F	E
Alkali Resistance	G-E	G-E	G-E	G-E	E	F	E	E
Gasoline, Kerosene, etc. (Aliphatic Hydrocarbons) Resistance	P	P-F	P-F	P-F	P-F	G	G	E
Benzol, Toluol, etc. (Aromatic Hydrocarbons) Resistance	P-F	P	P	P	P-F	P	G	E
Degreaser Solvents (Halogenated Hydrocarbons) Resistance	P-F	P	P	P	P	P	G	E
Alcohol Resistance	G-E	E	E	E	E	P	P	E

P = poor, F = fair, G = good, E = excellent, O = outstanding.

Table 6.9 Dielectric Constant

Dielectric Constant	Material	Note
1	Vacuum	By definition
1.0167	Air	Very close to 1
1.35	Foam, Air-filled Plastic	Current technological limit
2.1	Solid Teflon™	Best solid plastic
2.3	Solid Polyethylene	Most common plastic
3.5–6.5	Solid Polyvinyl Chloride	Low price, easy to work with

Wire insulation characteristics

The key difference between rubber compounds and plastic compounds is their recyclability. Plastic materials can be ground up, and re-melted into other objects. Polyethylene, for instance, can be recycled into plastic bottles, grocery bags, or even park benches. And, should the need arise, these objects could themselves be ground up and turned back into wire insulation, or many other uses. The term *thermoplastic* means changed by heat and is the source of the common term *plastic.*

Rubber compounds, on the other hand, are thermoset. That is, once they are made, they are set, and the process cannot be reversed. Rubber, and its family, is cured in a process sometimes called *vulcanizing.* These compounds cannot be ground up and recycled into new products. There are natural rubber compounds (such as latex-based rubber) and artificial, chemical-based rubber compounds such as EPDM (ethylenepropylene-diene monomer).

The vast majority of wire and cable insulations are plastic-based compounds. Rubber, while it is extremely rugged, is considerably more expensive than most plastics, so there are fewer and fewer manufacturers offering rubber-based products. These materials, both rubber and plastic, are used in two applications with cable. The first application is insulation of the conductor(s) inside the cable. The second is as a jacket material to protect the contents of the cable.

JACKETS

The jacket characteristics of cable have a large effect on its ruggedness and the effect of environment. A key consideration is often flexibility, especially at low temperatures. Audio and broadcast cables are manufactured in a wide selection of standard jacketing materials. Special compounds and variations of standard compounds are used to meet critical audio and broadcast application requirements and unusual environmental conditions. Proper matching of cable jackets to their working environment can prevent deterioration due to intense heat and cold, sunlight, mechanical abuse, impact, and crowd or vehicle traffic.

PLASTICS

Plastic is a shortened version of the term *thermoplastic. Thermo* means heat, *plastic* means change. Thermoplastic materials can be changed by heat. They can be melted and extruded into other shapes. They can be extruded around wires, for instance, forming an insulative (non-conductive) layer. There are many forms of plastic. Below is a list of the most common varieties used in the manufacture of wire and cable.

Vinyl

Vinyl is sometimes referred to as PVC or *polyvinyl chloride,* and is a chemical compound invented in 1928 by Dr. Waldo Semon (USA). Extremely high or low temperature properties cannot be found in one formulation, therefore formulations may have −55°C to +105°C (−67°F to +221°F) rating while other common vinyls may have −20°C to +60°C (−4°F to +140°F). The many varieties of vinyl also differ in pliability and electrical properties fitting a multitude of applications. The price range can vary accordingly. Typical dielectric constant values can vary from 3.5 at 1000 Hz to 6.5 at 60 Hz, making it a poor choice if high performance is required. PVC is one of the least expensive compounds, and one of the easiest to work with. Therefore, PVC is used with many cables that do not require high performance, or where cost of materials is a major factor. PVC is easy to color, and can be quite flexible, although it is not very rugged. In high-performance cables, PVC is often used as the jacket material, but not inside the cable.

Polyethylene

Polyethylene, invented by accident in 1933 by E.W. Fawcett and R.O. Gibson (Great Britain), is a very good insulation in terms of electrical properties. It has a low dielectric constant value over all frequencies and very high insulation resistance. In terms of flexibility, polyethylene can be rated stiff to very hard depending on molecular weight and density. Low density is the most flexible and high density high molecular weight formulations are very hard. Moisture resistance is rated excellent. Correct brown and black formulations have excellent sunlight resistance. The dielectric constant is 2.3 for solid insulation and as low as 1.35 for gas-injected foam cellular designs. Polyethylene is the most common plastic worldwide.

Teflon®

Invented in 1937 by Roy Plunkett (USA) at DuPont, Teflon has excellent electrical properties, temperature range, and chemical resistance. It is not suitable where subjected to nuclear radiation, and it does not have good high voltage characteristics. FEP (fluorinated ethylenepropylene) Teflon is extrudable in a manner similar to vinyl and polyethylene, therefore long wire and cable lengths are available. TFE (tetrafluoroethylene) Teflon is extrudable in a hydraulic ram-type process and lengths are limited due to amount of material in the ram, thickness of the insulation, and core size. TFE must be extruded over silver-coated or nickel-coated wire. The nickel- and silver-coated designs are rated +260°C and +200°C maximum (500°F and 392°F), respectively, which is the highest temperature for common plastics. The cost of Teflon is approximately 8 to 10 times more per pound than vinyl insulations. The dielectric constant for solid Teflon is 2.1, the lowest of all solid plastics. Foam Teflon (FEP) has a dielectric constant as low as 1.35. Teflon is produced by and a trademark of DuPont Corporation.

Polypropylene

Polypropylene is similar in electrical properties to polyethylene and is primarily used as an insulation material. Typically, it is harder than polyethylene, which makes it suitable for thin wall insulations. UL maximum temperature rating may be 60°C or 80°C (140°F or 176°F). The dielectric constant is 2.25 for solid and 1.55 for cellular designs.

THERMOSET COMPOUNDS

As the name implies, *thermoset* compounds are produced by heat (thermo) but are set. That is, the process cannot be reversed as in thermoplastics. They cannot be recycled into new products as thermoplastic materials can.

Silicone

Silicone is a very soft insulation which has a temperature range from −80°C to +200°C (−112°F to +392°F). It has excellent electrical properties plus ozone resistance, low moisture absorption, weather resistance, and radiation resistance. It typically has low mechanical strength and poor scuff resistance. Silicone is seldom used because it is very expensive.

Neoprene

Neoprene has a maximum temperature range from −55°C to +90°C (−67°F to +194°F). The actual range depends on the formulation used. Neoprene is both oil and sunlight resistant making it ideal for many outdoor applications. The most stable colors are black, dark brown, and gray. The electrical properties are not as good as other insulation material; therefore thicker insulation must be used for the same insulation.

Rubber

The description of *rubber* normally includes natural rubber and styrene-butadiene rubber (SBR) compounds. Both can be used for insulation and jackets. There are many formulations of these basic materials and each formulation is for a specific application. Some formulations are suitable for −55°C (−67°F) minimum while others are suitable for +75°C (+167°F) maximum. Rubber jacketing compounds feature exceptional durability for extended cable life. They withstand high-impact and abrasive conditions better than PVC and are resistant to degradation or penetration by water, alkali, or acid. They have excellent heat resistant properties, and also provide greater cable flexibility in cold temperatures.

EPDM

EPDM stands for *ethylene-propylene-diene monomer.* It was invented by Dr. Waldo Semon in 1927. It is extremely rugged, like natural rubber, but can be created from petroleum byproducts ethylene and propylene gas.

SINGLE CONDUCTOR

Single conductor wire starts with a single wire, either solid or stranded. It can be bare, sometimes called *buss bar,* or can be jacketed. There is no actual limit to how small, or how large, a conductor could be.

Table 6.10 Color Code for Nonpaired Cables per ICEA #2 and #2R

Conductor	Color	Conductor	Color	Conductor	Color	Conductor	Color
1st	Black	14th	Green/White	27th	Blue/Blk/Wht	40th	Red/Wht/Grn
2nd	White	15th	Blue/White	28th	Blk/Red/Grn	41st	Grn/Wht/Blue
3rd	Red	16th	Black/Red	29th	Wht/Red/Grn	42nd	Org/Red.Grn
4th	Green	17th	White/Red	30th	Red/Blk/Grn	43rd	Blue/Red/Grn
5th	Orange	18th	Orange/Red	31st	Grn/Blk/Org	44th	Blk/Wht/Blue
6th	Blue	19th	Blue/Red	32nd	Org/Blk/Grn	45th	Wht/Blk/Blue
7th	White/Black	20th	Red/Green	33rd	Blue/Wht/Org	46th	Red/Wht/Blue
8th	Red/Black	21st	Orange/Green	34th	Blk/Wht/Org	47th	Grn/Orn/Red
9th	Green/Black	22nd	Blk/Wht/Red	35th	Wht/Red/Org	48th	Org/Red/Blue
10th	Orange/Black	23rd	Wht/Blk/Red	36th	Org/Wht/Blue	49th	Blue/Red/Org
11th	Blue/Black	24th	Red/Blk/Wht	37th	Wht/Red/Blue	50th	Blk/Org/Red
12th	Black/White	25th	Grn/Blk/Wht	38th	Blk/Wht/Grn		
13th	Red/White	26th	Org/Blk/Wht	39th	Wht/Blk/Grn		

(Courtesy Belden.)

Choice of size (AWG) will be based on application and the current or wattage delivery required. If jacketed, the choice of jacket can be based on performance, ruggedness, flexibility, or any other requirement.

There is no single conductor plenum rating because the NEC (National Electrical Code) only applies to cables, more than one conductor. However, Articles 300 and 310 of the NEC are sometimes cited when installing single conductor wire for grounds and similar applications.

MULTICONDUCTOR

Bundles of two or more insulated wires are considered multiconductor cable. Besides the requirements for each conductor, there is often an overall jacket, chosen for whatever properties would be appropriate for a particular application.

There are specialized multiconductor cables, such as power cordage, used to deliver AC power from a wall outlet (or other source) to a device. There are UL safety ratings on such a cable to assure users will not be harmed.

There are other multiconductor applications such as VFD (variable frequency drive) cables, specially formulated to minimize standing waves and arcing discharge when running variable frequency motors. Since a multiconductor cable is not divided into pairs, resistance is still the major parameter to be determined, although reactions between conductors (as in VFD) can also be considered.

Multiconductor insulation color codes

The wire insulation colors help trace conductors or conductor pairs. There are many color tables; Table 6.10 is one example.

PAIRS AND BALANCED LINES

Twisting two insulated wires together makes a twisted pair. Since two conductive paths are needed to make a circuit, twisted pairs give users an easy way to connect power or signals from point to point. Sometimes the insulation color is different to identify each wire in each pair. Pairs can have dramatically better performance than multiconductor cables because pairs can be driven as a balanced line.

A balanced line is a configuration where the two wires are electrically identical. The electrical performance is referred to ground, the zero point in circuit design. Balanced lines reject noise, from low frequencies, such as 50/60 Hz power line noise, up to radio frequency signals in the Megahertz, or even higher.

When the two conductors are electrically identical, or close to identical, there are many other parameters, besides resistance, that come into play. These include capacitance, inductance, and impedance. And when we get to high-frequency pairs, such as data cables, we even measure the variations in resistance (resistance unbalance), variations in capacitance (capacitance unbalance,) or even variations in impedance (return loss). Each of these has a section farther on in this chapter.

Balanced lines work because they have a transformer at each end, a device made of two coils of wire wound together. Many modern devices now use circuits that act electrically the same as a transformer, an effect called *active balancing*. The highest-quality transformers can be extremely expensive, so high-performing balanced-line chips have been improving, some getting very close to the coils-of-wire performance.

It should be noted that virtually all professional installations use twisted pairs for audio because of their noise rejection properties. In the consumer world, the cable has one hot connection and a grounded shield around it and is called an *unbalanced cable*. These cables are effective for only short distances and have no other inherent noise rejection besides the shield itself.

Multipair

As the name implies, multipair cables contain more than one pair. Sometimes referred to as *multicore* cables, these can just be grouped bare pairs, or each pair could be individually jacketed, or each pair could be shielded (shielding is outlined below), or the pairs could even be individually shielded and jacketed. All of these options are easily available. Where there is an overall jacket, or individual jackets for each pair, the jacket material for each pair is chosen with regard to price, flexibility, ruggedness, color, and any other parameter required.

It should be noted that the jackets on pairs, or the overall jacket, have almost no effect on the performance of the pairs. One could make a case that, with individually jacketed pairs, the jacket moves the pairs apart and therefore improves crosstalk between pairs. It is also possible that poorly extruded jackets could leak the chemicals that make up the jacket into the pair they are protecting, an effect called *compound migration*, and therefore affect the performance of the pair.

Table 6.11 shows a common color code for paired cables where they are simply a bundle of pairs. The color coding is only to identify the pair and the coloring of the insulation has no effect on performance. If this cable were individually jacketed pairs, it would be likely that the two wires in the pair would be identical colors such as all black-and-red, and the jackets would use different colors to identify them as shown in Table 6.12.

Analog multipair snake cable

Originally designed for the broadcast industry, hard-wire multipair audio snake cables feature individually shielded pairs, for optimum noise rejection, and sometimes with individual jackets on each pair for improved physical protection. These cables are ideal, carrying multiple line-level or microphone-level signals. They will also interconnect audio components such as multichannel mixers and consoles for recording studios, radio and television stations, post-production facilities, and sound system installations. Snakes offer the following features:

- A variety of insulation materials, for low capacitance, ruggedness, or fire ratings.
- Spiral/serve, braid, French Braid™, or foil shields.
- Jacket and insulation material to meet ruggedness or NEC flame requirements.
- High temperature resistance in some compounds.
- Cold temperature pliability in some compounds.
- Low-profile appearance, based mostly on the gage of the wires, but also on the insulation.

Table 6.11 Color Codes for Paired Cables (Belden Standard)

Pair No.	Color Combination	Pair No.	Color Combination	Pair No.	Color Combination	Pair No.	Color Combination
1	Black/Red	11	Red/Yellow	21	White/Brown	31	Purple/White
2	Black/White	12	Red/Brown	22	White/Orange	32	Purple/Dark Green
3	Black Green	13	Red/Orange	23	Blue/Yellow	33	Purple/Light Blue
4	Black/Blue	14	Green/White	24	Blue/Brown	34	Purple/Yellow
5	Black/Yellow	15	Green/Blue	25	Blue/Orange	35	Purple/Brown
6	Black/Brown	16	Green/Yellow	26	Brown/Yellow	36	Purple/Black
7	Black/Orange	17	Green/Brown	27	Brown/Orange	37	Gray/White
8	Red/White	18	Green/Orange	28	Orange/Yellow		
9	Red/Green	19	White/Blue	29	Purple/Orange		
10	Red/Blue	20	White/Yellow	30	Purple/Red		

(Courtesy Belden.)

Table 6.12 Color Codes for Snake Cables

Pair No.	Color Combination	Pair No.	Color Combination	Pair No.	Color Combination	Pair No.	Color Combination	Pair No.	Color Combination
1	Brown	13	Lt. Gray/Brown stripe	25	Lt. Blue/Brown stripe	37	Lime/Brown stripe	49	Aqua/Brown stripe
2	Red	14	Lt. Gray/Red stripe	26	Lt. Blue/Red stripe	38	Lime/Red stripe	50	Aqua/Red stripe
3	Orange	15	Lt. Gray/Orange stripe	27	Lt. Blue/Orange stripe	39	Lime/Orange stripe	51	Aqua/Orange stripe
4	Yellow	16	Lt. Gray/Yellow stripe	28	Lt. Blue/Yellow stripe	40	Lime/Yellow stripe	52	Aqua/Yellow stripe
5	Green	17	Lt. Gray/Green stripe	29	Lt. Blue/Green stripe	41	Lime/Green stripe	53	Aqua/Green stripe
6	Blue	18	Lt. Gray/Blue stripe	30	Lt. Blue/Blue stripe	42	Lime/Blue stripe	54	Aqua/Blue stripe
7	Violet	19	Lt. Gray/Violet stripe	31	Lt. Blue/Violet stripe	43	Lime/Violet stripe	55	Aqua/Violet stripe
8	Gray	20	Lt. Gray/Gray stripe	32	Lt. Blue/Gray stripe	44	Lime/Gray stripe	56	Aqua/Gray stripe
9	White	21	Lt. Gray/White stripe	33	Lt. Blue/White stripe	45	Lime/White stripe	57	Aqua/White stripe
10	Black	22	Lt. Gray/Black stripe	34	Lt. Blue/Black stripe	46	Lime/Black stripe	58	Aqua/Black stripe
11	Tan	23	Lt. Gray/Tan stripe	35	Lt. Blue/Tan stripe	47	Lime/Tan stripe	59	Aqua/Tan stripe
12	Pink	24	Lt. Gray/Pink stripe	36	Lt. Blue/Pink stripe	48	Lime/Pink stripe	60	Aqua/Pink stripe

(Courtesy Belden.)

- Some feature overall shields to reduce crosstalk and facilitate star grounding.
- Allows easier and cheaper installs than using multiple single channel cables.

Snakes come with various terminations and can be specified to meet the consumer's needs. Common terminations are male or female XLR (microphone) connectors and ¼ inch male stereo connectors on one end, and either a junction box with male or female XLR connectors and ¼ inch stereo connectors or pigtails with female XLR connectors and ¼ inch connectors on the other end.

For stage applications, multipair individually shielded snake cables feature lightweight and small diameter construction, making them ideal for use as portable audio snakes. Individually shielded and jacketed pairs are easier to install with less wiring errors. In areas that subscribe to the NEC guidelines, the need for conduit in studios is eliminated when CM-rated snake cable is used through walls between rooms. Vertically between floors, snakes rated CMR (riser) do not need conduit. In plenum areas (raised floors, drop ceilings) CMP, plenum rated snake cables can be used without conduit. Color codes for snakes are given in Table 6.12.

High-frequency pairs

Twisted pairs were original conceived to carry low-frequency signals, such as telephone audio. Beginning in the 1970s, research and development was producing cables such as twinax that had reasonable performance to the megahertz. IBM Type 1 was the breakthrough product that proved that twisted pairs could indeed carry data. This led directly to the Category premise/data cable of today.

There are now myriad forms of high-frequency, high-data rate cable including DVI, USB, HDMI, IEEE 1394 FireWire, and others. All of these are commonly used to transport audio and video signals, Table 6.13.

DVI

DVI (Digital Visual Interface) is used extensively in the computer-monitor interface market for flat panel LCD monitors.

Table 6.13 Comparing Twisted-Pair High-Frequency Formats

Standard	Format	Intended Use	Connector Style	Cable Type	Transmission Distance[1]	Sample Rate	Data Rate (Mbps)	Guiding Document
D1 component	parallel	broadcast	multipin D	multipairs	4.5 m/15 ft	27 MHz	270	ITU-R BT.601-5
DV	serial	professional/ consumer		(see IEEE 1394)	4.5 m/15 ft	20.25 MHz	25	IEC 61834
IEEE 1394 (FireWire)	serial	professional/ consumer	1394	6 conductors, 2-STPs/2 pwr	4.5 m/15 ft	n/a	100, 200, 400	IEEE 1394
USB 1.1	serial	consumer	USB A & B	4 conductors, 1 UTP/2 pwr	5 m/16.5 ft	n/a	12	USB 1.1 Promoter Group
USB 2.0	serial	professional/ consumer	USB A & B	4 conductors, 1-UTP/2 pwr	5 m/16.5 ft	n/a	480	USB 2.0 Promoter Group
DVI	serial/ parallel	consumer	DVI (multipin D)	Four STPs	10 m/33 ft	To 165 MHz	1650	DDWG; DVI 1.0
HDMI	parallel	consumer	HDMI (19 pin)	Four STPs + 7 conductors	Unspecified	To 340 MHz	To 10.2 Gbps	HDMI LLC
DisplayPort	parallel	consumer	20 pin	Four STPs + 8 conductors	15 m	To 340 MHz	To 10.8 Gbps	VESA

STP = shielded twisted pair, UTP = unshielded twisted pair, n/a = not applicable
[1]Transmission distances may vary widely depending on cabling and the specific equipment involved.

The DVI connection between local monitors and computers includes a serial digital interface and a parallel interface format, somewhat like combining the broadcast serial digital and parallel digital interfaces.

Transmission of the TMDS (transition minimized differential signaling) format combines four differential, high-speed serial connections transmitted in a parallel bundle. DVI specifications that are extended to the dual mode operation allow for greater data rates for higher display resolutions. This requires seven parallel, differential, high-speed pairs. Quality cabling and connections become extremely important. The nominal DVI cable length limit is 4.5 m (15 ft). Electrical performance requirements are signal rise-time of 0.330 ns, and a cable impedance of 100 Ω. FEXT is less than 5%, and signal rise-time degradation is a maximum of 160 ps (picoseconds). Cable for DVI is application specific since the actual bit rate per channel is 1.65 Gbps.

Picture information or even the entire picture can be lost if any vital data is missing with digital video interfaces. DVI cable and its termination are very important and the physical parameters of the twisted pairs must be highly controlled as the specifications for the cable and the receiver are given in fractions of bit transmission.

Requirements depend on the clock rate or signal resolution being used. Transferring the maximum rate of 1600 × 1200 at 60 Hz for a single link system means that 1 bit time or 10 bits per pixel is 0.1 (1/165 MHz) or 0.606 ns.

The DVI receiver specification allows 0.40 × bit time, or 0.242 ns intrapair skew within any twisted pair. The pattern at the receiver must be very symmetrical. The interpair skew, which governs how bits will line up in time at the receiving decoder, may only be 0.6 × pixel time, or 3.64 ns. These parameters control the transmission distances for DVI.

Also, the cable should be evaluated on its insertion loss for a given length. DVI transmitter output is specified into a cable impedance of 100 Ω with a signal swing of ± 780 mV with a minimum signal swing of ± 200 mV. When determining DVI cable, assume minimum performance by the transmitter—i.e., 200 mV—and best sensitivity by the receiver which must operate on signals ± 75 mV. Under these conditions the cable attenuation can be no greater than 8.5 dB at 1.65 GHz (10 bits/pixel × 165 MHz clock) which is relatively difficult to maintain on twisted-pair cable.

DVI connections combine the digital delivery, described above, with legacy analog component delivery. This allows DVI to be the transition delivery scheme between analog and digital applications.

HDMI

HDMI (high definition multimedia interface) is similar to DVI except that it is digital-only delivery. Where DVI has found its way into the commercial space as well as consumer applications, HDMI is almost entirely consumer-based. It is configured into a 19 pin connector which contains four shielded twisted pairs (three pairs data, one pair clock) and seven wire for HDCP (copy protection), devices handshaking, and power. The standard versions of HDMI are nonlocking connector, attesting to its consumer-only focus.

IEEE – 1394 OR FIREWIRE SERIAL DIGITAL

FireWire, or IEEE 1394, is used to upload DV, or digital video, format signals to computers, etc. DV, sometimes called *DV25*, is a serial digital format of 25 Mbps. IEEE 1394 supports up to 400 Mbps. The specification defines three signaling rates: S100 (98.304 Mbps), S200 (196.608 Mbps), and S400 (393.216 Mbps).

IEEE 1394 can interconnect up to 63 devices in a peer-to-peer configuration so audio and video can be transferred from device to device without a computer, D/A, or A/D conversion. IEEE 1394 is not plugable from the circuit while the equipment is turned on.

The IEEE 1394 system uses two shielded twisted pairs and two single wires, all enclosed in a shield and jacket, Figure 6.2. Each pair is shielded with 100% coverage foil and a minimum 60% coverage braid. The outer shield is 100% coverage foil and a minimum 90% coverage braid. Each pair is shielded with aluminum foil and is equal to or greater than 60% braid. The twisted pairs handle the differential data and strobe (assists in clock regeneration) while the two separate wires provide the power and ground for remote devices. Signal level is 265 mV differential into 110 Ω.

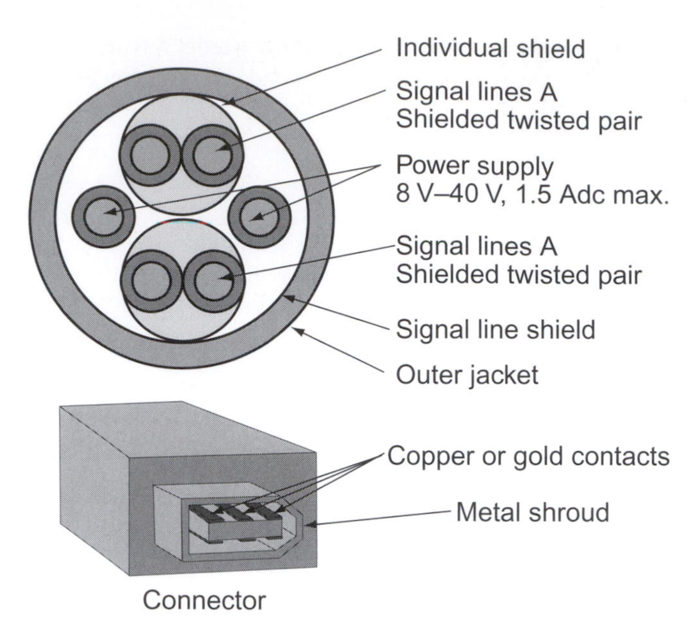

FIGURE 6.2
IEEE 1394 cable and connector.

Table 6.14 Critical IEEE 1394 Timing Parameters

Parameter	100 Mbps	200 Mbps	400 Mbps
Max Tr/Tf	3.20 ns	2.20 ns	1.20 ns
Bit Cell Time	10.17 ns	5.09 ns	2.54 ns
Transmit Skew	0.40 ns	0.25 ns	0.20 ns
Transmit Jitter	0.80 ns	0.50 ns	0.25 ns
Receive End Skew	0.80 ns	0.65 ns	0.60 ns
Receive End Jitter	1.08 ns	0.75 ns	0.48 ns

The IEEE 1394 specification cable length is a maximum of 4.5 m (15 ft). Some applications may run longer lengths when the data rate is lowered to the 100 Mbps level. The typical cable has #28 gage copper twisted pairs and #22 gage wires for power and ground. The IEEE 1394 specification provides the following electrical performance requirements:

- Pair-to-pair data skew is 0.40 ns.
- Crosstalk must be maintained below −26 dB from 1 MHz to 500 MHz.
- Velocity of propagation must not exceed 5.05 ns/m.

Table 6.14 gives details of the physical interface system for IEEE 1394.

USB

The USB, universal serial bus, simplifies connection of computer peripherals. USB 1.1 is limited to a communications rate of 12 Mbps, while USB 2.0 supports up to 480 Mbps communication. The USB cable consists of one twisted pair for data and two untwisted wires for powering downstream appliances. A full-speed cable includes a #28 gage twisted pair, and an untwisted pair of #28 to #20 gage power conductors, all enclosed in an aluminized polyester shield with a drain wire.

Nominal impedance for the data pair is 90 Ω. The maximum cable length is determined by the signal propagation delay which must be less than 26 ns from end to end. Table 6.15 lists some common

Table 6.15 Dielectric Constant, Delay, and Transmission Distance of Various Plastics

Material	Dielectric Constant	Delay ns/ft	Maximum USB Distance
Foam, Air-Filled Plastic	1.35	1.16	22.4 ft
Solid Teflon™	2.1	1.45	18 ft
Solid Polyethylene	2.3	1.52	17 ft
Solid Polyvinyl Chloride	3.5–6.5	1.87–2.55	10–14 ft

plastics and the theoretical distance each could go based on 26 ns. With an additional allowance of 4 ns, which is split between the sending device connection and the receiver connection/response function, the entire one-way delay is a maximum of 30 ns. The cable velocity of propagation must be less than 5.2 ns/m and the length and twist of the data pair must be matched so time skew is no more than 0.10 ns between bit polarities. The nominal differential signal level is 800 mV.

DISPLAYPORT

DisplayPort is an emerging protocol for digital video. Its original intention was the transfer of images from a PC or similar device to a display. It has some significant advantages over DVI and HDMI. DisplayPort is by design backward-compatible to single link DVI and HDMI. Those are both severely distance-limited by the delay skew of the three data pairs when compared with the clock pair. With DisplayPort the clock is embedded with the video, much as the clock is embedded with the audio bit stream in AES digital audio, so the distance limitations on DisplayPort are less likely to involve clock timing problems.

However, DisplayPort is also a nonlocking connector, of 20 pins, and is intended for maximum distance 15 m (50 ft). These cables are, like HDMI and DVI, only available in assemblies. Raw cable and connectorization in the field do not currently look like an option for the professional installer. All these factors make it less likely to be embraced by the professional broadcast video arena.

PREMISE/DATA CATEGORY CABLES

While premise/data category cables were never intended to be audio or video cables, their high performance and low cost, and their ubiquitous availability, have seen them pressed into service carrying all sorts of nondata signals.

It should also be noted that high-speed Ethernet networks are routinely used to transport these audio and video signals in data networks. The emergence of 10GBase-T, 10 gigabit networks, will allow the transport of even multiple uncompressed 1080p/60 video images. The digital nature of most entertainment content, with the ubiquitous video server technology in use today, makes high-bandwidth, high-data-rate networks in audio, video, broadcast, and other entertainment facilities, an obvious conclusion.

Cabling definitions

- *Telcom Closet* (TC). Location where the horizontal cabling and backbone cabling are made.
- *Main Cross-Connect* (MXC). Often called the *equipment room* and is where the main electronics are located.
- *Intermediate Cross-Connect* (IXC). A room between the TC and the MXC are terminated. Rarely used in LANs.
- *Horizontal Cabling.* The connection from the telcom closet to the work area.
- *Backbone Cabling.* The cabling that connect all of the hubs together.
- *Hub.* The connecting electronic box that all of the horizontal cables connect to which are then connected to the backbone cable.
- *Ethernet.* A 10, 100, or 1000 Mb/s LAN. The 10 Mbps version is called *10Base-T*. The 100 Mbps version is called *Fast Ethernet* and 1000 Mbps version is called *Gigabit Ethernet*.

Structured cabling

Structured cabling, also called *communications cabling, data/voice, low voltage,* or *limited energy* is the standardized infrastructure for telephone and local area network (LAN) connections in most commercial installations. The architecture for the cable is standardized by the Electronic Industries Association and Telecommunications Industry Association (EIA/TIA), an industry trade association. EIA/TIA 568, referred to as 568, is the main document covering structured cabling. IEEE 802.3 also has standards for structured cabling.

The current standard, as of this writing, is EIA/TIA 568-B.2-10 that covers all active standards up to 10GbaseT, 10 gigabit cabling.

Types of structured cables

Following are the types of cabling, Category 1 though Category 7, often referred to as Cat 1 through Cat 7. The standard TIA/EIA 568A no longer recognizes Cat 1, 2, or 4. As of July 2000, the FCC mandated the use of cable no less than Cat 3 for home wiring. The naming convention specified by ISO/IEC 11801 is shown in Figure 6.3.

Table 6.16 gives the equivalent TIA and ISO classifications for structured cabling.

Category 1. Meets the minimum requirements for analog voice or plain old telephone service (POTS). This category is not part of the EIA/TIA 568 standard.

Category 2. Defined as the IBM Type 3 cabling system. IBM Type 3 components were designed as a higher grade 100 Ω UTP system capable of operating 1 Mb/s Token Ring, 5250, and 3270 applications over shortened distances. This category is not part of the EIA/TIA 568 standard.

Category 3. Characterized to 16 MHz and supports applications up to 10 Mbps. Cat 3 conductors are 24 AWG. Applications range from voice to 10Base-T.

Category 4. Characterized to 20 MHz and supports applications up to 16 Mb/s. Cat 4 conductors are 24 AWG. Applications range from voice to 16 Mbps Token Ring. This category is no longer part of the EIA/TIA 568 standard.

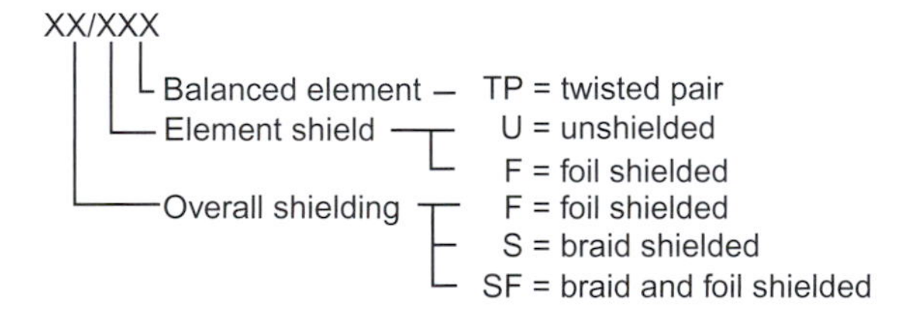

FIGURE 6.3
ISO/IEC 11801 cable naming convention.

Table 6.16 TIA and ISO Equivalent Classifications

Frequency bandwidth	TIA		ISO	
	Components	Cabling	Components	Cabling
1–100 MHz	Cat 5e	Cat 5e	Cat 5e	Class D
1–250 MHz	Cat 6	Cat 6	Cat 6	Class E
1–500 MHz	Cat 6a	Cat 6a	Cat 6a	Class E_A
1–600 MHz	n/s	n/s	Cat 7	Class F
1–1000 MHz	n/s	n/s	Cat 7A	Class F_A

Category 5. Characterized to 100 MHz and supports applications up to 100 Mbps. Cat 5 conductors are 24 AWG. Applications range from voice to 100 Base-T. This category is no longer part of the EIA/TIA 568 standard.

Category 5e. Characterized to 100 MHz and supports applications up to 1000 Mbps/1 Gbps. Cat 5e conductors are 24 AWG. Applications range from voice to 1000Base-T. Cat 5e is specified under the TIA standard ANSI/TIA/EIA-568-B.2. Class D is specified under ISO standard ISO/IEC 11801, 2nd Ed.

Category 6. Characterized to 250 MHz, in some versions bandwidth is extended to 600 MHz, and supports 1000 Mbps/1 Gbps and future applications and is backward compatible with Cat 5 cabling systems. Cat 6 conductors are 23 AWG. This gives improvements in power handling, insertion loss, and high-frequency attenuation. Figure 6.4 shows the improvements of Cat 6 over Cat 5e. Cat 6 is specified under the TIA standard ANSI/TIA/EIA-568-B.2-1. Class E is specified under ISO standard ISO/IEC 11801, 2nd Ed. Cat 6 is available most commonly in the United States as UTP.

Category 6 F/UTP. Cat 6 F/UTP (foiled unshielded twisted pair) or ScTP (screened twisted pair) consists of four twisted pairs enclosed in a foil shield with a conductive material on one side. A drain wire runs adjacent to the conductive side of the shield, Figure 6.5. When appropriately connected, the shield reduces ANEXT, RFI, and EMI. Cat 6 FTP can only be designed to 250 MHz per TIA/EIA 568B.2-1.

Category 6a. Cat 6a (Augmented Category 6) is characterized to 500 MHz, and in special versions to 625 MHz, has lower insertion loss, and has more immunity to noise. Cat 6a is often larger than the other cables. 10GBase-T transmission uses digital signal processing (DSP) to cancel out some of the internal noise created by NEXT and FEXT between pairs. Cat 6a is specified under the TIA standard ANSI/TIA/EIA 568-B.2-10. Class EA is specified under ISO standard ISO/IEC 11801, 2nd Ed. Amendment 1. Cat 6a is available as UTP or FTP.

Category 7 S/STP. Cat 7 S/STP (foil shielded twisted pair) cable is sometimes called *PiMF* (pairs in metal foil). Shielded-twisted pair 10GBase-T cable dramatically reduces alien crosstalk. Shielding reduces electromagnetic interference (EMI) and radio-frequency interference (RFI). This is particularly important as the airways are getting more congested. The shield reduces signal leakage and makes it harder to tap by an outside source. Shield termination at 14.16 Class F will be specified under ISO standard ISO/IEC 11801, 2nd Ed. Class FA will be specified under ISO standard ISO/IEC 11801, 2nd Ed. Amendment 1.

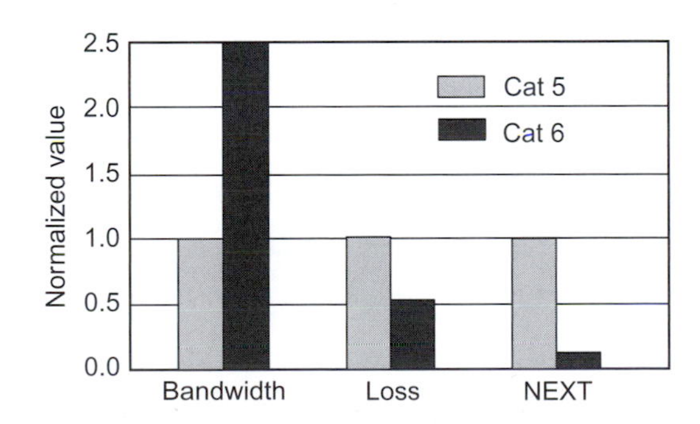

FIGURE 6.4
Normalized comparison of Cat 5e and Cat 6.

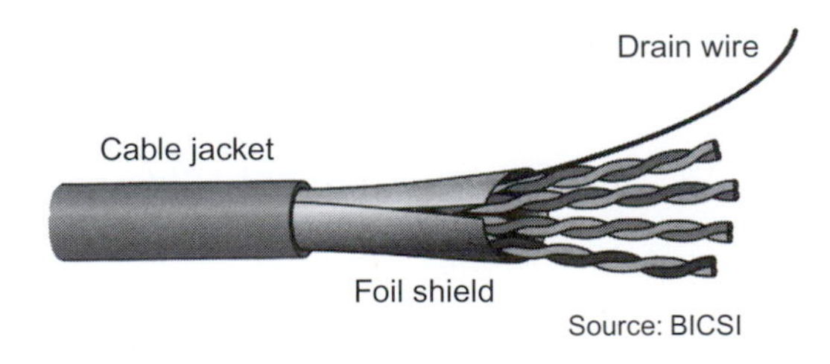

FIGURE 6.5
Cat 6 F/UTP.

Table 6.17 Network Data Rates, Supporting Cable Types, and Distance

Minimum Performance	Token Ring	Ethernet	Maximum Distance
Cat 3	4 Mb/s	10 Mbps	100 m/328 ft
Cat 4	16 Mb/s	–	100 m/328 ft
Cat 5	–	100 Mbps	100 m/328 ft
Cat 5e		1000 Mbps	100 m/328 ft
Cat 6	–	10 Gbps	55 m/180 ft
Cat 6a		10 Gbps	100 m/328 ft

Table 6.18 Characteristics of Cat 5e, Cat 6, and Cat 6a

Cabling Type	Cat 5e	Cat 6	Cat 6a
Relative Price (%)	100	135–150	165–180
Available Bandwidth	100 MHz	250 MHz	500 MHz
Data Rate Capability	1.2 Gbps	2.4 Gbps	10 Gbps
Noise Reduction	1.0	0.5	0.3
Broadband Video Channels 6 MHz/channel	17	42	83
Broadband Video Channels rebroadcast existing channels	6	28	60+
No. of Cables in Pathway 24 inches × 4 inches	1400	1000	700

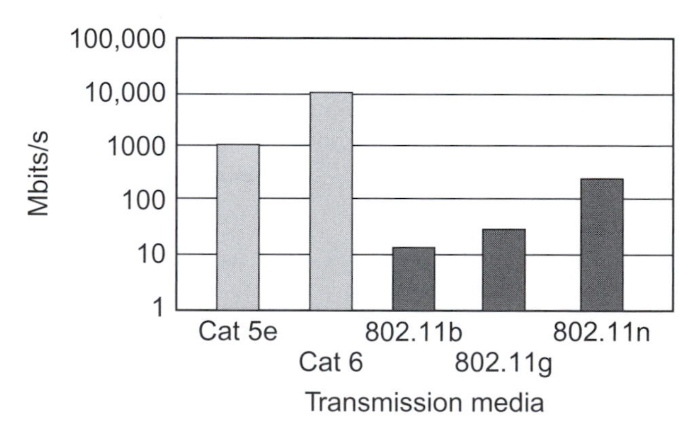

FIGURE 6.6
Comparison of media distance to bandwidth.

Comparisons

Table 6.17 compares network data rates for Cat 3 through Cat 6a and Table 6.18 compares various characteristics of Cat 5e, 6, and 6a. Figure 6.6 compares the media distance-bandwidth product of Cat 5e and Cat 6a with 802.11 (a, b, g, n) wireless media, often called *Wi-Fi*.

New cable designs can affect size and pathway load so consult the manufacturer. Note that cable density is continually changing with newer, smaller cable designs. Numbers in Table 6.18 should be considered worst case. Designers and installers of larger systems should get specific dimensional information from the manufacturer.

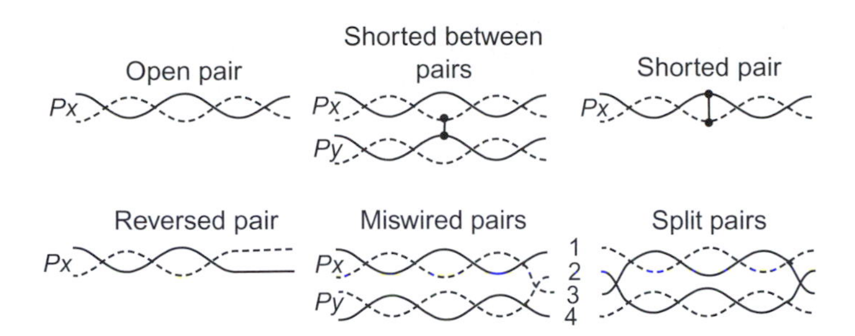

FIGURE 6.7
Paired wiring faults. (Courtesy Belden.)

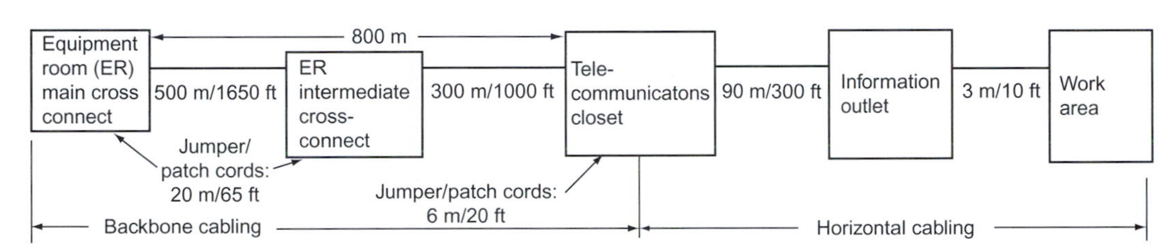

FIGURE 6.8
Maximum distances between areas for UTP cable.

Table 6.19 Color Code for UTP Cable

Pair No.	1st Conductor Base/Band	2nd Conductor
1	White/Blue	Blue
2	White/Orange	Orange
3	White/Green	Green
4	White/Brown	Brown

Figure 6.7 shows various problems that can be found in UTP cabling. Figure 6.8 gives the maximum distances for UTP cabling as specified by ANSI/TIA.

Four (4) pair 100 Ω ± 15% UTP Cat 5e cabling is the recommended minimum requirement for residential and light commercial installations because it provides excellent flexibility. Pair counts are four pair for desktop and 25 pair for backbone cabling. The maximum length of cable is 295 ft (90 m) with another 33 ft (10 m) for patch cords.

Unshielded twisted pairs (UTP) and shielded twisted pairs (STP) are used for structured cabling. Unshielded twisted pairs (UTP) are the most common today. These cables look like the POTS cable, however their construction makes them usable in noisy areas and at high frequencies because of the short even twisting of the two wires in each pair. The twist must be even and tight so complete noise cancellation occurs along the entire length of the cable. To best keep the twist tight and even, better cable has the two wires bonded together so they will not separate when bent or flexed. Patch cable is flexible so twist and impedance are not as well controlled. The color codes for the pairs are given in Table 6.19.

Cable diameter varies for the different types of cable. TIA recommends that two Cat 6 cables but only one Cat 6a cable can be put in a ¾ inch (21 mm) conduit at 40% fill. The diameter and the stiffness of the cables determine their bend radius and therefore the bend radius of conduits and trays, Table 6.20.

Figure 6.9 shows the construction of UTP and screened UTP cable.

Table 6.20 Diameter and Bend Radius for 10GbE Cabling

Cable	Diameter	Bend Radius
Category 6	0.22 inch (5.72 mm)	1.00 inch (4 × OD)
Category 6a	0.35 inch (9 mm)	1.42 inch (4 × OD)
Category 6 FTP	0.28 inch (7.24 mm)	2.28 inch (8 × OD)
Category 7 STP	0.33 inch (8.38 mm)	2.64 inch (8 × OD)

Cable diameters are nominal values.

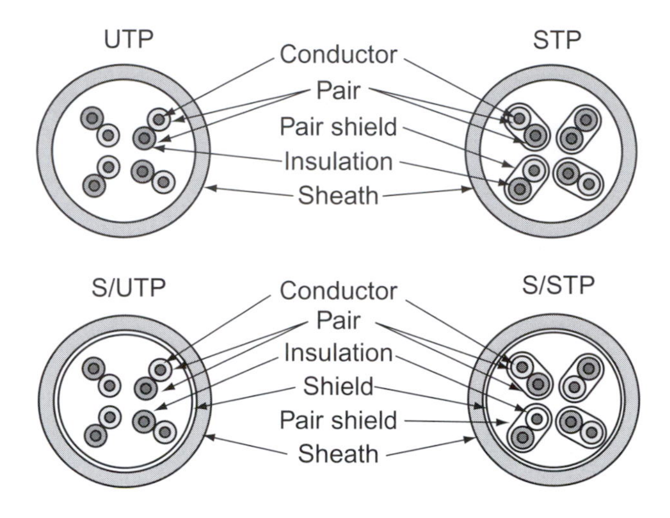

FIGURE 6.9
UTP and S/UTP cable design.

Critical parameters

Critical parameters for UTP cable are: NEXT, PS-NEXT, FEXT, EL-FEXT, PS-ELFEXT, RL, ANEXT.

NEXT. *NEXT*, or near-end crosstalk, is the unwanted signal coupling from the near end of one sending pair to a receiving pair.

PS-NEXT. *PS-NEXT*, or power-sum near-end crosstalk, is the crosstalk between all of the sending pairs to a receiving pair. With four-pair cable, this is more important than NEXT.

FEXT. *FEXT*, or far-end crosstalk, is the measure of the unwanted signal from the transmitter at the near end coupling into a pair at the far end.

EL-FEXT. *EL-FEXT*, or equal level far-end crosstalk, is the measure of the unwanted signal from the transmitter end to a neighboring pair at the far end relative to the received signal at the far end. The equation is:

$$EL - FEXT = FEXT - Attenuation \tag{6.3}$$

PS-ELFEXT. *PS-ELFEXT*, or power sum equal-level far-end crosstalk is the computation of the unwanted signal coupling from multiple transmitters at the near end into a pair measured at the far end relative to the received signal level measured on the same pair.

RL. *RL*, or return loss is a measure of the reflected energy from a transmitted signal and is expressed in −dB, the higher the value, the better. The reflections are caused by impedance mismatch caused by connectors, improper installation, such as stretching the cable or too sharp a bend radius, improper manufacturing, or improper load.

Broadcasters are very familiar with return loss, calling it by a different name, SWR (standing wave ratio) or VSWR (voltage standing wave ratio). In fact, return loss measurements can easily be converted into VSWR values, or vice versa. Return loss can be found with the equation:

$$RL = 20\log\frac{Difference}{Sum} \tag{6.4}$$

where,

Difference is the difference (absolute value) between the desired impedance and the actual measured impedance,

Sum is the desired impedance and the actual measured impedance added together.

- The desired impedance for all UTP data cables (Cat 5, 5e, 6, 6a) is 100 Ω.
- The desired impedance for all passive video, HD, HD-SDI or 1080p/60 components is 75 Ω.
- The desired impedance for all digital audio twisted pairs is 110 Ω.
- The desired impedance for all digital audio on coaxial cable is 75 Ω.

With 1000Base-T systems, the pairs simultaneously transmit and receive. As the transmitter sends data, it is also listening for data being sent from the opposite end of the same pair. Any reflected signal from the sending end that reflects back to the sending end mixes with the sending signal from the far end, reducing intelligibility. With 10Base-T or 100Base-T data networks, one pair transmits while another receives, so reflections (RL, return loss) are not a major consideration and were not required to be measured. Now, with pairs simultaneously transmitting and receiving, called *duplex mode*, RL is a critical measurement for data applications.

Delay Skew. Since every pair (and every cable) takes a specific amount of time to deliver a signal from one end to the other, there is a delay. Where four pairs must deliver data to be recombined, the delay in each pair should, ideally, be the same. However, to reduce crosstalk, the individual pairs in any Category cable have different twist rates. This reduces the pair-to-pair crosstalk but affects the delivery time of the separate parts. This is called *delay skew.*

While delay skew affects the recombining of data, in 1000Base-T systems, for instance, the same delay skew creates a problem when these UTP data cables are used to transmit component video or similar signals, since the three colors do not arrive at the receiving end at the same time, creating a thin bright line on the edge of dark images. Some active baluns have skew correction built in.

ANEXT. *ANEXT*, or alien crosstalk, is coupling of signals between cables. This type of crosstalk cannot be canceled by DSP at the switch level. Alien crosstalk can be reduced by overall shielding of the pairs, or by inserting a nonconducting element inside that cable to push away the cables around it.

Terminating connectors

All structured cabling use the same connector, an RJ-45. In LANs (local area networks) there are two possible pin-outs, 568A and 568B. The difference is pair 2 and pair 3 are reversed. Both work equally well as long as they are not intermixed. The termination is shown in Figure 6.10.

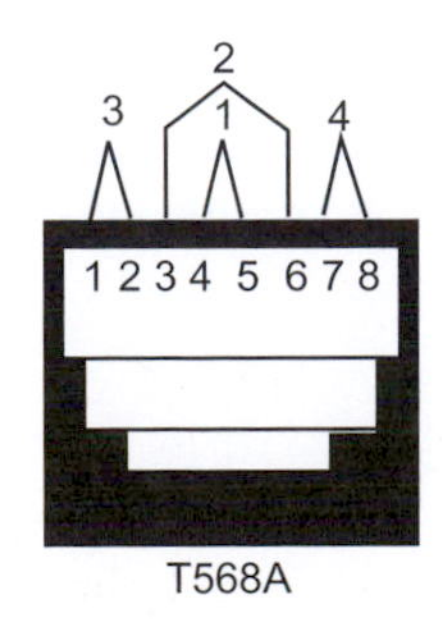

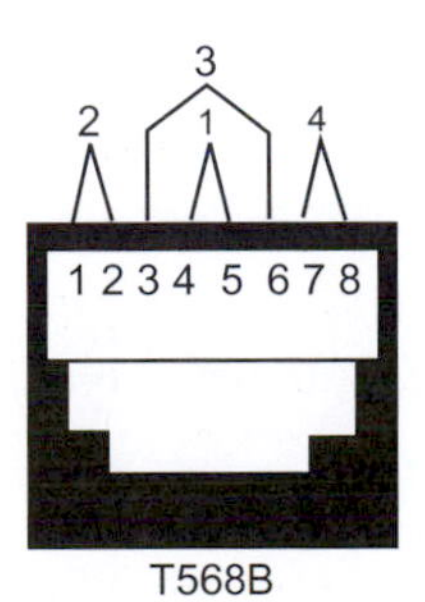

FIGURE 6.10
Termination layout for EIA/TIA 568-B.2 cable.

In the past decade, the B wiring scheme has become the most common. However, if you are adding to or extending and existing network, you must determine which wiring scheme was used and continue with that scheme. A mixed network is among the most common causes of network failure.

It is very important that the pairs be kept twisted as close to the connector as possible. For 100Base-T (100 MHz, 100 Mbps) applications, a maximum of ½ inch should be untwisted to reduce crosstalk and noise pickup. In fact, with Cat 6 (250 MHz) or Cat 6a (500 MHz) it is safe to say that any untwisting of the pairs will affect performance. Therefore there are many connectors, patch panels, and punch-down blocks that minimize the untwisting of the pairs.

Baluns

Baluns (*Bal*anced-*Un*balanced) networks are a method of connecting devices of different impedance and different formats. Baluns have been commonly used to convert unbalanced coax to balanced twin-lead for television antennas, or to match coaxial data formats (coaxial Ethernet) to balanced line systems (10Base-T, 100Base-T, etc.). Other balun designs can allow unbalanced sources, such as video or consumer audio, for instance, to be carried on balanced lines, such as UTP Cat 5e, 6, etc.

Since there are four pairs in a common data cable, this can carry four channels. Since category cables are rarely tested below 1 MHz, the audio performance was originally suspect. Crosstalk at audio frequencies in UTP has been measured and is consistently better than −90 dB even on marginal Cat 5. On Cat 6, the crosstalk at audio frequencies is below the noise floor of most network analyzers.

Baluns are commonly available to handle such signals as analog and digital audio, composite video, S-video, RGB or other component video (VGA, Y/R-Y/B-Y, Y/Cr/Cb), broadband RF/CATV, and even DVI and HDMI. The limitations to such applications are the bandwidth specified on the cable and the performance of the cable (attenuation, return loss, crosstalk, etc.) at those higher frequencies.

Passive baluns can also change the source impedance in audio devices. This dramatically extends the effective distance of such signals from only a few feet to many hundreds of feet. Consult the balun manufacturer for the actual output impedance of their designs.

Some baluns can include active amplification, equalizations, or skew (delivery timing) compensation. While more expensive, these active baluns can dramatically increase the effective distance of even marginal cable.

Adaptors

Users and installers should be aware there are adaptors, often that fit in wall plates, where keystone data jacks are intended to be snapped in place. These adaptors often connect consumer audio and video (RCA connectors) to 110 blocks or other twisted pair connection points. However, there is no unbalanced-to-balanced device in these, so the noise rejection inherent in twisted pairs when run as a balanced line is not provided. These adaptors simply unbalance the twisted pair and offer dramatically short effective distances. Further, baluns can change the source impedance and extend distance. Adaptors with no transformers or similar components cannot extend distance and often reduce the effective distance. These devices should be avoided unless they contain an actual balun.

Power over ethernet (PoE)

PoE supplies power to various Ethernet services as VoIP (Voice over Internet Protocol) telephones, wireless LAN access points, Blu-tooth access points, and Web cameras. Many audio and video applications will soon use this elegant powering system. IEEE 802.3af-2003 is the IEEE standard for PoE. IEEE 802.3af specifies a maximum power level of 15.4 W at the power sourcing equipment (PSE) and a maximum of 12.95 W of power over two pairs to a powered device (PD) at the end of a 100 m (330 ft) cable.

The PSE can provide power by one of two configurations:

1. Alternative A, sometimes called *phantom powering*, supplies the power over pairs 2 and 3.
2. Alternative B supplies power over pairs 1 and 4, as shown in Figure 6.11.

The voltage supplied is nominally 48 Vdc with a minimum of 44 Vdc, a maximum of 57 Vdc, and the maximum current per pair is 350 mAdc, or 175 mAdc per conductor. For a single solid 24 AWG wire,

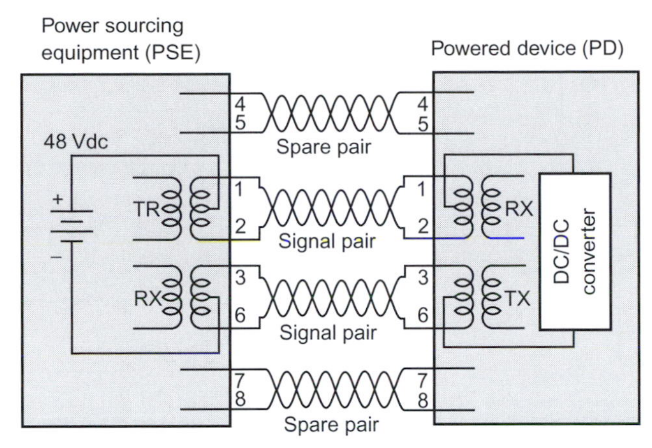

A. Power supplied via data pairs

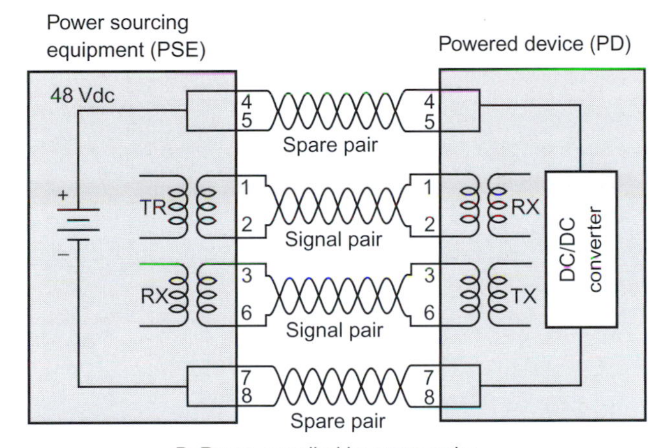

B. Power supplied by spare pairs

FIGURE 6.11
Two methods of supplying power via the Ethernet. (Courtesy of Panduit Int'l Corp.)

common to many category cable designs, of 100 m length (328 ft) this would be a resistance of 8.4 Ω. Each conductor would dissipate 0.257 W or 1.028 W per cable (0.257 W × 4 conductors). This causes a temperature rise in the cable and conduit which must be taken into consideration when installing PoE.

Power over ethernet plus (PoE Plus)

PoE Plus is defined in IEEE 802.3at and is capable of delivering up to 30 W. Work is being done to approach 60 W or even greater. This requires the voltage supply to be 50 to 57 Vdc. Assuming a requirement of 42 W of power at the endpoint at 50 Vdc, the total current would be 0.84 A, or 0.21 A per pair, or 0.105 A (105 mA) per conductor, or a voltage drop of only 0.88 V in one 24 AWG wire.

COAXIAL CABLE

Coaxial cable is a design in which one conductor is accurately centered inside another with both conductors carrying the desired signal currents (source to load and return), as shown in Figure 6.12. Coaxial cable is so called because if you draw a line through the center of a cross-sectional view, you will dissect all parts of the cable. All parts are on the same axis, or coaxial.

History of coaxial cable

It has been argued that the first submarine telegraph cable (1858) was coaxial, Figure 6.13. While this did have multiple layers, the outer layer was not part of the signal-carrying portion. It was a protective layer.

Modern coaxial cable was invented on May 23, 1929 by Lloyd Espenscheid and Herman Affel of Bell Laboratories. Often called *coax*, it is often used for the transmission of high-frequency signals.

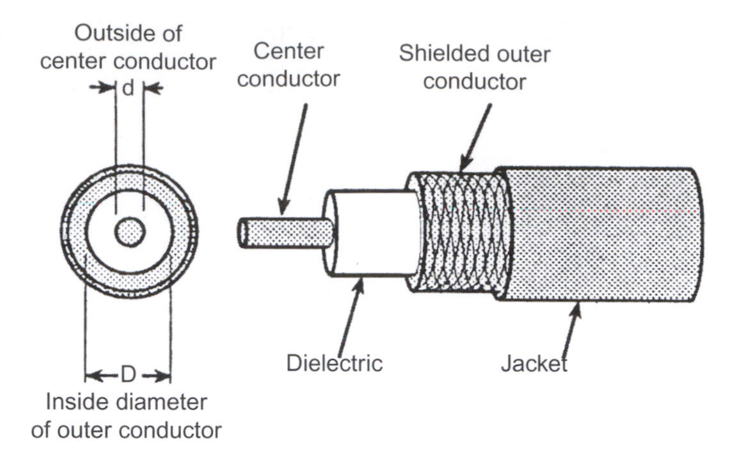

FIGURE 6.12
Construction of a coaxial cable.

FIGURE 6.13
First submarine cable.

At high frequencies, above 100 kHz, coax has a dramatically better performance than twisted pairs. However, coax lacks the ability to reject noise that twisted pairs can do when configured as balanced lines. Coaxial cable was first installed in 1931 to carry multiple telephone signals between cities.

Coaxial cable construction

The insulation between the center conductor and the shield of a coaxial cable affects the impedance and the durability of the cable. The best insulation to use between the center conductor and the shield would be a vacuum. The second best insulation would be dry air, the third, nitrogen. The latter two are familiar insulators in hard-line transmission line commonly used to feed high-power antenna in broadcasting.

A vacuum is not used, even though it has the lowest dielectric constant of "1," because there would be no conduction of heat from the center conductor to the outer conductor and such a transmission line would soon fail. Air and nitrogen are commonly used under pressure in such transmission lines. Air is occasionally used in smaller, flexible cables.

Polyethylene (PE) was common as the core material in coaxial cables during WW II. Shortly after the war, polyethylene was declassified and most early cable designs featured this plastic. Today most high-frequency coaxial cables have a chemically formed foam insulation or a nitrogen gas injected foam. The ideal foam is high-density hard cell foam, which approaches the density of solid plastic but has a high percentage of nitrogen gas. Current state-of-the-art polyethylene foam velocity is 86% (dielectric constant: 1.35) although most digital video cables are 82–84% velocity of propagation. High-density

Table 6.21 Transmission Distances for CCTV Cable

RG-59	1000 ft
RG-6	1500 ft
RG-11	3000 ft

foam of this velocity resists conductor migration when the cable is bent, keeping impedance variations to a minimum. This high velocity improves the high-frequency response of the cable.

A problem with soft foam is it easily deforms, which changes the distance between the center conductor and the shield, changing the cable impedance. This can be caused by bending the cable too sharply, or running over it, or pulling it too hard, or any other possibility. To reduce this problem, a hard cell foam is used. Some cable that is rated as having a very high velocity of propagation might use very soft foam. A simple test can be performed where the user squeezes the foam dielectric of various cables. It will be immediately apparent that some cables have a density (crush resistance) double that of other designs. Soft foam can lead to conductor migration over time which will change timing, impedance, return loss, and bit errors over distance.

Coaxial cable is used quite extensively with various types of test equipment. When such cable is replaced, the capacitance per foot, which is determined by the dielectric constant of the insulator, must be taken into consideration, particularly for oscilloscope probes.

CCTV CABLE

CCTV (closed circuit television) cable has a 75 Ω characteristic impedance. CCTV is a baseband signal comprised of low-frequency vertical and horizontal sync pulse information and high-frequency video information. Since the signal is broadband, only cable with a center conductor of solid copper should be used.

If the cable is constantly in motion as in pan and tilt operation, a stranded center conductor should be used as a solid conductor will work-harden and break. There are also robotic coaxes designed to flex millions of times before failure for intense flexing applications.

Shielding for CCTV cable should have a copper or tinned-copper braid of at least 80% coverage, for low-frequency noise rejection. If an aluminum foil shield is used in conjunction with a braid, either tinned copper or aluminum only may be used for the shield. A bare copper braid will result in a galvanic reaction.

CCTV distances

For common CCTV 75 Ω cables, their rule-of-thumb transmission distances are shown in Table 6.21. These distances can be extended by the use of in-line booster amplifiers.

CATV BROADBAND CABLE

For higher-frequency applications, such as carrying radio frequencies or television channels, only the skin of the conductor is working (see Skin Effect above). Television frequencies in the United States, for instance, start with Channel 2 (54 MHz) which is definitely in the area of skin effect. So these cables can use center conductors that have a layer of copper over a steel wire, since only the copper layer will be working.

If one uses a copper-clad steel conductor for applications below 50 MHz, the conductor has a DC resistance from four to seven times that of a solid copper conductor. If a copper-clad cable is used on a baseband video signal, for instance, the sync pulses may be attenuated too much. If such a cable is used to carry audio, almost the entire audio signal will be running down the steel wire.

CATV/broadband cable should have a foil shield for good high-frequency noise rejection. CATV cable should also have a braid shield to give the connector something to grab onto, 40% to 60% aluminum braid being the most common. Multiple layer shields are also available such as tri-shielded (foil-braid-foil) and quad shields (foil-braid-foil-braid). Assumptions that quad shields give the best shield effectiveness are erroneous, there being single foil/braid and tri-shield configurations that are measurably superior.

Modern CATV/broadband cable will use a foamed polyethylene or foamed FEP dielectric, and preferably one with gas injected foam. This will reduce the losses in the cable. The jacket material is determined by the environment that the cable will be working in.

Coaxial cable installation considerations

INDOOR INSTALLATION

Indoor environments are the most common for coaxial cable installations. A few tips on installing coaxial cable are as follows:

1. First and foremost, follow all NEC requirements when installing coaxial cables.
2. Distribute the pulling tension evenly over the cable and do not exceed the minimum bend radius of 10 times the diameter. Exceeding the maximum pulling tension or the minimum bend radius of a cable can cause permanent damage both mechanically and electrically to the cable.
3. When pulling cable through conduit, clean and deburr the conduit completely and use proper lubricants in long runs.

OUTDOOR INSTALLATION

Outdoor installations require special installation techniques that will enable the cable to withstand harsh environments. When using cable in an aerial application, lash the cable to a steel messenger, or buy cable with a built-in steel messenger. This will help support the cable and reduce the stress on the cable during wind, snow, and ice storms. When direct burying a cable, lay the cable without tension so it will not be stressed when earth is packed around it. When burying in rocky soil, fill the trench with sand. Lay the cable and then place pressure-treated wood or metal plates over the cable. This will prevent damage to the cable from rocky soil settling; in cold climate areas, bury the cable below the frost line. Buy direct burial cable designed to be buried.

Coaxial cable termination techniques

SOLDERING

Soldering offers several advantages as it can be used on solid or stranded conductors and it creates both a solid mechanical and electrical connection. The disadvantage is that it takes more time to terminate than other methods and cold solder joints can cause problems if the connector is not soldered to the cable properly. The use of lead-based solder might also be a consideration if RoHS (reduction of hazardous substances) requirements are part of the installation. Soldering is not recommended for high-frequency applications, such as HD-SDI or 1080p/60 as the variations in dimensions will show up as variations in impedance and contribute to return loss.

CRIMPING

Crimping is probably the most popular method of terminating BNC and F connectors on coax cable. Like the solder method, it can be used on solid or stranded conductors and provides a good mechanical and electrical connection. This method is the most popular because there is no need for soldering so installation time is reduced. It is very important to use the proper size connector for a tight fit on the cable. Always use the proper tool. Never use pliers as they are not designed to place the pressure of the crimp evenly around the connector. Pliers will crush the cable and can degrade the electrical properties of the cable.

TWIST-ON CONNECTORS

Twist-on connectors are the quickest way of terminating a coaxial cable; however, they do have some drawbacks. When terminating the cable with this type of connector, the center conductor is scored by the center pin on the connector, thus too much twisting can cause damage to the center conductor. It is not recommended for pan and tilt installations as the constant movement of the cable may work the connector loose. Because there is no mechanical or electrical crimp or solder connection, this connector is not as reliable as the other methods.

COMPRESSION CONNECTORS

There are connectors, often a one-piece connector, that fit over the stripped cable and fasten by having two parts squeeze or compress together. This is a very simple and reliable way of connecting cable. However, the very high-frequency performance (beyond 500 MHz) has yet to be proven and so these connectors are not recommended for professional digital applications. A compression connector that is measured with a return loss of −20 dB at 2 GHz would be acceptable for professional broadcast HD applications.

Return loss

At high frequencies, where cable and connectors are a significant percentage of a wavelength, the impedance variation of cable and components can be a significant source of signal loss. When the signal sees something other than 75 Ω, a portion of the signal is reflected back to the source. Table 6.22 shows the wavelength and quarter wavelength at various frequencies. One can see that this was a minor problem with analog video (quarter wave 59 ft) since the distances are so long. However, with HD-SDI and higher signals, the quarter wave can be 1 inch or less, meaning that everything in the line is critical: cable connectors, patch panels, patch cords, adaptors, bulkhead/feedthrough connectors, etc.

In fact, Table 6.22 above is not entirely accurate. The distances should be multiplied by the velocity of propagation of the cable or other component, to get the actual length, so they are even shorter still.

Since everything is critical at high frequencies, it is appropriate to ask the manufacturers of the cable, connectors, patch panels, and other passive components, how close to 75 Ω their products are. This can be established by asking for the return loss of each component. Table 6.23 will allow the user to roughly translate the answers given.

Most components intended for HD can pass −20 dB return loss. In fact, −20 dB return loss at 2 GHz is a good starting point for passive components intended for HD-SDI. Better components will pass −30 dB at 2 GHz.Better still (and rarer still) would be −30 dB at 3 GHz. There are currently no components that are consistently −40 dB return loss at any reasonable frequency. In Table 6.22, it can be seen that 1080p/60 signals need to be tested to 4.5 GHz. This requires expensive custom-built matching networks. As of this writing, only one company (Belden) has made such an investment.

Note that the number of nines in the Signal Received column in Table 6.23 is the same as the first digit of the return loss (i.e., −30 dB = 3 nines = 99.9%). There are similar tests, such as SRL (structural return loss). This test only partially shows total reflection. Do not accept values measured in any way except return loss. The SMPTE maximum amount of reflection on a passive line (with all components measured and added together) is −15 dB or 96.84% received, 3.16% reflected. A line with an RL of −10 dB (10% reflected) will probably fail.

Video triaxial cable

Video triaxial cable is used to interconnect video cameras to their related equipment. Triaxial cable contains a center conductor and two isolated shields, allowing it to support many functions on the one cable. The center conductor and outer shield carry the video signals plus intercoms, monitoring devices, and camera power. The center shield carries the video signal ground or common. Triax cable is usually of the RG-59 or RG-11 type.

Table 6.22 Wavelength and Quarter Wavelength of Various Signals at Various Frequencies

Signal	Clock Frequency	Third Harmonic	Wavelength	Quarter Wavelength
Analog video (4.2 MHz)	analog	analog	234 ft	59 ft
SD-SDI	135 MHz	405 MHz	2.43 ft	7.3 inches
HD-SDI	750 MHz	2.25 GHz	5.3 inches	1.3 inches
1080p/60	1.5 GHz	4.5 GHz	2.6 inches	0.66 inches

Table 6.23 Return Loss versus % of Signal Received and Reflected

Return Loss	% of Signal Received	% Reflected
−50 dB	99.999%	0.001%
−40 dB	99.99%	0.01%
−30 dB	99.9%	0.1%
−20 dB	99.0%	1.0%
−10 dB	90.0%	10.0%

S-Video

S-video requires a duplex (dual) coaxial cable to allow separate transmission of the luminance (Y) and the chrominance (C). The luminance signal is black or white or any gray value while the chrominance signal contains color information. This transmission is sometimes referred to as Y-C. Separating signals provides greater picture detail and resolution and less noise interference.

S-video is sometimes referred to as S-VHS™ (Super-Video Home System). While its intention was for improved consumer video quality, these cameras were also used for the lower end of the professional area, where they were used for news, documentaries, and other less-critical applications.

RGB

RGB stands for red-green-blue, the primary colors in color television. It is often called *component video* since the signal is split up to its component colors. When these analog signals are carried separately, much better image resolution can be achieved. RGB can be carried on multiple single video cables, or in bundles of cables made for this application. With separate cables, all the cables used must be precisely the same electrical length. This may or may not be the same as the physical length. Using a vectorscope, it is possibly to determine the electrical length and compare the RGB components. If the cables are made with poor quality control, the electrical length of the coaxes may be significantly different (i.e., one cable may have to be physically longer than the others to align the component signals). Cables made with very good quality control can simply be cut at the same physical length.

Bundles of RGB cables should be specified by the amount of timing error, the difference in the delivery time on the component parts. For instance, all Belden bundled coax cables are guaranteed to be 5 ns (nanosecond) difference per 100 ft of cable. Other manufacturers should have a similar specification and/or guarantee. The de facto timing requirement for broadcast RGB is a maximum of 40 ns. Timing cables by hand with a vectorscope allows the installer to achieve timing errors of >1 ns. Bundled cables made for digital video can also be used for RGB analog, and similar signals (Y, R–Y, B–Y or Y, Pb, Pr or YUV or VGA, SVGA, XGA, etc.) although the timing requirements for VGA and that family of signals has not been established.

These bundled coaxes come in other version besides just three coax RGB. Often the horizontal and vertical synchronizing signals (H and V) are carried with the green video signal on the green coax. For even greater control, these signals can be carried by a single coax (often called *RGBS*) or five coaxes, one for each signal (called *RGBHV*). These cables are becoming more common in the home, where they are often referred to as five-wire video. There are also four-pair UTP data cables made especially to run RGB and VGA signals. Some of these have timing tolerance (called *delay skew* in the UTP world) that is seriously superior to bundled coaxes. However, the video signals would have to be converted from 75 Ω to 100 Ω, and the baluns to do this, one for each end of the cable, would be added to the cost of the installation. Further, the impedance tolerance of coax, even poorly made coax, is dramatically superior to twisted pairs. Even bonded twisted pairs are, at best, ±7 Ω, where most coaxial cables are ±3 Ω, with precision cables being twice as good as that, or even better.

Table 6.24 Resolution of Various VGA and Family Formats

Signal Type	Resolution
VGA	640 × 480
SVGA	800 × 600
XGA	1024 × 768
WXGA	1280 × 720
SXGA	1280 × 1024
SXGA-HD	1600 × 1200
WSXGA	1680 × 1050
QXGA	2048 × 1536
QUSXG	3840 × 2400

VGA and family

VGA stands for video graphics array. It is an analog format to connect progressive video source to displays, such as projectors and screens. VGA comes in a number of formats, based on resolution. These are shown in Table 6.24.

There are many more variations in resolution and bandwidth than the ones shown in Table 6.24.

DIGITAL VIDEO

There are many formats for digital video, for both consumer, commercial and professional applications. This section concentrates on the professional applications, mainly SD-SDI (standard definition—serial digital interface) and HD-SDI (high-definition—serial digital interface.) There are sections on related consumer standards such as DVI and HDMI.

Digital signals and digital cable

Control communications, or data communications, uses digital signals. Digital video signals require wide bandwidth cabling. Control communications and data communications use lower-performance cabling because they carry less information, requiring less bandwidth. High-speed data communications systems have significant overhead added to handle error correction so if data is lost, it can be re-sent. Digital video has some error correction capabilities, however if all of the data bits required to make the system work are not received, picture quality is reduced or lost completely. Table 6.25 compares various digital formats.

Coax and SDI

Most professional broadcast formats (SDI and HD-SDI) are in a serial format and use a single coaxial cable with BNC connectors. Emerging higher resolution formats, such as 1080p/60, are also BNC based. Some work with smaller connectors for dense applications, such as patch panels and routers, which use subminiature connectors such as LCC, DIN 1.0/2.3 or DIN 1.0/2.5. Proprietary miniature BNC connectors are also available.

Cables and SDI

The most common form of SDI, component SDI, operates at data rates of 270 Mbps (clock 135 MHz). Cable loss specifications for standard SDI are specified in SMPTE 259M and ITUR BT.601. The maximum cable length is specified as 30 dB signal loss at one-half the clock frequency and is acceptable because serial digital receivers have signal recovery processing.

HD-SDI, whose cable loss is governed by SMPTE 292M, operates at a data rate of 1.5 Gbps (clock 750 MHz). The maximum cable length is specified at 20 dB signal loss at one-half the clock frequency. These are Manchester Coded signals and the bit rate is therefore double the clock rate. Emerging 1080p/60 applications are covered under SMPTE 424M. The data rate is 3 Gbps (clock 1.5 GHz).

Table 6.25 Comparing Coaxial Digital Formats

Standard	Format	Intended Use	Connector Style	Cable Type	Transmission Distance[2]	Sample Rate	Data Rate (Mbps)	Guiding Document
SDI	serial	broadcast	one BNC	coax[1]	300 m/1000 ft	27 MHz	270	SMPTE 259
SDTI	serial	data transport	one BNC	coax[1]	300 m/1000 ft	variable	270 or 360	SMPTE 305
SDTV	serial	broadcast	one BNC	coax[1]	300 m/1000 ft	27 MHz	3 to 8	ATSC; N53
HDTV	serial	broadcast	one BNC	coax[1]	122 m/400 ft	74.25 MHz	19.4	ATSC; A/53
HD-SDI	serial	broadcast	one BNC	coax[1]	122 m/400 ft	74.25 MHz	1500	SMPTE 292M
1080p/60	serial	Master format	one BNC	coax[1]	80 m/250 ft	148.5 MHz	3000	SMPTE 424M

[1]Also implemented over fiber systems.
[2]Transmission distances may vary widely depending on cabling and the specific equipment involved.

Table 6.26 SMPTE Serial Digital Performance Specifications

	SMPTE 259				SMPTE 292M	
Parameter	Level A NTSC 4fsc Composite	Level B PAL 4fsc Composite	Level C 525/625 Component	Level D 525/625 Component	Level D 1920 × 1080 Interlaced	Level L 1280 × 720 Progressive
Data Rate in Mbps (clock)	143	177	270	360	1485	1485
½ Clock Rate in MHz	71.5	88.5	135	180	742.5	742.5
Signal Amplitude (p-p)	800 mV	800 mV	800 mV	800 mV	800 mV	800 mV
DC Offset (volts)	0 ± 0.5	0 ±0.5	0 ±0.5	0 ±0.5	0 ±0.5	0 ±0.5
Rise/Fall Time Max. (ns)	1.50	1.50	1.50	1.5	0.27	0.27
Rise/Fall Time Min. (ns)	0.40	0.40	0.40	0.40	–	–
Rise/Fall Time Differential (ns)	0.5	0.5	0.5	0.5	0.10	0.10
% Overshoot Max.	10	10	10	10	10	10
Timing Jitter (ns)	1.40	1.13	0.74	0.56	0.67	0.67
Alignment Jitter (ns)	1.40	1.13	0.74	0.56	0.13	0.13

Receiver quality

The quality of the receiver is important in the final performance of a serial digital system. The receiver has a greater ability to equalize and recover the signal with SDI signals. SMPTE 292M describes the minimum capabilities of a type A receiver and a type B receiver. SDI receivers are considered adaptive because of their ability to amplify, equalize, and filter the information. Rise-time is significantly affected by distance, and all quality receivers can recover the signal from a run of HD-SDI RG-6 (such as Belden 1694A) for a minimum distance of 122 m (400 ft). The most important losses that affect serial digital are rise-time/fall-time degradation and signal jitter. Serial digital signals normally undergo reshaping and reclocking as they pass through major network hubs or matrix routers.

Table 6.26 gives the specifications mandated in SMPTE 259M and SMPTE 292M in terms of rise- /fall-time performance and jitter. If the system provides this level of performance at the end of the cable run, the SDI receiver should be able to decode the signal.

Serial digital video

Serial digital video (SDI) falls under standards by the Society of Motion Picture and Television Engineers (SMPTE) and ITU, and falls under the following categories:

SMPTE 259M	Digital video transmissions of composite NTSC 143 Mb/s (Level A) and PAL 177 Mb/s (Level B). It also covers 525/625 component transmissions of 270 Mb/s (Level C) and 360 Mb/s (Level D).
SMPTE 292M	HDTV transmissions at 1.485 Gb/s
SMPTE 344M	Component widescreen transmission of 540 Mb/s
ITU-R BT.601	International standard for PAL transmissions of 177 Mb/s

These standards can work with standard analog video coax cables, however the newer digital cables provide the more precise electrical characteristics required for high-frequency transmission.

SDI cable utilizes a solid bare-copper center conductor which improves impedance stability and reduced return loss (RL). Digital transmissions contain both low-frequency and high-frequency signals so it is imperative that a solid-copper center conductor is used rather than a copper-clad steel center conductor. This allows the low frequencies to travel down the center of the conductor and the high frequencies to travel on the outside of the conductor due to the skin effect. Since digital video consists of both low and high frequencies, foil shields work best. All SDI cable should be sweep tested for return loss to the third harmonic of the fundamental frequency. For HD-SDI which is 1.485 Gb/s or has a 750 MHZ bandwidth, the cable is this frequency.

BNC 50 Ω connectors are often used to terminate digital video lines. This is probably acceptable if only one or two connectors are used. However, if more connectors are used, 75 Ω connectors are required to eliminate RL. Connectors should exhibit a stable 75 Ω impedance out to 2.25 GHz, the third harmonic of 750 MHz.

RADIO GUIDE DESIGNATIONS

From the late 1930s the U.S. Army and Navy began to classify different cables by their constructions. Since the intent of these high-frequency cables, both coaxes and twisted pairs, was to guide radio-frequency signals, they carried the designation RG for radio guide.

There is no correlation between the number assigned and any construction factor of the cable. Thus an RG-8 came after an RG-7 and before an RG-9, but could be completely different and unrelated designs. For all intents and purposes, the number simply represents the page number in a book of designs. The point was to get a specific cable design, with predictable performance, when ordered for military applications.

As cable designs changed, with new materials and manufacturing techniques, variations on the original RG designs began to be manufactured. Some of these were specific targeted improvement, such as a special jacket on an existing design. These variations are noted by an additional letter on the designation. Thus RG-58C would be the third variant on the design of RG-58.

The test procedure for many of these military cables is often long, complicated, and expensive. For the commercial user of these cables, this is a needless expense. So many manufacturers began to make cables that were identical to the original RG specification except for testing. These were then designated utility grade and a slash plus the letter U is placed at the end. RG-58C/U is the utility version of RG-58C, identical in construction but not in testing.

Often the word *type* is included in the RG designation. This indicates that the cable under consideration is based on one of the earlier military standards but differs from the original design in some significant way. At this point, all the designation is telling the installer is that the cable falls into a family of cables. It might indicate the size of the center conductor, the impedance, and some aspects of construction, with the key word being *might*.

By the time the RG system approached RG-500, with blocks of numbers abandoned in earlier designs, the system became so unwieldy and unworkable that the military abandoned it in the 1970s and instituted MIL-C-17 (Army) and JAN C-17 (Navy) designations that continue to this day. RG-6, for instance, is found under MIL-C-17G.

VELOCITY OF PROPAGATION

Velocity of propagation, abbreviated V_p, is the ratio of the speed of transmission through the cable versus the speed of light in free space, about 186,282 miles per second (mi/s) or 299,792,458 meters per second (m/s). For simplicity, this is usually rounded up to 300,000,000 meters per second (m/s). Velocity of propagation is a good indication of the quality of the cable. Solid polyethylene has a V_p of 66%. Chemically formed foam has a V_p of 78%, and nitrogen gas injected foam has a V_p up to 86%, with current manufacturing techniques. Some hardline, which is mostly dry air or nitrogen dielectric, can exceed 95% velocity.

Velocity of propagation is the velocity of the signal as it travels from one end of the line to the other end. It is caused because a transmission line, like all electrical circuits, possesses three inherent properties: resistance, inductance, and capacitance. All three of these properties will exist regardless of how the line is constructed. Lines cannot be constructed to eliminate these characteristics.

Under the foregoing conditions, the velocity of the electrical pulses applied to the line is slowed down in its transmission. The elements of the line are distributed evenly and are not localized or present in a lumped quantity.

The velocity of propagation (V_p) in flexible cables will vary from 50% to a V_p of 86%, depending on the insulating composition used and the frequency. V_p is directly related to the dielectric constant (DC) of the insulation chosen. The equation for determining the velocity of propagation is:

$$V_p = \frac{100}{\sqrt{DC}} \tag{6.5}$$

where,

V_p is the velocity of propagation,

DC is the dielectric constant.

Velocity can apply to any cable, coax or twisted pairs, although it is much more common to be expressed for cables intended for high-frequency applications. The velocity of propagation of coaxial cables is the ratio of the dielectric constant of a vacuum to the square root of the dielectric constant of the insulator, and is expressed in percent.

$$\frac{V_L}{V_S} = \frac{1}{\sqrt{\varepsilon}} \tag{6.6}$$

or

$$V_L = \frac{V_S}{\sqrt{\varepsilon}} \tag{6.7}$$

where,

V_L is the velocity of propagation in the transmission line,

V_S is the velocity of propagation in free space,

ε is the dielectric constant of the transmission line insulation.

Various dielectric constants (ε) are as follows:

Material	Dielectric Constant
Vacuum	1.00
Air	1.0167
Teflon	2.1
Polyethylene	2.25
Polypropylene	2.3
PVC	3.0 to 6.5

SHIELDING

From outdoor news gathering to studios and control rooms to sound reinforcement systems, the audio industry faces critical challenges from EM/RF interference (EMI and RFI). Shielding cable and twisting pairs insures signal integrity and provides confidence in audio and video transmissions, preventing downtime and maintaining sound and picture clarity.

Cables can be shielded or unshielded, except for coaxial cable which is, by definition, a precise constructions of a shielded single conductor. There are a number of shield constructions available. Here are the most common.

Serve or spiral shields

Serve or spiral shields are the simplest of all wire-based shields. The wire is simply wound around the inner portions of the cable. Spiral shields can be either single or double spirals. They are more flexible than braided shields and are easier to terminate. Since spiral shields are, in essence, coils of wire, they can exhibit inductive effects which make them ineffective at higher frequencies. Therefore, spiral/serve shields are relegated to low frequencies and are rarely used for frequencies above analog audio. Serve or spiral shields tend to open up when the cable is bent or flexed. So shield effectiveness is less than ideal, especially at high frequencies.

Double serve shields

Serve or spiral shields can be improved by adding a second layer. Most often, this is run at a 90° angle to the original spiral. This does improve coverage although the tendency to open up is not significantly improved and so this is still relegated to low-frequency or analog audio applications. This double serve or spiral construction is also called a *Reussen* shield (pronounced roy-sen).

French braid™

The French Braid shield by Belden is an ultraflexible double spiral shield consisting of two spirals of bare or tinned copper conductors tied together with one weave. The shield provides the long flex life of spiral shields and greater flexibility than braided shields. It also has about 50% less microphonic and triboelectric noise. Because the two layers are woven along one axis, they cannot open up as dual spiral/serve constructions can. So French Braid shields are effective up to high frequencies, and are used up to the Gigahertz range of frequencies.

Braid

Braid shields provide superior structural integrity while maintaining good flexibility and flex life. These shields are ideal for minimizing low-frequency interference and have lower DC resistance than foil. Braid shields are effective at low frequencies, as well as RF ranges. Generally, the higher the braid coverage, the more effective the shield. The maximum coverage of a single braid shield is approximately 95%. The coverage of a dual braid shield can be as much as 98%. One hundred percent coverage with a braid is not physically possible.

Foil

Foil shields can be made of bare metal, such as a bare copper shield layer, but more common is an aluminum-polyester foil. Foil shields can offer 100% coverage. Some cables feature a loose polyester-foil layer. Other designs can bond the foil to either the core of the cable or to the inside of the jacket of the cable. Each of these presents challenges and opportunities.

The foil layer can either face out, or it can be reversed and face in. Since foil shields are too thin to be used as a connection point, a bare wire runs on the foil side of the shield. If the foil faces out, the drain wire must also be on the outside of the foil. If the foil layer faces in, then the drain wire must also be inside the foil, adjacent to the pair.

Unbonded foil can be easily removed after cutting or stripping. Many broadcasters prefer unbonded foil layers in coaxial cable to help prevent thin slices of foil that can short out BNC connectors. If the foil is bonded to the core, the stripping process must be much more accurate to prevent creating a thin slice of core-and-foil.

However, with F connectors, which are pushed onto the end of the coax, unbonded foil can bunch up and prevent correct seating of these connectors. This explains why virtually all coaxes for broadband/CATV applications have the foil bonded to the core—so F connectors easily slip on.

In shielded paired cables, such as analog or digital audio paired cables, the foil shield wraps around the pair. Once the jacket has been stripped off, the next step is to remove the foil shield. These cables are also available where the foil is bonded (glued) to the inside of the jacket. When the jacket is removed, the foil is also removed, dramatically speeding up the process.

A shorting fold technique is often used to maintain metal-to-metal contact for improved high-frequency performance. Without the shorting fold, a slot is created through which signals can leak. An isolation fold also helps prevent the shield of one pair contacting the shield of an adjacent pair in a multipair construction. Such contact significantly increases crosstalk between these pairs.

An improvement on the traditional shorting fold used by Belden employs the Z-Fold™, designed for use in multipair applications to reduce crosstalk, Figure 6.14. The Z-Fold combines an isolation fold and a shorting fold. The shorting fold provides metal-to-metal contact while the isolation fold keeps shields from shorting to one another in multipair, individually shielded cables.

Since the wavelength of high frequencies can eventually work through the holes in a braid, foil shields are most effective at those high frequencies. Essentially, foil shields represent a skin shield at high frequencies, where skin effect predominates.

Combination shields

Combination shields consist of more than one layer of shielding. They provide maximum shield efficiency across the frequency spectrum. The combination foil-braid shield combines the advantages of 100% foil coverage and the strength and low DC resistance of a braid. Other combination shields available include various foil-braid-foil, braid-braid, and foil-spiral designs.

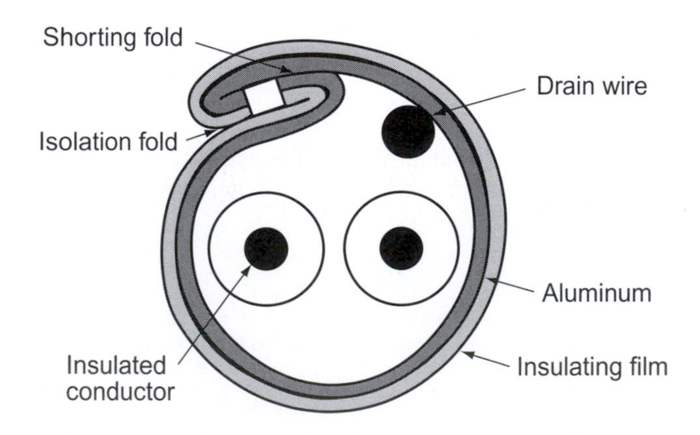

FIGURE 6.14
Z-Fold foil-type shielded wire improves high-frequency performance. (Courtesy Belden.)

FOIL + SERVE

Because of the inductive effects of serve/spiral shields, which relegate them to low-frequency applications, this combination is rarely seen.

FOIL + BRAID

This is the most common combination shield. With a high-coverage braid (95%) this can be extremely effective over a wide range of frequencies, from 1 kHz to many GHz. This style is commonly seen on many cables, including precision video cable.

FOIL + BRAID + FOIL

Foil-braid-foil is often called a *tri-shield*. It is most commonly seen in cable television (CATV) broadband coaxial applications. The dual layers of foil are especially effective at high frequencies. However, the coverage of the braid shield in between is the key to shield effectiveness. If it is a reasonably high coverage (>80%) this style of braid will have excellent shield effectiveness.

One other advantage of tri-shield coax cable is the ability to use standard dimension F connectors since the shield is essentially the same thickness as the common foil + braid shield of less expensive cables.

FOIL + BRAID + FOIL + BRAID

Foil-braid-foil-braid is often called *quad-shield* or just *quad* (not to be confused with starquad microphone cable or old POTS quad hookup cable). Like tri-shield above, this is most common in cable television (CATV) broadband coaxial applications. Many believe this to be the ultimate in shield effectives. However, this is often untrue.

If the two braids in this construction are high coverage braids (>80%) then, yes, this would be an exceptional cable. But most quad-shield cable uses two braids that are 40% and 60% coverage, respectively. With that construction, the tri-shield with an 80%+ braid is measurably superior. Further, quad-shield coaxial cables are considerably bigger in diameter and therefore require special connectors.

Table 6.27 shows the shield effectiveness of different shield constructions at various frequencies. Note that all the braids measured are aluminum braids except for the last cable mentioned. That last cable is a digital precision video (such as Belden 1694A) and is many times the cost of any of the other cables listed.

SHIELD CURRENT-INDUCED NOISE

There is significant evidence that constructions that feature bonded foil with an internal drain wire may affect the performance of the pairs, especially at high frequencies. Since an ideal balanced line is one where the two wires are electrically identical, having a drain wire in proximity would certainly seem to affect the symmetry of the pair. This would be especially critical where strong RF fields are around audio cables.

Despite this evidence, there are very few cables made with appropriate symmetry. This may be based on lack of end-user demand, as manufacturers would be glad to redesign their cables should the demand arise. The drain wire could be easily substituted with a symmetrical low-coverage braid, for instance.

Table 6.27 Shield Effectiveness of Different Shield Constructions

Shield Type (Aluminum Braid)	5 MHz	10 MHz	50 MHz	100 MHz	500 MHz
60% braid, bonded foil	20	15	11	20	50
60% braid, tri-shield	3	2	0.8	2	12
60%/40% quad shield	2	0.8	0.2	0.3	10
77% braid, tri-shield	1	0.6	0.1	0.2	2
95% copper braid, foil	1	0.5	0.08	0.09	1

GROUNDS OF SHIELDS

With any combination shield, the braid portion is the part that is making the connection. Even if we are shielding against high-frequency noise, in which case the foil is doing the actual work, the noise gets to ground by way of the braid which is much lower in resistance than the foil.

Where the foil uses a drain wire, it is that drain wire that is the shield connection. Therefore, that drain wire must be bare so it can make contact with the foil. If the foil is floating, not glued or bonded to the core of the cable, then another plastic layer is used to carry the foil. The foil itself is much too thin and weak to even be applied in the factory by itself. The second plastic layer adds enough strength and flex-life (flexes until failure) to allow the foil to be used.

The drain wire, therefore, must be in contact with the foil. In some cables, the foil faces out, so the drain wire must be on the outside of the foil, between the foil and the jacket. If the foil faces in, then the drain wire must be on the inside of the foil, adjacent to the pair (or other components) inside the cable.

With an internal drain wire, there are a number of additional considerations. One is SCIN, shield current-induced noise, mentioned earlier. Another is the ability to make a multipair shielded cable where the shields are facing in and the plastic facing out. This allows the manufacturer to color code the pairs by coloring the plastic holding the foil.

If you have a multipair cable, with individual foil shields, it is important that these foil shields do not touch. If the shields touch, then any signal or noise that is on one foil will be instantly shared by the other. You might as well put a foil shield in common around both pairs. Therefore, it is common to use foil shields facing in which will help prevent them from touching. These can then be color coded by using various colors of plastic with each foil to help identify each pair.

However, simply coiling the foil around the pair still leaves the very edge of the foil exposed. In a multipair cable with many individual foils, where the cable is bent and flexed to be installed, it would be quite easy for the edge of one foil to touch the edge of another foil, thus compromising shield effectiveness. The solution for this is a Z-fold invented by Belden in 1960, shown in Figure 6.14. This does not allow any foil edge to be exposed no matter how the cable is flexed.

Ground loops

In many installations, the ground potential between one rack and another, or between one point in a building and another, may be different. If the building can be installed with a star ground, the ground potential will be identical throughout the building. Then the connection of any two points will have no potential difference.

When two points are connected that do have a potential difference, this causes a ground loop. A ground loop is the flow of electricity down a ground wire from one point to another. Any RF or other interference on a rack or on an equipment chassis connected to ground will now flow down this ground wire, turning that foil or braid shield into an antenna and feeding that noise into the twisted pair. Instead of a small area of interference, such as where wires cross each other, a ground loop can use the entire length of the run to introduce noise.

If one cannot afford the time or cost of a star ground system, there are still two options. The first option is to cut the ground at one end of the cable. This is called a *telescopic ground*.

TELESCOPIC GROUNDS

Where a cable has a ground point at each end, disconnecting one end produces a telescopic ground. Installers should be cautioned to disconnect only the destination (load) end of the cable, leaving the source end connected.

For audio applications, the effect of telescopic grounds will eliminate a ground loop, but at a 50% reduction in shield effectiveness (one wire now connected instead of two). If one disconnects the source end, which in analog audio is the low-impedance end, and maintains the destination (load) connection, this will produce a very effective R-L-C filter at audio frequencies.

At higher frequencies, such as data cables, even a source-only telescopic shield can have some serious problems. Figure 6.15 shows the effect of a telescopic ground on a Cat 6 data cable. The left column

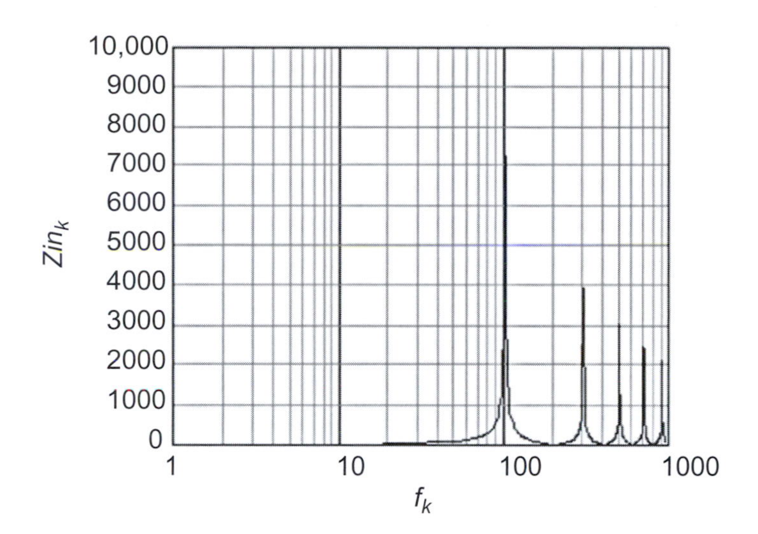

FIGURE 6.15
Effect of a telescopic ground on a Cat 6 cable.

shows the input impedance, the impedance presented to any RF traveling on the shield, at frequency F_k (bottom scale) in MHz.

You will note that at every half-wavelength, the shield acts like an open circuit. Since most audio cables are foil shielded, and the foil is effective only at high frequencies, this means that even a correctly terminated telescopic shield is less effective at RF frequencies.

UTP AND AUDIO

One other solution for ground loops is to have no ground connection. For the seasoned audio and video professional, this solution may require a leap of faith. In can clearly be seen that, with a cable that has no shield, no drain wire, and no ground wire, no ground loop can develop. This is a common form of data cable called *UTP*, unshielded twisted pairs.

With such a cable, having no shield means that you are totally dependent on the balanced line to reject noise. This is especially true, where you wish to use the four pairs in a Cat 5e, 6, or 6a cable to run four unrelated audio channels. Tests were performed on low-performance (stranded) Cat 5e patch cable (Belden 1752A) looking at crosstalk between the pairs. This test shows the average of all possible pair combinations, the worst possible case, and covered a bandwidth of 1 kHz to 50 kHz. The results are shown in Figure 6.16.

You will note that the worst case is around 40 kHz where the crosstalk is slightly better than −95 dB. In the range of common audible frequencies (20 kHz) the pair-to-pair crosstalk approaches −100 dB. Since a noise floor of −90 dB is today considered wholly acceptable, a measurement of −95 dB or −100 dB is even better still.

A number of data engineers questioned these numbers based on the fact that these measurements were FEXT, far-end crosstalk, where the signals are weakest in such a cable. So measurements were also taken of NEXT, near-end crosstalk, where the signals are strongest. Those measurements are shown in Figure 6.17.

The NEXT measurements are even better than the previous FEXT measurements. In this case, the worst case is exactly −95 dB at just under 50 kHz.At 20 kHz and below, the numbers are even better than the previous graph, around −100 dB or better.

There were attempts made to test a much better cable (Belden 1872A MediaTwist). This unshielded twisted-pair cable is now a Cat 6 bonded-pair design. After weeks of effort, it was determined that the pair-to-pair crosstalk could not be read on an Agilent 8714ES network analyzer. The crosstalk was somewhere below the noise floor of the test gear. The noise floor of that instrument is −110 dB. With a good cable, the crosstalk is somewhere below −110 dB.

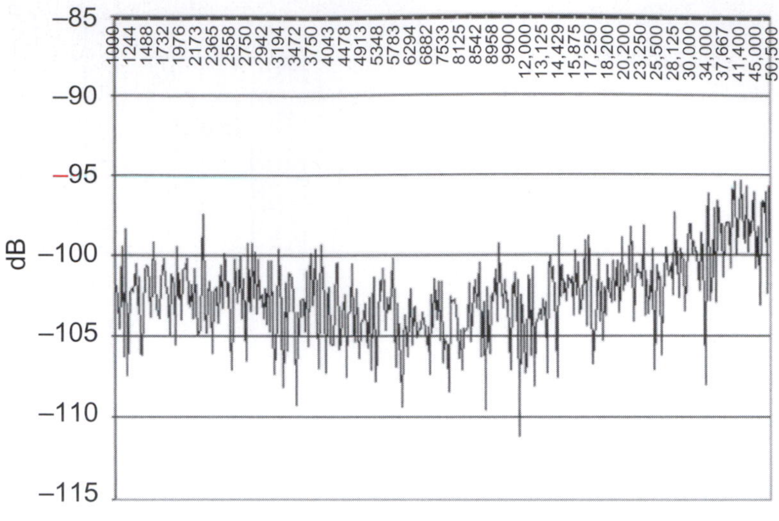

FIGURE 6.16
Crosstalk between Cat 5e patch cable.

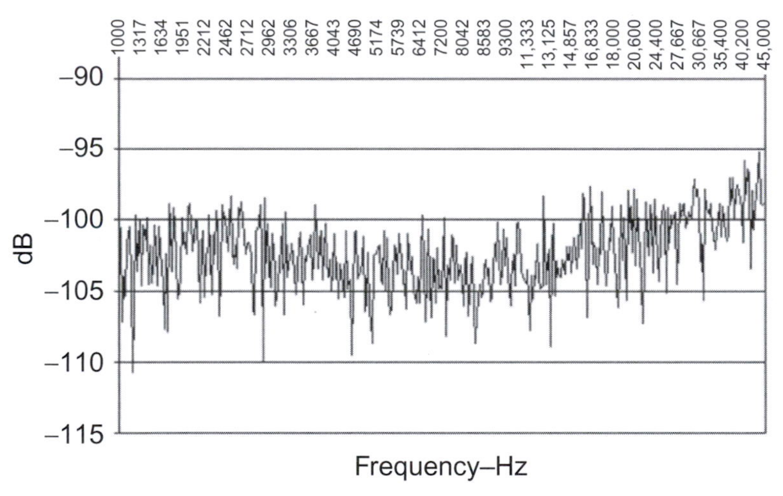

FIGURE 6.17
NEXT crosstalk.

So why shields?

These experiments with unshielded cable beg the question, why have a shield? In fact, the answer is somewhat reversed. The pairs in data cables are dramatically improved over the historic audio pairs. The bandwidth alone, 500 MHz for Cat 6a, for instance, indicates that these are not the same old pairs but something different. In fact, what has happened is that the wire and cable (and data) industries have fixed the pairs.

Before, with a poorly manufactured pair, a shield would help prevent signals from getting into, or leaking out of, a pair. The fact that either effect, ingress or egress, occurred indicated the poor balance, the poor performance of the pair.

This does not mean shields are dead. There are data cables with overall shields (FTP), even individually shielded pairs (Cat 7) common in Europe. However, these are subject to the same problems as all shielded, grounded cables in terms of ground loops and wavelength effects.

The truth to the efficacy of unshielded twisted pairs running audio, video, data and many other signals is commonplace today. Many audio devices routinely use UTP for analog and digital connections.

Table 6.28 Sampling Rate versus Bandwidth

Sampling Rate kHz	Bandwidth MHz	Sampling Rate kHz	Bandwidth MHz
32.0	4.096	48.0	6.144
38.0	4.864	96.0	12.228
44.1	5.6448	192.0	24.576

Where the source is not a balanced line, a device must change from balanced (UTP) to unbalanced (coax, for instance). Such a device matches *Balanced*-to-*Unbalanced* and is therefore called a *balun*. There is more on baluns later in this chapter.

AES/EBU DIGITAL AUDIO CABLE

Digital audio technology has been around for many years, even decades, but until recently it has not been used much for audio. This has now changed and digital audio is overtaking analog audio. For this reason it is important that the cable used for digital signals meet the digital requirements. To set a standard, the Audio Engineering Society (AES) and the European Broadcast Union (EBU) have set standards for digital audio cable. The most common sampling rates and equivalent bandwidth are shows in Table 6.28.

It is important that the line impedance be maintained to eliminate reflections that degrade the signal beyond recovery. Standard analog cable can be used for runs under 50 ft (15 m) but beyond that reliability decreases. The impedance and capacitance of analog cable is 40 to 70 Ω and 20 to 50 pF/ft. The impedance and capacitance for digital cable is 110 Ω and 13 pF/ft with a velocity of propagation of 78%. Proper impedance match and low capacitance are required so the square wave signal is not distorted, reflected, or attenuated.

Broadcast cable is most often #24 (7 × 32) tinned copper wire with short overall twist lengths, low-loss foam insulation, and 100% aluminum polyester foil shield for permanent installations. Braided shields are also available for portable use. If required, #22 to #26 wire can be obtained. Digital audio cable also comes in multiple pairs with each pair individually shielded, and often jacketed, allowing each pair and its shield to be completely isolated from the others. One pair is capable of carrying two channels of digital audio. Cables are terminated with either XLR connectors or are punched down or soldered in patch panels.

AES/EBU digital coaxial cable

Digital audio requires a much wider bandwidth than analog. As the sampling rate doubles, the bandwidth also doubles, as shown in Table 6.28.

Digital audio can be transmitted farther distances over coax than over twisted pairs. The coax should have a 75 Ω impedance, a solid copper center conductor, and have at least 90% shield coverage. When transmitting audio over an unbalanced coax line, the use of baluns may be required to change from balanced to unbalanced and back unless the device contains AES/EBU unbalanced coax inputs and outputs. The baluns change the impedance from 110 Ω balanced to 75 Ω unbalanced and back.

TRIBOELECTRIC NOISE

Noise comes in a variety of types such as EMI (electromagnetic interference) and RFI (radio-frequency interference). There are also other kinds of noise problems that concern cables. These are mechanically generated or mechanically-induced noise, commonly called *triboelectric noise.*

Triboelectric noise is generated by mechanical motion of a cable causing the wires inside the shield to rub against each other. Triboelectric noise is actually small electrical discharges created when conductors' position changes relative to each other. This movement sets up tiny capacitive changes that eventually pop. Highly amplified audio can pick this up.

Where signals are split up and recombined, the different cables supplying the components will each have a measurable delay. The trick is for all the component cables to have the *same delay* to deliver their portions at the same time. The de facto maximum timing variation in delay for RGB analog is delivery of all components within 40 ns. Measuring and adjusting cable delivery is often called *timing*. By coincidence, the maximum delay difference in the data world is 45 ns, amazingly close. In the data world, this is called *skew* or *delay skew*, where delivery does not line up.

In the RGB world, where separate coax cables are used, they have to be cut to the same electrical length. This is not necessarily the same physical length. Most often, the individual cables are compared by a Vectorscope, which can show the relationship between components, or a TDR (time domain reflectometer) that can establish the electrical length (delay) of any cable.

Any difference in physical versus electrical length can be accounted for by the velocity of propagation of the individual coaxes, and, therefore, the consistency of manufacture. If the manufacturing consistency is excellent, then the velocity of all coaxes would be the same, and the physical length would be the same as the electrical length. Where cables are purchased with different color jackets, to easily identify the components, they are obviously made at different times in the factory. It is then a real test of quality and consistency to see how close the electrical length matches the physical length.

Where cables are bundled together, the installer then has a much more difficult time in reducing any timing errors. Certainly in UTP data cables, there is no way to adjust the length of any particular pair. In all these bundled cables, the installer must cut and connectorize.

This becomes a consideration when four-pair UTP data cables (category cables) are used to deliver RGB, VGA, and other nondata component delivery systems. The distance possible on these cables is therefore based on the attenuation of the cables at the frequency of operation, and on the delay skew of the pairs. Therefore, the manufacturers measurement and guarantee (if any) of delay skew should be sought if nondata component delivery is the intended application.

ATTENUATION

All cable has attenuation and the attenuation varies with frequency. Attenuation can be found with the equation:

$$A = 4.35\frac{R_t}{Z_o} + 2.78pf\sqrt{\varepsilon} \tag{6.10}$$

where,

A is the attenuation in dB/100 ft,

R_t is the total DC line resistance in Ω/100 ft,

ε is the dielectric constant of the transmission line insulation,

p is the power factor of the dielectric medium,

f is the frequency,

Z_o is the impedance of the cable.

Table 6.29 gives the attenuation for various 50 Ω, 52 Ω, and 75 Ω cables. The difference in attenuation is due to either the dielectric of the cable or center conductor diameter.

CHARACTERISTIC IMPEDANCE

The *characteristic impedance* of a cable is the measured impedance of a cable of infinite length. This impedance is an AC measurement, and cannot be measured with an ohmmeter. It is frequency-dependent, as can be seen in Figure 6.19. This shows the impedance of a coaxial cable from 10 Hz to 100 MHz.

At low frequencies, where resistance is a major factor, the impedance is changing from a high value (approximately 4000 Ω at 10 Hz) down to a lower impedance. This is due to skin effect, where the signal

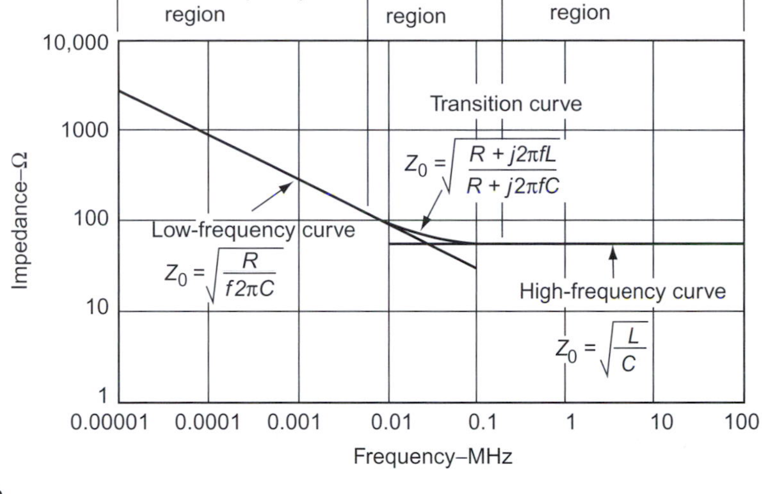

FIGURE 6.19
Impedance of coaxial cable from 10 Hz to 100 MHz.

is moving from the whole conductor at low frequencies to just the skin at high frequencies. Therefore, when only the skin is carrying the signal, the resistance of the conductor is of no importance. This can be clearly seen in the equations for impedance; Eq. 6.13, for low frequencies, shows *R*, the resistance, as a major component. For high frequencies, Eq. 6.14, there is no *R*, no resistance, even in the equation.

Once we enter that high-frequency area where resistance has no effect, around 100 kHz as shown in Figure 6.19, we enter the area where the impedance will not change. This area is called the *characteristic impedance* of the cable.

The characteristic impedance of an infinitely long cable does not change if the far end is open or shorted. Of course, it would be impossible to test this as it is impossible to short something at infinity. It is important to terminate coaxial cable with its rated impedance or a portion of the signal can reflect back to the input, reducing the efficiency of the transmission. Reflections can be caused by an improper load, using a wrong connector—i.e., using a 50 Ω video BNC connector at high frequencies rather than a 75 Ω connector—a flattened cable, or too tight a bend radius, which changes the spacing between the conductors. Anything that affects the dimensions of the cable will affect the impedance and create reflective losses. It would just be a question of how much reflection is caused. Reflections thus caused are termed *return loss.*

The characteristic impedance of common coaxial cable can be between 30 Ω and 200 Ω. The most common values are 50 Ω and 75 Ω. The characteristic Z_o is the average impedance of the cable equal to:

$$Z_o = \frac{138}{\sqrt{\varepsilon}}\log\frac{D}{d} \tag{6.11}$$

where,

ε is the dielectric constant,

D is the diameter of the inner surface of the outer coaxial conductor (shield) in inches,

d is the diameter of the center conductor in inches.

The true characteristic impedance, at any frequency, of a coaxial cable is found with the equation:

$$Z_o = \sqrt{\frac{R + j2\pi fL}{G + j2\rho C}} \tag{6.12}$$

where,

R is the series resistance of the conductor in ohms per unit length,

f is the frequency in hertz,

L is the inductance in henrys,

G is the shunt conductance in mhos per unit length,

C is the capacitance in farads.

At low frequencies, generally below 100 kHz, the equation for coaxial cable simplifies to:

$$Z_o = \sqrt{\frac{R}{j2\rho\pi C}} \tag{6.13}$$

At high frequencies, generally above 100 kHz, the equation for coaxial cable simplifies to:

$$Z_o = \sqrt{\frac{L}{C}} \tag{6.14}$$

CHARACTERISTIC IMPEDANCE

The *characteristic impedance* of a transmission line is equal to the impedance that must be used to terminate the line in order to make the input impedance equal to the terminating impedance. For a line that is longer than a quarter-wavelength at the frequency of operation, the input impedance will equal the characteristic impedance of the line, irrespective of the terminating impedance.

This means that low-frequency applications often have quarter-wavelength distance way beyond common practical applications. Table 6.30 shows common signals, with the wavelength of that signal and the quarter-wavelength. To be accurate, given a specific cable type, these numbers would be multiplied by the velocity of propagation.

The question is very simple: Will you be going as far as the quarter-wavelength, or farther? If so, then the characteristic impedance becomes important. As that distance gets shorter and shorter, this distance

Table 6.30 Characteristics of Various Signals

Signal Type	Bandwidth	Wavelength	Quarter-Wavelength	Quarter-Wavelength
Analog audio	20 kHz	15 km	3.75 km	12,300 ft
AES 3—44.1 kHz	5.6448 MHz	53.15 m	13.29 m	44 ft
AES 3—48 kHz	6.144 MHz	48.83 m	12.21 m	40 ft
AES 3—96 kHz	12.288 MHz	24.41 m	6.1 m	20 ft
AES 3—192 kHz	24.576 MHz	12.21 m	3.05 m	10 ft
Analog video (U.S.)	4.2 MHz	71.43 m	17.86 m	59 ft
Analog video (PAL)	5 MHz	60 m	15 m	49.2 ft
SD-SDI	135 MHz clock	2.22 m	55.5 cm	1 ft 10 in
SD-SDI	405 MHz third harmonic	74 cm	18.5 cm	7.28 in
HD-SDI	750 MHz clock	40 cm	10 cm	4 in
HD-SDI	2.25 GHz third harmonic	13 cm	3.25 cm	1.28 in
1080P/50-60	1.5 GHz clock	20 cm	5 cm	1.64 in
1080P/50-60	4.5 GHz third harmonic	66 mm	16.5 mm	0.65 in

becomes critical. With smaller distances, patch cords, patch panels, and eventually the connectors themselves become just as critical as the cable. The impedance of these parts, especially when measured over the desired bandwidth, becomes a serious question. To be truly accurate, the quarter-wavelength numbers in Table 6.28 need to be multiplied by the velocity of propagation of each cable. So, in fact, the distances would be even shorter than what is shown.

It is quite possible that a cable can work fine with lower-bandwidth applications and fail when used for higher-frequency applications. The characteristic impedance will also depend on the parameters of the pair or coax cable at the applied frequency. The resistive component of the characteristic impedance is generally high at the low frequencies as compared to the reactive component, falling off with an increase of frequency, as shown in Figure 6.19. The reactive component is high at the low frequencies and falls off as the frequency is increased.

The impedance of a uniform line is the impedance obtained for a long line (of infinite length). It is apparent, for a long line, that the current in the line is little affected by the value of the terminating impedance at the far end of the line. If the line has an attenuation of 20 dB and the far end is shortcircuited, the characteristic impedance as measured at the sending end will not be affected by more than 2%.

TWISTED-PAIR IMPEDANCE

For shielded and unshielded twisted pairs, the characteristic impedance is:

$$Z_o = \frac{101670}{C(V_P)} \tag{6.15}$$

where,

Z_o is the average impedance of the line,

C is found with Eqs. 6.16 and 6.17,

V_p is the velocity of propagation.

For unshielded pairs:

$$C = \frac{3.68\varepsilon}{\log\left[\frac{2(ODi)}{DC(Fs)}\right]}. \tag{6.16}$$

For shielded pairs:

$$C = \frac{3.68\varepsilon}{\log\left[\frac{1.06(ODi)}{DC(Fs)}\right]} \tag{6.17}$$

where,

ε is the dielectric constant,

ODi is the outside diameter of the insulation,

DC is the conductor diameter,

Fs is the conductor stranding factor (solid = 1, 7 strand = 0.939, 19 strand = 0.97.

The impedance for higher-frequency twisted-pair data cables is:

$$Z_o = 276\left(\frac{VP}{100}\right) \times \log\left[2\left(\frac{h}{DC \times Fs}\right) \times \left(\frac{1 - \frac{h}{DC + Fb}^2}{1 + \frac{h}{DC + Fb)}^2}\right)\right] \tag{6.18}$$

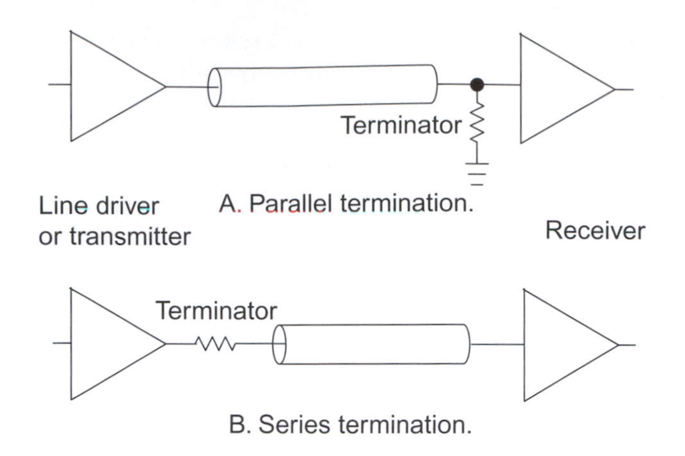

FIGURE 6.20
Basic termination of transmission lines.

where,

h is the center to center conductor spacing,

Fb is very near 0. Neglecting *Fb* will not introduce appreciable error.

Transmission line termination

All lines do not need to be terminated. Knowing when to terminate a transmission line is a function of the frequency/wavelength of the signal and the length of the transmission line. Table 6.30 can be a guideline, especially where the signal is long compared with the length of the line. If the wavelength of the signal is small compared with the transmission-line length, for instance a 4.5 GHz signal, a terminator is required to prevent the signal from reflecting back toward the source and interfering with forward traveling signals. In this case the line must be terminated for any line longer than a quarter of a wavelength.

Transmission-line termination is accomplished using parallel or series termination. Parallel termination connects a resistor between the transmission line and ground at the receiving end of the transmission line while series termination connects a resistor in series with the signal path near the beginning of the transmission line, Figure 6.20.

Resistive termination requires a resistor value that matches the characteristic impedance of the transmission line, most commonly a 50 Ω or 75 Ω characteristic impedance. The termination resistance is matched to the transmission-line characteristic impedance so the electrical energy in the signal does not reflect back from the receiving end of the line to the source. If the resistor is perfectly matched to the characteristic impedance, at all frequencies within the desired bandwidth, all of the energy in the signal dissipates as heat in the termination resistor so no signal reflects backwards down the line to the source causing cancellations.

LOUDSPEAKER CABLE

Much has been said about wire for connecting loudspeakers to amplifiers. Impedance, inductance, capacitance, resistance, loading, matching, surface effects, etc. are constantly discussed.

Most home and studio loudspeaker runs are short (less than 50 ft, or 15 m) and therefore do not constitute a transmission line. When runs are longer, it is common to connect the loudspeakers as a 70 V, or 100 V, system to reduce line loss caused by the wire resistance so the power lost in the line does not appreciably affect the power delivered to the loudspeaker. For instance, if a 4 Ω loudspeaker is connected to an amplifier with a cable which measures 4 Ω resistance, 50% of the power will be dissipated in the cable. If the loudspeaker was connected to a 70 V system, and the loudspeaker was taking 50 W from the amplifier, the loudspeaker/transformer impedance would be 100 Ω; therefore, the 4 Ω line resistance would dissipate 4% of the power.

When using a 70.7 V loudspeaker system, the choice of wire size for loudspeaker lines is determined by an economic balance of the cost of copper against the cost of power lost in the line. Power taken from the amplifier is calculated from the equation:

$$P = \frac{V^2}{Z} \tag{6.19}$$

where,

P is the power delivered by the amplifier,

V is the voltage delivered by the amplifier,

Z is the impedance of the load.

For a 70 V system:

$$P = \frac{5000}{Z} \tag{6.20}$$

If the voltage is 70.7 V and the load is 50 Ω, the power would be 100 W. However, if the amplifier was connected to a 50 Ω load with 1000 ft of #16 wire (2000 ft round trip), or 8 Ω of wire resistance, the power from the amplifier would be:

$$\begin{aligned} P &= \frac{5000}{50\,\Omega + 8\,\Omega} \\ &= 86.2\,\text{W} \end{aligned}$$

The current through the system is found with:

$$I = \sqrt{\frac{P}{R}} \tag{6.21}$$

or in this case

$$\begin{aligned} I &= \sqrt{\frac{P}{R}} \\ &= \sqrt{\frac{86.2}{58}} \\ &= 1.21\text{ A} \end{aligned}$$

The power to the 50 Ω load would be found with:

$$P = I^2R \tag{6.22}$$

or in this case

$$\begin{aligned} P &= I^2R \\ &= 1.21^2R \\ &= 74.3\text{ W} \end{aligned}$$

or 26% less power than assumed.

Only 11.7 W are lost to the line, the other 14 W cannot be taken from the amplifier because of the impedance mismatch. While high-power amplifiers are relatively inexpensive, it is still practical to use heavy enough wire so the amplifier can output almost its full power to the loudspeakers. Table 6.31 shows the characteristics of various cables which could be used for loudspeaker wire, and Table 6.32 is a cable selection guide for loudspeaker cable.

Table 6.31 Frequency Limitations for 33 ft (10 m) Lengths of Cable with Various Loads

Cable Type	Upper Corner Frequency, kHz		Resonant Measurement Phase (°)	
	2 Ω Load	4 Ω Load	4 μF Load	4 Ω Load
No. 18 zip cord	75	136	35	3
No. 16 zip cord	61	114	32	2
No. 14 speaker cable	82	156	38	2
No. 12 speaker cable	88	169	40	2
No. 12 zip cord	55	106	32	4
Welding cable	100	200	44	2
Braided cable	360	680	80	1
Coaxial dual cylindrical	670	1300	112	
Coaxial RG-8	450	880	92	

Table 6.32 Loudspeaker Cable Selection Guide (Courtesy Belden.)

Power (%) Loss (dB) Wire Size	11%	21%	50%
	0.5	1.0	3.0
	Maximum Cable Length–ft		
	4 Ω Loudspeaker		
12 AWG	140	305	1150
14 AWG	90	195	740
16 AWG	60	125	470
18 AWG	40	90	340
20 AWG	25	50	195
22 AWG	15	35	135
24 AWG	10	25	85
	8 Ω Loudspeaker		
12 AWG	285	610	2285
14 AWG	185	395	1480
16 AWG	115	250	935
18 AWG	85	190	685
20 AWG	50	105	390
22 AWG	35	70	275
24 AWG	20	45	170
	70 V Loudspeaker		
12 AWG	6920	14,890	56,000
14 AWG	4490	9650	36,300
16 AWG	2840	6100	22,950
18 AWG	2070	4450	16,720
20 AWG	1170	2520	9500
22 AWG	820	1770	6650

The following explains how to use Table 6.32:

1. Select the appropriate speaker impedance column.
2. Select the appropriate power loss column deemed to be acceptable.
3. Select the applicable wire gage size and follow the row over to the columns determined in steps one and two. The number listed is the maximum cable run length.
4. The maximum run for 12 AWG in a 4 Ω loudspeaker system with 11% or 0.5 dB loss is 140 ft.

Damping factor

The damping factor of an amplifier is the ratio of the load impedance (loudspeaker plus wire resistance) to the amplifier internal output impedance. The damping factor of the amplifier acts as a short circuit to the loudspeaker, controlling the overshoot of the loudspeaker. Present day amplifiers have an output impedance of less than 0.05 Ω which translates to a damping factor over 150 at 10 kHz, for instance, so they effectively dampen the loudspeaker as long as the loudspeaker is connected directly to the amplifier. Damping factor is an important consideration when installing home systems, studios, or any system where high-quality sound, especially at the low frequencies, is desired. As soon as wire resistance is added to the circuit, the damping factor reduces dramatically, reducing its effect on the loudspeaker. For instance, if a #16 AWG 50 ft loudspeaker cable (100 ft round trip) is used, the wire resistance would be 0.4 Ω, making the damping factor only 18, considerably less than anticipated.

It is not too important to worry about the effect the damping factor of the amplifier has on the loudspeakers in a 70 V system as the 70 V loudspeaker transformers wipe out the effects of the wire resistance.

Consider the line as a lump sum, Figure 6.21. The impedance of the line varies with wire size and type. Table 6.33 gives typical values of *R*, *C*, and *L* for 33 ft (10 m) long cables. Note, the impedance at 20 kHz is low for all but the smallest wire and the −3 dB upper frequency is well above the audio range. The worst condition is with a capacitive load. For instance, with a 4 μF load, resonance occurs around 35 kHz.

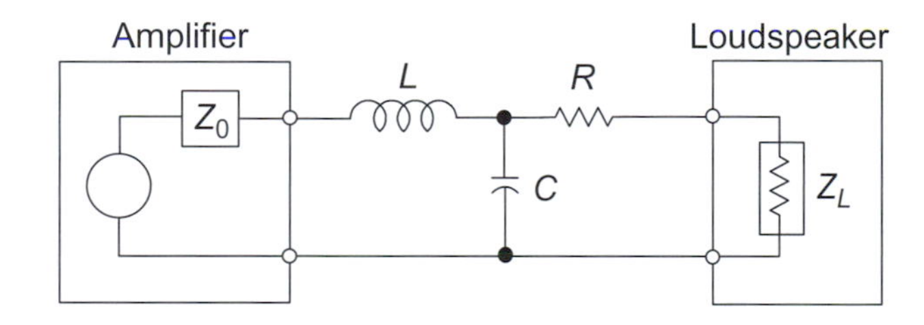

FIGURE 6.21
Amplifier, cable, loudspeaker circuit using lumped circuit elements to represent the properties of the cable.

Table 6.33 Lumped Element Values for 33 ft (10 m) Lengths of Cable

Cable Type	*L*–μH	C–pF	R_{dc}–Ω	Z–Ω@20 kHz
No. 18 zip cord	5.2	580	0.42	0.44
No. 16 zip cord	6.0	510	0.26	0.30
No. 14 speaker cable	4.3	570	0.16	0.21
No. 12 speaker cable	3.9	760	0.10	0.15
No. 12 zip cord	6.2	490	0.10	0.15
Welding cable	3.2	880	0.01	0.04
Braided cable	1.0	16,300	0.26	0.26
Coaxial dual cylindrical	0.5	58,000	0.10	0.10
Coaxial RG-8	0.8	300	0.13	0.13

The results of the above are as follows:

1. Make the amplifier to loudspeaker runs as short as possible.
2. Use a wire gage that represents less than 5% of the loudspeaker impedance at any frequency.
3. Use twisted pairs on balanced 70 or 100V distributed systems to reduce crosstalk (amplifier output is often fed back into the amplifier as negative feedback).
4. Use good connectors to reduce resistance.

Table 6.34 gives the length of cable run you can have for various loudspeaker impedances.

Crosstalk

When a plurality of lines, carrying different programs or signals, are run together in the same conduit, or where multiple pairs or multiple coax cables are bundled, they tend to induce crosstalk currents into each other. Crosstalk is induced by two methods:

1. Electromagnetically through unbalanced coupling between one circuit and others.
2. Electrostatically through unbalanced capacitance to other circuits, or to the conduit if it carries current. This develops a voltage difference between one circuit and the others, or to its own or other shields carrying current.

Table 6.34 Loudspeaker Cable Selection Guide

Power (%)	11%	21%	50%
Loss (dB)	0.5	1.0	3.0
Wire Size		Maximum Cable Length-ft	
		4 Ω Loudspeaker	
12 AWG	140	305	1150
14 AWG	90	195	740
16 AWG	60	125	470
18 AWG	40	90	340
20 AWG	25	50	195
22 AWG	15	35	135
24 AWG	10	25	85
		8 Ω Loudspeaker	
12 AWG	285	610	2285
14 AWG	185	395	1480
16 AWG	115	250	935
18 AWG 85 190 685			
20 AWG	50	105	390
22 AWG	35	70	275
24 AWG	20	45	170
		70V Loudspeaker	
12 AWG	6920	14,890	56,000
14 AWG	4490	9650	36,300
16 AWG	2840	6100	22,950
18 AWG	2070	4450	16,720
20 AWG	1170	2520	9500
22 AWG	820	1770	6650

(Courtesy Belden.)

If the line is less than a quarter-wavelength at the frequency of operation, then the cable does not have to have a specific impedance, or be terminated in a specific impedance. The terminating impedance could then be small compared to the open line characteristic impedance. The net coupling with unshielded pairs would then be predominantly magnetic. If the terminating impedance is much larger than the characteristic impedance of the wires, the net coupling will be predominantly electric.

Two wires of a pair must be twisted; this insures close spacing and aids in canceling pickup by transposition. In the measurements in Figure 6.22, all pickup was capacitive because the twisting of the leads effectively eliminated inductive coupling.

One application that is often ignored regarding crosstalk is speaker wiring, especially 70V distributed loudspeaker wiring. You will note in the first drawing that the two wires are not a balanced line. One is hot, the other is ground. Therefore, that pair would radiate some of the audio into the adjoining pair, also unbalanced. Twisting the pairs in this application would do little to reduce crosstalk.

The test was made on a 250ft twisted pair run in the same conduit with a similar twisted pair, the latter carrying signals at 70.7V. Measurements made for half this length produced half the voltages, therefore the results at 500ft and 1000ft were interpolated.

The disturbing line was driven from the 70V terminals of a 40W amplifier and the line was loaded at the far end with 125Ω, thus transmitting 40W. The crosstalk figures are for 1kHz. The voltages at 100Hz and 10kHz are 1/10 and 10 times these figures, respectively.

There are two ways to effectively reduce crosstalk. One is to run signals only on balanced-line twisted pairs. Even shielding has a small added advantage compared with the noise and crosstalk rejection of a balanced line. The second way to reduce crosstalk is to move the two cables apart. The inverse-square

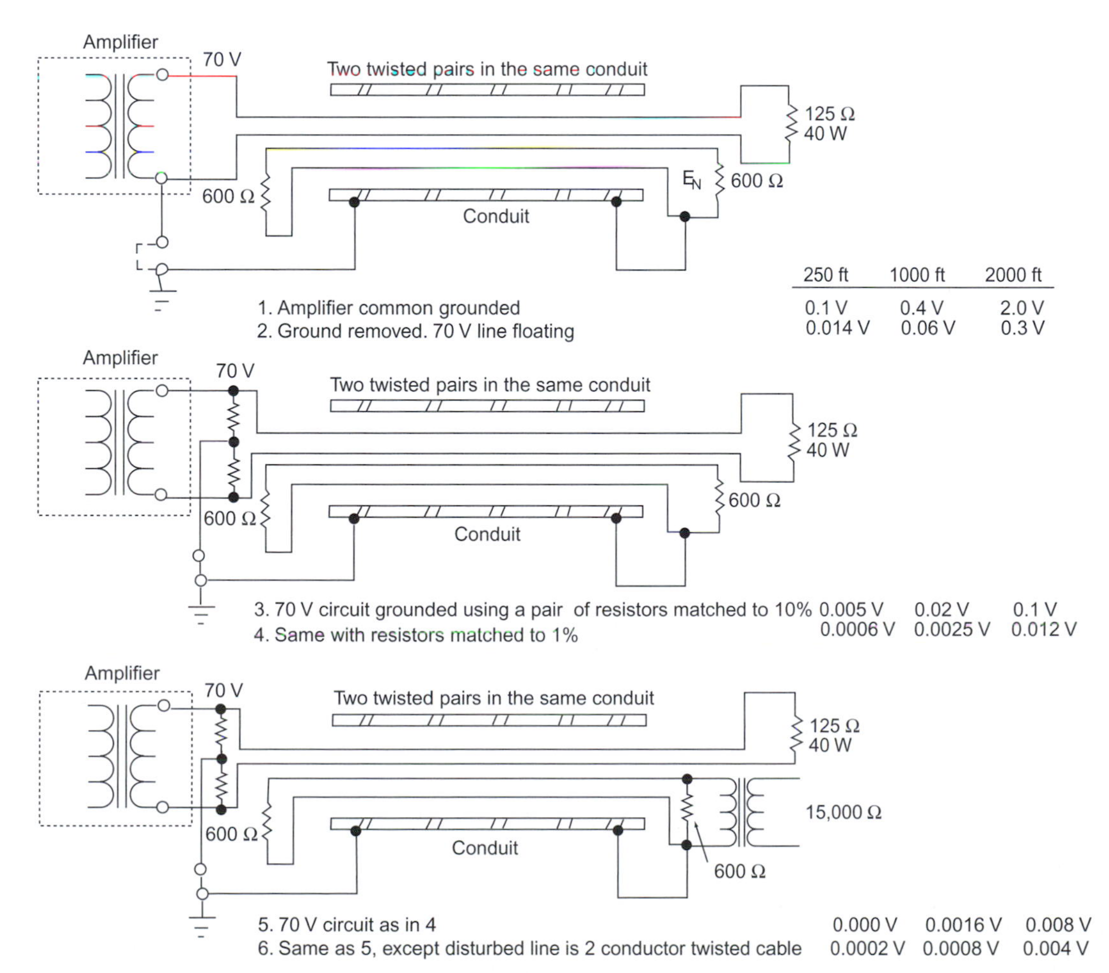

FIGURE 6.22
Effects of grounding on crosstalk. (Courtesy Altec Lansing Corp.)

law tells us that doubling the distance will produce four times less interference. Further, if cables cross at right angles, this is the point where the magnetic fields have minimum interaction. Of course, the latter solution is not an option in a prebundled cable, or in cable trays or installations with multiple cables run from point to point.

NATIONAL ELECTRICAL CODE

The *National Electrical Code* (NEC) is a set of guidelines written to govern the installation of wiring and equipment in commercial buildings and residential areas. These guidelines were developed to insure the safety of humans as well as property against fires and electrical hazards. Anyone involved in specifying cable for installation should be aware of the basics of the code.

The NEC code book is made up of nine chapters, with each chapter divided into separate articles pertaining to specific subjects. Five articles pertain to communication and power-limited cable. The NEC book is written by and available from the NFPA (National Fire Protection Association), 11 Tracy Drive, Avon, MA 02322. They can be reached at 1-800-344-3555 or www.nfpa.org.

Article 725—Class 1, Class 2, Class 3, Remote-Control, Signaling, and Power-Limited Circuits. Article 725 covers Class 1, Class 2, and Class 3 remote control and signaling cables as well as power-limited tray cable. Power-limited tray cable can be used as a Class 2 or Class 3 cable. Cable listed multi-purpose, communications, or power-limited fire protective can be used for Class 2 and Class 3 applications. A Class 3 listed cable can be used as a Class 2 cable.

Article 760—Fire Protective Signaling Systems. Article 760 covers power-limited fire-protective cable. Cable listed as power-limited fire-protective cable can also be used as Class 2 and Class 3 cable. Cable listed as communications and Class 3 can be used as power-limited fire protective cable with restrictions to conductor material and type gage size and number of conductors.

Article 770—Fiber Optic Systems. Article 770 covers three general types of fiber optic cable: nonconductive, conductive, and composite. Nonconductive type refers to cable containing no metallic members and no other electrically conductive materials. Conductive type refers to cable containing noncurrent carrying conductive members such as metallic strength members, etc. Composite type refers to cable containing optical fibers and current carrying electrical conductors. Composite types are classified according to the type of electrical circuit that the metallic conductor is designed for.

Article 800—Communication Circuits. Article 800 covers multipurpose and communication cable. Multi-purpose cable is the highest listing for a cable and can be used for communication, Class 2, Class 3, and power-limited fire-protective cable. Communication cable can be used for Class 2 and Class 3 cable and also as a power-limited fire protective cable with restrictions.

Article 820—Community Antenna Television. Article 820 covers community antenna television and RF cable. CATV cable may be substituted with multi-purpose or communication listed coaxial cable.

Designation and environmental areas

The NEC has designated four categories of cable for various environments and they are listed from the highest to the lowest listing. A higher listing can be used as a substitute for a lower listing.

Plenum—Suitable for use in air ducts, plenums, and other spaces used for environmental air without conduit and has adequate fire-resistant and low-smoke producing characteristics. It can also be substituted for all applications below.

Riser—Suitable for use in a vertical run, in a shaft, or from floor to floor, and has fire-resistant characteristics capable of preventing the spread of fire from floor to floor. It can also be substituted for all applications below.

General Purpose—Suitable for general-purpose use, with the exception of risers, ducts, plenums, and other space used for environmental air, and is resistant to the spread of fire. It can be substituted for the applications below.

Restricted Applications—Limited use and suitable for use in dwellings and in raceways and is flame retardant. Restricted use is limited to nonconcealed spaces of 10 ft or less, fully enclosed in conduit or raceway, or cable with diameters less than 0.25 inches for a residential dwelling.

Cable types

Signal cable used for audio, telephone, video, control applications, and computer networks of less than 50 V is considered low-voltage cabling and is grouped into five basic categories by the NEC, Table 6.35.

All computer network and telecommunications cabling falls into the CM class. The A/V industry primarily uses CM and CL2 cabling.

Table 6.36 defines the cable markings for various applications. Note plenum rated cable is the highest level because it has the lowest fire load, which means it does not readily support fire.

NEC substitution chart

NEC cable hierarchy, Figure 6.23, defines which cables can replace other cables. The chart starts with the highest listed cable on the top and descends to the lowest listed cable on the bottom. Following the arrows defines which cable can be substituted for others. Figure 6.24 defines the Canadian Electrical Code (CEC) substitution chart.

Final considerations

The National Electrical Code is widely accepted as the suggested regulations governing the proper installation of wire and cable in the United States. The code is revised every three years to keep safety in the forefront in wire and cable manufacturing and installation. Even though the code is generally accepted, each state, county, city, and municipality has the option to adopt all of the code, part of the code, or develop one of its own. The local inspectors have final authority of the installation. Therefore, the NEC is a good reference when questions arise about the proper techniques for a particular installation, but local authorities should be contacted for verification.

When choosing cable for an installation, follow these three guidelines to keep problems to a minimum:

1. The application and environment determine which type of cable to use and what rating it should have. Make sure the cable meets the proper ratings for the application.
2. If substituting a cable with another, the cable must be one that is rated the same or higher than what the code calls for. Check with the local inspector as to what is allowed in the local area.
3. The NEC code is a general guideline that can be adopted in whole or in part. Local state, county, city, or municipal approved code is what must be followed. Contact local authorities for verification of the code in the area.

Table 6.35 The Five Basic NEC Cable Groups

Cable Type	Use
CM	Communications
CL2, CL3	Class 2, Class 3 remote-control, signaling, and power-limited cables
FPL	Power-limited fire-protective signaling cables
MP	Multi-purpose cable
PLTC	Power-limited tray cable

Table 6.36 Cable Applications Designations Hierarchy

	Cable Family					
Application	MP	CM	CL2	CL3	FPL	PLTC
Plenum	MPP	CMP	CL2P	CL3P	FPLP	–
Riser	MPR	CPR	CL2R	CL3R	FPLR	–
General Purpose	MP, MPG	CM, CMG	CL2	CL3	FPL	PLTC
Dwelling	–	CMX	CL2X	CL3X	–	–

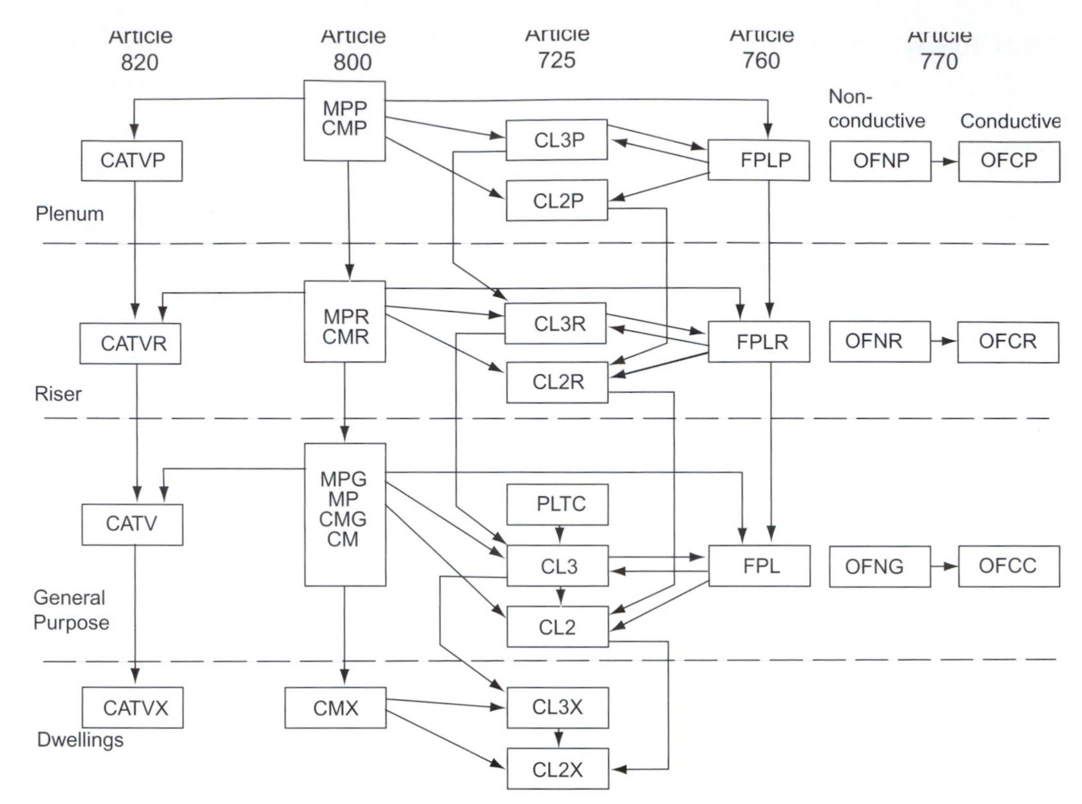

MPP, MPR, MPG, MP—Multipurpose Cables
CMP, CMR, CMG, CM, CMX—Communications Cables
CL3P, CL3R, CL3, CL3X, CL2P, CL2R, CL2, CL2X—Class 2 and Class 3 Remote-Control, Signaling and Power Limited Cables
FPLP, FPLR, FPL—Power Limited Fire Alarm Cables
CATVP, CATVR, CATV, CATVX—Community Antenna Television and Radio Distribution Cables
OFNP, OFNR, OFNG, OFN—Nonconductive Optical Fiber Cables
OFCP, OFCR, OFCG, OFC—Conductive Optical Fiber Cables
PLTC—Power Limited Tray Cables

FIGURE 6.23
National Electrical Code substitution and hierarchy. (Courtesy Belden.)

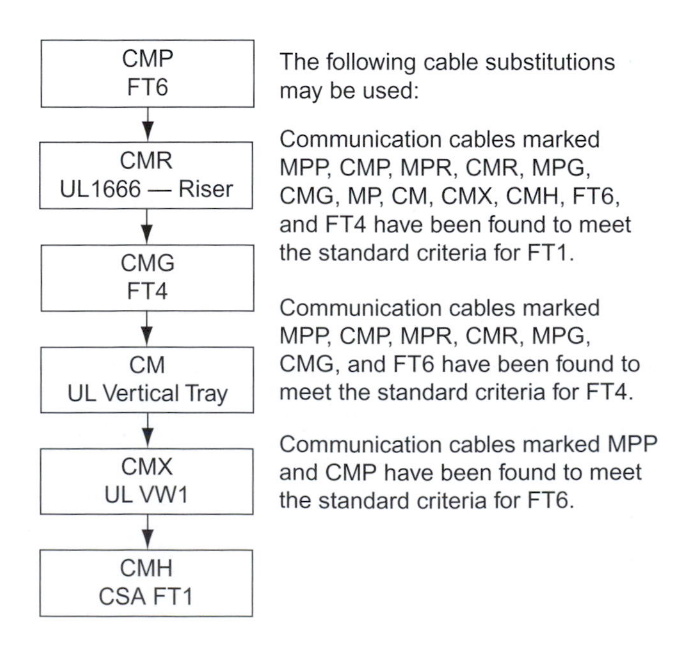

FIGURE 6.24
Canadian Electrical Code cable substitution hierarchy per C22.2 #214—Communication Cables.

The local inspector or fire marshal has the final authority to approve or disapprove any installation of cable based on the National Electric Code or on the local code.

Plenum cable

Plenum cable is used in ceilings where the air handling system uses the plenum as the delivery or the return air duct. Because of its flame-resistant and low smoke-emission properties, the special compound used in plenum cable jackets and insulations has been accepted under the provisions of the NEC and classified by Underwriters Laboratories Inc. (UL) for use without conduit in air plenums.

In a typical modern commercial building, cables are installed in the enclosed space between drop ceilings and the floors from which they are suspended. This area is also frequently used as a return air plenum for a building's heating and cooling system. Because these air ducts often run across an entire story unobstructed, they can be an invitation to disaster if fire breaks out. Heat, flames, and smoke can spread rapidly throughout the air duct system and building if the fire is able to feed on combustible materials (such as cable insulations) in the plenum. To eliminate this problem and to keep fumes from entering the air handling system, the NEC requires that conventional cables always be installed in metal conduit when used in plenums.

Plenums, with their draft and openness between different areas, cause fire and smoke to spread, so the 1975 NEC prohibited the use of electrical cables in plenums and ducts unless cables were installed in metal conduit. In 1978, Sections 725-2(b) (signaling cables), 760-4(d) (fire-protection cable), and 800-3(d) (communication/telephone cables) of the NEC allowed that cables "listed as having adequate fire-resistance and low-smoke producing characteristics shall be permitted for ducts, hollow spaces used as ducts, and plenums other than those described in Section 300-22(a)."

While plenum cable costs more than conventional cable, the overall installed cost is dramatically lower because it eliminates the added cost of conduit along with the increased time and labor required to install it.

In 1981 the jacket and insulation compound used in plenum cables was tested and found acceptable under the terms of the NEC and was classified by UL for use without conduit in air return ducts and plenums. Figure 6.25 shows the UL standard 910 plenum flame test using a modified Steiner tunnel equipped with a special rack to hold test cables.

Virtually any cable can be made in a plenum version. The practical limit is the amount of flammable material in the cable and its ability to pass the Steiner Tunnel Test, shown in Figure 6.25. Originally plenum cable was all Teflon inside and out. Today most plenum cables have a Teflon core with a special PVC jacket which meets the fire rating. But there are a number of compounds such as Halar® and Solef® that can also be used.

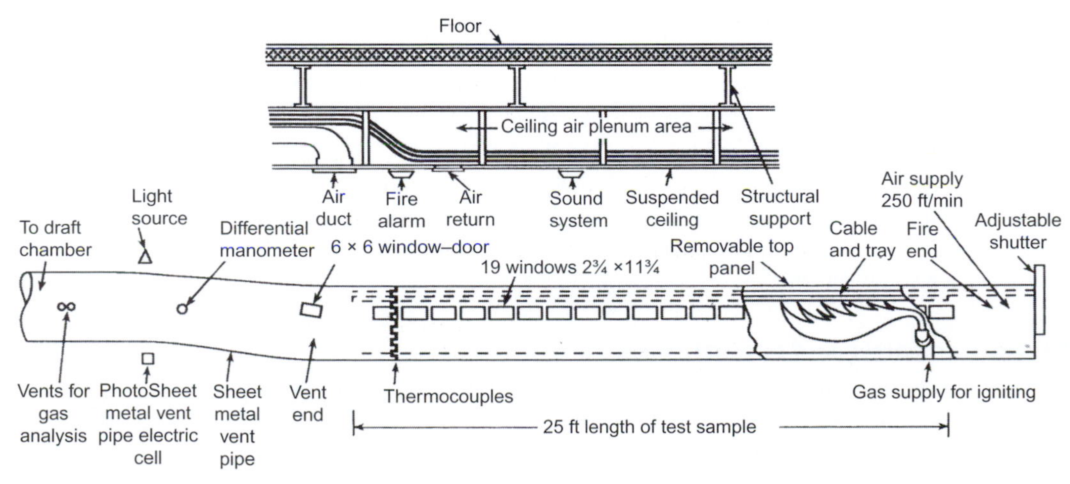

FIGURE 6.25
Plenum cable flame test, UL standard 910.

Table 6.37 Physiological Effects of Shock Current on Humans (From Amundson)

Shock Current in mArms	Circuit Resistance at 120 Vac	Physiological Effects
0.5–7 mA	240,000 Ω down to 17,000 Ω	**Threshold of Perception:** Large enough to excite skin nerve endings for a tingling sensation. Average thresholds are 1.1 mA for men and 0.7 mA for women.
1–6 mA	120,000 Ω down to 20,000 Ω	**Reaction Current:** Sometimes called the *surprise current*. Usually an involuntary reaction causing the person to pull away from the contact.
6–22 mA	20,000 Ω down to 5400 Ω	**Let-Go Current:** This is the threshold where the person can voluntarily withdraw from the shock current source. Nerves and muscles are vigorously stimulated, eventually resulting in pain and fatigue. Average let-go thresholds are 16 mA for men and 10.5 mA for women. Seek medical attention.
15 mA and above	8000 Ω and below	**Muscular Inhibition:** Respiratory paralysis, pain and fatigue through strong involuntary contractions of muscles and stimulation of nerves. Asphyxiation may occur if current is not interrupted.
60 mA–5 A	2000 Ω down to 24 Ω	**Ventricular Fibrillation:** Shock current large enough to desynchronize the normal electrical activity in the heart muscle. Effective pumping action ceases, even after shock cessation. Defibrillation (single pulse shock) is needed or death occurs.
1 A and above	120 Ω and below	**Myocardial Contraction:** The entire heart muscle contracts. Burns and tissue damage via heating may occur with prolonged exposure. Muscle detachment from bones possible. Heart may automatically restart after shock cessation.

Power distribution safety

Electricity kills! No matter how confident we are we must always be careful around electricity. Fibrillation is a nasty and relatively slow death so it is important that defibrillators are accessible when working around electricity. Table 6.37 displays the small amounts of current required to hurt or kill a person.

GROUND-FAULT INTERRUPTERS

Ground-fault circuit interrupters (GFCIs) are sometimes called *earth leakage* or *residual-current circuit breakers*. GFCIs sense leakage current to earth ground from the hot or neutral leg and interrupt the circuit automatically within 25 ms if the current exceeds 4 to 6 ma. These values are determined to be the maximum safe levels before a human heart goes into ventricular fibrillation. GFCIs do not work when current passes from one line to the other line through a person, for instance. They do not work as a circuit breaker.

One type of GFCI is the core-balance protection device, Figure 6.26. The hot and neutral power conductors pass through a toroidal (differential) current transformer. When everything is operating properly, the vector sum of the currents is zero. When the currents in the two legs are not equal, the toroidal transformer detects it, amplifies it, and trips an electromagnetic relay. The circuit can also be tested by depressing a test button which unbalances the circuit.

AC power cords and receptacles

AC power cords, like other cables, come with a variety of jacket materials for use in various environments. All equipment should be connected with three-wire cords. Never use ground-lift adapters to remove the ground from any equipment. This can be dangerous, even fatal, if a fault develops inside the equipment, and there is no path to ground.

A common European plug, with a rating of 250 Vac and 10 A, is shown in Figure 6.27.

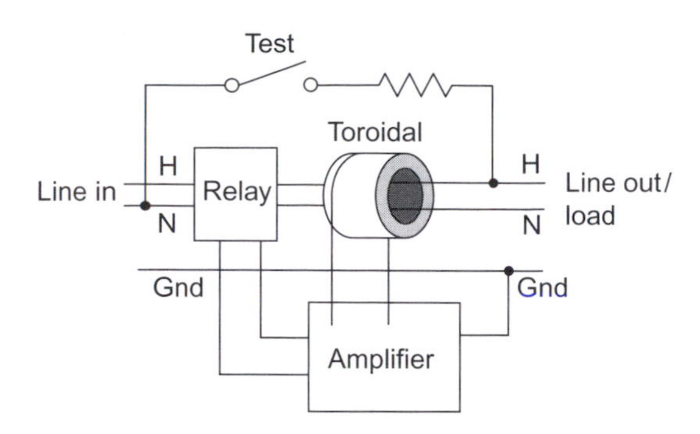

FIGURE 6.26
Typical ground-fault circuit interrupter.

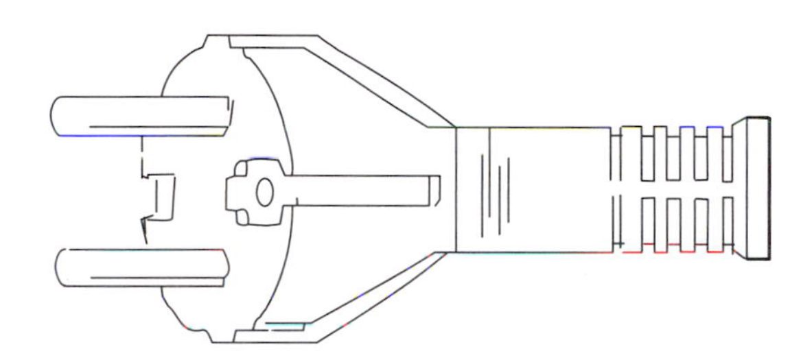

FIGURE 6.27
Standard international plug.

Table 6.38 Color Codes for Power Supply Cords

Function	North America	CEE and SAA Standard
N-Neutral	White	Light Blue
L-Live	Black	Brown
E-Earth or Ground	Green or Green/Yellow	Green/Yellow

Table 6.39 Approved Electrical Standards

Country		Standard
United States	UL	Underwriters Laboratory
Canada	cUL	Canadian Underwriters Laboratory
Germany	GS/TUV	German Product Certification Organization
International	IEC	International Electrotechnical Commission

The color codes used in North America and Europe for three conductors are shown in Table 6.38.

Cables should be approved to a standards shown in Table 6.39.

The UL listing signifies that all elements of the cords and assembly methods have been approved by the Underwriters Laboratories, Inc. as meeting their applicable construction and performance standards. UL listed has become a symbol of safety to millions of Americans and their confidence in it results in easier sales of electrical products.

The U.S. NEMA configurations for various voltage and current general-purpose plugs and receptacles are shown in Figure 6.28.

Table 6.40 Continued

Country	Type of Current	Phases	Voltage	# Wires
Morocco	AC 50	1,3	127/220	2,4
Mozambique	AC 50	1,3	220/380	2,3,4
Namibia	AC 50	1,3	220/380	2,4
Nassau	AC 60	1,3	120/240	2,3,4
Nepal	AC 50	1,3	220/440	2,4
Netherlands	AC 50	1,3	220/380	2,3
Netherlands, Antilles	AC 50	1,3	127/220	2,3,4
New Caledonia	AC 50	1,3	220/380	2,3,4
New Zealand	AC 50	1,3	230/400	2,3,4
Nicaragua	AC 60	1,3	120/240	2,3,4
Niger	AC 50	1,3	220/380	2,3,4
Nigeria	AC 50	1,3	230/415	2,4
Northern Ireland	AC 50	1,3	220/380 240/480	2,4
Norway	AC 50	1,3	230	2,3
Okinawa	AC 60	1	120/240	2,3
Oman	AC 50	1,3	240/415	2,4
Pakistan	AC 50	1,3	220/380	2,3,4
Panama	AC 60	1,3	110/220	2,3
Papua New Guinea	AC 50	1,3	240/415	2,4
Paraguay	AC 50	1,3	220/380	2,4
Peru	AC 50	1,3	220	2,3
Philippines	AC 60	1,3	110/220	2,3
Poland	AC 50	1,3	220/380	2,4
Portugal	AC 50	1,3	220/380	2,3,4
Puerto Rico	AC 60	1,3	120/240	2,3,4
Qatar	AC 50	1,3	240/415	2,3,4
Romania	AC 50	1,3	220/380	2,4
Russia	AC 50	1,3	220/380	
Rwanda	AC 50	1,3	220/380	2,4
Saudi Arabia	AC 60	1,3	127/220	2,4
Scotland	AC 50	1,3	240/415	2,4
Senegal	AC 50	1,3	127/220	2,3,4
Seychelles	AC 50	1,3	240	2,3
Sierra Leone	AC 50	1,3	230/400	2,4
Singapore	AC 50	1,3	230/400	2,4
Somalia	AC 50	1,3	230	2,3
South Africa	AC 50	1,3	220/380	2,3,4

(Continued)

Table 6.40 Continued

Country	Type of Current	Phases	Voltage	# Wires
Spain	AC 50	1,3	127/220 220/380	2,3,4
Sri Lanka	AC 50	1,3	230/400	2,4
St. Kitts and Nevis	AC 60	1,3	230/400	2,4
St. Lucia	AC 50	1,3	240/416	2,4
St. Vincent	AC 50	1,3	230/400	2,4
Sudan	AC 50	1,3	240/415	2,4
Suriname	AC 60	1,3	127/220	2,3,4
Swaziland	AC 50	1,3	230/400	2,4
Sweden	AC 50	1,3	220/380	2,3,4,5
Switzerland	AC 50	1,3	220/380	2,3,4
Syria	AC 50	1,3	220/380	2,3
Tahiti	AC 60	1,3	127/220	2,3,4
Taiwan	AC 60	1,3	110/220	2,3,4
Tanzania	AC 50	1,3	230/400	2,3,4
Thailand	AC 50	1,3	220/380	2,3,4
Togo	AC 50	1,3	220/380	2,4
Tonga	AC 50	1,3	240/415	2,3,4
Trinidad and Tobago	AC 60	1,3	115/230 230/400	2,3,4
Tunisia	AC 50	1,3	127/220 220/380	2,4
Turkey	AC 50	1,3	220/380	2,3,4
Uganda	AC 50	1,3	240/415	2,4
United Arab Emirates	AC 50	1,3	220/415	2,3,4
United Kingdom	AC 50	1	240/480	2,3
United States	AC 60	1,3	120/240	3,4
Uruguay	AC 50	1,3	220	2,3
Venezuela	AC 60	1,3	120/240	2,3,4
Vietnam	AC 50	1,3	220/380	2,4
Virgin Islands (American)	AC 60	1,3	120/240	2,3,4
Wales	AC 50	1,3	240/415	2,4
Western Samoa	AC 50	1,3	230/400	2,4
Yemen Arab Republic	AC 50	1,3	230/400	2,4
Yugoslavia	AC 50	1,3	220/380	2,4
Zaire, Rep. of	AC 50	1,3	220/380	2,3,4
Zambia	AC 50	1,3	220/380	2,4
Zimbabwe	AC 50	1,3	220/380	2,3,4

Everything physical is made of atoms whose outermost components are electrons. An electron carries a negative electric charge and is the smallest quantity of electricity that can exist. Some materials, called *conductors* and most commonly metals, allow their outer electrons to move freely from atom to atom. Other materials, called *insulators* and most commonly air, plastic, or glass, are highly resistant to such movement. This movement of electrons is called *current flow*. Current will flow only in a complete circuit consisting of a connected source and load. **Regardless of how complex the path becomes, all current leaving a source must return to it!**

Circuit theory

An electric potential or *voltage*, sometimes called *emf* for electromotive force, is required to cause current flow. It is commonly denoted *E* (from emf) in equations and its unit of measure is the *volt*, abbreviated *V*. The resulting rate of current flow is commonly denoted *I* (from intensity) in equations and its unit of measure is the ampere, abbreviated *A*. How much current will flow for a given applied voltage is determined by circuit resistance. Resistance is denoted *R* in equations and its unit of measure is the ohm, symbolized Ω.

Ohm's Law defines the quantitative relationship between basic units of voltage, current, and resistance:

$$E = I \times R$$

which can be rearranged as

$$R = \frac{E}{I}$$

$$I = \frac{E}{R}$$

For example, a voltage *E* of 12 V applied across a resistance *R* of 6 Ω will cause a current flow *I* of 2 A.

Circuit elements may be connected in series, parallel, or combinations of both, Figures 7.1 and 7.2.

Although the resistance of wires that interconnect circuit elements is generally assumed to be negligible, we will discuss this later.

In a parallel circuit, the total source current is the sum of the currents through each circuit element. The highest current will flow in the lowest resistance, according to Ohm's Law. The equivalent single resistance seen by the source is always lower than the lowest resistance element and is calculated as:

$$R_{EQ} = \frac{1}{\frac{1}{R1} + \frac{1}{R2} + \frac{1}{R} \cdots + \frac{1}{n}} \qquad (7.1)$$

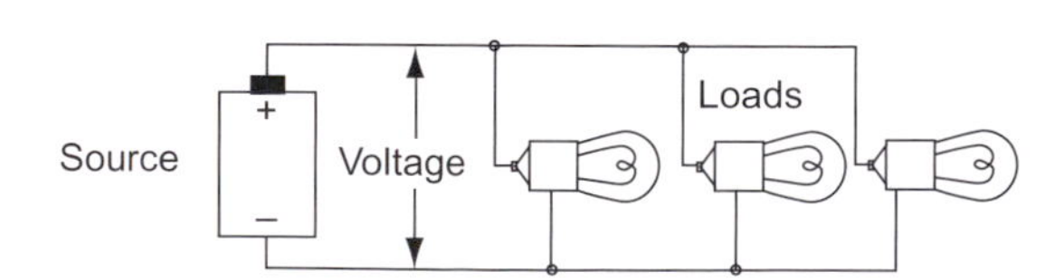

FIGURE 7.1
The voltage is the same across all elements in a parallel circuit.

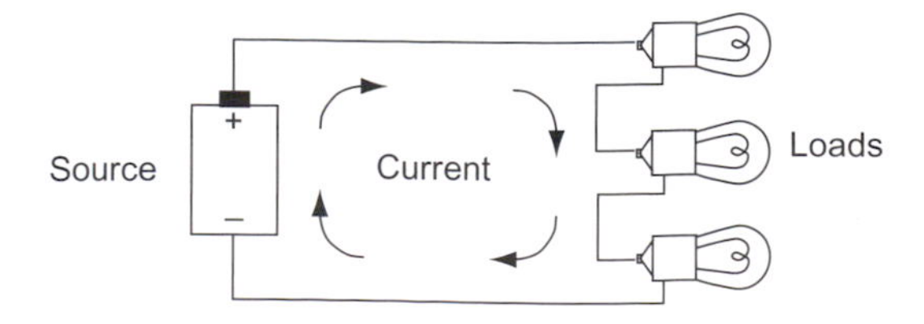

FIGURE 7.2
The current is the same through all elements in a series circuit.

In a series circuit, the total source voltage is the sum of the voltages across each circuit element. The highest voltage will appear across the highest resistance, according to Ohm's Law. The equivalent single resistance seen by the source is always higher than the highest resistance element and is calculated as:

$$R_{EQ} = R1 + R2 + R3 \cdots + Rn \tag{7.2}$$

Voltages or currents whose value (magnitude) and direction (polarity) are steady over time are generally referred to as *DC*. A battery is a good example of a DC voltage source.

AC circuits

A voltage or current that changes value and direction over time is generally referred to as AC. Consider the voltage at an ordinary 120 V, 60 Hz AC receptacle.

Since it varies over time according to a mathematical sine function, it is called a *sine wave*. Figure 7.3 shows how it would appear on an oscilloscope where time is the horizontal scale and instantaneous voltage is the vertical scale with zero in the center. The instantaneous voltage swings between peak voltages of +170 V and −170 V. A cycle is a complete range of voltage or current values that repeat themselves periodically (in this case every 16.67 ms). Phase divides each cycle into 360° and is used mainly to describe instantaneous relationships between two or more AC waveforms. Frequency indicates how many cycles occur per second of time. Frequency is usually denoted f in equations, and its unit of measure is the hertz, abbreviated Hz. Audio signals rarely consist of a single sine wave. Most often they are complex waveforms consisting of many simultaneous sine waves of various amplitudes and frequencies in the 20 Hz to 20,000 Hz (20 kHz) range.

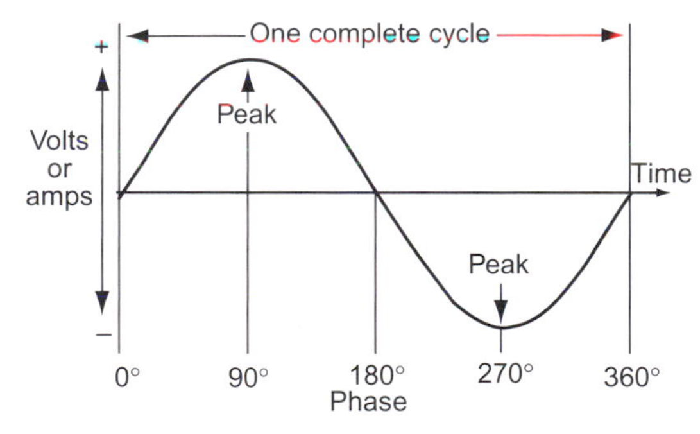

FIGURE 7.3
Sine wave as displayed on an oscilloscope.

Capacitance, inductance, and impedance

An electrostatic field exists between any two conductors having a voltage difference between them. Capacitance is the property that tends to oppose any change in the strength or charge of the field. In general, capacitance is increased by larger conductor surface areas and smaller spacing between them. Electronic components expressly designed to have high capacitance are called *capacitors*. Capacitance is denoted C in equations and its unit of measure is the Farad, abbreviated F. It's very important to remember that unintentional or parasitic capacitances exist virtually everywhere. As we will see, these parasitic capacitances can be particularly significant in cables and transformers!

Current must flow in a capacitor to change its voltage. Higher current is required to change the voltage rapidly and no current will flow if the voltage is held constant. Since capacitors must be alternately charged and discharged in AC circuits, they exhibit an apparent AC resistance called *capacitive reactance*. Capacitive reactance is inversely proportional to both capacitance and frequency since an increase in either causes an increase in current, corresponding to a decrease in reactance.

$$X_C = \frac{1}{2\pi fC} \tag{7.3}$$

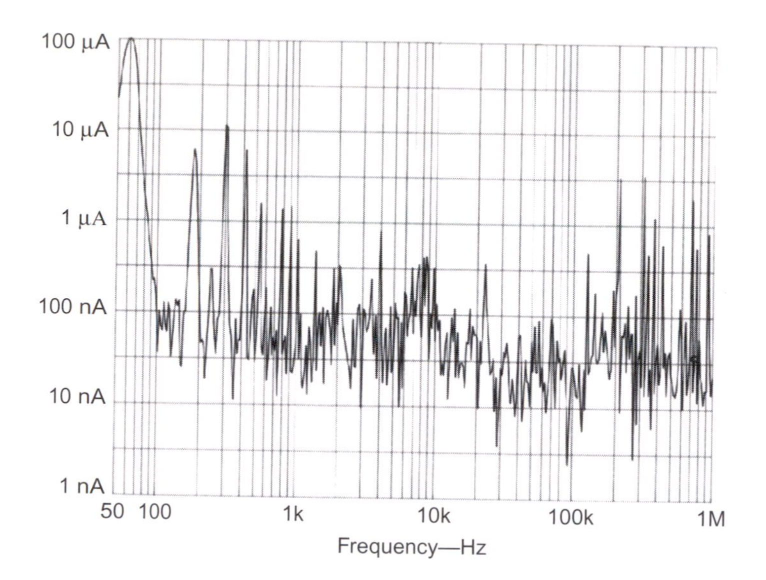

FIGURE 7.20
Typical leakage current from line to safety ground coupled via 3000 pF capacitance into a 75 Ω spectrum analyzer input.

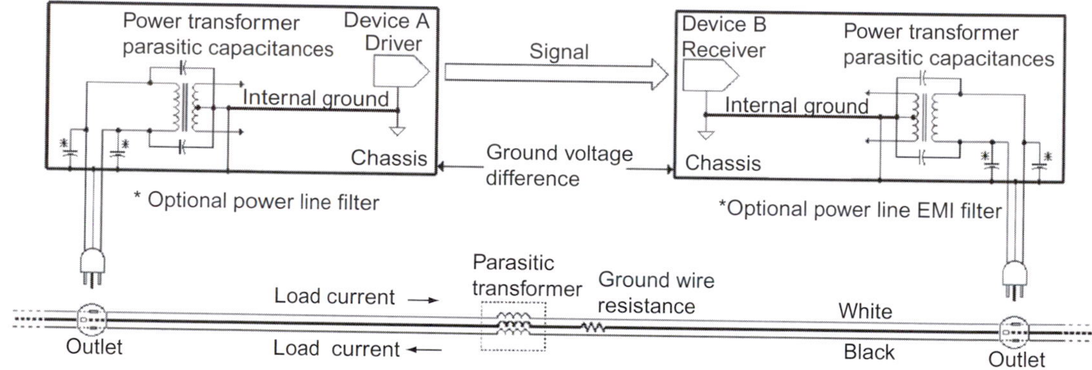

FIGURE 7.21
Voltage difference is magnetically induced over length of safety-ground premises wiring.

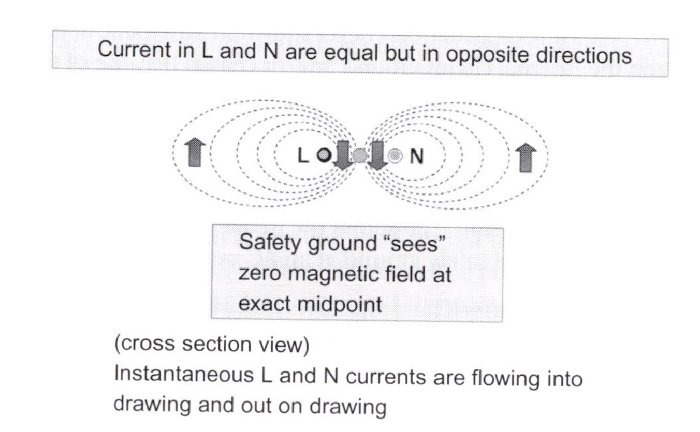

FIGURE 7.22
Magnetic fields surrounding line and neutral can induce voltage into safety ground.

The voltage induced in any transformer is directly proportional to the rate of change of load current in the circuit. With an ordinary phase-control light dimmer the peak voltages induced can become quite high. When the dimmer triggers current on 120 times per second, it switches on very quickly (a few microseconds) as shown in Figure 7.23. Since the magnetic induction into safety ground favors high frequencies, noise coupling problems in a system will likely become most evident when a light dimmer is involved. The problems are usually worst at about half-brightness setting of the dimmer.

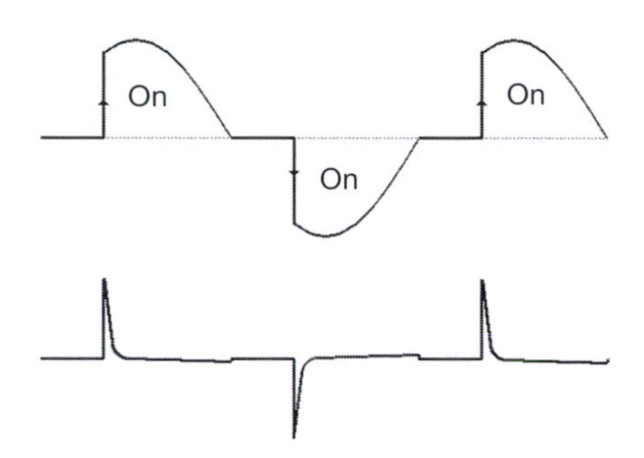

FIGURE 7.23
Lamp current (upper) versus induced voltage (lower) for phase-controlled dimmer.

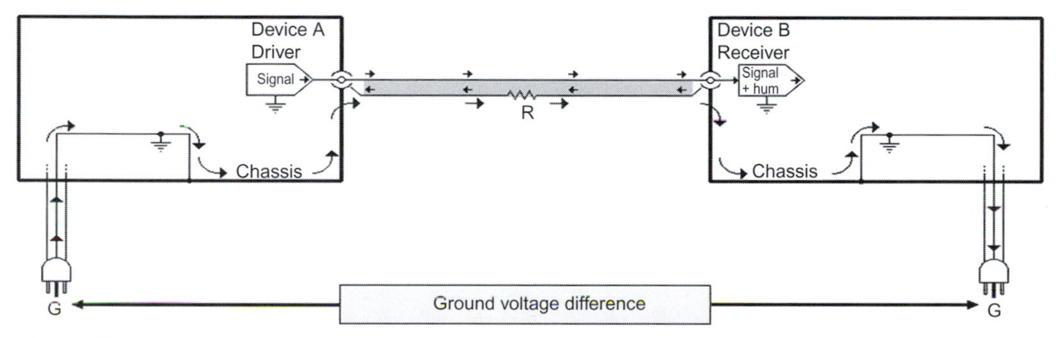

FIGURE 7.24
For grounded equipment, interconnect cables complete a wired loop.

This parasitic transformer action generates small ground voltage differences, generally under 1 V, between AC outlets. The voltage differences tend to be higher between two outlets on different branch circuits, and higher still if a device on the branch circuit is also connected to a remote or alien ground such as a CATV feed, satellite dish, or an interbuilding tie line. *We must accept interoutlet ground noise voltage as a fact of life.*

GROUND LOOPS

For our purposes, a ground loop is formed when a signal cable connects two pieces of equipment whose connections to the power line or other equipment causes a power-line-derived current to flow in the signal cable.

The first, and usually worst, kind of ground loop occurs between *grounded* devices—those with three-prong AC plugs. Current flow in signal cables, as shown in Figure 7.24, can easily reach 100 mA or more.

The second kind of ground loop occurs between floating devices—those with two-prong AC plugs. Each pair of capacitances CF (for EMI filter) and CP (for power transformer parasitic) in the schematic form a capacitive voltage divider between line and neutral, causing some fraction of 120 Vac to appear between chassis and ground. For UL-listed ungrounded equipment, this leakage current must be under 0.75 mA (0.5 mA for office equipment). This small current can cause an unpleasant, but harmless, tingling sensation as it flows through a person's body. More relevant is the fact that these noisy leakage currents will flow in any wire connecting such a floating device to safety ground, or connecting two floating devices to each other as shown in Figure 7.25.

INTERFACE PROBLEMS IN SYSTEMS

If properly designed balanced interfaces were used throughout an audio system, it would theoretically be noise-free. Until about 1970, equipment designs allowed real-world system to come very close to this ideal. But since then, balanced interfaces have fallen victim to two major design problems—and both can properly be blamed on equipment manufacturers. Even careful examination of manufacturers' specifications and data sheets will not reveal either problem—the devil is in the details. These problems

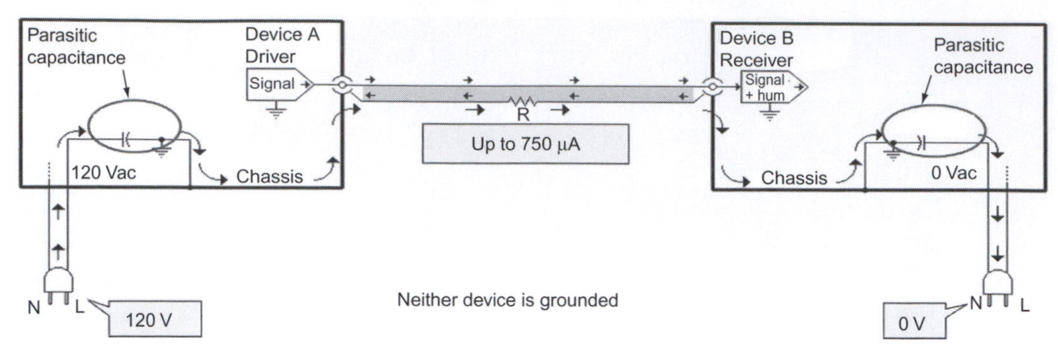

FIGURE 7.25
For ungrounded equipment, interconnect cables complete a capacitive loop.

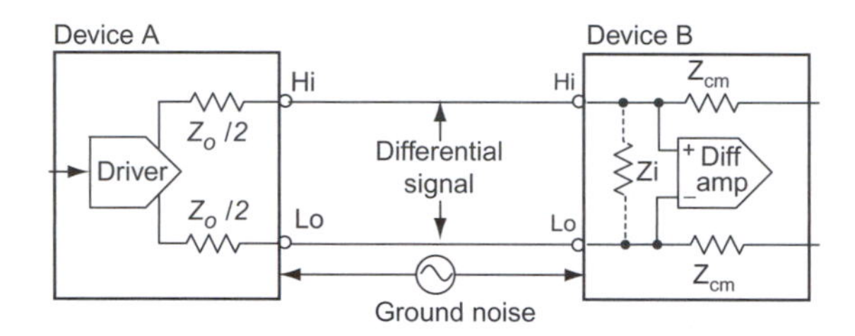

FIGURE 7.26
Simplified balanced interface.

are effectively concealed because the marketing departments of most manufacturers have succeeded in dumbing down their so-called *specifications* over the same time period.

First is degraded noise rejection, which appeared when solid-state differential amplifiers started replacing input transformers. Second is the pin 1 problem that appeared in large numbers when PC boards and plastic connectors replaced their metal counterparts. Both problems can be avoided through proper design, of course, but in this author's opinion, part of the problem is that the number of analog design engineers who truly understand the underlying issues is dwindling and engineering schools are steering most students into the digital future where analog issues are largely neglected. Other less serious problems with balanced interfaces are caused by balanced cable construction and choices of cable shield connections.

On the other hand, unbalanced interfaces have an intrinsic problem that effectively limits their use to only the most electrically benign environments. Of course, even this problem can be solved by adding external ground-isolation devices, but the best advice is to avoid them whenever possible in professional systems!

Degraded common-mode rejection

Balanced interfaces have traditionally been the hallmark of professional sound equipment. In theory, systems comprised of such equipment are completely noise-free. However, an often overlooked fact is that the common-mode rejection of a complete signal interface does not depend solely on the receiver, but on how the receiver interacts with the driver and the line performing as a subsystem.

In the basic balanced interface of Figure 7.26, the output impedances of the driver $Z_o/2$ and the input impedances of the receiver Z_{cm} effectively form the Wheatstone bridge shown in Figure 7.27. If the bridge is not balanced or nulled, a portion of the ground noise V_{cm} will be converted to a differential signal on the line. This nulling of the common-mode voltage is critically dependent on the ratio matching of the pairs of driver/receiver common-mode impedances R_{cm} in the − and + circuit branches. The balancing or nulling is unaffected by impedance across the two lines, such as the signal input impedance Z_i in Figure 7.28 or the signal output impedance of the driver. It is the common-mode impedances that matter!

The bridge is most sensitive to small fractional impedance changes in one of its arms when all arms have the same impedance (Sams, 1972). It is least sensitive when the upper and lower arms have widely differing impedances—e.g., when upper arms are very low and lower arms are very high, or vice

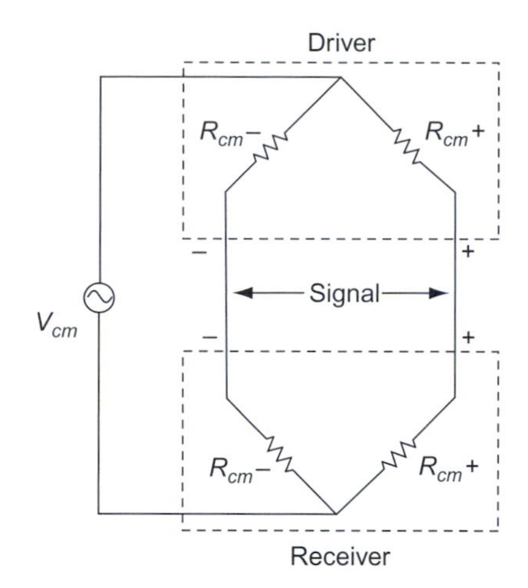

FIGURE 7.27
The balanced interface is a Wheatstone bridge.

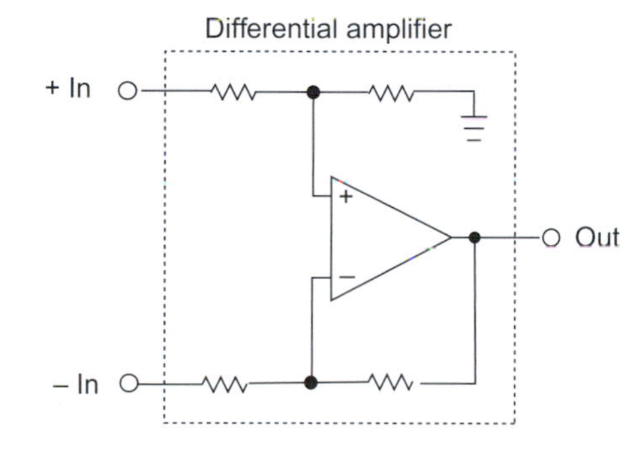

FIGURE 7.28
Basic differential amplifier.

versa. Therefore, we can minimize the sensitivity of a balanced system (bridge) to impedance imbalances by making common-mode impedances very low at one end of the line and very high at the other. This condition is consistent with the requirements for voltage matching discussed earlier.

Most active line receivers, including the basic differential amplifier of Figure 7.28, have common-mode input impedances in the 5 kΩ to 50 kΩ range, which is inadequate to maintain high CMRR with real-world sources. With common-mode input impedances of 5 kΩ, a source imbalance of only 1 Ω, which could arise from normal contact and wire resistance variations, can degrade CMRR by 50 dB. Under the same conditions, the CMRR of a good input transformer would be unaffected because of its 50 MΩ common-mode input impedances. Figure 7.29 shows computed CMRR versus source imbalance for different receiver common-mode input impedances. Thermal noise and other limitations place a practical limit of about 130 dB on most actual CMRR measurements.

How much imbalance is there in real-world signal sources? Internal resistors and capacitors determine the output impedance of a driver. In typical equipment, $Z_o/2$ may range from 25 to 300 Ω. Since the resistors are commonly ±5% tolerance and the coupling capacitors are ±20% at best, impedance imbalances up to about 20 Ω should be routinely expected. This defines a real-world source. In a previous paper, this author has examined balanced audio interfaces in some detail, including performance comparisons of various receiver types (Whitlock, 1995a, pp. 454–464). It was concluded that, regardless of their circuit topology, popular active receivers can have very poor CMRR when driven from such real-world sources. The poor performance of these receivers is a direct result of their low common-mode input impedances. If common-mode input impedances are raised to about 50 MΩ, 94 dB of ground noise rejection is attained from a completely unbalanced 1 kΩ source, which is typical of consumer outputs. When common-mode input impedances are sufficiently high, an input can be considered truly universal, suitable for any source—balanced or unbalanced. A receiver using either a good input transformer or the InGenius® integrated circuit (Whitlock, 1996a) will routinely achieve 90–100 dB of CMRR and remain unaffected by typical real-world output imbalances.

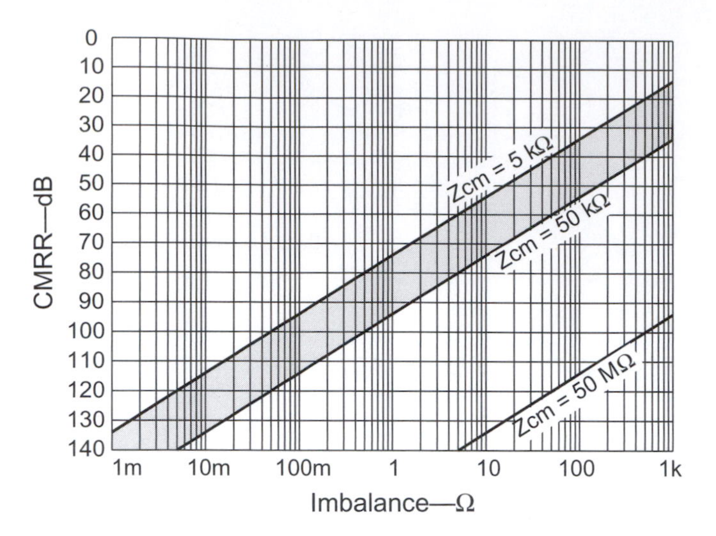

FIGURE 7.29
Noise rejection versus source impedance/imbalance.

The theory underlying balanced interfaces is widely misunderstood by audio equipment designers. Pervasive use of the simple differential amplifier as a balanced line receiver is evidence of this. And, as if this weren't bad enough, some have attempted to improve it. Measuring input X and Y input impedances of the simple differential amplifier individually leads some designers to alter its equal resistor values. However, as shown in Figure 7.30, if the impedances are properly measured simultaneously, it becomes clear that nothing is wrong. The fix grossly unbalances the common-mode impedances, which destroys the interface CMRR for any real-world source. This and other misguided *improvements* completely ignore the importance of common-mode input impedances.

The same misconceptions have also led to some CMRR tests whose results give little or no indication of how the tested device will actually behave in a real-world system. Apparently, large numbers of designers test the CMRR of receivers with the inputs either shorted to each other or driven by a laboratory precision signal source. The test result is both unrealistic and misleading. Inputs rated at 80 dB of CMRR could easily deliver as little as 20 dB or 30 dB when used in a real system. Regarding their previous test, the IEC had recognized that *test is not an adequate assurance of the performance of certain electronically balanced amplifier input circuits.* The old method simply didn't account for the fact that source impedances are rarely perfectly balanced. To correct this, this author was instrumental in revising IEC Standard: 60268-3 Sound System Equipment – Part 3: Amplifiers. The new method, as shown in Figure 7.31, uses typical $\pm 10\,\Omega$ source impedance imbalances and clearly reveals the superiority of input transformers and some new active input stages that imitate them. The new standard was published August 30, 2000. The Audio Precision APx520 and APx525, introduced in 2008, are the first audio instruments to offer the new CMRR test.

The pin 1 problem

In his now famous paper in the 1995 *AES Journal*, Neil Muncy says:

> This paper specifically addresses the problem of noise coupling into balanced line-level signal interfaces used in many professional applications, due to the unappreciated consequences of a popular and widespread audio equipment design practice which is virtually without precedent in any other field of electronic systems (Muncy, 1995, p. 436).

Common impedance coupling occurs whenever two currents flow in a shared or common impedance. A noise coupling problem is created when one of the currents is ground noise and the other is signal. The common impedance is usually a wire or circuit board trace having a very low impedance, usually well under an ohm. Unfortunately, common impedance coupling has been designed into audio equipment from many manufacturers. The noise current enters the equipment via a terminal, at a device input or output, to which the cable shield is connected via a mating connector. For XLR connectors, it's pin 1 (hence the name); for ¼ inch connectors, it's the sleeve; and for RCA/IHF connectors, it's the shell.

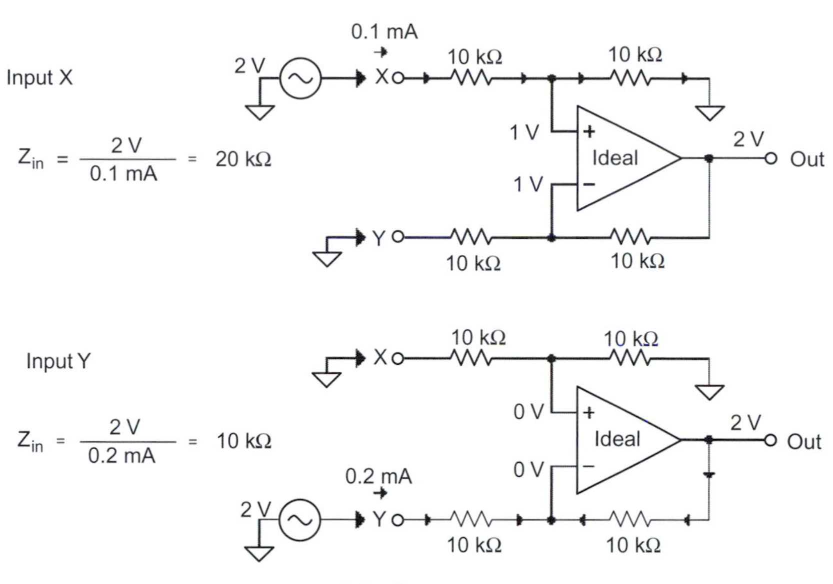

FIGURE 7.30
Common-mode impedances apply to a voltage applied to both inputs.

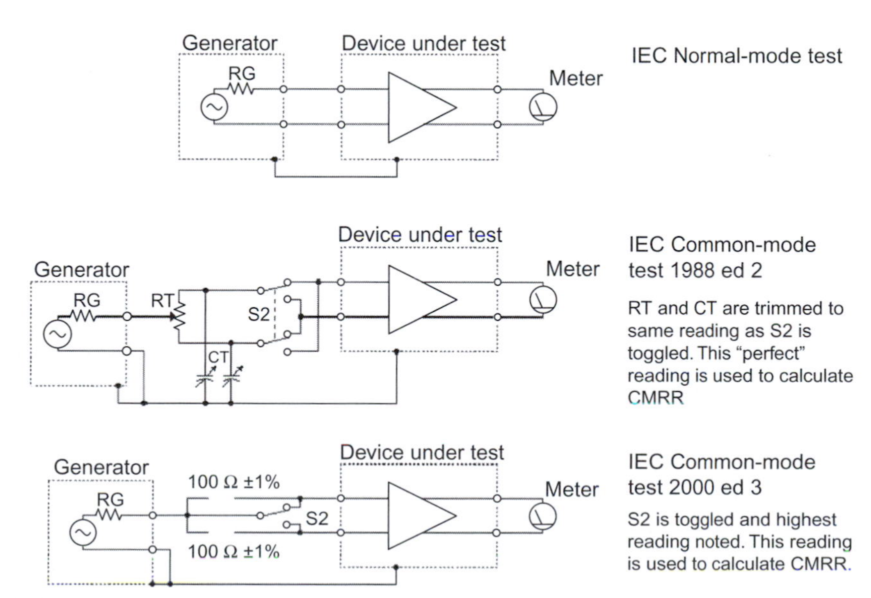

FIGURE 7.31
Old and new IEC tests for CMRR compared.

To the user, symptoms are indistinguishable from many other noise coupling problems such as poor CMRR. To quote Neil again:

> Balancing is thus acquiring a tarnished reputation, which it does not deserve. This is indeed a curious situation. Balanced line-level interconnections are supposed to ensure noise-free system performance, but often they do not.

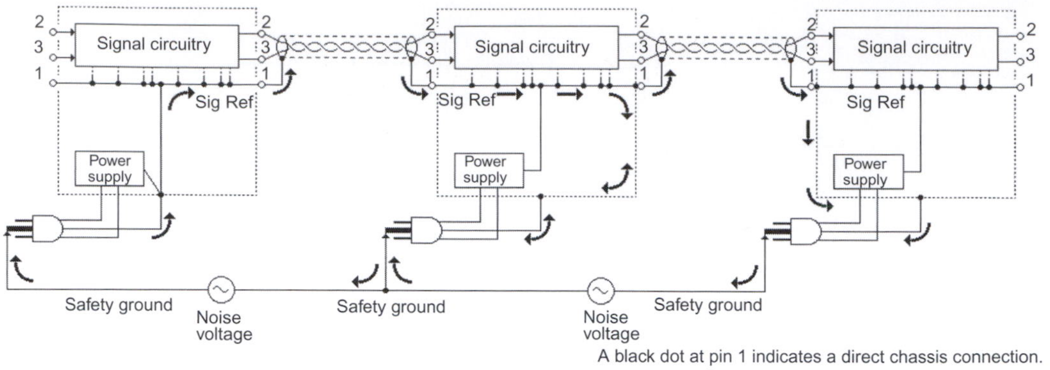

FIGURE 7.32
How poor routing of shield currents produces the pin 1 problem.

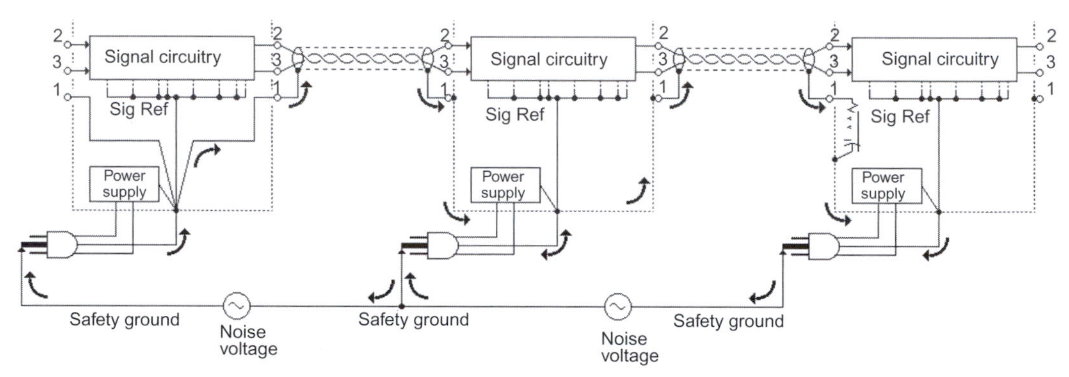

FIGURE 7.33
Equipment with proper internal grounding.

In balanced interconnections, it occurs at line inputs and outputs where interconnecting cables routinely have their shields grounded at both ends. Of course, grounding at both ends is required for unbalanced interfaces.

Figure 7.32 illustrates several examples of common impedance coupling. When noise currents flow in signal reference wiring or circuit board traces, tiny voltage drops are created. These voltages can couple into the signal path, often into very high gain circuitry, producing hum or other noise at the output. In the first two devices, pin 1 current is allowed to flow in internal signal reference wiring. In the second and third devices, power line noise current (coupled through the parasitic capacitances in the power transformer) is also allowed to flow in signal reference wiring to reach the chassis/safety ground. This so-called *sensitive equipment* will produce additional noise independent of the pin 1 problem. For the second device, even disconnecting its safety ground (not recommended) won't stop current flow through it between input and output pin 1 shield connections.

Figure 7.33 shows three devices whose design does not allow shield current to flow in signal reference conductors. The first uses a star connection of input pin 1, output pin 1, power cord safety ground, and power supply common. This technique is the most effective prevention. Noise currents still flow, but not through internal signal reference conductors. Before there were printed circuit boards, a metal chassis served as a very low-impedance connection (effectively a ground plane) connecting all pins 1 to each other and to safety ground. Pin 1 problems were virtually unknown in those vintage designs. Modern printed circuit board–mounted connectors demand that proper attention be paid to the routes taken by ground noise currents. Of course, this same kind of problem can and does exist with RCA connectors in unbalanced consumer equipment, too.

Fortunately, tests to reveal such common impedance coupling problems are not complex. Comprehensive tests using lab equipment covering a wide frequency range have been described by Cal Perkins (1995)

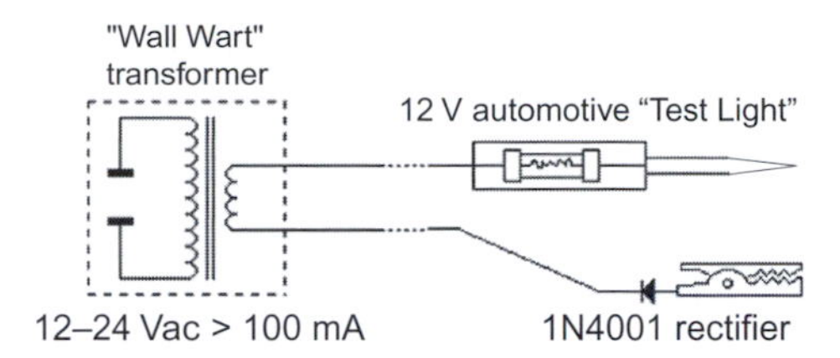

FIGURE 7.34
The Hummer II. (Courtesy Jensen Transformers, Inc.)

and simple tests using an inexpensively built tester called the *hummer* have been described by John Windt (1995). Jensen Transformers, Inc. variant of the Hummer is shown in Figure 7.34. It passes a rectified AC current of 60–80 mA through the potentially troublesome shield connections in the device under test to determine if they cause the coupling.

The glow of the automotive test lamp shows that a good connection has been made and that test current is indeed flowing. The procedure:

1. Disconnect all input and output cables, except the output to be monitored, as well as any chassis connections (due to rack mounting, for example) from the device under test.
2. Power up the device.
3. Meter and, if possible, listen to the device output. Hopefully, the output will simply be random noise. Try various settings of operator controls to familiarize yourself with the noise characteristics of the device under test without the hummer connected.
4. Connect the hummer clip lead to the device chassis and touch the probe tip to pin 1 of each input or output connector. If the device is properly designed, there will be no output hum or change in the noise floor.
5. Test other potentially troublesome paths, such as from an input pin 1 to an output pin 1 or from the safety ground pin of the power cord to the chassis (a three-to-two-prong AC adapter is handy to make this connection).

Note: Pin 1 might not be connected directly to ground in some equipment—hopefully, this will be at inputs only! In this case, the hummer's lamp may not glow—this is OK.

Balanced cable issues

At audio frequencies, even up to about 1 MHz, cable shields should be grounded at one end only, where the signal is ground referenced. At higher frequencies, where typical system cables become a small fraction of a wavelength, it's necessary to ground it at more than one point to keep it at ground potential and guard against RF interference (Ott, 1988, p. 10; Morrison, 1992, p. 55). Based on my own work, there are two additional reasons that there should always be a shield ground at the driver end of the cable, whether the receiver end is grounded or not, see Figures 7.35 and 7.36. The first reason involves the cable capacitances between each signal conductor and shield, which are mismatched by 4% in typical cable. If the shield is grounded at the receiver end, these capacitances and driver common-mode output impedances, often mismatched by 5% or more, form a pair of low-pass filters for common-mode noise. The mismatch in the filters converts a portion of common-mode noise to differential signal. If the shield is connected only at the driver, this mechanism does not exist. The second reason involves the same capacitances working in concert with signal asymmetry. If signals were perfectly symmetrical and capacitances perfectly matched, the capacitively coupled signal current in the shield would be zero through cancellation. Imperfect symmetry and/or capacitances will cause signal current in the shield. This current should be returned directly to the driver from which it came. If the shield is grounded at the receiver, all or part of this current will return via an undefined path that can induce crosstalk, distortion, or oscillation (Whitlock, 1995a, pp. 460–462).

With cables, too, there is a conflict between the star and mesh grounding methods. But this low-frequency versus high-frequency conflict can be substantially resolved with a hybrid approach involving grounding the receive end of cables through an appropriate capacitance (shown in the third device of Figure 7.33) (Ott, 1988, p. 105; Morrison, 1992, p. 55). Capacitor values in the range of 10 nF to 100 nF are most appropriate for the purpose. Such capacitance has been integrated into the Neutrik EMC

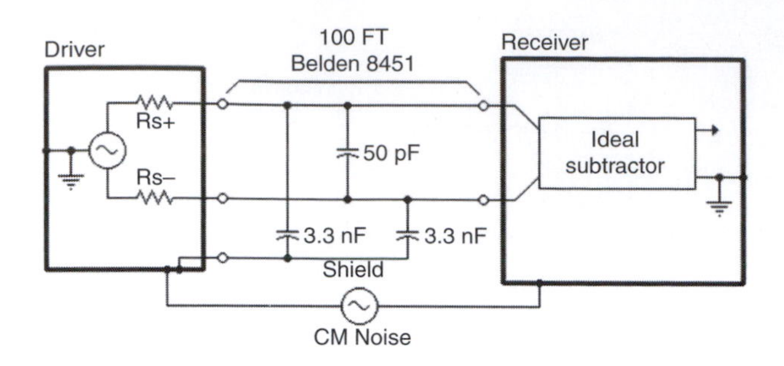

FIGURE 7.35
Shield grounded only at driver.

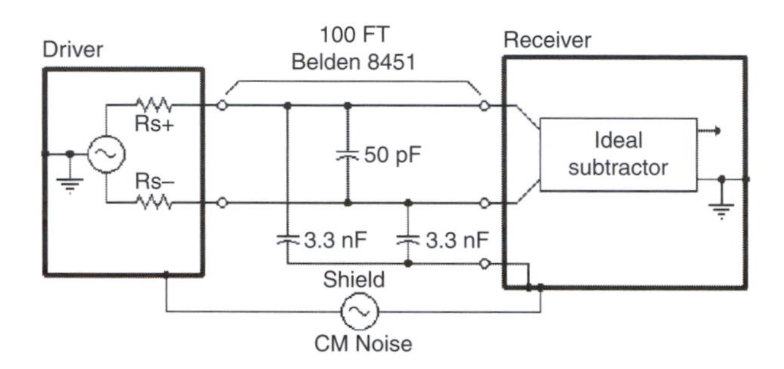

FIGURE 7.36
Shield grounded only at receiver.

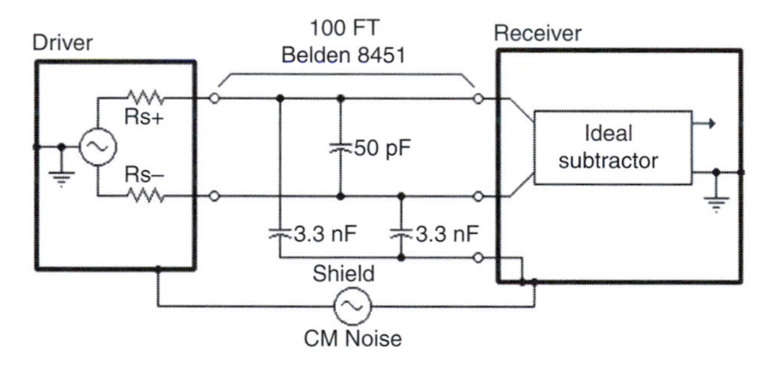

FIGURE 7.37
Shield of a shielded twisted pair cable is magnetically coupled to inner conductors.

series connectors. The merits of this scheme were the subject of several years of debate in the Audio Engineering Society Standards Committee working group that developed AES48.

Twisting essentially places each conductor at the same average distance from the source of a magnetic field and greatly reduces differential pickup. Star quad cable reduces pickup even further, typically by about 40 dB. But the downside is that its capacitance is approximately double that of standard shielded twisted pair.

SCIN, or shield-current-induced noise, may be one consequence of connecting a shield at both ends. Think of a shielded twisted pair as a transformer with the shield acting as primary and each inner conductor acting as a secondary winding, as shown in the cable model of Figure 7.37. Current flow in the shield produces a magnetic field which then induces a voltage in each of the inner conductors. If these voltages are identical, and the interface is properly impedanc balanced, only a common-mode voltage is produced that can be rejected by the line receiver. However, subtle variations in physical construction of the cable can produce unequal coupling in the two signal conductors. The difference voltage, since it appears as signal to the receiver, results in noise coupling. See Muncy, 1995, pp. 441–442 for test results on six commercial cable types. In general, braided shields perform better than foil shields and drain wires.

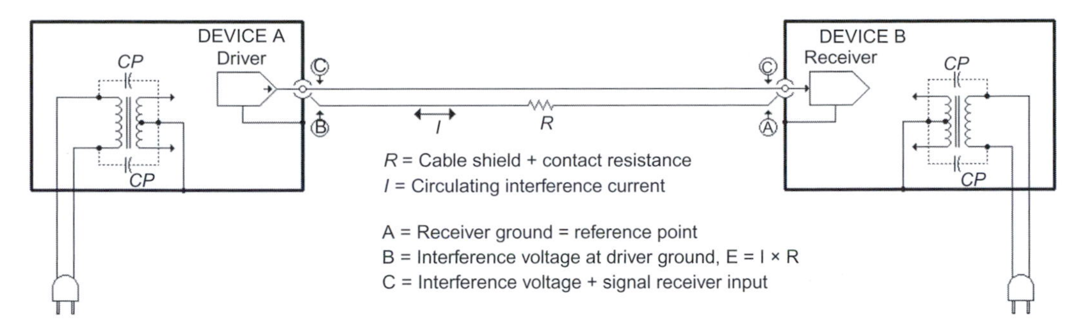

FIGURE 7.38
Common impedance coupling in an unbalanced audio, video, or data interface.

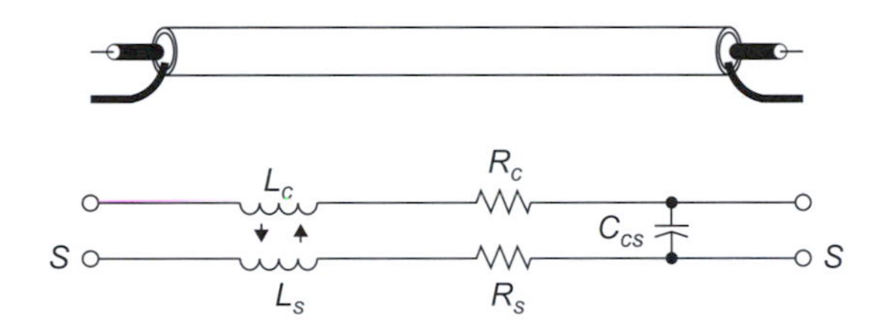

FIGURE 7.39
Magnetic coupling between shield and center conductor is 100%.

And, to make matters even worse, grounding the shield of balanced interconnect cables at both ends also excites the pin 1 problem if it exists. Although it might appear that there's little to recommend grounding at both ends, it is a widely accepted practice. As you can see, noise rejection in a real-world balanced interface can be degraded by a number of subtle problems and imperfections. But, it is virtually always superior to an unbalanced interface!

Coupling in unbalanced cables

The overwhelming majority of consumer as well as high-end audiophile equipment still uses an audio interface system introduced over 60 years ago and intended to carry signals from chassis to chassis inside the earliest RCA TV receivers! The ubiquitous RCA cable and connector form an unbalanced interface that is extremely susceptible to common impedance noise coupling.

As shown in Figure 7.38, noise current flow between the two device grounds or chassis is through the shield conductor of the cable. This causes a small but significant noise voltage to appear across the length of the cable. Because the interface is unbalanced, this noise voltage will be directly added to the signal at the receiver (Whitlock, 1996b). In this case, the impedance of the shield conductor is responsible for the *common impedance coupling*. This coupling causes hum, buzz, and other noises in audio systems. It's also responsible for slow-moving hum bars in video interfaces and glitches, lock-ups, or crashes in unbalanced—e.g., RS-232—data interfaces.

Consider a 25 ft interconnect cable with foil shield and a #26 AWG drain wire. From standard wire tables or actual measurement, its shield resistance is found to be 1.0 Ω. If the 60 Hz leakage current is 300 μA, the hum voltage will be 300 μV. Since the consumer audio reference level is about −10 dBV or 300 mV, the 60 Hz hum will be only $20\log(300\,\mu\text{V}/300\,\text{mV}) = -60\,\text{dB}$ relative to the signal. For most systems, this is a very poor signal-to-noise ratio! For equipment with two-prong plugs, the 60 Hz harmonics and other high-frequency power-line noise (refer to Figure 7.20) will be capacitively coupled and result in a harmonic-rich buzz.

Because the output impedance of device A and the input impedance of device B are in series with the inner conductor of the cable, its impedance has an insignificant effect on the coupling and is not represented here. Common-impedance coupling can become extremely severe between two grounded devices, since the voltage drop in the safety ground wiring between the two devices is effectively parallel connected across the length of the cable shield. This generally results in a fundamental-rich hum that may actually be larger than the reference signal!

Coaxial cables, which include the vast majority of unbalanced audio cables, have an interesting and underappreciated quality regarding common-impedance coupling at high frequencies, Figure 7.39. Any voltage appearing across the ends of the shield will divide itself between shield inductance L_s and resistance R_s according to frequency. At some frequency, the voltages across each will be equal (when reactance of L_s equals R_s). For typical cables, this frequency is in the 2 to 5 kHz range. At frequencies below this transition frequency, most of the ground noise will appear across R_s and be coupled into the audio signal as explained earlier. However, at frequencies above the transition frequency, most of the ground noise will appear across L_s. Since L_s is magnetically coupled to the inner conductor, a replica of the ground noise is induced over its length. This induced voltage is then subtracted from the signal on the inner conductor, reducing noise coupling into the signal. At frequencies 10 times the transition frequency, there is virtually no noise coupling at all—common-impedance coupling has disappeared. Therefore, common-impedance coupling in coaxial cables ceases to be a noise issue at frequencies over about 50 kHz. Remember this as we discuss claims made for power-line filters that typically remove noise only above about 50 kHz.

Unbalanced interface cables, regardless of construction, are also susceptible to magnetically-induced noise caused by nearby low-frequency AC magnetic fields. Unlike balanced interconnections, such noise pickup is not nullified by the receiver.

Bandwidth and RF interference

RF interference isn't hard to find—it's actually very difficult to avoid, especially in urban areas. It can be radiated through the air and/or be conducted through any cables connected to equipment. Common sources of radiated RF include AM, shortwave, FM, and TV broadcasts; ham, CB, remote control, wireless phone, cellular phone, and a myriad of commercial two-way radio and radar transmitters; and medical and industrial RF devices. Devices that create electrical sparks, including welders, brush-type motors, relays, and switches can be potent wideband radiators. Less obvious sources include arcing or corona discharge from power line insulators (common in seashore areas or under humid conditions) or malfunctioning fluorescent, HID, or neon lights. Of course, lightning, the ultimate spark, is a well-known radiator of momentary interference to virtually anything electronic.

Interference can also be conducted via any wire coming into the building. Because power and telephone lines also behave as huge outdoor antennas, they are often teeming with AM radio signals and other interference. But the most troublesome sources are often inside the building and the energy delivered through AC power wiring. The offending source may be in the same room as your system or, worse yet, it may actually be a part of your system! The most common offenders are inexpensive light dimmers, fluorescent lights, CRT displays, digital signal processors, or any device using a switching power supply.

Although cable shielding is a first line of defense against RF interference, its effectiveness depends critically on the shield connection at each piece of equipment. Because substantial inductance is added to this connection by traditional XLR connectors and grounding pigtails, the shield becomes useless at high radio frequencies. Common-mode RF interference simply appears on all the input leads (Morrison, 1992 pp. 136–137). Because the wire limitations discussed earlier apply to grounding systems. Contrary to widespread belief, grounding is not an effective way to deal with RF interference. To quote Neil Muncy:

> Costly technical grounding schemes involving various and often bizarre combinations of massive copper conductors, earth electrodes, and other arcane hardware are installed. When these schemes fail to provide expected results, their proponents are usually at a loss to explain why (Muncy, 1995, p. 436).

The wider you open the window, the more dirt flies in. One simple, but often overlooked, method of minimizing noise in a system is to limit the system bandwidth to that required by the signal (Ott, 1988, p. 134). In an ideal world, every signal-processing device in a system would contain a filter at each input and output connector to appropriately limit bandwidth and prevent out-of-band energy from ever reaching active circuitry. This RF energy becomes an audio noise problem because the RF is demodulated or detected by active circuitry in various ways, acting like a radio receiver that adds its output to the audio signal. Symptoms can range from actual reception of radio signals or a 59.94 Hz buzz from TV signals or various tones from cell phone signals to much subtler distortions, often described as a veiled or grainy audio quality (Jensen and Sokolich, 1988). The filters necessary to prevent these problems vary widely in effectiveness and, in some equipment, may not be present at all. Sadly, the performance of most commercial equipment will degrade when such interference is coupled to its input (Whitlock, 1999b).

SOLVING REAL-WORLD SYSTEM PROBLEMS

How much noise and interference are acceptable depends on what the system is and how it will be used. Obviously, sound systems in a recording studio need to be much more immune to noise and interference than paging systems for construction sites.

Noise perspective

The decibel is widely used to express audio-related measurements. For *power* ratios:

$$dB = 101\log\frac{P_1}{P_2} \tag{7.9}$$

For *voltage* or *current* ratios, because power is proportional to the square of voltage or current:

$$\begin{aligned} dB &= 20\log\frac{E_1}{E_2} \\ dB &= 20\log\frac{I_1}{I_2} \end{aligned} \tag{7.10}$$

Most listeners describe 10 dB level decreases or increases as halving or doubling loudness, respectively, and 2 dB or 3 dB changes as just noticeable. Under laboratory conditions, well-trained listeners can usually identify level changes of 1 dB or less. The dynamic range of an electronic system is the ratio of its maximum undistorted signal output to its residual noise output or noise floor. Up to 120 dB of dynamic range may be required in high-end audiophile sound systems installed in typical homes (Fielder, 1995).

Troubleshooting

Under certain conditions, many systems will be acceptably noise-free in spite of poor grounding and interfacing techniques. People often get away with doing the wrong things! But, notwithstanding anecdotal evidence to the contrary, logic and physics will ultimately rule.

Troubleshooting noise problems can be a frustrating, time-consuming experience but the method described on page 244 can relieve the pain. It requires no electronic instruments and is very simple to perform. Even the underlying theory is not difficult. The tests will reveal not only what the coupling mechanism is but also where it is.

OBSERVATIONS, CLUES, AND DIAGRAMS

A significant part of troubleshooting involves how you think about the problem. First, don't assume anything! For example, don't fall into the trap of thinking, just because you've done something a particular way many times before, it simply can't be the problem. Remember, even things that can't go wrong, do! Resist the temptation to engage in guesswork or use a shotgun approach. If you change more than one thing at a time, you may never know what actually fixed the problem.

Second, ask questions and gather clues! If you have enough clues, many problems will reveal themselves before you start testing. Be sure to write everything down—imperfect recall can waste a lot of time. Troubleshooting guru Bob Pease (1991) suggests these basic questions:

1. Did it ever work right?
2. What are the symptoms that tell you it's not working right?
3. When did it start working badly or stop working?
4. What other symptoms showed up just before, just after, or at the same time as the failure?

Operation of the equipment controls, and some elementary logic, can provide very valuable clues. For example, if a noise is unaffected by the setting of a gain control or selector, logic dictates that it must be entering the signal path after that control. If the noise can be eliminated by turning the gain down or selecting another input, it must be entering the signal path before that control.

Third, sketch a block diagram of the system. Figure 7.40 is an example diagram of a simple home theater system. Show all interconnecting cables and indicate approximate length. Mark any balanced inputs or outputs. Generally, stereo pairs can be indicated with a single line. Note any device that is

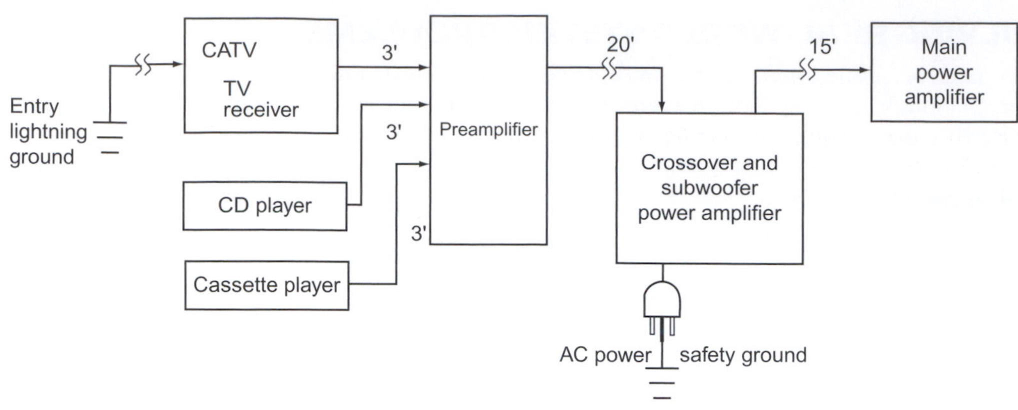

FIGURE 7.40
Block diagram of example system.

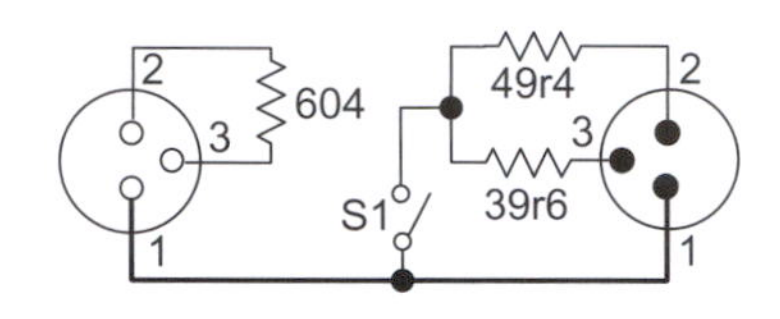

For balanced audio XLR
P1/J1 = Switchcraft S3FM adapter with QG3F and QG3M insert
All resistors 1%, ¼ W metal film
Close S1 for CMRR test only

For balanaced audio 3C phone
Use Switchcraft 383A and 387A adapters with XLR version

FIGURE 7.41
Balanced ground dummy.

grounded via a three-prong AC plug. Note any other ground connections such as equipment racks, cable TV connections, etc.

THE GROUND DUMMY PROCEDURE

An easily constructed adapter or ground dummy is the key element in this procedure. By temporarily placing the dummy at strategic locations in the interfaces, precise information about the nature and location of the problem is revealed. The tests can specifically identify:

1. Common-impedance coupling in unbalanced cables.
2. Shield-current-induced coupling in balanced cables.
3. Magnetic or electrostatic pickup of nearby magnetic or electrostatic fields.
4. Common-impedance coupling (the pin 1 problem) inside defective devices.
5. Inadequate CMRR of the balanced input.

The ground dummy can be made from standard connector wired as shown in Figures 7.41 and 7.42. Since a dummy does not pass signal, mark it clearly to help prevent it being accidentally left in a system.

Each signal interface is tested in four steps. As a general rule, always start at the inputs to the power amplifiers and work backward toward the signal sources. *Be very careful when performing the tests not to damage loudspeakers or ears!* The surest way to avoid possible damage is to turn off the power amplifier(s) before reconfiguring cables for each test step.

For unbalanced interfaces

STEP 1: Unplug the cable from the input of Box B and plug in only the dummy as shown below.

- Output quiet?

  No—The problem is either in Box B or farther downstream.
  Yes—Go to next step.

Unbalanced interfaces

For audio RCA
P1 = Switchcraft 3502 plug
J1 = Switchcraft 3503 jack
R = 1 kΩ, 5%, ¼ W resistor

For audio 2C phone
Use Switchcraft 336A and 345A adapters with RCA version

FIGURE 7.42
Unbalanced ground dummy.

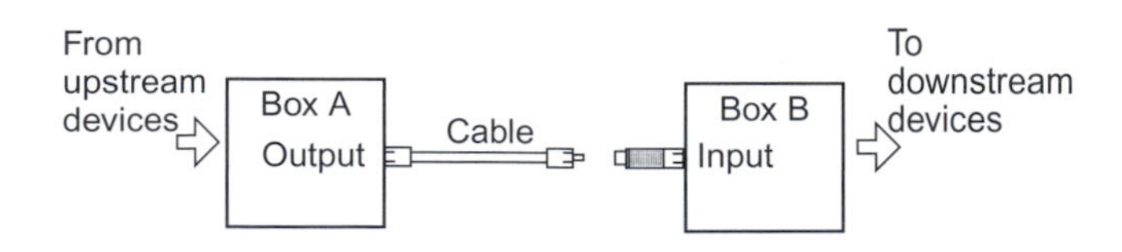

STEP 2: Leaving the dummy in place at the input of Box B, plug the cable into the dummy as shown below.

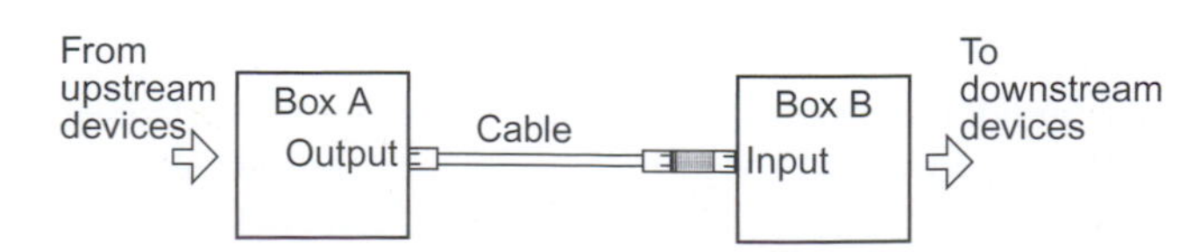

- Output quiet?

  No—Box B has a pin 1 problem.

  Yes—Go to next step.

STEP 3: Remove the dummy and plug the cable directly into the input of Box B. Unplug the other end of the cable from the output of Box A and plug it into the dummy as shown below. Do not plug the dummy into Box A or let it touch anything conductive.

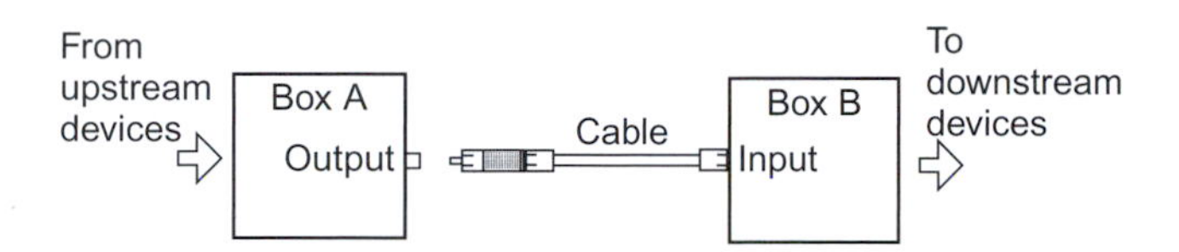

- Output quiet?

  No—Noise is being induced in the cable itself. Reroute the cable to avoid interfering fields.

  Yes—Go to next step.

STEP 4: Leaving the dummy in place on the cable, plug the dummy into the output of Box A as shown below.

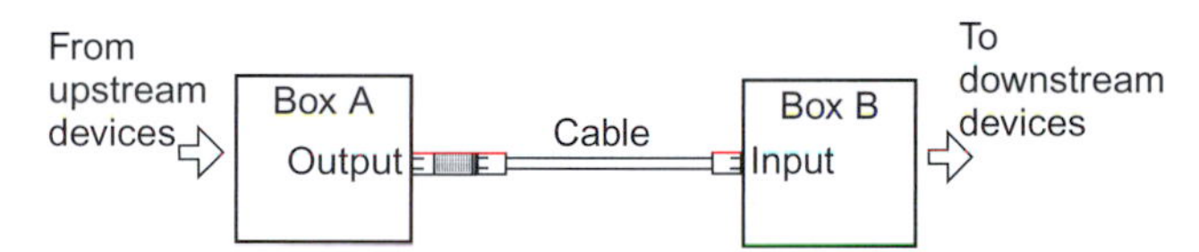

- Output quiet?

  No—The problem is common-impedance coupling. Install a ground isolator at the input of Box B.

  Yes—The noise is coming from (or through) the output of Box A. Perform the same test sequence on the cable(s) connecting Box A to upstream devices.

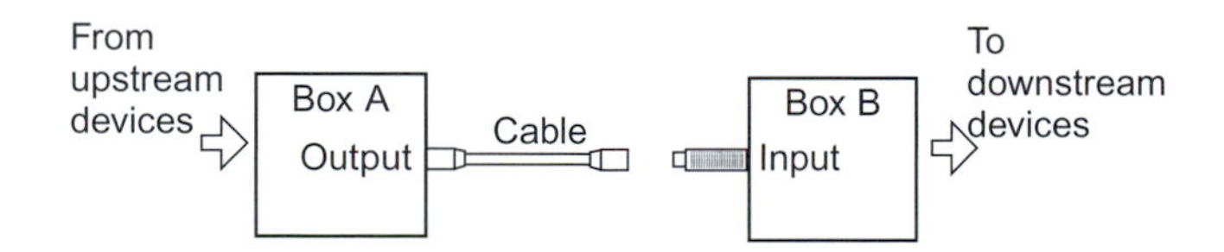

For balanced interfaces

STEP 1: Unplug the cable from the input of Box B and plug in only the dummy (switch open or NORM) as shown below.

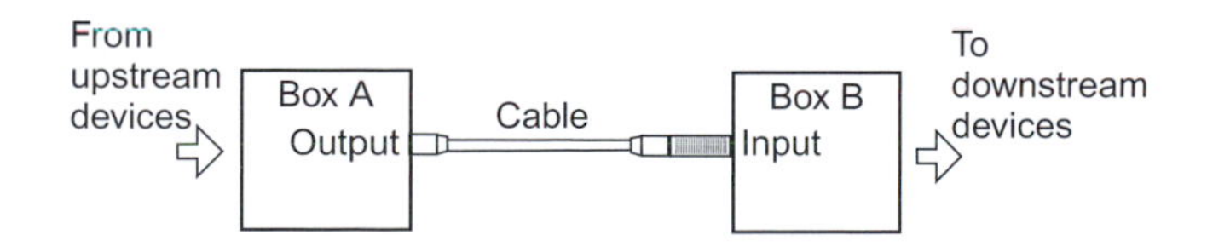

- Output quiet?

  No—The problem is either in Box B or farther downstream.

  Yes—Go to next step.

STEP 2: Leaving the dummy in place at the input of Box B, plug the cable into the dummy (switch open or NORM) as shown below.

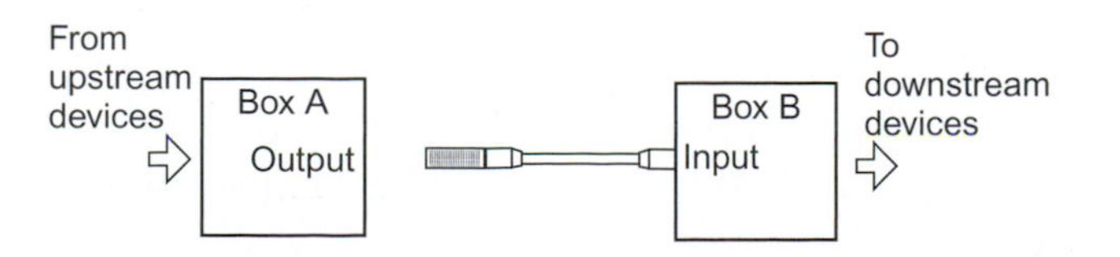

- Output quiet?

  No—Box B has a Pin 1 problem (see hummer test earlier to confirm this).
  Yes—Go to next step.

STEP 3: Remove the dummy and plug the cable directly into the input of Box B. Unplug the other end of the cable from the output of Box A and plug it into the dummy (switch open or NORM) as shown below. *Do not plug the dummy into Box A or let it touch anything conductive.*

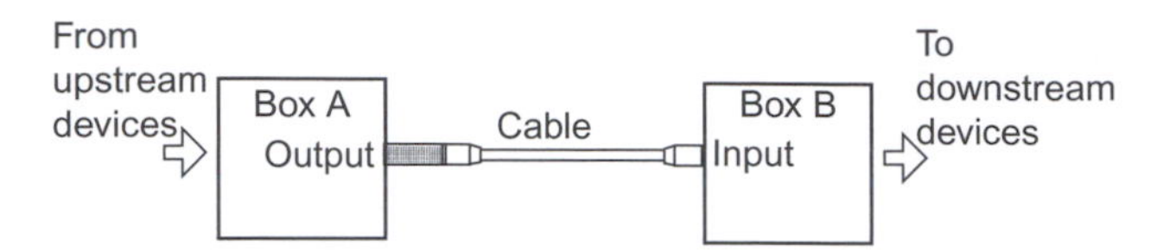

- Output quiet?

  No—Noise is being induced in the cable itself by an electric or magnetic field. Check the cable for an open shield connection, reroute the cable to avoid the interfering field, or replace the cable with a star quad type.
  Yes—Go to next step.

STEP 4: Leaving the dummy in place on the cable, plug the dummy (switch open or NORM) into the output of Box A as shown below.

- Output quiet?

  No—The problem is shield-current-induced noise. Replace the cable with a different type (without a drain wire) or take steps to reduce current in the shield.
  Yes—Go to next step.

STEP 5: Leave the dummy and cable as for step 4, but move the dummy switch to the CMRR (closed) position.

- Output quiet?

  No—The problem is likely inadequate common-mode rejection of the input stage of Box B. This test is based on the IEC common-mode rejection test but uses the actual common-mode voltage present in the system. The nominal 10 Ω imbalance may not simulate the actual imbalance at the output of Box A, but the test will reveal input stages whose CMRR is sensitive to source imbalances. Most often, adding a transformer-based ground isolator at the input of Box B will cure the problem.
  Yes—The noise must be coming from (or through) the output of Box A. Perform the same test sequence on the cable(s) connecting Box A to upstream devices.

Solving interface problems

GROUND ISOLATORS

A device called a *ground isolator* solves the inherent common-impedance coupling problem in unbalanced interfaces. Broadly defined, a ground isolator is a differential responding device with high common-mode rejection. It is not a filter that can selectively remove hum, buzz, or other noises when simply placed anywhere in the signal path. To do its job, it must be installed where the noise coupling would otherwise occur.

A transformer is a passive device that fits the definition of a ground isolator. Transformers transfer a voltage from one circuit to another without any electrical connections between the two circuits. It converts an AC signal voltage on its primary winding into a fluctuating magnetic field that is then converted back to an AC signal voltage on its secondary winding.

As shown in Figure 7.43, when a transformer is inserted into an unbalanced signal path, the connection between device grounds via the cable shield is broken. This stops the noise current flow in the shield conductor that causes the noise coupling. The highest noise rejection is achieved with input-type transformers containing Faraday shields. A transformer-based isolator for consumer audio signals using such transformers, the ISO-MAX® model CI-2RR, is shown in Figure 7.44. To avoid bandwidth loss, such isolators must be located at the receive end of interconnections, using minimum-length cables between isolator outputs and equipment inputs. Conversely, isolators using output-type transformers, such as

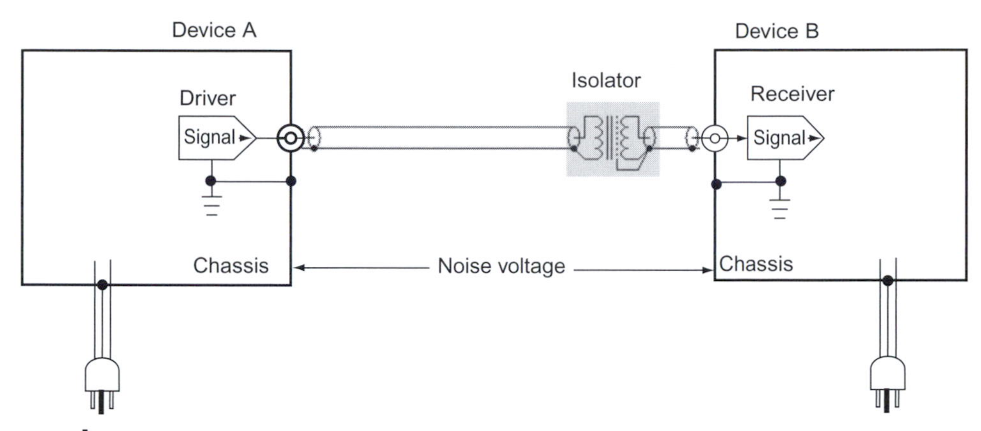

FIGURE 7.43
Ground isolator stops noise current in shield of unbalanced cable.

FIGURE 7.44
Stereo unbalanced audio isolator. (Courtesy of Jensen Transformers, Inc.)

FIGURE 7.45
Stereo balanced audio isolator. Courtesy of Jensen Transformers, Inc.

the ISO-MAX® model CO-2RR and most other commercial isolators, may be freely located but will achieve significantly less noise rejection.

Ground isolators can also solve most of the problems associated with balanced interfaces. The ISO-MAX® Pro model PI-2XX shown in Figure 7.45 often improves CMRR by 40 dB to 60 dB and provides excellent CMRR even if the signal source is unbalanced. Because it also features DIP switches to reconfigure cable shield ground connections, it can also solve pin 1 problems. Because it uses input-type transformers, it attenuates RF interference such as AM radio by over 20 dB. Again, to avoid bandwidth loss, it must be located at the receive end of long cable runs, using minimum-length cables between isolator outputs and equipment inputs. Other models are available for microphone signals and other applications. The vast majority of commercial hum eliminators and a few special-purpose ISO-MAX® models use output-type transformers, which may be freely located but offer significantly less CMRR improvement and have essentially no RF attenuation.

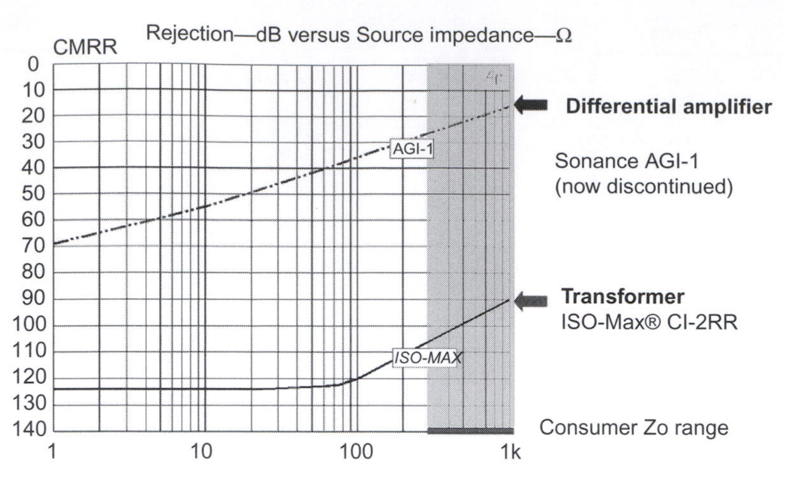

FIGURE 7.46
Measured hum rejection versus source impedance; active differential amplifier versus input transformer isolator.

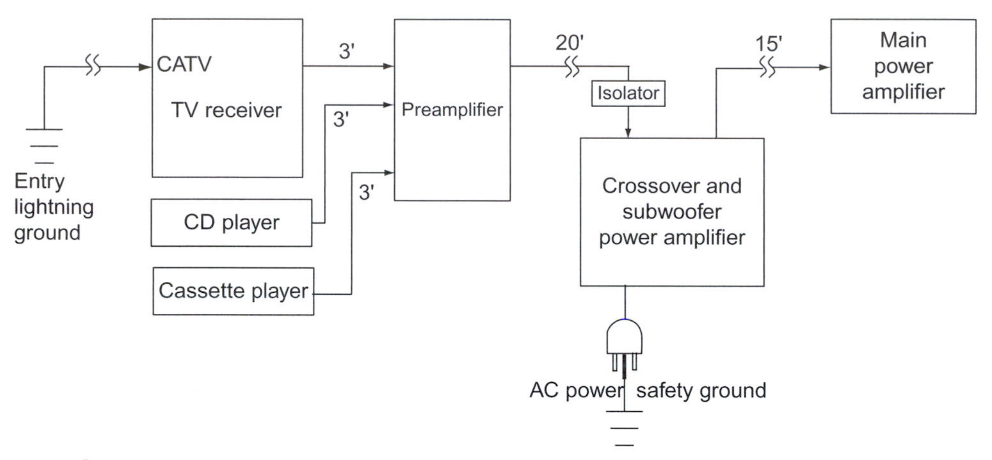

FIGURE 7.47
Using an audio ground isolator to break the loop.

Several manufacturers make active (i.e., powered) ground isolators using some form of the simple differential amplifier shown in Figure 7.31. Unfortunately, these circuits are exquisitely sensitive to the impedance of the driving source. Figure 7.46 compares the measured 60 Hz (hum) rejection of a typical active isolator to a transformer-based isolator. Over the typical range of consumer output impedances, 100 Ω to 1 kΩ, the transformer has about 80 dB more rejection!

Passive isolators based on input-type transformers have other advantages, too. They require no power, they inherently suppress RF interference, and they're immune to most overvoltages that can be sudden death to active circuitry.

MULTIPLE GROUNDING

When a system contains two or more grounded devices, such as the TV receiver and the subwoofer power amplifier in our example home theater system, a wired ground loop is formed as shown in Figure 7.47.

As discussed earlier, noise current flowing in the shaded path can couple noise into the signal as it flows in unbalanced cables or through the equipment's internal ground path. This system would likely exhibit a loud hum regardless of the input selected or the setting of the volume control because of noise current flow in the 20 ft cable. You might be tempted to break this ground loop by lifting the safety ground at the subwoofer.

One safe solution is to break the ground loop by installing a ground isolator in the audio path from preamp to subwoofer as shown in Figure 7.48. This isolator could also be installed in the path from

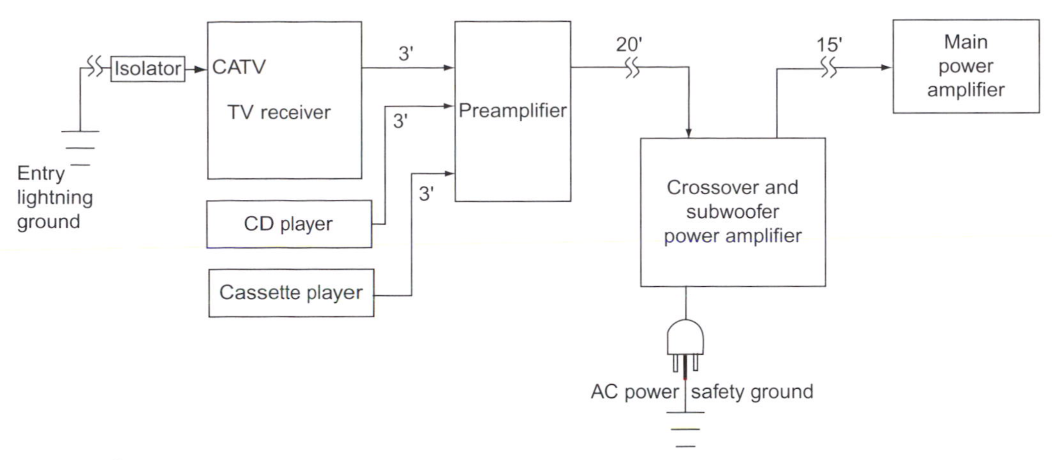

FIGURE 7.48
Using a CATV ground isolator to break the loop.

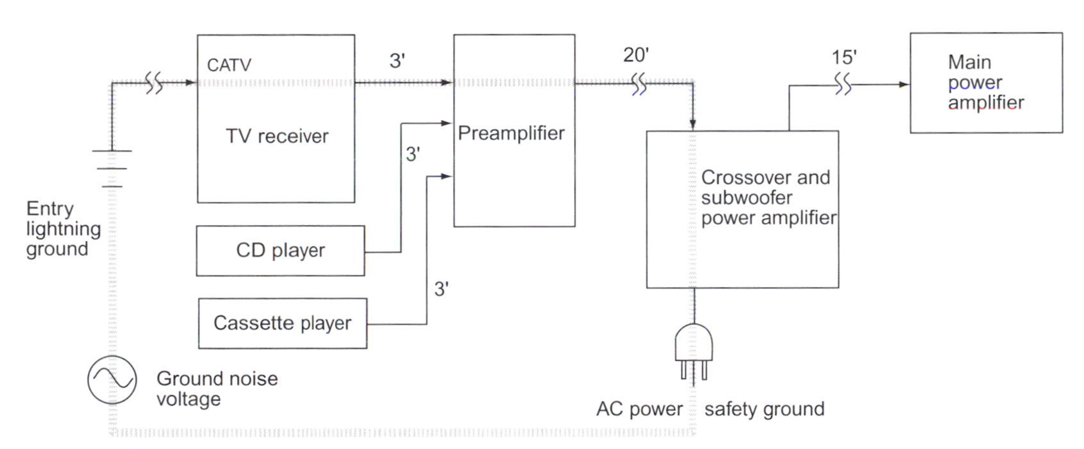

FIGURE 7.49
Loop created by two ground connections.

TV receiver to preamp, but it is generally best to isolate the longest lines since they are more prone to coupling than shorter ones.

Another safe solution is to break the ground loop by installing a ground isolator in the CATV signal path at the TV receiver as shown in Figure 7.49. These RF isolators generally should be installed where the cable connects to the local system, usually at a VCR or TV input. If an RF isolator is used at the input to a splitter, ground loops may still exist between systems served by the splitter outputs since the splitter provides no ground isolation. Although it can be used with a conventional TV or FM antenna, never install an RF isolator between the CATV drop or antenna and its lightning ground connection. Isolators will not pass DC operating power to the dish in DBS TV systems.

Since most unbalanced interfaces are made to consumer devices that have two-prong AC plugs, isolating the signal interfaces may leave one or more pieces of equipment with no ground reference whatsoever. This could allow the voltage between an isolator's input and output to reach 50 Vac or more. While this isn't dangerous (leakage current is limited in UL-listed devices), it would require unrealistically high (CMRR over 140 dB) performance by the isolator to reject it! The problem is solved by grounding any floating gear as shown in Figure 7.50. This is best done by replacing the two-prong AC plug with a three-prong type and adding a wire (green preferred) connected between the safety ground pin of the new AC plug and a chassis ground point.

A screw may be convenient as the chassis ground point. Use an ohmmeter to check for continuity between the screw and the outer contact of an RCA connector, which itself can be used if no other

PREFACE

The section on "Power Amplifiers" in this book is contributed by myself. I have spent a good deal of my time working on the apparently simple but actually highly complex business of turning small voltages and currents into large ones without damaging them in the process. I still find it a most absorbing subject of study. This chapter is taken from my *Audio Power Amplifier Design Handbook* which is now available in a greatly extended fifth edition.

This chapter from *The Handbook* begins by explaining the basic architecture of power amplifiers. While the detailed circuitry of the many audio power amplifiers that have been put on the market shows almost unending variation, most of them can be classified by the way that they are split up into stages, and can be quickly assigned to a two-stage, a three-stage, or a four-stage structure, in which a unity-gain power output configuration forms the final stage. Two-stage amplifiers are generally deficient in both open-loop gain and basic linearity, and they were only ever accepted for budget applications. Their place in the cheap-and-cheerful amplifier market sector has now been taken over by integrated power amplifiers.

The majority of power amplifiers use a three-stage architecture. The first stage is a transconductance stage that converts a voltage-difference to a current output. The second is a transadmittance stage, which takes a current input and gives a voltage output. This voltage then drives a unity-gain power output stage. The voltage-current-voltage conversions may appear to be a roundabout, if not actually perverse, way of doing things, but in fact this approach has a number of subtle advantages in terms of linearity and stability, and that is why it is so popular. It can easily provide enough open-loop gain for as much negative feedback as you can use.

Four-stage amplifiers keep the signal in a voltage format. Because of this, and the extra phase-shifts introduced by the extra stage, dependable stability requires some complicated compensation schemes. Linearity also tends to be inferior.

Audio power amplifiers are assigned to different classes, depending on how the output stage operates. These are called Class-A, Class-AB, Class-B, Class-D, Class-G and Class-S, though on examination most are simple combinations of the three fundamental classes, which are Class-A, Class-B, and Class-C. All are described here. Some less well known types such as error-correcting, Blomley, current-drive and non-switching power amplifiers are also covered.

I then look at the bridging of amplifiers, where two are connected in anti-phase to double the effective output voltage, and thus quadruple the power output. I describe my introduction of the "fractional bridging" principle, which can be extremely useful if you need more power from a pair of amplifiers, but not as much as four times more.

In the next section I examine the issue of AC coupling versus DC coupling at the amplifier output. Once all solid state amplifiers were coupled to their load via a big electrolytic capacitor. This component disappeared when differential pairs were introduced to the first stages of amplifiers, because they allowed small offset voltages at the output and thus direct connection, but reappeared in the power supply because two supply rails were needed. A serious drawback to AC coupling is that the output capacitor is likely to introduce distortion.

The final part of the chapter looks at negative feedback. The fact of the matter is that you need a good healthy amount of negative feedback to make a decent amplifier, and it is essential to understand how it works, and what are its advantages and limitations.

CHAPTER 8

Power Amplifier Architecture and Negative Feedback

Audio Power Amplifier Design Handbook by Douglas Self

AMPLIFIER ARCHITECTURES

This grandiose title simply refers to the large-scale structure of the amplifier; that is, the block diagram of the circuit one level below that representing it as a single white block labeled Power Amplifier. Almost all solid-state amplifiers have a three-stage architecture as described below, though they vary in the detail of each stage. Two-stage architectures have occasionally been used, but their distortion performance is not very satisfactory. Four-stage architectures have been used in significant numbers, but they are still much rarer than three-stage designs, and usually involve relatively complex compensation schemes to deal with the fact that there is an extra stage to add phase shift and potentially imperil high-frequency stability.

The three-stage amplifier architecture

The vast majority of audio amplifiers use the conventional architecture, shown in Figure 8.1, and so it is dealt with first. There are three stages, the first being a transconductance stage (differential voltage in, current out), the second a transimpedance stage (current in, voltage out), and the third a unity-voltage-gain output stage. The second stage clearly has to provide all the voltage gain and I have therefore called

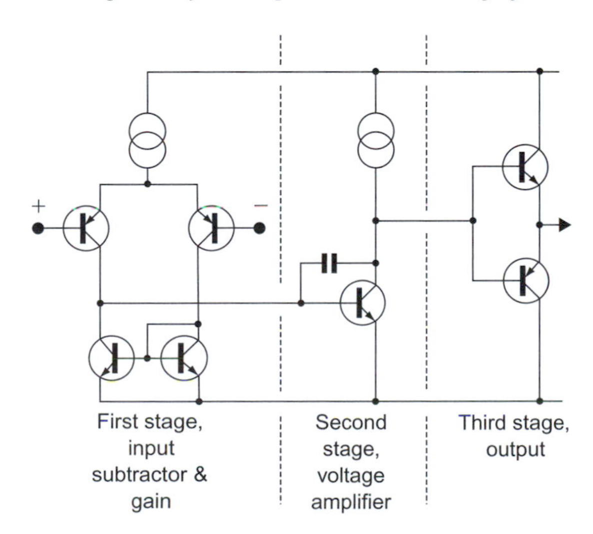

FIGURE 8.1
The three-stage amplifier structure. There is a transconductance stage, a transadmittance stage (the VAS), and a unity-gain buffer output stage.

Audio Engineering Explained

it the *voltage-amplifier stage* or VAS. Other authors have called it the *pre-driver stage* but I prefer to reserve this term for the first transistors in output triples. This three-stage architecture has several advantages, not least being that it is easy to arrange things so that interaction between stages is negligible. For example, there is very little signal voltage at the input to the second stage, due to its current-input (virtual-earth) nature, and therefore very little on the first stage output; this minimizes Miller phase shift and possible early effect in the input devices.

Similarly, the compensation capacitor reduces the second stage output impedance, so that the nonlinear loading on it, due to the input impedance of the third stage, generates less distortion than might be expected. The conventional three-stage structure, familiar though it may be, holds several elegant mechanisms such as this. They will be fully revealed in later chapters. Since the amount of linearizing global negative feedback (NFB) available depends upon amplifier open-loop gain, how the stages contribute to this is of great interest. The three-stage architecture always has a unity-gain output stage—unless you really want to make life difficult for yourself—and so the total forward gain is simply the product of the transconductance of the input stage and the transimpedance of the VAS, the latter being determined solely by the Miller capacitor C_{dom}, except at very low frequencies. Typically, the closed-loop gain will be between +20 and +30 dB. The NFB factor at 20 kHz will be 25–40 dB, increasing at 6 dB/octave with falling frequency until it reaches the dominant pole frequency *P*1, when it flattens out. What matters for the control of distortion is the amount of NFB available, rather than the open-loop bandwidth, to which it has no direct relationship. In my *Electronics World* Class-B design, the input stage g_m is about 9 mA/V, and C_{dom} is 100 pF, giving an NFB factor of 31 dB at 20 kHz. In other designs I have used as little as 26 dB (at 20 kHz) with good results.

Compensating a three-stage amplifier is relatively simple; since the pole at the VAS is already dominant, it can be easily increased to lower the HF negative-feedback factor to a safe level. The local NFB working on the VAS through C_{dom} has an extremely valuable linearizing effect.

The conventional three-stage structure represents at least 99% of the solid-state amplifiers built, and I make no apology for devoting much of this book to its behavior. I am quite sure I have not exhausted its subtleties.

The two-stage amplifier architecture

In contrast with the three-stage approach, the architecture in Figure 8.2 is a two-stage amplifier, the first stage being once more a transconductance stage, though now without a guaranteed low impedance to accept its output current. The second stage combines VAS and output stage in one block; it is inherent in this scheme that the VAS must double as a phase splitter as well as a generator of raw gain. There are then two quite dissimilar signal paths to the output, and it is not at all clear that trying to break this block down further will assist a linearity analysis. The use of a phase-splitting stage harks back to valve amplifiers, where it was inescapable, as a complementary valve technology has so far eluded us.

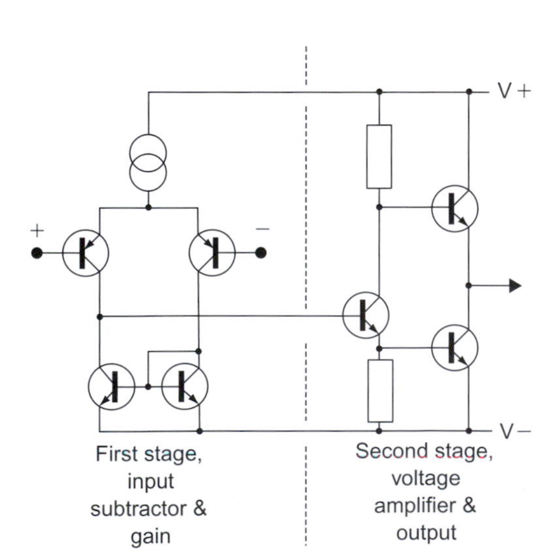

FIGURE 8.2
The two-stage amplifier structure. A voltage-amplifier output follows the same transconductance input stage.

Paradoxically, a two-stage amplifier is likely to be more complex in its gain structure than a three-stage one. The forward gain depends on the input stage g_m, the input stage collector load (because the input stage can no longer be assumed to be feeding a virtual earth) and the gain of the output stage, which will be found to vary in a most unsettling manner with bias and loading. Choosing the compensation is also more complex for a two-stage amplifier, as the VAS/phase splitter has a significant signal voltage on its input and so the usual pole-splitting mechanism that enhances Nyquist stability by increasing the pole frequency associated with the input stage collector will no longer work so well. (I have used the term Nyquist stability, or Nyquist oscillation, throughout this chapter to denote oscillation due to the accumulation of phase shift in a global NFB loop, as opposed to local parasitics, etc.)

The LF feedback factor is likely to be about 6 dB less with a 4 Ω load, due to lower gain in the output stage. However, this variation is much reduced above the dominant pole frequency, as there is then increasing local NFB acting in the output stage.

Here are two examples of two-stage amplifiers: (Linsley-Hood, 1969) and (Olsson, 1994). The two-stage amplifier offers little or no reduction in parts cost, is harder to design, and in my experience invariably gives a poor distortion performance.

The four-stage amplifier architecture

The best-known example of a four-stage architecture is probably that published by Lohstroh and Otala in their influential paper, which was confidently entitled 'An audio power amplifier for ultimate quality requirements' and appeared in December 1973 (Lohstroh and Otala, 1973). A simplified circuit diagram of their design is shown in Figure 8.3. One of their design objectives was the use of a low value of overall feedback, made possible by heavy local feedback in the first three amplifier stages, in the form of emitter degeneration; the closed-loop gain was 32 dB (40 times) and the feedback factor 20 dB, allegedly flat across the audio band. Another objective was the elimination of so-called transient intermodulation distortion, which after many years of argument and futile debate has at last been accepted to mean nothing more than old-fashioned slew-rate limiting. To this end dominant-pole compensation was avoided in this design. The compensation scheme that was used was complex, but basically the lead capacitors C1, C2 and the lead-lag network R19, C3 were intended to cancel out the internal poles of the amplifier. According to Lohstroh and Otala, these lay between 200 kHz and 1 MHz, but after compensation the open-loop frequency response had its first pole at 1 MHz. A final lag compensation network R15, C4 was located outside the feedback loop. An important point is that the third stage was heavily loaded by the two resistors R11, R12. The emitter-follower (EF)-type output stage was biased far into Class-AB by a conventional V_{be}-multiplier, drawing 600 mA of quiescent current, this gives poor linearity when you run out of the Class-A region.

You will note that the amplifier uses shunt feedback; this certainly prevents any possibility of common-mode distortion in the input stage, as there is no common-mode voltage, but it does have the frightening drawback of going berserk if the source equipment is disconnected, as there is then a greatly increased feedback factor, and high-frequency instability is pretty much inevitable. Input common-mode nonlinearity, where it is shown that in normal amplifier designs it is of negligible proportions, and certainly not a good reason to adopt overall shunt feedback.

Many years ago I was asked to put a version of this amplifier circuit into production for one of the major hi-fi companies of the time. It was not a very happy experience. High-frequency stability was very doubtful and the distortion performance was distinctly unimpressive, being in line with that quoted in the original paper as 0.09% at 50 W, 1 kHz (Lohstroh and Otala, 1973). After a few weeks of struggle the four-stage architecture was abandoned and a more conventional (and much more tractable) three-stage architecture was adopted instead.

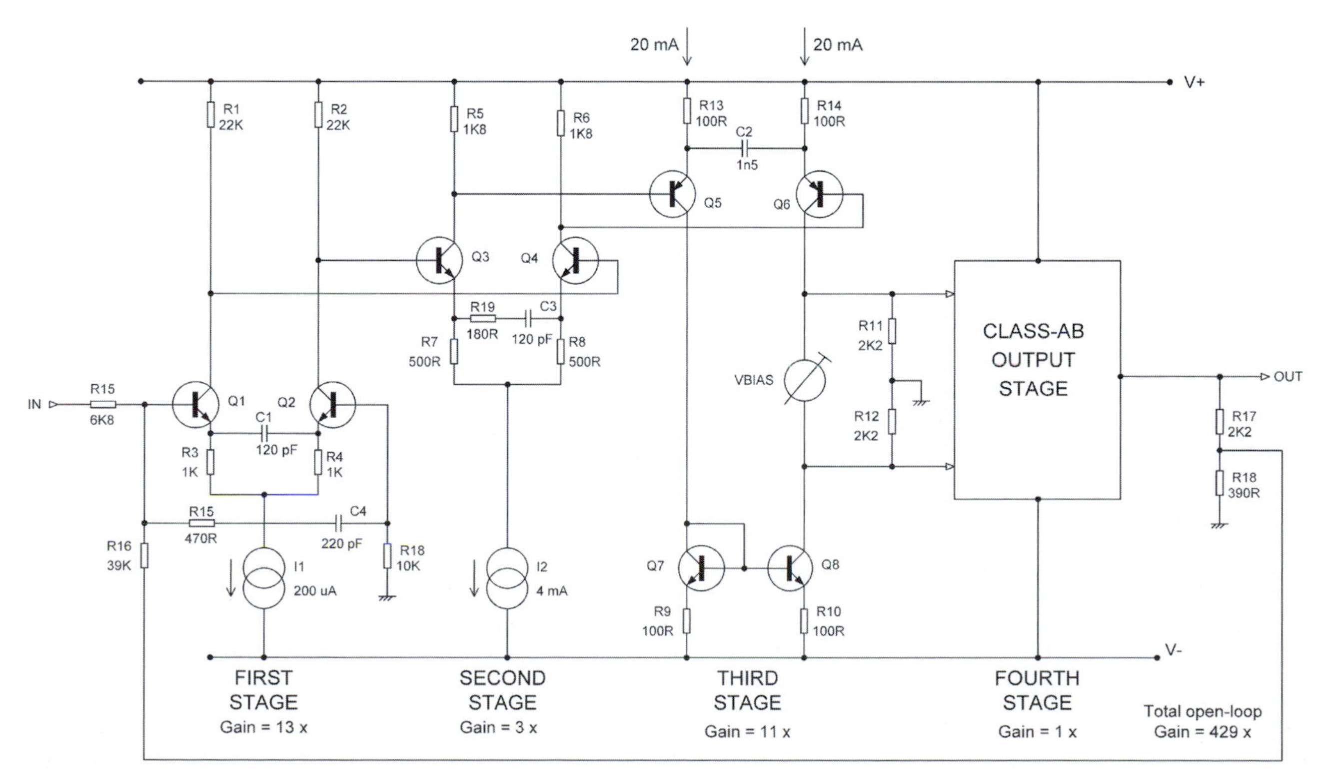

FIGURE 8.3
A simplified circuit diagram of the Lohstroh and Otala four-stage power amplifier. The gain figures for each stage are as quoted in the original paper.

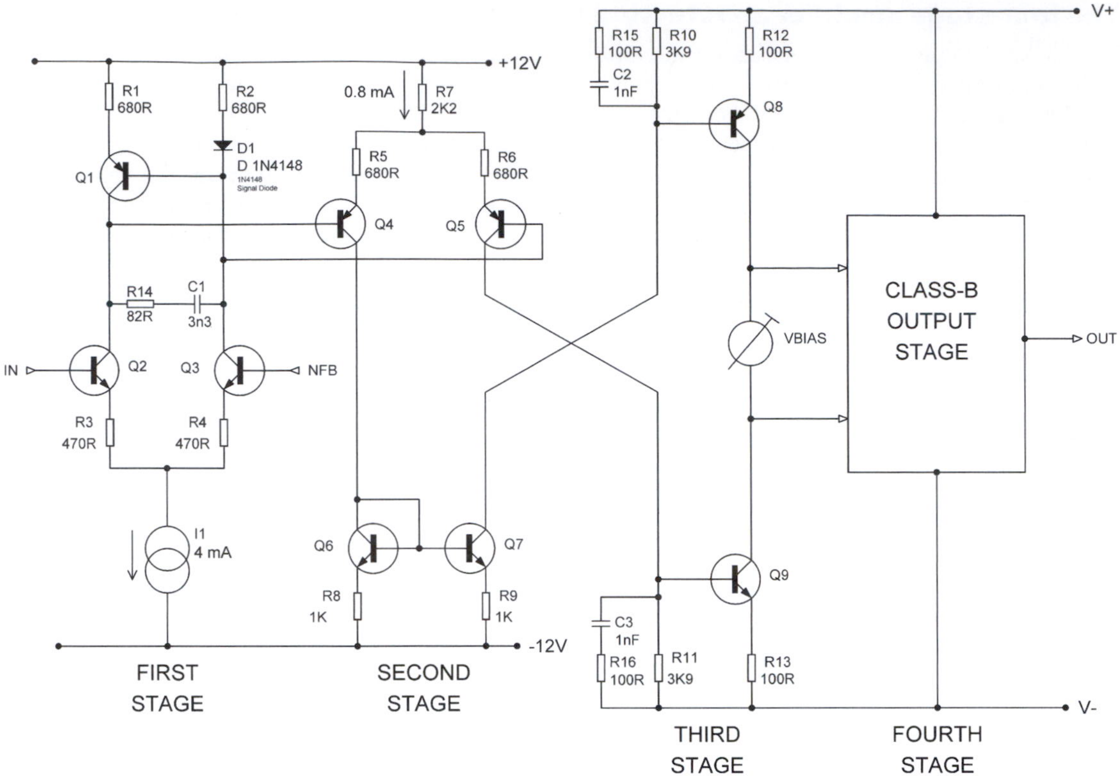

FIGURE 8.4
Four-stage amplifier architecture of a commercial amplifier.

Another version of the four-stage architecture is shown in Figure 8.4; it is a simplified version of a circuit used for many years by another of the major hi-fi companies. There are two differential stages, the second one driving a push–pull VAS Q8, Q9. Once again the differential stages have been given a large amount of local negative feedback in the form of emitter degeneration. Compensation is by the lead-lag network R14, C1 between the two input stage collectors and the two lead-lag networks R15, C2 and R16, C3 that shunt the collectors of Q5, Q7 in the second differential stage. Unlike the Lohstroh and Otala design, series overall feedback was used, supplemented with an op-amp DC servo to control the DC offset at the output.

Having had some experience with this design (no, it's not one of mine) I have to report that while in general the amplifier worked soundly and reliably, it was unduly fussy about transistor types and the distortion performance was not of the best.

The question now obtrudes itself: what is gained by using the greater complexity of a four-stage architecture? So far as I can see at the moment, little or nothing. The three-stage architecture appears to provide as much open-loop gain as can be safely used with a conventional output stage; if more is required then the Miller compensation capacitor can be reduced, which will also improve the maximum slew rates. A four-stage architecture does, however, present some interesting possibilities for using nested Miller compensation, a concept which has been extensively used in op-amps.

POWER AMPLIFIER CLASSES

For a long time the only amplifier classes relevant to high-quality audio were Class-A and Class-AB. This is because valves were the only active devices, and Class-B valve amplifiers generated so much distortion that they were barely acceptable even for public address purposes. All amplifiers with pretensions to high fidelity operated in push–pull Class-A.

Solid-state gives much more freedom of design; all of the amplifier classes below have been commercially exploited. This chapter deals in detail with Classes A, AB, B, D and G, and this certainly covers the vast majority of solid-state amplifiers. For the other classes plentiful references are given so that the intrigued can pursue matters further. In particular, my book *Self On Audio* (Self, 2006) contains a thorough treatment of all known audio amplifier classes, and indeed suggests some new ones.

Class-A

In a Class-A amplifier current flows continuously in all the output devices, which enables the nonlinearities of turning them on and off to be avoided. They come in two rather different kinds, although this is rarely explicitly stated, which work in very different ways. The first kind is simply a Class-B stage (i.e. two emitter-followers working back to back) with the bias voltage increased so that sufficient current flows for neither device to cut off under normal loading. The great advantage of this approach is that it cannot abruptly run out of output current; if the load impedance becomes lower than specified then the amplifier simply takes brief excursions into Class-AB, hopefully with a modest increase in distortion and no seriously audible distress.

The other kind could be called the controlled-current-source (VCIS) type, which is in essence a single emitter-follower with an active emitter load for adequate current-sinking. If this latter element runs out of current capability it makes the output stage clip much as if it had run out of output voltage. This kind of output stage demands a very clear idea of how low an impedance it will be asked to drive before design begins.

Valve textbooks will be found to contain enigmatic references to classes of operation called AB1 and AB2; in the former grid current did not flow for any part of the cycle, but in the latter it did. This distinction was important because the flow of output-valve grid current in AB2 made the design of the previous stage much more difficult.

AB1 or AB2 has no relevance to semiconductors, for in BJTs base current always flows when a device is conducting, while in power FETs gate current never does, apart from charging and discharging internal capacitances.

Class-AB

This is not really a separate class of its own, but a combination of A and B. If an amplifier is biased into Class-B, and then the bias further increased, it will enter AB. For outputs below a certain level both output devices conduct, and operation is Class-A. At higher levels, one device will be turned completely off as the other provides more current, and the distortion jumps upward at this point as AB action begins. Each device will conduct between 50% and 100% of the time, depending on the degree of excess bias and the output level.

Class-AB is less linear than either A or B, and in my view its only legitimate use is as a fallback mode to allow Class-A amplifiers to continue working reasonably when faced with a low-load impedance.

Class-B

Class-B is by far the most popular mode of operation, and probably more than 99% of the amplifiers currently made are of this type. My definition of Class-B is that unique amount of bias voltage which causes the conduction of the two output devices to overlap with the greatest smoothness and so generate the minimum possible amount of crossover distortion.

Class-C

Class-C implies device conduction for significantly less than 50% of the time, and is normally only usable in radio work, where an LC circuit can smooth out the current pulses and filter harmonics. Current-dumping amplifiers can be regarded as combining Class-A (the correcting amplifier) with Class-C (the current-dumping devices); however, it is hard to visualize how an audio amplifier using devices in Class-C only could be built. I regard a Class-B stage with no bias voltage as working in Class-C.

Class-D

These amplifiers continuously switch the output from one rail to the other at a supersonic frequency, controlling the mark/space ratio to give an average representing the instantaneous level of the audio signal; this is alternatively called pulse width modulation (PWM). Great effort and ingenuity has been devoted to this approach, for the efficiency is in theory very high, but the practical difficulties are severe, especially so in a world of tightening EMC legislation, where it is not at all clear that a 200 kHz high-power square wave is a good place to start. Distortion is not inherently low (Attwood, 1983), and the amount of global negative feedback that can be applied is severely limited by the pole due to the effective sampling frequency in the forward path. A sharp cut-off low-pass filter is needed between amplifier and speaker, to remove most of the RF; this will require at least four inductors (for stereo) and will cost money, but its worst feature is that it will only give a flat frequency response into one specific load impedance.

Important references to consult for further information are (Goldberg and Sandler, 1991) and (Hancock, 1991).

Class-E

This is an extremely ingenious way of operating a transistor so that it has either a small voltage across it or a small current through it almost all the time, so that the power dissipation is kept very low (Peters, 1975). Regrettably this is an RF technique that seems to have no sane application to audio.

Class-F

There is no Class-F, as far as I know. This seems like a gap that needs filling . . .

Class-G

This concept was introduced by Hitachi in 1976 with the aim of reducing amplifier power dissipation. Musical signals have a high peak/mean ratio, spending most of the time at low levels, so internal dissipation is much reduced by running from low-voltage rails for small outputs, switching to higher rails current for larger excursions (Feldman, 1976) (Raab, 1986).

The basic series Class-G with two rail voltages (i.e. four supply rails, as both voltages are ±) is shown in Figure 8.5. Current is drawn from the lower ±V1 supply rails whenever possible; should the signal exceed ±V1, TR6 conducts and D3 turns off, so the output current is now drawn entirely from the higher ±V2 rails, with power dissipation shared between TR3 and TR6. The inner stage TR3, TR4 is usually operated in Class-B, although AB or A are equally feasible if the output stage bias is suitably increased. The outer devices are effectively in Class-C as they conduct for significantly less than 50% of the time.

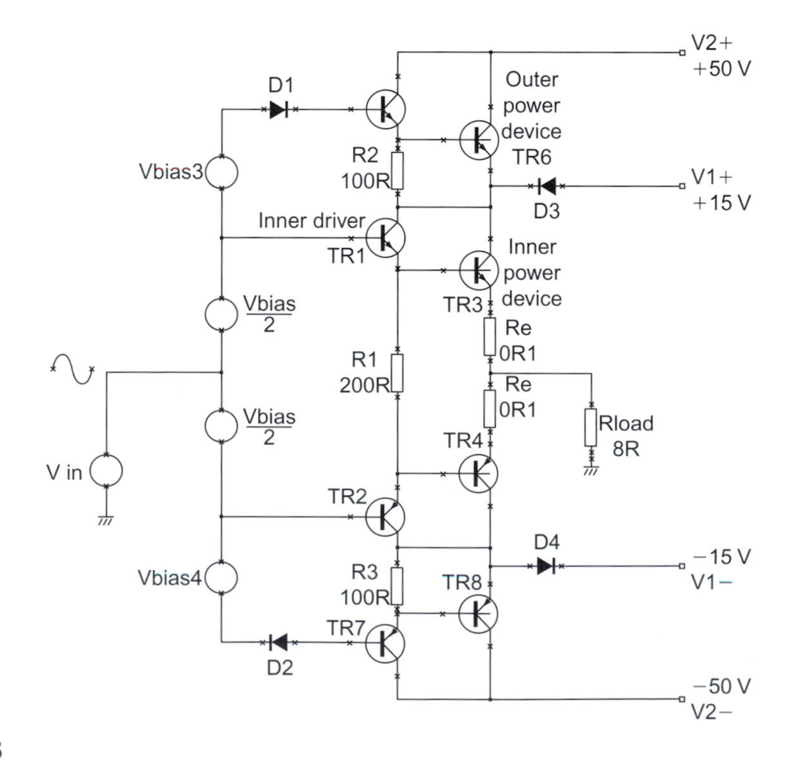

FIGURE 8.5
Class-G series output stage. When the output voltage exceeds the transition level, D3 or D4 turn off and power is drawn from the higher rails through the outer power devices.

In principle movements of the collector voltage on the inner device collectors should not significantly affect the output voltage, but in practice Class-G is often considered to have poorer linearity than Class-B because of glitching due to charge storage in commutation diodes D3, D4. However, if glitches occur they do so at moderate power, well displaced from the crossover region, and so appear relatively infrequently with real signals.

An obvious extension of the Class-G principle is to increase the number of supply voltages. Typically the limit is three. Power dissipation is further reduced and efficiency increased as the average voltage from which the output current is drawn is kept closer to the minimum. The inner devices operate in Class-B/AB as before, and the middle devices are in Class-C. The outer devices are also in Class-C, but conduct for even less of the time.

To the best of my knowledge three-level Class-G amplifiers have only been made in Shunt mode, as described below, probably because in Series mode the cumulative voltage drops become too great and compromise the efficiency gains. The extra complexity is significant, as there are now six supply rails and at least six power devices, all of which must carry the full output current. It seems most unlikely that this further reduction in power consumption could ever be worthwhile for domestic hi-fi.

A closely related type of amplifier is Class-G Shunt (Sampei et al., 1978). Figure 8.6 shows the principle; at low outputs only Q3, Q4 conduct, delivering power from the low-voltage rails. Above a threshold set by Vbias3 and Vbias4, D1 or D2 conduct and Q6, Q8 turn on, drawing current from the high-voltage rails, with D3, D4 protecting Q3, Q4 against reverse bias. The conduction periods of the Q6, Q8 Class-C devices are variable, but inherently less than 50%. Normally the low-voltage section runs in Class-B to minimize dissipation. Such shunt Class-G arrangements are often called *commutating amplifiers*.

Some of the more powerful Class-G Shunt PA amplifiers have three sets of supply rails to further reduce the average voltage drop between rail and output. This is very useful in large PA amplifiers.

Class-H

Class-H is once more basically Class-B, but with a method of dynamically boosting the single supply rail (as opposed to switching to another one) in order to increase efficiency (Buitendijk, 1991). The usual mechanism is a form of bootstrapping. Class-H is occasionally used to describe Class-G as above; this sort of confusion we can do without.

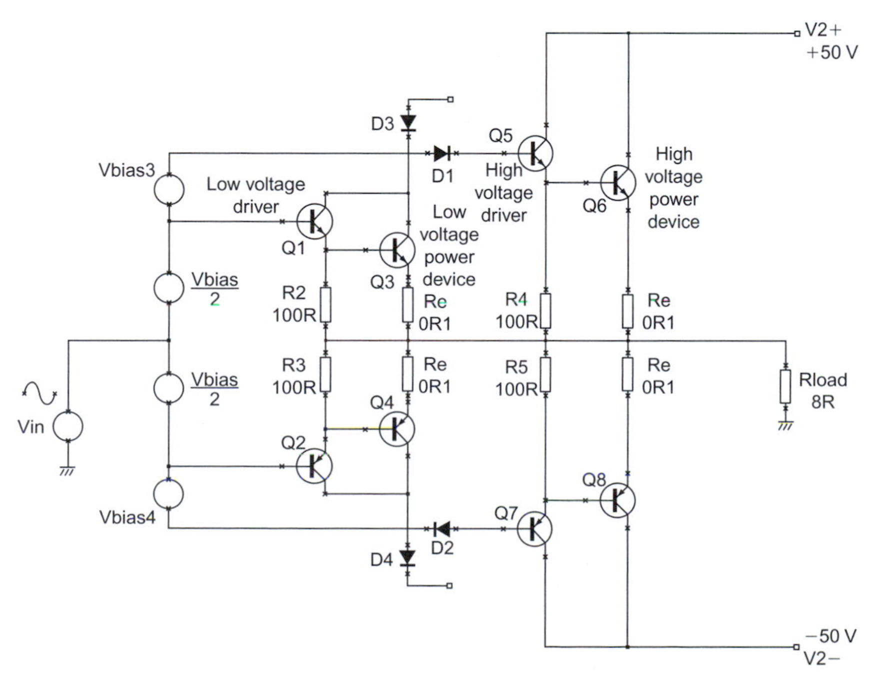

FIGURE 8.6
A Class-G Shunt output stage, composed of two EF output stages with the usual drivers. Vbias3, Vbias4 set the output level at which power is drawn from the higher rails.

Class-S

Class-S, so named by Dr Sandman (1982), uses a Class-A stage with very limited current capability, backed up by a Class-B stage connected so as to make the load appear as a higher resistance than is within the first amplifier's capability. The method used by the Technics SE-A100 amplifier is extremely similar (Sinclair, 1993). I hope that this necessarily brief catalog is comprehensive; if anyone knows of other bona fide classes I would be glad to add them to the collection. This classification does not allow a completely consistent nomenclature; for example, Quad-style current-dumping can only be specified as a mixture of Classes A and C, which says nothing about the basic principle of operation, which is error correction.

Variations on class-B

The solid-state Class-B three-stage amplifier has proved both successful and flexible, so many attempts have been made to improve it further, usually by trying to combine the efficiency of Class-B with the linearity of Class-A. It would be impossible to give a comprehensive list of the changes and improvements attempted, so I give only those that have been either commercially successful or particularly thought-provoking to the amplifier-design community.

Error-correcting amplifiers

This refers to error-cancelation strategies rather than the conventional use of negative feedback. This is a complex field, for there are at least three different forms of error correction, of which the best known is error feedforward as exemplified by the groundbreaking Quad 405 (Walker, 1975). Other versions include error feedback and other even more confusingly named techniques, some at least of which turn out on analysis to be conventional NFB in disguise. For a highly ingenious treatment of the feed-forward method see a design by Giovanni (Stochino, 1994). A most interesting recent design using the Hawksford correction topology has recently been published by Jan (Didden, 2008).

Non-switching amplifiers

Most of the distortion in Class-B is crossover distortion, and results from gain changes in the output stage as the power devices turn on and off. Several researchers have attempted to avoid this by ensuring that each device is clamped to pass a certain minimum current at all times (Tanaka, 1981). This approach has certainly been exploited commercially, but few technical details have been published. It is not intuitively obvious (to me, anyway) that stopping the diminishing device current in its tracks will give less crossover distortion.

Current-drive amplifiers

Almost all power amplifiers aspire to be voltage sources of zero output impedance. This minimizes frequency-response variations caused by the peaks and dips of the impedance curve, and gives a universal amplifier that can drive any loudspeaker directly.

The opposite approach is an amplifier with a sufficiently high output impedance to act as a constant-current source. This eliminates some problems – such as rising voice-coil resistance with heat dissipation— but introduces others such as control of the cone resonance. Current amplifiers therefore appear to be only of use with active crossovers and velocity feedback from the cone (Mills and Hawksford, 1989).

It is relatively simple to design an amplifier with any desired output impedance (even a negative one), and so any compromise between voltage and current drive is attainable. The snag is that loudspeakers are universally designed to be driven by voltage sources, and higher amplifier impedances demand tailoring to specific speaker types (Evenson, 1988).

The blomley principle

The goal of preventing output transistors from turning off completely was introduced by Peter Blomley in 1971 (Blomley, 1971); here the positive/negative splitting is done by circuitry ahead of the output stage, which can then be designed so that a minimum idling current can be separately set up in each output device. However, to the best of my knowledge this approach has not yet achieved commercial exploitation.

I have built Blomley amplifiers twice (way back in 1975) and on both occasions I found that there were still unwanted artefacts at the crossover point, and that transferring the crossover function from one part of the circuit to another did not seem to have achieved much. Possibly this was because the discontinuity was narrower than the usual crossover region and was therefore linearized even less effectively by negative feedback that reduces as frequency increases. I did not have the opportunity to investigate very deeply and this is not to be taken as a definitive judgment on the Blomley concept.

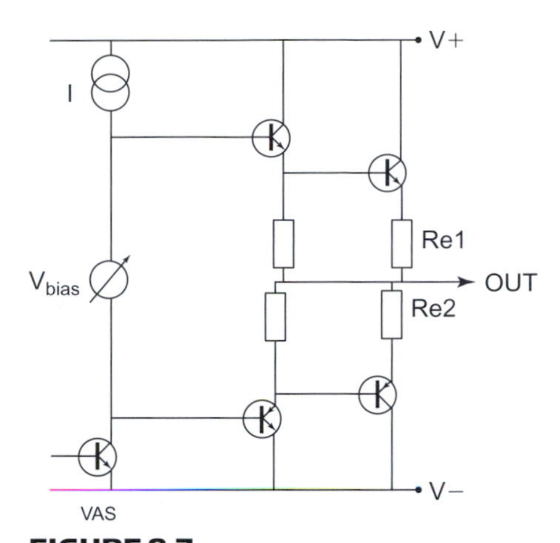

FIGURE 8.7
A conventional double emitter-follower output stage with emitter resistors Re shown.

Geometric mean Class-AB

The classical explanations of Class-B operation assume that there is a fairly sharp transfer of control of the output voltage between the two output devices, stemming from an equally abrupt switch in conduction from one to the other. In practical audio amplifier stages this is indeed the case, but it is not an inescapable result of the basic principle. Figure 8.7 shows a conventional output stage, with emitter resistors Re1, Re2 included to increase quiescent-current stability and allow current sensing for overload protection; it is these emitter resistances that to a large extent make classical Class-B what it is.

However, if the emitter resistors are omitted, and the stage biased with two matched diode junctions, then the diode and transistor junctions form a *translinear loop* (Gilbert, 1990), around which the junction voltages sum to zero. This links the two output transistor currents I_p, I_n in the relationship $I_n \cdot I_p$ = constant, which in op-amp practice is known as Geometric-Mean Class-AB operation. This gives smoother changes in device current at the crossover point, but this does not necessarily mean lower THD. Such techniques are not very practical for discrete power amplifiers; first, in the absence of the very tight thermal coupling between the four junctions that exists in an IC, the quiescent-current stability will be atrocious, with thermal runaway and spontaneous combustion a near certainty. Second, the output device bulk emitter resistance will probably give enough voltage drop to turn the other device off anyway, when current flows. The need for drivers, with their extra junction-drops, also complicates things.

A new extension of this technique is to redesign the translinear loop so that $1/I_n + 1/I_p$ = constant, this being known as Harmonic-Mean Class-AB operation (Thus, 1992). It is too early to say whether this technique (assuming it can be made to work outside an IC) will be of use in reducing crossover distortion and thus improving amplifier performance.

Nested differentiating feedback loops

This is a most ingenious but conceptually complex technique for significantly increasing the amount of NFB that can be applied to an amplifier. I wish I could tell you how well it works but I have never found the time to investigate it practically. For the original paper see (Cherry, 1982), but it's tough going mathematically. A more readable account was published in *Electronics Today International* in 1983, and included a practical design for a 60 W NDFL amplifier (Cherry, 1983).

AMPLIFIER BRIDGING

When two power amplifiers are driven with anti-phase signals and the load connected between their outputs, with no connection to ground, this is called bridging. It is a convenient and inexpensive way to turn a stereo amplifier into a more powerful mono amplifier. It is called bridging because if you draw the four output transistors with the load connected between them, it looks something like the four arms of a Wheatstone bridge (see Figure 8.8). Doubling the voltage across a load of the same resistance naturally quadruples the output power – in theory. In harsh reality the available power will be considerably less, due to the power supply sagging and extra voltage losses in the two output stages. In most cases you will get something like three times the power rather than four, the ratio depending on how seriously the bridge mode was regarded when the initial design was done. It has to be said that in many designs the bridging mode looks like something of an afterthought.

In Figure 8.8 an 8 Ω load has been divided into two 4 Ω halves, to underline the point that the voltage at their center is zero, and so both amplifiers are effectively driving 4 Ω loads to ground, with all that that implies for increased distortion and increased losses in the output stages. A unity-gain inverting

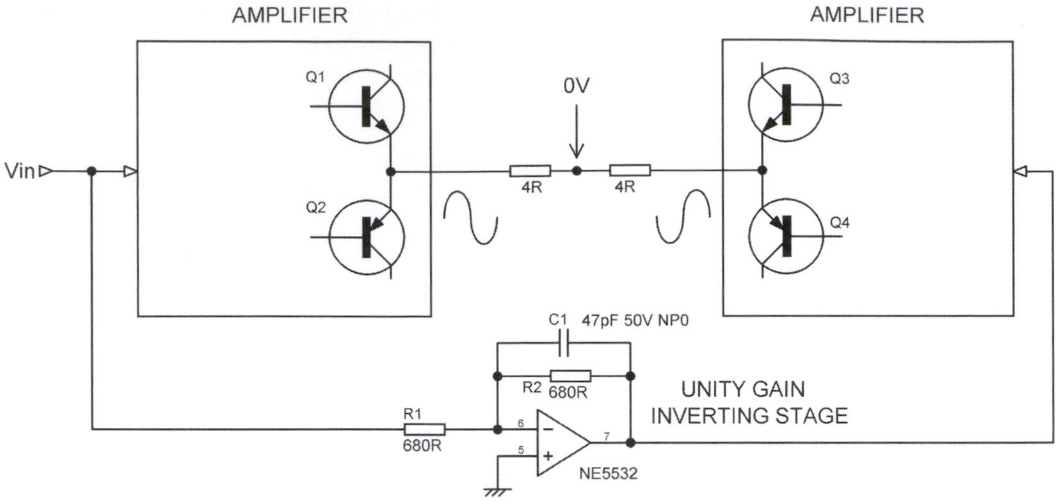

FIGURE 8.8
Bridging two power amplifiers to create a single, more powerful amplifier.

stage is required to generate the anti-phase signal; nothing fancy is required and the simple shunt-feedback stage shown does the job nicely. I have used it in several products. The resistors in the inverter circuit need to be kept as low in value as possible to reduce their Johnson noise contribution, but not of course so low that the op-amp distortion is increased by driving them; this is not too hard to arrange as the op-amp will only be working over a small fraction of its voltage output capability, because the power amplifier it is driving will clip a long time before the op-amp does. The capacitor assures stability – it causes a roll-off of 3 dB down at 5 MHz, so it does not in any way imbalance the audio frequency response of the two amplifiers.

You sometimes see the statement that bridging reduces the distortion seen across the load because the push–pull action causes cancelation of the distortion products. In brief, it is not true. Push–pull systems can only cancel even-order distortion products, and in a well-found amplifier these are in short supply. In such an amplifier the input stage and the output stage will both be symmetrical (it is hard to see why anyone would choose them to be anything else) and produce only odd-order harmonics, which will not be canceled. The only asymmetrical stage is the VAS, and the distortion contribution from that is, or at any rate should be, very low. In reality, switching to bridging mode will almost certainly increase distortion, because as noted above, the output stages are now in effect driving 4 Ω loads to ground instead of 8 Ω.

FRACTIONAL BRIDGING

I will now tell you how I came to invent the strange practice of "fractional bridging". I was tasked with designing a two-channel amplifier module for a multichannel unit. Five of these modules fitted into the chassis, and if each one was made independently bridgeable, you got a very flexible system that could be configured for anywhere between five and ten channels of amplification. The normal output of each amplifier was 85 W into 8 Ω, and the bridged output was about 270 W as opposed to the theoretical 340 W. And now the problem. The next unit up in the product line had modules that gave 250 W into 8 Ω unbridged, and the marketing department felt that having the small modules giving more power than the large ones was really not on; I'm not saying they were wrong. The problem was therefore to create an amplifier that only doubled its power when bridged. Hmm!

One way might have been to develop a power supply with deliberately poor regulation, but this implies a mains transformer with high-resistance windings that would probably have overheating problems. Another possibility was to make the bridged mode switch in a circuit that clipped the input signal before the power amplifiers clipped. The problem is that building a clipping circuit that does not exhibit poor distortion performance below the actual clipping level is actually surprisingly difficult—think about the nonlinear capacitance of signal diodes. I worked out a way to do it, but it took up an amount of PCB area that simply wasn't available. So the ultimate solution was to let one of the power amplifiers do the clipping, which it does cleanly because of the high level of negative feedback, and the fractional bridging concept was born.

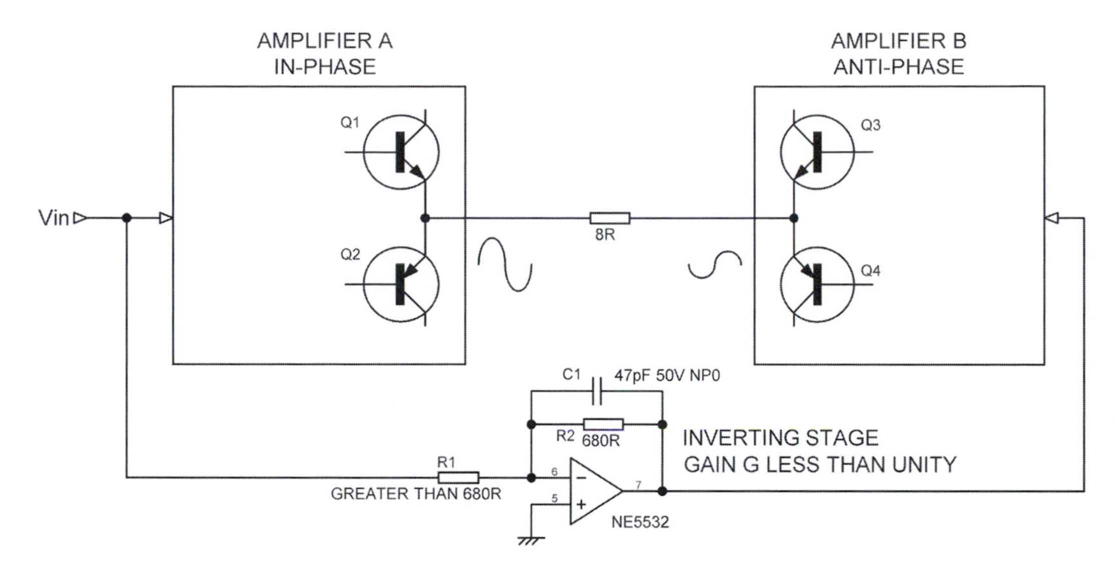

FIGURE 8.9
Fractional bridging of two power amplifiers to give doubled rather than quadrupled power output.

Figure 8.9 shows how it works. An inverter is still used to drive the anti-phase amplifier, but now it is configured with a gain G that is less than unity. This means that the in-phase amplifier will clip when the anti-phase amplifier is still well below maximum output, and the bridged output is therefore restricted. Double output power means an output voltage increased by root-2 or 1.41 times, and so the anti-phase amplifier is driven with a signal attenuated by a factor of 0.41, which I call the bridging fraction, giving a total voltage swing across the load of 1.41 times. It worked very well, the product was a considerable success, and no salesmen were plagued with awkward questions about power output ratings.

There are two possible objections to this cunning plan, the first being that it is obviously inefficient compared with a normal Class-B amplifier. Figure 8.10 shows how the power is dissipated in the pair of amplifiers; this is derived from basic calculations and ignores output stage losses. P_{dissA} is the power dissipated in the in-phase amplifier A, and varies in the usual way for a Class-B amplifier with a maximum at 63% of the maximum voltage output. P_{dissB} is the dissipation in anti-phase amplifier B that receives a smaller drive signal and so never reaches its dissipation maximum; it dissipates more power because it is handling the same current but has more voltage left across the output devices, and this is what makes the overall efficiency low. P_{tot} is the sum of the two amplifier dissipations. The dotted lines show the output power contribution from each amplifier, and the total output power in the load.

The bridging fraction can of course be set to other values to get other maximum outputs. The lower it is, the lower the overall efficiency of the amplifier pair, reaching the limiting value when the bridging fraction is zero. In this (quite pointless) situation the anti-phase amplifier is simply being used as an expensive alternative to connecting one end of the load to ground, and so it dissipates a lot of heat. Figure 8.11 shows how the maximum efficiency (which always occurs at maximum output) varies with the bridging fraction. When it is unity, we get normal Class-B operation and the maximum efficiency is the familiar figure of 78.6%; when it is zero the overall efficiency is halved to 39.3%, with a linear variation between these two extremes.

The second possible objection is that you might think it is a grievous offence against engineering ethics to deliberately restrict the output of an amplifier for marketing reasons, and you might be right, but it kept people employed, including me. Nevertheless, given the current concerns about energy, perhaps this sort of thing should not be encouraged. I have given semi-serious thought to writing a book called *How to Cheat with Amplifiers*.

AC- AND DC-COUPLED AMPLIFIERS

All power amplifiers are either AC-coupled or DC-coupled. The first kind have a single supply rail, with the output biased to be halfway between this rail and ground to give the maximum symmetrical voltage swing; a large DC-blocking capacitor is therefore used in series with the output. The second kind

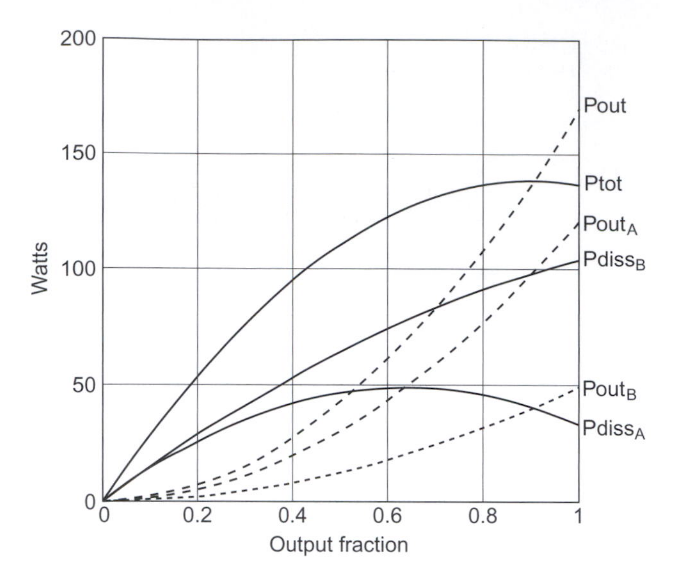

FIGURE 8.10
The variation of power output and power dissipation of two fractionally bridged power amplifiers, with a bridging fraction of 0.41 to give doubled rather than quadrupled power output.

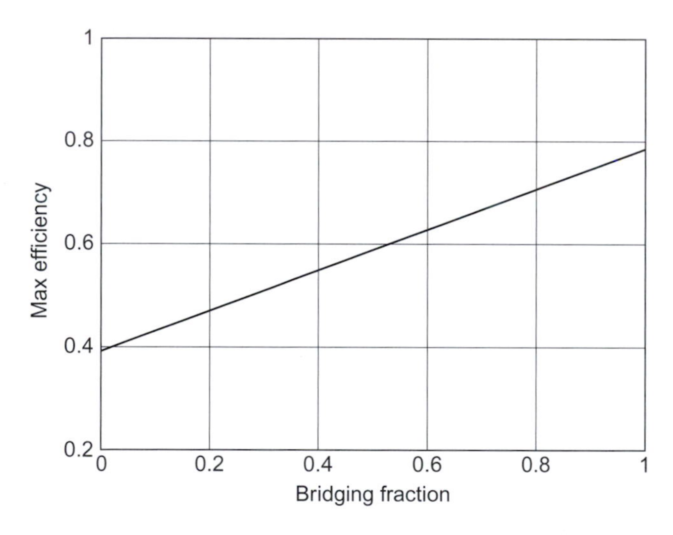

FIGURE 8.11
The variation of maximum efficiency of two fractionally bridged power amplifiers with bridging fraction.

have positive and negative supply rails, and the output is biased to be at 0 V, so no output DC-blocking is required in normal operation.

The advantages of AC-coupling

1. The output DC offset is always zero (unless the output capacitor is leaky).
2. It is very simple to prevent turn-on thump by purely electronic means; there is no need for an expensive output relay. The amplifier output must rise up to half the supply voltage at turn-on, but providing this occurs slowly there is no audible transient. Note that in many designs this is not simply a matter of making the input bias voltage rise slowly, as it also takes time for the DC feedback to establish itself, and it tends to do this with a snap action when a threshold is reached. The last AC-coupled power amplifier I designed (which was in 1980, I think) had a simple RC time-constant and diode arrangement that absolutely constrained the VAS collector voltage to rise slowly at turn-on, no matter what the rest of the circuitry was doing – cheap but very effective.

3. No protection against DC faults is required, providing the output capacitor is voltage-rated to withstand the full supply rail. A DC-coupled amplifier requires an expensive and possibly unreliable output relay for dependable speaker protection.
4. The amplifier should be more easy to make short-circuit proof, as the output capacitor limits the amount of electric charge that can be transferred each cycle, no matter how low the load impedance. This is speculative; I have no data as to how much it really helps in practice.
5. AC-coupled amplifiers do not in general appear to require output inductors for stability. Large electrolytics have significant equivalent series resistance (ESR) and a little series inductance.

   For typical amplifier output sizes the ESR will be of the order of 100 mΩ; this resistance is probably the reason why AC-coupled amplifiers rarely had output inductors, as it is often enough resistance to provide isolation from capacitive loading and so gives stability. Capacitor series inductance is very low and probably irrelevant, being quoted by one manufacturer as "a few tens of nanohenrys". The output capacitor was often condemned in the past for reducing the low-frequency damping factor (DF), for its ESR alone is usually enough to limit the DF to 80 or so. As explained above, this is not a technical problem because "damping factor" means virtually nothing.

The advantages of DC-coupling

1. No large and expensive DC-blocking capacitor is required. On the other hand, the dual supply will need at least one more equally expensive reservoir capacitor, and a few extra components such as fuses.
2. In principle there should be no turn-on thump, as the symmetrical supply rails mean the output voltage does not have to move through half the supply voltage to reach its bias point – it can just stay where it is. In practice the various filtering time-constants used to keep the bias voltages free from ripple are likely to make various sections of the amplifier turn on at different times, and the resulting thump can be substantial. This can be dealt with almost for free, when a protection relay is fitted, by delaying the relay pull-in until any transients are over. The delay required is usually less than a second.
3. Audio is a field where almost any technical eccentricity is permissible, so it is remarkable that AC-coupling appears to be the one technique that is widely regarded as unfashionable and unacceptable. DC-coupling avoids any marketing difficulties.
4. Some potential customers will be convinced that DC-coupled amplifiers give better speaker damping due to the absence of the output capacitor impedance. They will be wrong, but this misconception has lasted at least 40 years and shows no sign of fading away.
5. Distortion generated by an output capacitor is avoided. This is a serious problem, as it is not confined to low frequencies, as is the case in small-signal circuitry. For a 6800 μF output capacitor driving 40 W into an 8 Ω load, there is significant mid-band third harmonic distortion at 0.0025%, as shown in Figure 8.12. This is at least five times more than the amplifier generates in this part of the frequency range. In addition, the THD rise at the LF end is much steeper than in the small-signal case, for reasons that are not yet clear. There are two cures for output capacitor distortion. The straightforward approach uses a huge output capacitor, far larger in value than required for a good low-frequency response. A 100,000 μF/40 V Aerovox from BHC eliminated all distortion, as shown in Figure 8.13. An allegedly "audiophile" capacitor gives some interesting results; a Cerafine Supercap of only moderate size (4700 μF/63 V) gave the result shown in Figure 8.14, where the mid-band distortion is gone but the LF distortion rise remains. What special audio properties this component is supposed to have are unknown; as far as I know electrolytics are never advertised as "low mid-band THD", but that seems to be the case here. The volume of the capacitor case is about twice as great as conventional electrolytics of the same value, so it is possible the crucial difference may be a thicker dielectric film than is usual for this voltage rating.

   Either of these special capacitors costs more than the rest of the amplifier electronics put together. Their physical size is large. A DC-coupled amplifier with protective output relay will be a more economical option.

   A little-known complication with output capacitors is that their series reactance increases the power dissipation in the output stage at low frequencies. This is counter-intuitive as it would

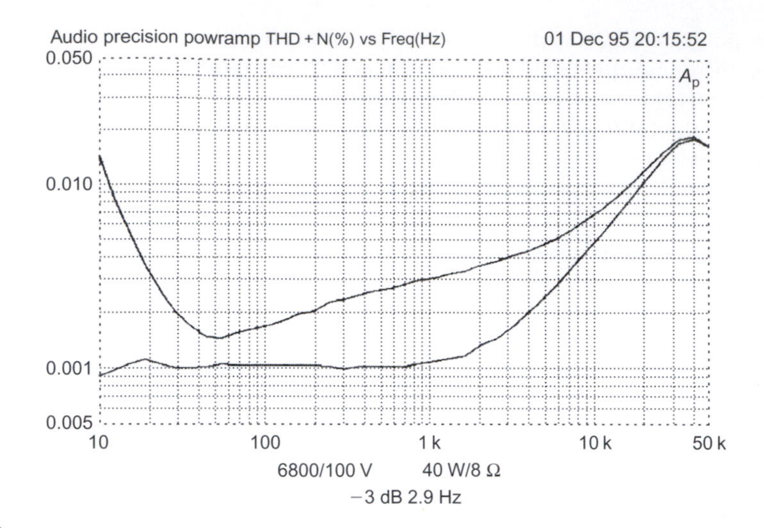

FIGURE 8.12
The extra distortion generated by a 6800 μF electrolytic delivering 40 W into 8 Ω. Distortion rises as frequency falls, as for the small-signal case, but at this current level there is also added distortion in the mid-band.

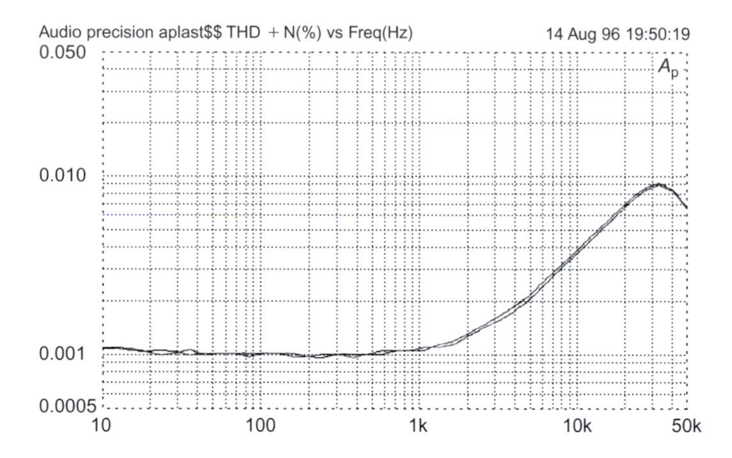

FIGURE 8.13
Distortion with and without a very large output capacitor, the BHC Aerovox 100,000 μF/40 V (40 W/8 W). Capacitor distortion is eliminated.

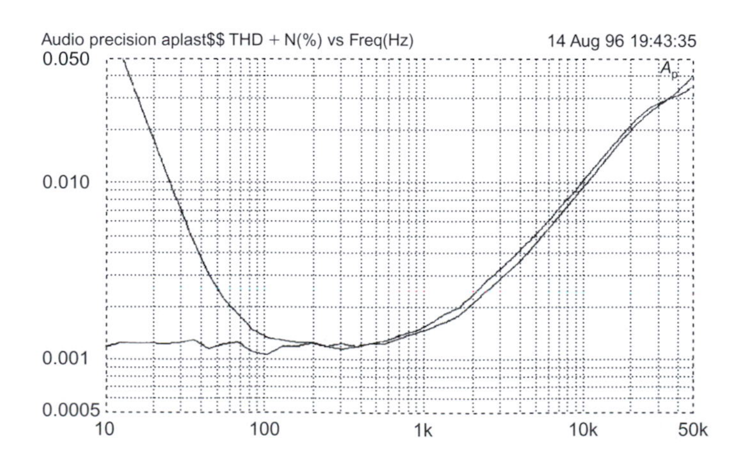

FIGURE 8.14
Distortion with and without an "audiophile" Cerafine 4700 μF/63 V capacitor. Mid-band distortion is eliminated but LF rise is much the same as the standard electrolytic.

seem that any impedance added in series must reduce the current drawn and hence the power dissipation. In fact it is the load phase shift that increases the amplifier dissipation.

6. The supply currents can be kept out of the ground system. A single-rail AC amplifier has half-wave Class-B currents flowing in the 0 V rail, and these can have a serious effect on distortion and crosstalk performance.

NEGATIVE FEEDBACK IN POWER AMPLIFIERS

It is not the role of this chapter to step through elementary theory that can be easily found in any number of textbooks. However, correspondence in audio and technical journals shows that considerable confusion exists on negative feedback as applied to power amplifiers; perhaps there is something inherently mysterious in a process that improves almost all performance parameters simply by feeding part of the output back to the input, but inflicts dire instability problems if used to excess. I therefore deal with a few of the less obvious points here.

The main use of NFB in power amplifiers is the reduction of harmonic distortion, the reduction of output impedance, and the enhancement of supply-rail rejection. There are also analogous improvements in frequency response and gain stability, and reductions in DC drift.

The basic feedback equation is dealt with in a myriad of textbooks, but it is so fundamental to power amplifier design that it is worth a look here. In Figure 8.15, the open-loop amplifier is the big block with open-loop gain A. The negative-feedback network is the block marked β; this could contain anything, but for our purposes it simply scales down its input, multiplying it by β, and is usually in the form of a potential divider. The funny round thing with the cross on is the conventional control theory symbol for a block that adds or subtracts and does nothing else.

Firstly, it is pretty clear that one input to the subtractor is simply V_{in}, and the other is $V_{out}\,\beta$, so subtract these two, multiply by A, and you get the output signal V_{out}:

$$V_{in} = A(V_{in} - \beta \cdot V_{out})$$

Collect the V_{out} values together and you get:

$$V_{in} = A(V_{in} - \beta \cdot V_{out})$$

$$\frac{V_{out}}{V_{in}} = \frac{A}{1 + A\beta} \qquad \text{Equation 2.1}$$

So:

This is the feedback equation, and it could not be more important. The first thing it shows is that negative feedback stabilizes the gain. In real-life circuitry A is a high but uncertain and variable quantity,

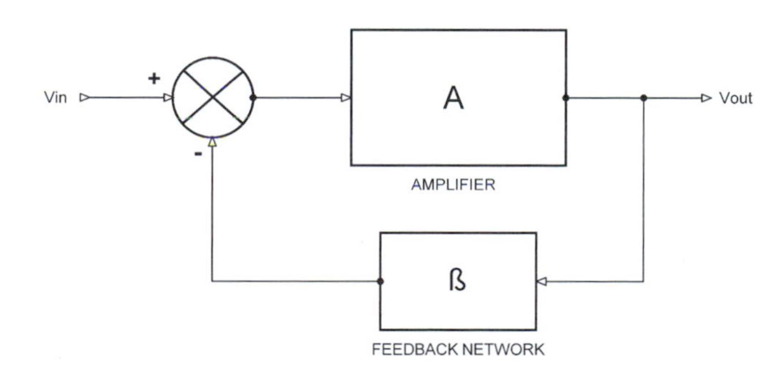

FIGURE 8.15
A simple negative-feedback system with an amplifier with open-loop gain A and a feedback network with a "gain", which is less than 1, of β.

while β is firmly fixed by resistor values. Looking at the equation, you can see that the higher A is, the less significant the 1 on the bottom is; the A values cancel out, and so with high A the equation can be regarded as simply:

$$\frac{V_{out}}{V_{in}} = \frac{1}{\beta}$$

This is demonstrated in Table 8.1, where β is set at 0.04 with the intention of getting a closed-loop gain of 25 times. With a low open-loop gain of 100, the closed-loop gain is only 20, a long way short of 25. But as the open-loop gain increases, the closed-loop gain gets closer to the target. If you look at the bottom two rows, you will see that an increase in open-loop gain of more than a factor of 2 only alters the closed-loop gain by a trivial second decimal place.

Negative feedback is, however, capable of doing much more than stabilizing gain. Anything untoward happening in the amplifier block A, be it distortion or DC drift, or any of the other ills that electronics is prone to, is also reduced by the negative-feedback factor (NFB factor for short). This is equal to:

$$\text{NFB factor} = \frac{1}{1 + A\beta} \qquad \text{Equation 2.2}$$

and it is tabulated in the fourth column in Table 8.1. To show why this factor is vitally important, Figure 8.16 shows the same scenario as Figure 8.11, with the addition of a voltage V_d to the output of A; this represents noise, DC drift, or anything that can cause a voltage error, but what is usually most interesting to the practitioners of amplifier design is its use to represent distortion.

Repeating the simple algebra we did before, and adding in V_d, we get:

$$V_{out} = A(V_{in} - \beta \cdot V_{out}) + V_d$$

$$V_{out}(1 + A\beta) = A \cdot V_{in} + V_d$$

$$\frac{V_{out}}{V_{in}} = \frac{A}{1 + A\beta} + \frac{V_d}{1 + A\beta}$$

So the effect of V_d has been decreased by the feedback factor:

$$\frac{1}{1 + A\beta}$$

In other words, the higher the open-loop gain A compared with the gain demanded by β, the lower the distortion. Since we are usually dealing with high values of A, the 1 on the bottom of the fraction has very little effect and doubling the open-loop gain halves the distortion. This effect is illustrated in the fifth and sixth columns of Table 8.1, which adds an error of magnitude 1 to the output of the amplifier; the closed-loop error is then simply the reciprocal of the NFB factor for each value of open-loop gain.

In simple circuits with low open-loop gain you just apply negative feedback and that is the end of the matter. In a typical power amplifier, which cannot be operated without NFB, if only because it would be saturated by its own DC offset voltages, there are several stages that may accumulate phase shift, and simply closing the loop usually brings on severe Nyquist oscillation at HF. This is a serious matter, as it will not only burn out any tweeters that are unlucky enough to be connected, but can also destroy the output devices by overheating, as they may be unable to turn off fast enough at ultrasonic frequencies.

The standard cure for this instability is compensation. A capacitor is added, usually in Miller-integrator format, to roll off the open-loop gain at 6 dB/octave, so it reaches unity loop-gain before enough phase shift can build up to allow oscillation. This means the NFB factor varies strongly with frequency, an inconvenient fact that many audio commentators seem to forget.

Table 8.1 How the closed-loop gain gets closer to the target as the open-loop gain increases

1 Desired C/L gain	2 NFB fraction	3 *A* O/L gain	4 NFB factor	5 C/L gain	6 O/L error	7 C/L error
25	0.04	100	5	20.00	1	0.2
25	0.04	1000	41	24.39	1	0.0244
25	0.04	10,000	401	24.94	1	0.0025
25	0.04	40,000	1601	24.98	1	0.0006
25	0.04	100,000	4001	24.99	1	0.0002

It is crucial to remember that a distortion harmonic, subjected to a frequency-dependent NFB factor as above, will be reduced by the NFB factor corresponding to its own frequency, not that of its fundamental. If you have a choice, generate low-order rather than high-order distortion harmonics, as the NFB deals with them much more effectively.

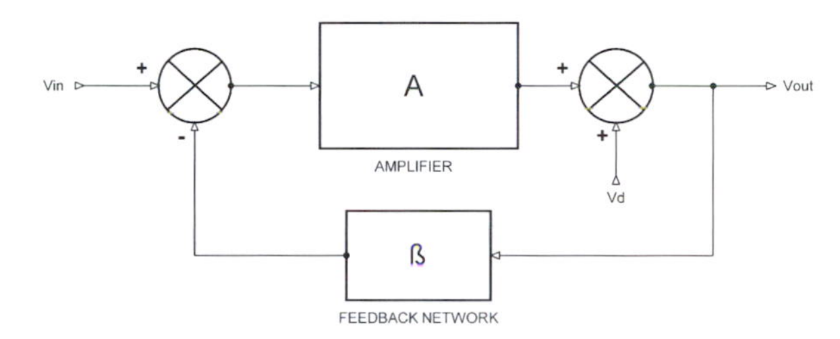

FIGURE 8.16
The negative-feedback system with an error signal V_d added to the output of the amplifier.

Negative feedback can be applied either locally (i.e. to each stage, or each active device) or globally, in other words right around the whole amplifier. Global NFB is more efficient at distortion reduction than the same amount distributed as local NFB, but places much stricter limits on the amount of phase shift that may be allowed to accumulate in the forward path (more on this later in this chapter).

Above the dominant-pole frequency, the VAS acts as a Miller integrator, and introduces a constant 90° phase lag into the forward path. In other words, the output from the input stage must be in quadrature if the final amplifier output is to be in phase with the input, which to a close approximation it is. This raises the question of how the 90° phase shift is accommodated by the negative-feedback loop; the answer is that the input and feedback signals applied to the input stage are then subtracted, and the small difference between two relatively large signals with a small phase shift between them has a much larger phase shift. This is the signal that drives the VAS input of the amplifier.

Solid-state power amplifiers, unlike many valve designs, are almost invariably designed to work at a fixed closed-loop gain. If the circuit is compensated by the usual dominant-pole method, the HF open-loop gain is also fixed, and therefore so is the important negative-feedback factor. This is in contrast to valve amplifiers, where the amount of negative feedback applied was regarded as a variable, and often user-selectable, parameter; it was presumably accepted that varying the negative-feedback factor caused significant changes in input sensitivity. A further complication was serious peaking of the closed-loop frequency response at both LF and HF ends of the spectrum as negative feedback was increased, due to the inevitable bandwidth limitations in a transformer-coupled forward path. Solid-state amplifier designers go cold at the thought of the customer tampering with something as vital as the NFB factor, and such an approach is only acceptable in cases like valve amplification where global NFB plays a minor role.

Some common misconceptions about negative feedback

All of the comments quoted below have appeared many times in the hi-fi literature. All are wrong.

Negative feedback is a bad thing. Some audio commentators hold that, without qualification, negative feedback is a bad thing. This is of course completely untrue and based on no objective reality. Negative feedback is one of the fundamental concepts of electronics, and to avoid its use altogether is virtually impossible; apart from anything else, a small amount of local NFB exists in every common-emitter transistor because of the internal emitter resistance. I detect here distrust of good fortune; the uneasy feeling that if something apparently works brilliantly then there must be something wrong with it.

One approach to appreciating negative feedback and its stability problems is SPICE simulation. Some SPICE simulators have the ability to work in the Laplace or s-domain, but my own experiences with this have been deeply unhappy. Otherwise respectable simulator packages output complete rubbish in this mode. Quite what the issues are here I do not know, but it does seem that s-domain methods are best avoided. The approach suggested here instead models poles directly as poles, using RC networks to generate the time-constants. This requires minimal mathematics and is far more robust. Almost any SPICE simulator – evaluation versions included should be able to handle the simple circuit used here.

Figure 8.17 shows the basic model, with SPICE node numbers. The scheme is to idealize the situation enough to highlight the basic issues and exclude distractions like nonlinearities or clipping. The forward gain is simply the transconductance of the input stage multiplied by the transadmittance of the VAS integrator. An important point is that with correct parameter values, the current from the input stage is realistic, and so are all the voltages.

The input differential amplifier is represented by G. This is a standard SPICE element—the VCIS, or voltage-controlled current source. It is inherently differential, as the output current from Node 4 is the scaled difference between the voltages at Nodes 3 and 7. The scaling factor of 0.009 sets the input stage transconductance (g_m) to 9 mA/V, a typical figure for a bipolar input with some local feedback. Stability in an amplifier depends on the amount of negative feedback available at 20 kHz. This is set at the design stage by choosing the input g_m and C_{dom}, which are the only two factors affecting the open-loop gain. In simulation it would be equally valid to change g_m instead; however, in real life it is easier to alter C_{dom} as the only other parameter this affects is slew rate. Changing input stage transconductance is likely to mean altering the standing current and the amount of local feedback, which will in turn impact input stage linearity.

The VAS with its dominant pole is modeled by the integrator E_{vas}, which is given a high but finite open-loop gain, so there really is a dominant pole *P*1 created when the gain demanded becomes equal to that available. With C_{dom} = 100 pF this is below 1 Hz. With infinite (or as near infinite as SPICE allows) open-loop gain the stage would be a perfect integrator. A explained elsewhere, the amount of open-loop gain available in real versions of this stage is not a well-controlled quantity, and *P*1 is liable to wander about in the 1–100 Hz region; fortunately this has no effect at all on HF stability. C_{dom} is the Miller capacitor that defines the transadmittance, and since the input stage has a realistic transconductance C_{dom} can be set to 100 pF, its usual real-life value. Even with this simple model we have a nested feedback loop. This apparent complication here has little effect, so long as the open-loop gain of the VAS is kept high.

The output stage is modeled as a unity-gain buffer, to which we add extra poles modeled by R1, C1 and R2, C2. Eout1 is a unity-gain buffer internal to the output stage model, added so the second pole does not load the first. The second buffer Eout2 is not strictly necessary as no real loads are being driven, but it is convenient if extra complications are introduced later. Both are shown here as a part of the output stage but the first pole could equally well be due to input stage limitations instead; the order in which the poles are connected makes no difference to the final output. Strictly speaking, it would be more accurate to give the output stage a gain of 0.95, but this is so small a factor that it can be ignored.

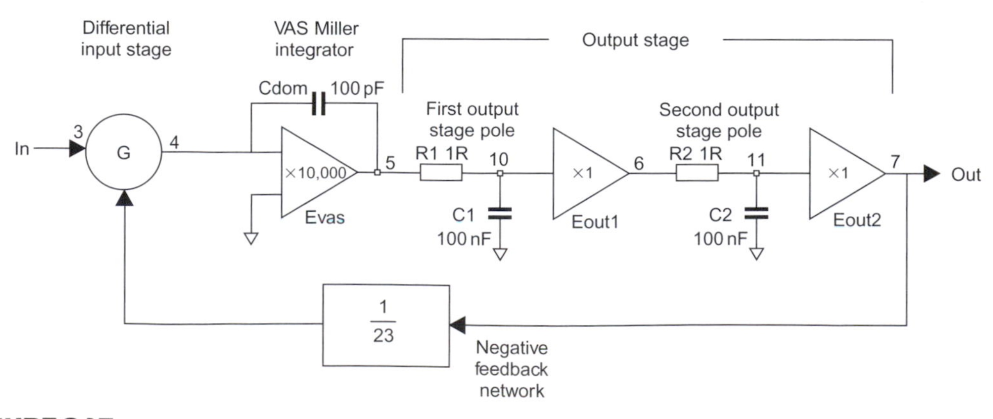

FIGURE 8.17
Block diagram of system for SPICE stability testing.

The component values used to make the poles are of course completely unrealistic, and chosen purely to make the maths simple. It is easy to remember that 1 Ω and 1 μF make up a 1 μs time-constant. This is a pole at 159 kHz. Remember that the voltages in the latter half of the circuit are realistic, but the currents most certainly are not.

The feedback network is represented simply by scaling the output as it is fed back to the input stage. The closed-loop gain is set to 23 times, which is representative of many power amplifiers.

Note that this is strictly a linear model, so the slew-rate limiting that is associated with Miller compensation is not modeled here. It would be done by placing limits on the amount of current that can flow in and out of the input stage.

Figure 8.18 shows the response to a 1 V step input, with the dominant pole the only time element in the circuit. (The other poles are disabled by making C1, C2 0.00001 pF, because this is quicker than changing the actual circuit.) The output is an exponential rise to an asymptote of 23 V, which is exactly what elementary theory predicts. The exponential shape comes from the way that the error signal that drives the integrator becomes less as the output approaches the desired level. The error, in the shape of the output current from G, is the smaller signal shown; it has been multiplied by 1000 to get mA onto the same scale as volts. The speed of response is inversely proportional to the size of C_{dom}, and is shown here for values of 50 and 220 pF as well as the standard 100 pF. This simulation technique works well in the frequency domain, as well as the time domain. Simply tell SPICE to run an AC simulation instead of a TRANS (transient) simulation. The frequency response in Figure 8.19 exploits this to show how the closed-loop gain in an NFB amplifier depends on the open-loop gain available. Once more elementary feedback theory is brought to life. The value of C_{dom} controls the bandwidth, and it can be seen that the values used in the simulation do not give a very extended response compared with a 20 kHz audio bandwidth.

In Figure 8.20, one extra pole *P*2 at 1.59 MHz (a time-constant of only 100 ns) is added to the output stage, and C_{dom} stepped through 50, 100 and 200 pF as before: 100 pF shows a slight overshoot that was not there before; with 50 pF there is a serious overshoot that does not bode well for the frequency response. Actually, it's not that bad; Figure 8.21 returns to the frequency-response domain to show that an apparently vicious overshoot is actually associated with a very mild peaking in the frequency domain.

From here on C_{dom} is left set to 100 pF, its real value in most cases. In Figure 8.22, *P*2 is stepped instead, increasing from 100 ns to 5 μs, and while the response gets slower and shows more overshoot, the system does not become unstable. The reason is simple: sustained oscillation (as opposed to transient ringing) in a feedback loop requires positive feedback, which means that a total phase shift of 180° must have accumulated in the forward path, and reversed the phase of the feedback connection. With only two poles in a system the phase shift cannot reach 180°. The VAS integrator gives a dependable 90° phase shift above *P*1, being an integrator, but *P*2 is instead a simple lag and can only give 90°

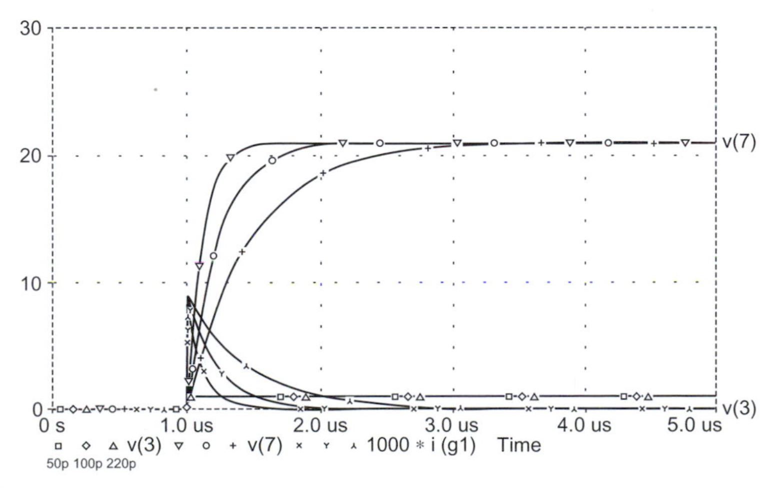

FIGURE 8.18
SPICE results in the time domain. As C_{dom} increases, the response V(7) becomes slower, and the error i(g1) declines more slowly. The input is the step-function V(3) at the bottom.

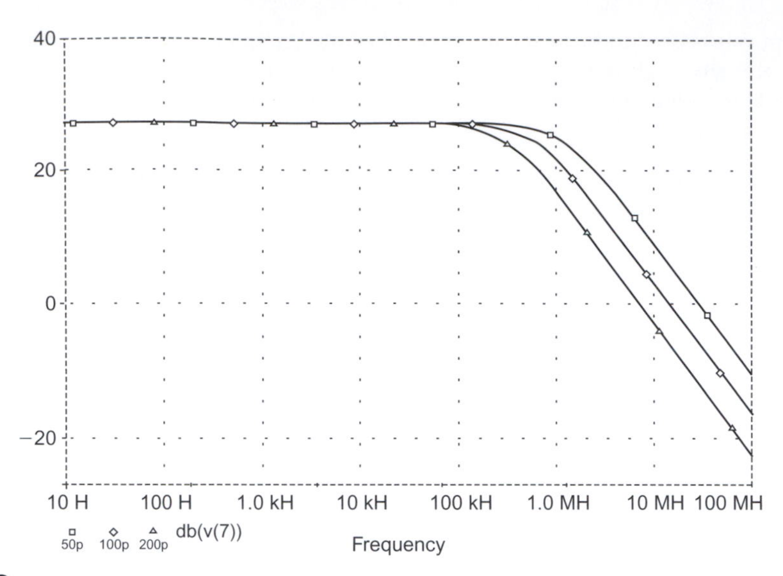

FIGURE 8.19
SPICE simulation in the frequency domain. As the compensation capacitor is increased, the closed-loop bandwidth decreases proportionally.

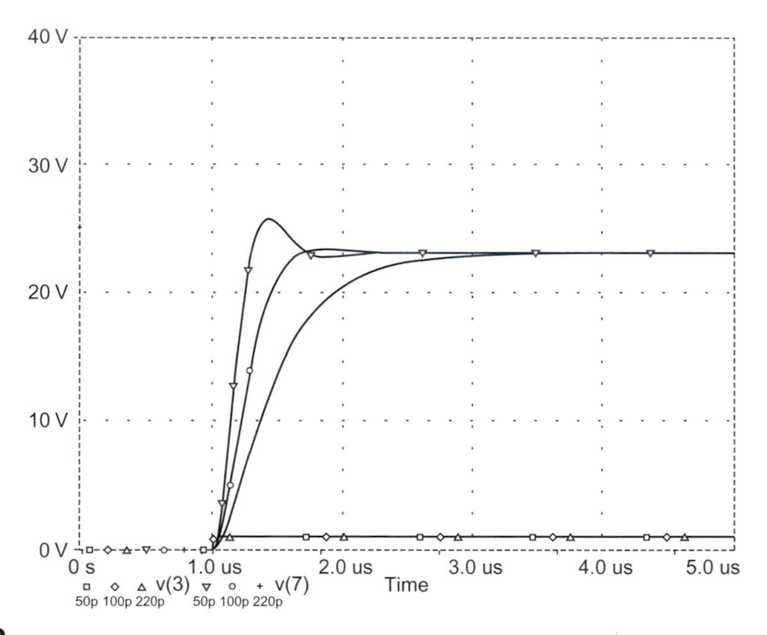

FIGURE 8.20
Adding a second pole *P*2 causes overshoot with smaller values C_{dom}, but cannot bring about sustained oscillation.

phase lag at infinite frequency. So, even this very simple model gives some insight. Real amplifiers do oscillate if C_{dom} is too small, so we know that the frequency response of the output stage cannot be meaningfully modeled with one simple lag.

As President Nixon is alleged to have said: "Two wrongs don't make a right—so let's see if three will do it!" Adding in a third pole *P*3 in the shape of another simple lag gives the possibility of sustained oscillation. This is case A in Table 8.2.

Stepping the value of *P*2 from 0.1 to 5 μs with *P*3 = 500 ns in Figure 8.23 shows that damped oscillation is present from the start. Figure 8.23 also shows over 50 μs what happens when the amplifier is made very unstable (there are degrees of this) by setting *P*2 = 5 μs and *P*3 = 500 ns. It still takes time for the oscillation to develop, but exponentially diverging oscillation like this is a sure sign of disaster. Even in the short time examined here the amplitude has exceeded a rather theoretical half a kilovolt. In

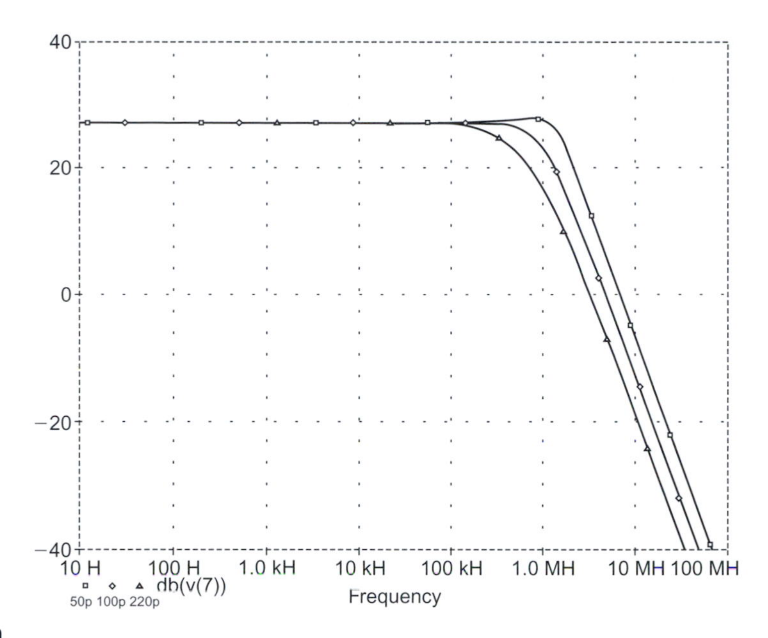

FIGURE 8.21
The frequency responses that go with the transient plots of Figure 8.20. The response peaking for $C_{dom} = 50$ pF is very small compared with the transient overshoot.

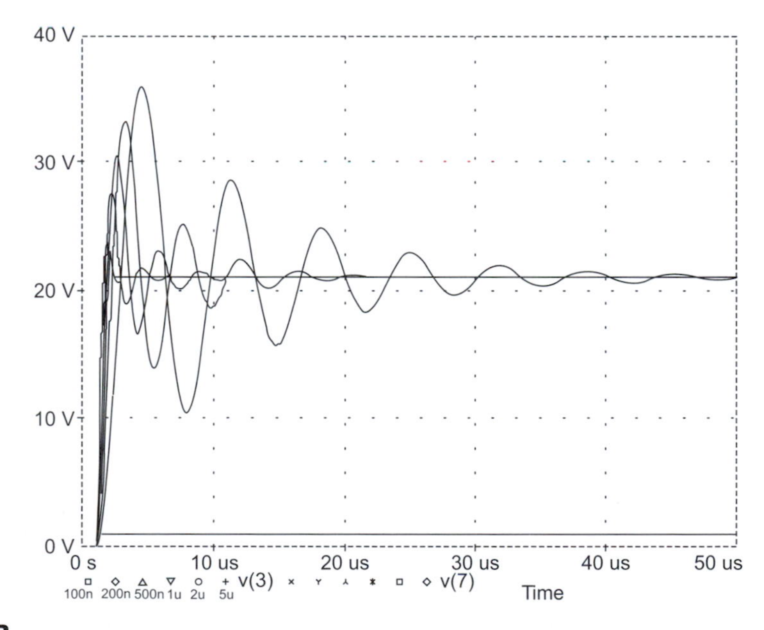

FIGURE 8.22
Manipulating the $P2$ frequency can make ringing more prolonged but it is still not possible to provoke sustained oscillation.

reality oscillation cannot increase indefinitely, if only because the supply rail voltages would limit the amplitude. In practice slew-rate limiting is probably the major controlling factor in the amplitude of high-frequency oscillation.

We have now modeled a system that will show instability. But does it do it right? Sadly, no. The oscillation is about 200 kHz, which is a rather lower frequency than is usually seen when an amplifier misbehaves. This low frequency stems from the low $P2$ frequency we have to use to provoke oscillation; apart from anything else this seems out of line with the known fT of power transistors. Practical amplifiers are likely to take off at around 500 kHz to 1 MHz when C_{dom} is reduced, and this seems to suggest that phase shift is accumulating quickly at this sort of frequency. One possible explanation is that there are a large number of poles close together at a relatively high frequency.

Table 8.2 Instability onset: *P2* is increased until sustained oscillation occurs

Case	C_{dom}	*P2*	*P3*	*P4*	*P5*	*P6*	
A	100p	0.45	0.5	–	–		200 kHz
B	100p	0.5	0.2	0.2	–		345 kHz
C	100p	0.2	0.2	0.2	0.01		500 kHz
D	100p	0.3	0.2	0.1	0.05		400 kHz
E	100p	0.4	0.2	0.1	0.01		370 kHz
F	100p	0.2	0.2	0.1	0.05	0.02	475 kHz

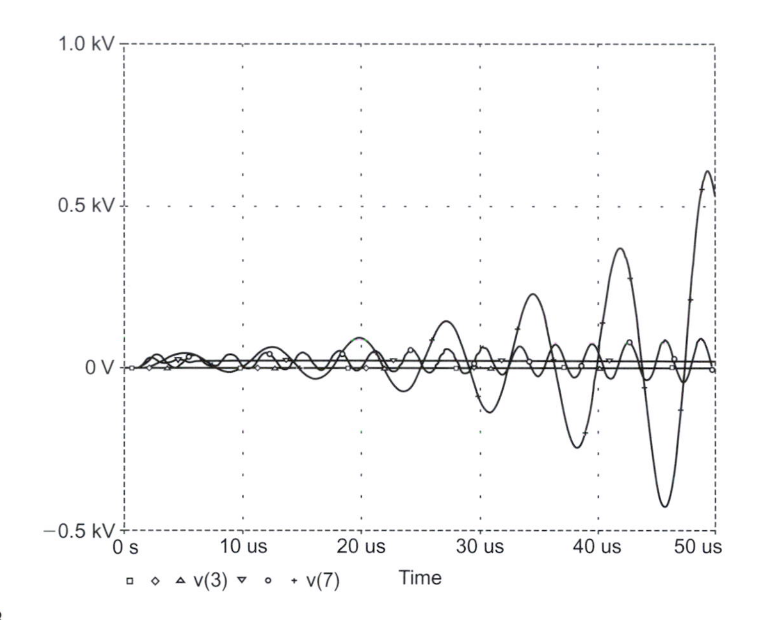

FIGURE 8.23
Adding a third pole makes possible true instability with exponentially increasing amplitude of oscillation. Note the unrealistic voltage scale on this plot.

A fourth pole can be simply added to Figure 8.17 by inserting another RC-buffer combination into the system. With *P2* = 0.5 µs and *P3* = *P4* = 0.2 µs, instability occurs at 345 kHz, which is a step towards a realistic frequency of oscillation. This is case B in Table 8.2.

When a fifth output stage pole is grafted on, so that *P3* = *P4* = *P5* = 0.2 µs, the system just oscillates at 500 kHz with *P2* set to 0.01 µs. This takes us close to a realistic frequency of oscillation. Rearranging the order of poles so *P2* = *P3* = *P4* = 0.2 µs, while *P5* = 0.01 µs, is tidier, and the stability results are of course the same; this is a linear system so the order does not matter. This is case C in Table 8.2.

Having *P2*, *P3*, and *P4* all at the same frequency does not seem very plausible in physical terms, so case D shows what happens when the five poles are staggered in frequency. *P2* needs to be increased to 0.3 µs to start the oscillation, which is now at 400 kHz. Case E is another version with five poles, showing that if *P5* is reduced *P2* needs to be doubled to 0.4 µs for instability to begin.

In the final case F, a sixth pole is added to see if this permitted sustained oscillation is above 500 kHz. This seems not to be the case; the highest frequency that could be obtained after a lot of pole twiddling was 475 kHz. This makes it clear that this model is of limited accuracy (as indeed are all models—it is a matter of degree) at high frequencies, and that further refinement is required to gain further insight.

Maximizing the NFB

Having hopefully freed ourselves from fear of feedback, and appreciating the dangers of using only a little of it, the next step is to see how much can be used. It is my view that the amount of negative feedback applied should be maximized at all audio frequencies to maximize linearity, and the only limit is the requirement for reliable HF stability. In fact, global or Nyquist oscillation is not normally a difficult design problem in power amplifiers; the HF feedback factor can be calculated simply and accurately, and set to whatever figure is considered safe. (Local oscillations and parasitics are beyond the reach of design calculations and simulations, and cause much more trouble in practice.)

In classical Control Theory, the stability of a servomechanism is specified by its *phase margin*, the amount of extra phase shift that would be required to induce sustained oscillation, and its *gain margin*, the amount by which the open-loop gain would need to be increased for the same result. These concepts are not very useful in audio power amplifier work, where many of the significant time-constants are only vaguely known. However, it is worth remembering that the phase margin will never be better than 90°, because of the phase lag caused by the VAS Miller capacitor; fortunately this is more than adequate.

In practice designers must use their judgment and experience to determine an NFB factor that will give reliable stability in production. My own experience leads me to believe that when the conventional three-stage architecture is used, 30 dB of global feedback at 20 kHz is safe, providing an output inductor is used to prevent capacitive loads from eroding the stability margins. I would say that 40 dB was distinctly risky, and I would not care to pin it down any more closely than that.

The 30 dB figure assumes simple dominant-pole compensation with a 6 dB/octave roll-off for the open-loop gain. The phase and gain margins are determined by the angle at which this slope cuts the horizontal unity-loop-gain line. (I am deliberately terse here; almost all textbooks give a very full treatment of this stability criterion.) An intersection of 12 dB/octave is definitely unstable. Working within this, there are two basic ways in which to maximize the NFB factor:

1. While a 12 dB/octave gain slope is unstable, intermediate slopes greater than 6 dB/octave can be made to work. The maximum usable is normally considered to be 10 dB/octave, which gives a phase margin of 30°. This may be acceptable in some cases, but I think it cuts it a little fine. The steeper fall in gain means that more NFB is applied at lower frequencies, and so less distortion is produced. Electronic circuitry only provides slopes in multiples of 6 dB/octave, so 10 dB/octave requires multiple overlapping time-constants to approximate a straight line at an intermediate slope. This gets complicated, and this method of maximizing NFB is not popular.
2. The gain slope varies with frequency, so that maximum open-loop gain and hence NFB factor is sustained as long as possible as frequency increases; the gain then drops quickly, at 12 dB/octave or more, but flattens out to 6 dB/octave before it reaches the critical unity loop-gain intersection. In this case the stability margins should be relatively unchanged compared with the conventional situation.

Overall feedback versus local feedback

It is one of the fundamental principles of negative feedback that if you have more than one stage in an amplifier, each with a fixed amount of open-loop gain, it is more effective to close the feedback loop around all the stages, in what is called an overall or global feedback configuration, rather than applying the feedback locally by giving each stage its own feedback loop. I hasten to add that this does not mean you cannot or should not use local feedback *as well* as overall feedback — indeed, one of the main themes of this book is that it is a very good idea, and indeed probably the only practical route to very low distortion levels.

It is worth underlining the effectiveness of overall feedback because some of the less informed audio commentators have been known to imply that overall feedback is in some way decadent or unhealthy, as opposed to the upright moral rigor of local feedback. The underlying thought, insofar as there is one, appears to be that overall feedback encloses more stages each with their own phase shift, and therefore requires compensation which will reduce the maximum slew rate. The truth, as is usual with this sort of moan, is that this could happen if you get the compensation all wrong; so get it right—it isn't hard.

It has been proposed on many occasions that if there is an overall feedback loop, the output stage should be left outside it. I have tried this, and believe me, it is not a good idea. The distortion produced

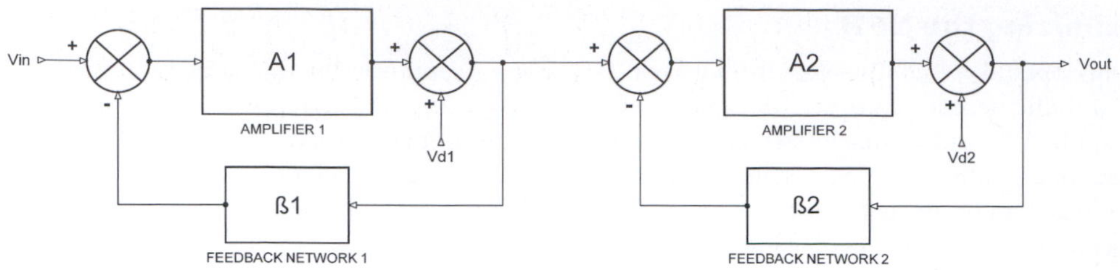

FIGURE 8.24
A negative-feedback system with two stages, each with its own feedback loop. There is no overall negative-feedback path.

Table 8.3 Open-loop gain and closed-loop errors in the two loops

1 Desired C/L gain	2 NFB fraction	3 A1 O/L gain	4 NFB factor	5 C/L gain	6 O/L error	7 C/L error	8 Total C/L gain	9 Total C/L error
5	0.2	10.00	3.00	3.333	0.5	0.1667	11.11	0.3333
5	0.2	31.62	7.32	4.317	0.5	0.0683	18.64	0.1365
5	0.2	100	21.00	4.762	0.5	0.0238	22.68	0.0476
5	0.2	200	41.00	4.878	0.5	0.0122	23.80	0.0244
5	0.2	316.23	64.25	4.922	0.5	0.0078	24.23	0.0156

by an output stage so operated is jagged and nasty, and I think no one could convince themselves it was remotely acceptable if they had seen the distortion residuals.

Figure 8.24 shows a negative-feedback system based on that in Figure 8.12, but with two stages. Each has its own open-loop gain A, its own NFB factor β, and its own open-loop error V_d added to the output of the amplifier. We want to achieve the same closed-loop gain of 25 as in Table 8.1 and we will make the wild assumption that the open-loop error of 1 in that table is now distributed equally between the two amplifiers A1 and A2. There are many ways the open- and closed-loop gains could be distributed between the two sections, but for simplicity we will give each section a closed-loop gain of 5; this means the conditions on the two sections are identical. The open-loop gains are also equally distributed between the two amplifiers so that their product is equal to column 3 in Table 8.1. The results are shown in Table 8.3: columns 1–7 show what's happening in each loop, and columns 8 and 9 give the results for the output of the two loops together, assuming for simplicity that the errors from each section can be simply added together; in other words there is no partial cancelation due to differing phases and so on.

This final result is compared with the overall feedback case of Table 8.1 in Table 8.4, where column 1 gives total open-loop gain, and column 2 is a copy of column 7 in Table 8.1 and gives the closed-loop error for the overall feedback case. Column 3 gives the closed-loop error for the two-stage feedback case, and it is brutally obvious that splitting the overall feedback situation into two local feedback stages has been a pretty bad move. With a modest total open-loop gain of 100, the local feedback system is almost twice as bad. Moving up to total open-loop gains that are more realistic for real power amplifiers, the factor of deterioration is between six and 40 times—an amount that cannot be ignored. With higher open-loop gains the ratio gets even worse. Overall feedback is totally and unarguably superior at dealing with all kinds of amplifier errors, though in this book distortion is often the one at the front of our minds.

While there is space here to give only one illustration in detail, you may be wondering what happens if the errors are not equally distributed between the two stages; the signal level at the output of the second stage will be greater than that at the output of the first stage, so it is plausible (but by no means automatically true in the real world) that the second stage will generate more distortion than the first. If this is so, and we stick with the assumption that open-loop gain is equally distributed between the two stages, then the best way to distribute the closed-loop gain is to put most of it in the first stage so

Table 8.4 Overall NFB gives a lower closed-loop error for the same total open-loop gain. The error ratio increases as the open-loop gain increases

1 *A* Total O/L gain	2 Overall NFB C/L error	3 Two-stage NFB C/L error	4 Error ratio
100	0.2000	0.3333	1.67
1000	0.0244	0.1365	5.60
10,000	0.0025	0.0476	19.10
40,000	0.0006	0.0244	39.05
100,000	0.0002	0.0156	62.28

we can get as high a feedback factor as possible in the second stage. As an example, take the case where the total open-loop gain is 40,000.

Assume that all the distortion is in the second stage, so its open-loop error is 1 while that of the first stage is zero. Now redistribute the total closed-loop gain of 25 so the first stage has a closed-loop gain of 10 and the second stage has a closed-loop gain of 2.5. This gives a closed-loop error of 0.0123, which is about half of 0.0244, the result we got with the closed-loop gain equally distributed. Clearly things have been improved by applying the greater part of the local negative feedback where it is most needed. But our improved figure is still about 20 times worse than if we had used overall feedback.

In a real power amplifier, the situation is of course much more complex than this. To start with, there are usually three rather than two stages, the distortion produced by each one is level-dependent, and in the case of the voltage-amplifier stage the amount of local feedback (and hence also the amount of overall feedback) varies with frequency. Nonetheless, it will be found that overall feedback always gives better results.

Maximizing linearity before feedback

Make your amplifier as linear as possible before applying NFB has long been a cliché. It blithely ignores the difficulty of running a typical solid-state amplifier without any feedback, to determine its basic linearity.

Virtually no dependable advice on how to perform this desirable linearization has been published. The two factors are the basic linearity of the forward path, and the amount of negative feedback applied to further straighten it out. The latter cannot be increased beyond certain limits or high-frequency stability is put in peril, whereas there seems no reason why open-loop linearity could not be improved without limit, leading us to what in some senses must be the ultimate goal—a distortionless amplifier.

REFERENCES

Attwood, B., November 1983. Design parameters important for the optimisation of PWM (class-D) amplifiers. JAES 31, 842.

Baxandall, P., December 1978. Audio Power Amplifier Design: Part 5. Wireless World 53. (This superb series of articles had six parts and ran on roughly alternate months, starting in Jan 1978.)

Blomley, P., February 1971. A New Approach to Class-B. Wireless World, p. 57.

Buitendijk, P. A 40 W integrated car radio audio amplifier," IEEE Conference Consumer Electronics, 1991 session, THAM 12.4, 174 (class-H).

Cherry, E., April/May 1983. Designing NDFL amps. Electronics Today International.

Cherry, E., May 1982. Nested differentiating feedback loops in simple audio power amplifiers. JAES 30 (5), 295.

Didden, J., April/May 2008. paX – A Power Amplifier with Error Correction. Elektor.

Evenson, R. Audio amplifiers with tailored output impedances. Preprint for November 1988 AES Convention, Los Angeles.

Feldman, L., August 1976. Class-G high-efficiency Hi-Fi amplifier. Radio Electron., 47.

Gilbert, B., 1990. In: Toumazou, C., Lidgey, F.G., Haigh, D.G. (Eds.), Current Mode Circuits from a Translinear Viewpoint, Chapter 2: Analogue IC Design: The Current-Mode Approach, IEEE.

Goldberg, J.M., Sandler, M.B., February 1991. Noise shaping and pulse-width modulation for all-digital audio power amplifier. JAES 39, 449.

Hancock, J.A., September 1991. Class-D amplifier using MOSFETS with reduced minority carrier lifetime. JAES 39, 650.

Linsley-Hood, J., April 1969. Simple Class-A Amplifier. Wireless World, p. 148.

Lohstroh, J., Otala, M., December 1973. An audio power amplifier for ultimate quality requirements. IEEE Trans Audio Electroacoustics AU-21 (6).

Mills, P.G.L., Hawksford, M.O.J., March 1989. Transconductance power amplifier systems for current-driven loudspeakers. JAES 37, 809.

Olsson, B., December 1994. Better Audio from Non-Complements? Electronics World, p. 988.

Peters, A., June 1975. Class-E RF amplifiers. IEEE J. Solid-State Circuits, 168.

Raab, F., May 1986. Average efficiency of class-G power amplifiers. IEEE Trans. Consum. Electron. CE-22, 145.

Sampei, T., et al., August 1978. Highest efficiency & super quality audio amplifier using mos-power fets in class-G. IEEE Trans. Consum. Electron. CE-24, 300.

Sandman, A., September 1982. Class S: A Novel Approach to Amplifier Distortion. Wireless World p. 38.

Self, D., 2006. Self On Audio, second ed. Newnes, Chapter 32.

Sinclair, R. (Ed.), 1993. Audio and Hi-Fi Handbook. Newnes, p. 541

Stochino, G., October 1994. Audio Design Leaps Forward? Electronics World, p. 818.

Tanaka, S., January/February 1981. A New Biasing Circuit for Class-B Operation. JAES, 27.

Thus, F., December 1992. Compact bipolar Class AB output stage. IEEE J. Solid-State Circuits, 1718.

Walker, P.J., December 1975. Current Dumping Audio Amplifier. Wireless World, p. 560.

PREFACE

The second chapter on power amplifiers is also from my *Audio Power Amplifier Design Handbook* and this one looks at the important business of controlling distortion in power amplifiers. Only the three-stage architecture is considered, but as you read through the chapter you will begin to understand why this form of amplifier is so popular.

The sources of distortion in a three-stage amplifier are, mercifully, not that numerous. They initially took me quite a lot of effort to disentangle, not least because changing one component value can change the functioning of two or more separate distortion mechanisms.

I initially classified the distortion mechanisms into seven different types, but further study and design experience has led me to increase this to eleven.

The importance of this is that once they are understood and appreciated, all of these eleven sources of distortion can be minimized, or in some cases completely eliminated, at the design stage. This means that it is possible to make Class-B amplifiers of extremely good linearity without using expensive components, complicated circuitry, or dangerously high levels of negative feedback. Typically the Total Harmonic Distortion can be reduced to less than 0.0006% at 1 kHz, which is barely detectable in the inherent circuit noise, and less than 0.004% at 10 kHz.

Such a design provides an extremely useful standard against which alternative amplifier concepts can be judged. I have called it a "blameless" amplifier, a name which indicates that its design avoids a series of possible errors that would worsen its linearity, but is not based on any new concepts or radical technologies.

It is of course possible to improve on this performance. For example, using a Class-A output stage instead of Class-B eliminates any possibility of crossover distortion and gives significantly better linearity, though at great cost in terms of efficiency and heatsinking requirements. The use of more sophisticated compensation arrangements than the standard dominant-pole Miller capacitor can also further reduce distortion, but care is always needed to make sure that stability is not compromised.

The eleven sources of amplifier distortion (I am not committing myself to that number, as I may discover some more) range from the inherent non-linearity of a bipolar transistor's Vbe/Ic characteristic, through some subtle but damaging possible component defects, to topological problems which can be completely solved simply by putting power supply connections in the right place. It is not possible to give a full explanation of eleven different mechanisms in one chapter, and the one reproduced here only gives a summary. Full explanations of each distortion mechanism, with various alternative methods of countering it, are given in the later chapters of the *Audio Power Amplifier Design Handbook*.

This chapter also looks at the use of what I call "model" amplifiers, where the normal high-current output stage is replaced by a small-signal linear output stage, usually running in Class-A, which can be made very linear indeed as it only has to drive the feedback network and the test instrumentation. This approach makes the study of the distortions produced by the first two stages of the amplifier very much simpler.

The important topic of the measurement of open-loop gain is also covered, with a way of doing it presented that is quite bullet-proof and requires nothing risky like tampering with the feedback connection.

CHAPTER 9

The General Principles of Power Amplifiers

Audio Power Amplifier Design Handbook by Douglas Self

HOW A GENERIC AMPLIFIER WORKS

Figure 9.1 shows a very conventional power amplifier circuit; it is as standard as possible. A great deal has been written about this configuration, though the subtlety and quiet effectiveness of the topology are usually overlooked, and the explanation below therefore touches on several aspects that seem to be almost unknown. The circuit has the merit of being docile enough to be made into a functioning amplifier by someone who has only the sketchiest of notions as to how it works.

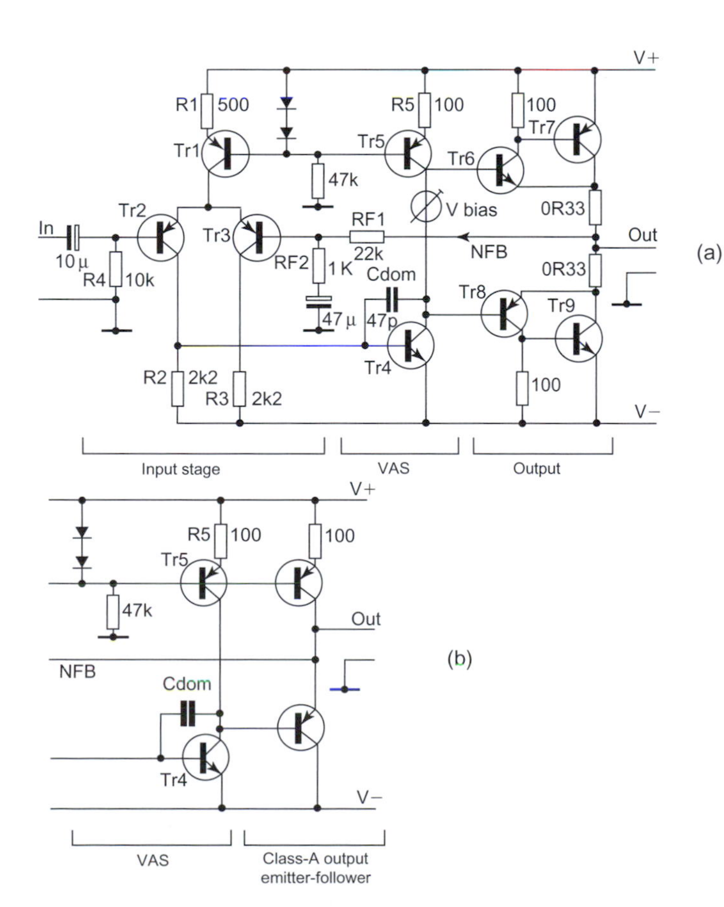

FIGURE 9.1
(a) A conventional Class-B power-amp circuit. (b) With small-signal Class-A output emitter-follower replacing Class-B output to make a model amplifier.

The input differential pair implements one of the few forms of distortion cancelation that can be relied upon to work reliably without adjustment—this is because the transconductance of the input pair is determined by the physics of transistor action rather than matching of ill-defined parameters such as beta; the logarithmic relation between I_c and V_{be} is proverbially accurate over some eight or nine decades of current variation.

The voltage signal at the voltage-amplifier stage (hereafter VAS) transistor base is typically a couple of millivolts, looking rather like a distorted triangle wave. Fortunately the voltage here is of little more than academic interest, as the circuit topology essentially consists of a transconductance amp (voltage-difference input to current output) driving into a transresistance (current-to-voltage converter) stage. In the first case the exponential V_{be}/I_c law is straightened out by the differential-pair action, and in the second the global (overall) feedback factor at LF is sufficient to linearize the VAS, while at HF shunt negative feedback (hereafter NFB) through C_{dom} conveniently takes over VAS linearization while the overall feedback factor is falling.

The behavior of Miller dominant-pole compensation in this stage is actually exceedingly elegant, and not at all a case of finding the most vulnerable transistor and slugging it. As frequency rises and C_{dom} begins to take effect, negative feedback is no longer applied globally around the whole amplifier, which would include the higher poles, but instead is seamlessly transferred to a purely local role in linearizing the VAS. Since this stage effectively contains a single gain transistor, any amount of NFB can be applied to it without stability problems.

The amplifier operates in two regions; the LF, where open-loop (O/L) gain is substantially constant, and HF, above the dominant-pole breakpoint, where the gain is decreasing steadily at 6 dB/octave. Assuming the output stage is unity gain, three simple relationships define the gain in these two regions:

$$LFgain = g_m \cdot \beta \cdot R_c \tag{9.1}$$

At least one of the factors that set this (beta) is not well controlled and so the LF gain of the amplifier is to a certain extent a matter of pot luck; fortunately this does not matter, so long as it is high enough to give a suitable level of NFB to eliminate LF distortion. The use of the word *eliminate* is deliberate, as will be seen later. Usually the LF gain, or HF local feedback factor, is made high by increasing the effective value of the VAS collector impedance R_c, either by the use of current-source collector load, or by some form of bootstrapping.

The other important relations are:

$$HFgain = \frac{g_m}{\omega \cdot C_{dom}} \tag{9.2}$$

Dominant-pole frequency

$$P1 = \frac{1}{C_{dom} \cdot \beta \cdot R_c} \tag{9.3}$$

Where:

$$\omega = 2 \cdot \pi \cdot frequency$$

In the HF region, things are distinctly more difficult as regards distortion, for while the VAS is locally linearized, the global feedback factor available to linearize the input and output stages is falling steadily at 6 dB/octave. For the time being we will assume that it is possible to define an HF gain (say, *N* dB at 20 kHz), which will assure stability with practical loads and component variations. Note that the HF gain, and therefore both HF distortion and stability margin, are set by the simple combination of the input stage transconductance and one capacitor, and most components have no effect on it at all.

It is often said that the use of a high VAS collector impedance provides a current drive to the output devices, often with the implication that this somehow allows the stage to skip quickly and lightly over the dreaded crossover region. This is a misconception—the collector impedance falls to a few kilohms at HF, due to increasing local feedback through C_{dom}, and in any case it is very doubtful if true current drive would be a good thing: calculation shows that a low-impedance voltage drive minimizes distortion due to beta-unmatched output halves (Oliver, 1971), and it certainly eliminates the effect of Distortion 4, described below.

THE ADVANTAGES OF THE CONVENTIONAL

It is probably not an accident that the generic configuration is by a long way the most popular, though in the uncertain world of audio technology it is unwise to be too dogmatic about this sort of thing. The generic configuration has several advantages over other approaches:

- The input pair not only provides the simplest way of making a DC-coupled amplifier with a dependably small output offset voltage, but can also (given half a chance) completely cancel the second-harmonic distortion that would be generated by a single-transistor input stage. One vital condition for this must be met; the pair must be accurately balanced by choosing the associated components so that the two collector currents are equal. (The *typical* component values shown in Figure 9.1 do *not* bring about this most desirable state of affairs.)
- The input devices work at a constant and near-equal V_{ce}, giving good thermal balance.
- The input pair has virtually no voltage gain so no low-frequency pole can be generated by Miller effect in the TR2 collector-base capacitance. All the voltage gain is provided by the VAS stage, which makes for easy compensation. Feedback through C_{dom} lowers VAS input and output impedances, minimizing the effect of input-stage capacitance, and the output-stage capacitance. This is often known as pole-splitting (Feucht, 1990); the pole of the VAS is moved downwards in frequency to become the dominant pole, while the input-stage pole is pushed up in frequency.
- The VAS Miller compensation capacitance smoothly transfers NFB from a global loop that may be unstable, to the VAS local loop that cannot be. It is quite wrong to state that *all* the benefits of feedback are lost as the frequency increases above the dominant pole, as the VAS is still being linearized. This position of C_{dom} also swamps the rather variable C_{cb} of the VAS transistor.

THE DISTORTION MECHANISMS

My original series of articles on amplifier distortion listed seven important distortion mechanisms, all of which are applicable to any Class-B amplifier, and do not depend on particular circuit arrangements. As a result of further experimentation and further thought, I have now increased this to ten.

In the typical amplifier, THD is often thought to be simply due to the Class-B nature of the output stage, which is linearized less effectively as the feedback factor falls with increasing frequency. This is, however, only true when all the removable sources of distortion have been eliminated. In the vast majority of amplifiers in production, the true situation is more complex, as the small-signal stages can generate significant distortion of their own, in at least two different ways; this distortion can easily exceed output stage distortion at high frequencies. It is particularly inelegant to allow this to occur given the freedom of design possible in the small-signal section.

If the ills that a Class-B stage is prone to are included then there are eight major distortion mechanisms. Note that this assumes that the amplifier is not overloaded in any way, and therefore is not suffering from:

- activation of any overload protection circuitry;
- overloading not affecting protection circuitry (for example, insufficient current to drive the output stage due to a VAS current source running set to too low a value);
- slew-rate limiting;
- defective or out-of-tolerance components.

It also assumes the amplifier has proper global or Nyquist stability and does not suffer from any parasitic oscillations; the latter, if of high enough frequency, cannot be seen on the average oscilloscope and tend to manifest themselves only as unexpected increases in distortion, sometimes at very specific power outputs and frequencies.

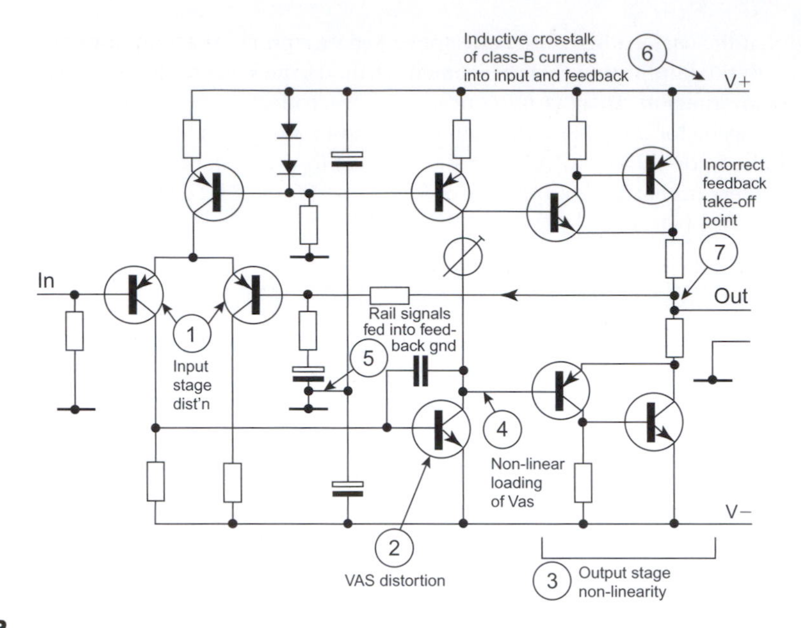

FIGURE 9.2
The location of the first seven major distortion mechanisms. The eighth (capacitor distortion) is omitted for clarity.

In Figure 9.2 an attempt has been made to show the distortion situation diagrammatically, indicating the location of each mechanism within the amplifier. Distortion 8 is not shown as there is no output capacitor.

The first four distortion mechanisms are inherent to any three-stage amplifier.

Distortion 1: Input Stage Distortion

This concerns nonlinearity in the input stage. If this is a carefully balanced differential pair then the distortion is typically only measurable at HF, rises at 18 dB/octave, and is almost pure third harmonic. If the input pair is unbalanced (which from published circuitry it usually is) then the HF distortion emerges from the noise floor earlier, as frequency increases, and rises at 12 dB/octave as it is mostly second harmonic.

Distortion 2: VAS Distortion

Nonlinearity in the voltage-amplifier stage (which I call the VAS for brevity) surprisingly does not always figure in the total distortion. If it does, it remains constant until the dominant-pole frequency $P1$ is reached, and then rises at 6 dB/octave. With the configurations discussed here it is always second harmonic.

Usually the level is very low due to linearizing negative feedback through the dominant-pole capacitor. Hence if you crank up the local VAS open-loop gain, for example by cascoding or putting more current-gain in the local VAS–C_{dom} loop, and attend to Distortion 4 below, you can usually ignore VAS distortion.

Distortion 3: Output Stage Distortion

Nonlinearity in the output stage, which is naturally the obvious source. This in a Class-B amplifier will be a complex mix of large-signal distortion and crossover effects, the latter generating a spray of high-order harmonics, and in general rising at 6 dB/octave as the amount of negative feedback decreases. Large-signal THD worsens with 4 Ω loads and worsens again at 2 Ω. The picture is complicated by dilatory switch-off in the relatively slow output devices, ominously signaled by supply current increasing in the top audio octaves.

Distortion 4: VAS-Loading Distortion

This is loading of the VAS by the nonlinear input impedance of the output stage. When all other distortion sources have been attended to, this is the limiting distortion factor at LF (say, below 2 kHz); it is simply cured by buffering the VAS from the output stage. Magnitude is essentially constant with frequency, though the overall effect in a complete amplifier becomes less as frequency rises and feedback through C_{dom} starts to linearize the VAS.

The next three distortion mechanisms are in no way inherent; they may be reduced to unmeasurable levels by simple precautions. They are what might be called topological distortions, in that they depend wholly on the arrangement of wiring and connections, and on the physical layout of the amplifier.

Distortion 5: Rail-Decoupling Distortion

Nonlinearity caused by large rail-decoupling capacitors feeding the distorted signals on the supply lines into the signal ground. This seems to be the reason that many amplifiers have rising THD at low frequencies. Examining one commercial amplifier kit, I found that rerouting the decoupler ground return reduced the THD at 20 Hz by a factor of 3.

Distortion 6: Induction Distortion

This is nonlinearity caused by induction of Class-B supply currents into the output, ground, or negative-feedback lines. This was highlighted by Cherry (1981) but seems to remain largely unknown; it is an insidious distortion that is hard to remove, though when you know what to look for on the THD residual it is fairly easy to identify. I suspect that a large number of commercial amplifiers suffer from this to some extent.

Distortion 7: NFB Take-Off Distortion

This is nonlinearity resulting from taking the NFB feed from slightly the wrong place near where the power-transistor Class-B currents sum to form the output. This may well be another very prevalent defect.

The next two distortion mechanisms relate to circuit components that are non-ideal or poorly chosen.

Distortion 8: Capacitor Distortion

In its most common manifestation this is caused by the non-ideal nature of electrolytic capacitors. It rises as frequency falls, being strongly dependent on the signal voltage across the capacitor.

The most common sources of nonlinearity are the input DC-blocking capacitor or the feedback network capacitor; the latter is more likely as it is much easier to make an input capacitor large enough to avoid the problem. It causes serious difficulties if a power amplifier is AC-coupled, i.e. has a series capacitor at the output, but this is rare these days.

It can also occur in ceramic capacitors that are nominally of the NP0/C0G type but actually have a significant voltage coefficient, when they are used to implement Miller dominant-pole compensation.

Distortion 9: Magnetic Distortion

This arises when a signal at amplifier output level is passed through a ferromagnetic conductor. Ferromagnetic materials have a nonlinear relationship between the current passing through them and the magnetic flux it creates, and this induces voltages that add distortion to the signal. The effect has been found in output relays and also speaker terminals. The terminals appeared to be made of brass but were actually plated steel.

Distortion 10: Input Current Distortion

This distortion is caused when an amplifier input is driven from a significant source impedance. The input current taken by the amplifier is nonlinear, even if the output of the amplifier is distortion free, and the resulting voltage drop in the source impedance introduces distortion.

This mechanism is purely a product of circuit design, rather than layout or component integrity, but it has been put in a category of its own because, unlike the inherent Distortions 1–4, it is a product of the interfacing between the amplifier and the circuitry upstream of it.

Distortion 11: Premature Overload Protection

The overload protection of a power amplifier can be implemented in many ways, but without doubt the most popular method is the use of VI limiters that shunt signal current away from the inputs to the output stage. In their simplest and most common form, these come into operation relatively gradually as their set threshold is exceeded, and introduce distortion into the signal long before they close it down entirely. It is therefore essential to plan a sufficient safety margin into the output stage so that the VI limiters are never near activation in normal use.

Other methods of overload protection that trigger and then latch the amplifier into a standby state cannot generate this distortion, but if this leads to repeated unnecessary shutdowns it will be a good deal more annoying than occasional distortion.

NONEXISTENT OR NEGLIGIBLE DISTORTIONS

Having set down what might be called the Eleven Great Distortions, we must pause to put to flight a few paper tigers ...

The first is common-mode distortion in the input stage, a specter that haunts the correspondence columns. Since it is fairly easy to make an amplifier with less than 0.00065% THD (1 kHz) without paying any attention at all to this issue it cannot be too serious a problem. It is perhaps a slight exaggeration to call it nonexistent, as under special circumstances it can be seen, but it is certainly unmeasurable under normal circumstances.

If the common-mode voltage on the input pair is greatly increased, then a previously negligible distortion mechanism is indeed provoked. This increase is achieved by reducing the C/L gain to between 1 and 2×; the input signal is now much larger for the same output, and the feedback signal must match it, so the input stage experiences a proportional increase in common-mode voltage.

The distortion produced by this mechanism increases as the square of the common-mode voltage, and therefore falls rapidly as the closed-loop gain is increased back to normal values. It therefore appears that the only precautions required against common-mode distortion are to ensure that the closed-loop gain is at least five times (which is no hardship, as it almost certainly is anyway) and to use a tail-current source for the input pair, which again is standard practice.

The second distortion conspicuous by its absence in the list is the injection of distorted supply-rail signals directly into the amplifier circuitry. Although this putative mechanism has received a lot of attention (Ball, 1990), dealing with Distortion 5 above by proper grounding seems to be all that is required; once more, if triple-zero THD can be attained using simple unregulated supplies and without paying any attention to the power-supply rejection ratio (PSRR) beyond keeping the amplifier free from hum (which it reliably can be) then there seems to be no problem. There is certainly no need for regulated supply rails to get a good performance. PSRR does need careful attention if the hum/noise performance is to be of the first order, but a little RC filtering is usually all that is needed.

A third mechanism of very doubtful validity is thermal distortion, allegedly induced by parameter changes in semiconductor devices whose instantaneous power dissipation varies over a cycle. This would surely manifest itself as a distortion rise at very low frequencies, but it simply does not happen. There are several distortion mechanisms that can give a THD rise at LF, but when these are eliminated the typical distortion trace remains flat down to at least 10 Hz. The worst thermal effects would be expected in Class-B output stages where dissipation varies wildly over a cycle; however, drivers and output devices have relatively large junctions with high thermal inertia. Low frequencies are of course also where the NFB factor is at its maximum.

To return to our list of the unmagnificent eleven, note that only Distortion 3 is directly due to output stage nonlinearity, though Distortions 4–7 all result from the Class-B nature of the typical output stage. Distortions 8–11 can happen in any amplifier, whatever its operating class.

THE PERFORMANCE OF A STANDARD AMPLIFIER

The THD curve for the standard amplifier is shown in Figure 9.3. As usual, distortion increases with frequency, and, as we shall see later, would give grounds for suspicion if it did not. The flat part of the curve below 500 Hz represents non-frequency-sensitive distortion rather than the noise floor, which for this case is at the 0.0005% level. Above 500 Hz the distortion rises at an increasing rate, rather than a constant number of dB/octave, due to the combination of Distortions 1–4. (In this case, Distortions 5–7 have been carefully eliminated to keep things simple; this is why the distortion performance looks good already, and the significance of this should not be overlooked.) It is often written that having distortion constant across the audio band is a good thing – a most unhappy conclusion, as the only practical way to achieve this with a normal Class-B amplifier is to *increase* the distortion at LF, for example by allowing the VAS to distort significantly.

It should now be clear why it is hard to wring linearity out of such a snake-pit of contending distortions. A circuit-value change is likely to alter at least two of the distortion mechanisms, and probably change the O/L gain as well.

OPEN-LOOP LINEARITY AND HOW TO DETERMINE IT

Improving something demands measuring it, and thus it is essential to examine the open-loop linearity of power-amp circuitry. This cannot be done directly, so it is necessary to measure the NFB factor and calculate open-loop distortion from closed-loop measurements. The closed-loop gain is normally set by input sensitivity requirements.

Measuring the feedback factor is at first sight difficult, as it means determining the open-loop gain. Standard methods for measuring op-amp open-loop gain involve breaking feedback loops and manipulating C/L gains, procedures that are likely to send the average power amplifier into fits. Nonetheless the need to measure this parameter is inescapable as a typical circuit modification – e.g., changing the value of R2 changes the open-loop gain as well as the linearity, and to prevent total confusion it is essential to keep a very clear idea of whether an observed change is due to an improvement in O/L linearity or merely because the O/L gain has risen. It is wise to keep a running check on this as work proceeds, so the direct method of open-loop gain measurement shown in Figure 9.4 was evolved.

DIRECT OPEN-LOOP GAIN MEASUREMENT

The amplifier shown in Figure 9.1 is a differential amplifier, so its open-loop gain is simply the output divided by the voltage difference between the inputs. If output voltage is kept constant by providing a constant swept-frequency voltage at the positive input, then a plot of open-loop gain versus frequency is obtained by measuring the error-voltage between the inputs, and referring this to the output level. This gives an upside-down plot that rises at HF rather than falling, as the differential amplifier requires more input for the same output as frequency increases, but the method is so quick and convenient that this can be lived with. Gain is plotted in dB with respect to the chosen output level (+16 dBu in this case) and the actual gain at any frequency can be read off simply by dropping the minus sign. Figure 9.5 shows the plot for the amplifier in Figure 9.1.

The HF-region gain slope is always 6 dB/octave unless you are using something special in the way of compensation, and by the Nyquist rules must continue at this slope until it intersects the horizontal line representing the feedback factor, if the amplifier is stable. In other words, the slope is not being

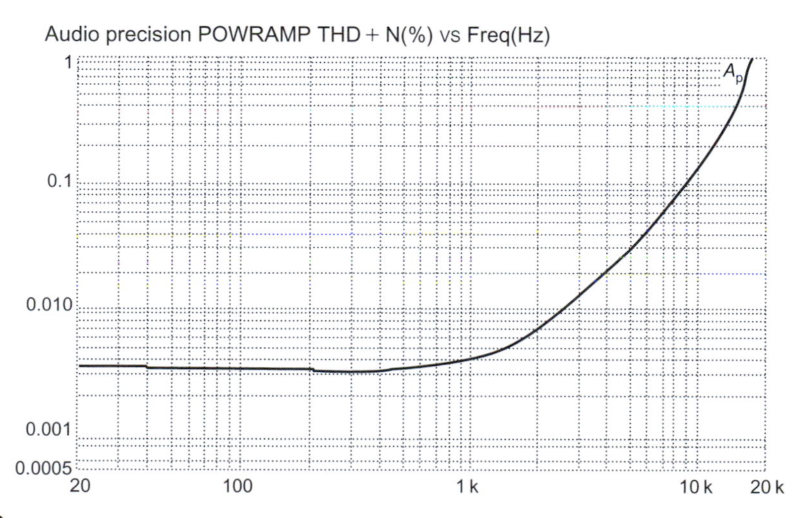

FIGURE 9.3
The distortion performance of the Class-B amplifier in Figure 9.1.

accelerated by other poles until the loop gain has fallen to unity, and this provides a simple way of putting a lower bound on the next pole *P*2; the important *P*2 frequency (which is usually somewhat mysterious) must be above the intersection frequency if the amplifier is seen to be stable.

Given test gear with a sufficiently high common-mode rejection ratio (CMRR) balanced input, the method of Figure 9.4 is simple; just buffer the differential inputs from the cable capacitance with TL072 buffers, which place negligible loading on the circuit if normal component values are used. In particular be wary of adding stray capacitance to ground to the negative input, as this directly imperils amplifier stability by adding an extra feedback pole. Short wires from power amplifier to buffer IC can usually be unscreened as they are driven from low impedances.

The test-gear input CMRR defines the maximum open-loop gain measurable; I used an Audio Precision System-1 without any special alignment of CMRR. A calibration plot can be produced by feeding the two buffer inputs from the same signal; this will probably be found to rise at 6 dB/octave, being set by the inevitable input asymmetries. This must be low enough for amplifier error signals to be above it by at least 10 dB for reasonable accuracy. The calibration plot will flatten out at low frequencies, and may even show an LF rise due to imbalance of the test-gear input-blocking capacitors; this can make determination of the lowest pole *P*1 difficult, but this is not usually a vital parameter in itself.

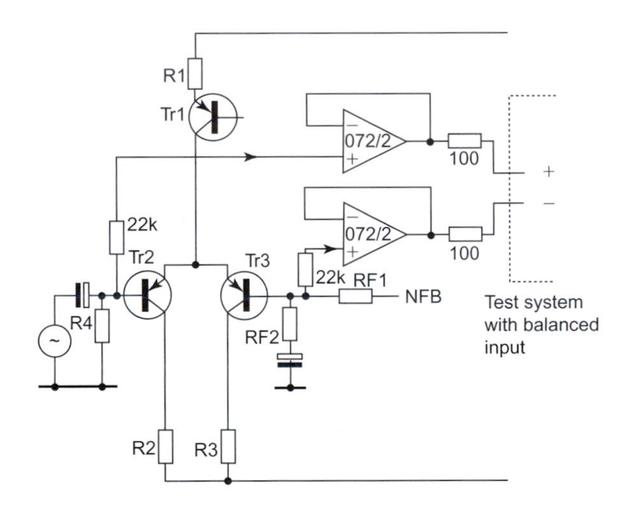

FIGURE 9.4
Test circuit for measuring open-loop gain directly. The accuracy with which high O/L gains can be measured depends on the test-gear CMRR.

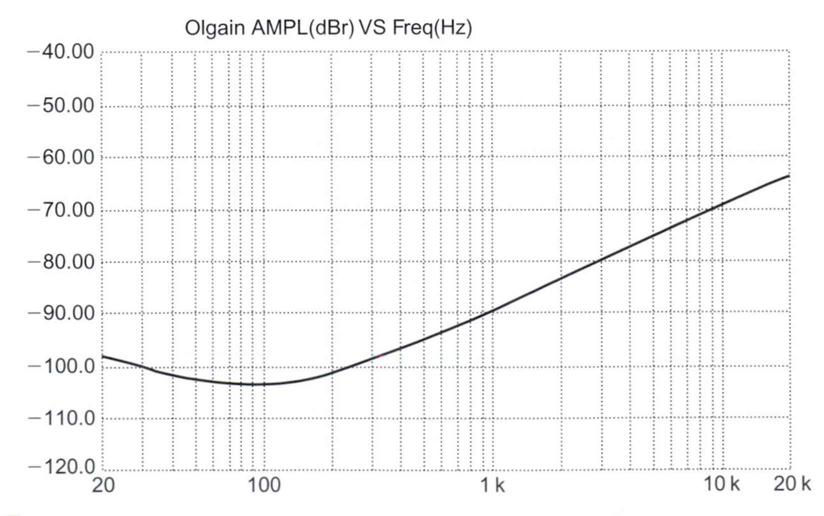

FIGURE 9.5
Open-loop gain versus frequency plot for Figure 9.1. Note that the curve rises as gain falls, because the amplifier error is the actual quantity measured.

USING MODEL AMPLIFIERS

Distortions 1 and 2 can dominate amplifier performance and need to be studied without the manifold complications introduced by a Class-B output stage. This can be done by reducing the circuit to a *model* amplifier that consists of the small-signal stages alone, with a very linear Class-A emitter-follower attached to the output to allow driving the feedback network; here *small signal* refers to current rather than voltage, as the model amplifier should be capable of giving a full power-amp voltage swing, given sufficiently high rail voltages. From Figure 9.2 it is clear that this will allow study of Distortions 1 and 2 in isolation, and using this approach it will prove relatively easy to design a small-signal amplifier with negligible distortion across the audio band, and this is the only sure foundation on which to build a good power amplifier.

A typical plot combining Distortions 1 and 2 from a model amp is shown in Figure 9.6, where it can be seen that the distortion rises with an accelerating slope, as the initial rise at 6 dB/octave from the VAS is contributed to and then dominated by the 12 dB/octave rise in distortion from an unbalanced input stage.

The model can be powered from a regulated current-limited PSU to cut down the number of variables, and a standard output level chosen for comparison of different amplifier configurations; the rails and output level used for the results in this work were ±15 V and + 16 dBu. The rail voltages can be made comfortably lower than the average amplifier HT rail, so that radical bits of circuitry can be tried out without the creation of a silicon cemetery around your feet. It must be remembered that some phenomena such as input-pair distortion depend on absolute output level, rather than the proportion of the rail voltage used in the output swing, and will be increased by a mathematically predictable amount when the real voltage swings are used.

The use of such model amplifiers requires some caution, and gives no insight into BJT output stages, whose behavior is heavily influenced by the sloth and low current gain of the power devices. As a general rule, it should be possible to replace the small-signal output with a real output stage and get a stable and workable power amplifier; if not, then the model is probably dangerously unrealistic.

THE CONCEPT OF THE BLAMELESS AMPLIFIER

Here I introduce the concept of what I have chosen to call a *Blameless* audio power amplifier. This is an amplifier designed so that all the easily defeated distortion mechanisms have been rendered negligible. (Note that the word *Blameless* has been carefully chosen *not* to imply perfection, but merely the avoidance of known errors.) Such an amplifier gives about 0.0005% THD at 1 kHz and approximately 0.003% at 10 kHz when driving 8 Ω. This is much less THD than a Class-B amplifier is normally

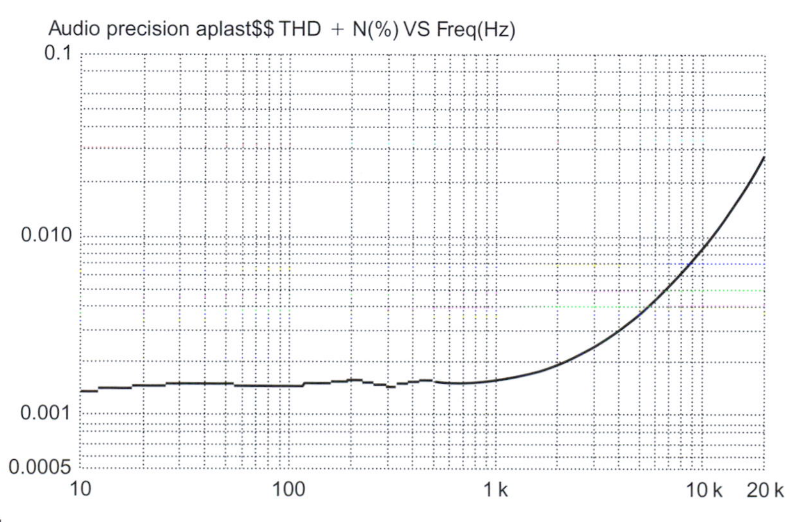

FIGURE 9.6
The distortion from a model amplifier, produced by the input pair and the voltage-amplifier stage. Note increasing slope as input pair distortion begins to add to VAS distortion.

expected to produce, but the performance is repeatable, predictable, and definitely does not require large global feedback factors.

Distortion 1 cannot be totally eradicated, but its onset can be pushed well above 20 kHz by the use of local feedback. Distortion 2 (VAS distortion) can be similarly suppressed by cascoding or beta-enhancement, and Distortions 4–7 can be made negligible by simple topological methods. All these measures will be detailed later. This leaves Distortion 3, which includes the intractable Class-B problems, i.e. crossover distortion (Distortion 3b) and HF switch-off difficulties (Distortion 3c). Minimizing 3b requires a Blameless amplifier to use a BJT output rather than FETs.

A Blameless Class-B amplifier essentially shows crossover distortion only, so long as the load is no heavier than 8 Ω; this distortion increases with frequency as the amount of global NFB falls. At 4 Ω loading an extra distortion mechanism (3a) generates significant third harmonic.

The importance of the Blameless concept is that it represents the best distortion performance obtainable from straightforward Class-B. This performance is stable and repeatable, and varies little with transistor type as it is not sensitive to variable quantities such as beta.

Blamelessness is a condition that can be defined with precision, and is therefore a standard other amplifiers can be judged against. A Blameless design represents a stable point of departure for more radical designs. This may be the most important use of the idea.

REFERENCES

Ball, G., 1990. Distorting power supplies. Electronics & Wireless World, 1084.

Cherry, E., 1981 May. A new distortion mechanism in Class-B amplifiers. JAES 6, 327.

Feucht, D., 1990. Handbook of Analog Circuit Design. Academic Press, p. 256 (pole-splitting).

Oliver, B. Distortion in complementary-pair Class-B amplifiers. Hewlett-Packard Journal, 1971 February, 11.

PREFACE

In the third chapter on power amplifiers, which is also taken from my *Audio Power Amplifier Design Handbook*, I explain in detail the operation and design of Class-G amplifiers. Music has a large peak-to-mean level ratio, with the output voltage usually a long way below the peak levels, and this is what makes possible the improved efficiency of Class-G. It is a clever technology that significantly economizes on power consumption and heat dissipation by taking the power for its output stage from low-voltage rails when reproducing small amplitude signals, and switching to high-voltage rails to accommodate the relatively rare signal peaks. Anyone acquainted with electronic design will at once raise an eyebrow at the word "switching" in there; it is hard enough to make a standard Class-B output stage perform the relatively gentle transition between the two halves of the output stage without introducing excessive crossover distortion, and a radical approach that involves stepping the output supply rail up and down at audio frequencies, including 20 kHz, sounds like it might be a nightmare to make work properly.

In fact, the operation of a two-rail series-mode Class-G amplifier, which is the most common kind, and the most useful for hi-fi, is not so fraught as it might sound. The rail-switching process certainly can introduce ugly glitches into the audio output, and this definitely happened in the early designs of the mid-1970s, causing class-G to get a reputation for rather mediocre quality despite its acknowledged efficiency, and it was at first not widely adopted.

When I took a look at Class-G technology in 1999, it had not changed significantly in the last 20 years or so. The major problem of glitching when transferring from one rail to another was caused by charge-carrier storage in the commutating diodes, and this was now straightforward to cure by adopting Schottky power diodes for the job. So far as I know I was the first person to do this; the technique was originally published in *Electronics World* at the end of 2001, and nobody wrote in to tell me they'd done it first. A more difficult problem was the increase in distortion caused by Early Effect in the output stage devices, which were affected by the abrupt changes in collector voltage. I halved the magnitude of this problem be rearranging the driver connections in the output stage, but it was not possible to eliminate it altogether, and today Class-G remains somewhat inferior in linearity to a Blameless Class-B design. I should point out, though, that the extra distortion is at a very low level (of the order of 0.002% at 1 kHz) and only manifests itself relatively infrequently, when the amplifier switches to make use of the higher-voltage supply rail. These improvements are detailed in the chapter here.

Class-G technology is of increasing importance today. It has found niches in driving the low-frequency units in multi-amped systems, and in powered subwoofers where relatively large amplifier power outputs are required to give extended bass through equalization. When it is properly implemented—as described in this chapter—it can give lower noise, a much better distortion performance, and an efficiency with real signals that is almost as good as the Class-D amplifiers which are sometimes used in this role.

CHAPTER 10

Class-G Power Amplifiers

Audio Power Amplifier Design Handbook by Douglas Self

Most types of audio power amplifier are less efficient than Class-B; for example, Class-AB is markedly less efficient at the low end of its power capability, while it is clear that Class-A wastes virtually all the energy put into it. Building amplifiers with higher efficiency is more difficult. Class-D, using ultrasonic pulse width modulation, promises high efficiency and indeed delivers it, but it is undeniably a difficult technology, and its linearity is still a long way short of Class-B. The practical efficiency of Class-D rests on details of circuit design and device characteristics. The apparently unavoidable LC output filter—second order at least—can only give a flat response into one load impedance, and its magnetics are neither cheap nor easy to design. There are likely to be some daunting EMC difficulties with emissions. Class-D is not an attractive proposition for high-quality domestic amplifiers that must work with separate speakers of unknown impedance characteristics.

There is, however, the Class-G method. Power is drawn from either high- or low-voltage rails as the signal level demands. This technology has taken a long time to come to fruition, but is now used in very-high-power amplifiers for large PA systems, where the savings in power dissipation are important, and is also making its presence felt in home theater systems; if you have seven or eight power amplifiers instead of two their losses are rather more significant. Class-G is firmly established in powered subwoofers, and even in ADSL telephone-line drivers. Given the current concern for economy in energy consumption, Class-G may well become more popular in mainstream areas where its efficiency can be used as a marketing point. It is a technology whose time has come.

THE PRINCIPLES OF CLASS-G

Music has a large peak-to-mean level ratio. For most of the time the power output is a long way below the peak levels, and this makes possible the improved efficiency of Class-G. Even rudimentary statistics for this ratio for various genres of music are surprisingly hard to find, but it is widely accepted that the range between 10 dB for compressed rock and 30 dB for classical material covers most circumstances.

If a signal spends most of its time at low power, then while this is true a low-power amplifier will be much more efficient. For most of the time lower output levels are supplied from the lowest-voltage rails, with a low-voltage drop between rail and output, and correspondingly low dissipation. The most popular Class-G configurations have two or three pairs of supply rails, two being usual for hi-fi, while three is more common in high-power PA amplifiers.

When the relatively rare high-power peaks do occur they must be handled by some mechanism that can draw high power, causing high internal dissipation, but which only does so for brief periods. These infrequent peaks above the transition level are supplied from the high-voltage pair of rails. Clearly the switching between rails is the heart of the matter, and anyone who has ever done any circuit design will immediately start thinking about how easy or otherwise it will be to make this happen cleanly with a high-current 20 k Hz signal.

Audio Engineering Explained

There are two main ways to arrange the dual-rail system: series and parallel (i.e. shunt). This chapter deals only with the series configuration, as it seems to have had the greatest application to hi-fi. The parallel version is more often used in high-power PA amplifiers.

INTRODUCING SERIES CLASS-G

A series configuration Class-G output stage using two rail voltages is shown in Figure 10.1. The so-called inner devices are those that work in Class-B; those that perform the rail-switching on signal peaks are called the outer devices—by me, anyway. In this design study the EF type of output stage is chosen because of its greater robustness against local HF instability, though the CFP configuration could be used instead for inner, outer, or both sets of output devices, given suitable care. For maximum power efficiency the inner stage normally runs in Class-B, though there is absolutely no reason why it could not be run in Class-AB or even Class-A; there will be more discussion of these intriguing possibilities later. If the inner power devices are in Class-B, and the outer ones conduct for much less than 50% of a cycle, being effectively in Class-C, then according to the classification scheme I have proposed (Self, 2006, p. 347), this should be denoted Class-B + C. The plus sign indicates the series rather than shunt connection of the outer and inner power devices. This basic configuration was developed by Hitachi to reduce amplifier heat dissipation (Sampei et al., 1978; Feldman, 1976). Musical signals

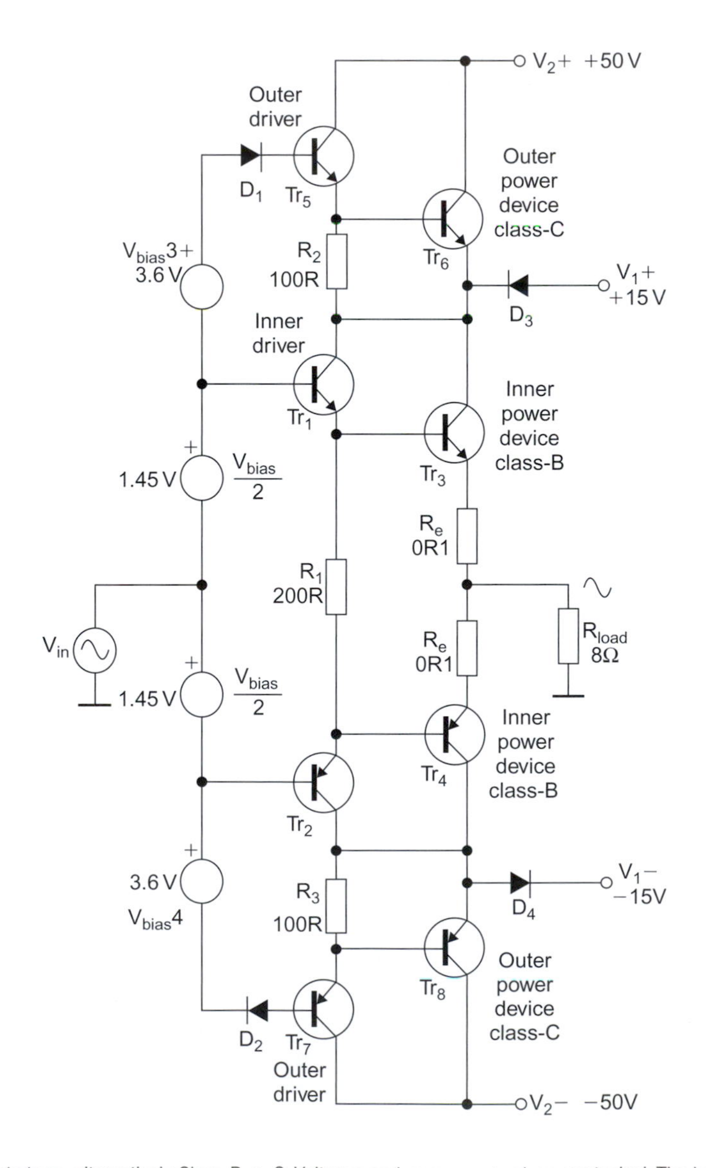

FIGURE 10.1
A series Class-G output stage, alternatively Class-B + C. Voltages and component values are typical. The inner stage is Class-B EF. Biasing by my method.

spend most of their time at low levels, having a high peak/mean ratio, and power dissipation is greatly reduced by drawing from the lower ±V1 supply rails at these times.

The inner stage TR3, TR4 operates in normal Class-B. TR1, TR2 are the usual drivers and R1 is their shared emitter resistor. The usual temperature-compensated V_{bias} generator is required, shown here theoretically split in half to maintain circuit symmetry when the stage is SPICE simulated; since the inner power devices work in Class-B it is their temperature that must be tracked to maintain quiescent conditions. Power from the lower supply is drawn through D3 and D4, often called the commutating diodes, to emphasize their rail-switching action. The word "commutation" avoids confusion with the usual Class-B crossover at zero volts. I have called the level at which rail-switching occurs the transition level.

When a positive-going instantaneous signal exceeds low rail +V1, D1 conducts, TR5 and TR6 turn on and D3 turns off, so the entire output current is now taken from the high-voltage +V2 rail, with the voltage drop and hence power dissipation shared between TR4 and TR6. Negative-going signals are handled in exactly the same way. Figure 10.2 shows how the collector voltages of the inner power devices retreat away from the output rail as it approaches the lower supply level.

Class-G is commonly said to have worse linearity than Class-B, the blame usually being loaded onto the diodes and problems with their commutation. As usual, received wisdom is only half of the story, if that, and there are other linearity problems that are not due to sluggish diodes, as will be revealed shortly. It is inherent in the Class-G principle that if switching glitches do occur they only happen at moderate power or above, and are well displaced away from the critical crossover region where the amplifier spends most of its time. A Class-G amplifier has a low-power region of true Class-B linearity, just as a Class-AB amplifier has a low-power region of true Class-A performance.

EFFICIENCY OF CLASS-G

The standard mathematical derivation of Class-B efficiency with sine-wave drive uses straightforward integration over a half-cycle to calculate internal dissipation against voltage fraction, i.e. the fraction of possible output voltage swing. As is well known, in Class-B the maximum heat dissipation is about 40% of maximum output power, at an output voltage fraction of 63%, which also delivers 40% of the maximum output power to the load.

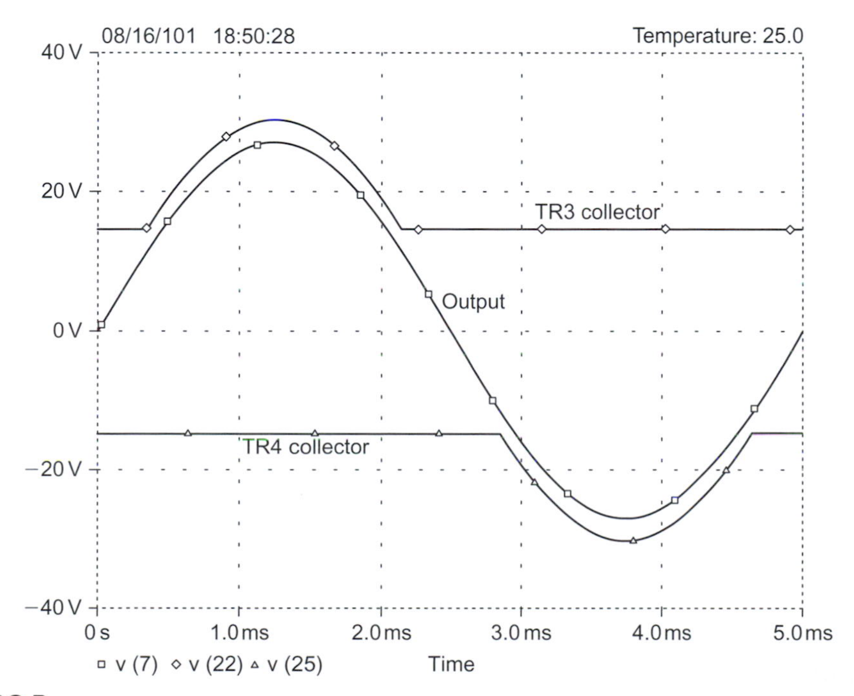

FIGURE 10.2
The output of a Class-G stage and the voltages on the collectors of the inner output devices.

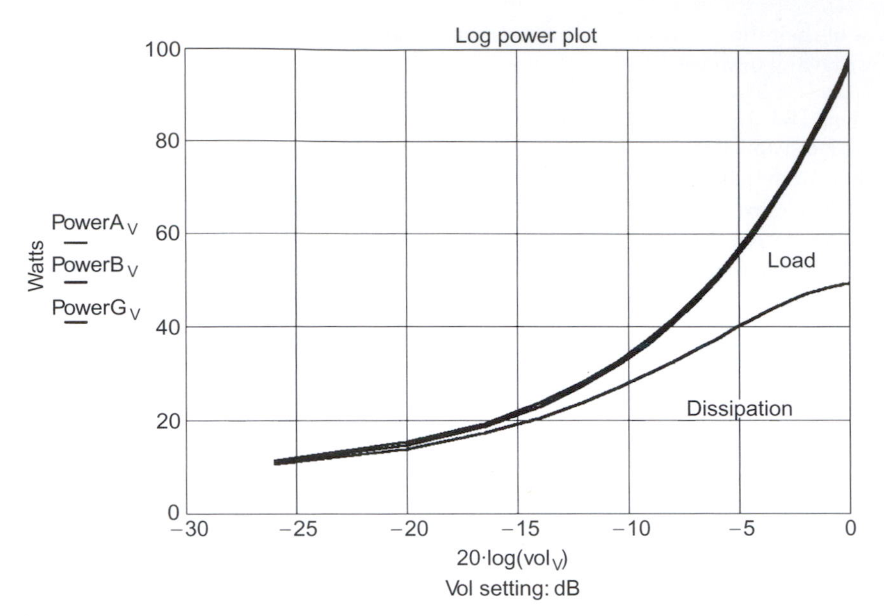

FIGURE 10.3
Power partition diagram for a conventional Class-B amplifier handling a typical music signal with a triangular probability density function. *X*-axis is volume.

The mathematics is simple because the waveforms do not vary in shape with output level. Every possible idealization is assumed, such as zero quiescent current, no emitter resistors, no $V_{ce(sat)}$ losses and so on. In Class-G, on the other hand, the waveforms are a strong function of output level, requiring variable limits of integration, and so on, and it all gets very unwieldy.

The SPICE simulation method described by Self (2006, p. 369) is much simpler, if somewhat laborious, and can use any input waveform, yielding a Power Partition Diagram (PPD), which shows how the power drawn from the supply is distributed between output device dissipation and useful power in the load.

No one disputes that sine waves are poor simulations of music for this purpose, and their main advantage is that they allow direct comparison with the purely mathematical approach. However, since the whole point of Class-G is power saving, and the waveform used has a strong effect on the results, I have concentrated here on the PPD of an amplifier with real musical signals or, at any rate, their statistical representation. The triangular probability density function (PDF) approach is described in Self (2006, p. 386).

Figure 10.3 shows the triangular PDF PPD for conventional Class-B EF, while Figure 10.4 is that for Class-G with ±V2 = 50 V and ±V1 = 15 V, i.e. with the ratio of V1/V2 set to 30%. The PPD plots power dissipated in all four output devices, the load, and the total drawn from the supply rails. It shows how the input power is partitioned between the load and the output devices. The total sums to slightly less than the input power, the remainder being accounted for as usual by losses in the drivers and Re resistors. Note that in Class-G power dissipation is shared, though not very equally, between the inner and outer devices, and this helps with efficient utilization of the silicon.

In Figure 10.4 the lower area represents the power dissipated in the inner devices and the larger area just above represents that in the outer devices; there is only one area for each because in Class-B and Class-G only one side of the amplifier conducts at a time. Outer device dissipation is zero below the rail-switching threshold at −15 dB below maximum output. The total device dissipation at full output power is reduced from 48 W in Class-B to 40 W, which may not appear at first to be a very good return for doubling the power transistors and drivers.

Figure 10.5 shows the same PPD but with ±V2 = 50 V and ±V1 = 30 V, i.e. with V1/V2 set to 60%. The low-voltage region now extends up to −6 dB ref. full power, but the inner device dissipation is higher due to the higher V1 rail voltages. The result is that total device power at full output is reduced from 4 in Class-B to 34 W, which is a definite improvement. The efficiency figure is highly sensitive to the way the ratio of rail voltages compares with the signal characteristics. Domestic hi-fi amplifiers are not

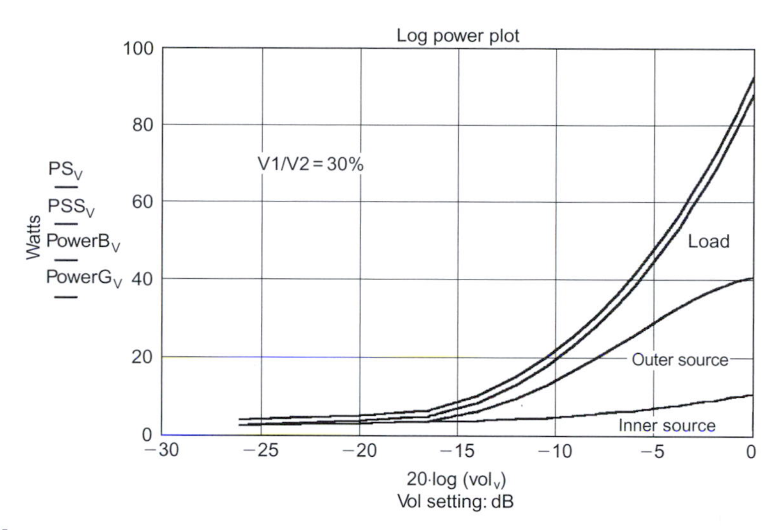

FIGURE 10.4
Power partition diagram for Class-G with V1/V2 = 30%. Signal has a triangular PDF. *X*-axis is volume; outer devices dissipate nothing until −15 dB is reached.

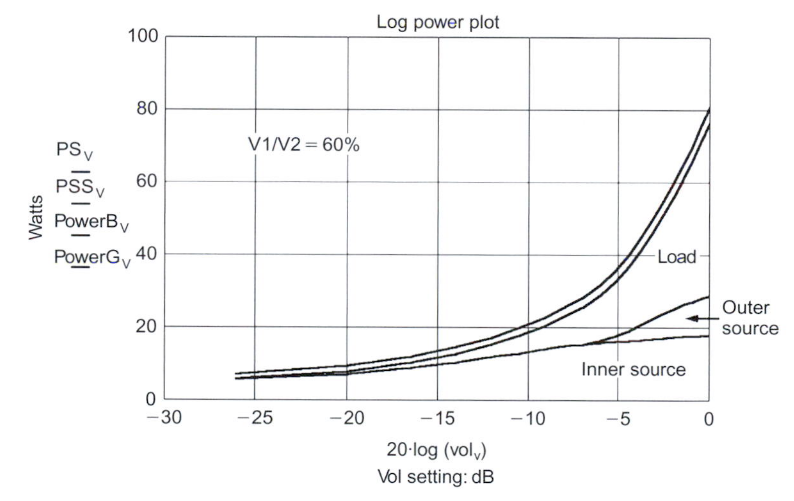

FIGURE 10.5
Power partition diagram for Class-G with V1/V2 = 60%. Triangular PDF. Compared with Figure 10.4, the inner devices dissipate more and the outer devices almost nothing except at maximum volume.

operated at full volume all the time, and in real life the lower option for the V1 voltage is likely to give lower general dissipation. I do not suggest that V1/V2 = 30% is the optimum lower-rail voltage for all situations, but it looks about right for most domestic hi-fi.

PRACTICALITIES

In my time I have wrestled with many "new and improved" output stages that proved to be anything but. When faced with a new and intriguing possibility, I believe the first thing to do is sketch out a plausible circuit such as Figure 10.1 and see if it works in SPICE simulation. It duly did.

The next stage is to build it, power it from low supply rails to minimize the size of any explosions, and see if it works for real at 1 kHz. This is a bigger step than it looks.

SPICE simulation is incredibly useful but it is not a substitute for testing a real prototype. It is easy to design clever and complex output stages that work beautifully in simulation but in reality prove impossible to stabilize at high frequencies. Some of the more interesting output-triple configurations seem to suffer from this.

The final step—and again it is a bigger one than it appears—is to prove real operation at 20 kHz and above. Again it is perfectly possible to come up with a circuit configuration that either just does not work at 20 kHz, due to limitations on power transistor speeds, or is provoked into oscillation or other misbehavior that is not set off by a 1 kHz testing.

Only when these vital questions are resolved is it time to start considering circuit details, and assessing just how good the amplifier performance is likely to be.

THE BIASING REQUIREMENTS

The output stage bias requirements are more complex than for Class-B. Two extra bias generators Vbias3, Vbias4 are required to make TR6 turn on before TR3 runs out of collector voltage. These extra bias voltages are not critical, but must not fall too low or become much too high. Should these bias voltages be set too low, so the outer devices turn on too late, then the V_{ce} across TR3 becomes too low, and its current sourcing capability is reduced. When evaluating this issue bear in mind the lowest impedance load the amplifier is planned to drive, and the currents this will draw from the output devices. Fixed Zener diodes of normal commercial tolerance are quite accurate and stable enough for setting Vbias3 and Vbias4.

Alternatively, if the bias voltage is set too low, then the outer transistors will turn on too early, and the heat dissipation in the inner power devices becomes greater than it need be for correct operation. The latter case is rather less of a problem so if in doubt this bias should be chosen to be on the high side rather than the low.

The original Hitachi circuit (Self, 2006, p. 347) put Zeners in series with the signal path to the inner drivers to set the output quiescent bias, their voltage being subtracted from the main bias generator, which was set at 10 V or so, a much higher voltage than usual (see Figure 10.6). SPICE simulation showed me that the presence of Zener diodes in the forward path to the inner power devices gave poor linearity, which is not exactly a surprise. There is also the problem that the quiescent conditions will be affected by changes in the Zener voltage. The 10 V bias generator, if it is the usual V_{be}-multiplier, will have much too high a temperature coefficient for proper thermal tracking.

I therefore rearrange the biasing as in Figure 10.1. The amplifier forward path now goes directly to the inner devices, and the two extra bias voltages are in the path to the outer devices; since these do not control the output directly, the linearity of this path is of lesser importance. The Zeners are out of the forward path and the bias generator can be the standard sort. It must be thermally coupled to the inner power devices; the outer ones have no effect on the quiescent conditions.

THE LINEARITY ISSUES OF SERIES CLASS-G

Series Class-G has often had its linearity called into question because of difficulties with supply-rail commutation. Diodes D3, D4 must be power devices capable of handling a dozen amps or more, and conventional silicon rectifier diodes that can handle such currents take a long time to turn off due to their stored charge carriers. This has the following unhappy effect: when the voltage on the cathode of D3 rises above V1, the diode tries to turn off abruptly, but its charge carriers sustain a brief but large reverse current as they are swept from its junction. This current is supplied by TR6, attempting as an emitter-follower to keep its emitter up to the right voltage. So far all is well.

However, when the diode current ceases, TR6 is still conducting heavily, due to its own charge-carrier storage. The extra current it turned on to feed D3 in reverse now goes through the TR3 collector, which accepts it because of TR3's low V_{ce}, and passes it onto the load via TR3 emitter and Re.

This process is readily demonstrated by a SPICE commutation transient simulation (see Figures 10.7 and 10.8). Note there are only two of these events per cycle—not four, as they only occur when the diodes turn off. In the original Hitachi design this problem was reportedly tackled by using fast transistors and relatively fast gold-doped diodes, but according to Sampei et al. (1978) this was only partially successful.

It is now simple to eradicate this problem. Schottky power diodes are readily available, as they were not in 1976, and are much faster due to their lack of minority carriers and charge storage. They have the added advantage of a low forward voltage drop at large currents of 10 A or more. The main snag is a relatively low reverse withstand voltage, but fortunately in Class-G usage the commutating diodes are only

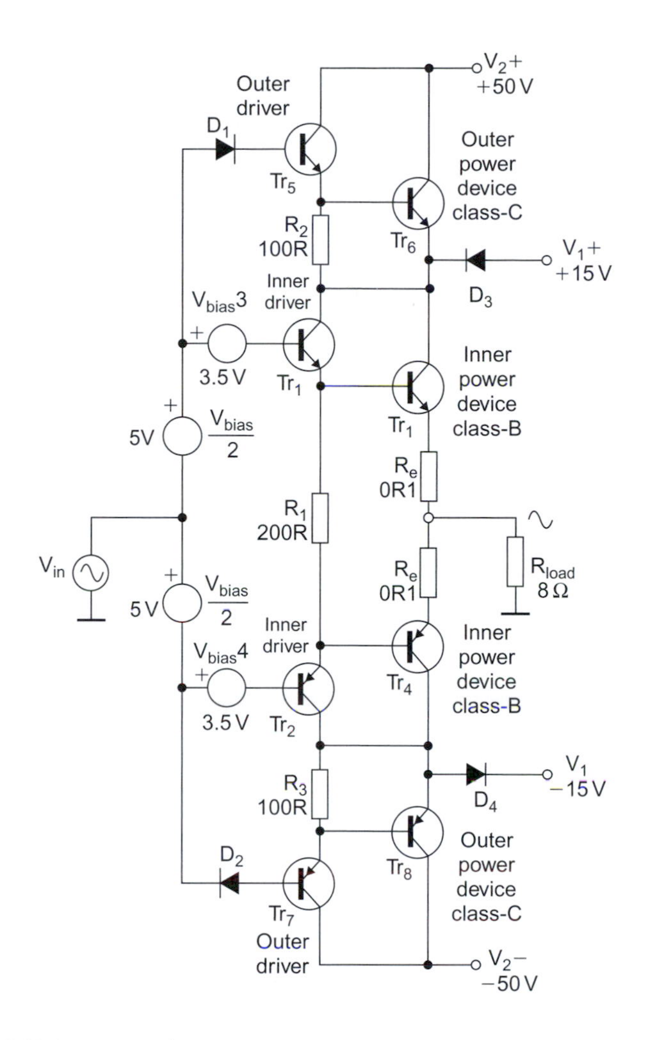

FIGURE 10.6
The original Hitachi Class-G biasing system, with inner device bias derived by subtracting Vbias3, Vbias4 from the main bias generator.

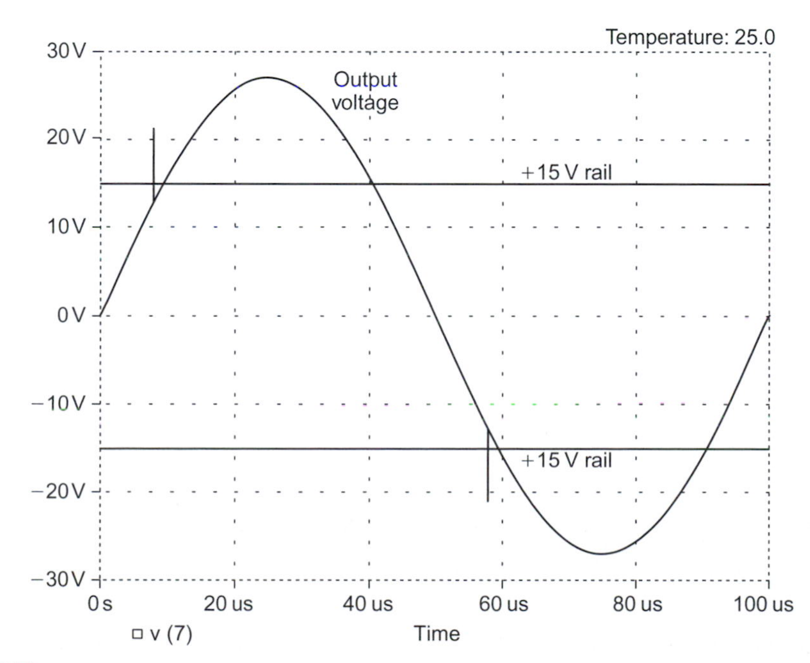

FIGURE 10.7
Spikes due to charge storage of conventional diodes, simulated at 10 kHz. They only occur when the diodes turn off, so there are only two per cycle. These spikes disappear completely when Schottky diodes are used in the SPICE model.

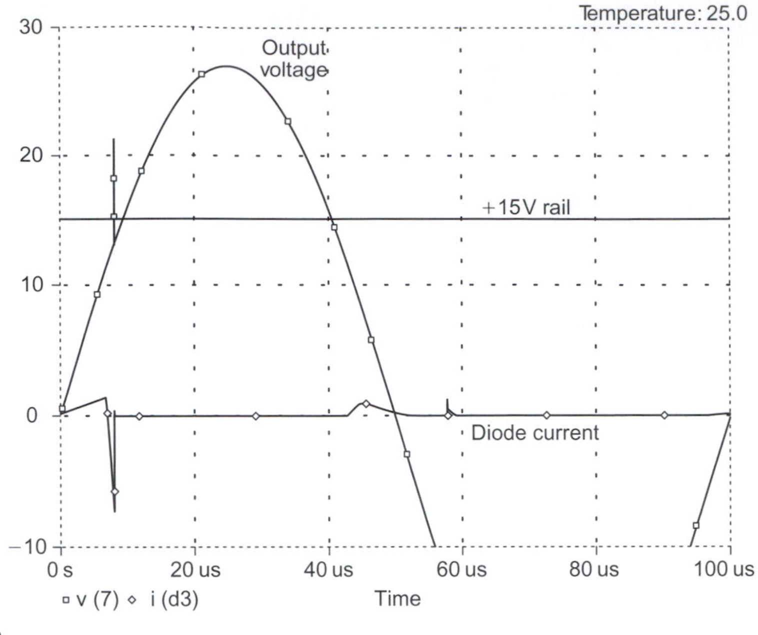

FIGURE 10.8
A close-up of the diode transient. Diode current rises as output moves away from zero, then reverses abruptly as charge carriers are swept out by reverse-biasing. The spike on the output voltage is aligned with the sudden stop of the diode reverse current.

exposed at worst to the difference between V2 and V1, and this only when the amplifier is in its low-power domain of operation. Another good point about Schottky power diodes is that they do appear to be robust; I have subjected 50 A Motorola devices to 60 A-plus repeatedly without a single failure. This is a good sign. The spikes disappear completely from the SPICE plot if the commutating diodes are Schottky rectifiers. Motorola MBR5025L diodes capable of 50 A and 25 PIV were used in simulation.

THE STATIC LINEARITY

SPICE simulation shows in Figure 10.9 that the static linearity (i.e. that ignoring dynamic effects like diode charge storage) is distinctly poorer than for Class-B. There is the usual Class-B gain wobble around the crossover region, exactly the same size and shape as for conventional Class-B, but also there are now gain-steps at ±16 V. The result with the inner devices biased into push–pull Class-A is also shown, and proves that the gain-steps are not in any way connected with crossover distortion. Since this is a DC analysis the gain-steps cannot be due to diode-switching speed or other dynamic phenomena, and the Early effect was immediately suspected (the Early effect is the increase in collector current when the collector voltage increases, even though V_{be} remains constant). When unexpected distortion appears in a SPICE simulation of this kind, and effects due to finite transistor beta and associated base currents seem unlikely, a most useful diagnostic technique is to switch off the simulation of the Early effect for each transistor in turn. In SPICE transistor models the Early effect can be totally disabled by setting the parameter VAF to a much higher value than the default of 100, such as 50,000. This experiment demonstrated in short order that the gain-steps were caused wholly by the Early effect acting on both inner drivers and inner output devices. The gain-steps are completely abolished with Early effect disabled. When TR6 begins to act, TR3 V_{ce} is no longer decreasing as the output moves positive, but substantially constant as the emitter of Q6 moves upwards at the same rate as the emitter of Q3. This has the effect of a sudden change in gain, which naturally degrades the linearity.

This effect appears to occur in drivers and output devices to the same extent. It can be easily eliminated in the drivers by powering them from the outer rather than the inner supply rails. This prevents the sudden changes in the rate in which driver V_{ce} varies. The improvement in linearity is seen in Figure 10.10, where the gain-steps have been halved in size. The resulting circuit is shown in Figure 10.11. Driver power dissipation is naturally increased by the increased driver V_{ce}, but this is such a small fraction of the power consumed that the overall efficiency is not significantly reduced. It is obviously not practical to apply the same method to the output devices, because then the low-voltage rail would

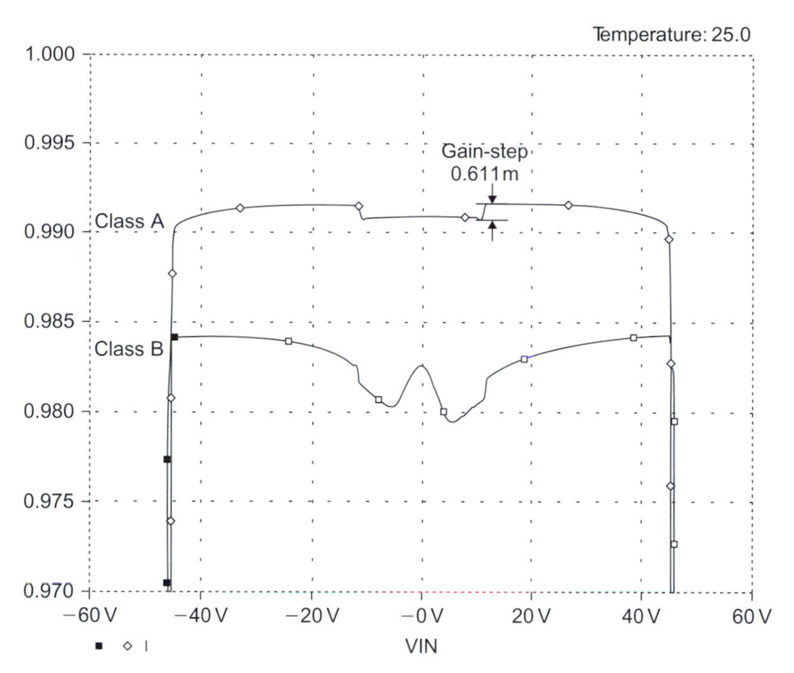

FIGURE 10.9
SPICE simulation shows variations in the incremental gain of an EF-type Class-G series output stage. The gain-steps at transition (at ±16V) are due to Early effect in the transistors. The Class-A trace is the top one, with Class-B optima below. Here the inner driver collectors are connected to the switched inner rails, i.e., the inner power device collectors, as in Figure 10.1.

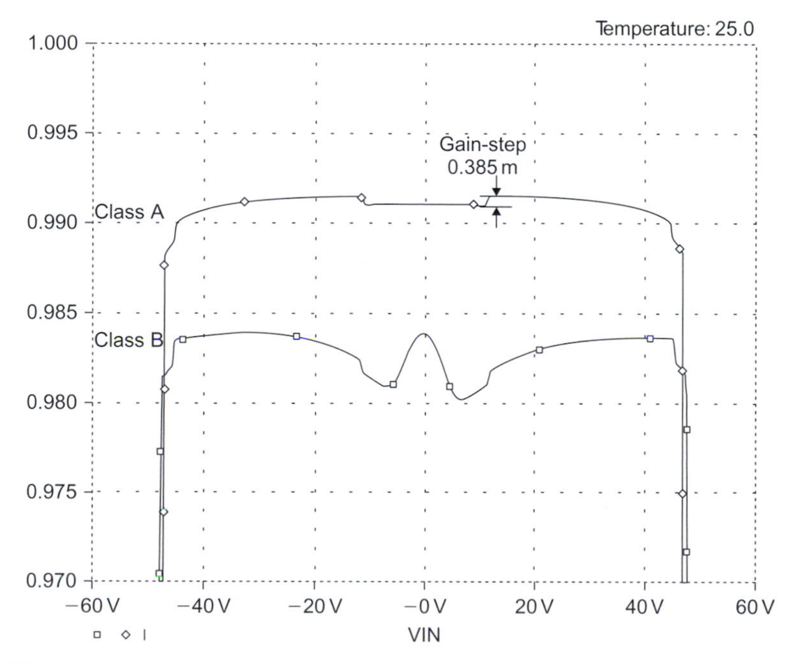

FIGURE 10.10
Connecting the inner driver collectors to the outer V2 rails reduces Early effect nonlinearities in them, and halves the transition gain-steps.

never be used and the amplifier is no longer working in Class-G. The small-signal stages naturally have to work from the outer rails to be able to generate the full voltage swing to drive the output stage.

We have now eliminated the commutating diode glitches and halved the size of the unwanted gain-steps in the output stage. With these improvements made it is practical to proceed with the design of a Class-G amplifier with mid-band THD below 0.002%.

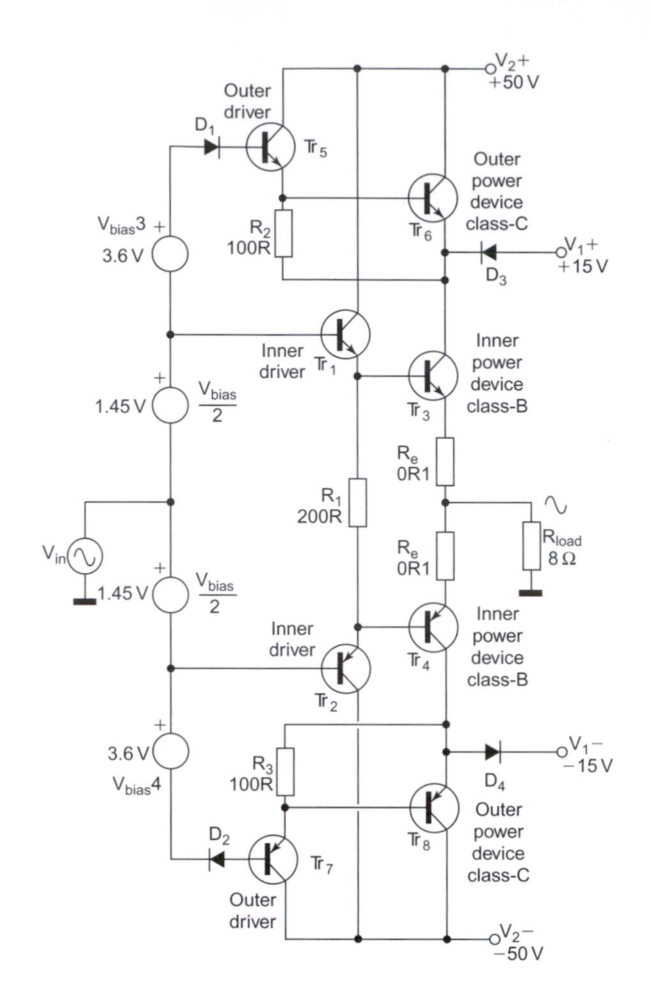

FIGURE 10.11
A Class-G output stage with the drivers powered from the outer supply rails.

PRACTICAL CLASS-G DESIGN

The Class-G amplifier design expounded here uses very similar small-signal circuitry to the Blameless Class-B power amplifier, as it is known to generate very little distortion of its own. If the specified supply voltages of ±50 and ±15 V are used, the maximum power output is about 120 W into 8 Ω, and the rail-switching transition occurs at 28 W.

This design incorporates various techniques, and closely follows the Blameless Class-B amp. A notable example is the low-noise feedback network, complete with its option of input bootstrapping to give a high impedance when required. Single-slope VI limiting is incorporated for overload protection; this is implemented by Q12, Q13. Figure 10.12 shows the circuit.

As usual in my amplifiers the global NFB factor is a modest 30 dB at 20 kHz.

CONTROLLING SMALL-SIGNAL DISTORTION

The distortion from the small-signal stages is only dealt with briefly here. The input stage differential pair Q1, Q2 is given local feedback by R5 and R7 to delay the onset of third-harmonic Distortion 1. Internal r_e variations in these devices are minimized by using an unusually high tail current of 6 mA Q3, Q4 are a degenerated current-mirror that enforces accurate balance of the Q1, Q2 collector currents, preventing the production of second-harmonic distortion. The input resistance (R3 + R4) and feedback resistance R16 are made equal and made unusually low, so that base-current mismatches stemming from input device beta variations give a minimal DC offset.

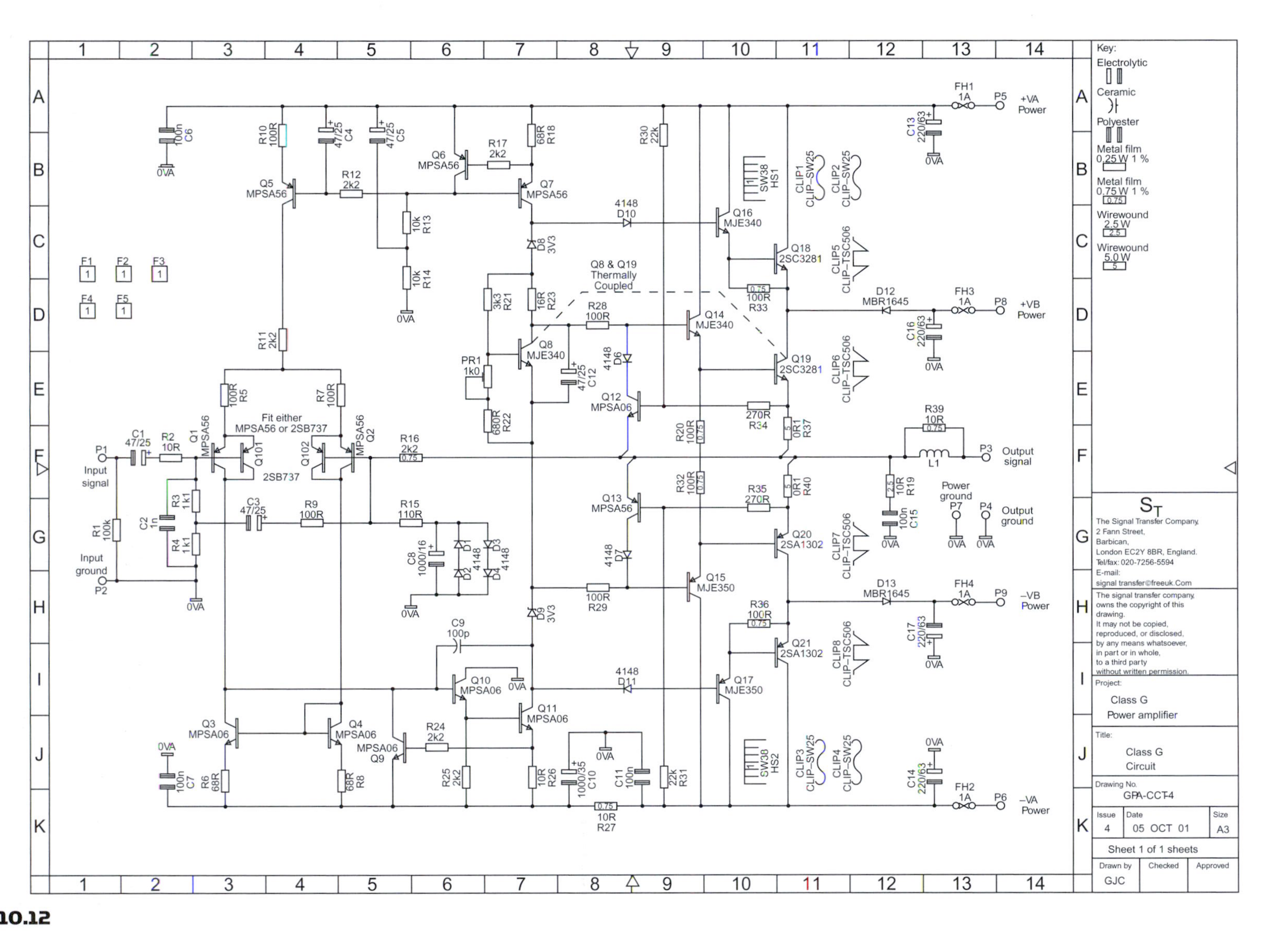

FIGURE 10.12
The circuit diagram of the Class-G amplifier.

V_{be} mismatches in Q1 and Q2 remain, but these are much smaller than the effects of I_b. Even if Q1 and Q2 are high-voltage types with relatively low beta, the DC offset voltage at the output should be kept to less than ±50 mV. This is adequate for all but the most demanding applications. This low-impedance technique eliminates the need for balance presets or DC servo systems, which is most convenient.

A lower value for R16 implies a proportionally lower value for R15 to keep the gain the same, and this reduction in the total impedance seen by Q2 improves noise performance markedly. However, the low value of R3 plus R4 at 2k2 gives an input impedance that is not high enough for many applications.

There is no problem if the amplifier is to have an additional input stage, such as a balanced line receiver. Proper choice of op-amp will allow the stage to drive a 2k2 load impedance without generating additional distortion. Be aware that adding such a stage—even if it is properly designed and the best available op-amps are used—will degrade the signal-to-noise ratio significantly. This is because the noise generated by the power amplifier itself is so very low—equivalent to the Johnson noise of a resistor of a few hundred ohms—that almost anything you do upstream will degrade it seriously.

If there is no separate input stage then other steps must be taken. What we need at the input of the power amplifier is a low DC resistance, but a high AC resistance; in other words we need either a 50 henry choke or recourse to some form of bootstrapping. There is to my mind no doubt about the way to go here, so bootstrapping it is. The signal at Q2 base is almost exactly the same as the input, so if the mid-point of R3 and R4 is driven by C3, so far as input signals are concerned R3 has a high AC impedance. When I first used this arrangement I had doubts about its high-frequency stability, and so added resistor R9 to give some isolation between the bases of Q1 and Q2. In the event I have had no trouble with instability, and no reports of any from the many constructors of the Trimodal and Load-Invariant designs, which incorporate this option.

The presence of R9 limits the bootstrapping factor, as the signal at the R3–R4 junction is thereby a little smaller than at Q2 base, but it is adequate. With R9 set to 100R, the AC input impedance is raised to 13 k, which should be high enough for almost all purposes. Higher values than this mean that an input buffer stage is required.

The value of C8 shown (1000 μF) gives an LF roll-off in conjunction with R15 that is −3 dB at 1.4 Hz. The purpose is not impossibly extended sub-bass, but the avoidance of a low-frequency rise in distortion due to nonlinearity effects in C8. If a 100 μF capacitor is used here the THD at 10 Hz worsens from $<$0.0006% to 0.0011%, and I regard this as unacceptable esthetically—if not perhaps audibly. This is not the place to define the low-frequency bandwidth of the system—this must be done earlier in the signal chain, where it can be properly implemented with more accurate non-electrolytic capacitors. The protection diodes D1–D4 prevent damage to C2 if the amplifier suffers a fault that makes it saturate in either direction; it looks like an extremely dubious place to put diodes but since they normally have no AC or DC voltage across them no measurable or detectable distortion is generated.

The voltage-amplifier stage (VAS) Q11 is enhanced by emitter-follower Q10 inside the Miller compensation loop, so that the local negative feedback that linearizes the VAS is increased. This effectively eliminates VAS nonlinearity. Thus increasing the local feedback also reduces the VAS collector impedance, so a VAS buffer to prevent Distortion 4 (loading of VAS collector by the nonlinear input impedance of the output stage) is not required. Miller capacitor C_{dom} is relatively big at 100 pF, to swamp transistor internal capacitances and circuit strays, and make the design predictable. The slew rate calculates as 40 V/μs use in each direction. VAS collector load Q7 is a standard current source.

Almost all the THD from a Blameless amplifier derives from crossover distortion, so keeping the quiescent conditions optimal to minimize this is essential. The bias generator for an EF output stage, whether in Class-B or Class-G, is required to cancel out the V_{be} variations of four junctions in series; those of the two drivers and the two output devices. This sounds difficult, because the dissipation in the two types of devices is quite different, but the problem is easier than it looks. In the EF type of output stage the driver dissipation is almost constant as power output varies, and so the problem is reduced to tracking the two output device junctions. The bias generator Q8 is a standard V_{be}-multiplier, with R23 chosen to minimize variations in the quiescent conditions when the supply rails change. The bias generator should be in contact with the top of one of the inner output devices, and not the heat-sink itself. This position gives much faster and less attenuated thermal feedback to Q8. The VAS collector circuit incorporates not only bias generator Q8 but also the two Zeners D8, D9, which determine how early rail-switching occurs as the inner device emitters approach the inner (lower) voltage rails.

The output stage was selected as an emitter-follower (EF) type as this is known to be less prone to parasitic or local oscillations than the CFP configuration, and since this design was to some extent heading into the unknown it seemed wise to be cautious where possible. R32 is the usual shared emitter resistor for the inner drivers. The outer drivers Q16 and Q17 have their own emitter resistors R33 and R36, which have their usual role of establishing a reasonable current in the drivers as they turn on, to increase driver transconductance, and also in speeding up turn-off of the outer output devices by providing a route for charge carriers to leave the output device bases.

As explained above, the inner driver collectors are connected to the outer rails to minimize the gain-steps caused by the abrupt change in collector voltage when rail transition occurs.

Deciding the size of heat-sink required for this amplifier is not easy, mainly because the heat dissipated by a Class-G amplifier depends very much on the rail voltages chosen and the signal statistics. A Class-B design giving 120 W into 8 Ω would need a heat-sink with thermal resistance of the order of 1 °C/W (per channel); a good starting point for a Class-G version giving the same power would be about half the size, i.e., 2 °C/W. The Schottky commutating diodes do not require much heat-sinking, as they conduct only intermittently and have a low forward voltage drop. It is usually convenient to mount them on the main heat-sink, even if this does mean that most of the time they are being heated rather than cooled.

C15 and R38 make up the usual Zobel network. The coil L1, damped by R39, isolates the amplifier from load capacitance. A component with 15–20 turns at 1 inch diameter should work well; the value of inductance for stability is not all that critical.

THE PERFORMANCE

Figure 10.13 shows the THD at 20 W and 50 W (into 8 Ω) and I think this demonstrates at once that the design is a practical competitor for Class-B amplifiers. Compare these results with the upper trace of Figure 10.14, taken from a Blameless Class-B amplifier at 50 W, 8 Ω. Note the lower trace of Figure 10.14 is for 30 kHz bandwidth, used to demonstrate the lack of distortion below 1 kHz; the THD data above 30 kHz is in this case meaningless as all the harmonics are filtered out. All the Class-G plots here are taken at 80 kHz to make sure any high-order glitching is properly measured.

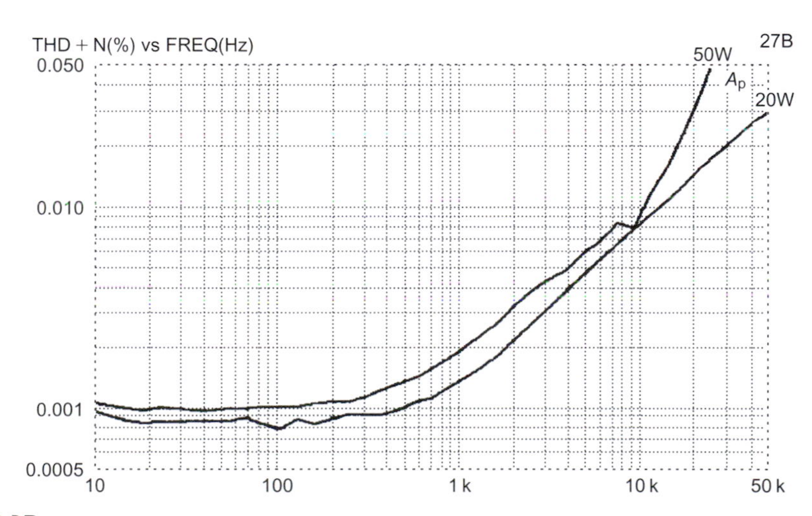

FIGURE 10.13
THD versus frequency, at 20 W (below transition) and 50 W into an 8 Ω load. The joggle around 8 kHz is due to a cancelation of harmonics from crossover and transition. Bandwidth 80 kHz.

DERIVING A NEW KIND OF AMPLIFIER: CLASS-A + C

A conventional Class-B power amplifier can be almost instantly converted to push–pull Class-A simply by increasing the bias voltage to make the required quiescent current flow. This is the only real circuit change, though naturally major increases in heat-sinking and power-supply capability are required for practical use. Exactly the same principle applies to the Class-G amplifier. In the book *Self On Audio* (2006, p. 293) I suggested a new and much more flexible system for classifying amplifier types and here it comes in very handy. Describing Class-G operation as Class-B + C immediately indicates that only a bias increase is required to transform it into Class-A + C, and a new type of amplifier is born. This amplifier configuration combines the superb linearity of classic Class-A up to the transition level, with only minor distortion artifacts occurring at higher levels, as demonstrated for Class-B + C above. Using Class-A means that the simple V_{be}-multiplier bias generator can be replaced with precise negative-feedback control of the quiescent current. There is no reason why an amplifier could not be configured as a Class-G Trimodal, i.e. manually switchable between Classes A and B. That would indeed be an interesting machine.

In Figure 10.19 is shown the THD plot for such an A + C amplifier working at 20 and 30 W into 8 Ω. At 20 W the distortion is very low indeed, no higher than a pure Class-A amplifier. At 30 W the transition gain-steps appear, but the THD remains very well controlled, and no higher than a Blameless Class-B design. Note that as in Class-B, when the THD does start to rise it only does so at 6 dB/octave. The quiescent current was set to 1.5 A.

Figure 10.20 reveals the THD residual during A + C operation. There are absolutely no crossover artifacts, and the small disturbances that do occur happen at such a high signal level that I really do think it is safe to assume they could never be audible. Figure 10.21 shows the complete absence of artifacts on the residual when this new type of amplifier is working below transition; it gives pure Class-A linearity. Finally, Figure 10.22 gives the THD when the amplifier is driving the full 50 W into 8 Ω; as before the A + C THD plot is hard to distinguish from Class-B, but there is the immense advantage that there is no crossover distortion at low levels, and no critical bias settings.

ADDING TWO-POLE COMPENSATION

Amplifier distortion can be very simply reduced by changes to the compensation, which means a scheme more sophisticated than the near-universal dominant-pole method. It must be borne in mind that any departure from the conventional 6 dB/octave all-the-way compensation scheme is likely to be a move away from unconditional stability. (I am using this phrase in its proper meaning; in Control Theory unconditional stability means that increasing open-loop gain above a threshold causes instability, but the system is stable for all lower values. Conditional stability means that lower open-loop gains can also be unstable.)

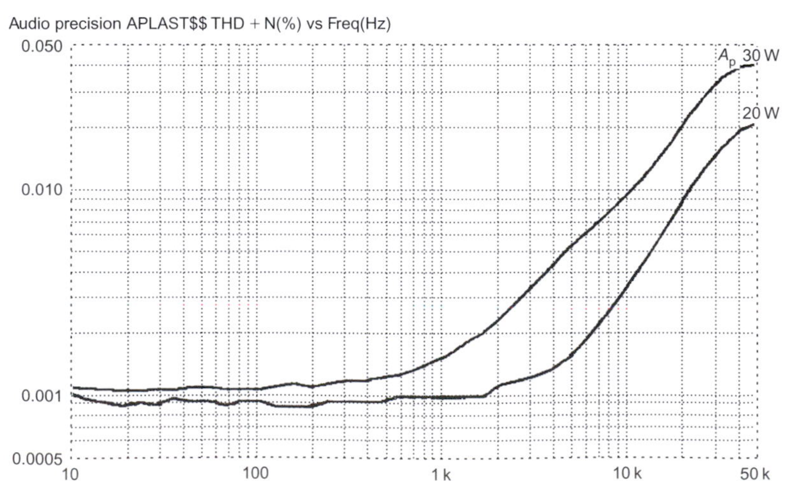

FIGURE 10.19
The THD plot of the Class-A + C amplifier (30 and 20 W into 8 Ω). Inner drivers powered from outer rails.

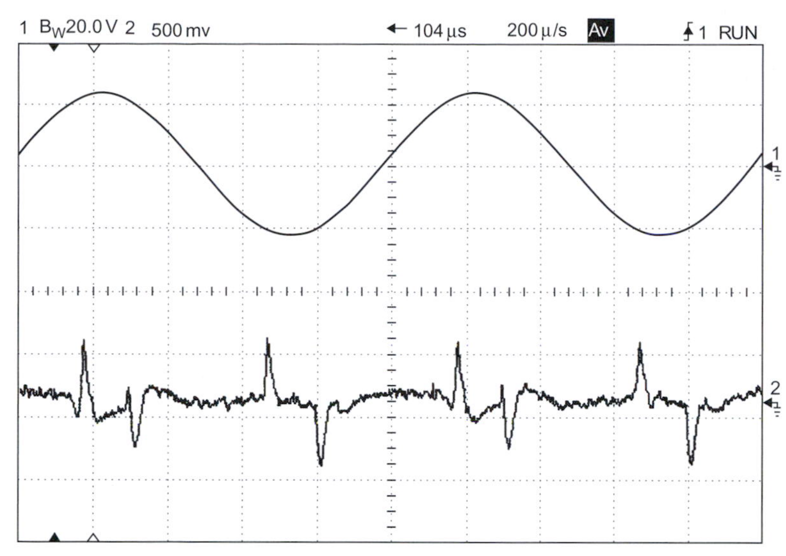

FIGURE 10.20
The THD residual waveform of the Class-A + C amplifier above transition, at 30W into 8 Ω. Switching artefacts are visible but not crossover distortion.

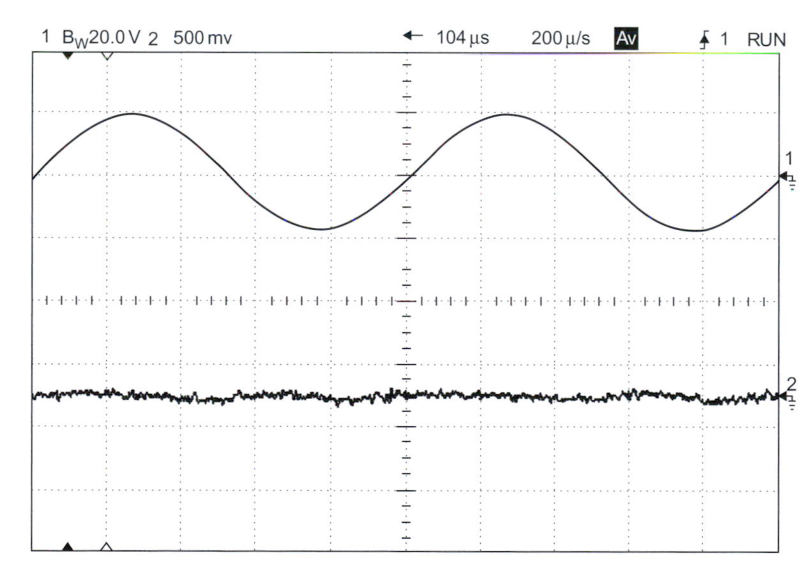

FIGURE 10.21
The THD residual waveform plot of the Class-A + C amplifier (20W into 8 Ω).

A conditionally stable amplifier may well be docile and stable into any conceivable reactive load when in normal operation, but shows the cloven hoof of oscillation at power-up and power-down, or when clipping. This is because under these conditions the effective open-loop gain is reduced.

Class-G distortion artifacts are reduced by normal dominant-pole feedback in much the same way as crossover nonlinearities, i.e. not all that effectively, because the artifacts take up a very small part of the cycle and are therefore composed of high-order harmonics. Therefore a compensation system that increases the feedback factor at high audio frequencies will be effective on switching artifacts, in the same way that it is for crossover distortion. The simplest way to implement two-pole circuit compensation is shown in Figure 10.23.

The results of two-pole compensation for B + C are shown in Figure 10.24; comparing it with Figure 10.13 (the normally compensated B + C amplifier) the above-transition (30W) THD at 10kHz has dropped from 0.008% to 0.005%; the sub-transition (20W) THD at 10kHz has fallen from 0.007% to 0.003%. Comparisons have to be done at 10kHz or thereabouts to ensure there is enough to measure.

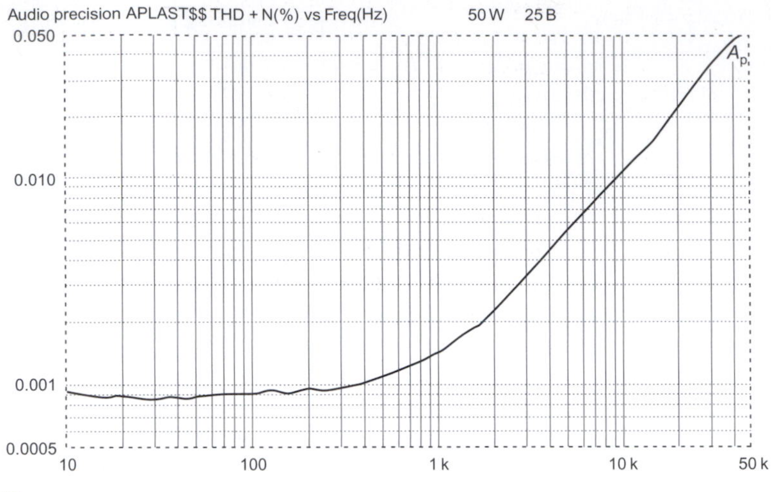

FIGURE 10.22
The THD plot of the Class-A + C amplifier (50 W into 8 Ω). Inner drivers powered from outer rails.

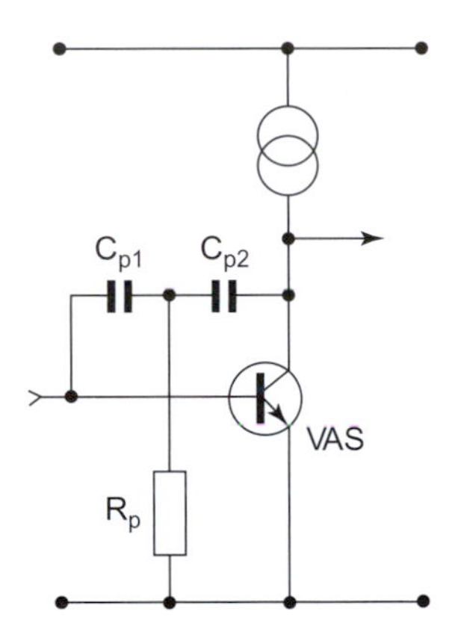

FIGURE 10.23
The circuit modification for two-pole compensation.

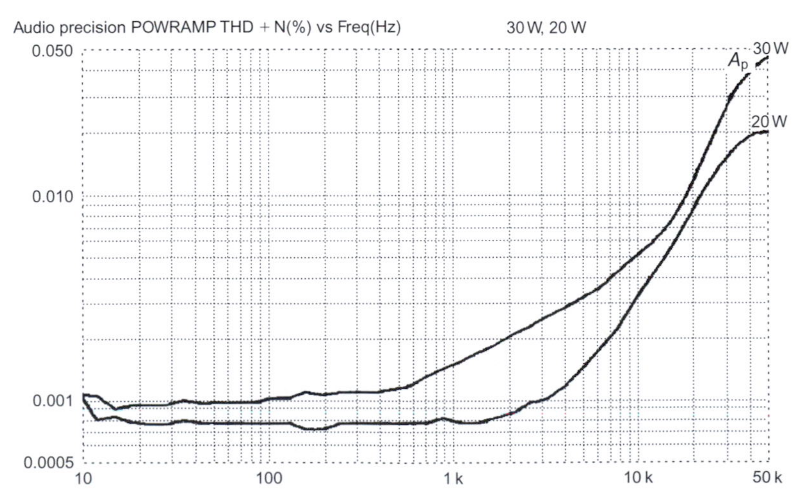

FIGURE 10.24
The THD plot for B + C operation with two-pole compensation (20 and 30 W into 8 Ω). Compare with Figures 10.13 (B + C) and 10.19 (A + C).

Now comparing the two-pole B + C amplifier with Figure 10.19 (the A + C amplifier) the above-transition (30 W) THD at 10 kHz of the former is lower at 0.005% compared with 0.008%. As I have demonstrated before, proper use of two-pole compensation can give you a Class-B amplifier that is hard to distinguish from Class-A—at least until you put your hand on the heat-sink.

FURTHER VARIATIONS ON CLASS-G

This by no means exhausts the possible variations that can be played on Class-G. For example, it is not necessary for the outer devices to operate synchronously with the inner devices. So long as they turn on in time, they can turn off much later without penalty except in terms of increased dissipation. In so-called syllabic Class-G, the outer devices turn on quickly but then typically remain on for 100 ms or so to prevent glitching (see Funada and Akiya, 1984 for one version). Given the good results obtained with straight Class-G, this no longer seems a promising route to explore.

With the unstoppable advance of multichannel amplifier and powered subwoofers, Class-G is at last coming into its own. It has recently even appeared in a Texas ADSL driver IC. I hope I have shown how to make it work, and then how to make it work better. I modestly suggest that this might be the lowest distortion Class-G amplifier so far.

REFERENCES

Feldman, L., (August 1976). Class-G high efficiency hi-fi amplifier, Radio Electron., p. 47.

Funada, S. and Akiya, H., (October 1984). A study of high-efficiency audio power amplifiers using a voltage switching method, JAES, 32(10), p. 755.

Sampei, T. et al, (August 1978). Highest efficiency and super quality audio amplifier using MOS power FETs in class-G operation, IEEE Trans. Consum. Electron. *CE-24*, 3, p. 300.

Self, D., 2006. Self On Audio, Newnes, 5, p. 347.

Loudspeakers

PREFACE

The fifth section of this book "Loudspeakers" grapples with the tricky business of turning electricity into sound in a faithful manner. It consists of four chapters, all taken from the book *Loudspeakers* by Philip Newell and Keith Holland.

This first chapter begins with a quick look at the history of moving-coil loudspeakers. This has some unexpected points of interest—for example, when Rice and Kellogg worked out the first practical cone loudspeaker in the early 1920s, they did it from fundamental physical principles, without going through the period of trial and error that characterizes the birth of most technologies. Hats off to them.

Newell & Holland then move on to the basics of loudspeaker operation, starting with acoustic wave propagation and going on to the concepts of mechanical and acoustic impedances. Mechanical impedance is a measure of how easy it is to get a loudspeaker cone moving with a given force, and acoustic impedance is a measure of how easy it is to move air with a given pressure. The underlying physics means that these impedances do not match up well, and that is why loudspeakers are so very inefficient, the radiated acoustic power rarely exceeding 1% of the electrical power that we so trustingly feed in.

The next section examines how loudspeaker radiation varies with direction. Considering how the directivity of that favorite theoretical device, the infinitely-rigid massless vibrating piston, alters as frequency increases, we find that it varies wildly, tending to focus in a forward direction as frequency goes up. We are already gaining an insight into why loudspeaker design is not a simple business.

The remainder of the chapter gives a thorough account of the electrical load impedance that a single loudspeaker unit presents to a power amplifier. A single speaker unit with a nominal 8 Ω impedance usually has a winding resistance of 6–7 Ω, and the impedance does not drop below this figure. There is a low-frequency peak at the resonant frequency of the cone which can easily rise to 40 Ω or more, and a steady rise at the high-frequency end due to the inductance of the voice-coil.

The load presented by a multi-way loudspeaker, which has not only multiple driver units but also a crossover unit to distribute the appropriate frequency bands to them, is much more complicated, and it is difficult to make any firm statements about what the frequency/impedance curve will look like. It varies widely from design to design, though the peak of the bass unit resonance is usually still clearly visible.

An audio power amplifier approaches theoretical perfection much more closely than any loudspeaker (and especially so, I assert, if it is one of my designs) and it is therefore standard practice in loudspeaker design to assume that the amplifier is perfect, which for the purposes in hand comes down to a flat frequency-response and zero output impedance. Swathes of stuff have been written about the alleged difficulty that amplifiers may or may not have driving a complex reactive load, much of it hopelessly misguided.

What is certain is that reactive loads can increase both the peak and average power dissipation in amplifier output devices, and this effect worsens rapidly as the phase angle of the load increases. Amplifier designers ignore this at their peril.

CHAPTER 11

What is a Loudspeaker?

Loudspeakers by Philip Newell and Keith Holland

A BRIEF LOOK AT THE CONCEPT

Before answering the question posed by the title of this chapter, perhaps we had better begin with the question "What is sound?" According to Fahy and Walker (1998) "sound may be defined as a time-varying disturbance of the density of a fluid from its equilibrium value, which is accompanied by a proportional local pressure, and is associated with small oscillatory movements of the fluid particles." The difference between the equilibrium (static) pressure and the local, oscillating pressure is known as the *sound pressure.*

Normally, for human beings, the fluid in which sound propagates is air, which is heavier than most people think—it has a mass of about 1.2 kg per cubic meter at a temperature of 20 °C at sea level. It is also interesting to note that sound propagation in air is by no means typical of its propagation in all substances, especially in that the speed that sound propagates in air is relatively slow, and is constant for all frequencies. For music lovers, this latter fact is quite fortunate, because it would be hard to enjoy a musical performance at the back of a concert hall if the notes arrived jumbled-up, with the harmonics arriving before the fundamentals, or vice versa. Conversely, as we shall see later, most of the materials from which loudspeakers are made do *not* pass all frequencies at the same speed of sound, a fact which can, at times, make design work rather complicated.

The speed of sound in air is about 343 meters per second (m/s or ms^{-1}) at 20 °C and varies proportionally with temperature at the rate of about 0.6 m/s for every degree Kelvin. (In fact, the speed of sound in air is *only* dependent upon temperature, because the changes that would occur due to changes in atmospheric pressure are equal and opposite to the accompanying changes in density, and the two serve to cancel each other out). Air therefore has some clearly defined characteristic properties, and our perception of sound in general, and music in particular, has developed around these characteristic properties.

The job of a loudspeaker is to set up vibrations in the air which are acoustic representations of the waveforms of the electrical signals that are being supplied to the input terminals. A loudspeaker is therefore an electro-mechanico-acoustic transducer. Loudspeakers transform the electrical drive signals into mechanical movements which, normally via a vibrating diaphragm, couple those vibrations to the air and thus propagate acoustic waves. Once these acoustic waves are perceived by the ear, we experience a sensation of sound.

To a casual observer, a typical moving-coil loudspeaker (or "driver" if you wish to restrict the use of "loudspeaker" to an entire system) seems to be a simple enough device. There is a wire "voice-coil" in a magnetic field. The coil is wound on a cylindrical former which is connected to a cardboard cone, and the whole thing is held together by a metal or plastic chassis. The varying electrical input gives rise to vibrations in the cone as the electromagnetic field in the voice-coil interacts with the static field of the (usually) permanent magnet. The cone thus responds to the electrical input, and there you have it, sound! It is all as simple as that! Or is it?

Well, if the aim is to make a sound from a small, portable radio that fits into your pocket, then maybe that concept will just about suffice, but if full frequency range, high fidelity sound is the object of the exercise, then things become fiendishly complicated at an alarming speed. In reality, in order to be able to reproduce the subtle structures of fine musical instruments, loudspeakers have a very difficult task to perform.

A LITTLE HISTORY AND SOME BACKGROUND

When Rice and Kellogg (1925) developed the moving coil cone loudspeaker in the early 1920s, (and no; they did not also invent Kellogg's Rice Krispies!) they were already well aware of the complexity of radiating an even frequency balance of sound from such a device. Although Sir Oliver Lodge had patented the concept in 1898 (following on from earlier work by Ernst Werner Siemens in the 1870s at the Siemens company in Germany), it was not until Rice and Kellogg that practical devices began to evolve. Sir Oliver had had no means of electrical amplification—the thermionic valve (or vacuum tube) had still not been invented, and the transistor was not to follow for 50 years. Remarkably, the concept of loudspeakers was worked out from fundamental principles; it was not a case of men playing with bits of wire and cardboard and developing things by trial and error. Indeed, what Rice and Kellogg developed is still the essence of the modern moving coil loudspeaker. Although they lacked the benefit of modern materials and technology, they had the basic principles very well within their understanding, but their goals at the time were not involved with achieving a flat frequency response from below 20 Hz to above 20 kHz at sound pressure levels in excess of 110 dB SPL. Such responses were not required because they did not even have signal *sources* of such wide bandwidth or dynamic range. It was not until the 1940s that microphones could capture the full frequency range, and the 1950s before it could be delivered commercially to the public via the microgroove, vinyl record.

Prior to 1925, the maximum output available from a radio set was in the order of milliwatts, normally only used for listening via earphones, so the earliest "speakers" only needed to handle a limited frequency range at low power levels. The six inch, rubber surround device of Rice and Kellogg used a powerful electro-magnet (not a permanent magnet), and as it could "speak" to a whole room-full of people, as opposed to just one person at a time via an earpiece, it became known as a *loud* speaker. The inventers were employed by the General Electric Company, in the USA, and they began by building a mains-driven power amplifier which could supply the then huge power of *one watt*. This massive increase in the available drive power meant that they no longer needed to rely on resonances and rudimentary horn loading, which typically gave very coloured responses. With a whole watt of amplified power, the stage was set to go for a flatter, cleaner response. The result became the Radiola Model 104, which with its built-in power amplifier sold for the then enormous price of 250 US dollars. (So there is nothing new about the concept of self-powered loudspeakers: they *began* that way!) Marconi later patented the idea of passing the DC supply current through the energizing coil of the loudspeaker, to use it instead of the usual, separate smoothing choke to filter out the mains hum from the amplifier. Therefore right from the early days it made sense to put the amplifier and loudspeaker in the same box.

Concurrently with the work going on at General Electric, Paul Voight was busy developing somewhat similar systems at the Edison Bell company. By 1924 he had developed a huge electro-magnet assembly weighing over 35 kg and using 250 watts of energizing power. By 1926 he had coupled this to his Tractrix horn, which rejuvenated interest in horn loudspeakers because it enormously improved the sensitivity and acoustic output of the moving coil loudspeakers, and when properly designed did not produce the "honk" sound associated with the older horns. Voight then moved on to use permanent magnets, with up to 3.5 kg of Ticonal and 9 kg of soft iron, paving the way for the permanent magnet devices and the much higher acoustic outputs that we have today.

Gilbert Briggs, the founder of Wharfedale loudspeakers, wrote in his book of 1955 (Briggs, 1990). "It is fairly easy to make a moving-coil loudspeaker to cover 80 to 8,000 cycles [Hz] without serious loss, but to extend the range to 30 cycles in the bass and 15,000 cycles in the extreme top presents quite a few problems. Inefficiency in the bass is due mainly to low radiation resistance, whilst the mass of the vibrating system reduces efficiency in the extreme top." The problem in the bass was, and still is, that with the cone moving so relatively slowly, the air in contact with it simply keeps moving out of the way, and then returning when the cone direction reverses, so only relatively weak, low efficiency pressure waves are being propagated. The only way to efficiently couple the air to a cone at low frequencies is to either make the cone very big, so that the air cannot get out of the way so easily, or to constrain the air in a gradually flaring horn, mounted directly in front of the diaphragm. Unfortunately, both of these methods can have highly detrimental effects on the high-frequency response of the loudspeakers. For a loudspeaker cone to vibrate at 20 kHz it must change direction 40,000 times a second. If the cone has the mass of a big diaphragm needed for the low frequencies, its momentum would be too great to respond to so many rapid accelerations and decelerations without enormous electrical input power—hence the loss of efficiency alluded to by Briggs. Large surfaces are also problematical in terms of the directivity of the high-frequency response, but we will come to that later. So, we can now begin to see how life becomes more complicated once we begin to extend the frequency range from 20 Hz to 20 kHz—the requirements for effective radiation become conflicting at the opposing frequency extremes.

The wavelength of a 20 Hz tone in air is about 17 meters, whereas the wavelength at 20 kHz is only 1.7 *centimeters*, a ratio of 1000 to 1, and for high quality audio applications we want our loudspeakers to produce all the frequencies in-between at a uniform level. We also need them to radiate the same waveforms, differing only in size (but not shape) over a power range of at least 10,000,000,000 to 1 *and* with no more than one part in a hundred of spurious signals (non-linear distortion). It is a tall order! Indeed, for a single drive unit, it still cannot be achieved at any realistic SPL (sound pressure level) if the full frequency range is required.

SOME OTHER PROBLEMS

There are also many mechanical concepts which must be considered in loudspeaker design. For example, the more that one pushes on a spring, the more one needs to push in order to make the same change in length. If the force is limited to less than what is necessary to fully compress the spring, equilibrium will be reached where the applied force and the reaction of the spring balance each other. This is useful when we go to bed, because it prevents the suitably chosen springs from bottoming out, and allows the mattress to adapt to our shape yet retain its springiness. The suspension systems of loudspeaker diaphragms are also springs, but they must try to maintain a consistent opposition to movement or they would compress the acoustical output. If, for example, the first volt of input moved the diaphragm x millimetres, and a second volt moved it only $0.8x$ mm further, this would be no good for high fidelity, because when we double the voltage we expect to see the same linear increase in motion. Otherwise, the acoustic output would not be linearly following the electrical input signal, and the non-linear movement would introduce distortion. Therefore, to keep the diaphragm well centered, but to still allow it to move linearly with the input signal, suspension systems must be used which do not exhibit a non-linear restorative force as the drive signal increases, at least not until the rated excursion limit is reached. (However, certain non-linear loudspeaker characteristics may actually be *desirable* for musical instrument amplification.) We tend to complicate this suspension problem further when we put a loudspeaker drive unit into a box, because the air in the box acts as an additional spring which is also not entirely linear.

Electrically we can also run into similar problems. Whenever an electric current flows through a wire, the wire will heat up. It is also a property of voice-coil wires that as they heat up their resistance increases. As the resistance increases, the signal voltage supplied by the amplifier will proportionally drive less current through the coil. So, as the drive force depends on the current flowing through the wire which is immersed in the magnetic field, if that current reduces, the movement of the diaphragm will reduce correspondingly. We therefore can encounter a situation where the amplifier sends out an accurate drive signal *voltage*, but how the loudspeaker diaphragm responds to it can, depending on level, change with the voice-coil temperature and the springiness of both the air and the suspension system. Even the very magnetic field of the permanent magnet, against which the drive force is developed, can be modulated by the magnetic field given rise to by the signal in the voice coil. Notwithstanding all of these effects, we still need our diaphragm to move exactly as instructed by the drive voltage from the amplifier, because in reality modern amplifiers are usually *voltage* sources. This is despite the fact that the loudspeaker motor is *current* driven, and that the voice-coil resistance and reactance (together they form the impedance) will not remain constant over the whole of the frequency range.

Clearly, things are beginning to get complicated, and already we have seen problems begin to pile up on each other. The concept of a loudspeaker being simply an electromagnet, coupled to a moveable cone and placed into a box, is obviously not going to produce the high fidelity sounds needed for music recording and reproduction. Good loudspeakers are complex devices which depend on the thorough application of some very rigid principles of electroacoustics in order to perform their very complex tasks. In the minds of most people the concept of a loudspeaker, if it exists at all, is usually grossly over-simplified, and unfortunately this is the case even with most *professional* users of loudspeakers.

Typical loudspeaker systems consisting of one or more vibrating diaphragms, either on one side of a rectangular cabinet or flush mounted into a wall, represent physical systems of sufficient complexity that accurate and reliable predictions of their sound radiation are rare, if not nonexistent, even with the aid of modern computer technology. Despite the fact that we inevitably have to deal with a degree of artistry and subjectivism in the final assessment, at the design and development stages we must stick close to the objective facts. So, to begin to understand the mechanisms of sound radiation it is necessary to establish the means by which a sound "signal" is transported from a source, through the air, and to our ears.

SOME BASIC FACTS

As explained in the opening paragraphs of this chapter, acoustic waves are essentially small local changes in the physical properties of the air which propagate through it at a finite speed. The mechanisms involved in the propagation of acoustic waves can be described in a number of different ways, depending upon the particular cause or source of the sound. With conventional loudspeakers that source is the movement of a diaphragm, so it is appropriate here to begin with a description of sound propagation away from a simple moving diaphragm.

Acoustic wave propagation

The process of sound propagation is illustrated in Figure 11.1. For simplicity, the figure depicts a diaphragm mounted in the end of a uniform pipe, the walls of which constrain the acoustic waves to propagate in one dimension only. Before the diaphragm moves (Figure 11.1(a)), the pressure in the pipe is the same everywhere and equal to the static (atmospheric) pressure P_0. As the diaphragm moves

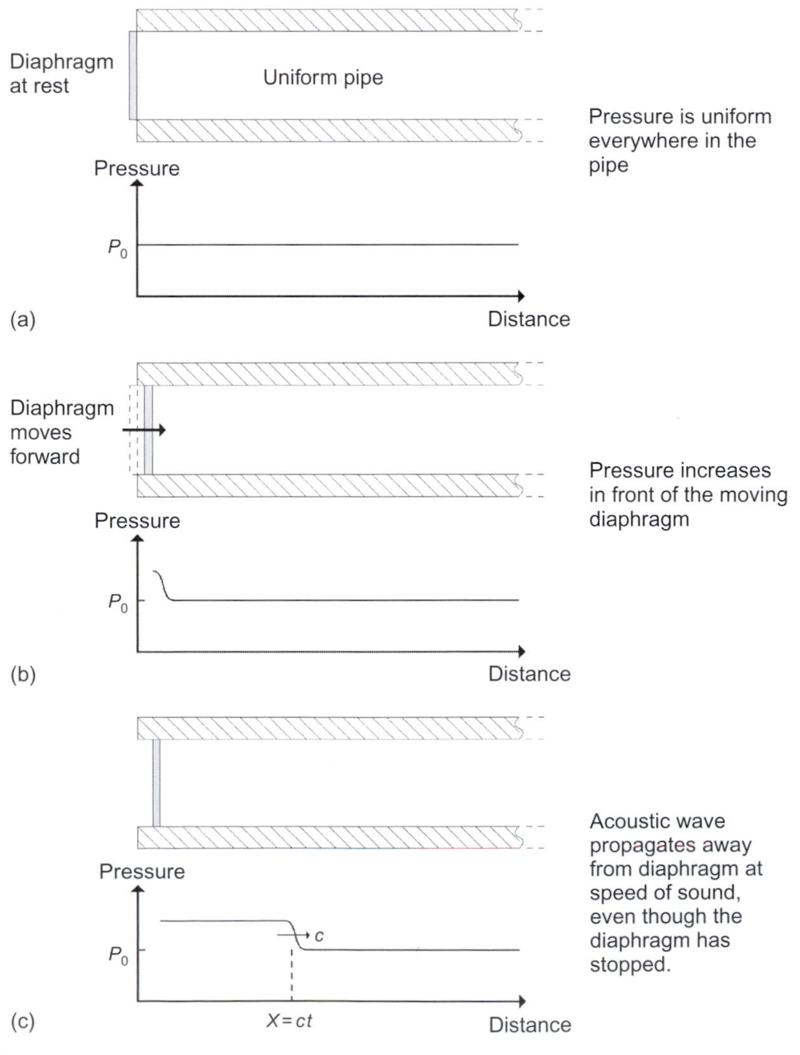

FIGURE 11.1
The generation and propagation of an acoustic wave in a uniform pipe.

forwards (Figure 11.1(b)), it causes the air in contact with it to move, compressing the air adjacent to it and bringing about an increase in the local air pressure and density. The difference between the pressure in the disturbed air and that of the still air in the rest of the pipe gives rise to a force which causes the air to move from the region of high pressure towards the region of low pressure. This process then continues forwards, and the disturbance is seen to propagate away from the source in the form of an acoustic wave. Because air has mass, and hence inertia, it takes a finite time for the disturbance to propagate through the air; a disturbance "leaves" a source and "arrives" at another point in space some time later (Figure 11.1(c)). The rate at which disturbances propagate through the air is known as the *speed of sound*, which has the symbol "c," and after a time of t seconds, the wave has propagated a distance of $x = ct$ meters. Note, though, that the speed of propagation is *not* related to the velocity of the diaphragm. In the great tsunami of December 2004, a five meter displacement of the ocean floor caused a tidal wave to travel at over 500 kilometers per hour, a speed much more rapid than that of the displacement which caused it. For most purposes, the speed of sound in air can be considered to be constant, and independent of the particular nature of the disturbance, although as previously stated it does vary with temperature. A one-dimensional wave, such as that shown in Figure 11.1, is known as a *plane wave.* A wave propagating in one direction only (e.g., left-to-right) is known as a *progressive wave.*

Mechanical and acoustic impedance

The description of sound propagation earlier mentioned the motion of the air in response to local pressure differences. This localized motion is often described in terms of acoustic particle velocity, where the term "particle" here refers to a small quantity of air that is assumed to move as a whole. Although we tend to think of a sound field as a distribution of pressure fluctuations, any sound field may be equally well described in terms of a distribution of particle velocity, and there is a relationship between the distribution of pressure in a sound field and the distribution of particle velocity. At a given frequency, the ratio of pressure to particle velocity at any point (and direction) in a sound field is known as acoustic impedance (strictly *specific* acoustic impedance), $Z_a = p/u$, and it is very important when considering the sound power radiated by a source. Acoustic impedance can be thought of as a quantity that expresses how difficult the air is to move. A low value of impedance tells us that the air moves easily in response to an applied pressure (low pressure, high velocity), and a high value of impedance tells us that it is hard to move (high pressure, low velocity). Mechanical impedance is directly equivalent to acoustic impedance, but with pressure replaced by force (pressure is force per unit area) and particle velocity replaced by velocity: $Z_m = F/u$.

With acoustic radiators, as well as electrical circuits and mechanical systems, there is a need to match impedances for good energy transfer. If a microphone needs to be connected to a 600 ohm input, then it will not sound as intended by its designers or exhibit its quoted sensitivity if connected to a 30 ohm or 10,000 ohm input. An amplifier which is optimized to function into a 4 ohm load will not produce its maximum power output capability into a load of 16 ohms. Loudspeakers are effectively "plugged into" the air, so if the air load impedance does not match the electro mechanical output impedance of the loudspeaker, the radiated power will be less than optimal. Impedance changes with frequency, so, for example, a resistor and capacitor in parallel have a frequency dependent impedance which is the combination of the purely resistive, frequency *independent* characteristic of the resistor, and the reactive, frequency *dependent* characteristic of the capacitor. This is the basis of electrical filter design.

Impedance (Z), whether it be electrical, mechanical or acoustic, can be divided into two components: resistance (R) and reactance (X). Reactive impedances represent systems which store input energy, but which later give it back, whereas resistive impedances represent systems which transfer energy away from the input, never to return. Figure 11.2 shows three mechanical systems which can be used to demonstrate the three different components of mechanical impedance and the way in which they relate to conventional loudspeakers. In Figure 11.2(a), the person applies a force to compress a spring. When the applied force is stopped, the spring returns to its original length and all of the (potential) energy applied to the spring is returned to the person as the spring pushes back. If the person pushes back and forth on the spring in an oscillatory manner, the energy flows from person to spring and back again in each half-cycle with an overall zero transfer of energy. A spring represents a purely reactive mechanical impedance. Figure 11.2(b) shows the person applying a force to a mass on a trolley. The force acts to accelerate the mass from rest, but when the force is stopped the mass tries to continue moving at a steady velocity. If the person then applies a force to slow the mass down, the (kinetic) energy possessed by the mass is returned as the mass pulls back on the person. Again, if the force is applied in an oscillatory manner, there is zero overall transfer of energy from the person to the mass. A mass also represents a purely reactive mechanical impedance, but with the opposite sign to that of a spring. (Note the different direction of the reactive force arrows in the figure.) Figure 11.2(c) shows the person pushing a block along a table. The applied force overcomes the friction between the block and the table and the block moves with a constant velocity. When the force is stopped the block also stops, and none of the energy supplied to the block is returned to the person. If the force is applied in an oscillatory manner in this case, the flow of energy is always from person to block, regardless of the direction of motion. All of the energy is "lost" to friction as heat, and none is returned to the person. The friction block represents a resistive mechanical impedance which is the mechanism by which power can be transferred from one system to another (in this case, from the person to heat). There are acoustic counterparts to each of these mechanical components, for example, a small sealed cavity driven by a piston is an acoustic spring. The electrical counterparts will be further discussed later in this chapter.

Impedance in loudspeakers

In a cone loudspeaker, we have all of these forms of impedance present at the same time. The diaphragm radiates useful sound power through its motion via an acoustic radiation *resistance*.

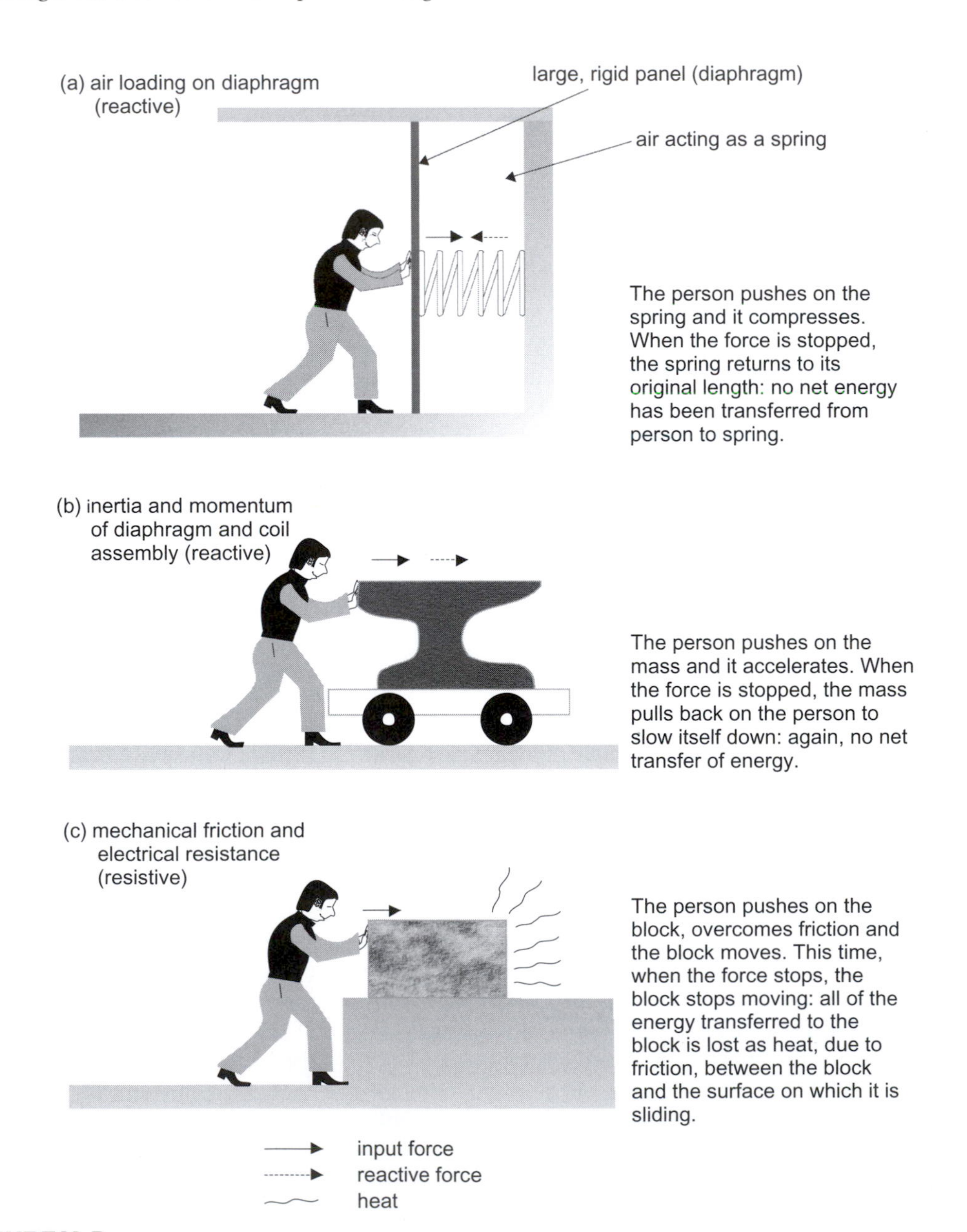

FIGURE 11.2
Three characteristic properties of a moving coil loudspeaker depicted as three components of its impedance. In electrical terms, these can be related to capacitance, inductance and resistance.

The mass of the voice coil and diaphragm, together with the stiffness of the suspension, produce a *reactive* impedance which merely serves to reduce the diaphragm motion. The *reactive* inductance of the voice coil reduces the current, and hence the applied force, at higher frequencies. And finally, the *resistive* frictional losses in the suspension and the electrical *resistance* of the voice coil simply waste power by turning it into heat.

THE PRACTICAL MOVING-COIL CONE LOUDSPEAKER

The majority of all loudspeakers are moving-coil devices employing a cone-shaped radiating diaphragm, but the mechanisms of sound radiation from these devices are not as straightforward as they may initially seem to the casual onlooker. This type of loudspeaker is essentially a "volume velocity" source. In other words it creates a pressure wave equivalent to injecting air from a point source at a rate of injection measured in cubic meters per second (or *extracting* the air on the rarefaction half-cycle). However, unlike the piston shown in Figure 11.1, most cabinet-mounted cone loudspeakers, direct radiating into a room (i.e., not radiating via a horn), do not couple effectively with the air. Instead, the cone finds itself punching into thin air, with much of the potential load being lost as most of the air adjacent to it simply moves out of the way, then returns when the direction of the cone movement reverses. Efficiency is therefore often extremely low, with less than 1% of the energy being supplied to the voice coil resulting in the radiation of sound. The remaining energy either gets lost by friction in the moving system, or by being burned up as heat by the resistance of the voice coil, or even by being reflected back into the output stages of the power amplifier. To complicate matters, many of these things are frequency dependent, so it is little wonder that many things must be considered and balanced before there is any chance of such a device having a flat frequency response.

As long as the circumference of the diaphragm remains small with respect to the wavelength, the radiation will be omnidirectional, but when the wavelength starts to become small compared with the circumference of the source, the radiation begins to beam directly ahead. (The wavelength, in meters, can be calculated simply by dividing the speed of sound in meters per second by the frequency in hertz.) This is a result of the interference field where the different parts of the diaphragm, radiating in phase, become significantly different in their path lengths to an off-axis listening position. The cause is depicted in Figure 11.3, and the effect is shown in Figure 11.4. The *ka* values referred to in the latter figure are derived from the wave number *k*, which is simply 2π divided by the wavelength (λ) in meters, and *a*, which is the radius of the radiating surface. In practice we can think of *ka* as being the number of wavelengths around the circumference of the diaphragm: $ka = 2\pi a/\lambda$.

We would tend instinctively to think of the whole diaphragm as being the radiating surface, however this is often not the case, and a large diaphragm driven at high frequencies will almost certainly *not* move uniformly. The outer parts would lag with respect to the movement of the central area, in which case the outer parts would radiate with a phase shift relative to the central area of the diaphragm. In fact they may even simply stay still, in which case they would not radiate at all. In either case the radiation may not correspond with the *ka* value taken from the *physical* measurements of the diaphragm.

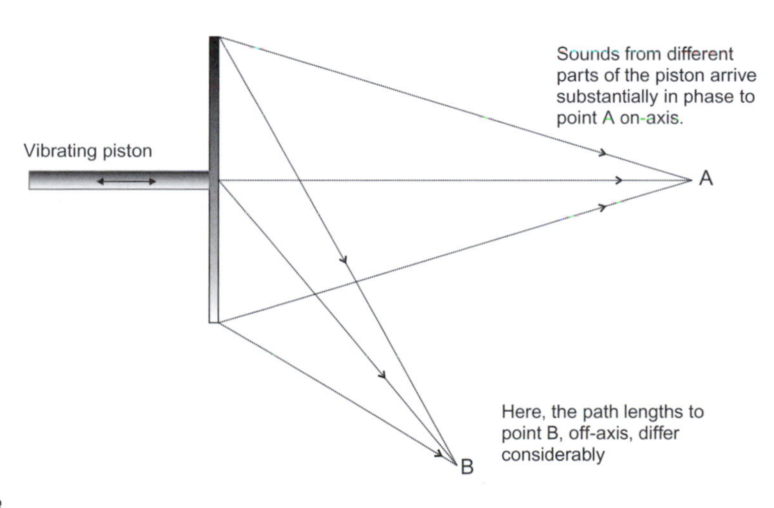

FIGURE 11.3
The cause of the off-axis interference effects that give rise to the directivity shown in Figure 11.4.

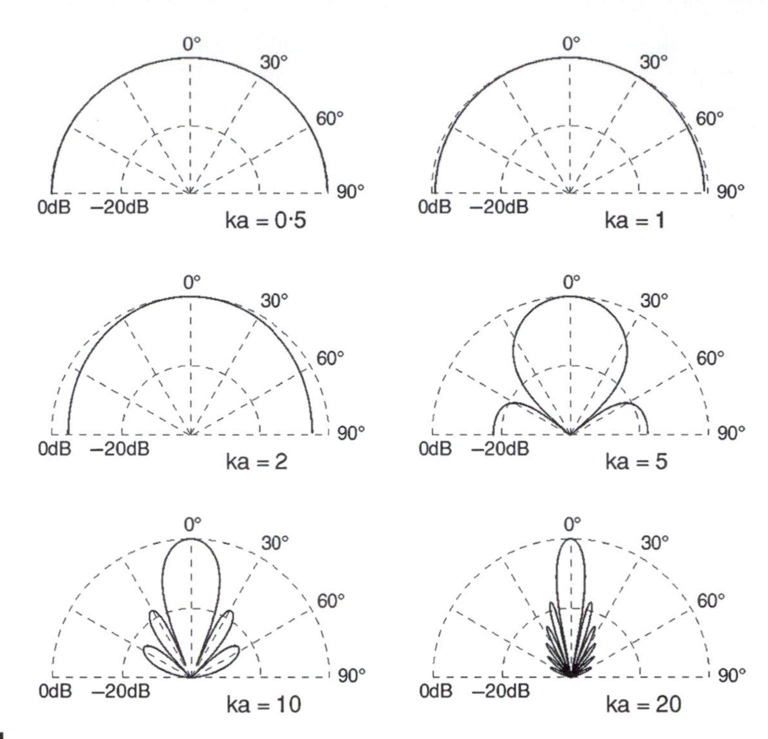

FIGURE 11.4
The directivity of a vibrating piston due to off-axis interference effects at different frequencies. Note the narrowing of the main lobe as the frequency increases.

Over the years, many "solutions" have been tried in the search for perfect pistonic motion in a diaphragm, such as using ultra-rigid cone materials, or solid, conical "plugs" as diaphragms, but internal losses and differential sound speeds *within* the materials have tended to confound all efforts. Remember, it was stated at the beginning of this chapter that air was not typical of all materials in terms of the speed with which sound propagates through it being equal for all frequencies. Many types of solid materials propagate high frequencies faster than low frequencies, a property known as phase dispersions, and exhibit sound speeds very different from that in air.

When the sound propagates from the voice coil *through* the material of the cone, in a radially outward direction, the different sound speeds may give rise to interference between the sound waves traveling to the edge of the cone and the forward radiation into the air by the pistonic action of the whole cone being driven by the voice coil. Resonances can also be set up within the cone itself, and one job of the cone surround is to absorb as much as possible of the waves propagating *radially* through the cone material, to prevent them from being reflected back and causing even more interference. Complex cone surrounds are often as much to do with suppressing these waves as with linearly suspending the cone itself.

The combined response

It was stated earlier that a loudspeaker was an electro-mechanico-acoustic transducer. Well, we have just looked at a few of the acoustic properties, but the electrical and mechanical aspects also present their own complications. The voice coil is a mixture of resistance and inductance, the latter being a form of frequency dependent reactance. The reactance, in ohms, of an inductor is given by $2\pi fL$, where f is the frequency in hertz and L is the inductance in henries. An 8 ohm voice coil is only nominally 8 ohms, and approximates that value only over a limited band of frequencies. The inductive part of the impedance (impedance here being resistance plus reactance) rises with frequency, whereas any capacitive reactance effects *decrease* with frequency. The reactance (c) of a capacitor, in ohms, is given by $1/2\pi fc$, so the frequency component, being in the denominator in this case, reduces the reactance as it rises. An overall impedance plot of a typical low-frequency loudspeaker with respect to frequency is shown in Figure 11.5. Fortunately, however, some effects counterbalance each other. Figure 11.6 shows how a flat frequency response can be achieved above the resonance frequency of a cone driver, mounted on a theoretical infinite baffle, where the roll-off in the diaphragm velocity compensates for its rising relationship with the pressure radiated to a point far away on the cone axis. However, at very close distances some rather more complex relationships manifest

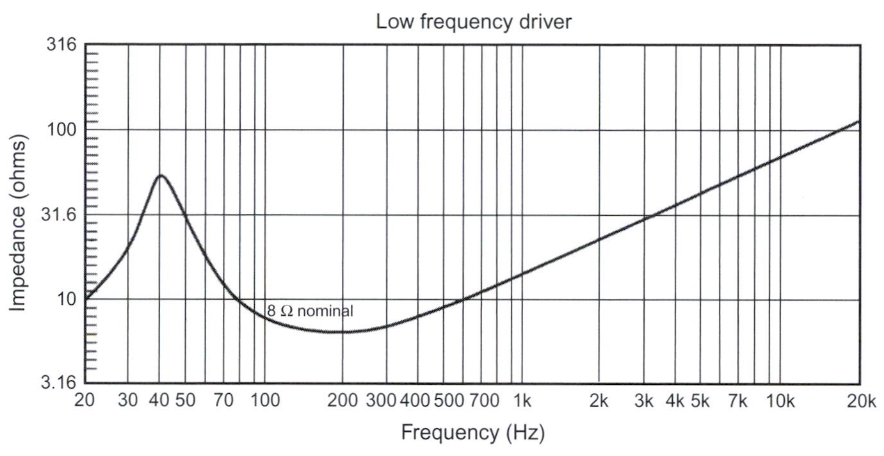

FIGURE 11.5
Magnitude of electrical impedance of a typical loudspeaker drive unit.

themselves. Academically speaking, this region is known as the near-field, but it should not be confused with the more colloquial "near-field" that many recording personnel speak of. In fact, it is better to use the term "close-field" for desk-top monitors, because the true near-fields are not good distances to listen within as the frequency balance can be strange in those regions.

For people conversant with electrical circuit diagrams, equivalent circuits can be devised which represent the electrical, mechanical and acoustic properties of a loudspeaker as a single electric circuit. This type of representation is very common in traditional textbooks on electroacoustics. In these circuits, the mechanical and acoustic components such as springs and masses are replaced by their electrical equivalents. One way in which this can be realized is to replace mechanical forces (and acoustic pressures) with electrical currents, and mechanical velocities with electrical voltages. It follows then that mechanical springs are replaced by electrical inductors, masses are replaced by capacitors, and mechanical resistances are replaced by resistors. This arrangement is known as the *mobility analogy,* although it should be noted that a different, but equally valid, *impedance analogy* exists, where voltages replace forces and currents replace velocities. Each analogy has its strengths and weaknesses and may better suit different aspects of loudspeaker analysis. The equivalent mobility analogy circuit of a typical moving-coil loudspeaker is shown in Figure 11.7. The transformers represent the coupling between the electrical and mechanical, and the mechanical and acoustical domains. This type of circuit is what an amplifier really "sees" when terminated by a typical moving-coil loudspeaker, which is a far cry from a resistive 8 ohm load.

FIGURE 11.6
How a flat frequency response results from a falling velocity and a rising radiated sound pressure.

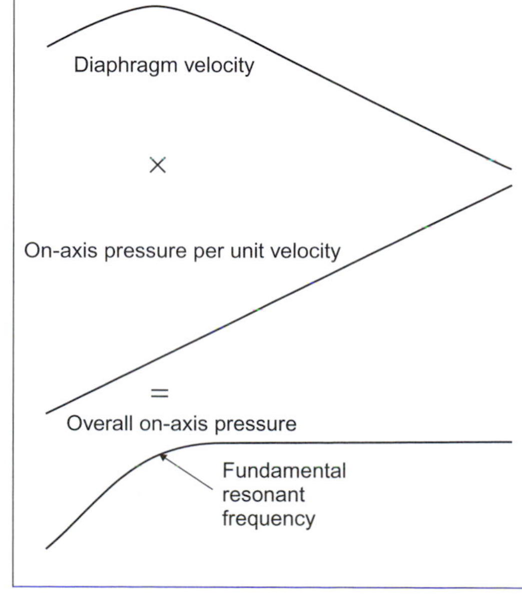

Getting a frequency-independent acoustic output from such a combination of components is no simple task. It can only be achieved by very careful balancing of the values of the components, and even then the perfect balance is usually only achievable over a limited bandwidth. It is apparent that any extra series resistance *increase,* or any parallel resistance *decrease,* will serve to reduce the power supplied to the load for any given input power, and so will reduce the efficiency of the system. What is more, with such a finely balanced system, any change to any component part(s) may require a counterbalancing change to many other component parts. As true perfection can never be achieved, there are an infinite number of close approximations which are possible, and this fact contributes to the diversity of loudspeaker designs that we have available in the market place.

RESISTIVE AND REACTIVE LOADS

Before ending this chapter, it may be worth looking a little more closely at the concepts of resistive and reactive loading. Figure 11.8 shows three potential divider networks. If we consider a constant voltage source to be applied between terminals A and C we can consider what voltages occur between points B and C, and how much power will be dissipated in each component. In the

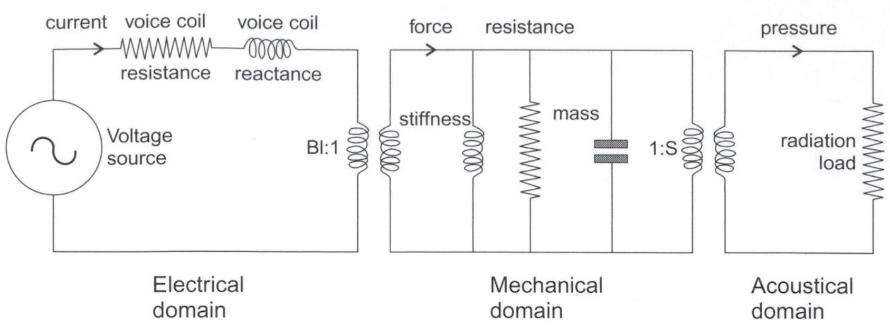

FIGURE 11.7
The equivalent electrical circuit of a moving-coil loudspeaker drive unit (mobility analogy): Bl = the product of the magnetic flux density and the length of wire in the magnetic field—the force factor—T_m (tesla × metres): S = area of the cone.

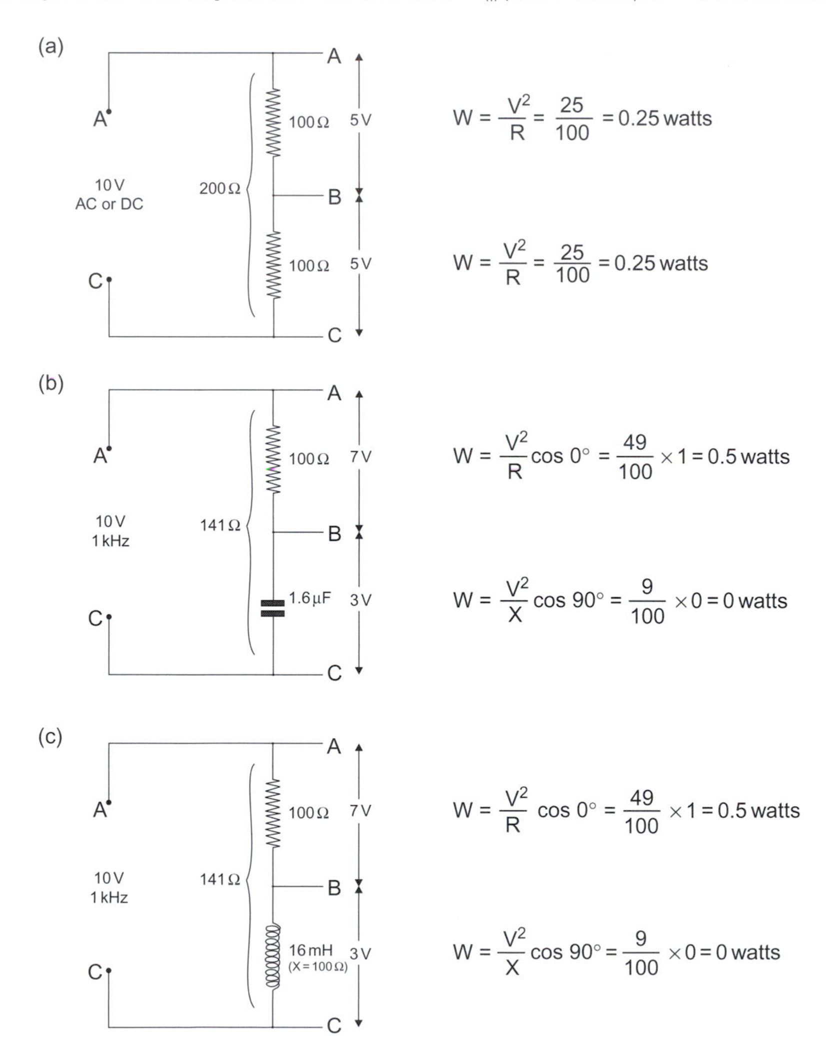

FIGURE 11.8
Three potential divider networks. W = watts, V = volts, R = resistance (ohms), X = reactance (ohms), Cos 0 degrees = 1, Cos 90 degrees = 0. In each case the total power dissipation is the same, at 0.5 watts, but it is only dissipated in the resistors. The inductors and capacitors merely store the energy then release it into the resistors. Note that a total of 0.5 watts is dissipated in each case.

case of the resistor-resistor network shown in Figure 11.8(a), the resistance of the two components R_1 and R_2 is equal. In a resistor, as the voltage is increased across its terminals the current will also increase in direct proportion, as given by Ohm's law, which may be written in the following ways:

$$I = \frac{V}{R} \tag{11.1}$$

$$V = IR \tag{11.2}$$

$$R = \frac{V}{I} \tag{11.3}$$

Where

I = current in amps
V = voltage in volts
R = resistance in ohms

The voltage and current are always in phase, so they will produce power, in watts (W), according to a simple multiplication:

$$V \times I = W \tag{11.4}$$

Strictly speaking, this should be multiplied by the cosine of the phase angle, but as this is 0 degrees in resistive circuits, and the cosine of 0 degrees is 1, then a multiplication by one has no effect, so it is traditionally omitted. Whether the current is direct or alternating is also of no consequence in resistors, and the resistance remains the same irrespective of frequency.

In the case of capacitors, there *is* a frequency dependent effect. Capacitors will not pass DC, because the impedance (reactance in this case) at 0 Hz is infinite, but, as the frequency rises, the reactance lowers. Reactance (X), like impedance (Z) and resistance (R), is measured in ohms, but unlike resistance it is frequency dependent. In the case of capacitors, the reactance is *inversely* proportional to the frequency: as the frequency rises the reactance lowers. The formula for the reactance (X) of a capacitor is given by:

$$X = \frac{1}{2\pi fc} \tag{11.5}$$

Where

f = frequency in hertz
c = capacitance in farads
π = 3.142

or,

$$X = \frac{1,00,000}{2\pi fc} \tag{11.6}$$

where c is in microfarads

In capacitors, the current and voltage are *not* in phase, but are shifted by 90 degrees. In the case of Figure 11.8(b) an application of a DC voltage across A-C will initially see an uncharged capacitor behaving like a short circuit (zero ohms). A current will then flow through R and the plates of the capacitor will begin to charge. As the voltage rises across B-C, the current will reduce until, once the plates are fully charged, the voltage will rise to a maximum (as across A-C) and the current will reduce to zero. All conduction will then cease. Thus it can be seen how the *current leads the voltage*: the current flows first, then it falls as the voltage rises. The *electrostatic* charge is a *voltage effect.*

In the case of Figure 11.8(c) the effect is the reverse. Inductors work on an *electromagnetic* principle, which is a *current effect.* The formula for the reactance (X) of an inductor is given by:

$$X = 2\pi fL \tag{11.7}$$

Where

f = frequency in hertz (Hz)
L = inductance in henries (L)

In this case, the reactance is *directly* proportional to the frequency. As the frequency rises, so does the reactance, and the *voltage leads the current.* In the cases of both the capacitor and the inductor, the voltage and current are 90 degrees out of phase. Equation 1.4 showed the formula for calculating the power from the voltage and the current, and it was noted that for AC currents and voltages there should be a phase angle multiplier, cos 6 (theta). The cosine of 90 degrees is *zero,* therefore whatever values of voltage and current exist in the circuit, the power dissipation (heating effect) in inductors and capacitors is always zero (except for losses due to imperfections). This is *wattless power,* and is why AC power circuits are measured in VA (volt-amps) and not in watts. Electricity meters measure kVA because in heavily inductive loads, such as electrical motors and machinery, the kW value would be less, and the electricity company would not be charging for all the current and voltage that they were supplying.

In Figure 11.8(a), a 10 volt, 1 kHz voltage at terminals A-C would give rise to a voltage of 5 volts across B-C if R_2 was equal to R_1, at, say, 100 ohms each, and the resistors would heat up with the power dissipation. In the case of Figure 11.8(b), we could select a capacitor to have a reactance of 100 ohms at 1 kHz, but the circuit would not behave in the same way. The capacitor would be selected from the formula

$$X = \frac{1,00,000}{2\pi fc}$$

which transposes to

$$c = \frac{1,00,000}{2\pi fx}$$

so for 100 ohms at 1 kHz

$$\begin{aligned} c &= \frac{1,000,000}{6.284(2\pi) \times 1,000 \times 100} \\ c &= \frac{1,000,000}{628400} \\ c &= 1.6\,\text{microfarads}\,(\mu F) \end{aligned}$$

However, despite the resistance and reactance in the circuit both being equal at 100 ohms, the total impedance (resistance *plus* reactance) would *not* be 200 ohms. The same current would flow through both components, but whereas the voltage across the resistor would be in phase with the current, the voltage across the capacitor would be 90 degrees out of phase with the current. We can draw a right-angled triangle, as in Figure 11.9, with one side representing the resistance and the other side, at 90 degrees, representing the reactance. The total impedance (Z) would be represented by the hypotenuse. From Pythagorus' theorem, the square of the hypotenuse is equal to the sum of the squares of the other two sides. Therefore:

$$\begin{aligned} 100^2 + 100^2 &= 10,000 + 10,000 = 20,000 \\ Z^2 &= 20,000 \\ Z &= \sqrt{20,000} \\ Z &= 141 \end{aligned}$$

total impedance = 141 ohms.

The resistor would dissipate power in the form of heat, but the capacitor would dissipate no power, and would not heat up. The inductive circuit in Figure 11.8(c) would behave similarly, except that the phase of the voltage across the inductor would be 90 degrees out with the voltage across the resistor in the opposite direction to the phase shift across the capacitor. The 90 degree differences in opposite directions leads to the 180 degree phase shift between capacitors and inductors, which gives rise to the resonance in tuned circuits. A capacitor and inductor in series are the electrical equivalent to the

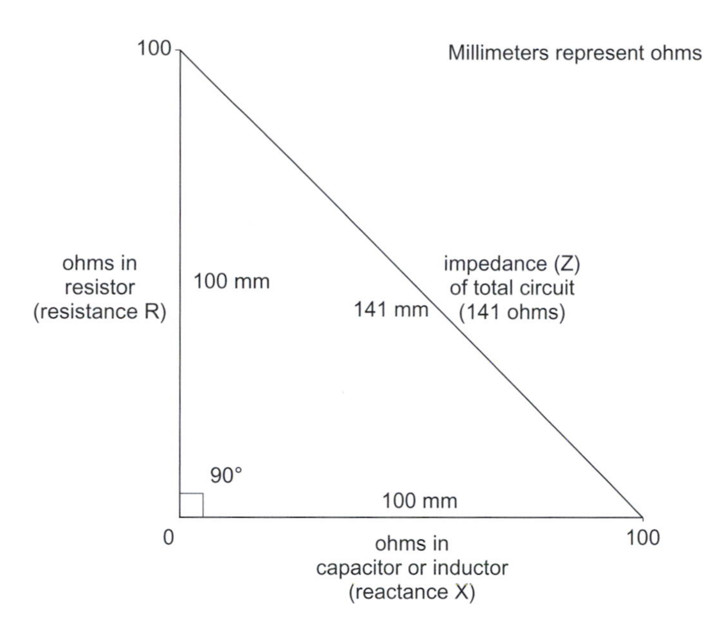

FIGURE 11.9
Phase angle vector triangle, using 1 mm to represent 1 ohm.

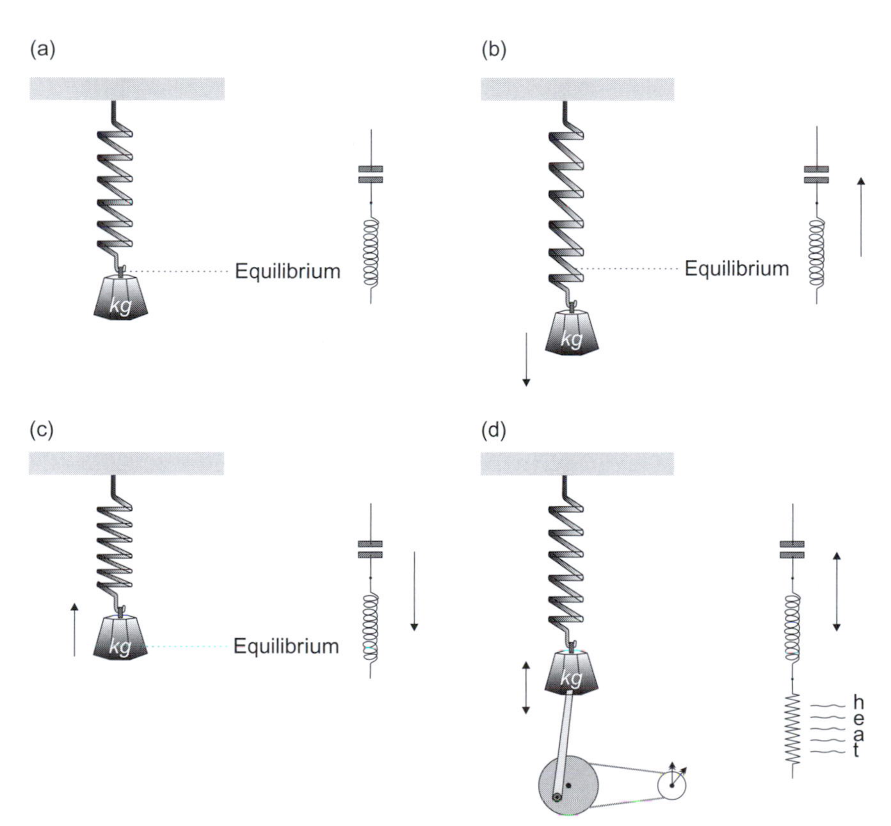

FIGURE 11.10
Masses, springs and resistances vis-à-vis inductors, capacitors and resistors.

mass and the spring shown in Figure 11.10. Both are tuned circuits, and both work in the same way, by transferring energy backwards and forwards and dissipating very little of it. The mass takes in energy when one tries to move it, but releases it again when one tries to stop it—this is why the brakes get hot when a car stops. A spring takes in energy when it is stretched or compressed, but releases it when it is released. A mass, basically, "wants" to stay free of accelerations and decelerations. A spring, basically, does not "want" to change its length. The two together, given their different phase relationships, take in and release energy alternately, and so remain in oscillation until frictional losses eventually turn the energy into heat.

At the risk of laboring the point (but it is really not well understood by most loudspeaker users) if we have a mass and a spring as shown in Figure 11.10(a) at equilibrium, the spring is "happy" because it is at its equilibrium length given the forces acting upon it. The mass is also "happy" because it is at rest; it is neither accelerating nor decelerating. If somebody then pulls down on the mass, and holds it there, the mass is still happy, but the spring is not, because it is stretched as in Figure 11.10(b). If the mass is released, the spring will act to overcome the inertia of the mass so that it can return to its equilibrium length. When the spring reaches its equilibrium length the mass is in motion, so the mass times velocity gives it momentum, which will carry it through the rest position of the spring, and will begin to compress it, as shown in Figure 11.10(c). The "unhappy" spring therefore begins to slow the mass, so that it (the spring) can once again return to its "happy" equilibrium position, but it overshoots once again, and the oscillation continues. This is a reactive system like a capacitor and inductor, and the only energy loss is due to its imperfection; which in the mechanical case is friction and air resistance, and in the electrical system is electrical resistance due to the less than perfect conductors—the coil of an inductor will always have a small resistance due to the wire.

Now, if we add some resistance to the mechanical system, we can make it do some work, like moving the hands of a clock, as in Figure 11.10(d), but the work takes energy, so the oscillations will be reduced more rapidly—the oscillations will be *damped*. The electrical circuit equivalent would be to put a resistor in series with the inductor and the capacitor, so that the current flowing through the circuit will produce some heat, and damp the electrical oscillation.

When we load a loudspeaker diaphragm with a horn, the air in front of the loudspeaker cannot escape to the side of the diaphragm, and the stretching pressure due to the sideways component of the wave is restricted. The reactive conditions are therefore minimized, and the air load in the horn provides a substantially resistive load, with the particle velocity and pressure in phase, so more useful work can be done by the diaphragm, such as producing sound instead of just flapping backwards and forwards.

These concepts are relative to horns and direct radiators with their predominantly resistive and reactive loadings respectively. The resistive horn loading tends to be efficient because it gives rise to *acoustic power* being radiated. The largely reactive loading on a direct radiator gives rise to little power being radiated, just as little heat is produced in a capacitor or an inductor.

THE BIGGER PICTURE

This chapter has tried to set out the fundamental characteristics of sound radiation from moving-coil loudspeakers, which represent 99% of all drive units manufactured, but up to now we have only been referring to single drive-units, which, as we have previously discussed, cannot be realistically expected to cover the full frequency range. Once we are forced to consider multi-driver systems with their obligatory crossover filters, the combined system can take on a further considerable degree of complexity. Many entire loudspeaker systems, some of great complexity, are made only from drivers using this motor technique. However, there are many other types of drive systems which will be discussed in the following chapter, and which we will need to look at in order to get a better appreciation of the characteristics which they can offer.

BIBLIOGRAPHY

Borwick, J., 2001. Loudspeaker and Headphone Handbook, Focal Press, third ed., Oxford, UK.

Borwick, J., 1994. Loudspeaker and Headphone Handbook, Focal Press, second ed., Oxford, UK, (significantly different in content from the aforementioned Third Edition).

Briggs, G.A., 1958. Loudspeakers, Rank Wharfedale Ltd, fifth ed., Bradford, UK, Reprinted until 1972.

Colloms, M., 2005. High Performance Loudspeakers, John Wiley and Sons, sixth ed., Chichester, UK.

Eargle, J.M., 1997. Loudspeaker Handbook, Chapman and Hall, New York, USA.

Jordan, E., 1963. Loudspeakers, Focal Press, Oxford, UK.

REFERENCES

Fahy, F., and Walker, J.,1998. Fundamentals of Noise and Vibration. Spon Press, Chapter 5, London, UK.

Rice, C., and Kellogg, E., 1925. Notes on the development of a new type of hornless loudspeaker, Transactions, American Institute of Electrical Engineers, 44, 461–475.

Briggs, G., Loudspeakers 1955. fourth ed., Wharfedale Wireless Works Ltd, Bradford, England. Reprinted by Audio Amateur Publications Inc (1990), Peterborough, NH, USA.

PREFACE

In our second chapter on loudspeakers, the great diversity in their design is explored, and the construction of both moving-coil and electrostatic loudspeakers is gone into in more detail.

We begin with the familiar moving-coil loudspeaker. The vast majority in use today are of this type, in which a voice-coil, fixed to the cone and carrying the audio signal, moves in a static magnetic field. Their development has continued steadily over the last 90 years or so, but they still retain some inherent limitations, as do all electroacoustic transducers. The earliest moving-coil speakers had their magnetic field generated by an electromagnet, as adequately powerful permanent magnets simply did not exist, and this meant that every loudspeaker needed a well-smoothed DC supply. Improvements in magnetic materials, giving rise to alloys like Alnico in the 1940s, meant that permanent-magnet speakers became practical, and made things a good deal more convenient.

Newell and Holland go on to consider the details of moving-coil loudspeaker construction, such as voice-coil centering, cone materials, cone break-up modes, and cone suspension methods.

We then move to consider dome-type moving-coil loudspeakers, where the voice-coil is the same size of the dome-shaped diaphragm, and there is no surrounding cone. This type is used extensively in middle and high-frequency drivers.

The next section deals with compression drivers, which are used with a horn structure that greatly improves their efficiency by matching the effective impedance of the air to the speaker diaphragm more closely, in a manner analogous to a matching transformer in the electrical domain. Efficiencies can go as high as 50% but this kind of driver requires very close mechanical tolerances and is not easy to manufacture.

Next we encounter ribbon loudspeakers, the Heil air-motion transformer, and flat-panel speakers such as those licensed by NXT. The most unusual type described is the obsolete ionic loudspeaker, or Ionophone, which produces sound by modulating a radio-frequency corona discharge. The lack of mechanical parts means that the sound can be stunning, but the fact that the ozone produced by the discharge is a lung irritant, poisonous, *and* carcinogenic would put off all but the most dedicated audiophile. I should also think that EMC testing would be hilarious.

The best-known alternative to the electromagnetic loudspeaker is the electrostatic loudspeaker, in which a very thin and light diaphragm is moved to and fro by high voltages, but small currents. The history of the electrostatic loudspeaker stretches back to the 1920s, but the first really practical design was not introduced until 1957, when the famous Quad ESL appeared. The first problem to overcome was that the attraction between two movable capacitor plates is inherently non-linear. This was dealt with by using constant-charge operation, in which the diaphragm coating is not metal, but a highly-resistive film that stops electric charges moving about at signal frequency. The development of the ESL crucially depended on this material becoming available.

While it is generally accepted that electrostatic loudspeakers can perform superbly, they have never swept the market because of some inherent snags. The flat panels need to be large and visually obtrusive to get a halfway decent bass response, and the need for polarizing voltages of the order of 3 kV puts the price up.

Electrostatic speakers present a highly capacitive load to a power amplifier, and there has been much hand-wringing over this issue in the past. The fact is that a properly designed solid-state amplifier has no problem at all in driving a capacitive load.

CHAPTER 12

Diversity of Design

Loudspeakers by Philip Newell and Keith Holland

Although the original moving-coil, cone loudspeaker of Rice and Kellogg was the first true loudspeaker of a type that we know today, it was, itself, a development of ideas which had gone before, principally relating to the design of telephone earpieces, which were *not* very loud speakers. The moving-coil direct radiator, along with amplifiers as great as 15 watts output—which was then huge—soon opened a door to room-filling sound levels, and, within only a couple of years, talking pictures at the cinema. The need to fill larger and larger theaters with sound led to horn designs, and the need for greater bandwidth led to the separation of the drive units into frequency ranges where they could operate more efficiently. Thus began a refinement and specialization of designs which continues to this day, with ever more ideas, magnet materials, diaphragm materials and radiator concepts all designed essentially to do the same thing—convert electrical energy into sound waves. What follows in this chapter is a discussion of some of the various ways in which this conversion can be made to take place, and the strengths and weaknesses of the various approaches.

MOVING-COIL CONE LOUDSPEAKERS

Of all types of drive units, there is probably none so varied in size, shape, materials of construction or performance as the moving-coil cone loudspeaker. They basically all follow the concept shown in Figure 12.1, and little has changed in the underlying principles of their operation in the 80 years of existence so far. They all need a magnet, which was often an electro-magnet in the early years before permanent magnets of sufficient strength were developed. In this case, a "field coil" was supplied with a DC current sufficient to generate the required strength of magnetic field for the "voice-coil" (which was fed with the output signal from the amplifier), to drive the cone with the required level of sensitivity.

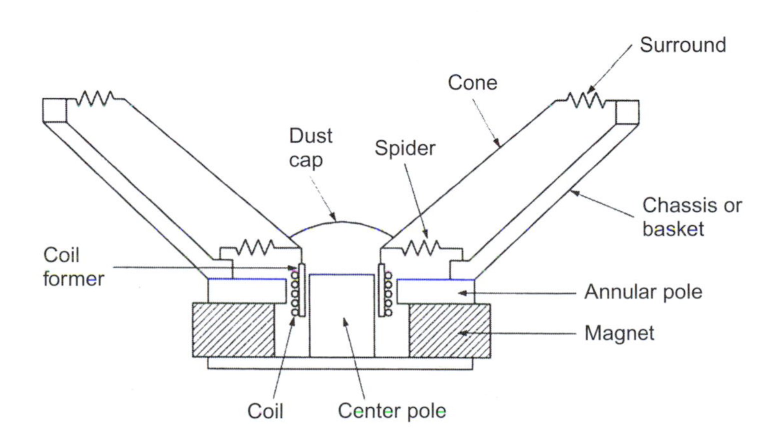

FIGURE 12.1
The components of a moving-coil loudspeaker.

Audio Engineering Explained

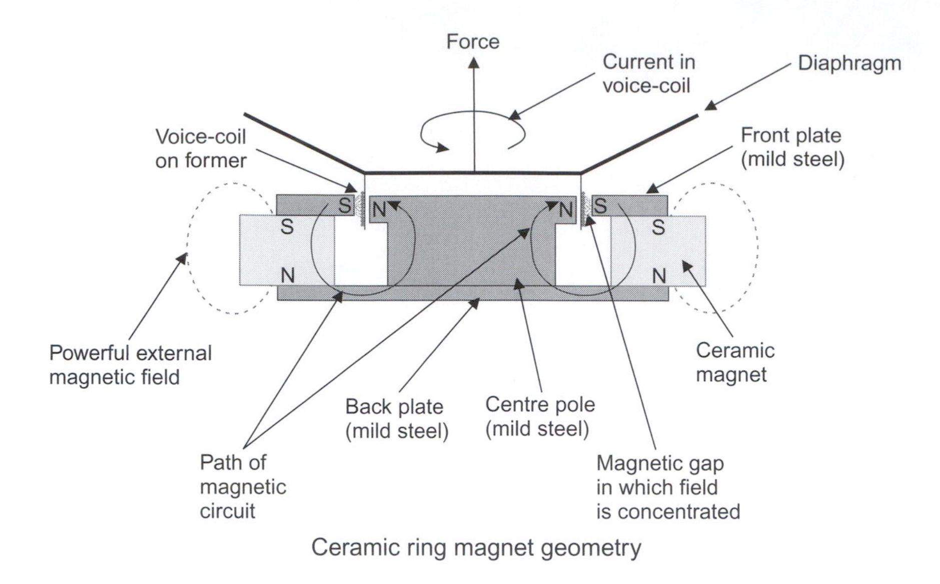

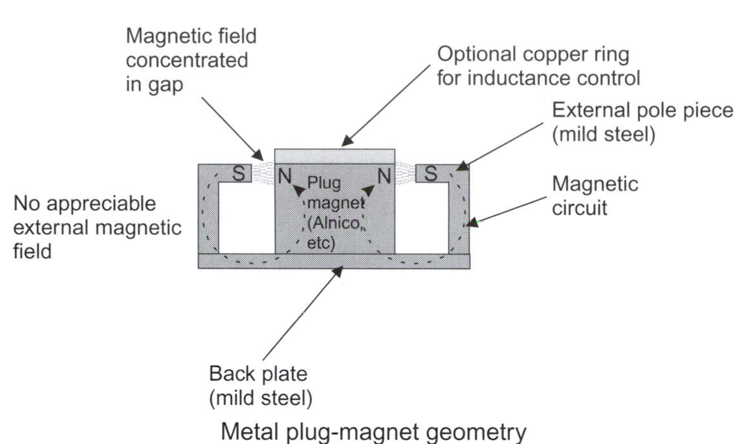

FIGURE 12.2
Typical motor topologies.

Early permanent magnets were often made from iron and chromium. Aluminium, nickel and cobalt were variously used in the early 1930s, alloyed with iron in different combinations, and the three together gave rise to the name Alnico. In the 1970s, the civil war in the Congo (then Zaire) created a big hole in the production of cobalt, whose price rose astronomically in a very short period of time. This led to the use of ferrite materials, known as ceramic magnets, which had their strengths and weaknesses. More recently, "rare earth" magnets, principally made from neodymium and samarium-based alloys, such as neodymium with iron and boron, or samarium and cobalt, have led to very lightweight magnets, and opened a door to new magnet shapes and magnetic field designs. The basic concept of two different magnet structures is shown in Figure 12.2.

The magnetic circuits are designed to concentrate the magnetic field in a circular gap, as shown in Figure 12.3. In this gap is inserted the voice-coil, which receives the electrical drive current from the power amplifier. This current produces its own alternating magnetic field, whose phase and amplitude depend on the drive signal. The variable field interacts with the static field in the circular gap, and creates a force which either causes the voice-coil to move into or out of the gap. Of course, a means is required to maintain the coil centralized in the gap, and this is achieved by the use of centering device, or inner suspension, which is still often referred to as a spider for reasons which should be clear from an inspection of Figure 12.4. A more typical modern device is shown in Figure 12.5. A chassis, also known as a frame or basket, supports the whole assembly and enables it to be mounted on a front baffle. The cone is connected rigidly to the former upon which the voice-coils is wound, and is also connected more or less at the same point to the inner suspension. At the chassis' outer edge the cone is attached via a flexible outer suspension, or surround, which may take the form of half-rolls, corrugations, or pleats. A dust cap is then normally placed in the apex of the cone in order to prevent the

(a)

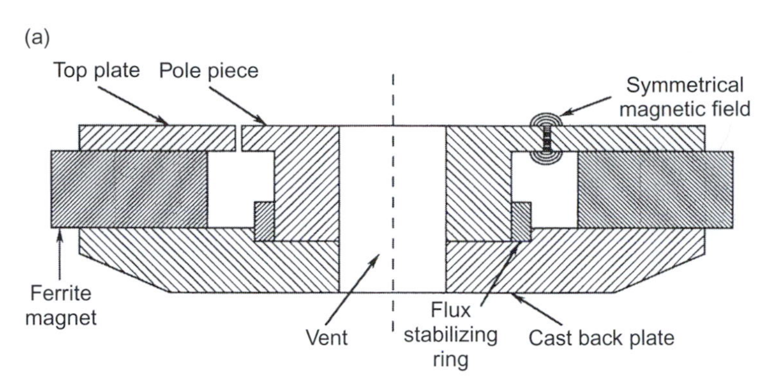

Magnet geometry for concentrating the flux in the voice-coil gap—section view. (Courtesy JBL Inc.)

(b)

The circular voice-coil gap—perspective view

FIGURE 12.3
The voice-coil gap.

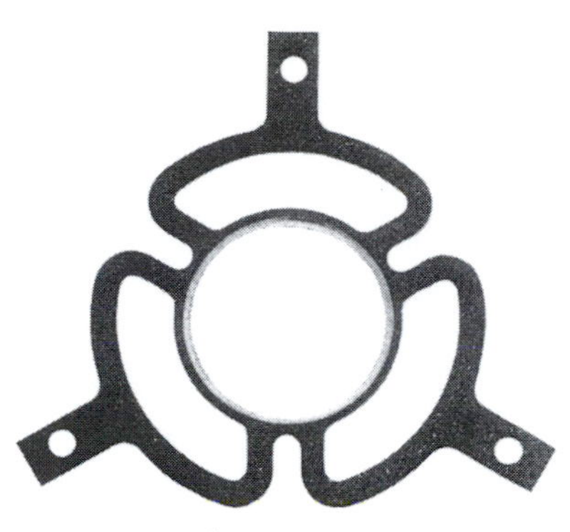

FIGURE 12.4
An early centering device—a "spider," some of which had more legs than the one shown.

Typical centering device in corrugated paper or fabric

FIGURE 12.5
A typical modern centering device, or inner suspension.

ingress of dust and any abrasive dirt, and may also be used as an air pump to cool the voice-coil and gap when the cone assembly moves in and out.

Cones

A three-way loudspeaker system consisting entirely of cone drivers is shown in Figure 12.6. Although the cone drivers all follow the above principles of construction, their designs are very different. In Chapter 11 it was explained how, at low frequencies, the electro-acoustic conversion efficiency is low, because the air tends to move out of the way of the vibrating cone. The result of this is that for a reasonable on-axis sensitivity, the cone needs to be quite large. In the loudspeaker shown in Figure 12.6, the low-frequency cone is of nominally 12 inch diameter (300 mm), although the effective radiating area is only just over 10 inches because the surround does not contribute much to the radiation. The cone needs to be rigid, because if it breaks up into non-uniform movement, phase cancellations will occur at some frequencies and the subsequent frequency response will not be flat. In some cases, cone break up can be used to extend the frequency response, and can be used in musical instrument loudspeakers to create desirable coloration in the sound, but, for flat uncolored low-frequency responses, the piston which is pumping the air needs to maintain its rigidity. Some of the ways in which a cone can break up are shown in Figure 12.7.

FIGURE 12.6
A full-range, all-cone driver loudspeaker system—the JBL L100.

Once cones exceed a diameter of about 18 inches (460 mm) it can become difficult to maintain their rigidity. The gain in efficiency due to the large radiating areas of big drivers can rapidly be offset by the greater proportional weight needed to keep them rigid. Many designers favor multiple smaller drivers, to the use of single larger drivers, partly because they feel that they can keep these better controlled. It is unusual to see drivers of greater than 18 inch diameter, although they do exist, as shown in Figure 12.8. Sandwich cones, honeycomb cones and Kevlar and carbon fiber and metal cones have all been employed in attempts to maintain cone rigidity, and consequently the pistonic movement. In each case, the cones exhibit different

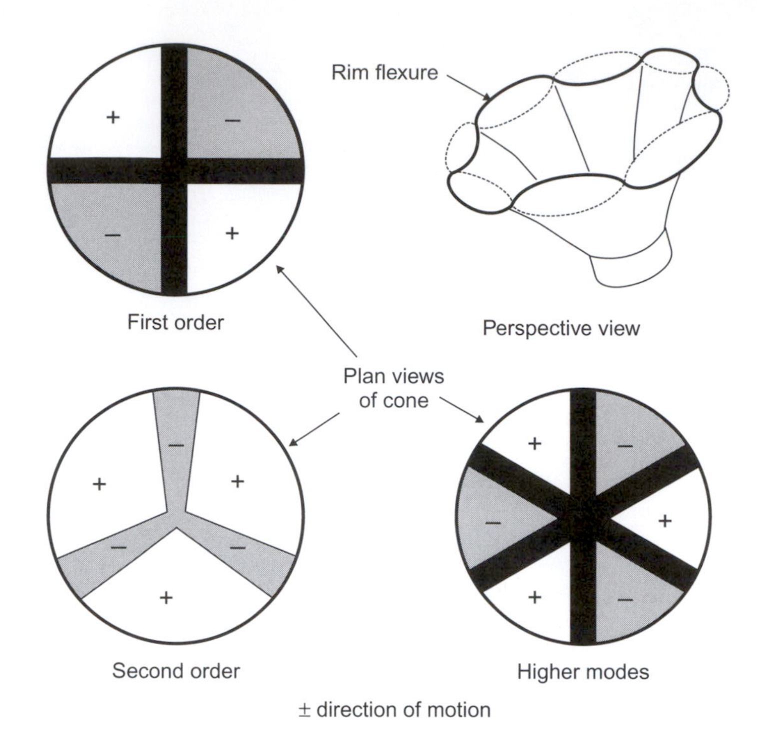

FIGURE 12.7
Bell-mode break-up in cones.

characteristics above certain frequencies, so the suitability of each material or construction may depend upon the upper frequency limit to which a driver will be used. Solid cones have also been used, but they can introduce as many problems as they solve, and they are not so obviously beneficial as they may at first appear to be. One problem with many of these approaches has been that the near-perfect rigidity has improved matters at low frequencies but has only pushed the resonances up in frequency, rather than eliminating them. When stiff structures do break up they tend to do so much more severely, so crossover frequencies must be chosen well away from the break-up frequencies if coloration is to be avoided. It has proved difficult to achieve uncolored mid-range responses from highly rigid low-frequency driver cones, so they are often best restricted to use at low frequencies only.

Bextrene, a mixture of polystyrene and neoprene, was pioneered as a cone material by the BBC, in the UK, as far back as the 1960s. This was originally researched largely to find a solution to the inconsistency problems encountered in the manufacture of paper pulp (cardboard) cones.

Paper, being made from wood, which is a natural material, can suffer from the problem of all natural organic materials; they are not homogeneous substances so they tend to vary from batch to batch. Bextrene was well damped and resisted break-up to relatively high frequencies. Designs have been employed using Bextrene cones which have used 12 inch (300 mm) bass drivers up to crossover frequencies beyond 1.5 kHz with little mid-range coloration. Polypropylene has since been developed for use as a cone material, offering even more consistency, long-term stability and sensitivity. However, opinions vary about the sonic neutrality of polypropylene-coned drivers.

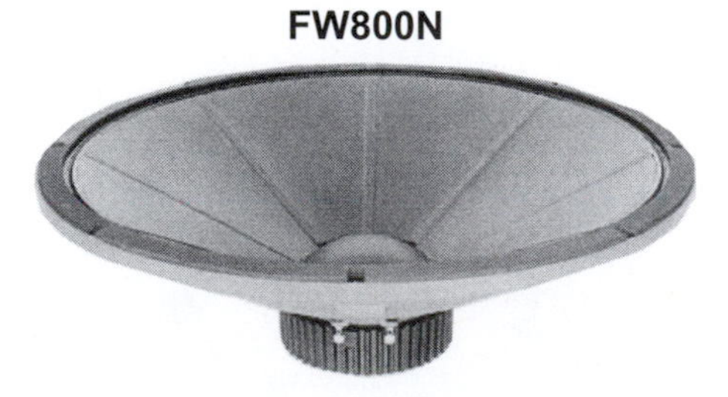

FIGURE 12.8
A very large loudspeaker—a 30 inch (800 mm) low-frequency loudspeaker with radial reinforcing ribs to augment the cone rigidity.

The original loudspeaker of Rice and Kellogg used a paper cone. Quite remarkably, despite all of the modern developments in materials and construction, paper pulp cones, and even folded and seamed paper cones, are still in use in all levels of performance ranges. Paper pulp cones are made by drawing a slurry (wet mix) of paper fibers through a fine screen in the required shape. The resulting cone is then cured and dried before being cut to the exact size required. This "old fashioned" material still exhibits excellent characteristics of high rigidity and high internal damping, and these are two things which normally are contradictory inasmuch as the augmentation of either one usually tends to reduce the other. At low frequencies the rigidity is necessary to maintain piston action, but once any break-up

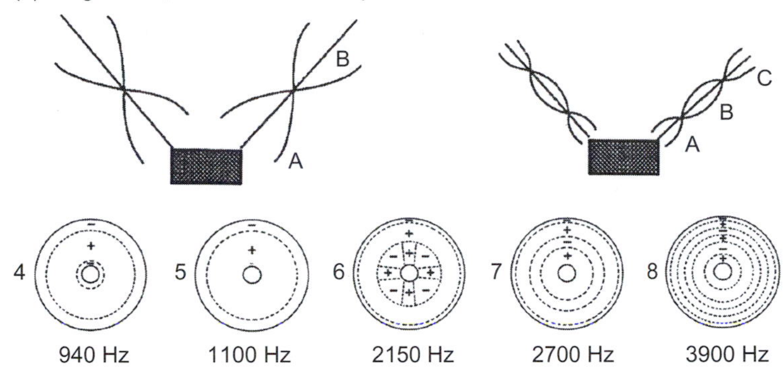

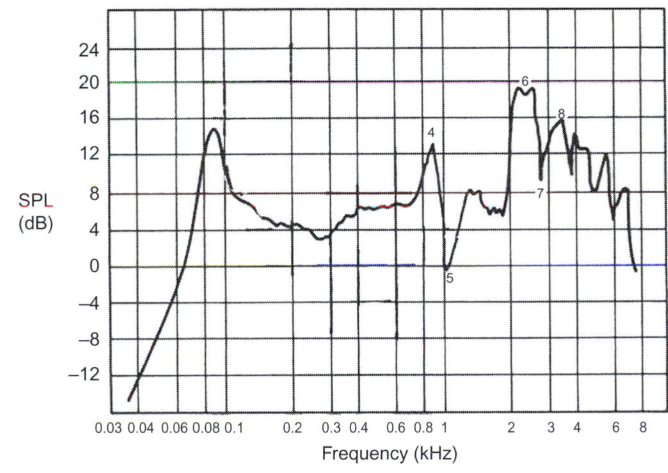

FIGURE 12.9
Concentric modes in loudspeaker cones.

does begin, the internal damping of the cone material needs to suppress the waves which travel as shown in Figure 12.9, which would cause peaks and dips in the frequency response. The bass cone shown in Figure 12.6 has been further treated on both sides with a damping material known as Aquaplas, which also adds some mass, and this lowers the free-air resonance of the driver.

Despite the fact that many drive units of similar specification which employ different cone materials may perform very closely in objective measurement, there is strong evidence that some very similarly performing drivers do not *sound* the same. In very high quality loudspeakers, paper pulp is still a favored material, despite its sensitivity to humidity and batch-to-batch variation. Colloms (1997) refers to the well-balanced characteristics of paper pulp, together with the fact that its properties and manufacturing techniques are well understood, as strong justification for its continued use. He notes how some high-loss materials also may tend to lose, or mask, fine musical detail. The authors of this chapter have noticed a loss of reverberation detail when substituting some synthetic cones for paper cones used up to 1 kHz, and have received comments from professional users about guitar strings not sounding as new when heard via the synthetic cones as when heard via high quality paper pulp cones. The Celestion loudspeakers company still produce the exact model of guitar loudspeaker which was made famous in the Vox AC30 guitar amplifier of the 1960s. This blue-chassised driver has resisted all efforts to update its construction, yet still maintain its highly desirable sound qualities. A great number of musicians claim to hear their guitars more "clearly" via paper cones.

The observations about guitar strings could suggest a harmonic enhancement due to non-linear distortion products enriching the sound, but harmonic distortion could not explain the increased sensation of low-level detail and reverberation. Synthesizing natural sounding reverberation is not something which one would expect from the addition of harmonic distortion. It seems probable that the guitar strings are benefiting from the same characteristics which are enabling the greater resolution of reverberation and room effects, and paper pulp seems to be a good performer in this respect. Work is currently under way to investigate the possibility of intermodulation distortion contributing to the low-level detail loss with certain materials, given rise to by non-linear hysteresis effects connected with the damping action. Intermodulation distortions result in harmonically *and* non-harmonically related products which together tend to produce a noise signal below the music.

As the frequency rises, two things begin to affect the performance of a cone driver. It was shown in Figures 11.3. and 11.4 how the directivity of a cone, or any pistonic radiating surface, narrows as the wavelength become small compared with the circumference of the radiating area. However, as frequencies rise, the mass of the moving assembly gradually tends to oppose more strongly the force which is trying to move it. There comes a point where the increasing efficiency of radiation due to the greater radiation resistance provided by the air as the frequency rises can no longer compensate for the mass effects of the moving parts, and the power response of the driver begins to roll off. For a 15 inch (380 mm) loudspeaker, 1 kHz is about the upper limit of either its flat response range or the acceptability of its narrowing directivity. Nevertheless, the directivity narrowing may not be too severe if the center section of the cone begins to decouple itself from the outer section, as often happens—either by design or accident. Eight inch loudspeakers (200 mm) can work well up to 2 kHz, or more, but the compromises which must be made if their responses are to be extended at the bottom end may begin to degrade the higher frequency performance. In the loudspeaker shown in Figure 12.6, the 12 inch (300 mm) low-frequency driver, having a free-air resonance of 25 Hz, is used up to a frequency of around 1 kHz, at which point the crossover begins to divert the signal towards the 5 inch (125 mm) cone. Five octaves is just about the limit of the bandwidth of a cone driver if very high quality, wide directivity, minimum-compromise sonic performance is required.

In smaller loudspeakers, cone rigidity is much easier to achieve, therefore much lighter moving assemblies can be employed which can exhibit good efficiency without the massive magnet assemblies needed for the low-frequency drivers. The low-frequency driver shown in Figure 12.6 has a sensitivity of 89 dB for 1 watt input at 1 meter distance, yet with a much smaller magnet but a lighter moving assembly, the 5 inch mid-range driver has a corresponding sensitivity of 94 dB. This sensitivity increase can be important, because smaller loudspeakers cannot lose the waste heat as easily as can large loudspeakers, which was one driving force behind the developments of domes. The tiny tweeter cone shown in Figure 12.6 is only 1½ inches (38 mm) in diameter, but handles frequencies from 4 kHz to almost 20 kHz, so in this design of loudspeaker cabinet all frequencies from below 40 Hz to almost 20 kHz are handled by paper cones.

Surrounds

The outer suspension, or cone surround, serves two functions. The most obvious is to maintain the outer edge of the cone stable during the rapid movements along the front-back axis, but surrounds also serve to absorb vibrational waves which propagate from the voice-coil in the manner shown in Figure 12.9. A selection of surround designs are shown in Figure 12.10. The surrounds are variously formed from a continuation of the cone material or from separate materials attached with adhesives. Polyurethane foam, butyl rubber, nitryl rubber, cambric (pronounced Kāymbrik—a woven linen or cotton fabric), or other treated fabrics are frequently used as cone surround materials. PVC is also sometimes used. Figure 12.10(a) shows a half-roll surround. These are usually made from synthetic foams or rubbers. They allow long travel because they tend to stretch in an elastic manner over a considerable range of movement, and hence give rise to little distortion which would be caused by restraining the cone travel. When the materials are carefully chosen they can effectively absorb the resonant modes which pass radially along the cone, thus avoiding standing waves within the materials of the cones. The *double* half-roll surrounds, shown in Figure 12.10(b) are usually made from treated cloths. They are more rugged than the single half-rolls, and find much use in sound reinforcement and musical instrument amplification, but tend in general to be used in less long-throw loudspeakers with higher resonant frequencies than those which usually employ the single half-rolls. The extra stiffness of the double half-roll surrounds both adds to their ruggedness and increases the resonant frequency of the moving system as compared with single half-roll devices. Figure 12.10(c) shows a concertina (accordion) surround. These are often pressed or moulded as part of the cone. They can allow extended travel, but unless very carefully damped can give rise to resonance problems. They also tend to be

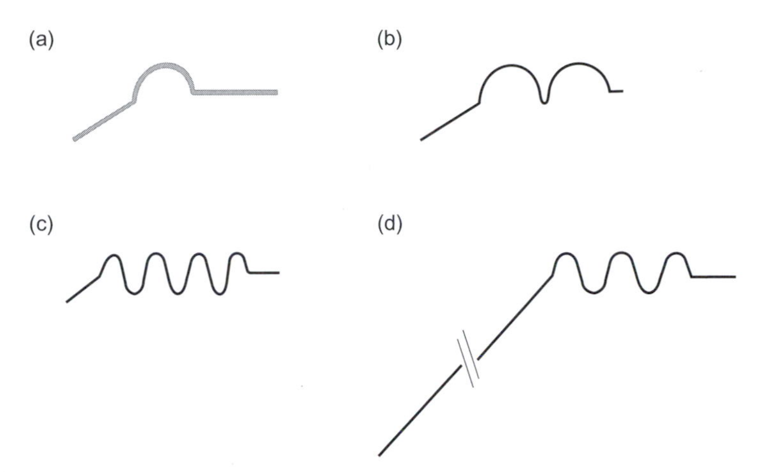

FIGURE 12.10
Cone surround variants. (a) Half-roll of polyurethane foam—low stiffness (high compliance) for long travel, but requires precise choice of centering device for controlled linearity. (b) Double half-roll cloth—shape of rolls can precisely tune the stiffness characteristic, (c) Multiple roll accordion pleat—long travel but prone to rim-resonance response dip problems (see Figure 12.11). (d) One-piece cone/surround with treated edge—stiff, non-linear suspension, provides HF resonance peak. Principally used for musical instrument loudspeakers to prevent over-excursions of the cone.

stiffer than the single or double half-roll surrounds, and are therefore not often found on low-resonance designs.

Surround *materials* are also a specialized subject. Some of the foams which are commonly used can deteriorate much quicker than expected in sunlight or polluted atmospheres, and can also suffer from insect damage. Nevertheless, in clean temperature-controlled environments such as exist in many sound control rooms they can easily last for 20 years or more, and can be replaced by skilled artisans without removing the cone from the chassis. Plasticized PVC is another material which has been employed with success, and is found on some mid-range cones where its properties efficiently absorb the waves reaching the cone edges. In sealed cabinets, the surrounds must also resist the differential pressures between the inside and the outside of the cabinet, and this fact can preclude the use of some foams in certain cases. In order to fulfil all of the demands made of them, surrounds are quite specialized devices. The surrounds not only need to be selected according to the density of the cone material, but also the principal frequency ranges over which a driver will work, and the excursion limits within which they must allow relatively unrestricted movement. For wide-range drive units, the compromise choices are not simple. Figure 12.11 shows a response dip due to an antiphase movement of a surround. The fact that the dip exists in the response of such a high quality driver suggests that solutions to the dip problem would have compromised the overall response to a greater degree.

Rear suspensions

The prime function of the rear suspension, or spider, is to maintain the coil centralized in the magnetic gap. On its inner edge it is connected to the voice-coil former, and at its outer edge it is glued to the chassis. Modern suspensions, such as the one shown in Figure 12.5, are usually made from phenolic resin impregnated cloth, hot pressed into shape. Care has to be taken in the design of the corrugations to ensure that movement in one direction is not favored over the other direction, because an asymmetrical movement would give rise to non-linear distortion. Some designs have employed double spiders, mirror imaged, in order to ensure symmetrical linear travel of the cone. A double spider arrangement is shown in Figure 12.12. These are fine in vented magnet designs, but a complication in double suspension designs arises if they must allow air to pass through to cool the voice-coil in designs that do not have vented magnet systems. As with the corrugated concertina surrounds, the corrugated inner suspensions can suffer from resonance problems unless they are carefully "tuned."

The stability of the inner suspensions needs to be very good because the gap between the voice-coil and the magnet can be less than half a millimeter, even with a relatively large cone and coil. With large heavy cone/coil assemblies, the suspensions can become stretched if the drivers are stored in a

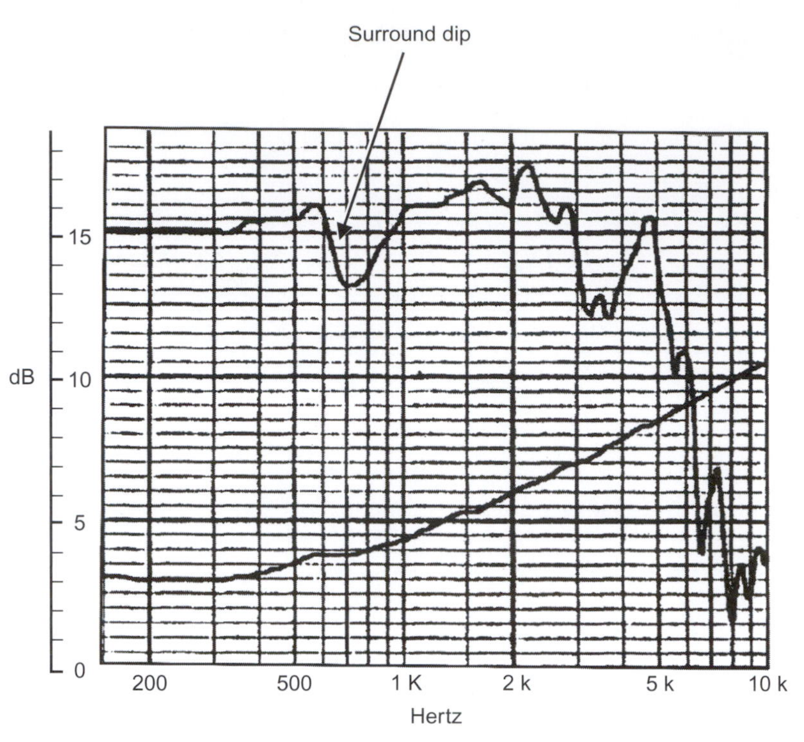

FIGURE 12.11
Response dip due to a surround resonance.

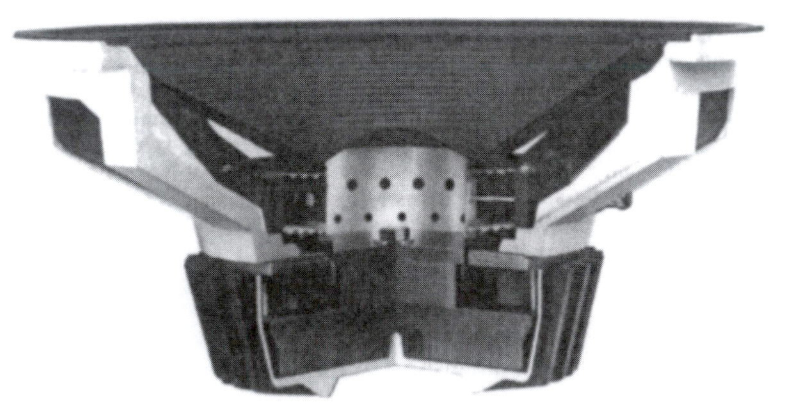

FIGURE 12.12
Cut-away view of a 380 mm loudspeaker with double spider (centering device) construction. (Cetec-Gauss Inc.)

horizontal position without adequate support for the cone (which effectively means in the manufacturers' shipping boxes). Likewise, complete loudspeaker systems should not be stored on their backs or "cone sag" is likely to result. Once mounted vertical again, the suspension may have "set" to a new equilibrium position which is not in the center of the cone's axial travel, hence the cone excursion will be prematurely limited in one direction. In many cases, this cone sag cannot be corrected by any simple means, and so storage conditions likely to give rise to it should be avoided.

The inner suspension is very critical because it usually provides the main restoring force for centralizing the cone in the axial as well as the radial directions. Over 50 years ago, Briggs recognized not only the third-harmonic distortion-producing mechanisms of badly designed suspensions, but also the fact that inadequate suspensions could give rise to distorted transient responses (Briggs, 1955). Over 40 years later, Colloms wrote about work done at KEF which correlated well with his own experiences that some inner suspensions could give excellent low distortion results on sine waves and the more open bass waveforms of orchestral music, but could be slow in responding to the more percussive bass

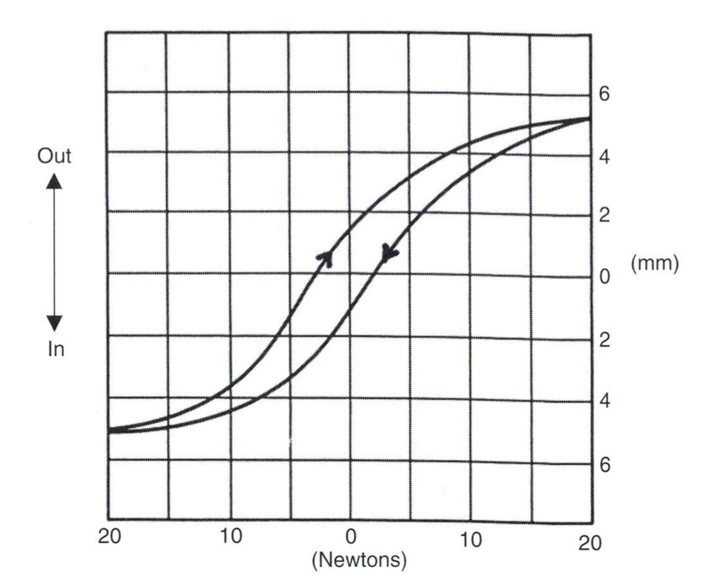

FIGURE 12.13
Hysteresis curves—the hysteresis curves represent processes which are cyclic but where the forward and backward processes do not follow the same path. "Hysteresis" is from the Greek word meaning "to lag behind."

sounds found in much modern music (Colloms, 1997). The KEF findings had emerged from investigations into the different low-frequency measurements obtained via the use of steady-state or impulsive signal sources. (The differences have been attributed to hysteresis in the suspensions, shown diagrammatically in Figure 12.13.)

In some very small cone loudspeakers, designed only for high frequency use, the additional complexity of the use of an inner suspension can often be omitted, the external surround being sufficient to maintain the cone in a central position, and thus avoiding all the inner-suspension-related problems which larger heavier cones must endure.

Although the suspension systems, both surrounds and spiders, are mechanically essential in low- and mid-frequency cone loudspeakers, towards their excursion limits they all begin to give rise to third-harmonic distortion due to the approaching elasticity limits where they become less compliant (i.e., more stiff). Once they no longer obey Hooke's Law (the law governing the relationship between force and compression [or expansion] of a simple spring) they can produce quite a number of undesirable artifacts.

FIGURE 12.14
A 15 inch (380 mm) loudspeaker manufactured by the British company Volt Loudspeakers Ltd, employing an external chassis for improved heat dissipation.

The chassis

Loudspeaker chassis provide a frame for the mounting of the magnet system and inner and outer suspensions. They can be made of plastic, or pressed or cast metal. Metal is preferred when power levels are high because it helps to conduct the heat away from the magnet assembly. Cast metal is usually preferable to pressed metal on grounds of stability and dimensional accuracy; cast aluminium being the material of choice for most large professional bass units. With magnet assemblies weighing 10 kg or more on some 18 inch loudspeakers, the chassis ("frames," "baskets") need to be strong to withstand shipping shocks without disturbing the centralization of the sub-half-millimeter coil clearances. Also, coil temperatures of 250 °C, or over, need to be withstood and dissipated without warping. Unfortunately, there is a conflicting requirement between the strength of the chassis and the need not to impede the free movement of air between the cone and the inside of the box (or outside the box in the case of external chassis designs, as shown in Figure 12.14). The chassis therefore needs to be as strong as necessary for support, whilst being as open as required for non obstruction of the air adjacent to the cone. It is also important that it should not suffer from resonance problems, and so needs to be well acoustically damped.

Once a cone and coil assembly is mounted in a chassis at the surround and the inner suppression attachment points, its mass will resonate with the compliance of the suspension systems to determine the free air resonance of the driver. The free air resonance normally defines the lower response limit of a loudspeaker system in any given volume or design of cabinet. However, there do exist some special designs of loudspeaker systems which drive the bass units through their resonances, but they are rare, and need electronic compensation.

The voice-coil assembly

The voice-coil former is normally attached to a cone at some point between its apex and a point mid-way between the apex and the perimeter. The coil former, of cylindrical design, must be mechanically stable under high degrees of vibration and temperature changes, neither deforming in circularity nor expanding or contracting to any significant degree. Without the required stability, the coil or the former could rub on the sides of the magnetic gap and make undesirable scraping noises. Although paper was used to great effect on early low-power loudspeakers, modern day, high-temperature voice-coils need to be bonded to thermally stable formers with thermally stable adhesives. Glass fiber has been used as a former material, but polyamides are now more normal (Nomex, Kapton, etc). Aluminium has also been used, but metal formers can suffer from eddy current problems by acting as a shorted turn in the alternating magnetic field, and can also conduct very high temperatures to the necks of the cones, where charring, melting or softening can occur, and adhesives may also be caused to fail.

The coils are almost exclusively made from copper or aluminium, although silver has also been used, reportedly to some sonic benefit (Colloms, 1997). Copper offers lower resistance, but aluminium offers lighter weight. Which one is the most appropriate depends on many other design factors, but both materials have been used at either frequency extreme—there are no hard and fast rules. Copper-clad aluminium is another option, which simplifies the soldering problems that may be encountered with pure aluminium. Round wire and rectangular section (ribbon) wire are also options. The ribbon wire packs more densely, eliminating the gaps between the adjacent round wires, but round wires offer much simpler winding processes. To prevent short circuits between turns, the copper wire is insulated with a heat resistant lacquer, and aluminium wires are anodized, which creates a layer of non-conducting aluminium oxide on the surface. In either case, the wires may also be coated with a thermosetting adhesive before winding, which, after curing, helps to render the entire assembly more rigid.

The most appropriate diameter of a voice-coil is also the subject of compromise. Larger diameter voice-coils have more surface area for any given length of wire than coils of less diameter, at least for any given gap depth, and can therefore lose heat more easily. They can also help to stiffen a cone by driving it in a more evenly distributed manner. However, if the coil is too big in diameter, the center of the cone can begin to decouple from the coil, which can cause strange frequency response and directivity problems. Proponents of small diameter voice-coils cite advantages of deeper coils in longer gaps giving them design advantages such as deeper gaps with less magnetic material at no thermal dissipation cost. As with so many things in loudspeaker design, the art of the science is finding the best compromise for any given situation. Low-frequency efficiency versus high-frequency extension, for example, can dictate the optimum choice of coil material as copper or aluminium. Some manufacturers have also great expertise in using particular design concepts or manufacturing processes which suit certain ways of doing things better than others. Electro-Voice, for decades, continued to make excellent 15 inch (380 mm) loudspeakers with 2½ inch (62 mm) voice-coils, whilst JBL had long since moved to 4 inch (100 mm) coils for their equivalent designs, but using different magnet topologies.

Magnet systems

The voice-coil and the magnet form the motor system of a moving-coil loudspeaker. As mentioned above, either one depends on the other in order to produce the required force to drive the cone in the required direction at the required speed, as instructed by the electrical input signal. Magnets are a huge and complex subject in themselves, so only some of the more fundamental aspects of their behavior can be dealt with here, but the Bibliography at the end of this chapter gives references to some excellent further reading. Cost, however, perhaps plays a bigger part in the choice of magnet systems than in any other aspect of the design of cone loudspeakers. Some of the best magnetic materials ever developed for loudspeaker use used cobalt, which, as mentioned p. 343, rose in price by over 2000% in a very short period of time when the civil war in Zaire (the former Belgian Congo) began in the 1970s, because Zaire was the world's largest producer of cobalt. This drove many manufacturers to use ferrite

(a)

Magnetic return path
Pole piece
Alnico ring magnet
Center vent path

(b)

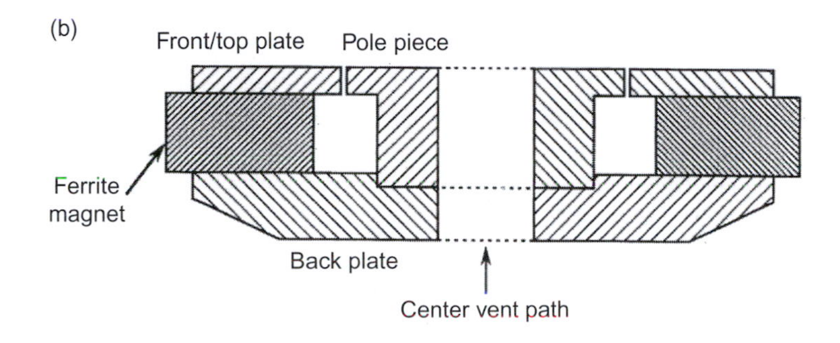

(c)

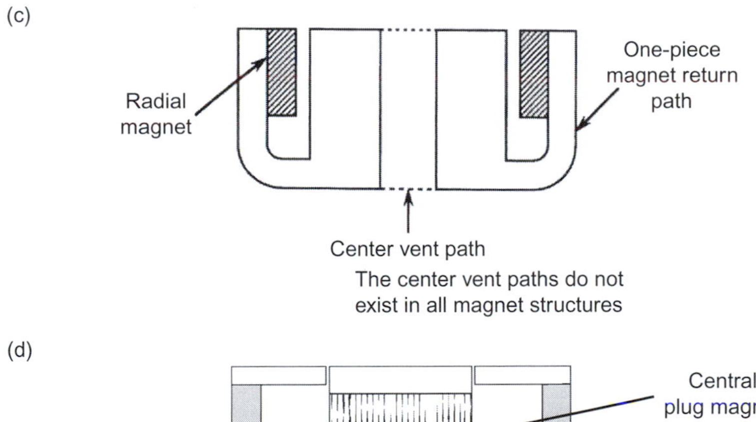

(d)

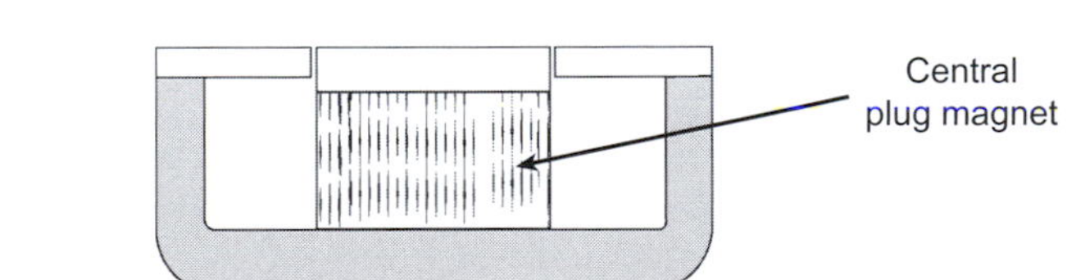

FIGURE 12.15
Basic magnet structures: (a) Alnico ring magnet; (b) Ferrite ring magnet; (c) Radial, high energy magnet (neodymium, etc); (d) Alcomax plug magnet; (a), (c) and (d) do not have any appreciable external magnetic fields.

materials, predominantly barium ferrite, which had been developed for deriving the static magnetic fields necessary around cathode ray tubes in television sets.

The design of typical Alnico (Aluminium, *nickel*, cobalt and iron), Ferrite (ceramic), neodymium and Alcomax magnets are shown in Figure 12.15, from which the geometrical differences are obvious (although the Alcomax and Alnico geometries shown are interchangeable). Many modern loudspeakers use neodymium alloys, and an alloy of samarium and cobalt is also finding use in loudspeaker designs. These materials give enormous magnetic strength for their weight, and they have given rise to further changes in magnetic geometry. The ferrite materials are very resistant to loss of magnetism due to time or heat stresses, but they exhibit powerful stray magnetic fields and can pose some difficulties in achieving the desired magnetic field geometries. Alnico is somewhat less durable, but can allow designs enabling very compact and concentrated magnetic fields. Strength for strength, neodymium magnets are much lighter than either ferrite or cobalt alloy magnets, but can be relatively easily demagnetized at relatively low temperatures, and cannot withstand 250 °C voice-coil temperatures without permanently losing some of their magnetic strength. The metal magnets are good conductors of electricity, but the ferrite magnets are ceramic materials, and hence are electrically non-conductive. The non-conducting nature of the ferrite materials can be a problem unless careful measures are taken to use other means to avoid unnecessary and undesirable flux modulation effects. Iron of high magnetic permeability is used in the magnetic structures shown in Figure 12.15 to complete the magnetic circuit, and to achieve the correct shape of field and density of magnetic flux in the gap in which the coil is positioned. The type of iron used is normally a mild steel of low carbon content, but when very high flux densities are required, a material known as Permendur is often used, especially in compression drivers. Permendur is an iron-cobalt-vanadium alloy, which is very hard and difficult to work, but its magnetic properties, when required, may demand its use.

There are many people who consider the metal magnets to be capable of better sonic performance than the ferrite magnets, citing better resolution of fine detail with materials such as Alnico and the neodymium alloys. It is hard to find evidence of tests which rigorously compare the differences in the magnetic materials only, because the required structural differences needed to get exactly the same magnetic field in the same gap can lead to other changes being necessary. Nevertheless, there is a tendency for many of the highest resolution devices to use metal magnets, and explanations have been put forward to suggest that the magnetic domain jumps which take place in non-conducting materials can give rise to effects not dissimilar to digital quantizing distortion. These jumps are smoothed out by large eddy currents flowing in the electrically conducting magnets. In some loudspeaker designs the central pole-piece of the magnetic assembly is fitted with a copper ring to provide a very low electrical resistance—less than that of steel—to effectively short out any flux-modulation currents.

As shown in Figure 12.15(a), (b) and (c), the entire magnet assemblies in many high-power low-frequency drivers have cylindrical holes through their central axes to allow air to be pumped through by the dust cover (dome) which caps the voice-coil former on the outer face of the cone. In some other designs, the spider (inner suspension) and the dust cover are of an open weave nature, to allow hot air to pass to the outside of the cabinet. As mentioned earlier, the voice-coils can get to temperatures above 250 °C in some cases, and this heat needs to be dissipated as quickly as possible, not only to prevent the burn-out of the coil but also to avoid overheating and weakening of the magnet. Furthermore, the resistance of a copper wire rises by about 0.6% for every degree of temperature rise, so a changing voice-coil resistance will affect the sensitivity of the motor system and may lead to signal compression.

Ferrofluids

The problem of the conduction of heat from the voice-coil to the pole piece of the magnet assembly can, in some instances, be augmented by the use of a ferrofluid. Air is not a good conductor of heat, so much of the heat is transferred from the voice-coil to the magnet assembly by radiation alone. In the 1970s, ferrofluids began to be introduced which were liquids with magnetic particles in colloidal suspension. The magnetic field holds the ferrofluid in the gap, and the good heat-conduction properties of the ferrofluids aids the cooling of the voice-coil by means of a thermal bridge to the magnet assembly. The viscosities of the ferrofluids can be adjusted according to the circumstances of use, a fast-moving tweeter needing lower viscosity than a mid-range driver if viscous damping effects are to be avoided. However, ferrofluids are rarely used in high-excursion low-frequency drivers, because the shearing effects of the large axial movements of the coil tend to create non-linear movement due to the non-laminar flow of the fluids. In high-frequency drive units, the ferrofluids can be advantageous in damping some mechanical resonances if the viscosities are chosen appropriately.

The complete system

The moving coil cone loudspeaker is quite remarkable in its degree of versatility of application, and has formed the backbone of loudspeaker system design since its first application in 1925. Despite all the technological developments and improvements in materials, Rice and Kellogg, if still alive today, would almost certainly be able to explain the workings of any modern moving-coil loudspeakers from simple inspection. When they filed their patent, they already had described, in principle, almost everything that we can find in a modern driver. Loudspeaker design is a science; but there is art in deciding about the best compromise points for what are imperfect devices. For example, whether a complete system should use wider range drivers and fewer crossover points, or narrow range drivers and more crossover points, is a question that may depend more on the circumstances of use than any single measurement at a fixed position. Things such as the room acoustics, the music, the required timbral fidelity and the listening distance may all influence a design, but these things will all be discussed later.

DOME LOUDSPEAKERS

In general principle, a dome loudspeaker is a cone loudspeaker with the voice-coil having the same diameter as the diaphragm. The diaphragm is also usually inverted, to be convex rather than concave to the exterior. A mid-range dome loudspeaker is shown in Figure 12.16. The development of dome loudspeakers largely grew out of the problems surrounding how to lose heat from the small voice-coils of mid- and high-frequency drivers at high power levels. A 1½ inch (38 mm) cone tweeter, with a coil of only 12 mm diameter simply cannot lose heat quickly enough to prevent itself from burning up at power levels much above 10 watts, because the heat production is all confined in a small space. However, if the coil were to be made the same diameter as the diaphragm, a much greater surface area would be available for heat loss. As light weight is important for sensitivity, the coil former can be kept to a minimum length if the dome diaphragm is convex because it will remain clear of the pole piece of the magnet assembly, as shown in Figure 12.17.

Despite the "common sense" belief of many people that domes radiate over a wider angle than cones, this is not so. It is important not to confuse domes with pulsating spheres. As shown in Figure 12.18, a pulsating sphere, which would radiate radially, moves by expanding and contracting in three dimensions, whereas a dome simply radiates as a piston, because it moves in one direction only—along its axis of movement. What is more, when a cone begins to decouple from its voice-coil, it does so from the outer parts first. The central part, nearest to the voice-coil, always remains under the control of the coil, so the radiating area concentrates towards the center of the diaphragm as the frequency rises, which is exactly what is needed to maintain its directivity. That is, the radiating area reduces in diameter as the frequency rises, which can be desirable. Conversely, with a dome, it is the *center* of the radiating area which is furthest from the coil, so as the frequency rises the tendency is for the voice-coil to keep control over the *outer perimeter* of the radiating area, whilst the center of the diaphragm decouples itself, as shown in Figure 12.19. This leads to a ring radiator, which has very peculiar directivity properties, so domes, if not applied very carefully and below their break-up frequencies, can actually have less smooth directivity than cones.

Hard and soft domes

The diaphragms of hard domes are usually either made from phenolic-resin impregnated cloth, aluminium, titanium, beryllium, carbon fiber or other similar materials with very high strength to weight ratios. Soft domes are typically made from moulded cloth which has been treated with a viscous damping material, usually a synthetic rubber. Other types of domes are also sometimes used. Hard domes

FIGURE 12.16
A cut-away view of an ATC 3 inch (75 mm) soft dome driver. The British company ATC pioneered the development of this type of high-output mid-range dome.

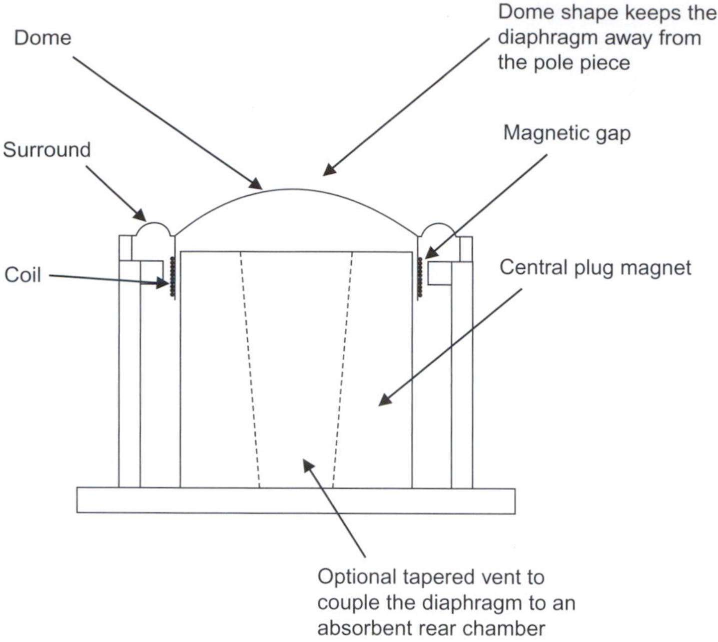

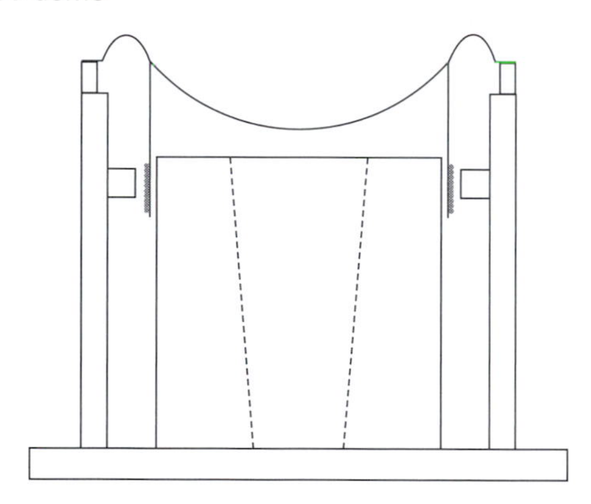

FIGURE 12.17
A dome as a piston-in principle, if the voice-coil former were to be extended, and the dome inverted as in "b" the sensitivity would drop due to the extra weight, but no material change would take place in terms of the radiation pattern (directivity).

are usually restricted to use at high frequencies, above 3 or 4 kHz, because their resonances are difficult to control in diaphragms of sufficient diameter to radiate useful power at much lower frequencies. In rare cases, hard domes of 3 inch (75 cm) or more can be found operating down to frequencies as low as 800 Hz, but in order to maintain piston action up to very high frequencies before the first break-up modes occur, materials such as titanium or beryllium need to be employed. As beryllium is difficult to work with and its vapor is highly toxic, the production of diaphragms out of this metal is an expensive process. Such units have found favor in the design of domestic high-fidelity loudspeakers, but are rarely to be seen in studios. At higher SPL, when they *do* break up into separately radiating sections, they tend to do so suddenly and in a sonically most unpleasant manner, and produce non-linear distortion products quite differently from soft domes.

Soft domes, on the other hand, can be used down to around 400 Hz, but many exhibit a hysteresis type of response, as discussed earlier, and the same comments apply. The lagging response shown in Figure 12.14 can tend to mask low level detail. The "rigid" domes have been developed partly in response to this problem.

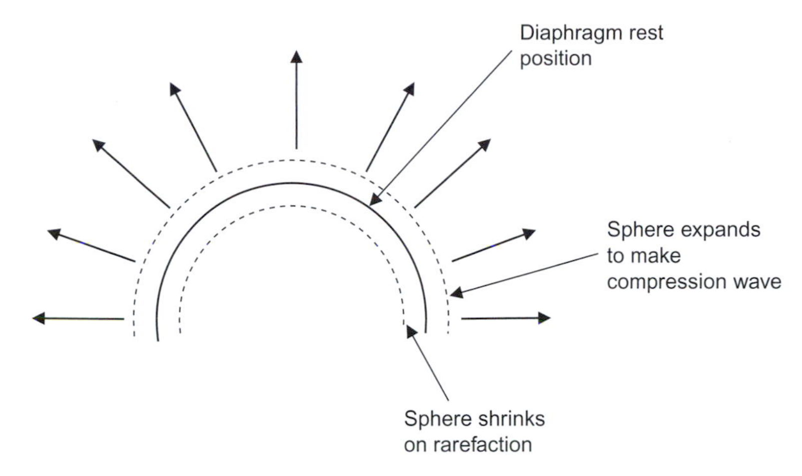

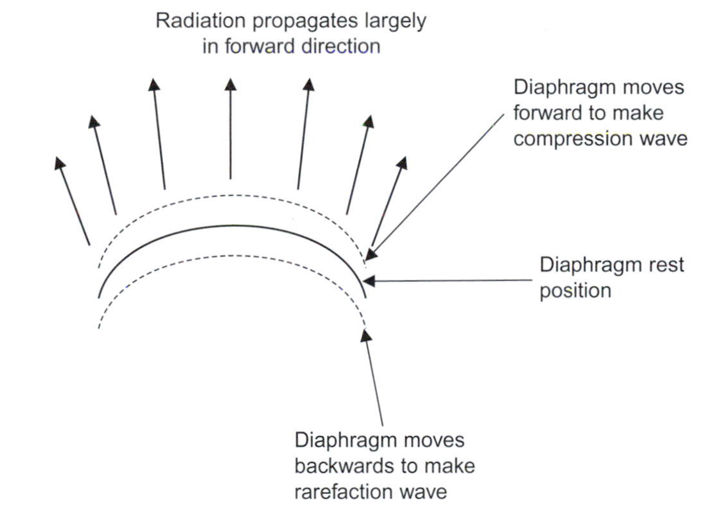

FIGURE 12.18
Comparison of radiation from pulsating spheres and domes.

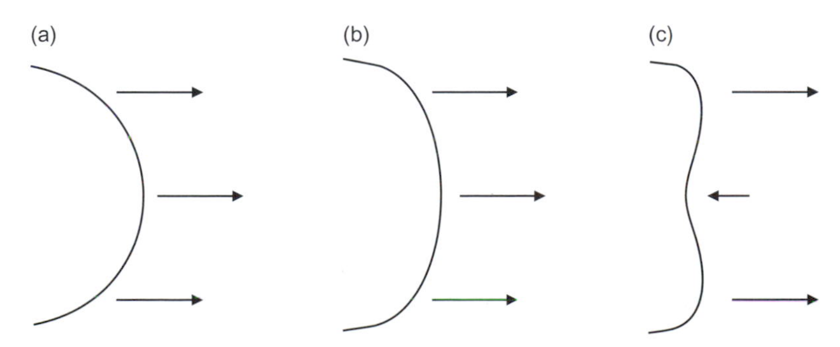

FIGURE 12.19
A dome in break-up: (a) Radiation equal in phase and amplitude from all parts of the dome; (b) First bending begins. The radiation amplitude increased from the dome edges and begins to reduce from the center; (c) The dome breaks up. Radiation from the center becomes out of phase with the edge radiation.

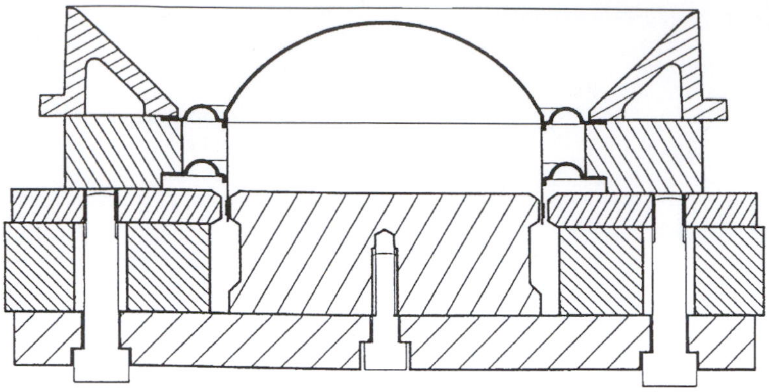

FIGURE 12.20
Sectional view of an ATC soft-dome driver showing the double suspension which helps to eliminate rocking motion.

Because of their nature, domes only have an outer suspension. The surround may be formed from the material of the dome in the case of soft diaphragms, but hard domes often employ a separate bonded suspension material. Rocking motion can be a problem in some designs, and solutions to remedy this are not always practical if they add weight to the moving system, and hence lower the sensitivity, because they may introduce other problems as a result. Lower sensitivity, for example, means more heat in the coil for the same SPL, and can lead to various problems. Ferrofluids can be beneficial in some cases by damping the rocking modes. On rare occasions, double outer suspensions are used. A cross-section of such a construction is shown in Figure 12.20. The choice of material for surrounds can be quite an arduous task in the mid frequencies because the ideal damping properties and compliance (the reciprocal of stiffness) may be conflicting for frequencies two or three octaves apart, yet which must all be radiated by the same diaphragm. At low mid-frequencies, where the diaphragm excursion may still be considerable, a cone driver may need to be in a separate enclosure with at least half a liter of air. Dome diaphragms are no different, but the magnetic assembly is an obstruction to the air that is trapped behind the diaphragm, and which tends to push up the resonant frequency. The hollow cavity between the diaphragm and the magnet center pole, clearly visible in Figure 12.20, is often filled with absorbent material to reduce the cavity resonances. However, if this cavity is too small, it can create problems by way of excessive back pressure on the diaphragm. Relieving the back pressure may require quite complex boring of the magnetic system if resonances in the tubes and cavities are to be kept out of the working range of the drivers. At high frequencies, these problems are less complex because low resonance frequencies are not required, so neither are such large air cavities required.

Dome tweeters have become very widespread in use, and now probably account for the majority of high-frequency drivers. Composite diaphragms are also not uncommon, such as polyester bonded to PVC. Unfortunately, dome tweeters tend to be rather low in sensitivity. This in some ways is ironic, because one of the driving forces behind the development of domes was to overcome the thermal failure problems due to the very small coil surface area in small cone loudspeakers when used at high SPLs. The lower sensitivity of an equivalent dome driver needs more power to drive it, and hence produces more heat. Nevertheless, in many cases, the balance is still in favor of the dome.

COMPRESSION DRIVERS

Compression drivers are almost always used with horns, except in some rare cases where their internal throats are sufficient to act as horns for very high frequency use. Essentially, the diaphragm, coil and surround assemblies of compression drivers are rather similar to those of dome drivers. The principle difference lies in the way in which the diaphragm couples to the outside air. In the case of the compression driver, the acoustic radiation passes through a restricted aperture, the ratio of its area to the area of the diaphragm being known as the compression ratio. This puts a highly resistive air load on the diaphragm, which is then passed through a horn of roughly exponential flare rate that prevents the air from moving out of the way of the gradually expanding radiated sound wave. Figure 12.21 shows the cross-section of a typical compression driver, and due to the very resistive load, electro-acoustic efficiencies of up to 50% can be achieved. This means that for every watt of electrical input, only half a watt will be dissipated as heat, and the other half a watt will be radiated as sound. It is thus not unusual for

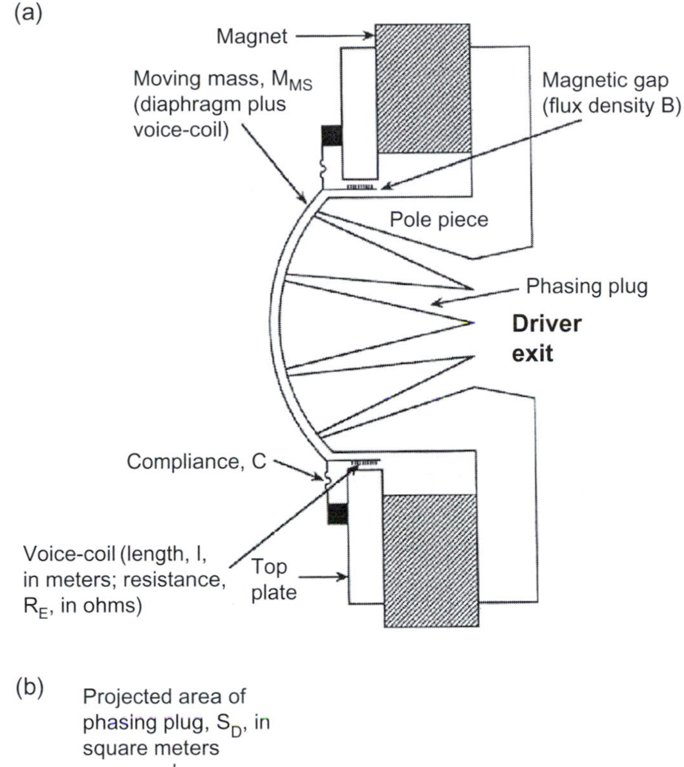

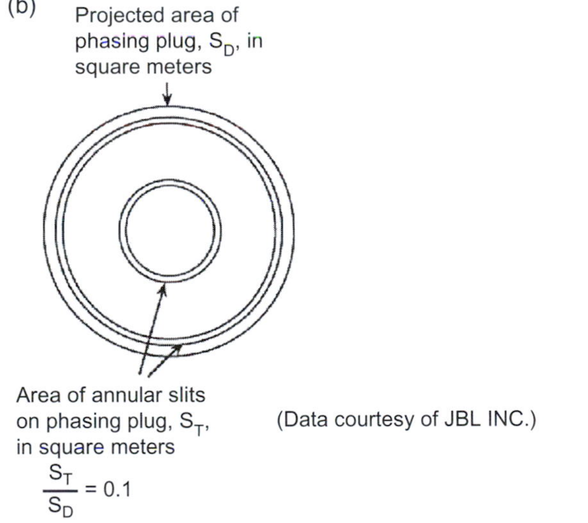

FIGURE 12.21
A high-frequency compression driver. (Data courtesy of JBL Inc.); (a) Section view of a JBL high-frequency compression driver; (b) Plan view of the diaphragm side of the phasing plug.

mid- and high-frequency compression driver/horn combinations to reach sensitivities of over 110 dB SPL for 1 watt input at 1 meter distance. Domes, on the other hand, can rarely convert more than 5% of the electrical input into radiated sound, so the other 95% serves only to heat up the voice-coil.

Horns are often shunned by many people who have not heard the best examples. Much of this negative attitude has arisen from the days when studio loudspeakers using compression drivers and horns were virtual transplants from the world of sound reinforcement and public address, and once a bad reputation sticks it can be very difficult to lose. To far too many people, because they have heard some bad horns, all horns must be bad. This is the absolute opposite to the general perception of soft dome mid-range drivers, where many people have heard some excellent ones and thus think that they all must be excellent. It is sometimes difficult to understand human reactions to these situations. In neither case is the point of view either logical or correct. It is also worth noting how so many people who state that they do not like horn loudspeakers in the mid-range will also say how they like the classic Tannoy Dual Concentric loudspeakers which, above 1 kHz, are exactly no more and no less than compression drivers and horns.

One problem which does always plague compression drivers is that the sound pressure levels within their throats can reach levels where the air itself cannot linearly propagate sound waves. Air at these SPLs does not compress and rarefy to the same degree under the same applied force in each direction, and so gives rise to harmonic distortion. However, at recording studio SPLs, which are way below live sound and cinema SPLs because of the much closer listening distances, air overload often does not become a problem until levels where direct radiating loudspeakers are suffering from mechanical and thermal non-linearities of their own. When a compression driver made from 5 kg of metal is dissipating half a watt of heat and producing 100 dB SPL for the people behind the mixing console, thermal problems absolutely do not exist, and neither do mechanical stress problems because the diaphragms are moving over such short distances. Non-linear distortions can remain remarkably low, and transient attack can be second to none. The problems with compression driver/horn combinations usually arise when designers fail to respect the physical realities of how to couple the horn to the air. High sensitivity loudspeakers, in general, also enjoy another benefit in that the lower current in the voice-coil for any given output SPL gives rise to much less disturbance of the static magnetic field in the gap, and thus avoid some intermodulation distortion products that are largely unavoidable in less sensitive drive systems when passing high currents through their voice-coils.

Unfortunately, good compression drivers are not cheap to manufacture, because they require precision, low tolerance engineering. It is therefore futile judging compression drivers in general by listening to cheap examples. Tolerances of less than 50 micrometers are not unusual in manufacturing specifications, and the magnetic flux density required in the gap can be so high that special materials may be needed which are difficult to cut and need heat treatment afterwards. Diaphragms also need to be made to very high standards of uniformity whilst often only being about 50 micrometers in thickness. Some of the finest diaphragms are made out of beryllium, because of its enormous strength-to-weight ratio, but its melting point of 1600 °C is too high to be accurately moulded in any practicable manner, and rigidity is such that it would shatter like glass if stamped to shape in a press. Some diaphragms are therefore made by a time-consuming and laborious *in-vacuo* vapor deposition process which is definitely not suited to mass production and is obviously expensive, but 2 inch (50 mm) diameter diaphragms can be made in this manner weighing less than 0.15 grammes, and with frequency responses from 500 Hz to well over 20 kHz.

In order to avoid phase cancellations at high frequencies in compression drivers, a phasing plug is usually incorporated which guides the pressure from different parts of the diaphragm down a series of tubes or concentric slits (as can be seen in Figure 12.21), in order to bring a phase-coherent wavefront to the driver exit, even at the highest frequencies of use. Alternatively, for very high-frequency horns used largely only above about 5 kHz, ring diaphragms may be used, as shown in Figure 12.22. These drivers also usually incorporate a short exponential horn as a part of the driver itself, and hence require no external horn. The diaphragms are clamped at the outer and inner edges, and radiate into a ring-shaped aperture which gradually flares into a single exit by means of some sort of central "nose." The sensitivities of these devices range up to about 108 dB SPL for 1 watt input at 1 meter distance. With a typical power handling capacity of 40 watts, they are virtually indestructible in recording studio or domestic playback use when operating only above about 7 kHz.

Compression drivers tend to work best from about 500 Hz upwards. Below that frequency they *can* be used, but the horns required to couple them optimally to the outside air tend to become impractically large.

One of the principal differences between the design of compression drivers and dome loudspeakers is that a soft diaphragm is not an option for compression devices. The diaphragms must be light and they must be rigid if high sensitivity and low distortion are required. The rear cavities in compression drivers are small, and the space between the diaphragm and the phasing plug is so small that any flexing of the diaphragm would be likely to make contact with the plug. Leaving more space between the diaphragm and the phasing plug would lead to a loss of sensitivity at high frequencies.

RIBBON LOUDSPEAKERS

The origin of the ribbon loudspeaker actually predates the moving-coil cone loudspeaker, with Schottky and Gerlack filing their patent two years before the moving-coil loudspeaker patent application. However, in practice it was rather disastrous, with a response of about two octaves between 250 Hz and 1000 Hz. Eight years later, Olson and Massa made use of the concept when they reversed

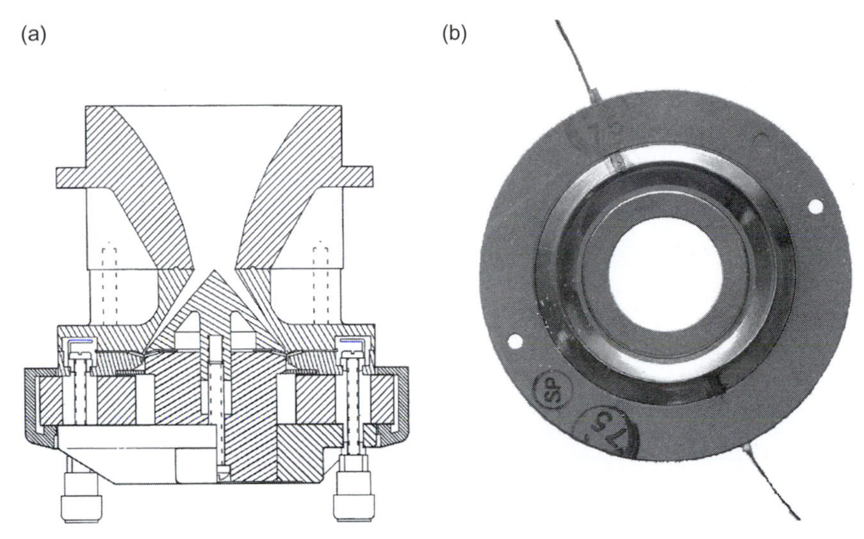

FIGURE 12.22
Ring diaphragm drivers: (a) Section view of a JBL high-frequency compression driver employing a centrally and peripherally clamped ring diaphragm (Drawing courtesy of JBL Inc.); (b) Photograph of a ring diaphragm.

it and turned it into a ribbon *microphone*. Nevertheless, the concept of a "current sheet" had emerged, with the diaphragm being suspended between the extended poles of a magnet system, and with the diaphragm itself passing the current. The idea is shown graphically in Figure 12.23. Stanley Kelly patented a much superior device in the 1950s, and, in his own words: "The ideal radiator is one which (a) vibrates in phase over its whole surface, (b) has a mass comparable to the air load, and (c) has only resonances which are outside the working frequency band. In order to meet these requirements, the radiator must be subject to a mechanical force equal in amplitude and phase over its whole surface. There are only two commercial systems which meet this requirement, viz, the constant-charge electrostatic and the ribbon electromagnetic loudspeakers." (Borwick, 1994).

In practice, the ribbon is corrugated to give it more rigidity. In order to avoid giving rise to resonances in its frequency band of operation it is supported at each end but it is not stretched. The support of the ribbon in the gaps is normally by means of elastomers and silicone rubber. As a current flows through the diaphragm, the corresponding magnetic field interacts with the static magnetic field from the magnets parallel to the plane of the ribbon. This generates a force at right angles to the plane of both the magnetic field and the ribbon, which moves the ribbon back and forth, but parallel to the direction of the magnets. To maintain adequate efficiency, the mass and electrical resistance must be kept as low as possible, the latter being typically as low as 0.2 ohms and requiring an impedance matching transformer in order to be useable with normal amplifiers. The low impedances also imply high currents, so a conflict can arise between the thickness of the diaphragm being low enough to keep the weight down but high enough to keep the electrical resistance down, otherwise the diaphragm might melt with the signal current. The classic Decca/Kelly "London" ribbon loudspeaker had a horn attached to it, had a sensitivity of 92 dB SPL for 1 watt at 1 meter distance, and covered a frequency range from 1 to 30 kHz with a power handling of 25 watts. Sonically, it was widely appreciated.

Modern technical advances have led to printed circuit current sheets, with the copper tracks on polyimide sheets of around 12 micrometers thickness. The sheets can be made with overlapping copper tracks on each surface, thus allowing the whole sheet area to be conductive. In conventional ribbons, however, the diaphragm material is usually aluminium, because it has the best compromise between resistance and mass. In recent years, the American company SLS Loudspeakers has made a big feature of the use of ribbon loudspeakers beyond about 2 kHz. Ribbons, traditionally, have been delicate, and difficult to manufacture, but sonically they have always had many friends.

HEIL AIR-MOTION TRANSFORMERS

These devices often get mistaken for ribbon loudspeakers when people see the folded diaphragms set in short horns, but they are definitely *not* ribbon loudspeakers. A ribbon radiates sound by the whole diaphragm moving backwards and forwards in a uniform manner, with the pleats never changing their

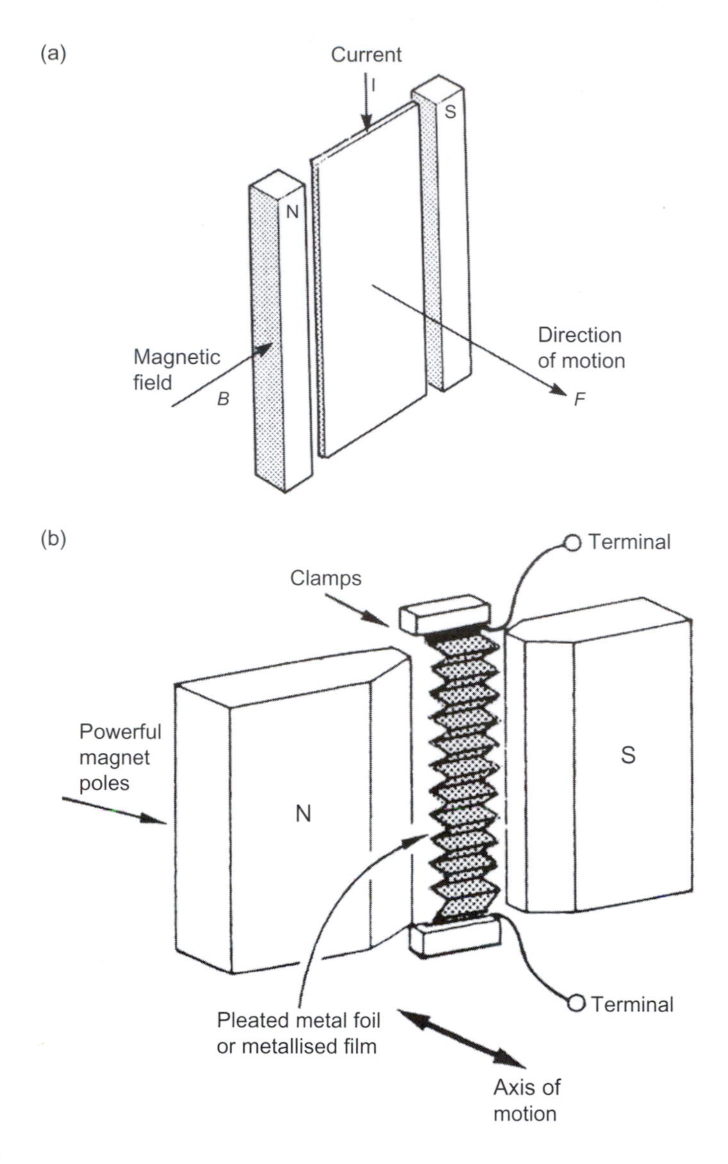

FIGURE 12.23
The ribbon driver: (a) The basic concept of a ribbon driver. The current flowing through the diaphragm reacts with the static magnetic field between the north (N) and south (S) poles, and gives rise to a force, and hence a movement of the diaphragm in the direction shown; (b) A more practical realization.

angles. The air-motion transformer, quite differently, moves its diaphragm in a concertina movement, drawing the air into the folds as they expand, and expelling the air as they contract. The German company A.D.A.M Audio have recently featured this technology rather in the way that SLS have made big use of ribbons. The air-motion transformer was designed by Dr. Oskar Heil and its general outline is shown in Figure 12.24. The current flows through a flat conducting track which is bonded to the diaphragm, and which is folded such that the conductive strip lies parallel to itself on the adjacent fold. When the current flows in one direction through the entire circuit, it travels in different directions in the conductors on adjacent folds, and the magnetic field is either attracted to or repelled from the nearby permanent magnets. When the current in the circuit reverses, the folds which were opened are then closed, and vice versa. It is called an air-motion transformer because there is a ratio of about four to one between the air particle velocity in and out of the folds relative to the velocity of movement of the pleated diaphragm. The magnet structures need to be very large because the entire pleated diaphragm must fit in the gap between the poles. The diaphragms are made from plastics such as p.t.f.e. or polyethylene, which have good damping. Current units can work from about 500 Hz to 20 kHz. Some models have been shown to produce quite high levels of second-harmonic distortion above about 5 kHz, but the subjective audibility of this does not seem to be significant as the distortion products are all above 10 kHz and about 30 dB down relative to the signal.

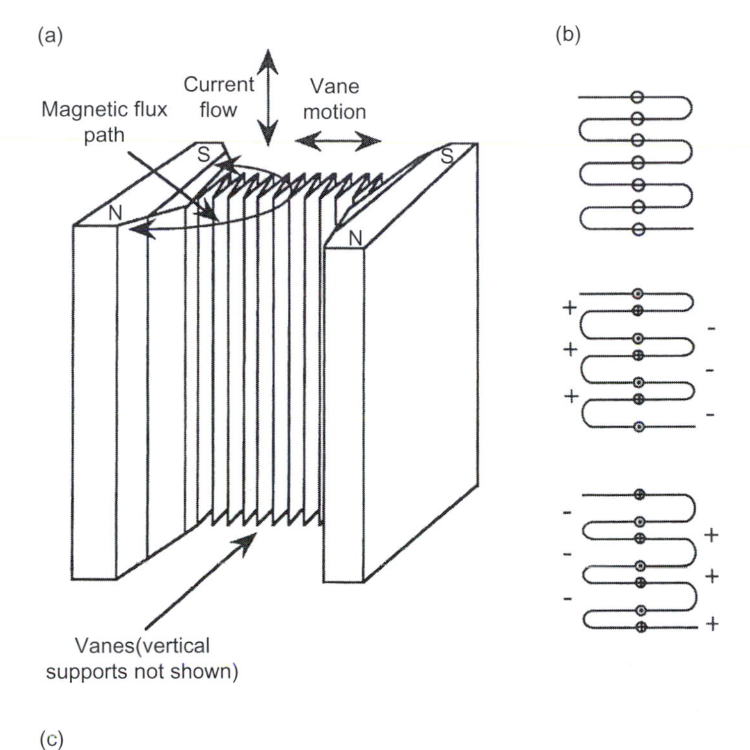

(c)

FIGURE 12.24
The Heil air-motion transformer: (a) Perspective view of the basic concept; (b) Polarity changes in the conductors cause the opening and closing of alternating folds, drawing in and expelling air on alternate half-cycles; (c) A full-range loudspeaker system using Heil air-motion transformers for the mid and high frequencies in a symmetrical D'Appolito layout—an A.D.A.M. S7.

DISTRIBUTED-MODE LOUDSPEAKERS

These are the flat panels developed under licenses from NXT and its subsidiary New Transducers Ltd in the UK. The principal patent is held by the British Ministry of Defence, on whose behalf Dr. Ken Heron was not actually trying to develop a loudspeaker at all—he was trying to build lighter helicopters and stumbled upon an aluminium honeycomb panel that radiated sound quite efficiently.

At first glance, the concept of a distributed-mode loudspeaker (DML) seems to be a total contradiction. It is a mess of resonances, when, in almost all other aspects of loudspeaker design diaphragm, resonances are taboo. The drive points where the electromagnetic exciters couple to the panel are chosen so that they couple to as many of the vibrational modes as possible, in order to excite the panel in the most uniform manner. The panels which are currently used are typically made from resin-impregnated glass-fiber sheets bonded to a 3 mm honeycomb core of "Nomex"—a polyamide which is often used to make loudspeaker voice-coil formers. The panels are not driven by a voice-coil attached to the panel which is then connected to the same frame/chassis as the magnet, but rather the lightweight panel and coil react against the mass of a much heavier, freely suspended magnet. Panels are commonly excited by two or four coils, to more evenly distribute the drive force. An example of a commercial panel is shown in Figure 12.25.

As the name "distributed-mode loudspeaker" suggests, they radiate all frequencies from all parts of the panel, and so are naturally diffuse sources. As such, they can work well in both studios and domestic circumstances as the rear channels of a surround system, where they can create excellent ambient diffusion effects. Their sonic coloration, though not unpleasant, tends to render them inappropriate as main front channels where serious listening is the goal, but their performance on the surround channels can be very involving, and numerous professional installations have used them in this role. It is also interesting to note that the radiation from both sides of the panel (if the rear is not enclosed) couples to the room neither in an omnidirectional way nor as a figure-of-eight pattern (or dipole) like electrostatic panel loudspeakers, but as something more akin to a bi-pole, where the radiation from each side is only partially correlated. As such, and if spaced away from a reflective wall, they can fill a room with reflexions which have very little tendency towards showing typical summation and cancellation effects (peaks and dips in the response) in different places in the room. This characteristic can add still further to the beneficial way that they can be used for ambience channels, where their inherent sonic coloration appears to be little disadvantage.

A constant problem for DML loudspeakers has been the lack of low-frequency output, but designs are now emerging, such as the Fane Minipro, which extend reasonably well down to 60 Hz. For surround use this is often quite adequate, especially since many of these systems will be used with bass management systems which will pass the low frequencies to a separate (sub-)woofer. Panels with dimensions of about 40 cm × 60 cm are sufficient for such purposes, but smaller panels suffer from much higher roll-off frequencies. The low-frequency response can be extended by mounting the panels on a shallow, absorbent-lined box, of about 10 cm depth, but care must be taken to avoid over-filling the box with absorbent material or the damping on the rear of the panel can become too great. When the open-backed panels are hung against walls they should be hung at an angle, in order to prevent resonances in the parallel cavity formed between the panels and the wall. As the general radiation directivity is very broad at all frequencies, as shown in Figure 12.26(a), it is not necessary to point the panel at a

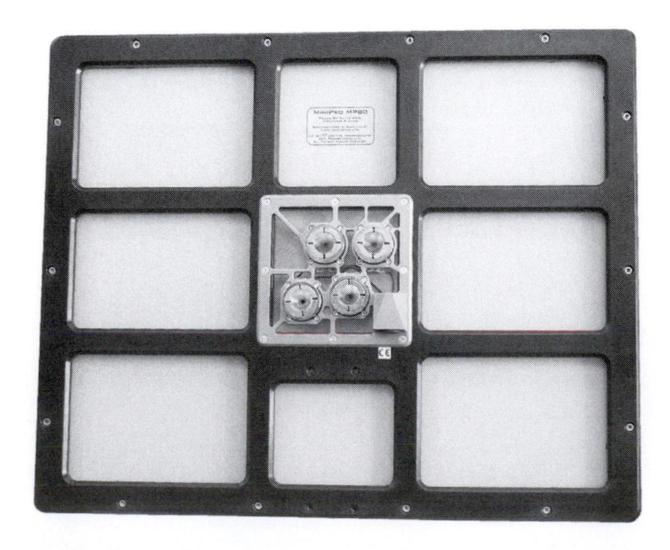

FIGURE 12.25
A distributed-mode loudspeaker (DML).

normal to the listening area. From almost whatever angle a panel is radiating, and no matter where it is in a room, it will excite the whole room with its full frequency range. Only from a position in line with the panel edge is there a region of a few degrees where low-frequency cancellation takes place. The high frequencies radiate in an almost omnidirectional manner, though from a spacially diffuse source.

As a result of the diffuse nature of the radiation, the fall-off of SPL with distance is initially less than from a conventional loudspeaker, as shown in Figure 12.26(b). It is more typical of larger, planar radiators. This can also be a benefit when filling a room with ambient sounds, because the left and right

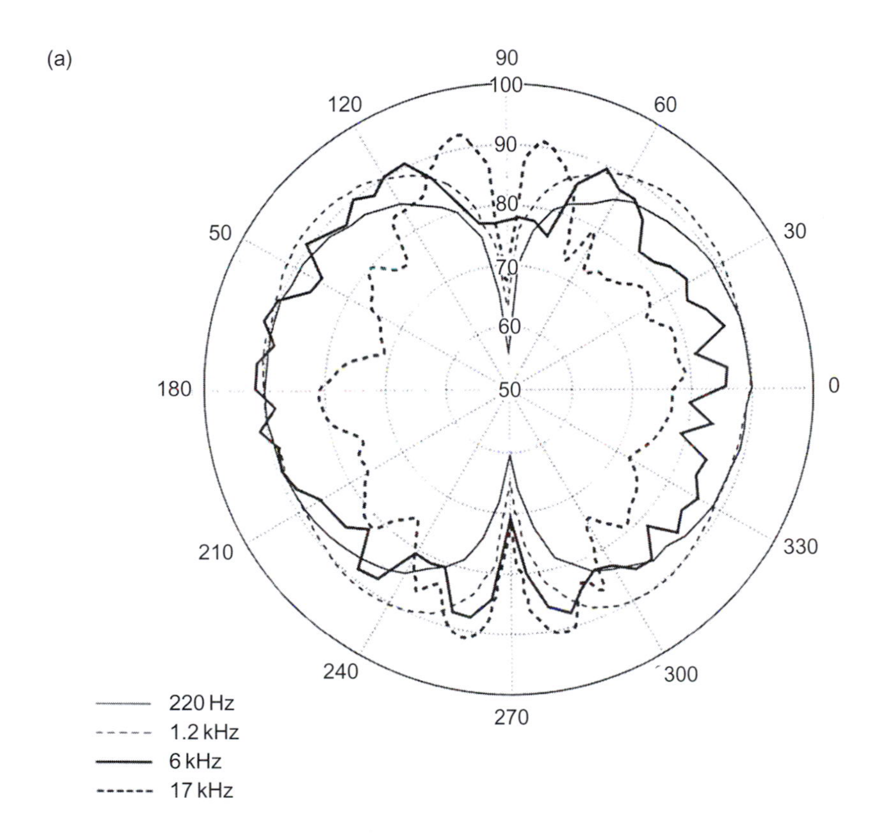

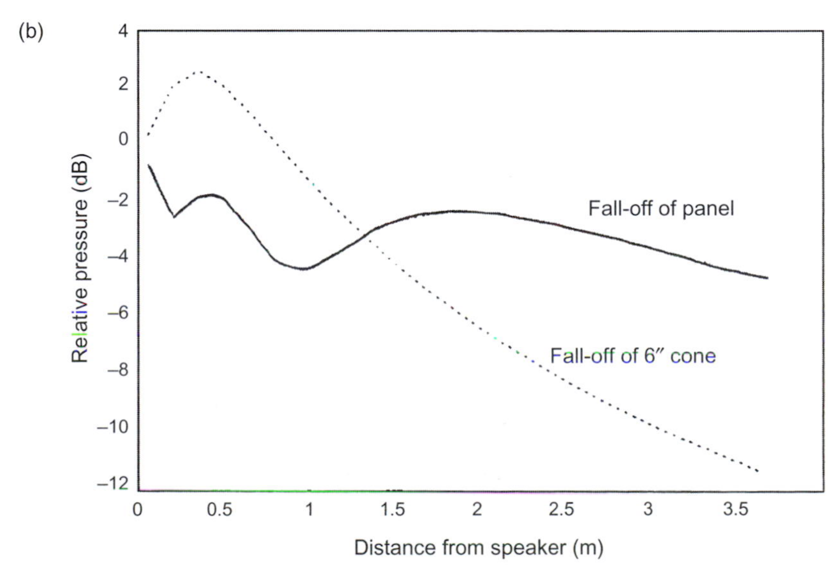

FIGURE 12.26
Radiation characteristics from a typical DML: (a) DML polar response at 220 Hz, 1.2 kHz, 6 kHz and 17 kHz; (b) Computed comparison of loudness with distance—a distributed mode panel versus a 6 inch (150 mm) piston. (From data published by NXT.)

signals, from positions more or less laterally alongside the listeners, are more evenly distributed across the room, even when the room acoustics are relatively dead. The coupling to the room modes is also accomplished in a different manner to the way in which conventional radiators couple with modes. This again has been shown to give rise to fewer peaks and dips in the room. When all the characteristics are taken together, the DMLs do offer some interesting opportunities for surround applications. In fact, the low-frequency response can be noticeably better if panels larger than those referred to above are used, but very large flat panels tend to become unwieldy and can introduce resonance and reflexion problems into the room when used in conjunction with other sources.

Panel/piston combinations

Since the mid-1990s, various efforts have been made to develop combinations of DMLs and conventional loudspeakers in such manners as to take advantage of their uncorrelated and correlated radiation characteristics respectively. The thinking behind these ideas is that in live music situations the direct propagation from the instruments to the ears tends to be highly correlated, whilst the reflected energy from walls, ceilings, floors and other hard surfaces tends towards being uncorrelated. In good concert halls it has been found that the ones with low levels of inter-aural cross-correlation tend to produce the generally most desirable sensations of spaciousness. In domestic situations, the reflexions from walls do not tend to be as diffuse and uncorrelated as in good concert halls, and their frequency response is inevitably affected by the directivity characteristics of the loudspeakers. To help to combat these deficiencies, the concept of creating a sensation of diffuse reflexions has been pursued by the use of relatively diffuse sources in combination with a conventional stereo pair of loudspeakers. The relative level of the two types of sources can be adjusted to taste.

The KEF company in the UK has patented a concept of using DMLs behind conventional loudspeakers with the axes of the DMLs at right angles to the axes of the conventional loudspeakers. The conventional loudspeakers generally point towards the listeners, whilst the DMLs work more omnidirectionally (although with their weak lateral nulls towards the listening position) in an attempt to excite the rooms with diffuse reflexions from the surround channel(s) of multichannel recordings.

Another system, marketed under the trade name of Layered Sound, was patented by Dr. Shelley Katz, a Canadian pianist, who initially researched the concept as a means to make electric pianos sound less "stiff" and more acoustic. This technology was licensed for research purposes to the Japanese company Korg. In the domestic reproduction or sound reinforcement formats, the panels are placed closely above or behind the conventional loudspeakers, but usually with the axes parallel to each other, and not necessarily at 90° as with KEF systems. The panels in this system are fed with the same signal as the conventional loudspeakers, and can also be fed via a delay, with the delay time and the relative SPLs from the different sources being used to control the overall effect.

By definition, such systems are not high-fidelity in the classical sense, because they seek to introduce artifacts which are not in the original drive signals. Nevertheless, that overall sense of realism which they can help to generate may be considered in many cases to be highly faithful to the *sensations* of the performance spaces or the wishes of the recording personnel or musicians. Proponents claim that as the current recording processes via conventional microphones and the reproduction via conventional loudspeakers are still limited and compromised by the inherent short-falls of their performance, then piston/panel combinations may be able to realistically add, globally, more than is lost in conventional reproduction, and so can be considered to be more than making up for those short-falls. However, all assessments of these types of loudspeaker systems currently need to be made subjectively, because there are still no measurement systems which can reliably define the performances of such combined systems in any meaningful manner.

Therefore, whilst technical accuracy in terms of conventional reproduction might be compromised, the developers of such combined systems can reasonably claim that the perceptual fidelity, in terms of overall realism, may be superior, at least on certain types of music, to reproduction on systems with more measurable fidelity in the conventional sense. Whilst it would seem to be unwise to use the combined systems for music recording quality monitoring, the beneficial effects of Layered Sound for electric piano amplification seems to be established. Of course, if recordings were being specially made for reproduction on these composite systems or their derivatives, then the monitoring of the recordings via such systems may also be justified. Development of DMLs is still in progress, however, and some new designs are already emerging with much less colored responses than have previously been achievable.

BEYOND MAGNETICS

All of the loudspeaker drive systems discussed in this chapter so far have been electromagnetic transducers. One way or another, all of them have employed the magnetic field generated by an alternating music signal current in a moveable conductor to react with a static magnetic field. The force generated at right angles to the current and the static magnetic field has then been applied to a diaphragm of some sort or other which has been designed to move air and radiate sound. They are all, basically, variations on the same theme. There are, however, various other means by which loudspeakers can be made.

Piezoelectric devices

There are certain materials which can be made to twist and bend when electrical signals are applied to opposing surfaces, and, in general, there is a useful proportional relationship between the applied voltage and the degree of movement of the material. Such transducers have found use as high-frequency loudspeakers of very robust design, which have in turn found use in guitar amplifiers and sound reinforcement systems where conventional tweeters have been considered to be too fragile. Quartz, Rochelle salt and some ceramics such as barium titanate have piezoelectric properties, as do some high polymer plastics, such as polyvinylidene fluoride. Direct radiator and horn-loaded piezoelectric radiators are available, the horn-loaded Motorola device being quite widespread and having an axial response within ±3 dB from 4 kHz to 20 kHz. Pioneer have also developed a cylindrical piezo radiator. In these, a thin film of high polymer plastic is made into a cylindrical shape which is then caused to pulsate with the applied signal voltage. The response is respectably flat from 2 kHz to 20 kHz, with 360° horizontal radiation.

The piezoelectric units are rugged largely because they are self-protecting. The impedance tends to rise as the frequency lowers, so driving low-frequency signals through them is difficult, even with no crossover. They also, effectively, have nothing to burn out and nothing to go off-center. Although they are very rarely encountered in loudspeakers used for music monitoring, they can be found in domestic system as well as in music amplification systems. The principles of their construction are shown in Figure 12.27. Piezoelectric drivers tend to be mid-sensitivity devices, offering the low 90s of decibels SPL for 1 watt (or at least its voltage equivalent—2.83 volts—into their varying impedance) at 1 meter distance.

Ionic loudspeakers

It is unlikely that anybody will find these in use today because production ceased around 1968, but the concept is interesting. Radio-frequency interference and the production of irritating ozone were unwanted side-effects that helped towards their demise, along with low output, but there is widespread agreement that the sound of these devices was true high fidelity. In ionic drivers a 27 MHz high voltage signal is fed to the electrodes of a quartz cell. A corona discharge results, giving off a blue light as the air is ionized. When the radio-frequency voltage is modulated by an applied audio-frequency signal the volume of ionized air will vary and produce pressure fluctuations in the air. The frequency response was given as 3 kHz to 50 kHz ± 2 dB, with only 0.5% distortion at 93 dB SPL. Absolute peak output was around 98 dB SPL for the "Ionofane" version. This was the cutting edge of high fidelity in the 1950s and 1960s, and is still good today; however, at higher SPLs, above about 96 dB at 1 meter, compression soon set in, seriously limiting the output. They also cost 28 guineas each (just under 30 pounds, sterling) in the early 1960s, which was an entire month's salary for many working people in those days.

ELECTROSTATIC LOUDSPEAKERS

Just as dynamic microphones, such as ribbons and moving coils, have their equivalent loudspeakers, and as piezoelectric loudspeakers relate to crystal microphones, electrostatic loudspeakers are the counterparts to condenser (capacitor) microphones. (Condenser is the old term, capacitor is the modern term, but the old terms often take root.) Within their limitations, electrostatic loudspeakers can produce a sound quality which, in microphone terms, we would only associated with the finest condenser microphones. Sonically they can be astounding. The limitations are size (because they need large surface areas of diaphragm), relatively low maximum output SPL, and their figure-of-eight radiation pattern which does not suit all room layouts. The typical radiation pattern from a full-range electrostatic loudspeaker (ESL) is shown in Figure 12.28. The absorption of the rear radiation is not a

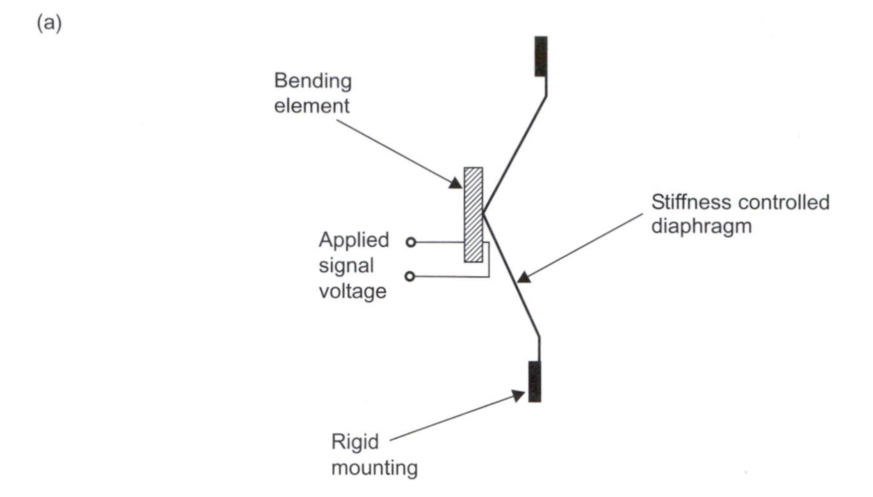

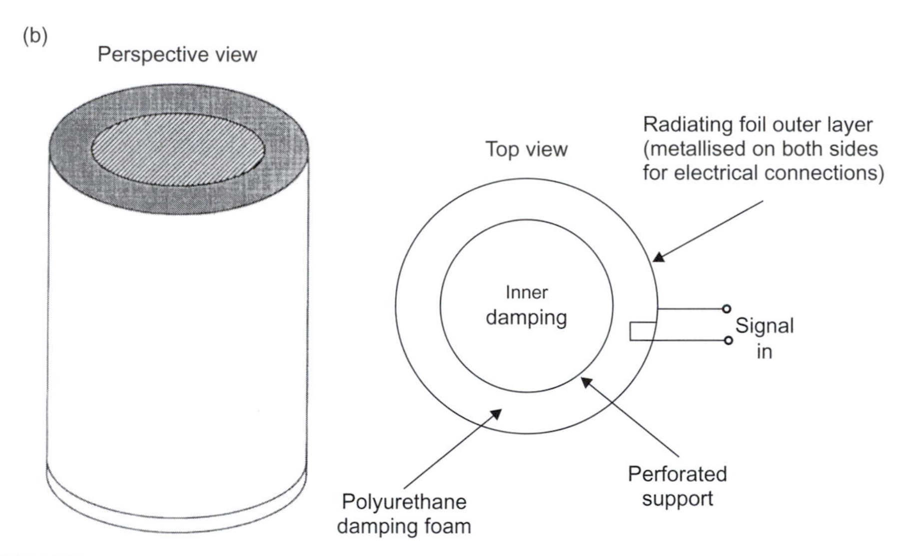

FIGURE 12.27
Piezoelectric radiators: (a) Section view of a typical piezoelectric HF radiator; (b) The Pioneer cylindrical piezoelectric driver—the High Polymer Radiator.

practicable solution unless the box is enormous, because the air load would inhibit the movement of the extremely light diaphragm, whose mass is critically chosen to match that of the free air surrounding it. The loudspeakers therefore need to be placed away from walls, and the nature of the walls close to the sides and behind them need to be duly taken into account, acoustically, when the response in front of the loudspeakers is being considered. Full-range electrostatic loudspeakers may therefore be less forgiving in terms of where they can be placed.

These devices do not act as volume-velocity pumps like cone loudspeakers, but radiate as pressure gradient sources. This means that they do not couple to the *pressure* anti-nodes of the room modes, but to the *velocity* anti-nodes, which are the pressure *nodes*. Other than in anechoic chambers, this means that the optimal siting of electrostatic loudspeakers will be different to that for moving-coil loudspeakers. The maximum SPL is limited (although 95 dB SPL should be no problem) because at higher SPLs the polarizing voltages would need to be so great that highly specialized materials and techniques would need to be used in the manufacture of the devices, and also because the air would reach its own electrical breakdown limits. This is rather similar to the situation in compression drivers, where the air itself begins to be the limiting factor at higher SPLs. Air is not just something that you can do what you want with. It has its own characteristic properties and it can impose its own limits on what it can be made to do.

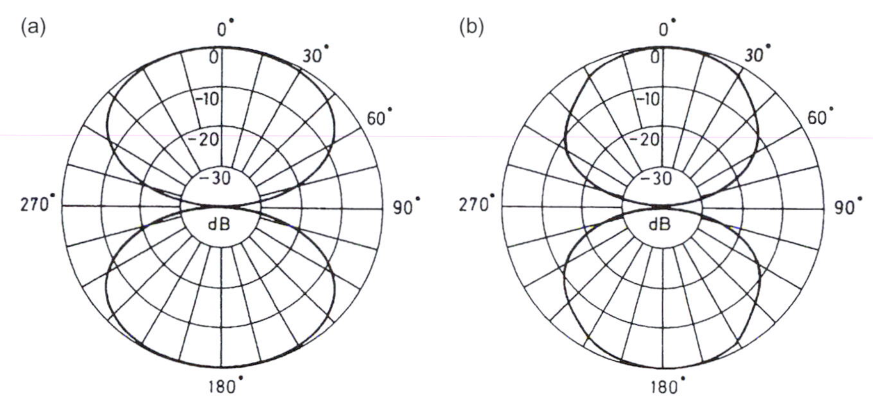

FIGURE 12.28
Typical directivity pattern from a dipole radiator: (a) Low- to mid-frequency response; (b) High-frequency response.

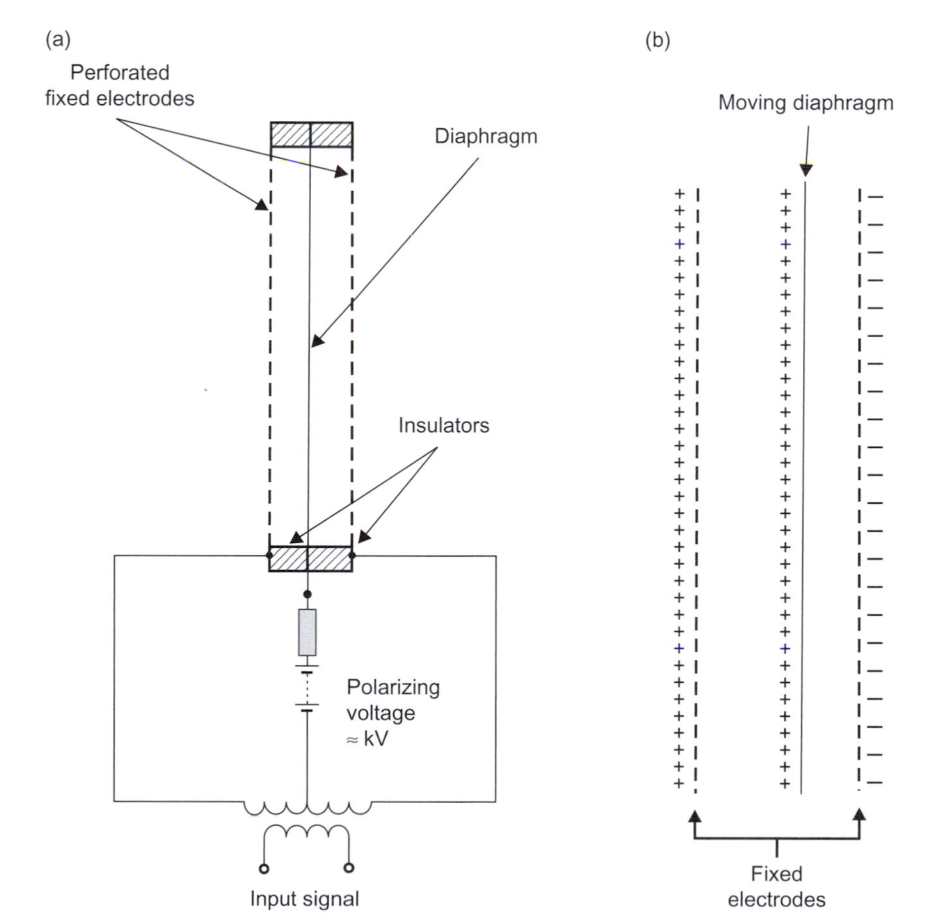

FIGURE 12.29
The electrostatic radiator: (a) The basic principle of operation of an electrostatic loudspeaker; (b) When the charge becomes opposite on the fixed electrodes the diaphragm moves to take up a new equilibrium position. If the charge between the fixed electrodes changes, the diaphragm position will change correspondingly.

The basic concept of an ESL is shown in Figure 12.29. A polarizing voltage of around 3000 volts is applied to the diaphragm whilst the two outer perforated electrodes are grounded via resistances, and spaced away by about 2 mm. The charge on the diaphragm keeps it centralized, in equilibrium, in the absence of signal. If, via an input transformer, a signal voltage is applied across the electrodes, the equilibrium point will shift and the diaphragm will move to chase it. The whole device operates on high voltages and high impedances, so the signals have to be fed from the power amplifiers via a step-up transformer, which itself requires careful design if it is to pass all frequencies equally. Nevertheless, this type of loudspeaker is largely a capacitor, so it still presents a predominantly capacitive load to the amplifier, which therefore needs to be able to supply high currents even when voltages are low, due to the phase angle difference between the voltage and the current (described previously). The choice of amplifier may therefore depend on its ability to supply current more than its ability to supply power.

Because of the inevitably small distances between the electrodes, necessary to maintain useful sensitivity with polarizing voltages which will not cause air breakdown, the distance that a low-frequency diaphragm can travel is severely limited. Therefore, the only way that the diaphragm can move quantities of air sufficient to generate the required SPLs is to be large. However, the diaphragm size of a single unit is limited by its ability to maintain an even tension, and not to sag in places, so this also restricts the SPL achievable by single units.

The large source area would, as with the DML, give a rather irregular directivity at high frequencies (but without the DMLs' ability to produce so many irregularities that they become almost regular again). Full-range ESLs therefore tend to be made as two- or three-way devices, as with moving-coil loudspeakers, using a much smaller radiating area for the higher frequencies. The Quad ESL63 uses a series of concentric rings, as shown in Figure 12.30, in order to mimic a point source situated some way behind the diaphragm. These loudspeakers have also been made available with dipole moving-coil subwoofers mounted beneath them to extend their rather limited low-frequency responses, or rather, their limited low-frequency *output* capability due to the limited maximum excursion and size of the diaphragms. Electrostatics are therefore not the ideal loudspeakers for monitoring a solo bass drum in a large control room, but they do find use in critical listening rooms and audiophile high fidelity applications, where their natural sound and resolution of low-level detail are highly valued. When heard in an anechoic room reproducing recordings of acoustic instruments recorded in the same room, and with the instruments alongside for reference, their ability to mimic the original sound can be quite startling.

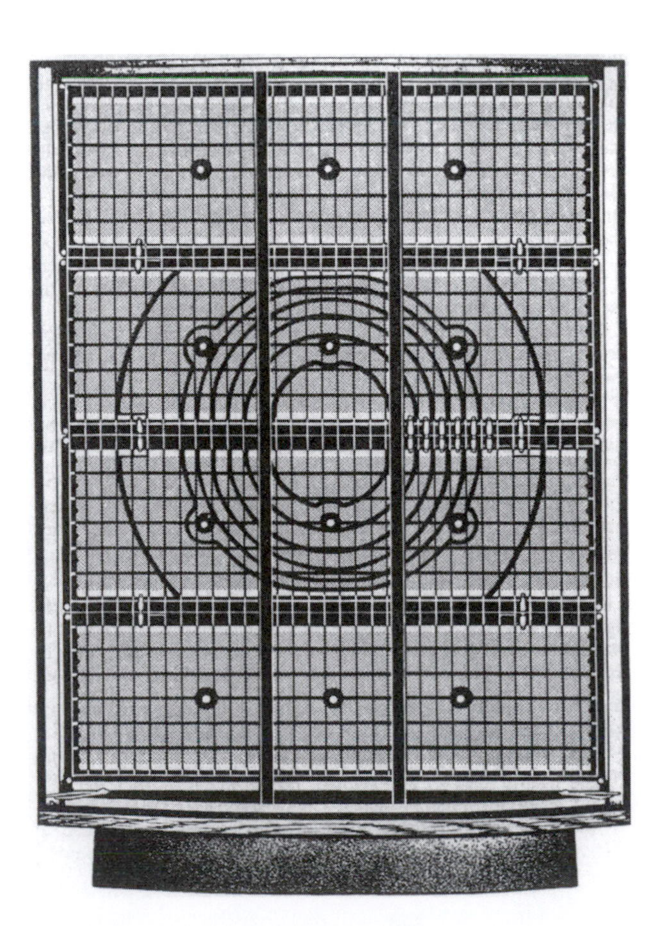

FIGURE 12.30
The Quad ESL 63 electrostatic loudspeaker with its concentric diaphragms. The higher frequencies radiate only from the central sections.

Granted, the anechoic response is not the be-all and end-all of loudspeaker reproduction, but the general tendency is for anything which can work so well in such circumstances to have a good start when transposed to other circumstances. Figure 12.31 shows the step function response of a quad electrostatic loudspeaker: the attack of the signal is exemplary, and very hard to beat with other loudspeakers.

Although the electrostatic loudspeaker principle was experimented with as early as the 1920s, it was not until 1957 that the first really viable design was put into production. It took the advent of the concept of a constant charge and the development of new plastic foils before it could be fully realized. However, once all the pieces of the jig-saw were in place, the progress was remarkable. A pair of ELSs from the 1950s can, even 50 years later, put many of the latest loudspeakers to shame in terms of low coloration, low distortion, transient response, frequency response flatness and, perhaps most of all, perceived sound quality. What Walker and Williamson did when they developed the Quad ESL was to take a step forwards to a degree that has rarely been equalled in the world of sound reproduction, and this is especially so considering all the technical difficulties which they had to overcome.

Whereas the moving-coil loudspeaker exhibits non-linearities in its inner suspension, outer suspension, magnetic flux disturbances, magnetic field asymmetries and various other sources, the electrostatics more or less only exhibit non-linearities due to very small asymmetries in construction, which can be minimized by careful quality control. In general, the non-linear distortion production by electrostatic loudspeakers is much lower than that produced by most moving-coil devices. The authors of this chapter have, for decades, used full-range electrostatic loudspeakers as benchmarks against which to judge other loudspeakers, both objectively and subjectively. This is not to say that they cannot be surpassed on individual aspects of their performance, but their global performance is hard to beat.

Occasionally, electrostatic mid-range drivers and/or high-frequency drivers can also be found in compound, electromagnetic/electrostatic designs of domestic loudspeaker systems.

ELECTROMAGNETIC PLANAR LOUDSPEAKERS

As if to rise to the electrostatic challenge, one of the electro-dynamic (electromagnetic) responses was the planar loudspeaker. These use light, thin, tensioned plastic film diaphragms which have voice-coil circuits printed on them, rather in the manner of a very thin printed circuit board. The diaphragms are stretched over frames with many openings and a large number of small magnets dispersed over their area. There is no attempt to concentrate the flux in any area, but just to set up a field of fringe flux in the vicinity of the magnets. In this way, a diaphragm is caused to move by the reaction of the signal current in the printed tracks with the static magnetic field, which results in the diaphragm being more or less uniformly driven over its entire surface. As with large electrostatic diaphragms, they must cross over at higher frequencies into drivers of smaller radiating areas if strange directivity problems are to be avoided. They do not have the "random" distribution of high-frequency sources as exhibited by the DMLs, but neither do they tend to have as much coloration.

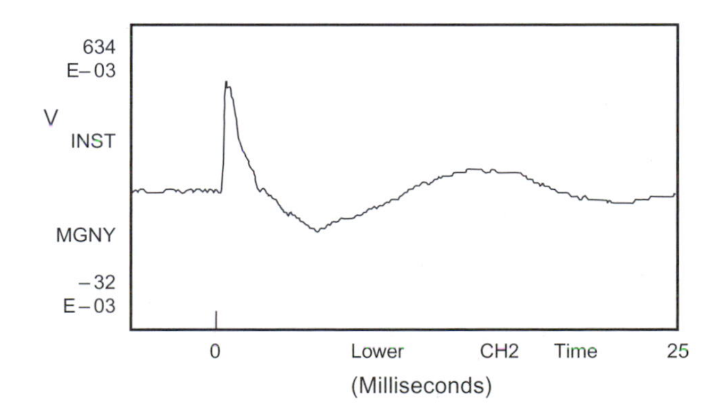

FIGURE 12.31
Step function response of an electrostatic loudspeaker, showing the exemplary attack (rise-time).

SUMMARY

There are a considerable number of different ways to transform electrical drive signals into sound waves, and no one system has all of the advantages to itself. In fact, all the drive systems are electro-mechanico-acoustic transducers; that is, they must first convert the electrical signals into mechanical forces which are then used to drive sound-radiating diaphragms of some sort or another. (The one exception perhaps, being the now defunct ionic tweeter.) The necessary double conversion tends to involve a number of non-linear processes, and it is largely the mechanical components which are the main offenders. It is therefore unfortunate that we often refer to loudspeakers as simply electro-acoustic transducers because it fails to recognize the existence of the principal culprit for our problems.

At the limit, the air itself is non-linear, so when we try to reproduce loud sounds from small sources that were originally produced by large sources, local concentrations of high air pressures close to the small sources will give rise to non-linearities, which our ears will recognize as not being the real thing. The reason for describing this wide range of loudspeaker drive unit concepts (and there are other, less common ones) is to establish the point that at the very heart of all loudspeaker systems are imperfect components, and, as mentioned earlier, that the art of the science of the designs is to find the best compromise for any individual requirement.

BIBLIOGRAPHY

Chapter 3 of Borwick (1994) contains what is perhaps the definitive work on electrostatic loudspeakers, written by the late Peter Baxandall. In the Third Edition of the book, published in 2001, Peter Walker somewhat modified the text. Either edition, in its entirety, is recommended reading for anybody wishing to delve deeper into the world of loudspeakers and headphones.

Eargle, J., 1997. Loudspeaker Handbook. Chapman and Hall, New York, USA and London, UK.

Borwick, J., 2001. Loudspeaker and Headphone Handbook. Focal Press, third ed., Oxford, UK.

REFERENCES

Briggs, G., 1955. Loudspeakers, The Why and How of Good Reproduction. Wharfedale Wireless Works Ltd, fourth ed., Bradford, UK.

Borwick, J., 1994. Loudspeaker and Headphone Handbook. Chapter 2 (By Stanley Kelly), Focal Press, second ed., Oxford, UK.

Colloms, M., 1997. High Performance Loudspeakers. John Wiley & Sons, fifth ed., Chichester, UK .

PREFACE

Our third chapter on loudspeakers describes their cabinets. These are not just enclosures to keep the dust off the drive units, but an integral part of the functioning of the loudspeaker, especially at the bass end of the response.

If you try to use a loudspeaker unit all by itself in free air, the first thing you will notice is that the bass response is very poor indeed. This is because when the cone moves, the air set in motion takes the path of least resistance and simply sidles round the edges of the unit to the back of the cone, and overall little sound is radiated.

The most obvious solution is to mount the speaker in the center of a large rigid board, so the air has a much greater distance to go on its round trip. This scheme is called a *finite* baffle, and it has obvious problems in that the baffle board needs to be large to get a good bass response, and there is only atmospheric loading on the rear of the speaker cone.

If you bend the edges of your baffle board around so they meet and form a sealed box around the rear of the speaker unit, you have what is called an *infinite* baffle. This is probably the most common sort of speaker enclosure in use. The air in the sealed box acts as a spring on the rear of the speaker cone, effectively stiffening its suspension and raising the resonant frequency below which roll-off occurs. The smaller the box the stiffer the air-spring, and the higher the resonant frequency; this is why small boxes give poor bass.

Newell and Holland then move on to consider the next most popular enclosures, known as reflex enclosures or vented boxes. These are not sealed boxes, but have a tube or port connecting to the outside world. The mass of the air in the port resonates with the air-spring inside the box, and exploits the output from the rear side of the bass driver cone to increase the efficiency of the system at low frequencies compared with a typical closed box loudspeaker. One problem is that there is no loading on the driver below the resonant frequency, which can cause mechanical over-excursions unless the audio signal has been suitably filtered to remove very low frequencies; the classic source of these low frequency disturbances is the playing of non-flat vinyl discs. This is why every phono preamplifier should include an effective subsonic filter.

Newell and Holland proceed from there to look at the other options for enclosures, such as acoustic labyrinths or so-called "transmission line" types, and Auxiliary Bass Radiator (ABR) systems that are a variation on the reflex enclosure, using the mass of a dummy speaker with no magnet system to replace the air mass in the reflex port.

Also covered are bandpass enclosures, normally only suitable for sub-woofers, and isobaric systems that use two bass drivers closely coupled by the air between them.

The shape of the cabinet is yet another factor in speaker design. A spherical cabinet gives a nice flat response, due to an absence of diffraction effects, but it is hard to manufacture and even harder to fit into most schemes of decoration. For practical reasons the rectangular shape is by far the most popular.

Finally, a word on front grilles. These are rarely seen on loudspeakers in professional situations, such as recording studios, but they are essential in domestic use to give some protection to the drivers. I can vouch for that.

CHAPTER 13

Loudspeaker Cabinets

Loudspeakers by Philip Newell and Keith Holland

THE CONCEPT OF THE INFINITE BAFFLE

When the diaphragm of an open-framed driver moves forwards, the compression of the air at the face of the diaphragm is accompanied by a rarefaction at the other side of the diaphragm, and the natural tendency is for the pressure difference to equalize itself by a movement of air around the sides of the driver. At frequencies whose wavelengths are large compared to the circumference of the diaphragm, the equalization is almost perfectly accomplished, and so almost no sound is radiated. It is therefore necessary to discourage this pressure equalization if low frequencies are to be radiated. The simplest means of accomplishing this is to mount the loudspeaker in a large rigid board, or baffle, as shown in Figure 13.1. If the board were to extend in all directions to infinity, it would be a true infinite baffle. It would cause no change in the air loading on each side of the diaphragm, it would exhibit no resonances, it could cause no diffraction, and, with a good quality driver (or drivers) would sound excellent. Unfortunately, its great drawback is that it is a rather impractical concept.

The two practical realizations of this idea are the finite baffle, where a baffle of perhaps a metre square is employed, or the so-called infinite baffle, which is, in fact, a sealed box. The radiation pattern of the finite baffle is shown in Figure 13.2(a). The cancellation around the sides of the extended plane of the driver causes response nulls to the sides, in the direction of the plane of the baffle, resulting in a three-dimensional figure-of-eight pattern in free space. The low frequency cut-off is determined by the size of the baffle. The final rate of low frequency roll-off is 18 dB per octave, but some measures can affect the nature of the entry to the roll-off. Varying the Q of the driver resonance, by mechanical and/or magnetic changes, can yield response shapes such as those shown in Figure 13.2(b). By placing the driver off-center, the cut-off can be made more gradual due to the distance from the driver to each edge of the baffle being different. Open baffles are rarely used in recording studio control rooms because of the problems of where to site them and how to control the rear radiation, but they find use in listening rooms and domestic high-fidelity systems. In these instances the baffles can be sited somewhat more flexibly than in an equipment-loaded control room, and the loudspeaker and listener positions can usually be found which give good results. Subjectively, open

FIGURE 13.1
An open baffle of Wharfedale design from the 1950s. The front panel was a sand-filled plywood sandwich, to damp resonances. The upward-pointing tweeter was to generate a more diffuse high frequency response.

Audio Engineering Explained

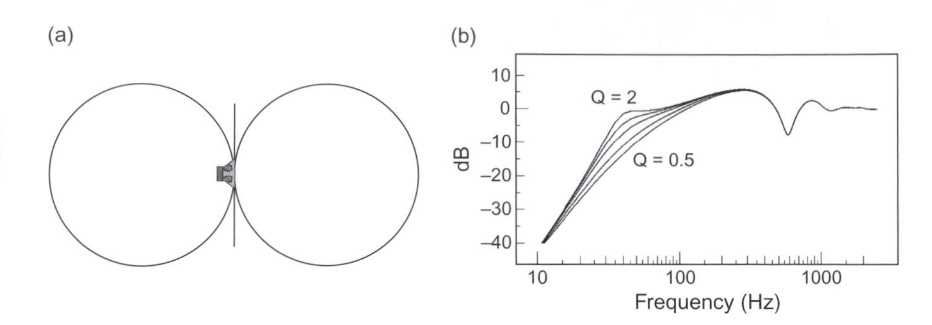

FIGURE 13.2
Directivity and roll-off of open baffles: (a) Radiation pattern polar plot of an open finite baffle; (b) Low frequency response roll-off of an open baffle, showing the effect of the Q (degree of sharpness of resonance) of the driver. The final roll-off tends towards 18 dB per octave below the driver resonance.

baffles tend to sound very clean and, not surprisingly, open. They are largely free of resonances, so their time-domain responses are limited only by the drivers and the rooms in which they are placed.

When mounted on the floor, the solid surface below the open baffle acts like an acoustic mirror, so a baffle of 1 square meter placed on the floor behaves like a baffle of 2 square meters in free space. This enables baffles of practical size to be useful down to frequencies of 40 Hz or below, but the lack of anything other than atmospheric loading on the rear of the diaphragms and poor efficiency of radiation may lead to over-excursion problems with high sound pressure levels at low frequencies. The resonance frequency of the driver on an open baffle will be that of its free-air resonance. Because the open baffle mounting does not push up the free air resonance of the driver, and the back-pressures are not augmented by any constraint of the air behind the diaphragm, lighter moving assemblies may be used. Driver cooling is also something that poses no problem with open baffles, so power compression problems are rarely encountered. The open baffle, in the hi-fi world, still enjoys a devoted following of aficionados.

THE SEALED BOX

The practical realization of an infinite baffle (the sealed box) is rarely large enough to avoid significantly loading the rear of the loudspeaker, so is best called what it really is, *a sealed box.* Just as open baffles tend to sound "open," sealed boxes often tend to sound "boxy." However, this need not be the case if the box and driver are of adequate size, well matched, and if sufficient attention is paid to the suppression of resonances within the box. The constraint of the air within a sealed box causes it to act like a spring, which reacts against the movement of the diaphragm in either direction. This effectively stiffens the suspension of the drive unit, and raises its resonant frequency. The smaller the box, the stiffer will be the spring, so the higher will be the resonant frequency of the driver/box system. As the system resonance defines the frequency at which the low frequency roll-off will begin, then for any given driver the low frequency response will become progressively more curtailed as the box size reduces. The only way to counteract this tendency is to use drivers of lower free-air resonance frequency, which means using a driver with a heavier moving assembly or, to a lesser extent, a more compliant suspension, but both of these characteristics have their drawbacks.

A heavier cone takes more energy to move it, so it will need more amplifier power to drive it to produce the same SPL as a lighter cone. A more compliant suspension will be much less rugged than a stiffer suspension, and will tend to be much more easily damaged in the event of an overload. What is more, a very loose flexible suspension may not be able to adequately resist the pressure changes inside the box at high SPLs, and may physically deform, giving rise to non-linearities in its travel and non-linear distortions in its radiation. A small sealed box must suffer from either poor system sensitivity (due to its poor overall electro-acoustic conversion efficiency) or a low frequency roll-off that begins well into the musical spectrum. The roll-off exhibits a rate of 12 dB per octave below its frequency of resonance, but considerable roll-off may begin well above this frequency, depending on the system Q. Some typical roll-off curves are shown in Figure 13.3. Nevertheless, the time responses (transient responses) of well-designed sealed boxes with correctly matched drivers and adequate damping can be very accurate. Largely for this reason, sealed boxes have a strong following, and large sealed boxes can be the bases of excellent loudspeaker systems.

A sealed box system is said to be critically damped when its size and the driver resonant frequency are matched such that the overall response is already 6 dB down at the resonant frequency. With this alignment, the transient response can be exemplary, with no perceptible ringing. The total system Q_{TC} (or quality factor of resonance) is 0.5. The Butterworth "B2" (maximally flat) alignment is very popular, with a Q_{TC} of 0.7. This exhibits a system response which is 3 dB down at the resonant frequency, and still has

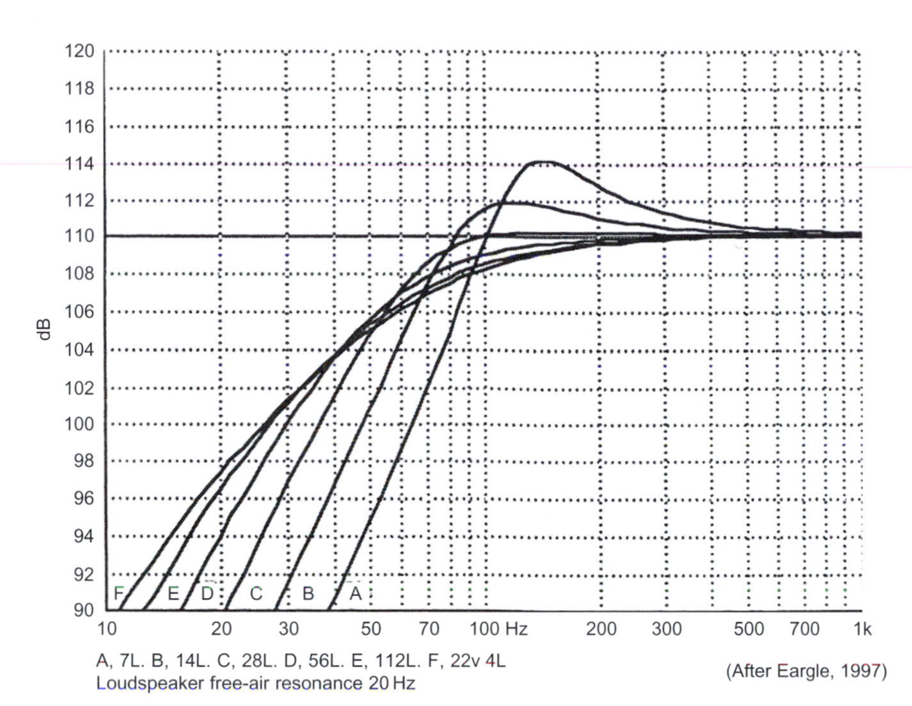

FIGURE 13.3
Typical low frequency roll-off behavior of a sealed box loudspeaker, showing the responses of the same driver in cabinets of six different volumes. All measurements in free-field conditions.

a transient response which is extremely well controlled. The low frequency responses can be extended downwards with alignments where the Q_{TC} is set at 1, or even up to 2, but, as the Q increases, so does the tendency for the transient response to become extended, and for audible ringing or "boominess" to become obtrusive. The outcome of these relationships is that if the low frequency −3 dB point is to be dropped to 30 Hz, and a fast well-damped transient response is required at the same time, then the box must be big. If high SPLs are required, then the only solution to the compromise of a low resonance driver with an adequately robust construction and a good sensitivity is that the driver must also be big.

A 15 inch (380 mm) driver, of high quality, with a 20 Hz free air resonance in a 500 liter enclosure can yield some very impressive bass. However, "impressive" in this context means full, flat, fast and low distortion—in other words—"accurate". Unfortunately, many sealed boxes get themselves a bad name by trying to use "boomy" alignments in forlorn efforts to keep the size down whilst seeking to extend the low frequency response to frequencies that the box size cannot really support. The penalty paid is in terms of low sensitivity and poor transient response. It must be thoroughly understood that there is no clever computer program which can solve this problem. The restrictions that we must accept are deeply entrenched in the physical laws of the universe in which we live. They are that fundamental!

Some manufacturers have tried to sacrifice system sensitivity by lowering the magnet flux in order to lower the system Q. There is a strong "amplifier power in cheap" lobby, who believe that lower efficiency systems can exhibit higher Qs, and hence can be extended in their low frequency range. What they often seem to fail to realize is that a heavier current in the voice coil and a lower power magnet will drastically alter the ratio of the fixed magnetic field to the variable magnet field. The much higher variable field due to the voice coil current can severely distort the position of the flux lines of the weak permanent magnet, and give rise to loss of low level detail in the sound and increased levels of intermodulation distortion. This highlights perhaps one of the worse aspects of the use of programmable calculators or computers in the wrong hands—they can lead to good results on paper, but they can give rise to unpleasant side-effects in practice. Figure 13.4 gives a graphic illustration of the connection between system Q (Q_{TC}) and the transient response. The Q_{TC} is derived from the electrical, magnetic, mechanical and acoustical properties of the total system—electro magnetic damping, mechanical stiffness and air loading. Note that as the Q_{TC} increases, the transient response becomes longer. This is perfectly logical because the transient response becomes more resonant as the system Q_{TC} becomes more resonant. The amplitude response is boosted and extended downwards by keeping the energy responding for a longer time, and not by instantaneously boosting the level. As stated before, in order to boost the level, and nothing else, a bigger box and driver are needed.

Small sealed boxes with relatively high rates of roll-off can be mounted near to room boundaries, where the constraint of their radiation angle can boost their low frequency output, acoustically, without

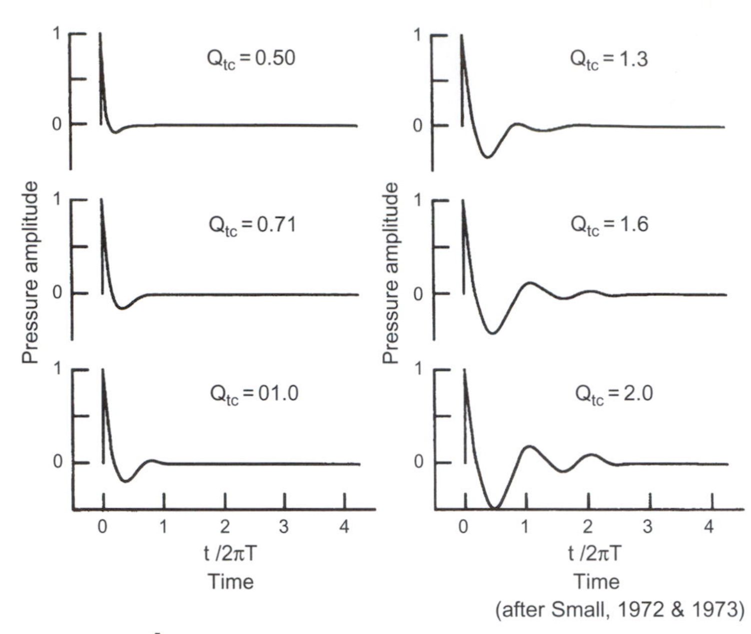

FIGURE 13.4
Transient response of a sealed box enclosure as a function of the total system Q (Q_TC). As the Q_{TC} increases, the transient decay time also increases.

suffering time penalties, but on pedestals in the center of a room, the low frequencies from small sealed boxes will be found to be either weak, resonant or both. On the meter bridge of a solidly-built mixing console they can also receive some low frequency support, but coloration problems due to the reflective surface being *between* the small loudspeaker and the listening position can be a problem. This "acoustic mirror" concept was discussed in the previous section, but when applied to floor standing open baffles, the mid and high frequency drive units are usually mounted well clear of the floor. When a small sealed box is placed on top of a mixing console, the sources of mid-range and high frequencies are inevitably close to the reflecting surface, so comb-filtering of the response is the likely result.

To put things into proportion with respect to size, a cabinet which is 3 dB down with a given low frequency driver at 80 Hz would need to be four times larger if it were to be 3 dB down at 40 Hz and 16 times larger to be 3 dB down at 20 Hz, so sealed box sizes do tend to get larger very quickly if lower roll-of frequencies are required.

One advantage of sealed box systems is that they are relatively self-protecting in terms of excessive cone excursions. Compared with the open baffle, which offers almost no protection, an input signal below the resonant frequency of a sealed box system will tend to drive the cone at a constant excursion for any given input level, independent of frequency. (See Note 1 at end of chapter.) The thermal overload of the coil is therefore the biggest risk factor in terms of driver integrity at input level extremes.

The lining materials in the boxes also have an effect on the low frequency response. Although they are primarily intended to prevent cabinet resonances at mid frequencies, which may color the sound by passing to the outside via a relatively acoustically transparent cone, the lining materials can also affect the low frequency damping and total system Q. They should not be too tightly packed, or effectively they will be more or less solid and will reduce the enclosure volume. Neither should they be able to move en masse, or they can introduce non-linear distortion due to their somewhat erratic movement. Given just the right quantity, however, they can not only reduce box resonances but can also make the boxes appear to be up to around 20% acoustically bigger due to their ability to act as heat sinks and slow down the speed of sound by absorbing heat on compression half cycles and releasing it on rarefaction half cycles. The tortuosity of the path through the pores or fibers also gives rise to sound absorption. The density and quantity of the absorbent material inside a sealed box are therefore chosen for the parts that they play in the air loading and damping calculations for the whole system.

Acoustic suspensions

Developed in the 1950s by Edgar Villchur, in the USA. The principal is to use a very low resonance loudspeaker in a small sealed box. The air in the box may push up the resonance by an octave, or more, and is the predominant restoring force for centralizing the diaphragm, because the low resonance suspension is very weak. Generally, although an acoustic suspension system is a sealed box, the specialized term is normally only used when the ratio of the cabinet (air) compliance to the driver's compliance exceeds a factor of about 4 to 1. In the late 1950s and early 1960s, the company Acoustic Research enjoyed very great success with these designs by making huge improvements in the low frequency fidelity of small loudspeaker systems.

REFLEX ENCLOSURES

Also known as ported enclosures, vented boxes or phase inverters, reflex enclosures use openings, or ports, to tune the cabinet resonance to a desired frequency. Effectively, the air in the port, which may be a simple hole or a tube, acts as a mass which resonates with the spring created by the air inside the cabinet. In Figure 13.5, a mass is shown suspended below a spring. Almost everybody will intuitively know what would happen if they were to pull down on the weight and then let go—the weight would spring back and the system would go into oscillation until the energy was finally dissipated. Adding more weight would cause the oscillation to slow down, as would using a weaker spring. Therefore:

more weight)
weaker spring) slower oscillation (lower frequency)

less weight)
stronger spring) faster oscillation (higher frequency)

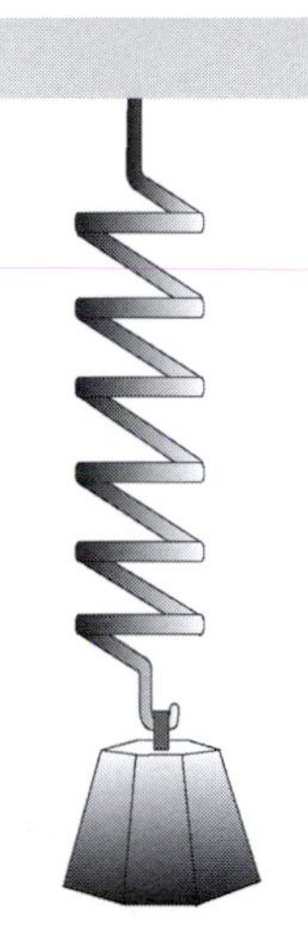

FIGURE 13.5
A mass/spring system. A mass suspended beneath a spring. It is easy to imagine how pulling down on the weight and releasing it would set up an oscillation due to the mass-spring interaction.

In the case of a reflex cabinet, a bigger box provides a weaker spring, because the enclosed air is compressed or rarefied proportionately less than in a small box for any given diaphragm displacement. For any given diameter of hole (port), extending it with a tube will lower the resonant frequency because a greater mass of air will be trapped within it. For any given cabinet volume and mass of air in the port, changing the *area* of the port will also change the resonant frequency. *Increasing* the area will *increase* the resonant frequency. This is because there is more surface area in contact with the air-spring, so more force acts upon the air mass, effectively stiffening the spring. There are therefore three variables in the equation—the cabinet volume, the length of the port, and the area of the port—the latter two defining the volume of air in the port, and hence its mass. Air weighs about 1.2 kg per cubic meter, and thus about 1.2 grams per liter.

The cabinet tuning frequency can therefore be calculated approximately from either of the two following equations, the first in imperial measure and the second in metric units:

$$f_v^2 = \frac{2700A}{V(L + \sqrt{A})} \tag{13.1}$$

where:

f_v = resonant frequency of box (Hz)
A = area of port in square inches
V = volume of box in cubic feet
L = length of port in inches

or

$$f = \frac{1}{2\pi} \times \sqrt{\frac{C^2A}{VL_e}} \tag{13.2}$$

where:

f = resonant frequency of box (Hz)
c = speed of sound in air – 340 m/s
A = area of port in *square meters*
V = volume of box in *cubic meters*
L_e = effective length of port in *meters*

Note: L_e allows for an end correction. The effective length of a port tube is, in reality, somewhat longer than the physical length, but for many calculations the actual physical length can be used.

The formulae are not precise, because there are always variables such as the quantity of air displaced by the drive units themselves, the air displaced by the port tubes, the air displaced by internal bracing, and the effect of the absorbent material inside the enclosure. Nevertheless, the formulae give good working approximations or starting points for calculations. Of course, the cabinet volume is calculated from the *interior* dimensions of the cabinet, not the exterior dimensions.

In practice, when the frequency of resonance of the driver in the cabinet is just above the resonant frequency of the box, the port resonance gives rise to a high load on the diaphragm and greatly reduces the diaphragm excursion. In this way, the ported cabinet can protect the driver from excessive travel while still maintaining a flat response. Below this frequency, the driver output falls, but the port, itself, begins to radiate, thus extending downwards the frequency response. At still lower frequencies the port and loudspeaker outputs occur in opposite polarity, so the response falls off rapidly at 24 dB per octave. Moreover, below the port resonance, air simply pumps in and out of the port under the influence of the driver. At these frequencies, the cabinet is just a box with a big air leak, and it can provide no loading on the driver diaphragm, which then behaves as if it were in an open baffle with no air loading protection, so over-excursions are easy to encounter in reflex enclosures unless the low frequency drive signal is filtered, or has no natural content, below the resonant frequency of the box.

A comparison of the performance of two different low frequency drivers in an open baffle, a sealed box, and a reflex enclosure is shown in Figure 13.6. In practice, a driver would be specifically designed for each type of loading, because the different cabinets or baffles match more optimally with drivers of specific Q_{TC} values. The Q_{TS}, which can be found in many formulae and reference texts, is the sum of the Q_{MS} and the Q_{ES} , which are the mechanical and electrical system quality factors (sharpness of resonance), respectively. The higher the Q, in each case, the more highly tuned is the resonance, as shown in Figure 13.7. For reference, the Q terms commonly found in loudspeaker texts are as follows:[2]

Q_{MS} is the mechanical system Q. It is the ratio of the electrical equivalent of the frictional resistance of the moving parts of the driver to the reflected motional reactance at the free-air resonance frequency of the driver.

Q_{ES} is the electrical system Q, which is given by the ratio of the voice coil DC resistance to the reflected motional reactance at the free-air resonance frequency of the driver.

Q_{TS} is the parallel combination of the Q_{MS} and Q_{ES}, and the equation takes the same form as that for two, parallel, electrical resistors:

$$Q_{TS} = \frac{Q_{MS} \times Q_{ES}}{Q_{MS} + Q_{ES}} \tag{13.3}$$

Q_{TC} is the total system Q of the driver and the cabinet.

The free-air resonant frequency of the drivers for reflex enclosures may also need to be different to the optimum resonant frequencies for sealed enclosures of similar size or covering a similar frequency range. The modern tendency is to tailor complete driver designs to given box sizes and system concepts, and the programs now available for computer analysis are very powerful and very accurate. However, careful listening tests are still a *sine qua non* because concentration on the optimization of one aspect of driver design may unexpectedly change for the worse some other aspect of performance that was not under such close scrutiny. Unfortunately, listening rooms are expensive to build and listening panels can be an expensive luxury. Computer time on the other hand is cheap, and quick, and there has developed a strong tendency to design systems ever more by computer and ever less by careful listening.

ACOUSTIC LABYRINTHS

These are sometimes referred to as transmission lines, but at low frequencies they are usually *not* transmission lines. A true transmission line needs a rear cavity, straight or folded, at least a quarter of a wavelength long. At 30 Hz, with a wavelength of about 11 meters, the line would need to be around 3 meters long, and lines of this length are rare indeed. A true transmission line works by presenting the correct acoustic impedance at the rear of a driver so that all the backwards radiation propagates

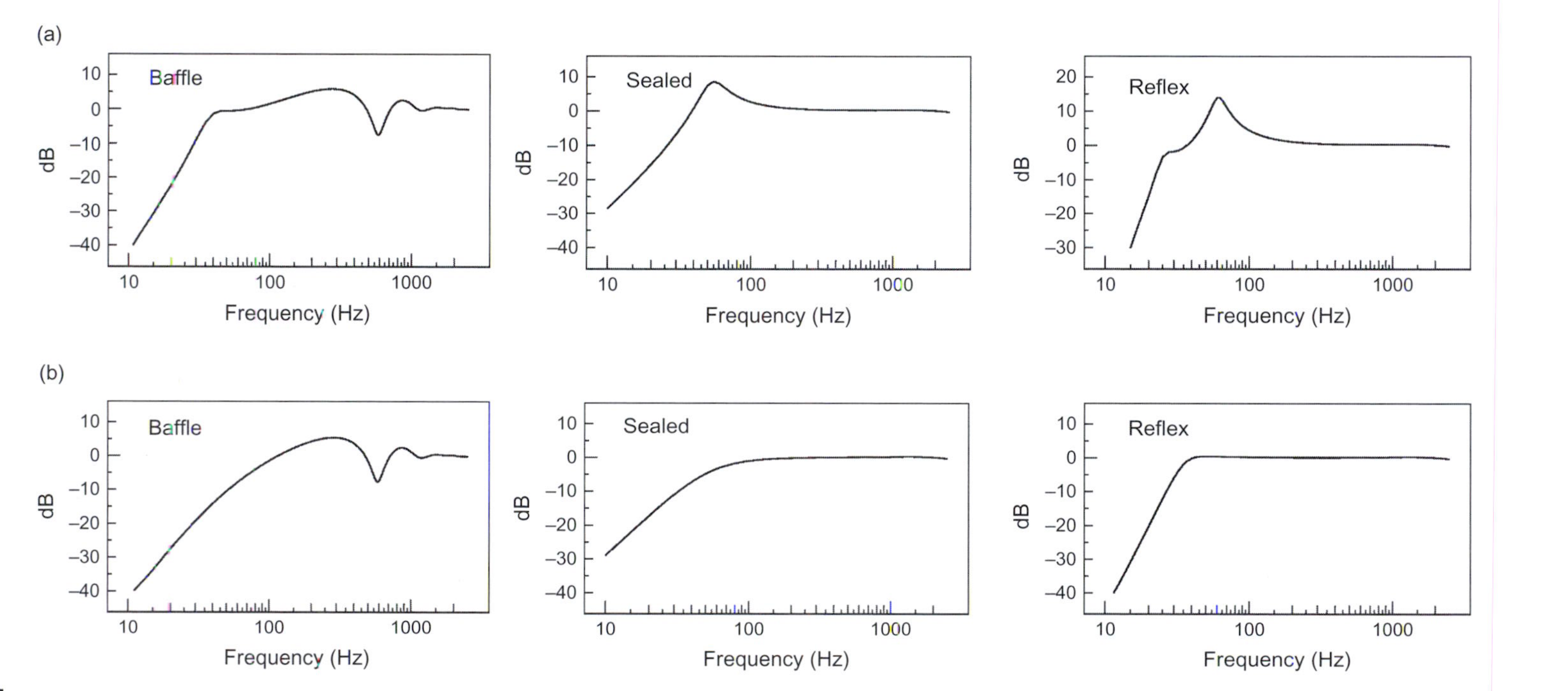

FIGURE 13.6
(a) Response of one driver mounted in an open baffle, a 50L sealed box, and a 50L reflex enclosure with the driver Q optimized for the open baffle response; (b) As in (a), but with the driver Q optimized for the reflex enclosure. Note how as the driver is optimized for one type of loading, the response may suffer with other types of loading.

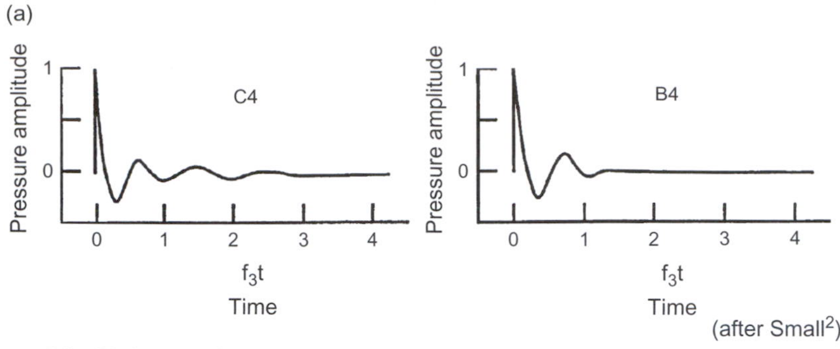

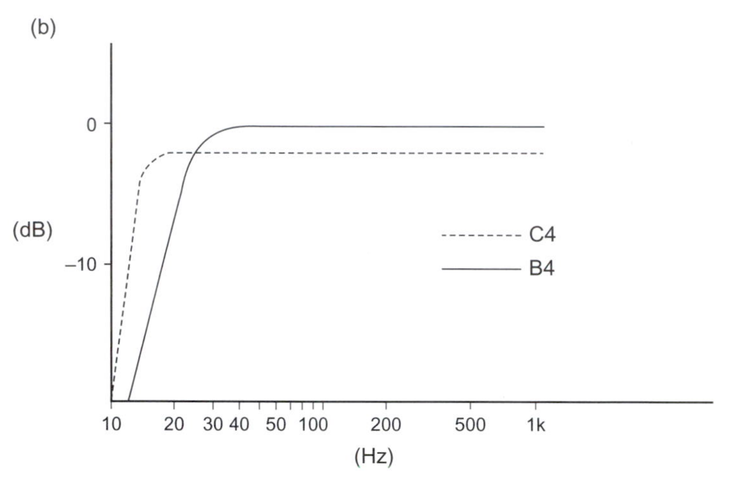

FIGURE 13.7
(a) Transient response of a reflex enclosure as a function of the total system Q (Q_{TC}). Compare with Figure 13.4. (b) Pressure amplitude response of the C4 and B4 alignments. Note how the response extension of the C4 alignment corresponds with an increase in the decay time, as shown in (a). Transient response is traded for response extension.

away from the driver, never to return. This can be achieved by loading the rear of the driver with an infinitely long pipe (which is rather impractical) or by some other system that absorbs all the sound energy. A finite length pipe can therefore be made to operate as a transmission line if it contains sound absorbing material strategically placed to give the correct, purely resistive acoustic impedance, such as an anechoic wedge. However, in order to work at low frequencies this pipe still needs to be very long, so some form of low frequency tuning is often employed.

If, instead of an infinite pipe we think of an organ pipe, it would exhibit a series of resonant frequencies determined by its length. If we attach it to the rear of the driver and fold it round to the front, there will be interference between the sound from the front of the driver and that from the open end of the pipe. When the pipe is one quarter wavelength long, there will be a high acoustic pressure at the rear of the driver and a high acoustic velocity at the open end, which combines with a phase difference of 90° with the acoustic velocity from the front of the driver, and this provides a useful boost in output. As the frequency is lowered, the output from the pipe increases in phase difference with respect to the direct output from the driver, and so tends towards cancelling the combined output. This yields a 24 dB per octave roll-off below the tuning frequency, which leads to transient problems not unlike those of a conventional reflex cabinet. A finite length open pipe, which is what the vast majority of so-called transmission lines certainly are, is clearly *not* a transmission line, as it works on a completely different principle and yields a different acoustic performance. A typical cross-section of such a design is shown in Figure 13.8(a).

In order to tame the strong resonant behavior exhibited by the open pipe, absorbent material is introduced into the pipe to add damping. As the amount of absorbent material is increased, the acoustic performance tends towards that of a transmission line except at very low frequencies, where there is insufficient absorption. A carefully lined, or filled, open-ended pipe may thus exhibit some of the properties of a transmission line, but may also rely on the quarter wavelength resonance to supplement the total low frequency sound output. Most commercial "transmission lines" are therefore really something between the two extremes of a transmission line and an open pipe, depending on the amount of absorption present at any given frequency.

Some versions of "transmission lines" are closed. In these designs the line is made to be as absorbent as possible. In reality it is a sealed box, but it differs from the sealed box in that there is no added air-spring effect, and therefore it does not raise the free-air resonance of the driver.

In practice, the low frequency "pipe," whether straight or folded, must not only be a quarter wavelength long at the lowest frequencies, but must also be wide enough so as not to obstruct the rear radiation from the driver. In some cabinets, as shown in Figure 13.8(b), the line narrows along its length, each section doubling the length of the previous section and halving its cross-section. Inevitably, in order to maintain close to ideal working, transmission line enclosures need to be large, therefore a small low frequency transmission line tends to be a contradiction in terms. The smaller cabinets at low frequencies may act as either sealed boxes or reflex enclosures, dependent upon the density and distribution of the absorbent material. The labyrinth enclosures with open-end terminations tend to

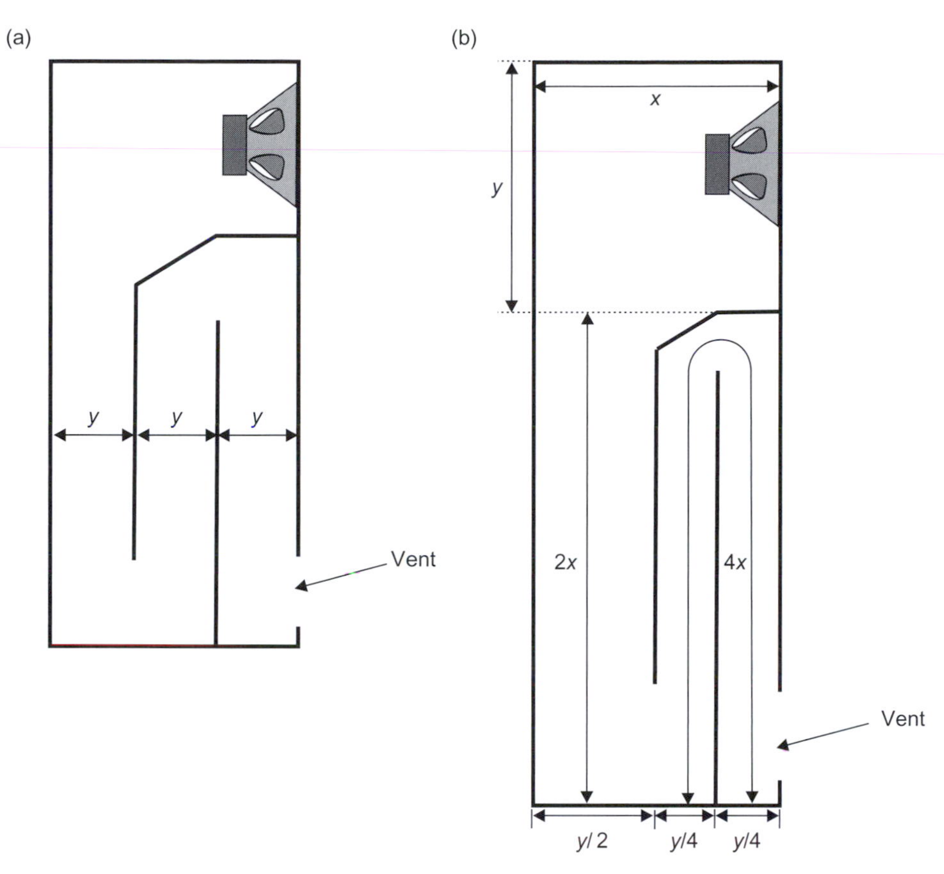

FIGURE 13.8
Acoustic labyrinths: (a) A parallel labyrinth; (b) A tapered labyrinth—each time the section of the labyrinth changes, the length (x) doubles and the width *(y)* halves. The entire labyrinth would be lined with absorbent material in both (a) and (b).

act like reflex enclosures in that the cone excursions are much reduced at the resonant frequency of the enclosure tuning, which may result in lower distortion than might be expected from an approximately equivalent sealed box at similar SPLs.

One other interesting aspect of acoustic labyrinths is that the highly resistive rear loading can actually *lower* the driver's free-air resonance due to the mass of air which acts directly on the cone, effectively making it heavier when in movement.

Labyrinths/transmission lines have their following, but many designers feel that they are a complex way of achieving rather little. In lines which almost totally lose the rear radiation, there is no out of phase output at low frequencies, so they exhibit 12 dB per octave roll-offs like sealed boxes.

Modern transmission lines

In the early 1990s, the British company PMC (Professional Monitor Company) began manufacturing a range of "transmission line" loudspeakers which have since achieved considerable acclaim and commercial success. The subject of what is and what is not a transmission line has been rather controversial in recent years, so Peter Thomas, the principal design engineer and managing director, was asked by the authors of this chapter to try to clarify the matter, and he subsequently supplied the following paragraphs of this section.

The birth of the modern transmission line speaker design came about in 1965 with the publication of A. R. Bailey's article in Wireless World, "A Non-resonant Loudspeaker Enclosure Design," (Bailey, 1965) detailing a working transmission line. Radford Audio took up this innovative design and briefly manufactured the first commercial transmission line loudspeaker. Shortly thereafter John Wright of IMF Electronics designed a range of transmission line designs and made them popular through his refinement and development of Bailey's theory. Although acknowledged as the father of the transmission line, Bailey's work drew on the work on labyrinth design, dating back as early as the 1930s. His design, however, differed significantly in the way in which he filled the cabinet with absorbent materials. Bailey hit upon the idea of absorbing all the energy generated by the bass unit inside the cabinet, providing an inert platform for the drive unit to work from. Unchecked, this energy produces spurious resonances in the cabinet and its structure, adding distortion to the original signal.

The transmission line (TL) is the theoretical ideal and most complex construction with which to load a moving-coil drive unit. The most practical implementation is to fit a drive unit to the end of a long duct that is open ended. In practice, the duct is folded inside a conventional-shaped cabinet with the open end of the duct usually appearing as a vent on the front of the cabinet. There are many ways in which the duct can be folded, and Figure 13.8 illustrates two typical forms. The line is often tapered in cross-section to avoid parallel internal surfaces that encourage standing waves. Depending upon the drive unit and quantity—and various physical properties—of absorbent material, the amount of taper will be adjusted during the design process to tune the duct to remove irregularities in its response. The internal partitioning provides substantial bracing for the entire structure, reducing cabinet flexing and coloration. The inside faces of the duct or line are treated with an absorbent material to provide the correct termination with frequency to load the drive unit as a TL. A theoretically perfect TL would absorb all frequencies entering the line from the rear of the drive unit, but remains theoretical as it would have to be infinitely long. The physical constraints of the real world demand that the length of the line must often be less than 4 meters before the cabinet becomes too large for any practical applications, so not all the rear energy can be absorbed by the line. In a realized TL, only the upper bass is TL loaded in the true sense of the term (i.e., fully absorbed); the low bass is allowed to freely radiate from the vent in the cabinet. The line therefore effectively works as a low-pass filter, another crossover point in fact, achieved acoustically by the line and its absorbent filling. Below this "crossover point" the low bass is loaded by the column of air formed by the length of the line. The length is specified to reverse the phase of the rear output of the drive unit as it exits the vent. This energy combines with the output of the bass unit, extending its response and effectively creating a second driver.

Phase inversion is achieved by selecting a length of line that is equal to the quarter wavelength of the target lowest frequency. The effect is illustrated in Figure 13.9(a), which shows a hard boundary at one end (the speaker) and the open-ended line vent at the other. The phase relationship between the bass driver and vent is in-phase in the pass band until the frequency approaches the quarter wavelength, when the relationship reaches 90° as shown. However by this time the vent is producing most of the output—as shown in Figure 13.9(b). Because the line is operating over several octaves with the drive unit, cone

(a)

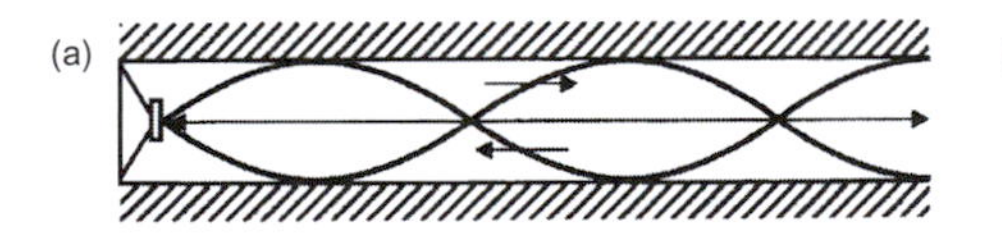

(b)

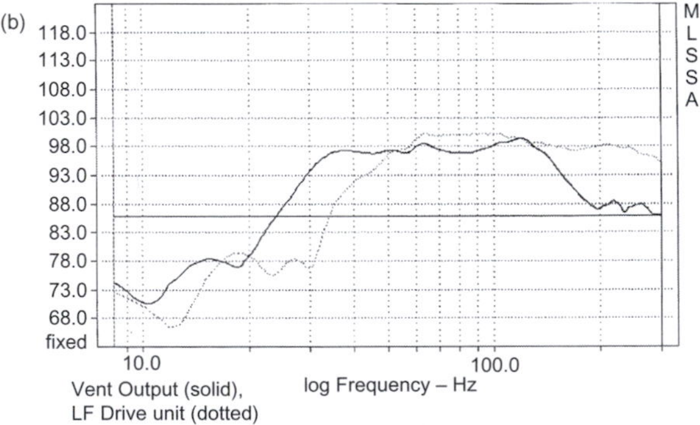

(c)

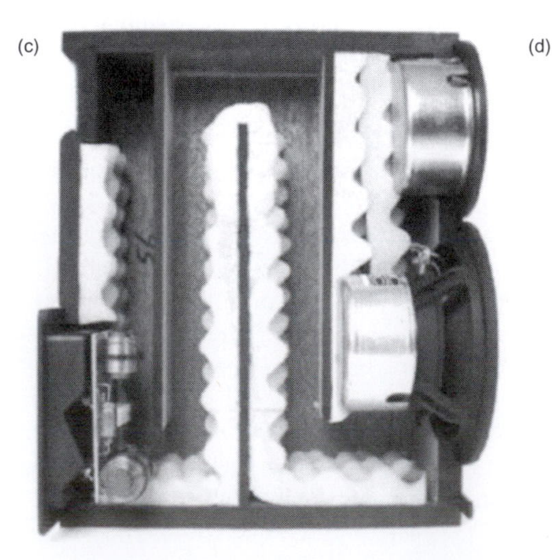

(d)

FIGURE 13.9
PMC transmission lines: (a) Phase relationship between the driver and the vent in the quarter wave transmission line; (b) The drive unit and vent contributions to the overall output; (c) Cut-away view of a PMC transmission line; (d) A pair of large PMC cabinets, with their designer.

excursion is reduced, providing higher SPLs and lower distortion levels compared with reflex and sealed box designs.

The calculation of the length of the line required for a certain bass extension appears to be straightforward, based on a simple formula:

$$\lambda = 344/4 \times f \qquad (13.4)$$

where:

f is the quarter wavelength frequency
344 ms is the speed of sound in air at 20 °C
λ is the length of the transmission line

However the introduction of the absorption materials reduces the velocity of sound through the line, as discovered by Bailey in his original work. Bradbury published his extensive tests to determine this effect in an AES Journal in 1976 (Bradbury, 1976) and his results agreed that heavily damped lines could reduce the velocity of sound by as much as 50%, although 35% is typical in medium damped lines. Bradbury's tests were carried out using fibrous materials, typically longhaired wool and glass fiber. These kinds of materials however produce highly variable effects that are not consistently repeatable for production purposes. They are also liable to produce inconsistencies due to movement, climatic factors and effects over time. High specification acoustic foams, developed by PMC with similar characteristics to longhaired wool, provide repeatable results for consistent production. The density of the polymer, the diameter of the pores and the sculptured profiling are all specified to provide the correct absorption for each speaker model. Quantity and position of the foam is critical to engineer a low-pass acoustic filter that provides adequate attenuation of the upper bass frequencies, whilst allowing an unimpeded path for the low bass frequencies.

There are therefore two distinct forms of bass loading employed in a TL, which historically and confusingly have been amalgamated in the TL description. Separating the upper and lower bass analysis reveals why the TL has so many advantages over reflex and sealed box designs. The upper bass is almost completely absorbed by the line allowing a clean and neutral response. The lower bass is extended effortlessly and distortion is lowered by the line's control over the drive unit's excursion. One great advantage of the low frequency extension provided by transmission lines is the perception of deep bass even at low listening levels, due to the extended flatness of the response.

The complex loading of the bass drive unit demands specific Thiele-Small driver parameters to realize the full benefits of a TL design. Most drive units in the market place are developed for the more common reflex and sealed box designs and are usually not suitable for TL loading. To design a high efficiency woofer with extended low frequency ability, one tends to need cones which are often extremely light and flexible with very compliant suspensions. Whilst performing well in a reflex design, these characteristics do not match the demands of a TL design. The drive unit is effectively coupled to a long column of air which has mass. This lowers the resonant frequency of the drive unit, negating the need for a highly compliant device. Furthermore, the control of this column of air requires an extremely rigid cone, to avoid deformation and consequent distortion. The lack of available suitable drive units created the necessity for PMC to design a series of drivers employing a flat, 6 mm thick diaphragm, manufactured from aerospace materials, that provide extraordinary stiffness whilst maintaining a relatively low mass.

The combination of extended frequency response, higher sound pressure levels and lower distortion afforded by TLs, separates them from reflex and sealed box models. In addition, phase accuracy is superior to many other moving-coil designs as a result of the absorption provided by the line in the upper bass range. The low frequency roll-off can be as low as 12 dB per octave in highly damped lines, matching the sealed box arrangement and avoiding the large phase changes inherent in reflex designs. A cutaway section of a PMC transmission line is shown in Figure 13.9(c), and a pair of complete cabinets in Figure 13.9(d).

ABR SYSTEMS

A variation on the use of reflex enclosures is to use an auxiliary bass radiator (ABR) instead of a port. The ABR often takes the form of a loudspeaker with no magnet assembly. They are often also referred to as passive radiators or drone cones, and are normally of approximately the same size as the cone of the driver. One of the primary advantages of ABRs is that there is no air-flow associated with them, so there

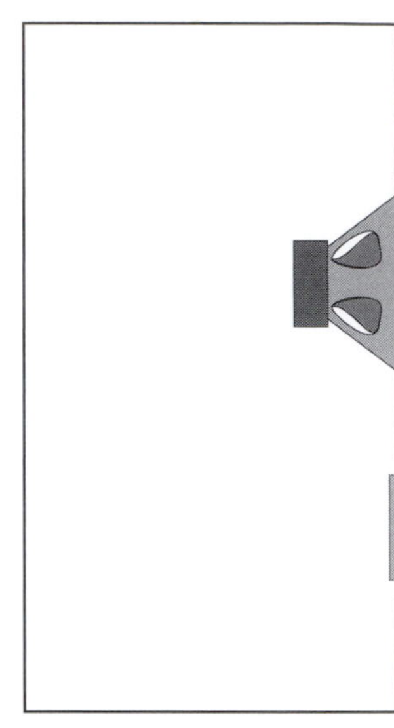

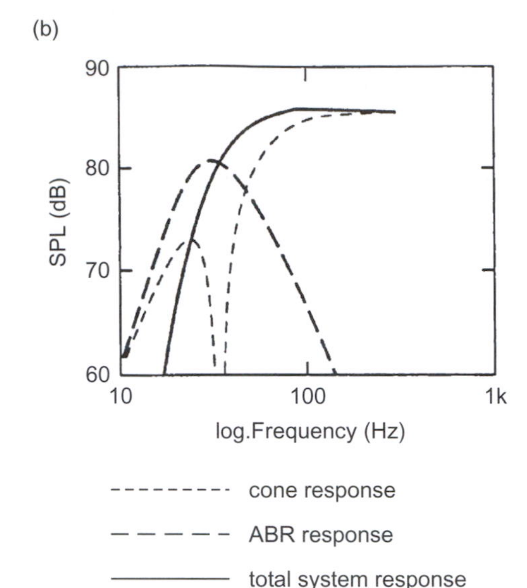

FIGURE 13.10
The ABR: (a) A passive radiator (ABR) loudspeaker system; (b) Cone and ABR contributions to the overall response.

is no turbulence or wind noise, which can be a problem when small diameter ports are the only means of tuning a cabinet. On the negative side, the compliance (springiness) of the suspension system of an ABR can be non-linear at high excursions, and hence can be a source of distortion which can be more subjectively noticeable than port distortion.

ABRs can be used in small boxes, giving them a reflex-type response when the tuning frequency would require ports which would be too small to be practical. In general, ABRs offer reflex-type advantages over sealed boxes, such as 4–6 dB more output on typical program material and a lower −3 dB point at low frequencies, but the cut-off is more steep once it begins, and, as with any resonant system, the transient response gets smeared. The transient response of a resonant system depends upon the Q of the resonance, as shown in Figure 13.7. There are some alignments (tunings) known as quasi-Butterworth third order (QB3) and sub-Chebychev fourth order (SC4) which use lower Qs. They exhibit transient characteristics rather like sealed boxes of Q = 1 (see Figure 13.4) but maintain the reduced diaphragm excursions of the reflex enclosures. However, the choice of musical program and room acoustics may well be the determining factor as to whether these alignments are beneficial or not. Electronic music in a highly damped control room is much more revealing of tuning resonances than would be romantic music in a relatively live domestic room. In many ways it is fortunate that we have at our disposal a range of loudspeaker performances, because none are perfect, and we have a similar range of musical and acoustical requirements.

FIGURE 13.11
A typical bandpass enclosure.

ABR systems have not had a totally continuous development. They tend to emerge from time to time as solutions to specific problems. When air is used in a resonator, its density is fixed, so if not enough volume is available in the cabinet for the necessary sized port, or if excessively long tubes are called for (which suffer from viscous losses), then low tuning frequencies cannot be accomplished. An ABR offers the use of a number of different materials of varying weights (masses, densities) and also offers the possibilities of lower box tuning when size is limited. Polystyrene diaphragms of the correct weight, suspended in a loudspeaker-type surround (usually a half-roll) are now often the choice of the designers who use ABRs, as shown in Figure 13.10(a). The total acoustic output is the sum of the volume velocity of the front of the driver diaphragm and the out of phase volume velocity of the ABR. The ABR is, of course, driven by the rear of the driver via the air in the cabinet. The relative contributions of the driver and the ABR to the combined output of the system is shown in Figure 13.10(b). ABRs are also sometimes chosen for use in systems where an air-tight cabinet is required, such as for outdoor use.

BANDPASS CABINETS

If the low frequency driver is enclosed with air volumes on each side of the cone, and the only radiation to the outside air is via a port in one of the air volumes, the result is a bandpass loudspeaker, as shown in Figure 13.11. Bandpass cabinets are usually restricted to use as sub-woofers because the pass-band is very limited—rarely more than one octave of flat response. The roll-offs at either end of their spectrum of use tend to be 12 dB per octave, because there is no direct radiation to give rise to the phase differences that lead to the 24 dB per octave lower roll-off rate of reflex enclosures, unterminated transmission lines or ABR systems. In the design shown in Figure 13.11, the lower roll-off frequency is governed largely by the inner sealed chamber, and the upper roll-off frequency is governed by the outer ported chamber.

However, there do exist some designs with both chambers vented, generally in an attempt to gain overall system efficiency in the pass-band.

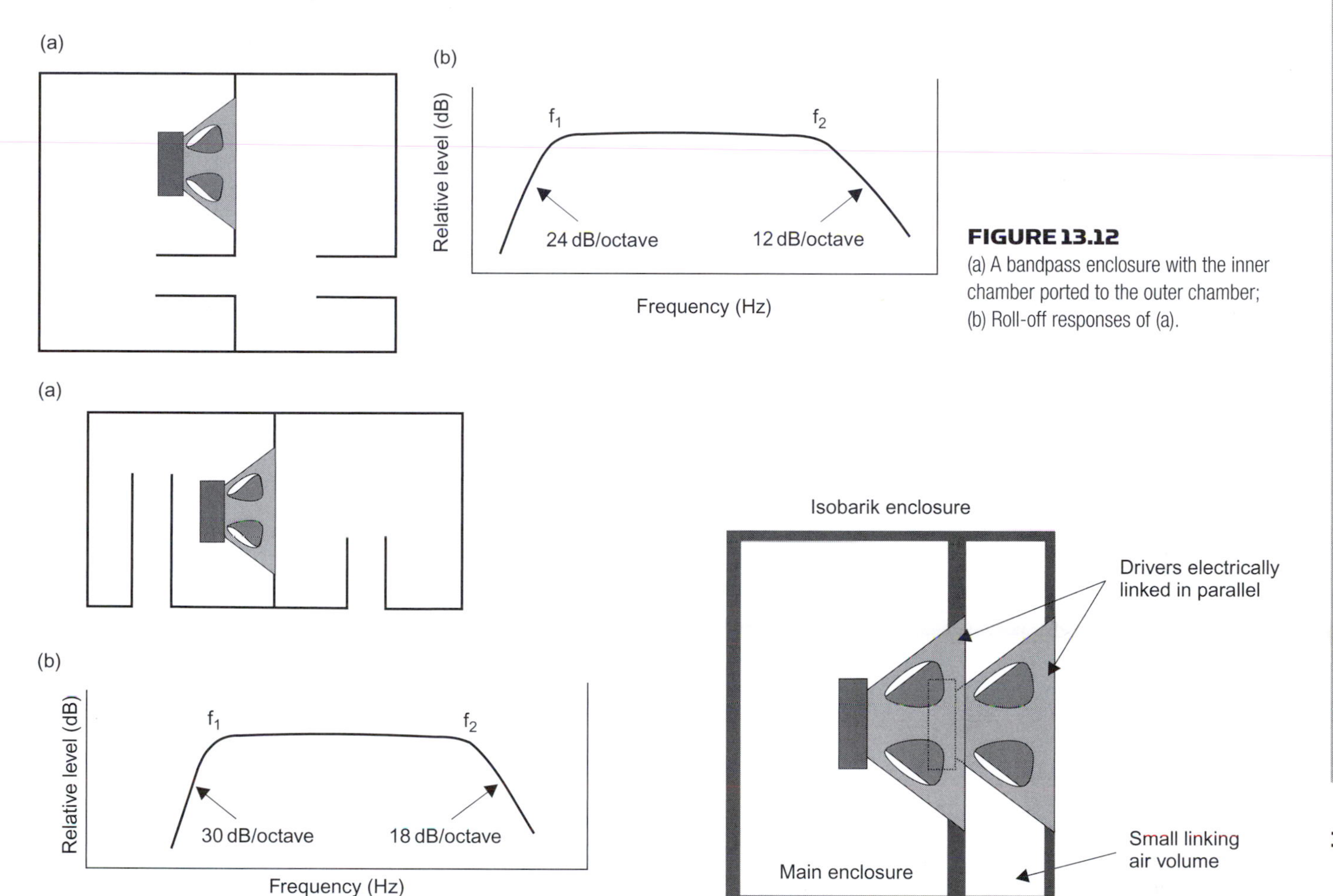

FIGURE 13.12
(a) A bandpass enclosure with the inner chamber ported to the outer chamber; (b) Roll-off responses of (a).

FIGURE 13.13
(a) A bandpass enclosure with both chambers ported to the outside. (b) Roll-off response of (a).

FIGURE 13.14
The Linn Isobarik enclosure concept. The loudspeaker drivers are connected in acoustic series (cascade), but electrically in parallel.

Bandpass enclosures can normally be much physically smaller than other configurations for any given output capability, but the responses, not surprisingly, tend to be resonant and thus exhibit poor transient responses. Figure 13.12 shows the design of a bandpass enclosure in which both chambers are ported, one to the outside and the other to the outer chamber. This results in a 24 dB per octave roll-off at the lower frequency. A further development is shown in Figure 13.13. In this instance, both chambers are ported to the outside, resulting in a 30 dB per octave roll-off at the lower frequencies and an 18 dB per octave roll-off at the higher frequencies. Both of the latter two designs were developed by the Bose Corporation. (Eargle, 1997).

Care must be taken in the acoustic treatment of the chambers which vent to the outside because internal resonances at high frequencies can escape through the ports if not dealt with internally. Care must also be taken when siting bandpass enclosures, because their proximity to solid boundaries may severely affect their response, and the high velocity ports must be free from obstructions. Some designers claim better response linearities due to reduced cone excursions for any given output SPL as compared with other low frequency loudspeaker systems, but their use with true high-fidelity systems is very limited due to transient response anomalies.

SERIES DRIVER OPERATION AND ISOBARIC LOUDSPEAKERS

If the port in the outer chamber shown in Figure 13.11 were to be replaced by another drive unit, a system would result as shown in Figure 13.14, with the two drivers in acoustic series (but with the drivers still connected in parallel, electrically). At low frequencies, the two drivers behave like a single driver with twice the moving mass. The result is a downward shift in resonant frequency to about 70% of that

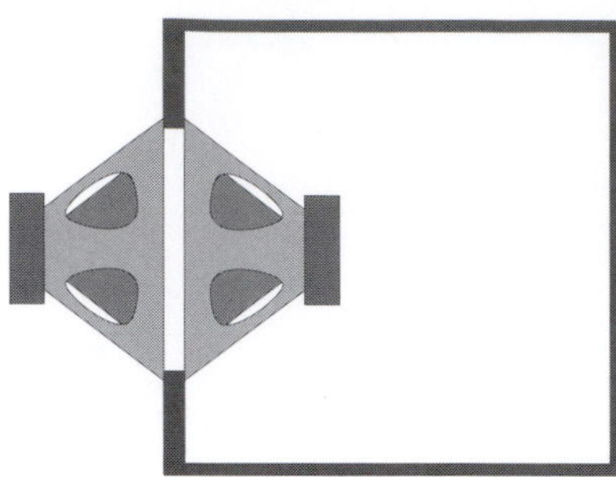

FIGURE 13.15
A variation on the theme of Figure 13.14.

of just one of the drivers in the sealed enclosure. The reduced back-pressure exerted by the enclosure on the externally radiating driver will result in lower distortion because the inner loudspeaker is tending to keep the pressure in the outer chamber constant. However, in practice, the two drivers also tend to function as one due to the strong coupling by the trapped air mass. Teifenbrun first used the term isobaric for this type of operation—"isobaric" meaning "same pressure" (UK patent 1,500,711). Another variation on the theme is shown in Figure 13.15.

GENERAL DISCUSSION

Although there are many esoteric designs of low frequency loudspeaker systems on the market, the ones described so far in this chapter are the ones which will cover 99.9% of the loudspeakers to be found in the mainstream music recording and domestic reproduction environments. In general, high levels of fast flat responses are not available from small enclosures and it seems obvious that nobody would choose to use large, expensive, unwieldy cabinets if more compact solutions were available. Nevertheless, despite the barrage of marketing claims about revolutionary low frequency sources, and a widespread tendency for many people to believe that computer control and signal processing can resolve any problems, the fact remains that the radiation of low frequencies is something that resides firmly in the domain of the laws of acoustics, and they tend to be somewhat inflexible. It should be noted that in nature, only large objects radiate low frequency sounds. The tendency towards using compact, single, mono low frequency sources is something that can greatly improve the overall response, both subjectively and objectively, in poor-to-quite-good circumstances of use. However, as the room acoustics become better controlled and the signal path, including the loudspeakers, becomes higher in resolution, the optimum choice for low frequency reproduction tends to return to favor stereo sources in integrated loudspeaker systems—i.e. without physically separated sub-woofers. What is optimum in the mid-to-reasonably-high quality range of loudspeaker systems may not necessarily be extrapolated as being best at the highest quality levels. One must be very careful not to generalize about things which are for specific applications. For example, an orchestral recording with phase differences in the left and right channels at low frequencies would benefit, in a good room, from stereo bass. It has now been determined that only by restricting the crossover frequency to less than 50 Hz can the bass be summed into mono without losing the spaciousness in the sound (Martens et al., 2004). However, in a poorly controlled room, a mono bass may lead to less general confusion if crossed over and combined into mono an octave higher. Such choices can be very circumstantially dependent.

CABINET LINING MATERIALS

Any hard-surfaced box will suffer from internal reflexions and resonances when excited by a loudspeaker drive unit mounted in one of its surfaces. The nature of cone loudspeakers is such that they are relatively transparent to sound at mid frequencies, so any reflexions and resonances occurring within the box are likely to pass to the outside via the cone, and combine with the directly radiating sound in a way that will give rise to undesirable coloration. One of the fundamental reasons for applying absorbent linings of foams or fibrous materials to the inside of loudspeaker cabinets is to reduce to inconsequentially low levels the coloration effects by rendering the boxes as acoustically non-reflective as is reasonably possible. A further advantage of the application of porous or fibrous materials is that they can slow down the speed of sound, and thus make the cabinets appear to be acoustically larger than they are physically. The practical limit to this size increase appears to be around 20% (which is still a useful gain) because the thermal transfer characteristics of the materials normally used are not sufficient to achieve the theoretical 41% maximum. Air heats up when compressed and cools on rarefaction. Both of these effects tend to augment the speed of sound on the successive half-cycles because the air itself is a poor conductor of heat, so the thermal changes are trapped within the waves. However, if the air is in close contact with another material, distributed throughout its volume, which *can* conduct the heat, the augmentation of the speed of sound due to the heat of compression and the cold of rarefaction will not be apparent. The ratio of the speed of sound with and without this augmentation is about 1.41 to 1, hence the approximately 40% difference in apparent box size if the heat conduction were total.

Partly for this reason, but also because fibrous materials are better absorbers where the particle velocities of the air movement are at their highest (they must be zero at rigid boundaries, so they are at their lowest close to the boundaries), the absorbent materials are best placed in the volume of the box, lightly packed, and not only against the sides of the box. Reticulated (open cell) foams, glass fiber, mineral wool, bonded acetate fibers, polyester fibers and cotton-waste felt are all common lining materials. The cut-away view of the "transmission-line" cabinet shown in Figure 13.9 illustrates the use of a synthetic foam lining which the manufactures specifically chose for its ability to maximally damp the line. Material types and densities are normally chosen with care for specific applications.

The KEF loudspeaker company, in the 1980s, noticed some differences in their loudspeaker frequency responses depending upon whether the excitation signal was of a steady state or transient nature. The discrepancy turned out to be due to non-uniform movement of some of the internal lining materials, which was somewhat uncontrolled after the shock excitation of a transient signal. The lining materials should therefore not be in panels which can vibrate en masse, or non-linear effects may be sufficient to be noticeable in the sound from the loudspeakers. Vibrating lining materials may settle into regular patterns on relatively steady signals, but can be excited rather unpredictably by transient shocks. At high SPLs, the linings can move in rather erratic manners, but the effects are usually swamped by the higher SPLs of the radiation direct from the driver. Nevertheless the ideal lining would be relatively inert. Colloms (2005) claims that unstable lining materials can impair the sense of "rhythm" from a loudspeaker system.

At low frequencies, the thicknesses of the lining materials are far too little to provide much absorption, but the cabinets are normally also far too small to support any resonant modes (100 Hz would require an internal dimension of at least 1.6 meters) so the lack of absorption rarely becomes a practical problem. However, at higher frequencies, the absorption is important in order to reduce internal resonances which could powerfully excite structural resonances in the cabinet walls, and which would then radiate into the listening rooms.

CABINET CONSTRUCTIONS

Above all, loudspeaker cabinets should be either rigid or heavy or both. A non-rigid (and/or lightweight) cabinet will be excited into vibration by the drive unit(s). In most cases, rigid materials tend to be heavy or expensive, so light cheap loudspeakers always tend to be sonically suspect because they will probably suffer from cabinet vibration coloration. Any part of a cabinet which vibrates in sympathy with the driver diaphragm will, itself, act as a diaphragm and interfere with the driver output, leading to coloration of the overall sound output. In order to prevent the structural vibration of cabinets, sandwich constructions can be used, with lead sheet or plasticized deadsheets between the layers. Internal bracing is also an option, which pushes up the resonant frequencies into regions that tend to be more easily damped. Phenolic materials are another common choice because they can produce wood composites of very high density and rigidity. Lighter weight, highly rigid materials are sometimes to be found, in the nature of honeycombs or matrices, but they tend to be expensive and difficult to manufacture. However, in all cases, the goals are similar—rigidity and high vibration damping. It is important that the materials do not ring when struck.

The panel radiation is proportional to the size of the cabinet, so the problem of resonance avoidance becomes greater as cabinet size increases. For this reason, a given thickness of material for the wall of a small cabinet may need to be significantly increased as the cabinet size increases. The weight therefore increases greatly, because not only are the panel sizes larger, but they must also be thicker if the same vibrational insensitivity is to be maintained. Large loudspeaker systems of high quality tend to be very heavy indeed.

In many cases, a stiff large panel, which is well behaved at low frequencies may still ring at mid frequencies, and a well-damped panel at mid frequencies may flex at low frequencies. Finding solutions which are dead at all frequencies is often difficult. In some of the more esoteric designs, mineral-loaded acrylics and melamine are used as panel materials, but they can be very difficult to work with.

CABINET SHAPES AND DIFFRACTION EFFECTS

In many cases, modern loudspeakers, although basically of rectangular shape, have rounded or chamfered edges. Figure 13.16 reproduces the classic work of Olson (1969) on the subject of the diffraction effects on the overall loudspeaker response due to cabinet shape. The responses are for an identical

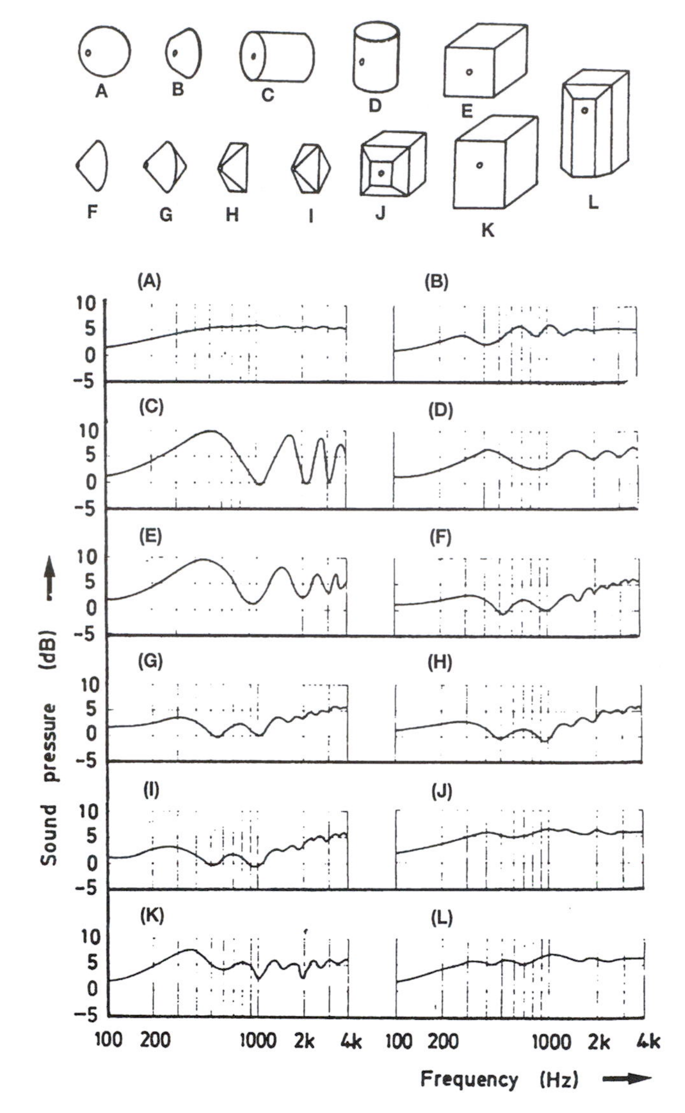

FIGURE 13.16
Olson's classic work on the effects of cabinet shapes on driver responses (Olson, 1969).

drive unit in each cabinet. The sphere looks attractive, but difficulty of manufacture and problems due to all the internal axial reflexion path lengths being the same lead to practical problems in its implementation. Figures 13.16 J and L are reminiscent of many modern designs, and their validity is borne out by the response plots. Of course, with built-in/flush-mounted loudspeakers, the diffraction problem is nullified, which is one reason why so many professional monitors are so mounted.

At low frequencies, the sound from a loudspeaker cabinet is radiated spherically if the loudspeaker is mounted in free space. At higher frequencies, where the wavelengths are small with respect to the front face of the cabinet, the *cabinet* will tend to act like an infinite baffle, and the radiation will be hemispherical. At still higher frequencies, where the wavelengths are small compared with the *radiating diaphragm*, the sound will be beamed forwards, regardless of whether it is mounted on a baffle, or not. The diffraction effects largely arise at the transition between the first two zones of radiation, i.e., between the spherical and hemispherical radiation zones.

As a sound wave in this transition zone radiates away from a source on a finite-sized cabinet wall, it spreads out as it propagates in the manner of half of a spherical wave. When the wave reaches the edge of the wall, it suddenly has to expand more rapidly to fill the space where there is no wall (see Figure 13.17). There are two consequences of this sudden expansion. First, some of the sound effectively "turns" the corner, around the edge, and carries on propagating into the region behind the plane of the source. Second, the sudden increase in expansion rate of the wave creates a lower sound pressure in front of the wall, near the edge, than would exist if the edge were not there. This drop in pressure then propagates away from the edge into the region in front of the plane of the source. The sound wave that propagates behind the plane of the source is in phase with the wave that is incident on the edge, but the one that propagates to the front is in phase opposition. These two "secondary" sound waves are known as diffracted waves and they "appear" to emanate from the edge; the total sound field may then be thought of as being the sum of the direct wave from the source (as if it were on an infinite baffle) and the diffracted waves.

The edges of the cabinets can be thought of as small loudspeaker radiating in antiphase to the real driver. The direct wave exists only in front of the baffle; the region behind is known as the "shadow" region where only the diffracted wave exists.

At low frequencies, the diffracted waves from all of the edges of the finite-sized cabinet sum to yield a sound field with almost exactly one half of the pressure radiated by the source of an infinite baffle. Thus behind the cabinet there is pressure due to the diffracted wave only, and in front of the cabinet there is the direct wave plus the negative-phased diffracted wave. Assuming that the edge is infinitely sharp (has no radius of curvature), there can be no difference between the strength of the diffracted wave at low frequencies and that at high frequencies (the edge remains sharp regardless of scale). The only difference, therefore, between the diffracted waves at low frequencies and those at higher frequencies is the effect that the path length differences between the source and different parts of the edge has on the radiated field. The diffracted waves from those parts of the edge further away from the source will be delayed relative to those from the nearer parts, giving rise to significant phase differences at high frequencies but not at low frequencies. The net result is a strong diffracted sound field at low frequencies and a weak diffracted sound field at high frequencies.

Figure 13.18 shows the results of a computer simulation of the typical effect that a finite-sized cabinet has on the frequency response of a loudspeaker. Figure 13.18(a) is the on-axis frequency response of an idealized loudspeaker drive-unit mounted in a true infinite baffle. The response is seen to be uniform over a wide range of frequencies. Figure 13.18(b) is the frequency response of the same drive-unit mounted on the front of a cabinet of dimensions 400 mm high by 300 mm wide by 250 mm deep. The 6 dB decrease in response at low frequencies, due to the change in radiation from baffled to unbaffled, is evident from a comparison between Figures 13.18(a) and (b). Also evident is an unevenness in the response in the mid-range of frequencies. These response irregularities are due to path length differences from the diaphragm to the different parts of the diffracting edges and on to the on-axis observation point. Unlike the low frequency behavior, these are dependent upon the detailed geometry of the driver and cabinet and the position of the observation point. Therefore, in order to try to ameliorate these response irregularities, many loudspeakers have contoured edges. Although this does not eliminate diffraction, it tends to make the transitions from the baffled to the unbaffled conditions occur in a less abrupt manner, and thus have a less disturbing effect on the axial frequency response. The off-axis responses are also improved.

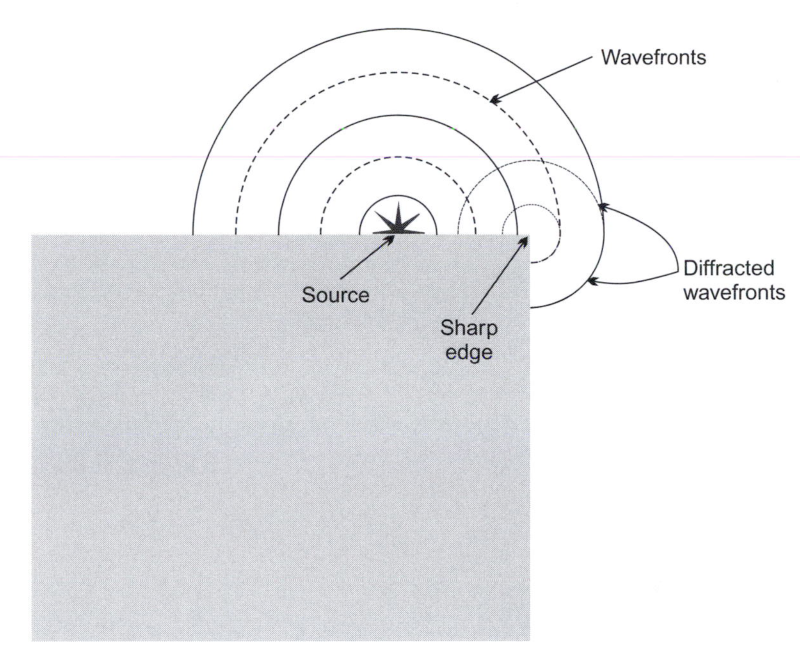

FIGURE 13.17
Graphical representation of the sudden increase in the rate of expansion of a wavefront at a sharp edge. The diffracted wave in the shadow region behind the source plane has the same effect as the wave incident on the edge; the diffracted wave in front of the source plane is phase reversed.

It is worth noting here the combined effect of cabinet diffraction and nearby surfaces. Figure 13.19 shows the effect on the axial response when the loudspeaker shown in Figure 13.18(b) is placed against, and close to, a wall. The different distances to the wall, behind the loudspeaker cabinet, give rise to different reflected paths for the diffracted waves, and hence different disturbance patterns in the on-axis forward response. The clear advantage of flush-mounting the loudspeakers into a wall can be seen by comparison with Figure 13.18(a). Nevertheless, it should be noted that loudspeakers designed for free-standing may have had their low frequency responses engineered for a higher low frequency output, and flush-mounting them may result in an excess of low frequencies. Active loudspeakers often have filter controls which can compensate for this, and reflex cabinets can sometimes have their ports reduced in size such that a flat response can easily be restored. In other cases, a slight bass boost may be deemed to be more acceptable than the irregular response resulting from the diffraction. However, the bass rise due to the flush mounting is much more easily equalized than the diffraction irregularities, which tend not to be equalizable because of the complexity of the response delays from the reflexions.

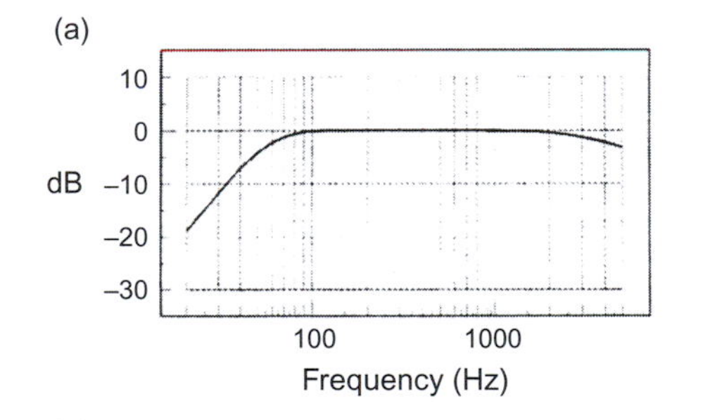

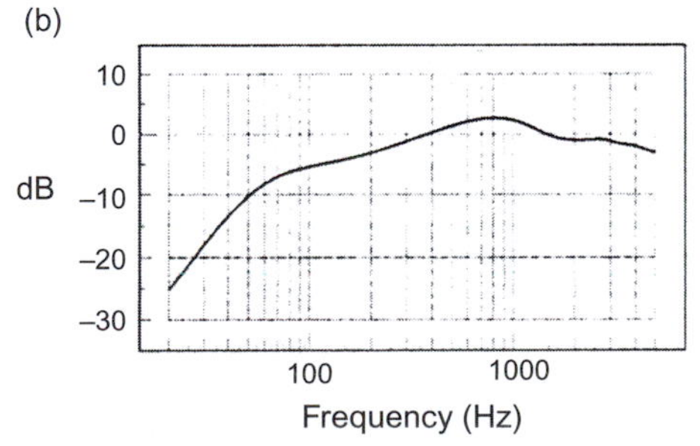

FIGURE 13.18
(a) On-axis frequency response of an idealized loudspeaker diaphragm mounted in an infinite baffle: the response is uniform over a wide range of frequencies. (b) On-axis frequency response of the same loudspeaker diaphragm mounted on the front face of a finite-sized cabinet (the rear enclosure size is assumed to be the same in both cases). The response has reduced by 6 dB at low frequencies and is uneven at higher frequencies. The differences between this and (a), above, are due to the diffraction from the edges of the cabinet.

In Figure 13.16, most of the drive unit positions are symmetrically placed. In practice, the diffraction effects can often be reduced by the non-symmetrical positions of the drivers with respect to the cabinet boundaries, but the necessity for this will depend upon the nature of the drive unit and the size of the cabinet. Some modern loudspeakers have the mid and high frequency drivers mounted in shallow horns, sometimes referred to as waveguides, which can project the wave in a more forward direction, and greatly reduce the effects of

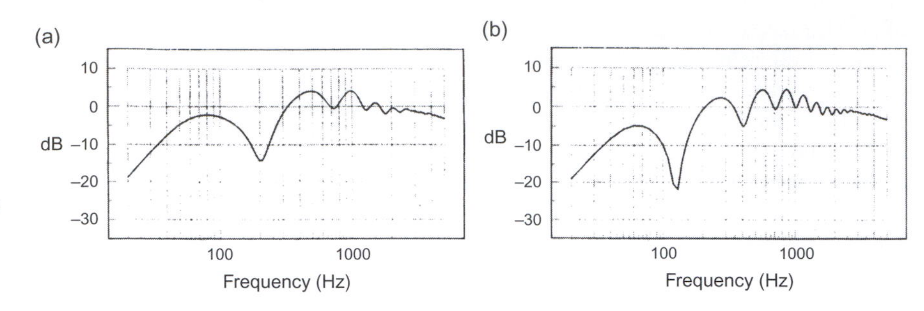

FIGURE 13.19
(a) On-axis frequency response of the same diaphragm and cabinet as in Figure 13.18(b) but with the rear of the cabinet against a rigid wall. Interference between the direct sound from the loudspeakers and that reflected from the wall produces a comb-filtered response, but the response level at low frequencies is restored to that for the infinite baffle case (Figure 13.18(a)); (b) As in (a), above, but with the rear of the cabinet 0.25 meters from the rigid wall.

edge diffraction. In general, it is also important to keep the front surface of the loudspeaker system as smooth as possible by recessing the drivers and screw heads.

FRONT GRILLES

It is somewhat rare, these days, to see front grilles on loudspeakers for professional use. The fact is that there are no truly transparent grille materials, but in domestic use their use is strongly justified, not only for aesthetic reasons but also for protection against children and over-zealous cleaners. Fabric grilles usually require wooden frames, and these can become diffraction sources as they provide a step at the edge of the baffle. Foam grilles can avoid this problem, but self-supporting foams of sufficient thickness will exhibit different distances through which the sound must pass, dependent upon the angle of radiation. They can also obstruct the air flow in the ports of reflex enclosures (and open transmission lines) where the exits are on the front face of the cabinets.

However, in some cases, the grille losses have been taken into account in the design of the loudspeaker system, so the removal of grilles should not be undertaken without careful listening to the effects. The diffusive effects of some grilles have been reported to impair the stereo imaging of some loudspeaker designs. In general, the best grille is no grille from a purely sonic viewpoint.

CABINET MOUNTING

Figure 13.19 shows the effects of placing a loudspeaker near to reflective surfaces. Obviously, a floor standing loudspeaker should have been designed for standing on the floor, but the intended mounting conditions for small cabinets is not always so obvious. The well-known Yamaha NS10 was originally designed as a bookshelf loudspeaker for domestic use. Its response was therefore tailored such that its bass/mid/treble response was best balanced when the loudspeaker was placed with its back against a wall. This fact was partly responsible for its success when mounted on top of the meter bridges of mixing consoles, because the flat surface below the loudspeaker tended to reinforce the low frequencies in the same way as a wall behind the cabinet. Mounting this type of loudspeaker on a stand in free space will lead to a reduction in the low frequency response, but mounting the loudspeaker on a mixing console will cause time-smearing and comb-filtering, the causes of which are highlighted in Figure 13.20.

Mounting loudspeakers on table tops or work surfaces is something that should be avoided, because the effects shown in Figure 13.20 will be exaggerated and coloration of the sound will be inevitable. Far too many people fail to realize that the mounting of a loudspeaker forms part of the loudspeaker system itself, and no loudspeaker's sound is independent of its mounting conditions. A poor loudspeaker, well mounted, may sound better than a good loudspeaker poorly mounted. The number of television and video studios in which the loudspeakers are really appallingly mounted indicates how little many of their users care about the sound, despite their protestations to the contrary.

Wall mounting of loudspeakers will need to take into account the nature of the wall, as not all walls can be considered to be rigid at low frequencies. The nature of the wall—plasterboard, hollow bricks, solid bricks, stone, concrete, etc.—will affect the loudspeaker response, so if the wall in the showroom was not the same as the wall at home, the same low frequency response will not be heard.

Placing loudspeakers on pieces of furniture is not recommended because of the vibrational coupling which will lead to sound being radiated by the furniture. Even so-called bookshelf loudspeakers should

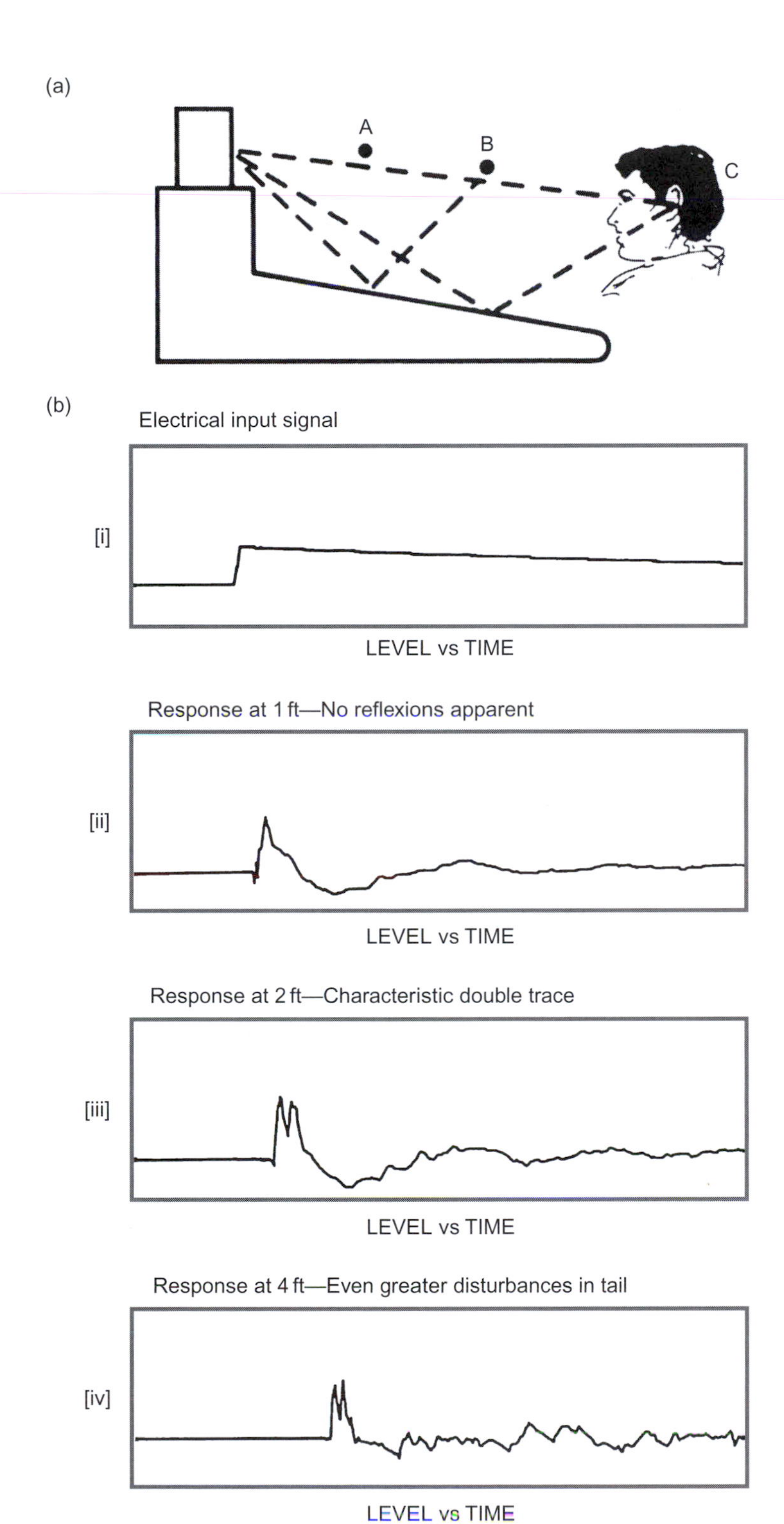

FIGURE 13.20
Effects of desk-top reflexions on transients: (a) Positions A, B and C relate to the 1 foot (30 cm) 2 foot (60 cm) and 4 foot (120 cm) responses, shown below; (b) Transient responses at the positions shown in (a), above.

only be mounted on substantial bookshelves, which should ideally also be rather full of books, to add mass, damping, and reduce diffraction effects.

Heavy narrow floor stands are recommended, with wide bases for stability, and coupled to the floors (either wood, or carpeted) by spikes. Soft rubber pads are not usually a good idea because they can give rise to rocking; hard rubber is a better option. Broad, columnar stands can induce floor reflexion problems by obstructing the free passage of the sound below the cabinets. Under all circumstances, wobbly mounting systems should be avoided, because the imaging can be severely impaired, even by small movements of the cabinets.

In effect, all of these different mounting regimes can be thought of as extensions to the loudspeaker cabinets, because they directly affect the loading on the diaphragms and the radiated sound output. They are not just simply different places to put a loudspeaker cabinet.

Note 1

In order to maintain a constant output SPL, a diaghragm needs to move four times the distance each time the frequency falls by an octave. The fact that the cone excursions are independent of frequency below the resonant frequency of a sealed box loudspeaker system is what gives rise to the 12dB/octave roll-off below that frequency.

REFERENCES

Bailey, A.R., 1965. A non-resonant loudspeaker enclosure design, Wireless World, 483–486.

Bradbury, L.J.S., 1976. The use of fibrous materials in loudspeaker enclosures, Journal of the Audio Engineering Society, 24, 404–412.

Colloms, M., 2005. High Performance Loudspeakers, sixth edition, John Wiley & Sons Chichester, UK.

Eargle, J.M., 1997. Loudspeaker Handbook, Chapman and Hall, New York, USA.

Martens, W., Braasch, J., & Woszczyk, W. October 2004. Identification and discrimination of listener envelopment percepts associated with multiple low-frequency signals in multi-channel sound reproduction. AES 117th Convention, Pre-print No 6229.

Olson, H.F., January 1969. Direct radiator loudspeaker enclosures, Journal of the Audio Engineering Society, 17(1), 22–29.

Small, R., 1972 and 1973. Direct radiator loudspeaker system analysis and synthesis—(Parts 1 and 2), Journal of the Audio Engineering Society, Vol 20, No 5, and Vol 21, No 1.

Finally, the fourth chapter in this section on loudspeakers deals with the crossover units that direct the appropriate frequencies to the high-frequency and low-frequency units of a multi-way loudspeaker system.

There are two kinds of crossover: passive and active. Passive crossovers deal with the signal after it has left a single power amplifier, and must therefore be able to handle large voltages and currents. An active crossover uses a separate power amplifier for each driver unit, and the division of the audio spectrum is done before these amplifiers, using circuitry working at preamp signal levels. A passive crossover is clearly a much simpler affair, using a relatively small number of components, but paradoxically its design is much more difficult; components such as inductors fall a long way short of the theoretical ideal, and minimizing power losses is a constant concern. Filters of order three or more are rare in passive crossovers because of the greater power loss in the extra inductors.

Active crossovers are much more complex in terms of hardware, but their design is much easier because almost any desired response can be produced by simple opamp circuitry and there is no need to worry about power losses.

However they are implemented, crossovers have the same basic job to do. The audio spectrum must be split up into two, three or (rarely) four bands, with appropriate crossover frequencies and filter slopes so that they recombine properly in the air in front of the loudspeaker. The roll-off rates must be fast enough to avoid exciting response inaccuracies that occur outside the intended frequency range of the unit; for example the bass filter output must drop fast enough as frequency increases to avoid exciting the resonance of the mid-frequency drive unit.

Practical drive units do not necessarily have a flat response in their band, and so crossover features that equalize response errors are often required. Drive units also do not act like purely resistive terminations to the crossover filters, having resonances and significant voice-coil inductance. All of these factors have to be taken into account, and I think you will agree that crossover design does not emerge as a pastime for the faint-hearted.

The crossover may also need to compensate for differences in drive unit mounting depth by the use of electrical delays. For example, if a tweeter and a bass driver are installed on a flat panel, the acoustic center of the tweeter can easily be 40 mm in front of that of the bass driver. Assume the crossover occurs at 3 kHz; at that frequency the sound wavelength is 115 mm, so a 40 mm misalignment is equivalent to a phase shift of 125 degrees, and this will increase to 180 degrees at 4.3 kHz, which will play merry hell with the polar response of the whole loudspeaker. The answer is to delay the signal to the tweeter. This can be realized very simply with opamp all-pass filters in an active crossover, but is awkward and expensive to do in a passive crossover.

Active crossovers have a long list of advantages, but the hard inescapable fact is that the vast majority of crossovers are passive. The most important reason is simply that customers want to be able to choose their own amplifiers.

CHAPTER 14

Crossovers

Loudspeakers by Philip Newell and Keith Holland

WHAT IS A CROSSOVER?

The term "crossover" appears to have been originally used to describe the relationship of the filter slopes—crossover filters—as shown in Figure 14.1. In reality, and in many languages other than English, they are better described as frequency dividing networks, or words to that effect, though the name crossover has generally stuck. The fact that no loudspeaker drive unit suitable for music monitoring or serious listening can provide a flat response over the entire musical frequency range requires that the multiple drivers in a system need to be fed by signals which are only appropriate to their designed performance range. The two normal ways to apply these filtered signals are via high level, passive crossovers—where the filter components are placed between the power amplifier and the loudspeaker drive units, or low level active crossovers—where the filters are placed in the line level signal circuits, ahead of the amplifier inputs. In the latter case, each filter output feeds a separate amplifier, which is then directly connected to the corresponding drive unit(s). In some cases, mixtures of the two concepts are applied to one system, such as an active crossover between the bass and mid drivers, and a high level passive crossover between the mid and high frequency drivers, as shown in Figure 14.2.

Other forms of crossover also exist, such as simple, low level passive crossovers, though they are rarely used because the filters can be more precisely tailored when the components are part of the feedback path in an electronic circuit. Mechanical crossovers are another type of filter. These can take the form of aluminium domes in the center of the cones, which decouple from the main cone at higher frequencies and radiate separately, extending the frequency response above that which could be achieved by the main cone, alone. However, the response tends to be somewhat irregular, but this type of high frequency extension can find use in loudspeakers for music *production*—as opposed to reproduction—and

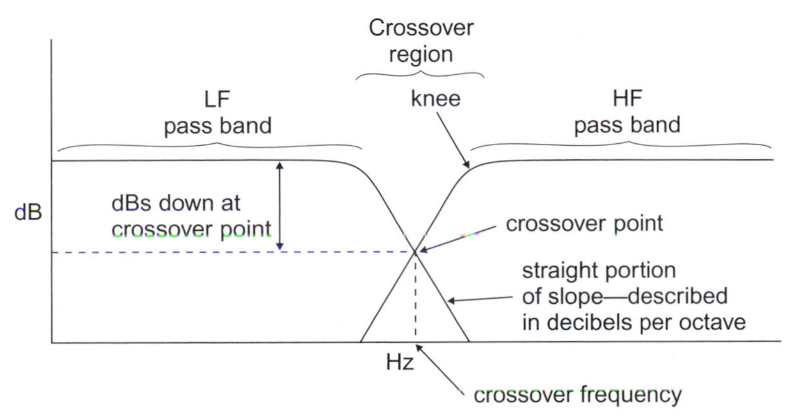

FIGURE 14.1
The basic concept of a pair of crossover filters.

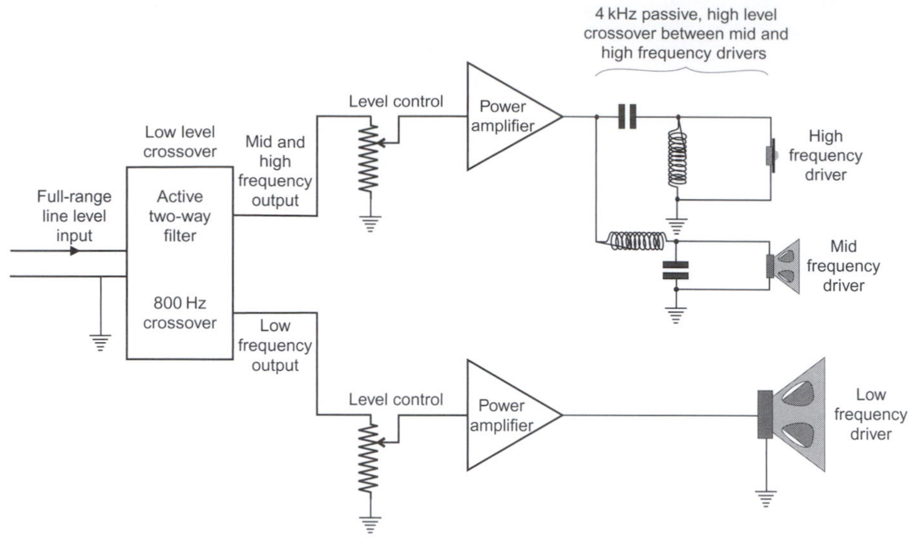

FIGURE 14.2
Example of a crossover system employing both active, low level and passive, high level filters.

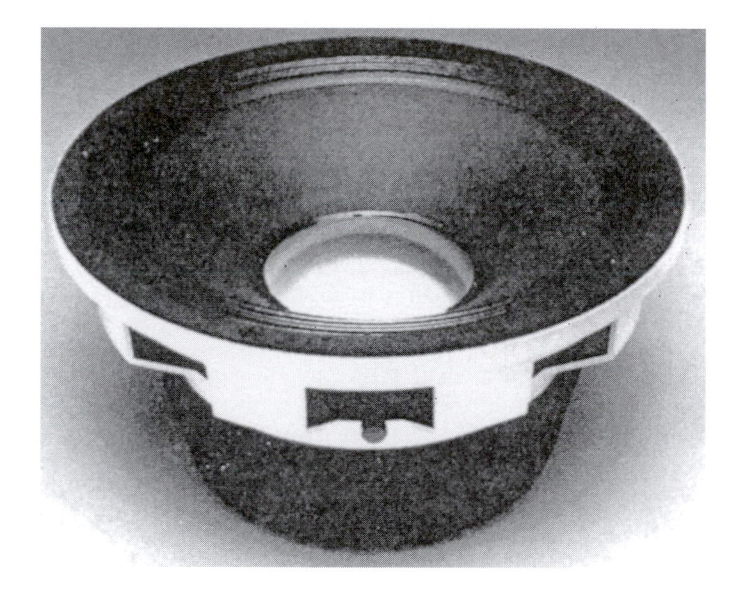

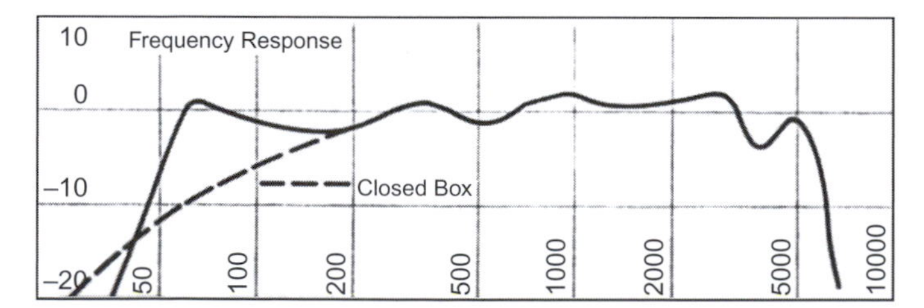

FIGURE 14.3
The Gauss model 4281, 12 inch (300 mm) drive unit for musical instrument use, which used an aluminium dome to extend the high frequency response to over 5 kHz.

the technique is extensively used in loudspeakers for musical instrument amplification, such as guitar amplifiers. An example is shown in Figure 14.3. Figure 14.4 shows a "parasitic cone" or "whizzer cone". The concept is generally the same in principle as that of the metal dome—the small cone decouples from the main cone at high frequencies—although the response of the parasitic cone tends to be more controlled, and a flatter frequency response can normally be achieved. Another concept, although not

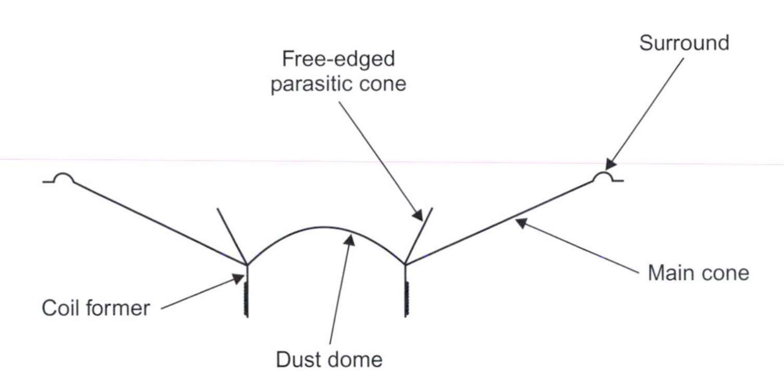

FIGURE 14.4
Parasitic, free-edged cone for high frequency extension.

widely used, is inductive coupling, where the high frequency cone is not electrically connected to the amplifier. In fact, the coil can be the single shorted turn formed by the metal dome itself. The dome and a former are simply placed over the center pole of the magnet assembly, sharing the same gap as the LF/MF cone assembly. Such inductively coupled transducers, or ICTs, are operated by the modulated magnetic coupling between the "coil" and the magnetic circuit. This type of "crossover" is neither electrical, electronic nor mechanical, but is simply a magnetic-inductive effect.

RECONSTRUCTION PROBLEMS

Unfortunately, the division of the frequencies is not all that a crossover must achieve. They must divide the frequencies in a way that the individual drive units can re-construct in the acoustic far-field of the loudspeaker a representation of the waveform which was electrically applied to the electronic amplification system, and it is not an easy task to do so. Figure 14.5 shows a representation of a typical, two-way loudspeaker system. Note how, due to the physical requirement of the radiation of the different frequency bands, the sizes of the drive units give rise to a displacement of the voice coils if the front faces of the drives share a plane common baffle. If the displacement were to be 10 cm, then a frequency with a wavelength of 20 cm would be received on the axis between the two drivers with its polarity reversed from either driver with respect to the other. The wavelength (λ) can be calculated by dividing the speed of sound (c), in meters per second, by the frequency *(f)* in Hz, so we arrive at the formula:

$$\lambda = \frac{c}{f} \tag{14.1}$$

To find the frequency with a wavelength of 20 cm, we can re-arrange the formula as:

$$f = \frac{c}{\lambda}$$

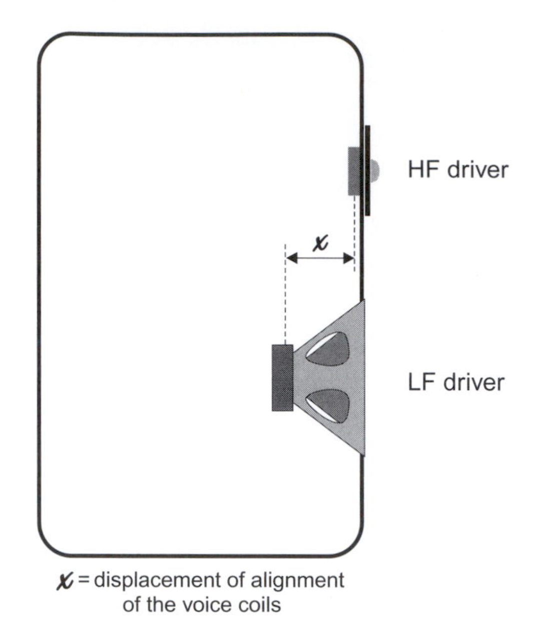

FIGURE 14.5
Elevation of a two-way loudspeaker system showing the typical offset of the voice coils in the axial plane.

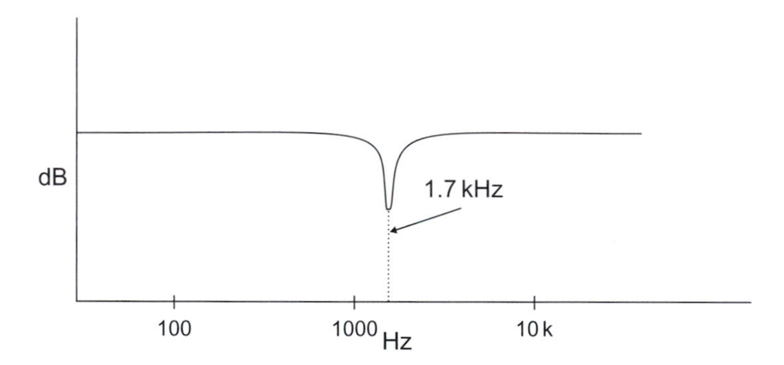

FIGURE 14.6
Composite response of the loudspeaker shown in Figure 14.5 on its central axis if the distance x were to be set at 10 cm.

Therefore: $f = \dfrac{340}{0.2} = 1700\ \text{Hz}$

At 1700 Hz we would have cancellation on-axis, producing a response as shown in Figure 14.6.

However, this is not the only complication which arises in the reconstruction. All conventional filters exhibit the property of "group delay." There is a finite time necessary for the information in a signal waveform to pass through a filter, which is a function of the slope of the filter and its cut-off frequency. As the frequency drops, the delay increases. As the steepness of the filter slope increases, so does the group delay. A filter of 24 dB/octave at 300 Hz would exhibit a group delay of around one millisecond. With a speed of sound of 340 m/s, one millisecond would represent 340/1000 m/s, or a 34 cm equivalent physical displacement. If the loudspeaker represented in Figure 14.5 were to be fed via such a crossover, the *real* radiation from the low frequency driver would be delayed by the equivalent of being mounted 44 cm behind the HF driver (34 cm due to the filter and 10 cm due to the physical misalignment). Figure 14.7 shows the step-function responses of four loudspeaker systems with different degrees of arrival synchronization from the drive units. Figure 14.7(d) shows the step response of a commercial three-way loudspeaker which clearly exhibits delays between the arrival times of the individual drive units.

In effect, if not compensated for, the flat-response-axis of the loudspeaker shown in Figure 14.5 would be tilted, as shown in Figure 14.8, for moderate low frequency signal delays. Conversely, though, if we engineer a flat response on-axis by compensating for the delays, it follows that the frequency

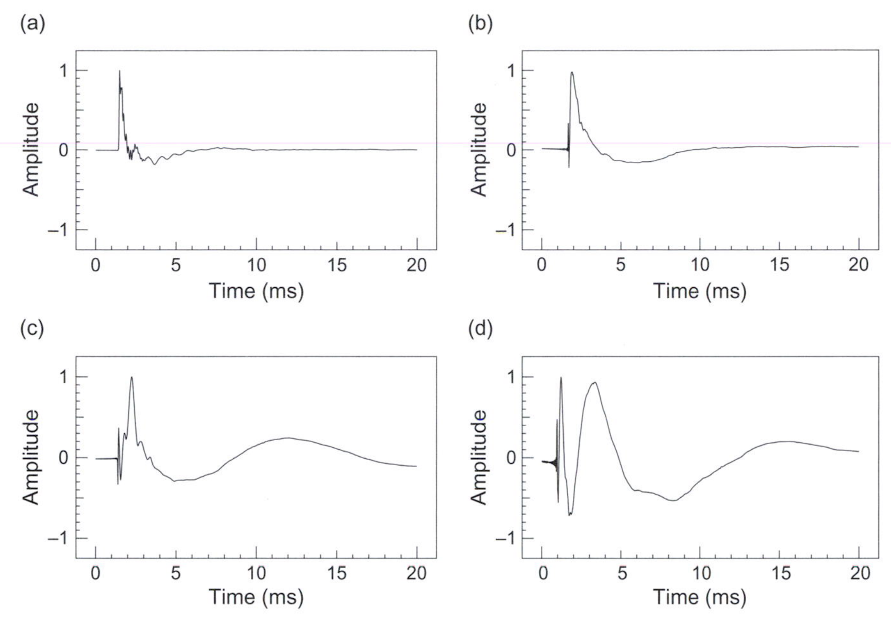

FIGURE 14.7
Step-function responses with one, two and three drivers: (a) Integrated attack of a relatively wideband, single driver; (b) Integrated attack of two-way system with excellent time alignment; (c) Separate rise-times visible from the two drivers of a system with slightly delayed low frequency driver response; (d) Three-way system with a clearly delayed response from the bass driver. In the latter case, the delay was a consequence of designing the system so that the main lobe of irregular response was forced upwards into a generally inoffensive direction. The trade-off was considered to be beneficial to the overall performance in typical situations in which is was expected to be used.

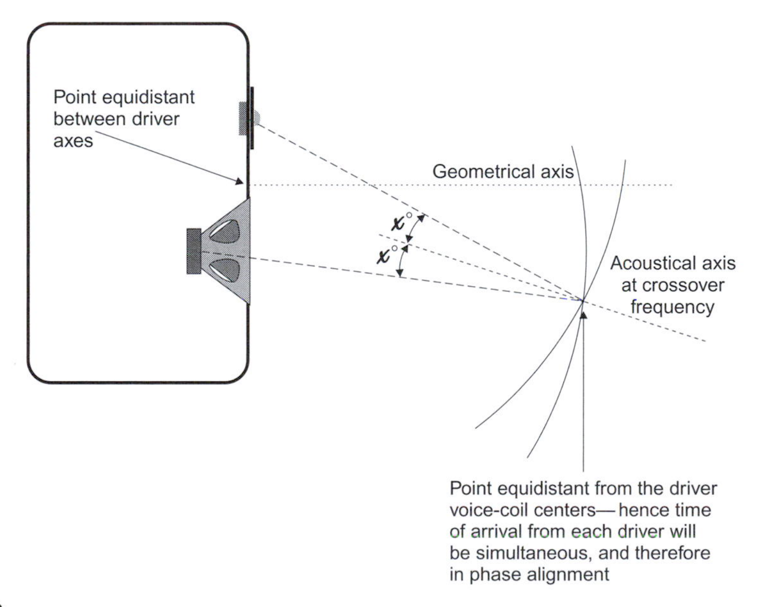

FIGURE 14.8
Tilting of the acoustic axis at the crossover frequency due to voice-coil physical displacement. On this axis, the response dip shown in Figure 14.6 would not be observed—the response would be flat.

responses *off-axis* must be incorrect, as shown in Figure 14.9. So, it can be appreciated that once the drive units have been physically separated, the problem of reconstructing the waveform of the original signal can become very difficult indeed. The problem can be partially solved by the use of concentric loudspeakers, but these concepts bring their own problems with them. For example, with the Tannoy, Dual Concentric approach, or the KEF Uni Q for that matter, the low frequency cone serves as a horn/waveguide for the high frequency driver. Modulating the LF cone with high levels of low frequencies can hardly be expected *not* to affect the high frequencies. In the case of the Tannoys, the LF coil gap also

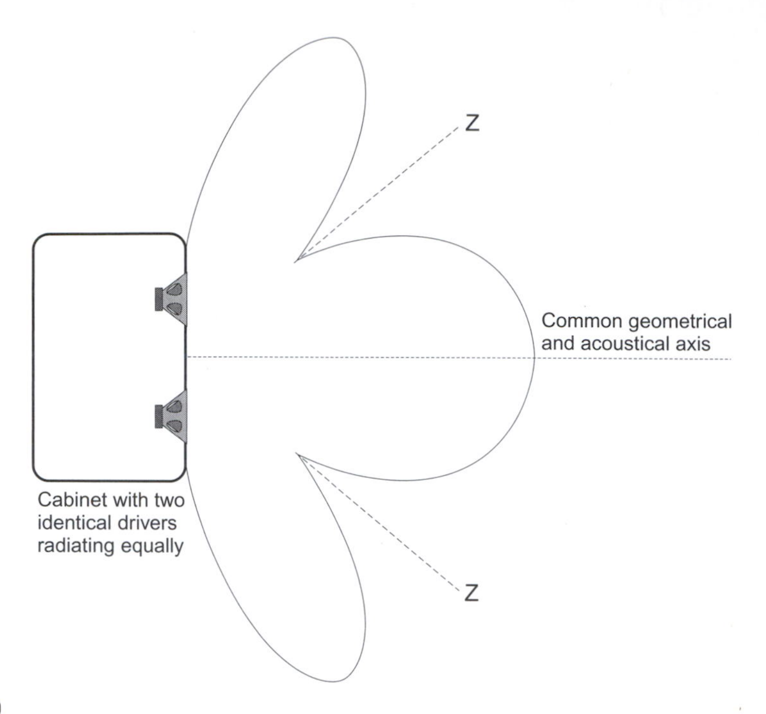

FIGURE 14.9
Lobing of the response when physically displaced drivers radiate a common frequency whose wavelength is close to, or smaller than, the distance between the drivers Z = axis of phase cancellation – varies with frequency – cancellation occurs whenever the distance to the two drive units varies by half a wavelength. The pattern shown therefore represents the situation at one frequency, only.

serves as an unwanted discontinuity in the HF horn. The general concepts of these drivers are shown in Figure 14.10, along with the Altec/UREI approach. In the latter case, the separate concentrically mounted horn is left hanging in free air, but this method of mounting is really too abrupt for proper mouth termination at the 1 kHz crossover frequency. The termination problem was discussed in detail in the previous chapter. Therefore, as so very often is the case with loudspeaker design, the tendency is to be trading one problem for another, rather than solving them—finding the best compromise for each situation—but that is often the reality of loudspeakers.

ORDERS, SLOPES AND SHAPES

Despite the different solutions on offer, electrical filters are overwhelmingly the most common manner of dividing the frequency bands. Whether this is done at high level, low level, actively or passively, the same basic filter concepts apply. Figure 14.11 shows a simple high pass filter, (a) to (d) showing first, second, third and fourth order roll-offs, respectively. Each inductor or capacitor adds 6 dB per octave of roll-off, and each 6 dB is known as an *order* of roll-off, a term which comes from the mathematical application of filter theory. An alternative approach to the inductor/capacitor (LC) design is a resistor/capacitor (RC) method shown in Figure 14.12. The LC approach is preferred for high-level crossover in the loudspeaker/amplifier interface because the power losses are much less, but the RC approach is preferred in low-level circuitry because of its simplicity (perfect inductors are not easy to make) and its relative insensitivity to drift and interference pickup. In the active circuits, where gain is plentifully available, the higher losses of the RC circuits are of little consequence.

First order crossovers are rarely used, because the low rate of roll-off requires the individual drivers to have respectably flat responses for at least two, if not three, octaves each side of the crossover point, which is usually not practicable. Nonetheless, when they *are* able to be used, they have the advantage that they are the only conventional crossovers whose combined outputs reconstruct the input wave-form. This is shown in Figure 14.13, and results from the fact that although each side of the filter is only 3 dB down at the crossover frequency (*voltage* summing would normally require that they should be 6 dB down to sum back to a flat response) the +45 degrees phase shift through one half of the

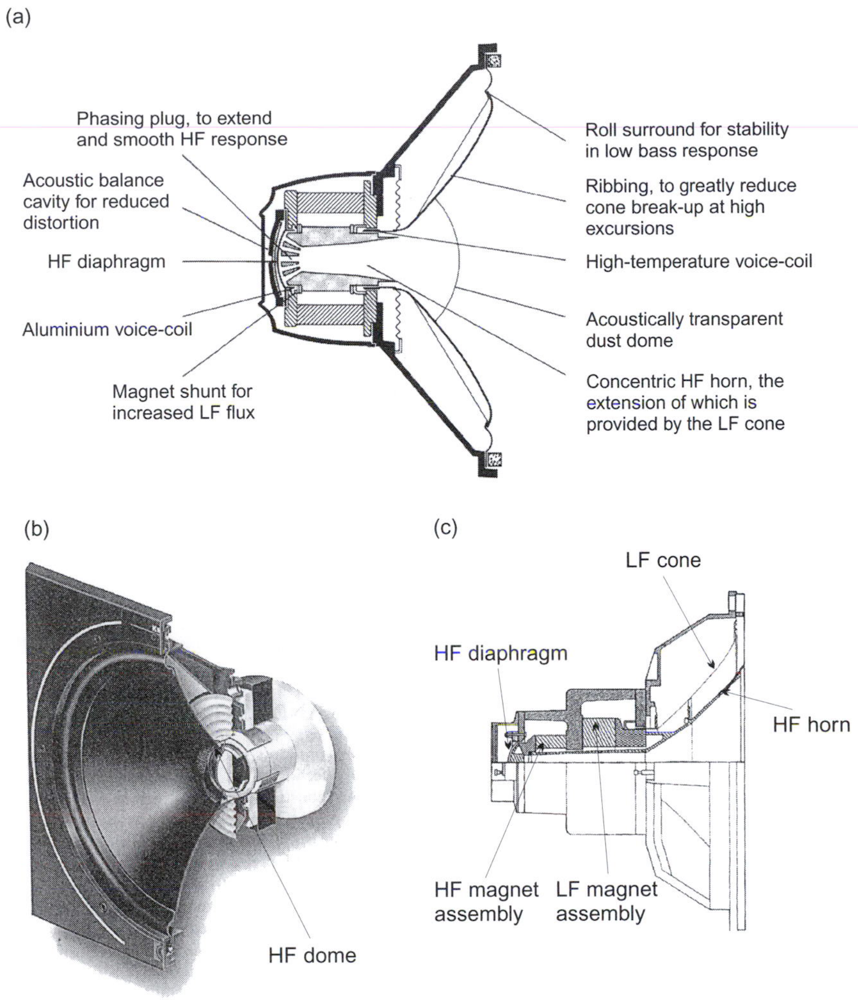

FIGURE 14.10
Some concentric drive units: (a) Tannoy Dual Concentric—with Alnico magnet structure (Courtesy of Tannoy Ltd); (b) The KEF Uni-Q; (c) The Altec 604.

crossover and the −45 degrees phase shift through the other half give rise to a 90 degrees combined shift which leads to another 3 dB of attenuation. The combination of amplitude roll-off and phase shifts leads to the perfect reconstruction shown in Figure 14.13.

Second order crossovers, with their 12 dB/octave roll-offs, are very popular with the manufacturers of small two-way cabinet loudspeakers. They are relatively cheap to construct and the power losses through the filters are quite small, but, as can be seen from Figure 14.14, they will *not* reconstruct the original waveform. The group delays associated with the phase shifts cause a temporal offset between the two halves, so the reconstruction is not summing the two outputs at the same instant. This "latency" in one half of the filter, relative to the other, creates a phase shift at the crossover frequency which does *not* compensate for the amplitude summation; hence the time and amplitude summations shown in Figure 14.14(a). Despite the fact that the reversing of the polarity of one of the outputs yields a flat amplitude response, the time response (waveform) becomes even more distorted. This is shown in Figure 14.14(b). There are ways of "juggling" this arrangement by offsetting the 3 dB down points so that the filter sections overlap, but this can introduce other irregularities unless it is very carefully implemented. However, in general, a standard second order crossover yields *either* a flat frequency response *or* a synchronous time response, but cannot exhibit both properties at the same time.

Third order crossovers, with slopes of 18 dB/octave are popular in more expensive passive loudspeaker systems, and are not infrequently used in active crossovers. They are more expensive in the passive form than the second order types, not only because they use more components, but also because the

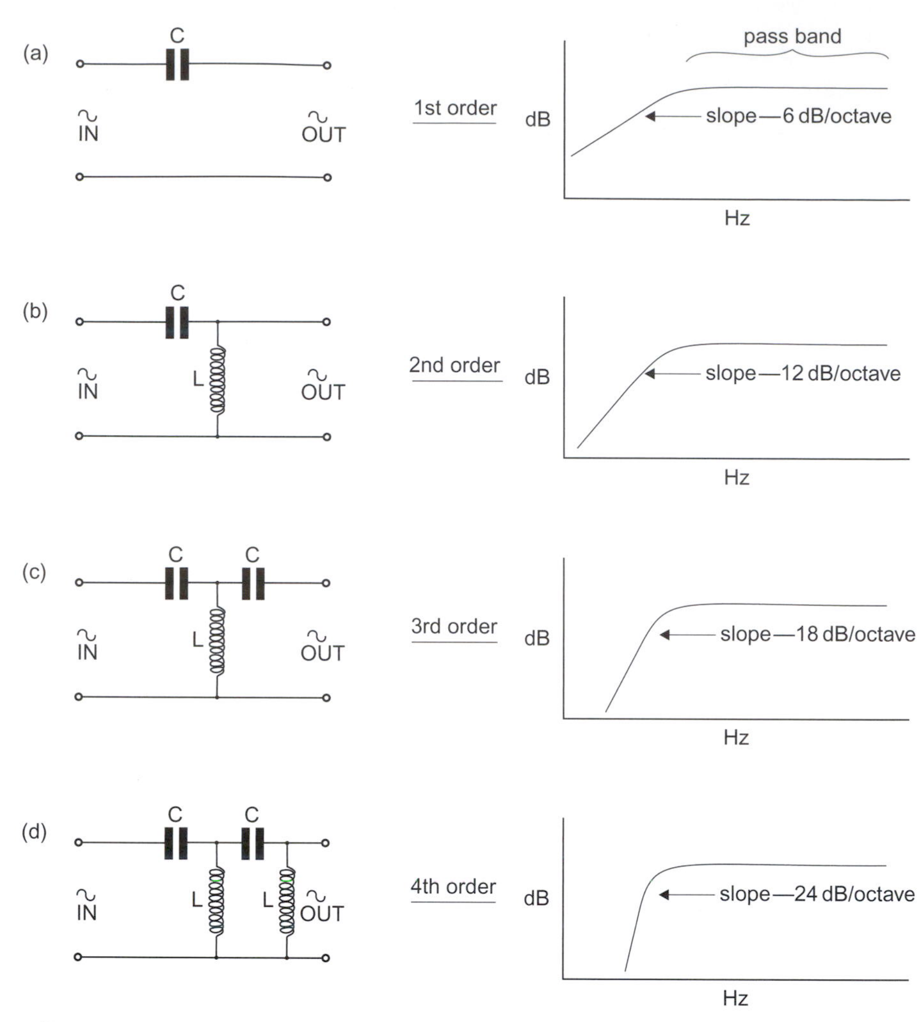

FIGURE 14.11
Capacitor/inductor filters—circuits and slopes. Values C and L depend upon the turnover frequency and the load impedance.

components may need to be less lossy. Otherwise, with so many components between the amplifier and the drive units, they would begin to sap considerable power from the signal. Typical circuit diagrams are shown in Figure 14.15, and a response summation is shown in Figure 14.16. The 18 dB/octave roll-off is useful in reducing the disturbance from out-of-band irregularity of the driver responses, because the driver responses can be almost 20 dB down an octave beyond the crossover frequency. This allows drivers to be used over almost all of their flat response region, but with third-order crossovers there is no "correct" polarity relationship between the outputs. The phase responses for normal and reversed polarity are shown in Figure 14.17, and the reverse polarity connection actually exhibits less phase shift through the crossover region after the summation of the outputs. However, this is the effect of simply summing the electrical outputs. Once those outputs are connected to physically displaced loudspeakers the story can be rather different.

One of the most popular types of crossover filter for use in active designs is the fourth order Linkwitz-Riley. The 24 dB/octave slope is achieved by cascading a pair of Butterworth (see next section) 12 dB/octave crossovers. The 24 dB/octave slopes are beneficial in high-power loudspeaker systems, where out-of-band energy is rapidly cut off, and for use with drivers whose out-of-band response is also irregular. The *power* response exhibits a dip at the crossover frequency, but the width of the dip is so narrow as to be, in many cases, almost inconsequential. The on-axis amplitude response is flat, because each section is normally designed to be 6 dB down at the crossover frequency. The two outputs sum to unity (−6 dB = half voltage [or half pressure], therefore $1/2 + 1/2 = 1$) because each section of the crossover is rotated 180 degrees, and the overall response therefore remains in-phase. However, due to the shape of the polar response, there are dips *off-axis* at the crossover frequency, hence the dip in

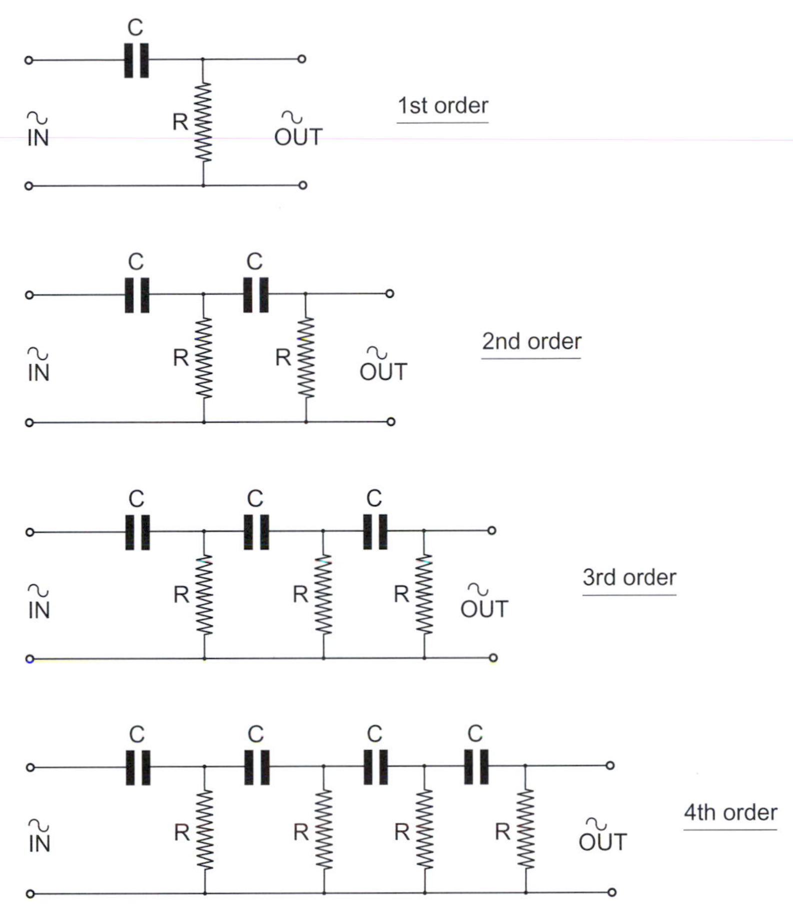

FIGURE 14.12
Resistor/capacitor equivalents of the filters shown in Figure 14.11. Except for the first order filter, and where the R is supplied by the loudspeaker load impedance, these filters are not suitable for use as high level filters because of the excessive power dissipated in the resistors. Values of C and R depend on the turnover frequency and the load impedance.

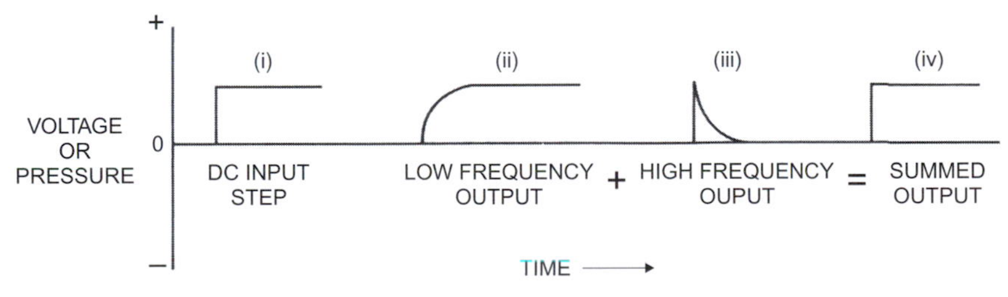

FIGURE 14.13
6 dB/octave crossover waveform reconstruction.

the total power response. The degree of audibility of this effect depends on how much reflected energy is returned to the listening position. In typical highly damped control rooms, the effect is usually imperceptible close to the axis and around the typical principal listening positions. With these crossovers, the narrowness of the band of frequencies over which the drivers on either side of the crossover frequency are simultaneously radiating ensures that the interference effects are kept small.

Filter orders higher than fourth are not normally used in crossover pairs, but they can sometimes be found in asymmetrical designs, such as a sixth order with a second order, to help to compensate for group delays or physical alignment delays. These individual filters sometimes also have their 3 dB or 6 dB down points offset in frequency, in order to achieve a flat on-axis response or flat power response, depending upon the circumstances. In practice, filters above sixth order are rarely used because they

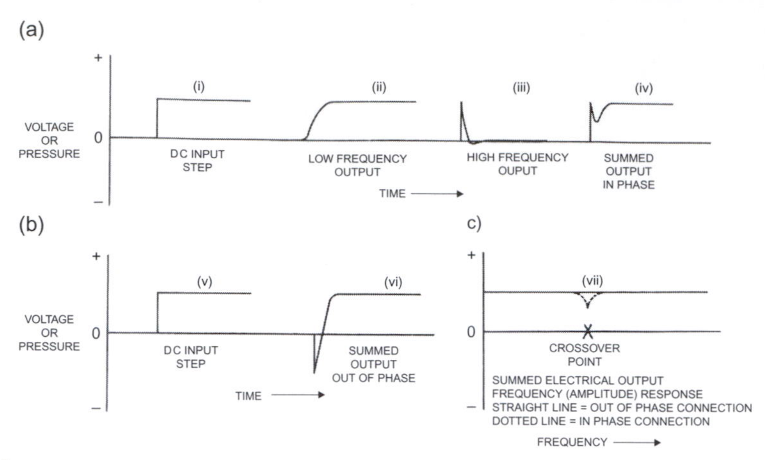

FIGURE 14.14
12 dB/octave crossover waveform reconstruction: (a) In-phase; (b) Reversed polarity; (c) Amplitude responses.
Note that whichever polarity is applied, the output waveform is not a true representation of the input waveform. (The waveform in Figure 14.13 reconstructed perfectly.)

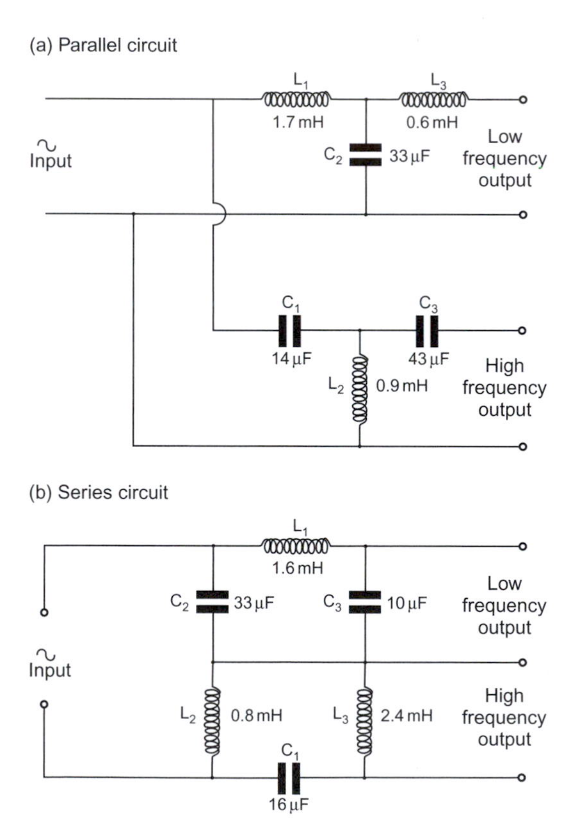

FIGURE 14.15
Typical two-way, 18 dB/octave, third order passive crossover circuits. Approximate values shown for 1 kHz crossover frequency and a uniform loading of 8 ohms.

tend to serve no useful purpose, and due to their tendency to introduce time response anomalies may actually create more problems than they can solve.

FILTER SHAPES

Until now we have been looking at standard filters, where, for generating higher orders, the filters are simply repeating the characteristics of the first order sections. However, different entry slopes and

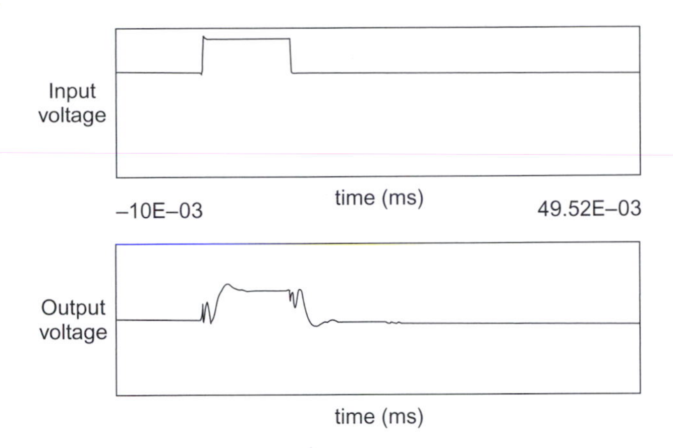

FIGURE 14.16
Summed output of a three-way, 18 dB/octave crossover. The signal path delays through the individual filter sections are clearly evident.

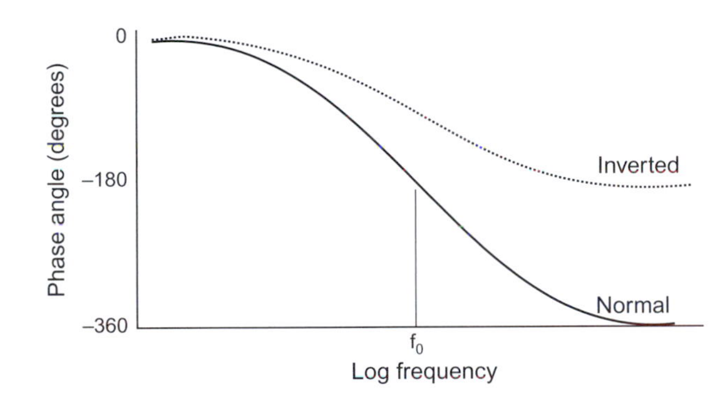

FIGURE 14.17
Phase relationships through the crossover region for the normal and inverted polarity connection of a third order Butterworth crossover—f_0 is the nominal crossover frequency.

different response shapes can be contoured by adjusting the components which are used to derive the higher orders. Therefore, by adjusting the Q of the filter, the "knee" of the curve—between the flat response and the uniform slope—can be adjusted in shape to produce either a more abrupt or more gradual entry to the final roll-off. The modified shapes can be used to help to achieve the desired overall electro-acoustic response when the individual responses of the drivers are taken into account, or when system summation is affected by physical displacement problems. Many different Q factors—or quality factors—are in use. Butterworth filters are "maximally flat" until the roll-off begins. Bessel filters are more abrupt in their transition from flat to sloping response, but exhibit alternating up and down responses before the main roll-off begins. Figure 14.18(a) shows a typical 6 dB/octave roll-off produced by a simple capacitor-resistor filter, but Figure 14.18(b) shows how the addition of an extra resistor can limit the rate of roll-off. There are therefore means at our disposal to modify the slopes of the curves of the filters, which is extremely useful when we have to tailor curves to mirror the responses of real drivers, which never behave like pure resistors, and so can never ideally terminate the standard filters which were discussed earlier.

Figure 14.19 shows an example of a conjugate network. These are used to flatten the impedance curves of drive units by compensating for the reactive properties of their electro-mechanico-acoustic characteristics. However, the resistors in these networks do dissipate power, so the overall efficiency of the system may reduce when conjugate networks are applied. Sonically, they can be questionable, and some complicated passive crossovers applying such technology for electrical and response flattening purposes have been considered to sound worse than simpler networks, but the overall effect may also depend on whether the power amplifiers with which they are used are capable of driving complex loads, or not. If they are not, then the conjugate networks may be helpful, but in professional situations, the choice of a more load-tolerant amplifier is often the preferred solution. Complex passive crossovers are not entirely distortion free, because the inductors, in particular, can produce non-linear

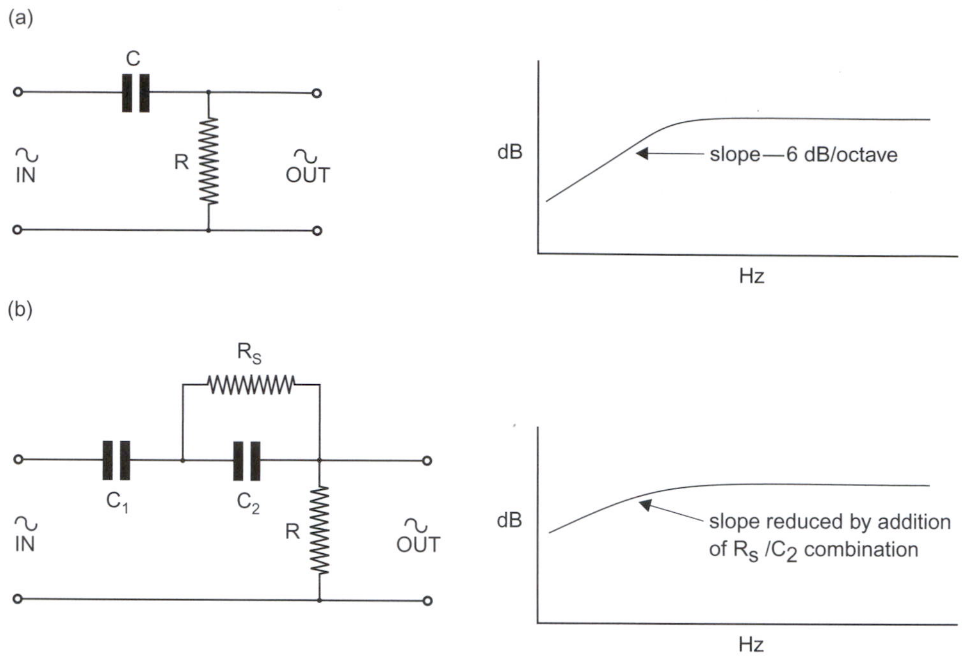

(a)

C
IN
R
OUT
dB
slope—6 dB/octave
Hz

(b)

R_S
C_1
C_2
IN
R
OUT
dB
slope reduced by addition of R_S /C_2 combination
Hz

FIGURE 14.18
Response contouring: (a) A simple high-pass filter of first order characteristic roll-off. The response will be 3 dB down at the frequency where the reactance of the crossover equals the resistance (R). Note that a resistor having the same value as R, substituted for C, would give rise to a 6 dB reduction of level at the output. The 90 degree phase shift through the capacitor is the reason for the 3 dB difference; (b) Added components for reducing the slope.

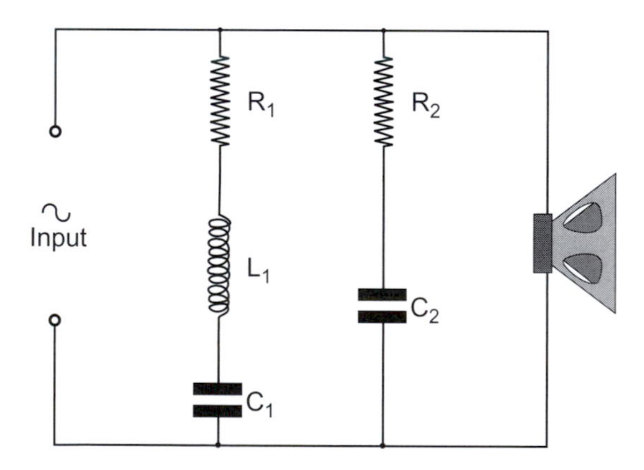

FIGURE 14.19
A conjugate network. In the circuit above, R_1, C1, and L1 provide compensation for the impedance rise at the resonant frequency of the driver, whilst R_2 and C2 compensate for the rising impedance at higher frequencies due to the voice-coil inductance. Opinions vary about the sonic effects of this type of impedance equalization, but such circuits are undoubtedly useful to flatten the response of systems using hi-order passive crossovers.

distortion. Cases tend to need to be judged individually, partly because, amplifier outputs stages and power supplies vary so widely in design concept.

Clearly, where complex circuitry is involved, it is essential to use components whose values remain stable if the performance of the overall system is not to change as time passes. Unfortunately, the large electrolytic capacitors which tend to be required by high-power low frequency passive crossovers are notoriously prone to changing their capacitance as they age. Likewise, changes in mechanical compliances of the drive units with age can also cause a drift in the circuit parameters. These factors tend to put high level passive crossovers at a disadvantage when compared with low level active circuitry, especially when precise tailoring of the response is required. The higher impedance active circuitry of the

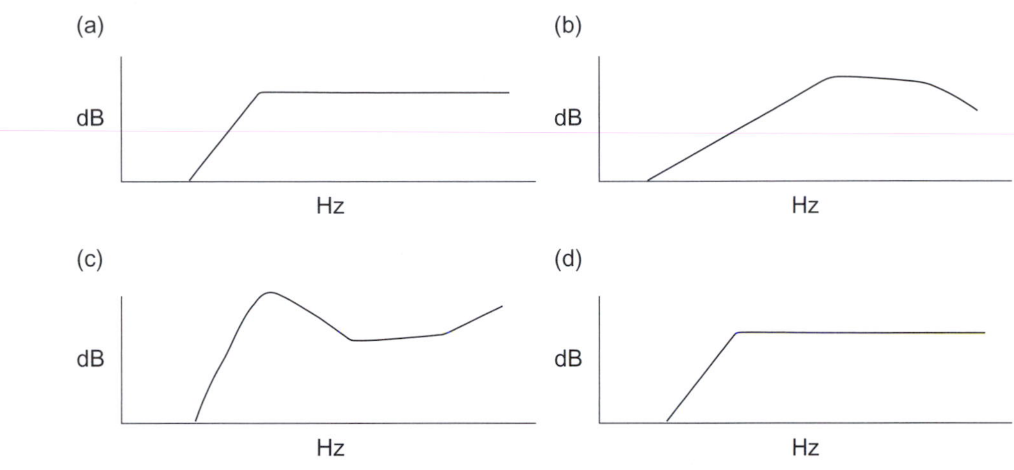

FIGURE 14.20
Target functions—practical realization: (a) Desired target function; (b) Measured driver response; (c) Electrical filter response—allows for a non-flat driver response; (d) System response as measured—equal to (a), i.e., it achieves its target.

latter case can avoid the use of large value capacitors, and can therefore employ smaller value components of much greater long-term stability. This is an important concern when we need to produce very precise response curves. Active circuitry can also eliminate the need for lossy, and potentially non-linear, inductors.

TARGET FUNCTIONS

Until as late as the 1970s it was commonplace to treat a moving-coil loudspeaker as if it were a resistor when designing crossover circuitry, but, as we saw in Chapter 11, this is usually a long way from reality. In fairness, before the 1970s, and the seminal work of Thiele and Small, (see references) it was not always very easy to find the necessary electro-mechanical information about drive unit characteristics, so design and development were often two separate processes—design it ideally, and then select components during tests to modify the response. Since the 1980s, with much more data available, it has been customary to look at a drive unit as a complex impedance, and to view its response for what it actually is, and not as an ideal response. There has subsequently emerged the concept of the "target function".

Whereas in Figure 14.1 the plots represent the responses of the filter circuits which presume that the driver is of constant impedance and has a flat acoustic response, we can also view the same plots as the *target* response for our combined electro-mechanico-acoustic system. In order to achieve this, the desired *filter* response becomes whatever is required to combine with the actual *driver* response to realize the *target* response. Figure 14.20(a) shows a target response; (b) shows the response of an actual driver, and (c) shows the response of a filter which, when used with (b) will yield the response shown in (a). This "new" approach obviously calls into question the value of purchasing ready-made crossovers as standalone devices, be they active or passive. They may be of value for sound reinforcement systems, where each loudspeaker box has been engineered to be more or less flat, and multi-band system equalization is de rigueur, but their use may be over-simplistic when applied to many other loudspeaker systems. Also, many modern drive units are not necessarily *engineered* for flat responses if other benefits can be gained by sacrificing the flatness. Computer-aided filter design can then be applied in order to design a crossover which will achieve the required target function from the complete system. In fact, in the current cost-conscious world, it is often less expensive to use electronic means to flatten a response rather than electro-mechanical means. Nonetheless, despite having said that, there is still in the experience of many listeners a certain *je ne sais quoi* about the purity of sound of an inherently flat driver. It should also be acknowledged that the more complex filter shapes such as that shown in Figure 14.20(c) are much more easily implemented with active crossover designs.

Minimum and non-minimum phase effects

The selection of the target function is somewhat more complicated than it may initially appear, because the phase shifts and group delays associated with the filters, *and* the physically induced delays due to the different drive units occupying different points in space, lead to non-minimum-phase responses. A minimum-phase response is one where the correction of the amplitude towards a flat response also leads to a corresponding flattening of the phase response, or vice versa. In the case of a non-minimum-phase response, the correction of either the amplitude or the phase response does not automatically correct the other. Non-minimum-phase responses give rise to situations where a flat amplitude response cannot be accompanied by an accurate transient (time) response. The amplitude and phase responses are defined by the time response, and vice versa, which is why the Fourier Transform and Inverse Fourier Transform can be used to derive the frequency response from the impulse response or the impulse response from the frequency response. The frequency response, in this case, is referring to the *complete* frequency response, i.e., the amplitude *and* the phase. Non-minimum-phase effects are typically associated with the recombination of non-time-synchronous signals, such as a recombination of a reflexion with a direct signal, or the summation of signals where different group delays or digital latency have been incurred.

Unfortunately, for loudspeaker designers and users, the non-minimum-phase effects often prevent the perfect reconstruction of a waveform from a multi-drive unit loudspeaker system. When this fact is coupled with the fact that no *single* drive unit can cover the whole frequency range in a manner which is flat in frequency response and adequate in its directivity pattern (at least not at useful sound pressure levels) we are condemned to compromise. The time-shifted sections of the overall response cannot be equalized flat in a minimum-phase manner, and so the transient responses and directivity patterns will be altered. Even when using concentric drivers, where the vertical and lateral displacements can be avoided, the problem of the crossover group delays still dogs the design process, although, if crossover frequencies are carefully matched to distances on the front/back plane, they can sometimes almost be eliminated.

Corrective measures and side-effects

In some cases, the electrical crossover filters can be overlapped to some degree, in order to take into account certain physical aspects of the relative driver positions and in order to vary the polar pattern of the complete system. Lobes in certain directions may or may not be problematical dependent upon the intended position in which the loudspeaker is expected to be used, for example, see Figure 14.21. If a loudspeaker system does exhibit a lobed radiation pattern with an aberration in the frequency response, then that lobe is best directed towards places where there is least likelihood of returning a reflexion to the listening position which would be detrimental to the overall perception of the music. Decisions must also be made about whether the most likely movements of the listener to the loudspeaker will be in the horizontal or vertical direction when the loudspeakers are in critical use, for example, either sitting in a chair whilst moving left or right along a mixing console, or working predominantly in one place but at various times either sitting down or standing up.

Earlier it was described how a first order crossover could reconstruct a perfect waveform. However, if the crossover is used with physically non-coincident drivers, then even a first order crossover will suffer from the above problems except on its acoustic axis, which may not always correspond with its physical axis. This means that off-axis reflexions would *not* have time-coincident origins, and there would be a lobe which was tilted, either upwards or downwards. One reason for the great popularity of the fourth order Linkwitz-Riley filters for high quality monitoring is that the in-phase relationship between the two drivers on either side of the crossover gives rise to the main lobe being symmetrical about the central axis between the drivers.

A very important point to emphasize here is that the design of crossover filters is not by any means the easy task which many people believe it to be. Many things must be taken into consideration before the filter functions can be decided upon, and the design of suitable filters can be work of a very specialized nature. Computer aided design of filters has been a great boon to loudspeaker engineers.

ACTIVE VERSUS PASSIVE CROSSOVERS

For high quality loudspeaker applications, the consensus is almost universally in favor of active crossovers. By virtue of their feedback loops they can remain remarkably stable over very many years, and complex filter shapes can be devised without any loss of power efficiency. Conversely, passive crossovers rely entirely on the long-term stability of each component part for their overall stability, which is

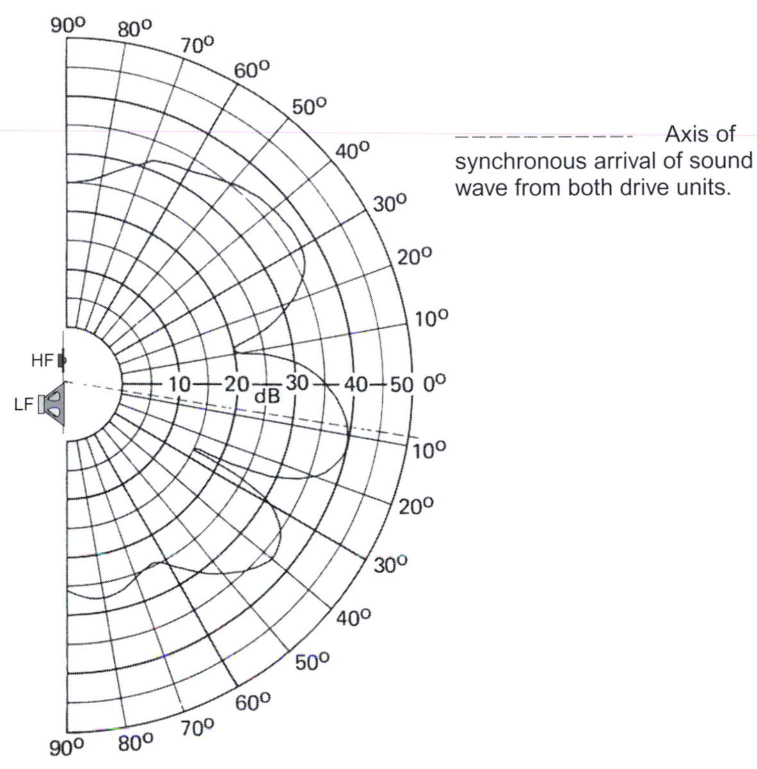

FIGURE 14.21
This type of lobing could be problematical if an irregular frequency response returns towards the listener from a reflective floor. Some designers choose to invert the cabinet in order to direct the irregular response above the listeners' heads, especially if the ceiling is high, or more absorbent than the floor.

not easy to achieve when the low impedances of loudspeaker circuits call for high value capacitors—both in terms of capacitance *and* working voltage—which in turn call for capacitors of types which may not be able to provide good long-term stability. Complex filter shapes may need to use many components, which when placed in the power circuitry will probably lead to lower system efficiency, wasting many watts of amplifier output. If large electrolytic capacitors are needed, their stability can be questionable, but, if the much larger solid dielectric capacitors are used, their physical construction and large size can lead to them having considerable unwanted inductance, which can upset the crossover operation.

Active filters are free from these problems, and in fact require no inductors at all. What is more, if state-variable filters are used, any drift which does occur can reflect equally in both halves of the crossover response, thus only slightly varying the crossover frequency as opposed to opening a gap or causing an overlap. The overall frequency response of the system is therefore unlikely to be affected. There can be no equivalent to this type of stability or self-correction with passive crossovers, and neither can passive crossovers easily compensate for group delays. Figure 14.22 shows a passive all-pass circuit (an analog delay circuit) of a type which has been applied commercially to loudspeaker systems, but there are people who feel that this type of circuitry between an amplifier and a loudspeaker can again cause as many problems as it solves—if not more!

Conversely, active filters can *easily* incorporate delay compensation for driver mounting offsets, such as when a horn driver is set behind the woofers. Response tailoring is independent of the loudspeaker impedance complexities. The list of advantages in favor of active crossovers and multi-amplification is impressive:

1) Loudspeaker drive units of different sensitivities may be used in one system without the need for lossy resistive networks or transformers. This can be advantageous because drive units of *sonic* compatibility may be electronically incompatible in passive systems.
2) Distortions due to overload in any one band are captive within that band, and cannot affect any of the other drivers.

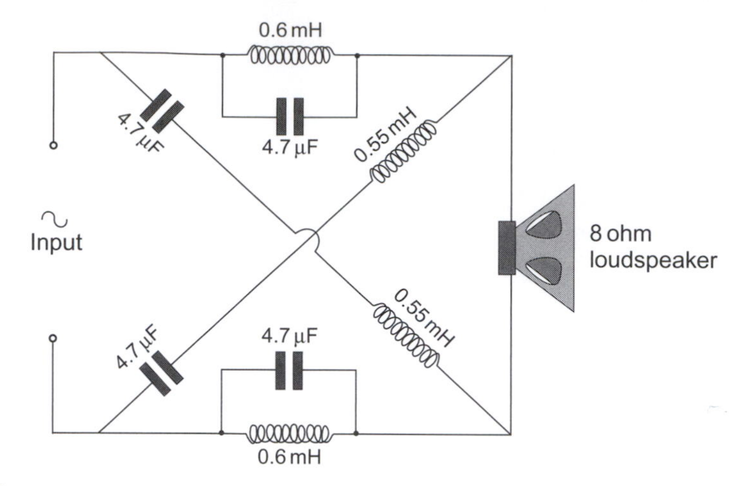

FIGURE 14.22
An all-pass delay circuit. Tannoy developed a delay circuit similar to the one shown, but unwanted phase shifts, response ripples and mistermination problems tend to prevent the theoretical benefits of its approximately 150 microsecond delay from being fully realized. One hundred and fifty microseconds of delay would ideally compensate for the distance of about 5 cm between the voice coil planes.

3) Occasional low frequency overloads do not pass distortion products into the high frequency drivers, and instead of being objectionable may, if slight, be inaudible.
4) Amplifier power and distortion characteristics can be optimally matched to the drive unit sensitivities and frequency ranges.
5) Driver protection, if required, can be precisely tailored to the needs of each driver.
6) Complex frequency response curves can easily be realized in the electronics to deliver flat (or as required) acoustic responses in front of the loudspeakers. Driver irregularities can, except if too sharp, be easily regularized.
7) There are no complex load impedances as found in passive crossovers, making amplifier performance (and the whole system performance) more dynamically predictable.
8) System intermodulation distortion can be significantly reduced.
9) Cable problems can be dramatically reduced.
10) If *mild* low frequency clipping or limiting can be tolerated, much higher SPLs can be generated from the same drive units (vis-à-vis their use in passive systems) without subjective quality impairment. (See 2) and 3) above.)
11) Modeling of thermal time constants can be incorporated into the drive amplifiers, helping to compensate for thermal compression in the drive units, although they cannot totally eliminate its effects.
12) Low source impedances at the amplifier outputs can damp out-of-band resonances in drive units, which otherwise may be uncontrolled due to the passive crossover effectively buffering them away from the amplifier.
13) Drive units are essentially voltage-controlled, which means that when coupled directly to a power amplifier (most of which act like voltage sources), they can be more optimally driven than when impedances are placed between the source and load, such as by passive crossover components. When "seen" from the point of view of a voice coil, the crossover components represent an irregularity in the amplifier output impedance.
14) Direct connection of the amplifier and loudspeaker is a useful distortion-reducing system. It can eliminate the strange currents which can often flow in complex passive crossovers.
15) Higher order filter slopes can easily be achieved without loss of system efficiency.
16) Low frequency cabinet/driver alignments can be made possible which, by passive means, would be more or less out of the question.
17) Drive unit production tolerances can easily be trimmed out.
18) Driver ageing drift can easily be trimmed out.
19) Subjectively, clarity and dynamic range are generally considered to be better on an active system compared to the passive equivalent (i.e. same box, same drive units). (See also Figure 14.23.)

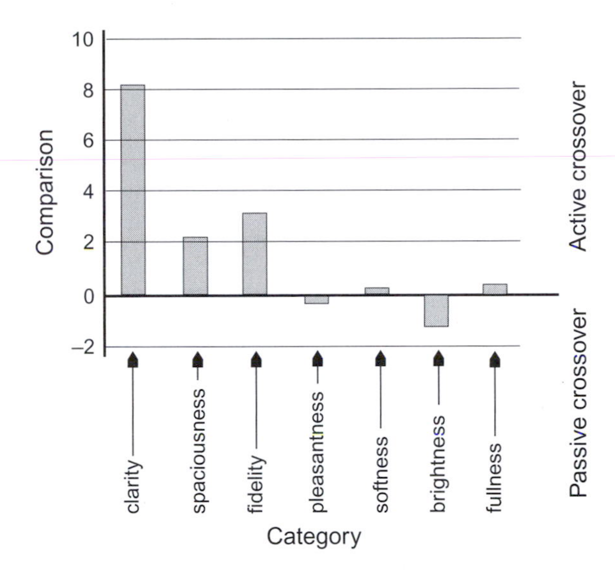

FIGURE 14.23
Active versus passive crossovers—subjective data (after Campbell and Holland, 2004). The results show a clear overall favor for the active crossover. The assessment of the extra brightness from the passive crossover could well be due to greater levels of non-linear distortion, which may in fact *not* be a result in its favor.

20) Out-of-band filters can easily be accommodated, if required.
21) Amplifier design may be able to be simplified, sometimes to sonic benefit.
22) In passive loudspeakers used at high levels, voice-coil heating will change the impedance of the drive units, which in turn will affect the crossover termination. Crossover frequencies, as well as levels, may dynamically shift. Actively crossed-over loudspeakers are immune to such crossover frequency changes.
23) Problems of inductor siting (to minimize interaction with drive unit voice coils at high current levels) do not occur.
24) Active systems have the potential for the relatively simple application of motional feedback, which may come more into vogue as time passes.

Conversely, the list of benefits for the use of passive high level crossovers for studio monitors would typically consist of:

1) Reduced cost? Not necessarily, because several limited bandwidth amplifiers may be cheaper to produce than one large amplifier capable of driving complex loads. What is more, the passive crossovers for 1000 watt Kinoshita studio monitors cost over 3000 euros each.
2) Passive crossovers are less prone to being misadjusted by misinformed users, who think that crossovers are some sort of "adjust to taste" tone controls. On the other hand, passive systems have a tendency to mis-adjust themselves with age.
3) Simplicity? Not really, because very high quality passive high level crossovers can be hellishly complicated to implement, not to mention the amplifiers which are needed to drive them.
4) Ruggedness? No, because the electrolytic capacitors (necessary for the large values) are notorious for ageing, and gradually changing their values.

However, it must be stated that in less demanding circumstances than studio monitor loudspeakers passive crossovers obviously have their appropriate applications, but the above lists highlight the benefits of active designs where the *highest* system performance levels are required.

Clearly, the advantages of active low level crossovers completely eclipse those of passive high level crossovers, yet it was only around the late 1980s that dedicated active crossovers began to be seriously used on large-scale monitor systems. Prior art used stock electronic crossovers, and perhaps these caused some delay in the acceptance of totally active designs because they were usually made with fixed slopes on all the filter bands. This tended to necessitate the use of multi-band equalizers, many of which were of dubious sonic quality. It took some time before people generally began to accept the need to buy a specific crossover with a monitor system, which was relatively useless in any other application. There was still a mix-and-match mentality towards components parts, each of which was expected to function as a "stand alone" device. Once attitudes like this become established it can be very difficult to introduce new concepts. In fact, it took a long time before self-powered actively crossed-over *small* monitors could establish their place in studio use, but domestic resistance to their acceptance has been even more pronounced. Established practices die hard, and they can be remarkably difficult to change, even in the face of clearly superior technology.

In 2003 and 2004, Alex Campbell worked on a performance comparison between active and passively crossed-over *domestic* loudspeaker systems at the Institute of Sound and Vibration Research, in the UK. He had an identical amount of money to spend on each design. His findings were presented to an international conference of the UK's Institute of Acoustics (Campbell and Holland, 2004). Even at this modest level of engineering, aimed at the *retail* price range of £400–£500 (€600–€750) per pair, the subjective assessment, made under ISVR control and using a panel of 30 subjects, came out heavily in favor of the active design. The results are reproduced in Figure 14.23, with the "clarity" and "fidelity" ratings being strongly in favor of the active designs (probably also as a result of greatly reduced intermodulation distortion.) In fact, the only tendency for the passive design to show a more positive result than the active design was in "brightness," which could perhaps be a result of higher non-linear distortion levels. Perhaps the only real block to the general acceptance of the superiority of active crossovers and multi-amplification in the world of domestic hi-fi is the fact that so many hi-fi enthusiasts want to select their own favorite amplifiers and loudspeakers as separate items, but this perhaps has more to do with human psychology rather than audio engineering. Also, of course, choosing your own system with future up-grades in mind, as extra money becomes available, is a fun part of building up a hi-fi system, and perhaps a necessity in some smaller studios.

PHYSICAL DERIVATION OF CROSSOVER DELAY

Crossovers are more than simple filters. As we have seen, due to the group delays which exist in filter circuits, the outputs of the various filter sections are time-shifted by virtue of the phase shifts which are inextricably linked to the roll-offs. The result is that when the outputs are re-combined, either electrically or acoustically, there are often non-minimum-phase response irregularities in either the amplitude or phase responses. Essentially, they cannot be corrected by analog, *electrical* means. Digital crossovers *can* be made to provide summing outputs, but the sonic benefits are not necessarily worth the efforts. For example, digital crossovers, unless they employ high sampling rates (96 kHz or more) may introduce limitations in such a way that would make it difficult to accurately monitor analog or high sampling rate digital recordings. What is more, a three-way stereo crossover would require six D to A converters on the outputs, and two A to D converters if they were being fed from analog sources. If these were to be of the highest quality (and in a high quality monitoring system they should be expected to be nothing less) the cost of the unit could be exorbitant. Using anything less than the finest converters would make a mockery of trying to monitor *recordings* made through the best converters. This subject is discussed further in the following section.

There is, however, an analog means of deriving delays. The loudspeaker cabinets, themselves, can be stepped, or the different drivers can be mounted in separate enclosures which are then mounted at different distances from the listeners. These two systems are depicted in Figure 14.24, but care needs to be taken to ensure that diffraction effects do not become problematical due to the increased number of cabinet edges. Figure 14.25 shows how a composite system using direct radiating bass drivers and a horn-loaded mid/high frequency driver can compensate for the delayed output from the bass units. In this case, the "natural" position of the high frequency voice coil is *behind* the coil of the bass driver. All of these means achieve the same result by delaying the high frequency signals with respect to the low frequency signals. The distances between the voice coils of the different drivers can be further physically off-set to whatever degree necessary to compensate for the electrically derived group delays. Given the speed of sound at 20° C, each centimeter that a driver is mounted behind another (relative to their voice coils) would give rise to a delay of 29.4 microseconds. Ten centimeters would therefore give rise to a delay of 294 microseconds, which would be sufficient to reverse the polarity of a wave at 1.7 kHz. (Look again at Figure 14.6.) Incidentally, the voice coils are used as an approximate reference for the source of the sound propogation because even though the diaphragms are ahead of the voice coils there is a finite time of propogation from the coil to the diaphragm face which, with most loudspeaker constructions, is roughly of the same order as the speed of sound in air.

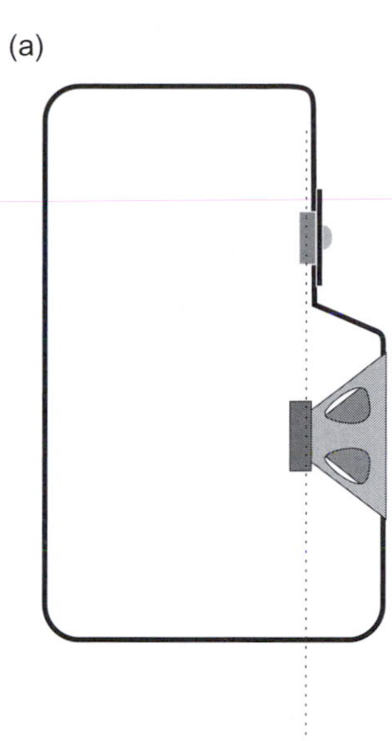

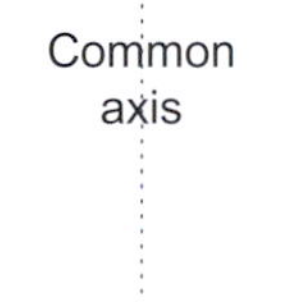

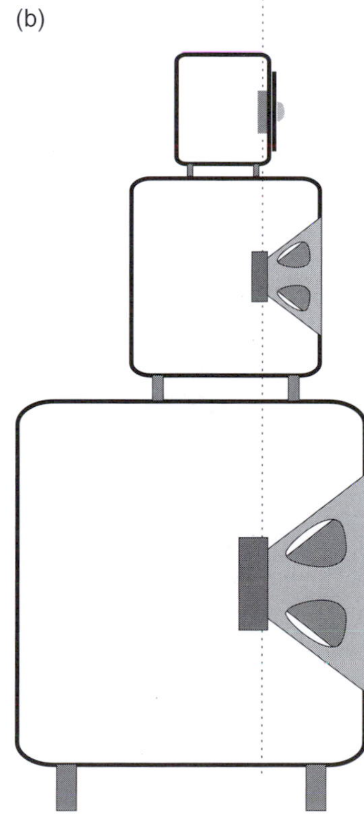

FIGURE 14.24
Methods for the physical compensation of propagation delays: (a) The stepped baffle; (b) Separate boxes.

DIGITAL CROSSOVERS

As time passes, digital crossovers have become more commonplace in professional loudspeaker systems, although their use in domestic circumstances is still largely restricted to home recording facilities. They are particularly attractive because of the easy implementation of almost any amplitude response, phase response, signal delay, driver compensation and even room compensation. However, all this flexibility comes at a considerable cost.

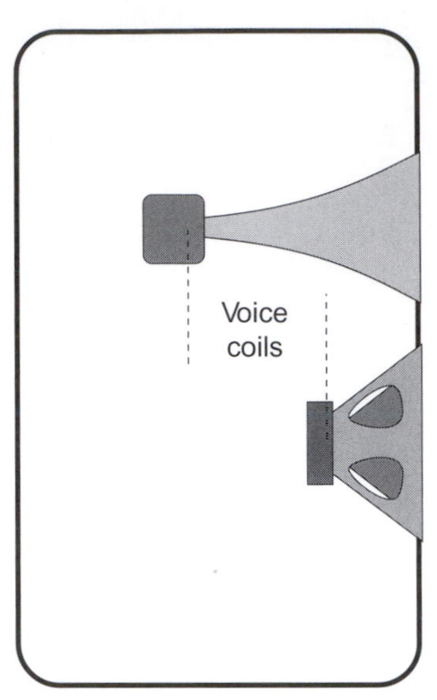

In this case the voice coil of the high frequency driver is *behind* the low frequency coil, which is rarely the case for direct radiating high frequency drivers sharing a common baffle with a low frequency driver—see, for example, Figure 14.8.

FIGURE 14.25
Flat baffle with a horn-loaded high frequency driver. In many cases it is possible for horn mounted compression drivers to align themselves very closely to the ideal whilst sharing a common, flat fronted baffle with a low frequency driver when the electrical group delays of the crossover filters are also taken into account, especially when steep-slope crossovers are used.

Sonically, in terms of "hi-end" hi-fi or high resolution studio monitoring, the highest fidelity can only be achieved if the sample rate and bit rate used in the crossovers at least equal those of the recording medium, or exceed the resolution of the ear and produce no audible artifacts. It may be difficult to hear the difference between a 20 bit/96 kHz recording and a 24 bit/192 kHz recording when listening to a crossover based on 16 bit/48 kHz processing. Even if a crossover has internal processing which seems higher than necessary, the resultant output after signal manipulation has taken place may not be as great as the marketing figures would suggest. The main problem, however, is concerned with the converters. As the finest amplifiers are still of the analog variety, D to A (digital to analog) conversion must take place in each output of the crossover, and, for either professional monitoring or "high-end" high fidelity, these crossover D to As must be of higher resolution than any other part of the signal chain. If they are not, then they will limit the quality and the resolution of the chain. As with the power amplifiers and loudspeaker cables the splitting of the frequency bands does offer some respite from the demands of handling the full audio bandwidth. Nonetheless, to achieve the highest levels of sonic quality, D to A converters are expensive, and a stereo three-way digital crossover would require six of them. At the time of writing, and whilst analog power amplifiers are the general order of the day, it would be reasonable to expect to pay 3000 to 5000 euros for those six converters. If analog *inputs* were required, then the A to D (analog to digital) converters may add another 1000 euros or so to the price of the crossover if the equivalent quality was to be expected.

Under less critical circumstances, digital crossovers can be very useful tools, but when they are user-programmable, they run the risk of being inappropriately applied. One must be very careful when trying to "solve" amplitude/phase problems that the solutions do not go against the laws of nature. Straightening out the phase associated with an amplitude roll-off may be very tempting, especially where the application is at the extremes of the frequency bands, but the results can sound unnatural because they *are* unnatural. On the other hand, in concert sound applications, where subtleties are by no means as important as solving the normal problems faced by such events, digital crossovers have been an enormous step forwards. Their real drawback is their cost when they must operate at the highest levels of sonic transparency. In many such cases, analog crossovers may simultaneously be simpler, cheaper, more robust, *and better*.

REFERENCES

Campbell, A.M. and Holland, K.R., 2004. Active vs Passive Crossovers for Mid-Priced Hi Fi Loudspeakers, Proceedings of the Institute of Acoustics, Vol 26, Part 8, pp 116–123, Reproduced Sound 20 conference, Oxford, UK.

Small, R.H., 1972. Direct Radiator Loudspeaker System Analysis and Synthesis, Journal of the Audio Engineering Society, 20(5).

Small, R.H., 1974. Passive Radiator Loudspeaker Systems. Journal of the Audio Engineering Society, Part I, 'Analysis', Vol 22, No 8 Part II, 'Synthesis', Vol 22, No 9.

Small, R.H., Closed Box Loudspeaker Systems, Journal of the Audio Engineering Society, Analysis,, 20(10), 1972. , Part II, 'Synthesis', Vol 21, No 1 (1973).

Small, R.H., Vented Box Loudspeaker System, Journal of the Audio Engineering Society, Part I. Small signal analysis, 21(5), 1973, 363–372. , Part II, 'Large signal analysis', Vol 21, No 6, pp 438–444 (1973) Part III, 'Synthesis', Vol 21, No 7, pp 549–554 (1973).

Theile, A.N., 1971. Loudspeakers in Vented Boxes, Part 1, Journal of the Audio Engineering Society, 19(5), 382–392.

Theile, A.N., 1971. Loudspeakers in Vented Boxes, Part 2, Journal of the Audio Engineering Society, 19(6), 471–483.

BIBLIOGRAPHY

Borwick, J., 2001. Loudspeaker and headphone handbook, [Chapters 5 and 6], third ed., Focal Press, Oxford, UK.

Colloms, M., 2005. High Performance Loudspeakers, [Chapter 6], sixth ed., John Wiley & Sons, Chichester, UK.

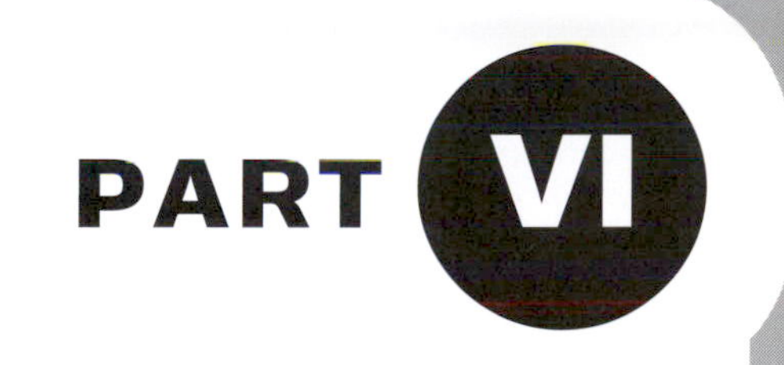

Digital Audio

PREFACE

This chapter by Dr Craig Richardson addresses the important and growing field of Digital Signal Processing (DSP). When digital audio began for most of us, with the introduction of the CD in 1982, it was certain that the data on the disc would be converted from the digital to the analog domain at once, that it would leave the CD player in an analog format which would be applied to the input of a conventional preamplifier, and that any processing such as tone control or level adjustment would also be carried in the analog domain.

Today these simple processes are often carried out in the digital domain, but there is also an important set of audio functions that can only be practically carried out there. A pure time-delay of almost any length is relatively straightforward in digital technology. In the analog world it can be done satisfactorily with all-pass filters, but only for short delays measured in tens of microseconds. Beyond this it would be necessary to resort to tape-loops, which, while just about adequate in their day, had an undeniably mediocre performance, and required regular replacement of the tape.

A more complicated application, such as the production of artificial reverberation, is by far best done in the digital domain, though the algorithms are naturally a good deal more complex than for a simple delay. Anyone who has attempted to get a satisfactory result out of spring-line reverb units will agree that digital is definitely the way to go.

The chapter begins by looking at the basics of sampling theory, Z-transforms, and sample-rate selection. It then moves on to considering Digital Signal Processor themselves. These processors are quite different from general-purpose processors or embedded microcontrollers, though they often incorporate some microcontroller features to make their integration into a system simpler. A DSP chip is optimized to perform very quickly and efficiently some very simple standard calculations on a flow of data; the calculations (typically floating-point multiplication and addition) may be simple but the volume that have to be performed in real-time is enormous. Very often they employ the Harvard Architecture, which has a program memory space/bus and a separate data memory space/bus, to speed up data throughput. Ordinary processors use the von Neumann architecture which has a single program and data memory structure; more flexible but not so fast.

A good many people, including myself, had their first introduction to DSP by working on the Motorola 56001 chip which used fixed-point arithmetic, with 24-bit program words and 24-bit data words. This was the first practical processor to offer this data length, which promised to be good enough for high-quality applications. In the late 1980s I was writing code for gain adjustment, EQ stages, filters, and sine-wave oscillators, as Soundcraft took its first steps towards a digital mixing console. More modern DSP chips now use floating-point arithmetic, which considerably simplifies the provision of adequate dynamic range in the calculations.

Every time a multiplication is performed on the audio data, the result is a word of twice the original length, and to prevent word-length growing out of control is necessary to truncate the result, by lopping digits off the end of it. If simple truncation is used, noise and distortion are added, and much research has gone into noise-shaping techniques that allow word-length to be controlled without losing quality.

CHAPTER 15

DSP Technology

Handbook for Sound Engineers by Dr. Craig Richardson

INTRODUCTION

Over the past 40 years, the field of digital signal processing (DSP) has grown from its origins as a collection of techniques for simulating the behavior of analog systems on digital computers into one of the most widely studied and universally used tools in modern technology. The use of DSP algorithms and implementations has become the rule rather than the exception, with applications in many areas such as music, communications, radar, sonar, image processing, robotics, seismology, meteorology, and applied physics. The remarkable growth of this discipline is largely due to two factors. First, DSP is a powerful problem-solving tool because it exploits the theoretical insights of discrete system theory to describe, analyze, and implement many interesting linear and nonlinear algorithms. Second, and more important, there is a special relationship between VLSI technology and DSP applications. The rapid development of digital integrated circuit technology has continually reduced the cost and increased the speed of the arithmetic operations necessary for DSP applications. In addition, DSP algorithms, which have demanding computational requirements but usually a very regular structure, are very well matched to the capabilities of VLSI. Integrated circuits are making complex DSP applications possible, and DSP applications have become a major motivating factor for building fast complex integrated circuits. Perhaps the most visible embodiments of this phenomenon are the families of DSP microprocessors commonly called DSP chips. These chips have already had an immense impact on technology and are currently in the process of revolutionizing much of our industrial and technological base.

This chapter will introduce some of the important aspects of DSP technology including the fundamentals of DSP, the sampling process for converting analog signals to digital signals, the algorithm development process, and an introduction to programmable DSP devices. References are provided for finding additional information.

DIGITAL SIGNAL PROCESSING

DSP is a technology and technique for analyzing and extracting information from signals, synthesizing signals, and manipulating signals. The acronym DSP is often used as both a noun and an adjective. DSP also often stands for digital signal processor—the actual microprocessor/computer that is used to implement the system. Common applications of DSP include cellular telephones, MP3 players, surround sound receivers, compact disc players, digital cameras, answering machines, and modems.

As with many disciplines, there are different perspectives and different layers of abstraction from which to explore DSP. For the purposes of this chapter, DSP will be approached and introduced from the theoretical, physical, and embedded software perspectives.

The theoretical perspective is concerned with the question "is something possible" and is built from fundamentals of DSP theory. This foundation includes linear system theory, complex number theory, and applied mathematics. The theoretical level provides a common language for DSP researchers to study and advance the state of the art.

Audio Engineering Explained

The physical perspective is concerned with the devices that are used to implement DSP systems. These devices include the programmable digital signal processors that perform mathematical operations at a very high speed, and the details of converting an analog signal into a digital signal and then back to an analog signal.

The embedded software perspective is concerned with the actual software that makes the digital signal processors perform the desired tasks. This software is called *embedded* because it is executed internally on the DSP device and is only user accessible through some user interface, effectively hidden or embedded in the product, hiding the implementation details from the user.

DSP SIGNALS AND SYSTEMS THEORY

The concepts of signals and systems are critical to an understanding of DSP. Signals can be a function of continuous time (i.e., analog) or of discrete time. Continuous-time signals have a signal value at any given instant of time while discrete-time signals only have a signal value at discrete instants of time. Values of discrete-time signals between the samples are determined by mathematically interpolating between the known sample values.

Signals represent the data that is to be processed. Examples include an audio file that needs to be compressed for low bit-rate storage or transmission or an image that will be searched for a particular object. A system is a transformation that maps an input signal (or multiple input signals) to an output signal (or multiple output signals)—i.e., the black box that maps inputs to outputs. In the music compression example, the output signal could be an MP3 file that was created by compressing an input signal. In the image example the output signal could simply be a yes/no decision along with positioning information. DSP systems are typically designed from simpler subsystems much like computer software is developed—subroutine by subroutine (one level of abstraction at a time). This section will introduce some fundamental systems and also introduce the useful properties that some systems possess.

Sequences

Discrete-time signals, also called sequences, are most often created by sampling analog, or continuous-time, signals. By sampling a continuous-time signal, a sequence of samples, really a sequence of numbers, can be processed and manipulated in a digital signal processor. Before going further into the sampling process, an introduction to signal and system theory will be presented, starting with discrete-time signals.

Discrete-time signals are represented mathematically as a sequence of numbers. The notation used will denote a sequence, x, as $x = \{x[n]\}$ where n is the index of the nth element in the sequence. In terms of notation, $x[n]$ represents both the nth sample in the sequence and the entire sequence that is a function of n. The index, n, can range over all values from $-\infty$ to $+\infty$.

From a programming perspective, a sequence can be thought of as an infinitely large array of data indexed by an integer variable. In reality, an infinitely long array is not practical, so a sequence is usually represented as a continuous stream of data. Often it is assumed that the sequence starts at time = 0 ($n = 0$) and ends some finite time later ($n = M$).

There are several sequences that are fundamental building blocks of DSP systems. These are the unit impulse, the unit step sequence, and the sinusoid (cosine or sine). The unit impulse is a signal that has a value of 1 at index $n = 0$ and is 0 everywhere else as shown in Figure 15.1. Mathematically this is denoted by

$$\delta[n] = \begin{cases} 0, n \neq 0 \\ 1, n = 0 \end{cases} \tag{15.1}$$

Having defined the unit impulse, it is possible to represent a sequence $x[n]$ as a sum of delayed impulses that have a value of $x[k]$ at $n = k$. Mathematically this is formulated as

$$x[n] = \sum_k x[k]\delta[n-k] \tag{15.2}$$

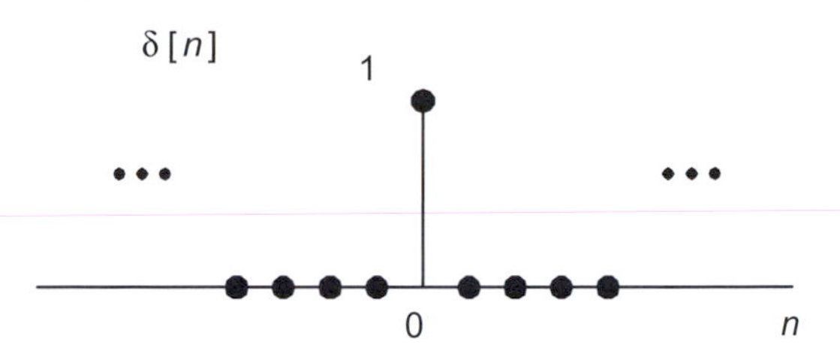

FIGURE 15.1
Unit impulse sequence has a value of 1 at $n = 0$ and is 0 everywhere else.

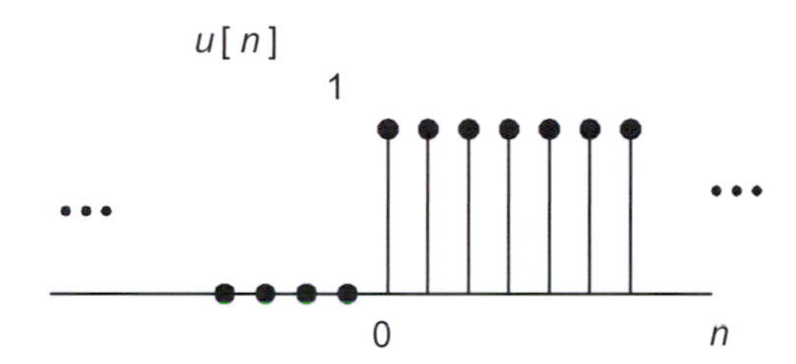

FIGURE 15.2
The unit step sequence has a value of 1 for $n \geq 0$ and is 0 everywhere else.

which simply says that the value of $x[n]$ is the collection of its individual samples at time $n = k$.

The unit step is a signal that starts at index 0 with value 1 and has value 1 for all positive indices as shown in Figure 15.2. Mathematically, this is denoted by

$$u[n] = \begin{cases} 0, n < 0 \\ 1, n \geq 0 \end{cases} \tag{15.3}$$

The cosine signal is a sinusoid of ω frequency and phase φ. An example of the cosine signal is shown in Figure 15.3. Mathematically, the cosine signal is denoted by

$$\cos[n] = \cos(\omega n + \phi) \tag{15.4}$$

All sequences can also be represented by the numbers that are the sample values $x[n]$. Table 15.1 shows the sample values for the sequence in Figure 15.4. Only the first 16 sample values are listed because the sequence repeats itself after the 16th value ($x[15]$).

Systems

Systems transform input signals into output signals. Some commonly used systems include the ideal delay system that delays the output relative to the input and the moving average system that performs some simple low-pass filtering. Systems operate on a signal by operating on each sample individually or groups of samples at a time. For instance, multiplying a sequence by a constant can be implemented by multiplying each sample of the sequence by the constant. Similarly, the addition of two sequences is performed by adding the signals together on a sample-by-sample basis. Other systems, such as an MPEG audio compression system, may operate on frames of data that have 1152 samples in each frame. The choice of whether to operate sample-by-sample or frame-by-frame is made by the system designer and algorithm developer.

A fundamental system is the ideal delay. The ideal delay system delays or advances a sequence by the delay amount. This system is defined by the equation

$$y[n] = x[n - n_d], -\infty < n < \infty \tag{15.5}$$

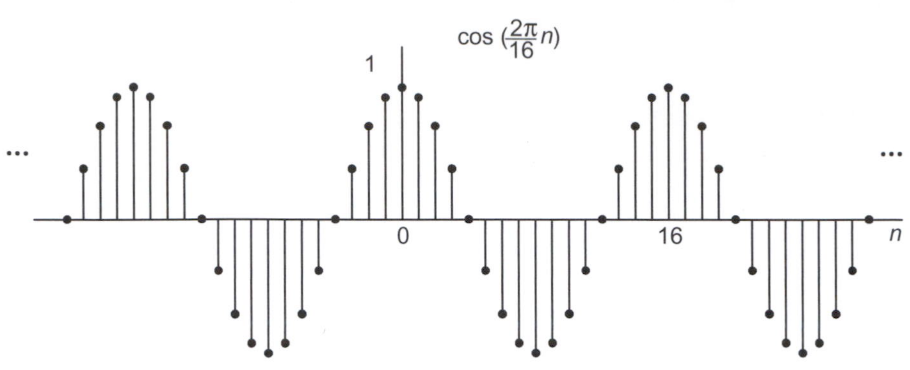

FIGURE 15.3
A cosine sequence of period 16. This particular cosine sequence is an infinite sequence of values that repeat with a period of 16 samples.

Table 15.1 The Values of the Signal $x[n]$ in Figure 15.4.

$x[0]$	1.0000	$x[6]$	−0.7071	$x[12]$	0.0000
$x[1]$	0.9239	$x[7]$	−0.9239	$x[13]$	0.3827
$x[2]$	0.7071	$x[8]$	−1.0000	$x[14]$	0.7071
$x[3]$	0.3827	$x[9]$	−0.9239	$x[15]$	0.9239
$x[4]$	0.0000	$x[10]$	−0.7071	...	
$x[5]$	−0.3827	$x[11]$	−0.3827		

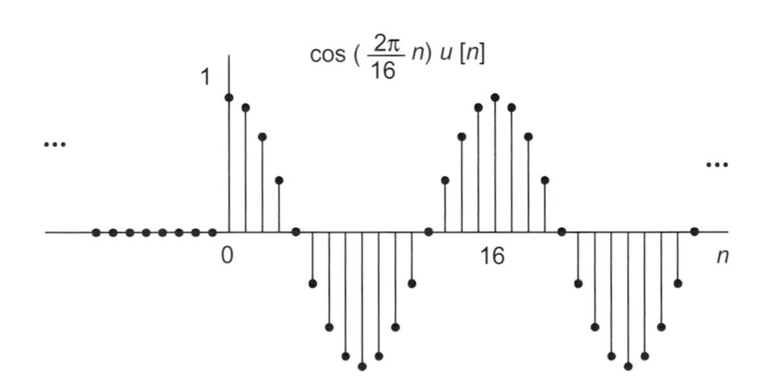

FIGURE 15.4
The product of the cosine sequence with the unit step sequence, $u[n]$. Notice that all the signal values for $n < 0$ are set to 0.

where,

n_d is an integer that is the delay of the signal.

The ideal delay system creates an output $y[n]$ by shifting the input signal, x, by n_d samples to the right when n_d is positive. This means that the value of the output signal $y[n]$ at a particular index n is the value of the input signal at index $n - n_d$. For example if the signal is delayed by three samples, then $n_d = 3$ and the output value $y[7]$ is equal to the value of $x[4]$—i.e., the value of $x[k]$ at $k = 4$ now appears at $y[j]$, $j = 7$. The system shifted the input signal three samples to the right as shown in Figure 15.5.

The moving average system takes an average of the input signal over some window and then moves to the next sample and takes an average over the new window, etc. The general moving average system is defined by the equation below, where M_1 and M_2 are positive integers. It is called a moving

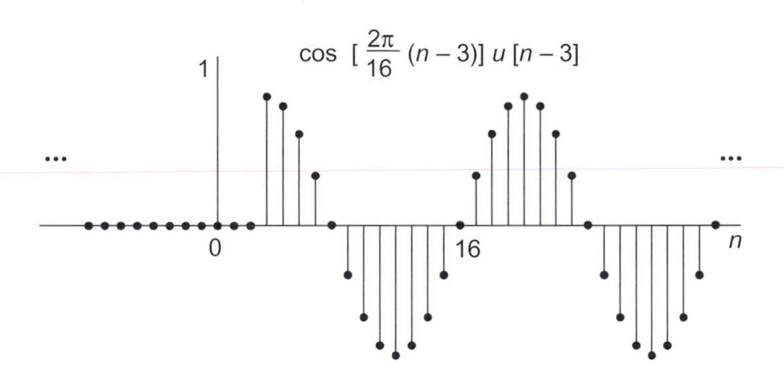

FIGURE 15.5
The cosine signal from Figure 15.4 delayed by $n_d = 3$ samples. This delay shifts the sequence to the right by three samples.

average because to compute each output, $y[n]$, the filter must be moved to the next index and the average recomputed.

$$y[n] = \frac{1}{M_1 + M_2 + 1}\sum_{k=-M_1}^{M_2} x[n-k] \tag{15.6}$$

The average sums values together starting M_1 samples forward from the current point and moving M_2 samples back from the current point and divides by the number of points that were summed together to form an average that smooths out the signal. The moving average is a digital filter that removes high frequency information through averaging.

System properties

System properties are a convenient way to describe broad classes of systems. Important system properties include linearity, shift invariance, causality, and stability. These properties are important because they lead to a representation of systems that can be readily analyzed.

LINEARITY

A linear system is one where the output of a sum of linearly scaled input signals is equal to the sum of the linearly scaled output signals. Mathematically, a system, $T\{\bullet\}$, is linear when

$$y_1[n] = T\{x_1[n]\}$$

and

$$y_2[n] = T\{x_2[n]\}$$

Then

$$\begin{aligned} T\{ax_1[n] + bx_2[n]\} &= T\{ax_1[n]\} + T\{bx_2[n]\} \\ aT\{x_1[n]\} + bT\{x_2[n]\} &= ay_1[n] + by_2[n] \end{aligned} \tag{15.7}$$

This means that when the input to a linear system is a sum of signals, the output is the sum of the signals transformed individually.

As an example, consider a system that performs a scalar multiply $y[n] = \alpha x[n]$ (when $\alpha > 1$, $y[n]$ is a louder version of $x[n]$, and when $\alpha < 1$, $y[n]$ is a quieter version of $x[n]$). This system is linear because

$$\begin{aligned} y[n] &= \alpha(ax_1[n] + bx_2[n]) \\ &= (\alpha ax_1[n] + \alpha bx_2[n]) \end{aligned}$$

An example of a nonlinear system would be a compressor/limiter because the output of a compressor/limiter to a sum of signals is generally not equal to the sum of the compressor/limiters applied to the signals individually.

TIME INVARIANCE

A time-invariant system is one where a delay in the input signal causes the output to be delayed by the same amount. Mathematically, a system, $T\{\bullet\}$, is time invariant if when $y[n] = T\{x[n]\}$ then

$$T\{x[n-N]\} = y[n-N] \tag{15.8}$$

When the input, $x[n]$, to a linear system is delayed, the output, $y[n]$, is delayed correspondingly. There is no absolute time reference associated with the system. The combination of time invariance and linearity makes the design and analysis of a large class of DSP theory and applications much simpler due to the convolution operation and Fourier analysis tools. (Oppenheim et al., 1999)

CAUSALITY

A causal system is one where the output of the system at a given time only depends on the present and past values of the input signal. No future data can be required to produce an output signal at the present time in a causal system. In the moving average system of Eq. 15.6, the system is causal only if $M_1 = 0$.

STABILITY

A system is bounded input/bounded output stable if and only if every bounded input sequence produces a bounded output sequence. A sequence is bounded if each value in the sequence is less than infinity. For real applications, stability is critically important because a system would stop operating properly should it ever become unstable.

Linear time-invariant systems

When the linearity property is combined with the time-invariance property to form a linear time-invariant (LTI) system, then the analysis of systems is very straightforward. Because a sequence can be represented as a sum of weighted delayed impulses as shown in Eq. 15.2, and an LTI system response is the sum of the component responses of the sequence components as shown in Eq. 15.7, the response of an LTI system is completely determined from its response to an impulse. Since an input signal can be represented as a collection of delayed and scaled impulses, the response to the full sequence is known. The response of a system to an impulse is commonly referred to as the *impulse response* of the system. Mathematically,

$$x[n] = \sum_k x[k]\delta[n-k]$$

i.e., the sequence $x[n]$ is a sum of scaled and delayed impulses. If $h_k[n] = T\{\delta[n-k]\}$, i.e., the system response to the delayed impulse at $n = k$, then the output $y[n]$ can be formed as

$$\begin{aligned} y[n] &= T\{x[n]\} \\ &= T\left\{\sum_k x[k]\delta[n-k]\right\} \\ &= \sum_k x[k]h_k[n] \end{aligned} \tag{15.9}$$

If the system is also time invariant, then $h_k[n] = h[n-k]$, and the output $y[n]$ is given by

$$\begin{aligned} y[n] &= \sum_k x[k]h[n-k] \\ &= \sum_k h[k]x[n-k] \end{aligned} \tag{15.10}$$

This representation is known as the convolution sum and is commonly written as $y[n] = x[n] \bullet h[n]$. The convolution system takes two sequences, $x[n]$ and $h[n]$, and produces a third sequence $y[n]$. For each value of $y[n]$, the computation requires multiplying $x[k]$ by $h[n-k]$ and summing over all valid indices for k where the signals are non-zero. To compute the output $y[n+1]$, move to the next point, $n+1$, and perform the same computation. The convolution is an LTI system and is a building block for many larger systems.

As an example, consider the convolution of the sequences in Figure 15.6 where $h[n]$ has only three non-zero sample values and $x[n]$ is a cosine sequence that has non-zero sample values for $n \geq 0$.

The computation of

$$y[n] = \sum_{k=0}^{2} h[k]x[n-k]$$

is performed as follows. Values of $x[n]$ for $n < 0$ are 0. Only the computation for the first three output samples are shown.

$$\begin{aligned} y[0] &= h[0]x[0] + h[1]x[-1] + h[2]x[-2] \\ &= 1.0 \\ y[1] &= h[0]x[1] + h[1]x[0] + h[2]x[-1] \\ &= 1.4239 \\ y[2] &= h[0]x[2] + h[1]x[1] + h[2]x[0] \\ &= 1.4190 \end{aligned}$$

The result of the convolution is shown in Figure 15.7 and has the sample values shown in Table 15.2.

FREQUENCY DOMAIN REPRESENTATION

Having defined an LTI system, it is possible to look at the signal from the frequency domain perspective and understand how a system changes the signals in the frequency domain. The frequency domain

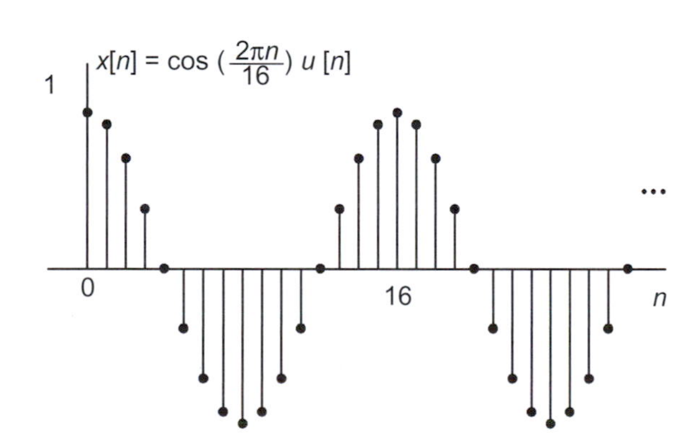

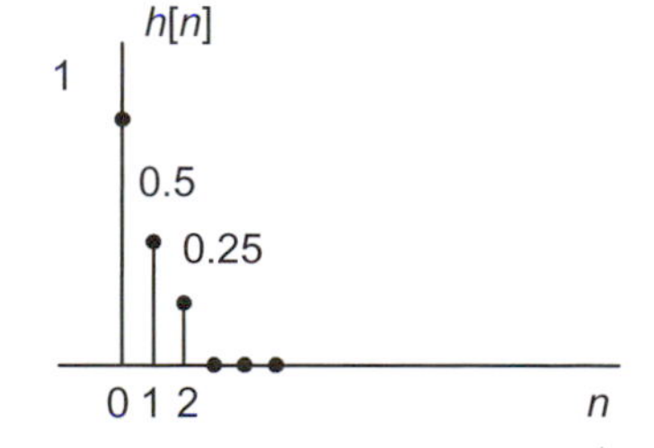

FIGURE 15.6

A convolution example with two sequences. $x[n]$ is the same signal from Figure 15.4 with values shown in Table 15.1, and $h[n]$ has the values shown above.

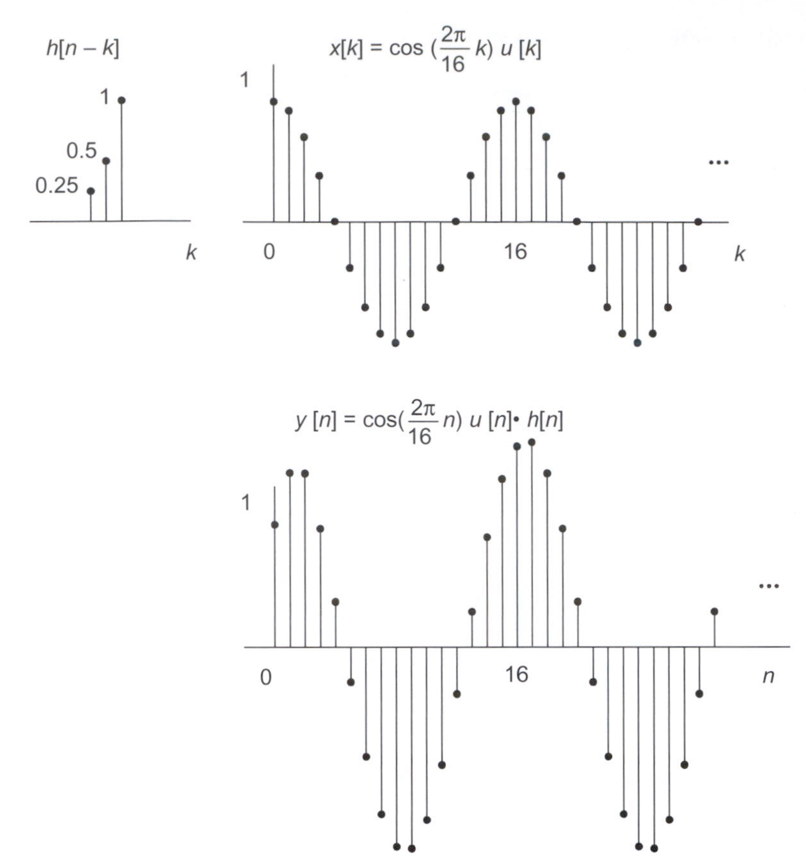

FIGURE 15.7
The output, $y[n]$, from the convolution of $x[n]$ and h[n] in Figure 15.6.

Table 15.2 The Result of the Convolution in Fig. 15.7.

y[0]	1.0000	y[11]	−0.9672	y[22]	−0.8984
y [1]	1.4239	y[12]	−0.3681	y [23]	−1.3731
y [2]	1.4190	y[13]	0.2870	y[24]	−1.6387
y[3]	0.9672	y[14]	0.8984	y[25]	−1.6548
y[4]	0.3681	y[15]	1.3731	y[26]	−1.4190
y[5]	−0.2870	y[16]	1.6387	y[27]	−0.9672
y[6]	−0.8984	y[17]	1.6548	y[28]	−0.3681
y[7]	−1.3731	y[18]	1.4190	y[29]	0.2870
y[8]	−1.6387	y[19]	0.9672	y[30]	0.8984
y[9]	−1.6548	y[20]	0.3681	...	
y[10]	−1.4190	y[21]	−0.2870		

represents signals as a combination of various frequencies from low frequency to high frequency. Each time-domain signal has a representation as a collection of frequency components where each frequency component can be thought of as sinusoids or tones. Sinusoids are important because a sinusoidal input to a linear time-invariant system generates an output of the same frequency but with amplitude and phase determined by the system. This property makes the representation of signals in terms of sinusoids very useful.

As an example, assume an input signal $x[n]$ is defined as $x[n] = e^{j}w^{n}$ —i.e., a complex exponential (Euler's relationship from complex number theory that states that $e^{j}w^{n} = \cos(\omega n) + j\sin(\omega n)$, where ω is the radian frequency that ranges from $0 \le \omega \le 2\pi$), then using the convolution sum of

$$y[n] = \sum_{k} h[k]x[n-k]$$

generates

$$y[n] = \sum_{k} h[k]e^{j\omega(n-k)} \quad (15.11)$$

$$y[n] = e^{j\omega n}\left(\sum_{k} h[k]e^{-j\omega k}\right) \quad (15.12)$$

By defining

$$H(e^{j\omega}) = \sum_{k} h[k]e^{-j\omega k}$$

we have

$$y[n] = H(e^{j\omega})e^{j\omega n}$$

where,

$H(e^{j}w)$ represents the phase and amplitude determined by the system.

This shows that a sinusoidal (or, in this case, the complex exponential) input to a linear time-invariant system will generate an output that has the same frequency but with an amplitude and phase determined by the system.

$H(e^{j}w)$ is known as the frequency response of the system and describes how the LTI system will modify the frequency components of an input signal. The transformation

$$H(e^{j\omega}) = \sum_{k} h[k]e^{-j\omega k}$$

is known as the Fourier transform of the impulse response, $h[n]$. If $H(e^{j}w)$ is a low-pass filter, then it has a frequency response that attenuates high frequencies but not low frequencies—hence it passes low frequencies. If $H(e^{j}w)$ is a high-pass filter, then it has a frequency response that attenuates low frequencies but not high frequencies.

In many instances it is more useful to process a signal or analyze a signal from the frequency domain than in the time domain either because the phenomenon of interest is frequency based or our perception of the phenomenon is frequency based.

An example of this is the family of MPEG audio compression standards that exploits the frequency properties of the human auditory system to dramatically reduce the number of bits required to represent the signal without significantly reducing the audio quality.

THE Z-TRANSFORM

The Z-transform is a generalization of the Fourier transform that permits the analysis of a larger class of systems than the Fourier transform. In addition, the analysis of systems is easier due to the convenient notation of the Z-transform.[1] The Fourier transform is defined as

$$X(e^{j\omega}) = \sum_{k} x[k]e^{-j\omega k}$$

while the Z-transform is defined as

$$X(z) = \sum_{k} x[k]z^{-k}$$

When working with linear time-invariant systems, an important relationship is that the Z-transform of the convolution of two sequences is equal to the multiplication of the Z-transforms of the two sequences, i.e., $y[n] = x[n]*h[n]7Y(z) = X(z)H(z)$. $H(z)$ is referred to as the system function (a generalization of the transfer function from Fourier analysis).

A common use of the Z domain representation is to analyze a class of systems that are defined as linear constant-coefficient difference equations that have the form of

$$\sum_{k=0}^{N} a_k y[n-k] = \sum_{k=0}^{M} b_k x[n-k] \tag{15.13}$$

where,

the coefficients a_k and b_k are constant (hence the name constant coefficient).

This general difference equation forms the basis for both finite impulse response (FIR) linear filters, and infinite impulse response (IIR) linear filters. Both FIR and IIR filters are used to implement frequency selective filters (e.g., high-pass, low-pass, bandpass, bandstop, and parametric filters) and other more complicated systems.

FIR filters are a special case of Eq. 15.13, where, except for the first coefficient, all the a_k are set to 0, leading to the equation

$$y[n] = \sum_{k=0}^{M} b_k x[n-k] \tag{15.14}$$

The important fact to notice is that each output sample $y[n]$ in the FIR filter is formed by multiplying the sequence of coefficients (also known as *filter taps*) by the input sequence values. There is no feedback in an FIR filter—i.e., previous output values are not used to compute new output values. A block diagram of this is shown in Figure 15.8 where the z^{-1} blocks are used to denote a signal delay of one sample (i.e., the Z-transform of the system $h[n] = \delta[n-1]$).

An IIR filter contains feedback in the computation of the output $y[n]$—i.e., previous output values are used to create current output values. Because of this feedback, IIR filters can be created that have a better frequency response (i.e., steeper slope for attenuating signals outside the band of interest) than FIR filters for a given amount of computation. However, most DSP architectures are optimized for computing FIR filters—i.e., multiplying and adding signals together continuously—so the choice of which filter style to use will depend on the particular application.

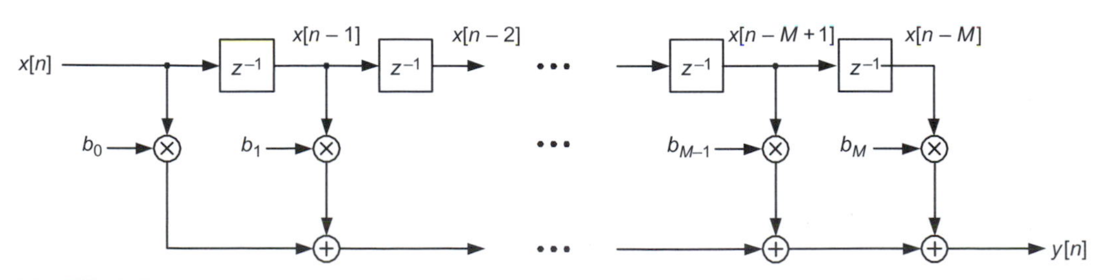

FIGURE 15.8
A block diagram of an FIR system where the input $x[n]$ is fed into a system that multiplies the delayed input signal with the filter coefficients b_k and sums the results together to form the output $y[n]$.

SAMPLING OF CONTINUOUS-TIME SIGNALS

The most common way to generate a digital sequence is to start with a continuous-time (analog) signal and create a discrete-time signal. For example, speech signals are continuous-time signals because they are continuous waves of acoustic pressure. A microphone is the transducer that converts the acoustic signal into a continuous-time electric signal. In order to process this signal digitally, it is necessary to convert this signal into the digital domain. Finally, after processing, it is often necessary to convert the discrete-time signal back into a continuous-time signal for playback through a loudspeaker system.

The process of converting an analog signal to a digital signal is often modeled as a two-step process, as shown in Figure 15.9, of converting a continuous-time signal to a discrete-time signal (with infinite resolution of the amplitude) and then quantizing the discrete-time signal into finite precision values (creating the digital sequence) that can be processed by a computer.[1] The process of converting the continuous-time signal into a discrete-time signal will be introduced, and then quantization will be reviewed. The quantization step is necessary to create a sample value that has a data word size that is compatible with the arithmetic capabilities of the target DSP. All real-world analog-to-digital converters (A/Ds) perform both the sampling and quantization process internal to the A/D device, but it is useful to discuss the subsystems separately because they have different significance and design trade-offs.

Continuous to discrete conversion

The most common method for converting a continuous-time signal, $x_c(t)$, into a discrete-time signal, $x[n]$, is to uniformly sample the signal every T seconds with the equation

$$x[n] = x_c(nT), -\infty < n < \infty \qquad (15.15)$$

This generates a sequence of samples, $x[n]$, where the value of $x[n]$ is the same as the value of $x_c(t)$ whenever $t = nT$—i.e., at each sampling interval T. $1/T$ is known as the *sampling frequency* and is usually expressed in Hertz or cycles per second.

Mathematically, when a continuous-time signal is sampled, the resulting signal has a frequency response that is related to the underlying continuous-time signal frequency response and the sampling rate. As shown next, this has significant ramifications for how often the signal must be sampled in order for the digital sequence to be reconstructed into an analog signal that accurately represents the original signal.

The sampling process will be analyzed in the frequency domain where it will be assumed that a band limited signal, $x_c(t)$, is to be sampled periodically with sample period T. A band-limited signal is one that has no signal energy higher than a particular frequency, Ω_N, as shown in Figure 15.10, where Ω represents the frequency axis of the signal. The reason the signal is assumed to be band limited is to prevent frequency aliasing, as will be evident shortly. The assumption of being band limited is significant although generally easily realizable in real-world systems.

The sampling of the continuous-time signal, $x_c(t)$, generates a signal, $x_s(t)$, from equation

$$x_s(t) = \sum_{n=-\infty}^{\infty} x_c(nT)\delta(t - nT) \qquad (15.16)$$

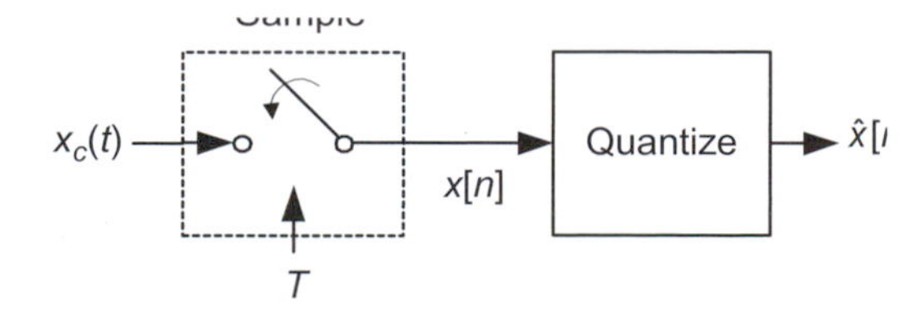

FIGURE 15.9
Analog-to-digital conversion can be thought of as a two-step process: converting a continuous-time signal to a discrete-time signal, $x[n]$, followed by quantizing the sample to create the digital sequence.

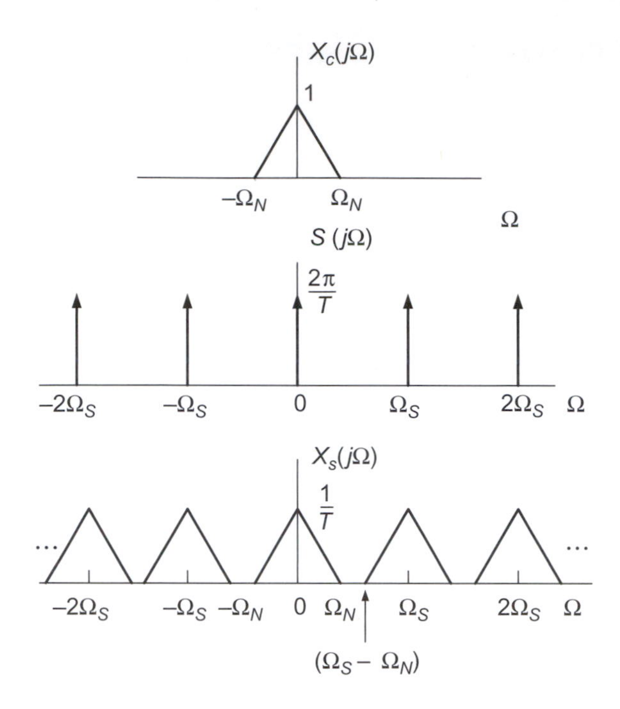

FIGURE 15.10
The frequency response of the analog signal, $X_c(j\Omega)$, the sampling function, $S(j\Omega)$, and the resulting frequency response of the sampled signal, $X_s(j\Omega)$.

$x_s(t)$ is the collection of values of $x_c(t)$ at the sampling interval of T. A convenient representation of this signal is as a collection of delayed and weighted impulse functions. The amplitude is the value at the sampling instant and the samples are spaced out by the sampling period T. The process can be analyzed in the frequency domain by first representing the Fourier transform of the impulse sequence as a sequence of impulses in the frequency domain (Oppenheim et al., 1997). This means that a sequence of equally spaced impulses in the time domain have a frequency representation that is a sequence of equally spaced impulses in the frequency domain, spaced by the sampling frequency $2\pi/T$. This is shown as

$$S(j\Omega) = \frac{2\pi}{T} \sum_{k=-\infty}^{\infty} \delta(\Omega - k\Omega_s) \tag{15.17}$$

where,

$\Omega_s = 2\pi/T$ is the sampling frequency in radians/second.

The Fourier transform of the sampled signal, $x_s(t)$, becomes

$$X_s(j\Omega) = \frac{1}{T} \sum_{k=-\infty}^{\infty} X_c(j(\Omega - k\Omega_s)) \tag{15.18}$$

Now the frequency response of the sampled continuous-time signal becomes a collection of shifted copies of the original frequency response of the analog signal $X_c(j\Omega)$. Figure 15.10 shows the frequency response of $X_c(j\Omega)$, the impulse train, $S(j\Omega)$, and the resulting frequency response of the sampled signal, $X_s(j\Omega)$.

This frequency response, $X_s(j\Omega)$, can also be interpreted as the convolution in the frequency domain between the frequency response of the continuous-time signal and the frequency response of the impulse train, $S(j\Omega)$.

$$X_s(j\Omega) = \frac{1}{2\pi} X_c(j\Omega) * S(j\Omega) \tag{15.19}$$

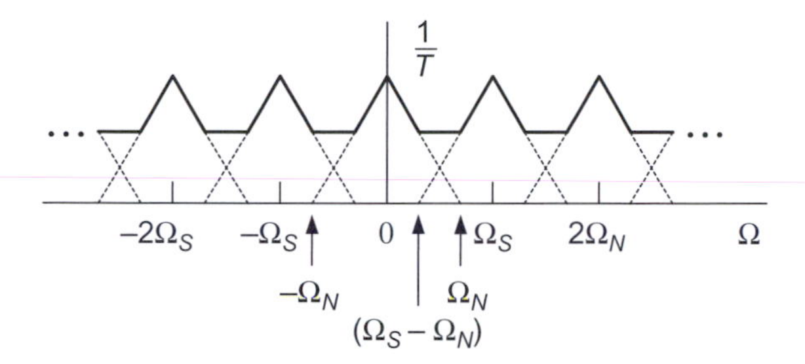

FIGURE 15.11
Sampling where the sampling frequency, Ω_S, is less than twice the highest frequency, Ω_N.

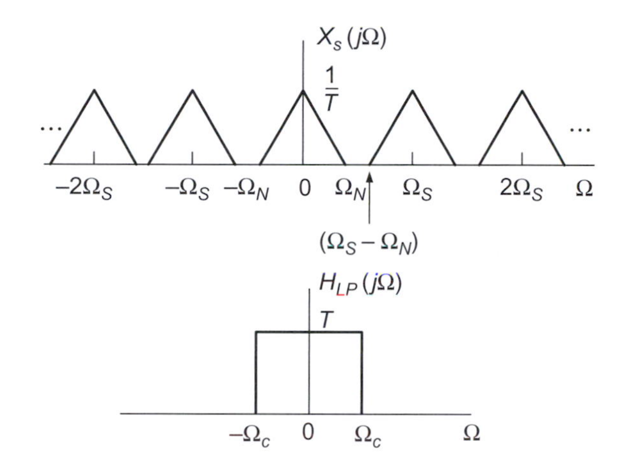

FIGURE 15.12
The spectrum replicas and the ideal low-pass filter that will remove the copies except for the desired baseband spectrum.

From Figure 15.10 it can be seen that as long as the sampling frequency minus the highest frequency is greater than the highest frequency, $\Omega_S - \Omega_N > \Omega_N$, the frequency copies do not overlap. This condition can be rewritten as $\Omega_N > 2\Omega_N$, which means that the sampling frequency must be at least twice as high as the highest frequency in the signal. If the sampling frequency is less than the highest frequency in the signal, $\Omega_S < 2\Omega_N$, then the frequency copies overlap as shown in Figure 15.11. This overlap causes the frequencies of the adjacent spectral copies to be added together, which results in the loss of spectral information. It is impossible to remove the effects of aliasing once aliasing has happened. The overlap is caused because the sampling frequency, Ω_S, is not high enough relative to the highest frequency in the continuous-time signal $X_c(j\Omega)$. As shown above, the sampling frequency must be at least twice as high as the highest frequency in the continuous-time signal in order to prevent this overlap, or aliasing, of frequencies.

Reconstructing the continuous-time signal

As seen from sampling a continuous-time signal, if the signal is not sampled fast enough, then the resulting frequency response of the sampled signal will have overlapping copies of the frequency response of the original signal. Assuming the signal is sampled fast enough (at least twice the bandwidth of the signal), the continuous-time signal can be reproduced by simply removing all of the spectral copies except for the desired one. This frequency separation can be performed with an ideal low-pass filter with gain, T, and cut-off frequency, Ω_C, where the cut-off frequency is higher than the highest frequency in the signal as well as the frequency where the first frequency replica starts,—i.e., $\Omega_N < \Omega_C < \Omega_S < -\Omega_N$. Figure 15.12 shows the repeated frequency spectrum and the ideal low-pass filter. Figure 15.13 shows the result of applying the low-pass filter to $X_S(j\Omega)$.

Sampling theory

The requirements for sampling are summarized by the Nyquist sampling theorem.[1] Let $x_c(t)$ be a band-limited signal with $X_c(j\Omega) = 0$ for $|\Omega| \geq \Omega_N$. Then $x_c(t)$ is uniquely determined by its samples,

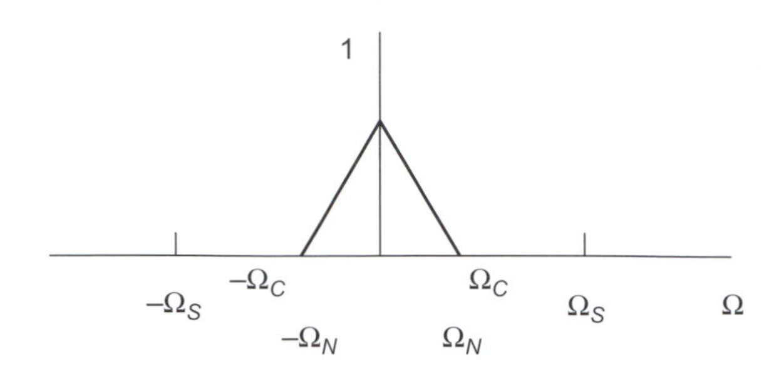

FIGURE 15.13
The final result of reconstructing the analog signal from the sampled signal.

$x[n] = x_c(nT)$, if $\Omega_S = 2\pi/T2\Omega_N$. The frequency Ω_N is referred to as the *Nyquist frequency*, and the frequency $2\Omega_N$ is referred to as the *Nyquist rate*. This theory is significant because it states that as long as a continuous-time signal is band-limited and sampled at least twice as fast as the highest frequency, then it can be exactly reproduced by the sampled sequence.

The sampling analysis can be extended to the frequency response of the discrete time sequence, $x[n]$, by using the relationships $x[n] = x_c(nT)$ and

$$X(e^{j\omega}) = \frac{1}{T}\sum_{k=-\infty}^{\infty} x[n]e^{-j\omega n}$$

The result is that

$$X(e^{j\omega}) = \frac{1}{T}\sum_{k=-\infty}^{\infty} X_c\left(j\left(\frac{\omega}{T} - \frac{2\pi k}{T}\right)\right) \tag{15.20}$$

$X(e^jw)$ is a frequency-scaled version of the continuous-time frequency response, $X_s(j\Omega)$, with the frequency scale specified by $\omega = \Omega T$. This scaling can also be thought of as normalizing the frequency axis by the sample rate so that frequency components that occurred at the sample rate now occur at 2π. Because the time axis has been normalized by the sampling period T, the frequency axis can be thought of as being normalized by the sampling rate $1/T$.

Quantization

The discussion up to this point has been on how to quantify the effects of periodically sampling a continuous-time signal to create a discrete-time version of the signal. As shown in Figure 15.9, there is a second step—namely, mapping the infinite-resolution discrete-time signal into a finite precision representation (i.e., some number of bits per sample) that can be manipulated in a computer. This second step is known as *quantization*. The quantization process takes the sample from the continuous-to-discrete conversion and finds the closest corresponding finite precision value and represents this level with a bit pattern. This bit pattern code for the sample value is usually a binary 2s-complement code so that the sample can be used directly in arithmetic operations without the need to convert to another numerical format (which takes some number of instructions on a DSP processor to perform). In essence, the continuous-time signal must be both quantized in time (i.e., sampled), and then quantized in amplitude.

The quantization process is denoted mathematically as

$$x[n] = Q(x[n])$$

where,

$Q(\bullet)$ is the nonlinear quantization operation,

$x[n]$ is the infinite precision sample value.

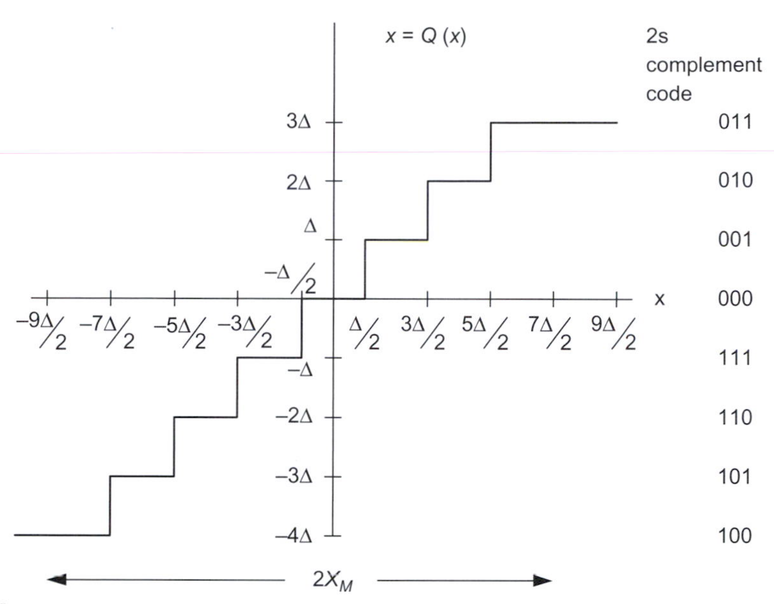

FIGURE 15.14
The quantization of an input signal, x, into $Q(x)$.

Quantization is nonlinear because it does not satisfy Eq. 15.7 —i.e., the quantization of the sum of two values is not the same as the sum of the quantized values due to how the nearest finite precision value is generated for the infinite-precision value.

To properly quantize a signal, it is required to know the expected range of the signal— i.e., the maximum and minimum signal values. Assuming the signal amplitude is symmetric, the most positive value can be denoted as X_M. The signal then ranges from $+X_M$ to $-X_M$ for a total range of $2X_M$. Quantizing the signal to B bits will decompose the signal into 2^B different values. Each value represents $2X_M/2^B$ in amplitude and is represented as the step size $\delta = 2X_M 2^{-B} = X_M 2^{-(B-1)}$. As a simplified example of the quantization process, assume that a signal will be quantized into eight different values which can be conveniently represented as a 3-bit value. Figure 15.14 shows one method of how an input signal, $x[n]$, can be converted into a 3-bit quantized value, $Q(x[n])$. In this figure, values of the input signal between $-\Delta/2$ and $\Delta/2$ are given the value 0. Input signal values between $\Delta/2$ and 3/2 are represented by their average value Δ, and so forth. The eight output values range from -4Δ to 3Δ for input signals between $-9\Delta/2$ and $7\Delta/2$. Values larger than $7\Delta/2$ are set to 3Δ and values smaller than $-9\Delta/2$ are set to -4Δ—i.e., the numbers saturate at the maximum and minimum values, respectively.

The step size, Δ, has an impact on the resulting quality of the quantization. If Δ is large, fewer bits will be required for each sample to represent the range, $2X_M$, but there will be more quantization errors. If Δ is small, more bits will be required for each sample, although there will be less quantization error. Normally, the system design process determines the value of X_M and the number of bits required in the converter, B. If X_M is chosen too large, then the step size, Δ, will be large and the resulting quantization error will be large. If X_M is chosen too small, then the step size, Δ, will be small, but the signal may clip the A/D converter if the actual range of the signal is larger than X_M.

This loss of information during quantization can be modeled as noise signal that is added to the signal as shown in Figure 15.15. The amount of quantization noise determines the overall quality of the signal. In the audio realm, it is common to sample with 24 bits of resolution on the A/D converter. Assuming a ± 15V swing of an analog signal, the granularity of the digitized signal is $30\text{V}/2^{24}$, which comes to $1.78\,\mu$V.

With certain assumptions about the signal, such as the peak value being about four times the rms signal value, it can be shown that the signal to noise ratio (SNR) of the A/D converter is approximately 6 dB per bit (Oppenheim et al., 1999). Each additional bit in the A/D converter will contribute 6 dB to the SNR. A large SNR is usually desirable, but that must be balanced with overall system requirements, system cost, and possibly other noise issues inherent in a design that would reduce the value of having a high-quality A/D converter in the system. The dynamic range of a signal can be defined as the range of the signal levels over which the SNR exceeds a minimum acceptable SNR.

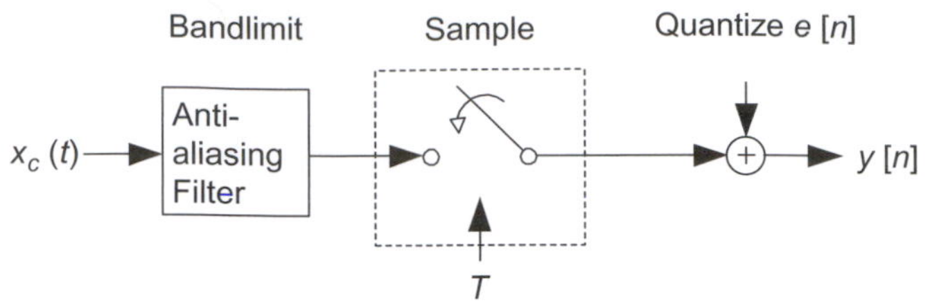

FIGURE 15.15
The sampling process of Figure 15.9 with the addition of an antialiasing filter and modeling the quantization process as an additive noise signal.

There are cost-effective A/D converters that can shape the quantization noise and produce a high-quality signal. Sigma-Delta converters, or noise-shaping converters, use an oversampling technique to reduce the amount of quantization noise in the signal by spreading the fixed quantization noise over a bandwidth much larger than the signal band (Aziz et al., 1996). The technique of oversampling and noise shaping allows the use of relatively imprecise analog circuits to perform high-resolution conversion. Most digital audio products on the market use these types of converters.

Sample rate selection

The sampling rate, $1/T$, plays an important role in determining the bandwidth of the digitized signal. If the analog signal is not sampled often enough, then high-frequency information will be lost. At the other extreme, if the signal is sampled too often, there may be more information than is needed for the application, causing unnecessary computation and adding unnecessary expense to the system.

In audio applications it is common to have a sampling frequency of 48 kHz = 48,000 Hz, which yields a sampling period of 1/48,000 = 20.83 μs. Using a sample rate of 48 kHz is why, in many product data sheets, the amount of delay that can be added to a signal is an integer multiple of 20.83 μs.

The choice of which sample rate to use depends on the application and the desired system cost. High-quality audio processing would require a high sample rate while low bandwidth telephony applications require a much lower sample rate. A table of common applications and their sample rate and bandwidths are shown in Table 15.3. As shown in the sampling process, the maximum bandwidth will always be less than ½ the sampling frequency. In practice, the antialiasing filter will have some roll-off and will band-limit the signal to less than ½ the sample rate. This band-limiting will further reduce the bandwidth, so the final bandwidth of the audio signal will be a function of the filters implemented in the specific A/D and the sample rate of the system.

ALGORITHM DEVELOPMENT

Once a signal is digitized, the next step in a DSP system is to process the signal. The system designer will begin the design process with some goal in mind and will use the algorithm development phase to develop the necessary steps (i.e., the algorithm) for achieving the goal.

Table 15.3 Common Sample Rates Found in Typical Applications and the Practical Bandwidths Realized at Each Sample Rate.

Application	Sample Rate	Bandwidth
Telephony applications	8 kHz	3.5 kHz
Videoconferencing	16 kHz	7 kHz
FM radio	32 kHz	15 kHz
CD audio	44.1 kHz	20 kHz
Professional audio	48 kHz	22 kHz
Future audio	96 kHz	45 kHz

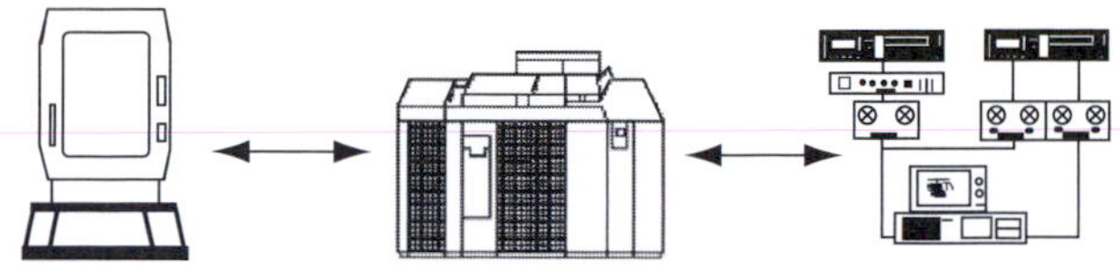

FIGURE 15.16
The three phases of DSP application development.

The design cycle for a DSP system generally has three distinct phases as shown in Figure 15.16: an abstract algorithm conceptualization phase, in which various mathematical algorithms and systems are explored; an algorithm development phase, where the algorithms are tested on large amounts of data; and a system implementation phase, where specific hardware is used to realize the system.

Traditionally, the three phases in the DSP design cycle have been performed by three entirely different groups of engineers using three entirely different classes of tools, although this process has converged as development tools have improved. The algorithm conceptualization phase is most often performed by researchers in a laboratory environment using highly interactive, graphically oriented DSP simulation and analysis tools. In this phase, the researcher begins with the concept of what to accomplish and creates the simulation environment that will enable changes and reformulations of the approach to the problem. No consideration is given, at this point, to the computational performance issues. The focus is on proof of concept issues—proving that the approach can solve the problem (or a temporarily simplified version of the problem).

In the algorithm development phase, the algorithms are fine-tuned by applying them to large databases of signals, often using high-speed workstations to achieve the required throughput. During this step it is often necessary to refine the high-level conceptualization in order to address issues that arose while running data through the system. Simulations are characterized by having many probes on the algorithm to show intermediate signal values, states, and any other useful information to aid in troubleshooting both the algorithm and the implementation of the simulation.

Once the simulation performs as desired, the next step is to create a real-time implementation of the simulation. The purpose of the real-time implementation is to better simulate the final target product, to begin to understand what the real-time memory and computational requirements will be, and to run real-time data through the system. There is no substitute for running real-time data through the system because real-time data typically exhibits characteristics that were either not anticipated or have unintended consequences in the simulation environment. Real-time data is generally more stressful to an algorithm than simulated, or non-real-time, data.

Often, with the introduction of real-time data, it may be necessary to go to the conceptual level again and further refine the algorithm.

Although advanced development tools and high-speed processors have blurred the distinction between simulation and real-time implementation, the goal of the real-time implementation is to "squeeze as much algorithm as possible" into the target processor (or processors). Squeezing more into the target processor is a desirable goal because it is usually much less expensive to use a single signal processor than to use multiple processors.

DIGITAL SIGNAL PROCESSORS

Programmable digital signal processors are microprocessors with particular features suited to performing arithmetic operations such as multiplication and addition very efficiently (Lee, 1988 and 1989). Traditionally, these enhancements have improved the performance of the processor at the expense of ease of programmability.

A typical microprocessor will have an arithmetic and logic unit for performing arithmetic operations, a memory space, I/O pins, and possible other peripherals such as serial ports and timers. A DSP processor will often have fewer peripherals, but will include a hardware multiplier, often a high-speed internal memory space, more memory addressing modes, an instruction cache and a pipeline, and even a separation of the program and data memory spaces to help speed program execution. The hardware multiplier allows the DSP processor to perform a multiplication in a single clock cycle while microprocessors typically take multiple clock cycles to perform this task. With clock cycles easily exceeding 100 MHz, up to 100 million multiples can occur every second. At this rate, 2083 multiplies can occur in the time span required to collect one sample of data at a 48 kHz sample rate (100 M/48,000).

A high-speed internal memory bank can be used to speed the access to the data and/or program memory space. By making the memory high speed, the memory can be accessed twice within a single clock cycle, allowing the processor to run at maximum performance. This means that proper use of internal memory enables more processing to take place within a given speed processor when compared with using external memory.

The instruction cache is also used to keep the processor running more efficiently because it stores recently used instructions in a special place in the processor where they can be accessed quickly, such as when looping program instructions over signal data.

The pipeline is a sequential set of steps that allow the processor to fetch an instruction from memory, decode the instruction, and execute the instruction. By running these subsystems in parallel, it is possible for the processor to be executing one instruction while it is decoding the next one and fetching the instruction after that. This streamlines the execution of instructions.

DSP Arithmetic

Programmable DSPs offer either fixed-point or floating-point arithmetic. Although floating-point processors are typically more expensive and offer less performance than fixed-point processors, VLSI hardware advances are minimizing the differences. The main advantage of a floating-point processor is the ability to be free of numerical scaling issues, simplifying the algorithm development and implementation process.

Most people naturally think in terms of fractions and decimal points, which are examples of floating-point numbers. Typically, floating-point DSPs can represent very large and very small numbers and use 32-bit (or longer) words composed of a 24-bit mantissa and an 8-bit exponent, which together provide a dynamic range from 2^{-127} to $2^{128.}$ This vast range in floating-point devices means that the system developer does not need to spend much time worrying about numerical issues such as overflow (a number too large to be represented) or underflow (a number too small to be represented). In a complicated system, there is enough to worry about without having to worry about numerical issues as well.

Fixed-point arithmetic is called fixed-point because it has a fixed decimal point position and because the numbers have an implicit scale, depending on the range that must be represented. This scale must be tracked by the programmer when performing arithmetic on fixed-point numbers. Most DSPs use the fixed-point 2s-complement format, in which a positive number is represented as a simple binary value and a negative value is represented by inverting all the bits of the corresponding positive value and then adding 1. Assuming a 16-bit word, there are $2^{16} = 65{,}536$ possible combinations or values that can be represented which allows the representation of numbers ranging from the largest positive number of $2^{15} - 1 = 32{,}767$ to the smallest negative (e.g., most negative) number of $-2^{15} = -32{,}768$.

There are many times when it is important to represent fractions in addition to integer numbers. To represent fractions, the implied position of the decimal point must be moved. When using 16-bit arithmetic to represent fractions only, with no integer component, a Q15 arithmetic format with an implied decimal point and 15 bits of fraction data to the right of the decimal point could be used. In this case, the largest number that can be represented is still $2^{15} - 1$, but now this number represents $32{,}767/32{,}768 = 0.999969482$, and the smallest negative number is still -2^{15}, but this number represents $-32{,}768/32{,}768 = -1$. Using Q15 arithmetic, it is possible to represent numbers between 0.999969482 and -1. As another example, representing numbers that range between 16 and -16 would require Q11 arithmetic (4 bits before the implied decimal point). An implementation may use different implied decimal positions for different variables in a system.

Because of the smaller word size and simpler arithmetic operations when compared with floating-point processors, fixed-point DSPs typically use less silicon area than their floating-point counterparts, which translates into lower prices and less power consumption. The trade-off is that, due to the limited dynamic range and the rules of fixed-point arithmetic, an algorithm designer must play a more active role in the development of a fixed-point DSP system. The designer has to decide whether the given word width (typically 16 or 24 bits) will be interpreted as integers or fractions, apply scale factors if required, and protect against possible register overflows at potentially many different places in the code. Overflow occurs in two ways in a fixed-point DSP (Lee, 1988). Either a register overflows when too many numbers are added to it or the program attempts to store N bits from the accumulator and the discarded bits are important. A complete solution to the overflow problem requires the system designer to be aware of the scaling of all the variables so that overflow is sufficiently unlikely. An underflow occurs if a number is smaller than the smallest number that can be represented. Floating-point arithmetic keeps track of the scaling automatically in order to simplify the programmer's job. The exponent keeps track of where the decimal point should be. Checking for overflow/underflow and preventing these conditions makes changing a DSP algorithm more difficult because, not only are algorithmic changes required, there are also numeric issues to contend with. Usually, once an implementation for a particular application has matured past the development stage, the code (which may have begun as floating-point code) may be ported to a fixed-point processor to allow the cost of the product to be reduced.

The dynamic range supported in a fixed-point processor is a function of the bit width of the processor's data registers. As with A/D conversion, each bit adds 6 dB to the SNR. A 24-bit DSP has 48 dB more dynamic range than a 16-bit DSP.

Implementation issues

The implementation of an algorithm into a real system is often much more complicated than using a compiler to automatically optimize the code for maximum performance. Real-time systems have constraints such as limited memory, limited computational performance, and most importantly, need to handle the real-time data that is continuously sent from the A/D converter to the DSP and the real-time data that must be sent from the DSP back to the D/A converter. Interruptions in this real-time data are typically not acceptable because, for example, in an audio application, these interruptions will cause audible pops and clicks in the audio signal.

Real-time programming requires that all of the computation required to produce the output signal must happen within the amount of time it takes to acquire the input signal from the A/D converter. In other words, each time an input sample is acquired, an output sample must be produced. If the processing takes too long to produce the output, then, at some point, incoming data from the A/D will not be able to be processed, and input samples will be lost. As an example, assume a system samples at 48 kHz and performs parametric equalization on a signal. Assuming that each band of parametric equalization requires 5 multiplies and 4 adds, which can be implemented in 9 clock cycles, then a 100 MHz DSP has 2083 instructions that can be executed in the time between samples. These instructions would allow a maximum of 231 bands of parametric equalization ($2083/9 = 231$). Now, realistically, the system is performing other tasks such as collecting data from the A/D converter, sending data to the D/A converter, handling overhead from calling subroutines and returning from subroutines, and is possibly responding to interrupts from other subsystems. So the actual number of bands of equalization could be significantly less than the theoretical maximum of 231 bands.

DSPs will have a fixed amount of internal memory and a fixed amount of external memory that can addressed. Depending on the system to be designed, it can be advantageous to minimize the amount of external memory that is required in a system because that can lead to reduced parts costs, reduced manufacturing expense, and higher reliability. However, there is usually a trade-off between computational requirements and memory usage. Often, it is possible to trade memory space for increased computational power and vice versa. A simple example of this would be the creation of a sine wave. The DSP can either compute the samples of a sine wave, or look-up the values in a table. Either method will produce the appropriate sine wave, but the former will require less memory and more CPU while the latter will require more memory and less CPU. The system designer usually makes a conscious decision regarding which trade-off is more important.

System delay

Depending on the application, one of the most important issues in an implementation is the amount of delay or latency that is introduced into the system by the sampling and processing. Figure 15.17 shows the typical digital system. The analog signal comes into the A/D converter that digitizes and quantizes

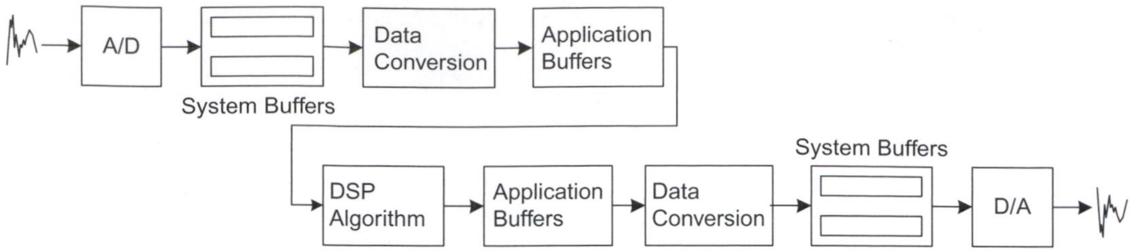

FIGURE 15.17
A block diagram of the typical DSP system.

the signal. Once digitized, the signal is typically stored in some data buffers or arrays of data. The data buffers could be one sample long or could be longer depending on whether the algorithm operates on a sample-by-sample basis or requires a buffer of data to perform its processing. The system buffers are usually configured in a ping-pong fashion so that while one buffer is filling up with new data from the A/D, the other is being emptied by the DSP as it pulls data from the buffer to process the data.

Following the system buffer may be a data conversion block that converts the data from a fixed-point integer format provided by the A/D to either some other fixed-point format or a floating-point processor, depending on the DSP and the numerical issues. Following this, there may be some application buffers that store buffers of data to give the DSP some flexibility in how much time it takes to process a single block of data. The application buffers can be viewed as a rubber band that allows the DSP to use more time for some frames of data and less time for other frames of data. As long as the average amount of time required to process a buffer of data is less than the amount of time required to acquire that buffer of data, the DSP will make real-time. If the amount of time required to process a buffer takes longer than the time to acquire the buffer, then the system will be unable to process all buffers and will have to drop buffers because there will not be any processing time left over to collect the next buffer from the A/D converter. In this case the system will not make real time and the missing buffers will produce audible pops in an audio signal. The application buffers can be used to compensate for some frames that may require more processing (more CPU time) than others. By providing more frames over which to average the computation, the DSP will more likely make real time. Of course, if the DSP cannot perform the required amount of computation on average during the time that a buffer of data is acquired, then averaging over more and more frames will not help. The system will eventually miss real-time and have to drop samples.

After the application buffers, the DSP algorithm performs the operations that are desired and then passes the data to possibly another set of application buffers that in turn can be converted from the numerical format of the DSP to the format required by the D/A converter. Finally the data will be sent to the D/A converter and converted back into an analog signal.

An accounting of the delay of the system should include all delays beginning when the analog signal comes in contact with the A/D converter to when the analog signal leaves the D/A converter. Table 15.4 shows the potential delays in each of the blocks of Figure 15.17. For this exercise, it is assumed that a frame of data consists of N samples, where $N \geq 1$. Each frame of delay adds $N \bullet 1/T$ seconds of delay to the system. For example, a delay of 16 samples at 48 kHz corresponds to $16/48{,}000 = 333.3\,\mu s$.

Further complicating the delay measurements is the possible requirement of sending information to an external system. This could be in the form of sending a bitstream to a remote decoder, receiving a bitstream from a remote encoder, and also any error detection and/or correction on a bitstream that may be required.

Choosing a DSP

The choice of which DSP to use for a particular application depends on a collection of factors including:

- **Cost.** DSPs range in price from several dollars to hundreds of dollars. Low-cost DSP processors are typically 16-bit fixed-point devices with limited amounts of internal memory and few peripherals. Low-cost DSPs are typically suited for extremely high volume applications, where the exact capabilities required, and no more, are built into the chip. High-cost DSPs typically are newer

Table 15.4 A Summary of Delay Issues in a Typical DSP System.

Block	Delay	Description
A/D	From 1 to 16 samples	Most A/D converters have some amount of delay built in due to the processing that is done. Oversampling A/Ds in particular have more delay than other types of A/Ds.
System Buffers	Adds at least 1 frame of delay	In the ping-pong buffer scheme, the system is always processing the last frame of data while the A/D is supplying the data from the next frame of data.
Data conversion	Possibly none	The conversion of the data format may be lumped with the algorithm processing delay.
Application buffers	Adds M-1 frames of delay for M buffers	Generalizing the ping-pong buffer scheme to M buffers, the system is always processing the oldest buffer, which is M-1 buffers behind the most recent buffer.
DSP algorithm	Variable, although usually at least 1 frame	There are two primary ways a DSP algorithm adds delay. One is processing delay and the other is algorithmic delay. Processing delay occurs because the processor is not infinitely fast, so it takes some amount of time to perform all of the computation. If the DSP has no extra CPU cycles after performing the computation, then the processing time adds a full frame of delay to the system. If it takes more than a frame of delay to perform the computation, then the system will not make real time.
		The algorithmic delay comes from any requirement to use data from future frames of data (i.e., buffer the data) in order to make decisions about the current frames of data and other delays inherent in the algorithm process.
D/A	From 1 to 16 samples	As with the A/D converter there is some delay associated with converting a digital signal into an analog signal. Current converters typically have no more than 16 samples of delay.

processors that have a great deal of internal memory or other architectural features including floating-point arithmetic and high-speed communication ports.

- **Computational Power: MHz, MIPs, MFLOPs.** Computational power is measured in several different ways including processor speed (MHz), millions of instructions per second (MIPS), and millions of floating-point operations per second (MFLOPS). The computational power of a processor is usually directly related to cost. An MIP means that one million instructions can be executed per second. The instructions that can be executed could include memory loads and stores or perhaps arithmetic operations. An MFLOP means one million floating-point operations can be executed per second. A floating-point operation includes multiplies and/or adds. Often the architecture of the DSP allows the DSP to execute two (or more) floating-point operations per instruction. In this case the MFLOPs would be twice (or more) the MIPs rating of the processor.

Higher-speed processors allow the user to pack more features into a DSP product, but with a higher cost.

Power Consumption. Depending on the application, low power may be important for long battery life or low heat dissipation. DSPs will have a power rating and, often, a watt/MIP rating to estimate power consumption.

Architecture. Different manufacturers' DSPs have different features and trade-offs. Some processors may allow extremely high-speed computational rates but at the expense of being difficult to program. Some may offer ease of multiprocessing, multiple arithmetic processors, or other features.

Arithmetic Precision. The use of floating-point arithmetic simplifies arithmetic operations. Fixed-point processors often have lower cost but often require additional instructions to maintain the level of numerical accuracy that is often required. The final production volume of the end product often dictates whether the added development time is worth the cost savings.

Peripherals. Certain features of processors such as the ability to share processor resources among linked processors or access to external memory/devices can have a significant impact on which processor to use for a particular application. Integrated timers, serial ports, and other features can reduce the number of additional parts required in a design.

Code Development. The amount of code already developed for a particular processor family may dictate the choice of processors. Real-time code development takes significant time and the investment can be substantial. The ability to reuse existing code is a significant time saver in getting products to market.

Development Tools. The development tools are critical to the timely implementation of an algorithm on a particular processor. If the tools are not available or are not functional, the development process will most likely be extended beyond any reasonable time estimate.

Third Party Support. DSP processor manufacturers have a network of companies that provide tools, algorithm implementations, and hardware solutions for particular problems. It is possible that some company has already implemented, and makes a living out of implementing, the type of solution that is required for a given application.

PROGRAMMING A DSP

DSPs, like many other processors, are only useful if they can input and output data. The software system used to input and output data is called an I/O system. As shown in Figure 15.17, a DSP application program typically processes an input stream of data to produce some output data. The processing of this data is performed under the direction of the application program, which usually includes one or more algorithms programmed on the DSP. The DSP application program consists of acquiring the input stream data, using the algorithms to process the data, and then outputting the processed data to the output data stream. An example of this is a speech data compression system where the input stream is a data stream representing uncompressed speech. The output stream, in this case, is the compressed speech data and the application consists of getting the uncompressed input speech data, compressing the data, and then sending the compressed data to the output stream.

One of the most important factors that a DSP I/O system must address is the idea of real-time. An extremely important aspect of these real-time A/D and D/A systems is that the samples must be produced and consumed at a fixed rate in order for the system to work in real-time. Although an A/D or D/A converter is a common example of a real-time device, other devices not directly related to real-time data acquisition can also have real time constraints. This is particularly true if they are being used to supply, collect, or transfer real-time information from devices such as disk drives and inter-processor communication links. In the speech compression example, the output stream might be connected to a modem that would transmit the compressed speech to another DSP system that would uncompress the speech. The I/O system should be designed to interface to these devices (or any other) as well.

Another important aspect of a real-time I/O system is the amount of delay imposed from input to output. For instance, when DSPs are used for in-room reinforcement or two-way speech communication (i.e., telecommunications), the delay must be minimized. If the DSP system causes a noticeable delay, the conversation would be awkward and the system would be considered unacceptable. Therefore, the DSP I/O system should be capable of minimizing I/O delay to a reasonable value.

Programming a DSP is usually accomplished in a combination of C and assembly languages. The C code provides a portable implementation that can potentially be run on multiple different platforms. Assembly language allows for a more computationally efficient implementation at the expense of increased development time and decreased portability. By starting in C, the developer can incrementally optimize the implementation by benchmarking which subroutines are taking the most time, optimizing these routines, and then finding the next subroutine to optimize.

The typical C code shell for implementing a DSP algorithm is shown in Figure 15.18. Here, the C code allocates some buffer memory to store signal data, opens an I/O signal stream, and then gets data, processes the data, and then sends the data to the output stream. The input and output streams typically have lower level device drivers for talking directly to the A/D and D/A converters, respectively.

CONCLUSION

This chapter has introduced the fundamentals of DSP from a theoretical perspective (signal and system theory), and a practical perspective. The concepts of real-time systems, data acquisition, and digital signal processors have been introduced. DSP is a large and encompassing subject and the interested reader is encouraged to learn more through the exhaustive treatment given to this material in the references.

```
#include <stdio.h>
#include <aspi_io.h>
#include <malloc.h>

#define LEN 800

void main(argc,argp)
char **argp;
int argc;
{
   SIG_Stream input, output;
   SIG_Attrs sig_attrs;
   BUF_Buffer buffer;

   buffer = BUF_create(SEG_DRAM,LEN,0);

   input = SIG_open(argp[1],SIG_READ,buffer,0);

   SIG_getattrs(input,&sig_attrs);

       output =
          SIG_open(argp[2],SIG_WRITE,buffer,&sig_attrs);

   while (SIG_get(input,buffer))
            {
      /* data processing of buffer */
      my_DSP_algorithm(buffer);

      SIG_put(output,buffer);
            }
      return(0);
}
```

FIGURE 15.18
An example C program for collecting data from an A/D using an input signal stream created with SIG_open and sending data to the D/A using the output signal stream and processing the data with the function my_DSP_algorithm ().

REFERENCES

Aziz, P.M., Sorensen, H.V. and Van Der Spiegel, J., 1996. An Overview of Sigma-Delta Converters, IEEE Signal Processing Magazine, 13(1), 61–84.

Lee, E.A., 1988. Programmable DSP Architectures: Part I, IEEE ASSP Magazine, 5(4), 4–19.

Lee, E.A., 1989. Programmable DSP Architectures: Part II, IEEE ASSP Magazine, 6(1), 4–14.

McClellan, J.H., Schafer, R.W. and Yoder, M., 1999. DSP First: A Multimedia Approach, Prentice Hall, Upper Saddle River, NJ.

Oppenheim, A.V., 1999. Schafer, R.W. and Buck, J.R., Discrete-Time Signal Processing, 2nd Edition., Prentice Hall, Upper Saddle River, NJ.

Oppenheim, A.V., Willsky, A.S. and Nawab, S.H., 1997. Signals and Systems, 2nd Edition., Prentice Hall, Upper Saddle River, NJ.

PREFACE

In Chapter 16 David Huber gives an overview of MIDI technology, including MIDI sequencing software and MIDI time code.

MIDI stands for Musical Instrument Digital Interface. It is *not* a means of carrying digital audio as a signal. It is a technology, and a very well worked-out and robust technology, for sending instructions to synthesizers, samplers and drum machines by transmitting short digital codes, which typically mean something like "play this note on synthesizer 2 at this volume." Such a command is transmitted in only three bytes, which allows incredible economy compared with encoding and transmitting the actual musical event as sampled audio. It means that MIDI data can be transmitted over a digital connection that is relatively slow but electrically bullet-proof, while still maintaining the time precision that is essential for the satisfactory production of complex musical parts.

There is however a great deal more to MIDI than just sending note messages, and David covers these comprehensively. He begins by looking at the structure of the MIDI message. Most messages consist of an initial status byte, with a channel number in the low 4 bits, and an instruction code in the high 4 bits, followed by one or two data bytes. The channel number is used to address one of 16 instruments, which can be all on the same daisy-chained MIDI connection. The data rate of MIDI has its limits, though, and in practice trying to run more than a few instruments simultaneously is likely to lead to timing problems. Given the inherent synchronization provided by a MIDI sequencer putting out the data, this problem is easily solved by multi-tracking.

The MIDI standard uses various techniques to permit the fastest possible data flow despite the low bandwidth. For example, the Running Status facility means that the initial status byte can be omitted if it would be the same as that of the previous message.

David goes on to look at the additional capabilities of MIDI, apart from basic note-on note-off messages. These include key pressure (for controlling note volume), program change (selecting the voice on a synthesizer) and pitch-bend data generated by the control wheels on keyboard instruments.

David then moves on to examine more complex features such as MIDI time code, song pointer positions, and MIDI machine control, which can operate hard disk recorders and tape and video transports.

At the physical level MIDI is a one-way serial digital current loop running at 31,250 bits per second, with one start bit, eight data bits, no parity bit and one stop bit. The transmitting end drives fixed resistors that define the signalling current, while at the receiving end an opto-isolator prevents the formation of ground loops.

When I had my own eight-track recording studio, I made very extensive use of MIDI, and I was always greatly impressed by its robustness and its flexibility. There were sometimes issues with the early DOS-based sequencing software and patch editors that were available at the time, but I don't recall that MIDI itself ever gave any trouble at all. I even found it possible to design and build a MIDI-to-control-voltage converter that did not need a microprocessor—if you picked the right UART you could run MIDI data straight into it and take a parallel output to a D/A converter. I have no idea if this possibility was foreseen by those who wrote the MIDI specification, but if it was then I have an even higher regard for their cleverness.

CHAPTER 16

MIDI

Handbook for Sound Engineers by David Huber

INTRODUCTION TO MIDI

Simply stated, Musical Instrument Digital Interface (MIDI) is a digital communications language and compatible specification that allows multiple hardware and software electronic instruments, performance controllers, computers, and other related devices to communicate with each other over a connected network. MIDI is used to translate performance- or control-related events (such as playing a keyboard, selecting a patch number, varying a modulation wheel, triggering a staged visual effect, etc.) into equivalent digital messages and then transmit these messages to other MIDI devices where they can be used to control sound generators and other performance parameters. The beauty of MIDI is that its data can be easily recorded into a hardware device or software program (known as a sequencer), where it can be edited and transmitted to electronic instruments or other devices to create music or control any number of parameters.

In artistic terms, this digital language is an important medium that lets artists express themselves with a degree of flexibility and control that wasn't possible at an individual level beforehand. Through the use of this performance language, an electronic musician can create and develop a song or composition in a practical, flexible, affordable, and fun production environment.

The word *interface* refers to the actual data communications link and software/hardware systems in a connected MIDI network. Through MIDI, it's possible for all of the electronic instruments and devices within a network to communicate real-time performance and control-related MIDI data messages throughout a system to multiple instruments and devices via MIDI, USB, or FireWire networked data lines. Given that MIDI data can simultaneously transmit performance and control messages over multiple channels (usually in groupings of 16 channels per port), an electronic musician can record, overdub, mix, and play back their performances in a building-block fashion that resembles the multitrack recording process. In fact, the true power of MIDI lies in its ability to edit, control, alter and automate parts of a composition after the original performance has been recorded, allowing performance parameters to be easily altered in ways that are unique to the medium.

What MIDI isn't

For starters, let's dispel one of MIDI's greatest myths: MIDI doesn't communicate audio—it cannot create sounds! It is a digital language protocol that can only be used to trigger and/or control a device (which, in turn generates, reproduces, or controls the sound). Thus, the MIDI data and the audio routing paths are kept entirely separate from each another, Figure 16.1. Even if they digitally share the same transmission cable (such as through USB or FireWire), the actual data paths and formats are distinct.

In short, MIDI's control-related language can be thought of as the dots on a player-piano roll—when we put the paper roll up to our ears, we hear nothing. However, when the cutout dots pass over the sensors on a player piano, the instrument itself begins to make beautiful music. The analogy is

Audio Engineering Explained

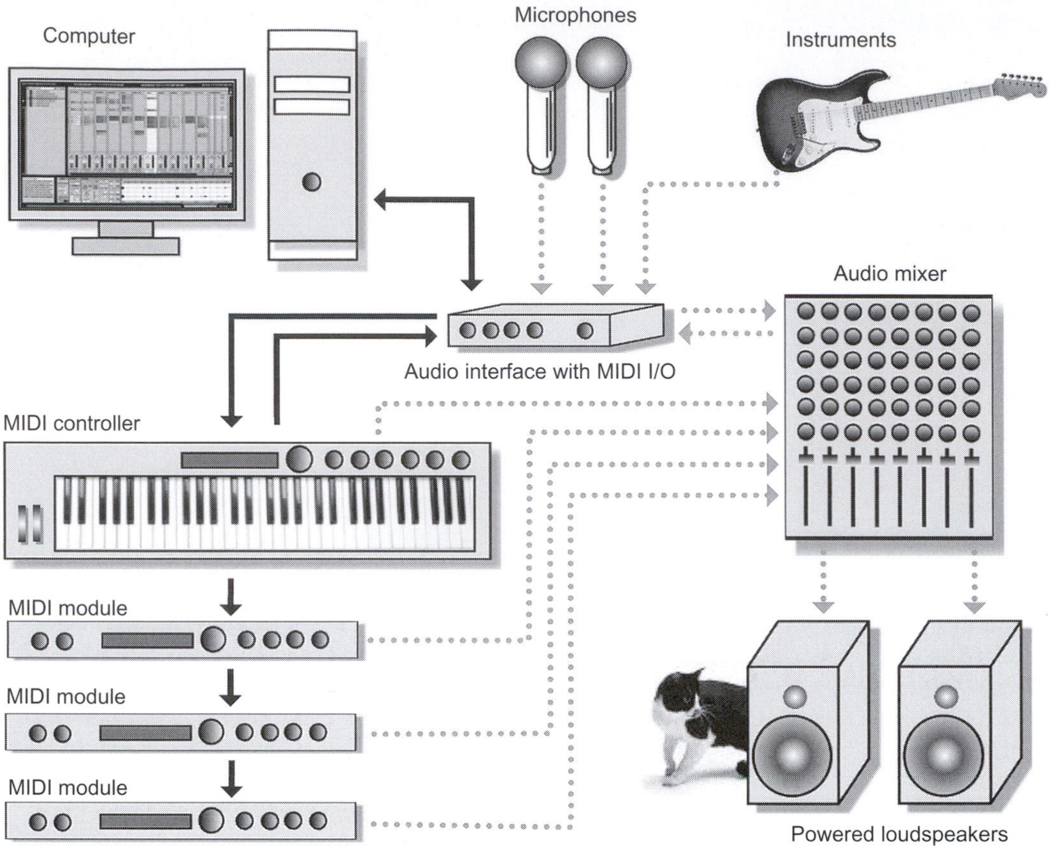

FIGURE 16.1
Example of a typical MIDI system with the MIDI network connections being shown in solid lines and audio connections shown using dotted lines.

pretty much the same with MIDI. A MIDI file or data stream is simply a set of instructions that pass through wires in a serial fashion, but when an electronic instrument interprets the data, we then hear sound.

As a performance-based control language, MIDI complements modern music production, by allowing a performance track to be edited, layered, altered, spindled, mutilated, and improved with relative ease under completely automated computer control and after the fact, during post-production. If you played a bad note, fix it. If you want to change the key or tempo of a piece, change it. If you want to change the expressive volume of a phrase in a song, just do it! Even its sonic character (timbre) can be changed! These capabilities merely hint at the power of this medium that widely affects the project studio, professional studio, audio or visual and film, live performance, multimedia, and even your cell phone!

THE MIDI MESSAGE

From its inception in the early 80s, the MIDI 1.0 spec (which is still the adopted version to this day) must be strictly adhered to by those who design and manufacture MIDI-equipped instruments and devices. As such, users needn't worry about whether the MIDI Out of one device will be understood by the MIDI In of a device that's made by another manufacturer (at least the basic performance level). We need only consider the day-to-day dealings that go hand-in-hand with using electronic instruments, without having to be concerned with data compatibility between devices.

MIDI messages are communicated through a standard MIDI line in a serial fashion at a speed of 31,250 bits/s. These messages are made up of groups of 8-bit words (known as bytes), which are used to convey instructions to one or all MIDI devices within a system. Only two types of bytes are defined by the MIDI specification: the status byte and the data byte.

Table 16.1 Status and Data Byte Interpretation.

	Status Byte	Data Byte 1	Data Byte 2
Description	Status/Channel #	Note #	Attack Velocity
Binary Data	(1001.0100)	(0100.0000)	(0101.1001)
Numeric Value	(Note On/Ch #5)	(64)	(89)
0000 = CH#1	0100 = CH#5	1000 = CH#9	1100 = CH#13
0001 = CH#2	0101 = CH#6	1001 = CH#10	1101 = CH#14
0010 = CH#3	0110 = CH#7	1010 = CH#11	1110 = CH#15
0011 = CH#4	0111 = CH#8	1011 = CH#12	1111 = CH#16

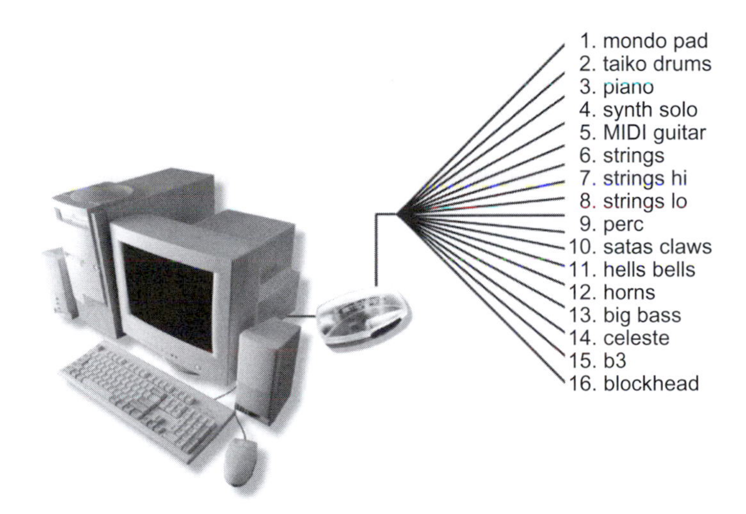

FIGURE 16.2
Up to 16 channels can be transmitted through a single MIDI cable.

A status byte is used to identify what type of MIDI function is to be performed by a device or program. It's also used to encode channel data (allowing the instruction to be received by a device that's set to respond to a specific channel). A data byte is used to associate a value to the event that's given by the accompanying status byte.

The most significant bit (MSB), the leftmost binary bit within a digital word within a MIDI byte, is used solely to identify the data's particular function. The MSB of a status byte is always 1, while the MSB of a data byte is always 0. For example, a 3 byte MIDI note-on message (which is used to signal the beginning of a MIDI note) in binary form might read as shown in Table 16.1. Thus, a 3 byte note-on message of (10010100) (01000000) (01011001) will transmit instructions that would be read as "Transmitting a note-on message over MIDI channel #5, using keynote #64, with an attack velocity (volume level of a note) of 89."

MIDI channels

Just as a public speaker might single out and communicate a message to one individual in a crowd, MIDI messages can be directed to communicate information to a specific device or series of devices within a MIDI system. This is done by imbedding a channel-related nibble (4 bits) within the status byte, allowing data to be conveyed to any of 16 channels over a single MIDI data cable line, Figure 16.2. This makes it possible for performance or control information to be communicated to a specific device or a sound generator within a device that's assigned to a particular channel.

Whenever a MIDI device, sound generator, or program function is instructed to respond to a specific channel number, it will only respond to messages that are transmitted on that channel (i.e., it ignores

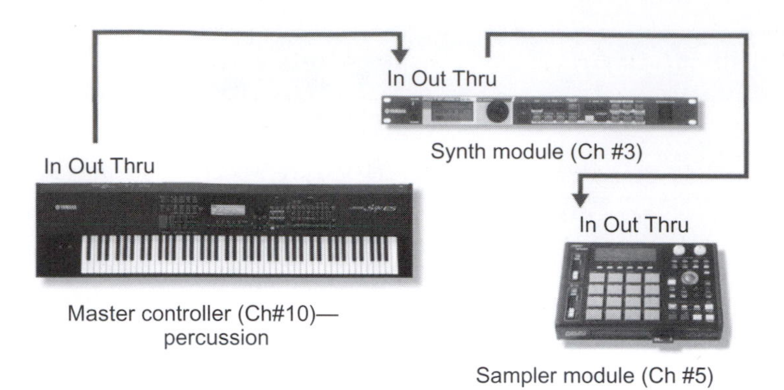

FIGURE 16.3
MIDI setup showing a set of MIDI channel assignments.

channel messages that are transmitted on any other channel). For example, let's assume that we're going to create a short song using a synthesizer that has a built-in sequencer (a device or program that's capable of recording, editing, and playing back MIDI data) and two other synths, Figure 16.3.

1. We could start off by recording a drum track into the master synth using channel 10 (many synths are pre-assigned to output drum/percussion sounds on this channel).
2. Once recorded, the sequence will then transmit the notes and data over channel 10, allowing the synth's percussion section to be heard.
3. Next, we could set a synth module to channel 3, and instruct the master synth to transmit on the same channel (since the synth module is set to respond to data on channel 3, its generators will sound whenever the master keyboard is played). We can now begin recording a melody line into the sequencer's next track.
4. Playing back the sequence will then transmit data to both the master synth (percussion section) and the module (melody line) over their respective channels. At this point, our song is beginning to take shape.
5. Now, we can set a sampler (or other instrument type) to respond to channel 5, and instruct the master synth to transmit on the same channel, allowing us to further embellish the song.
6. Now that the song's complete, the sequencer can then play the musical parts to the synths on their respective MIDI channels, all in an environment that allows us to have complete control of volume, edit, and a wide range of functions over each instrument. In short, we've created a true multichannel working environment.

It goes without saying that the above example is just but one of the infinite setup and channel possibilities that can be encountered in a production environment. It's often true, however, that even the most complex MIDI and production rooms will have a system, a basic channel and overall layout that makes the day-to-day operation of making music easier. This layout and the basic decisions in your own room are, of course, up to you. Streamlining a system to work both efficiently and easily will come over time with experience and practice.

MIDI modes

Electronic instruments often vary in the number of sounds and/or notes that can be simultaneously produced by their internal sound-generating circuitry. For example, certain instruments can only produce one note at a single time (known as a *monophonic* instrument), while others can generate 16, 32, and even 64 notes at once (these are known as *polyphonic* instruments). The latter type is easily capable of playing chords and/or more than one musical line on a single instrument.

In addition, some instruments are only capable of producing a single generated sound patch (often referred to as a voice) at any one time. Its generating circuitry could be polyphonic, allowing the player to lay down chords and bass/melody lines), but it can only produce these notes using a single characteristic sound at any one time (e.g., an electric piano, or a synth bass, or a string patch). However, the vast majority of newer synths differ from this in that they're multitimbral in nature, meaning that they can generate numerous sound patches at any one time (e.g., an electric piano, and a synth bass, and a string patch). That is, it's common to run across electronic instruments that can simultaneously generate a number of voices, each offering its own control over parameters (such as volume, panning, modulation, etc.) and—best of all—it's also common for different sounds to be assigned to their own MIDI channels, allowing multiple patches to be internally mixed within the device (often top a stereo output bus), or to independent outputs.

As a result of these differences between instruments and devices, a defined set of guidelines (known as *MIDI reception modes*) has been specified that allows a MIDI instrument to transmit or respond to MIDI channel messages in several ways. For example, one instrument might be programmed to respond to all 16 MIDI channels at one time, while another might be polyphonic in nature, with each voice being programmed to respond to only a single MIDI channel.

POLY/MONO

An instrument or device can be set to respond to MIDI data in either the poly mode or the mono mode. Stated simply, an instrument that's set to respond to MIDI data polyphonically will be able to play more than one note at a time. Conversely, an instrument that's set to respond to MIDI data monophonically will only be able to play a single note at any one time.

OMNI ON/OFF

Omni on/off refers to how a MIDI instrument will respond to MIDI messages at its input. When Omni is turned on, the MIDI device will respond to all channel messages that are being received regardless of its MIDI channel assignment. When Omni is turned off, the device will only respond to a single MIDI channel or set of assigned channels (in the case of a multitimbral instrument).

The following list and figures explain the four modes that are supported by the MIDI spec in more detail.

- **Mode 1—Omni On/Poly:** In this mode, an instrument will respond to data that's being received on any MIDI channel, and then redirect this data to the instrument's base channel, Figure 16.2a. In essence, the device will play back everything that's presented at its input in a polyphonic fashion... regardless of the incoming channel designations. As you might guess, this mode is rarely used.
- **Mode 2—Omni On/Mono:** As in Mode 1, an instrument will respond to all data that's being received at its input, without regard to channel designations. However, this device will only be able to play one note at a time, Figure 16.2b. Mode 2 is used even more rarely than Mode 1, as the device can't discriminate channel designations and can only play one note at a time.
- **Mode 3—Omni Off/Poly:** In this mode, an instrument will only respond to data that matches its assigned base channel in a polyphonic fashion, Figure 16.2c. Data that is assigned to any other channel will be ignored. This mode is by far the most commonly used because it allows the voices within a multitimbral instrument to be individually controlled by messages that are being received on different MIDI channels. For example, each of the 16 channels in a MIDI line could be used to independently play each of the parts in a 16-voice, multitimbral synth.
- **Mode 4—Omni Off/Mono:** As with Mode 3, an instrument will be able to respond to performance data that's transmitted over a single dedicated channel; however, each voice will only be able to generate one MIDI note at a time, Figure 16.2d. A practical example of this mode is often used in MIDI guitar systems, where MIDI data is monophonically transmitted over six consecutive channels (one channel/voice per string).

Channel messages

Channel-voice messages are used to transmit real-time performance data throughout a connected MIDI system. They're generated whenever a MIDI instrument's controller is played, selected, or varied by the performer. Examples of such control changes could be the playing of a keyboard, pressing of program selection buttons, or movement of modulation or pitch wheels. Each channel-voice message contains a MIDI channel number within its status byte, meaning that only devices that are assigned to the same channel number will respond to these commands. There are seven channel-voice message types: note-on, note-off, polyphonic-key pressure, channel pressure, program change, pitch-bend change and control change.

Note-On Messages. A note-on message is used to indicate the beginning of a MIDI note. It is generated each time a note is triggered on a keyboard, controller, or other MIDI instrument (i.e., by pressing a key, hitting a drum pad, or by playing a sequence).

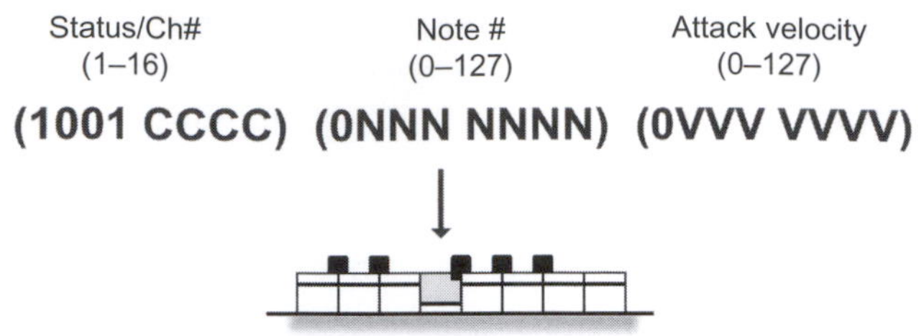

Status/Ch#
(1–16)
Note #
(0–127)
Attack velocity
(0–127)
(1001 CCCC) (0NNN NNNN) (0VVV VVVV)

FIGURE 16.4
Byte structure of a MIDI note-on message.

A note-on message consists of 3 bytes of information, Figure 16.4: Note-on status/MIDI channel number, MIDI pitch number and attack velocity value.

The first byte in the message specifies a note-on event and a MIDI channel (1–16). The second byte is used to specify which of the possible 128 notes (numbered 0–127) will be sounded by an instrument. In general, MIDI note number 60 is assigned to the middle C key of an equally tempered keyboard, while notes 21 to 108 correspond to the 88 keys of an extended keyboard controller. The final byte is used to indicate the velocity or speed at which the key was pressed (over a value range that varies from 0 to 127). Velocity is used to denote the loudness of a sounding note, which increases in volume with higher velocity values (although velocity can also be programmed to work in conjunction with other parameters such as expression, control over timbre, sample voice assignments, etc).

Note-Off Messages. A note-off message is used as a command to stop playing a specific MIDI note. Each note-on message will continue to play until a corresponding note-off message for that note has been received. In this way, the bare basics of a musical composition can be encoded as a series of MIDI note-on and note-off events. It should also be pointed out that a note-off message wouldn't cut off a sound; it'll merely stop playing it. If the patch being played has a release (or final decay) slope, it will begin this stage upon receiving the message.

A note-off message consists of three bytes of information, Figure 16.5: Note-off status/MIDI channel number, MIDI pitch number and attack velocity value.

In contrast to the dynamics of attack velocity, the release velocity value (0–127) indicates the velocity or speed at which the key was released. A low value indicates that the key was released very slowly, whereas a high value shows that the key was released quickly. Although not all instruments generate or respond to MIDI's release velocity feature, instruments that are capable of responding to these values can be programmed to vary a note's speed of decay, often reducing the signal's decay time as the release velocity value is increased.

A note-on message that contains an attack velocity of 0 (zero) is generally equivalent to the transmission of a note-off message. This common implementation tells the device to silence a currently sounding note by playing it with a velocity (volume) level of 0.

All Notes Off. On the odd occasion (often when you least expect it), a MIDI note can get stuck! This can happen when data drops out or a cable gets disconnected, creating a situation where a note receives a note-on message, but not a note-off message, resulting in a note that continues to

Status/Ch#
(1–16)
Note #
(0–127)
Attack velocity
(0–127)
(1000 CCCC) (0NNN NNNN) (0VVV VVVV)

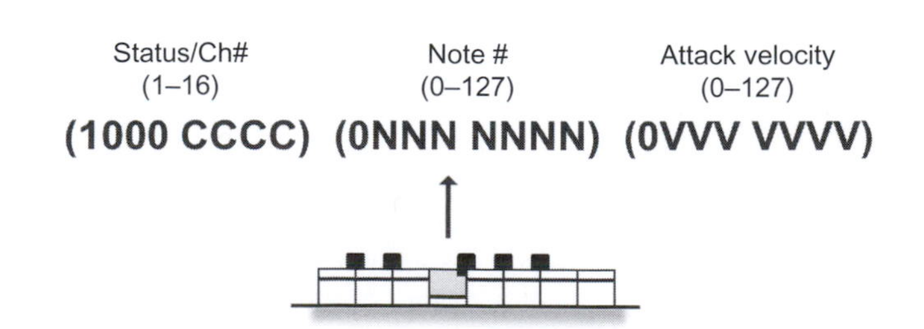

FIGURE 16.5
Byte structure of a MIDI note-off message.

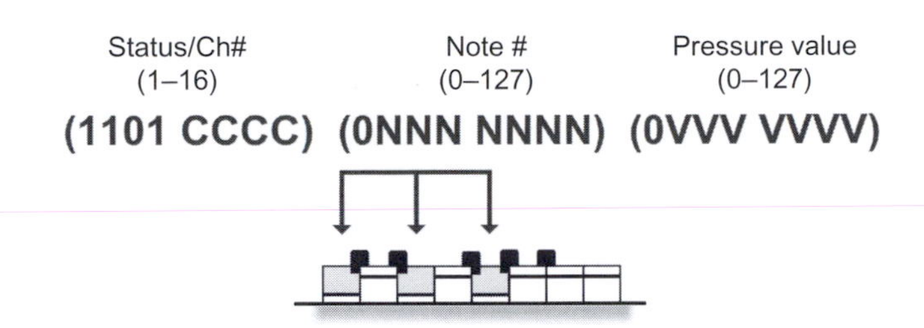

FIGURE 16.6
Byte structure of a MIDI channel-pressure message.

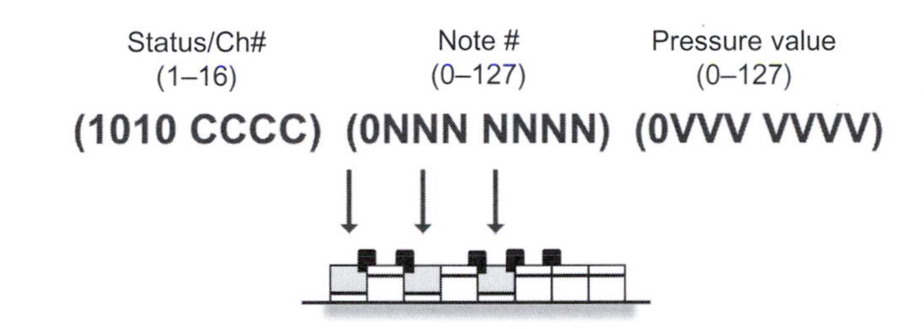

FIGURE 16.7
Byte structure of a MIDI polyphonic-key pressure message.

plaaaaaaaaaaayyyyyyyyyy! Since you're often too annoyed or under pressure to take the time to track down which note is the offending sucka… it's generally far easier to transmit an all notes off message that silences everything on all channels and ports. If it exists, this can easily be done by pressing a Panic Button that's built into the sequencer or hardware MIDI interface.

Pressure (Aftertouch) Messages. Pressure-related messages (often referred to as aftertouch) occur after you've pressed a key and then decide to press down harder to gain a particular effect. For devices that can respond to (and therefore generally transmit) these messages, aftertouch can often be assigned to such parameters as vibrato, loudness, filter cut-off, and pitch. Two types of pressure messages are defined by the MIDI spec:

- Channel-pressure.
- Polyphonic-pey pressure.

Channel-pressure messages are commonly transmitted by instruments that only respond to a single overall pressure, regardless of the number of keys that are being played at any one time, Figure 16.6. For example, if six notes are played on a keyboard and additional aftertouch pressure is applied to just one key, the assigned parameter would be applied to all six notes.

A channel-pressure message consists of 3 bytes of information, Figure 16.6: Channel-pressure status/MIDI channel number, MIDI note number, and pressure value.

Polyphonic-key pressure messages respond to pressure changes that are applied to the individual keys of a keyboard. That's to say that a suitably equipped instrument can transmit or respond to individual pressure messages for each key that's depressed.

How a device responds to these messages will often vary from manufacturer to manufacturer (or can be assigned by the user). However, pressure values are commonly assigned to such performance parameters as vibrato, loudness, timbre, and pitch. Although controllers that are capable of producing polyphonic pressure are generally more expensive, it's not uncommon for an instrument to respond to these messages.

A polyphonic-key pressure message consists of 3 bytes of information, Figure 16.7: Polyphonic-key pressure status/MIDI channel number, MIDI note number, and pressure value.

Program-Change Messages. Program-change messages are used to change a MIDI instrument or device's active program or preset number. A preset is a user- or factory-defined number that actively

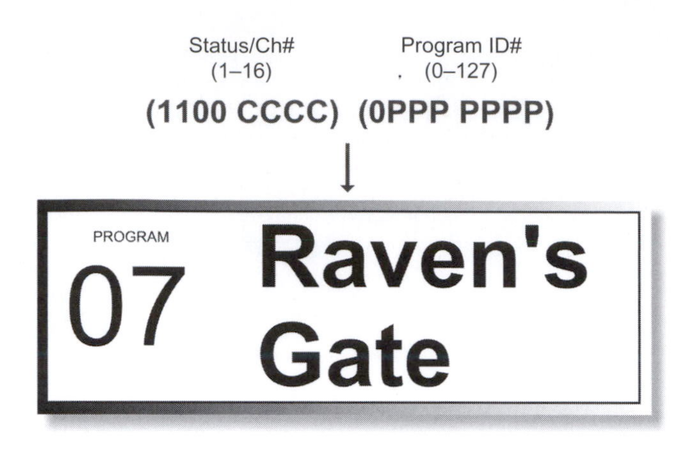

FIGURE 16.8
Byte structure of a MIDI program-change message.

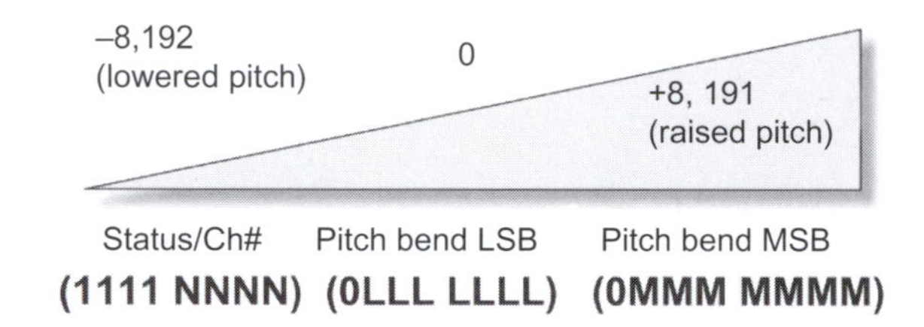

FIGURE 16.9
Byte structure of a pitch-bend message.

selects a specific sound patch or system setup. Using this extremely handy message, up to 128 presets can be remotely selected from another device or controller. For example:

- A program-change message can be transmitted from a remote keyboard or controller to an instrument, allowing sound patches to be remotely switched, Figures 16.8 and 16.9.
- Program-change messages could be programmed at the beginning of a sequence, so as to instruct the various instruments or voice generators to set to the correct sound patch before playing.
- It could be used to alter patches on an effects device, either in the studio or on stage. The list goes on.

A program-change message, Figure 16.8, consists of 2 bytes of information: program-change status/MIDI channel number and program ID number.

Pitch-bend Messages. Pitch-bend sensitivity refers to the response sensitivity (in semitones) of a pitch-bend wheel or other pitch-bend controlle, which, as you'd expect, is used to bend the pitch of a note upward or downward. Since the ear can be extremely sensitive to changes in pitch, this control parameter is encoded using 2 data bytes, yielding a total of 16,384 steps. Since this parameter is most commonly affected by varying a pitch wheel, Figure 16.9, the control values range from −8,192 to +8,191, with 0 being the instrument's or part's unaltered pitch.

Control-Change Messages. Control-change messages are used to transmit information to a device (either internally or through a MIDI line/network) that relates to real-time control over its performance parameters.

Three types of control-change messages can be transmitted via MIDI:

1. Continuous controllers: Controllers that relay a full range of variable control settings (often ranging in value between 0–127 although, in certain cases, two controller messages can be combined in tandem to achieve a greater resolution).
2. Switch controllers: Controllers that have either an off or an on state with no intermediate settings.

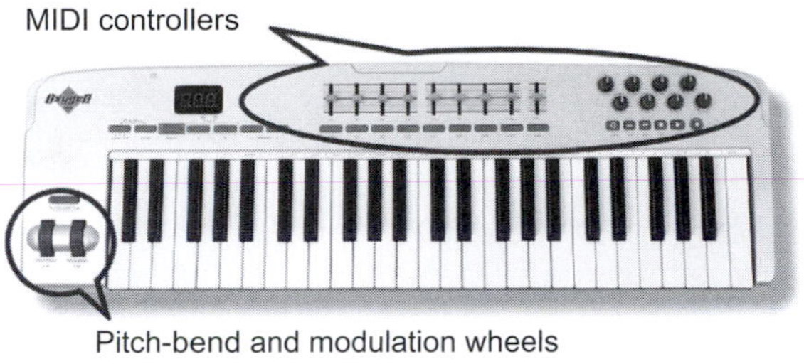

MIDI controllers

Pitch-bend and modulation wheels

FIGURE 16.10
M-audio controller. (Courtesy of M-Audio, a division of Avid Technology, Inc., www.m-audio.com.)

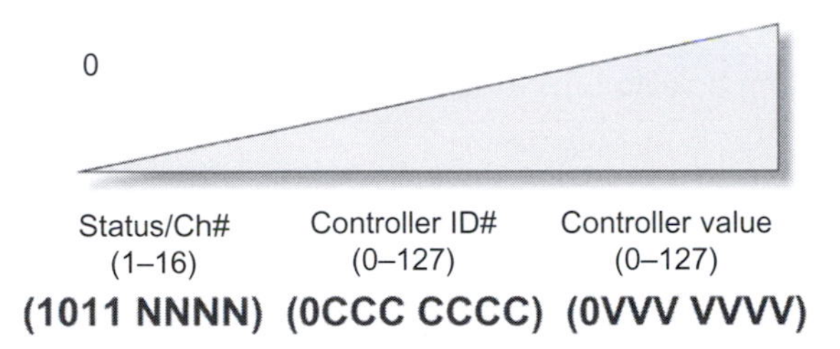

FIGURE 16.11
Byte structure of a control-change message.

3. Channel-mode message controllers: The final set of control change messages range between controller numbers 120 through 127, and are used to set the note sounding status, instrument reset, local control on/off, all notes off, and MIDI mode status of a device or instrument.

A single control-change message or a stream of such messages is transmitted whenever controllers (such as foot switches, foot pedals, pitch-bend wheels, modulation wheels, breath controllers, etc.) are varied in real-time. Newer controllers and software editors often offer up a wide range of switched and variable controllers, allowing for extensive, user-programmable control over any number of device, voice, and mixing parameters in real-time, Figure 16.10.

A control-change message, Figure 16.11, consists of 3 bytes of information: control-change status/MIDI channel number, controller ID number, and corresponding controller value.

As you can see, the second byte of the control-change message is used to denote the controller ID number. This all-important value is used to specify which of the device's program or performance parameters are to be addressed.

Table 16.2 details the general categories and conventions for assigning controller numbers to an associated parameter, as specified by the 1995 update of the MMA (MIDI Manufacturers Association, www.midi.org). This is definitely an important section to earmark, as these numbers will be an important guide towards knowing and/or finding the right ID number that can help you on your path towards finding that perfect variable for making it sound right.

The third byte of the control-change message is used to denote the controller's actual data value. This value is used to specify the position, depth, or level of a parameter. Here are a few examples as to how these values can be implemented to vary control and mix parameters.

In certain cases, greater resolutions than can be given by a single 7-bit course message (128 steps) might be available to increase a controller's resolution. This is simply accomplished by adding an additional fine controller value message to the data stream, resulting in an overall resolution that yields an overall total of 16,384 discrete steps!

System messages

System Messages. As the name implies, system messages are globally transmitted to every MIDI device in the MIDI chain. This is accomplished because MIDI channel numbers aren't addressed within the byte structure of a system message. Thus, any device will respond to these messages, regardless of its

Table 16.2 Listing of Controller ID Numbers, Outlining Both the Defined Format and Convention and Controller Assignments.

Control Number	Parameter
14 Bit Controllers Coarse/MSB (most significant bit)	
0	Bank Select 0–127 MSB
1	Modulation Wheel or Lever 0–127 MSB
2	Breath Controller 0–127 MSB
3	Undefined 0–127 MSB
4	Foot Controller 0–127 MSB
5	Portamento Time 0–127 MSB
6	Data Entry MSB 0–127 MSB
7	Channel Volume (formerly Main Volume) 0–127 MSB
8	Balance 0–127 MSB
9	Undefined 0–127 MSB
10	Pan 0–127 MSB
11	Expression Controller 0–127 MSB
12	Effect Control 1 0–127 MSB
13	Effect Control 2 0–127 MSB
14	Undefined 0–127 MSB
15	Undefined 0–127 MSB
16–19	General Purpose Controllers 1–4 0–127 MSB
20–31	Undefined 0–127 MSB
14-bit Controllers Fine/LSB (least significant bit)	
32	LSB for Control 0 (Bank Select) 0–127 LSB
33	LSB for Control 1 (Modulation Wheel or Lever) 0–127 LSB
34	LSB for Control 2 (Breath Controller) 0–127 LSB
35	LSB for Control 3 (Undefined) 0–127 LSB
36	LSB for Control 4 (Foot Controller) 0–127 LSB
37	LSB for Control 5 (Portamento Time) 0–127 LSB
38	LSB for Control 6 (Data Entry) 0–127 LSB
39	LSB for Control 7 (Channel Volume, formerly Main Volume) 0–127 LSB
40	LSB for Control 8 (Balance) 0–127 LSB
41	LSB for Control 9 (Undefined) 0–127 LSB
42	LSB for Control 10 (Pan) 0–127 LSB
43	LSB for Control 11 (Expression Controller) 0–127 LSB
44	LSB for Control 12 (Effect control 1) 0–127 LSB
45	LSB for Control 13 (Effect control 2) 0–127 LSB
46–47	LSB for Control 14–15 (Undefined) 0–127 LSB
48–51	LSB for Control 16–19 (General Purpose Controllers 1–4) 0–127 LSB
52–63	LSB for Control 20–31 (Undefined) 0–127 LSB

7-bit Controllers	
64	Damper Pedal On/Off (Sustain) <63 off, >64 on
65	Portamento On/Off <63 off, >64 on
66	Sustenuto On/Off <63 off, >64 on
67	Soft Pedal On/Off <63 off, >64 on
68	Legato Footswitch <63 Normal, >64 Legato
69	Hold 2 <63 off, >64 on
70	Sound Controller 1 (Default: Sound Variation) 0–127 LSB
71	Sound Controller 2 (Default: Timbre/Harmonic Intens.) 0–127 LSB
72	Sound Controller 3 (Default: Release Time) 0–127 LSB
73	Sound Controller 4 (Default: Attack Time) 0–127 LSB
74	Sound Controller 5 (Default: Brightness) 0–127 LSB
75	Sound Controller 6 (Default: Decay Time—see MMA RP-021) 0–127 LSB
76	Sound Controller 7 (Default: Vibrato Rate—see MMA RP-021) 0–127 LSB
77	Sound Controller 8 (Default: Vibrato Depth—see MMA RP-021) 0–127 LSB
78	Sound Controller 9 (Default: Vibrato Delay—see MMA RP-021) 0–127 LSB
79	Sound Controller 10 (Default undefined—see MMA RP-021) 0–127 LSB
80–83	General Purpose Controller 5–8 0–127 LSB
84	Portamento Control 0–127 LSB
85–90	Undefined
91	Effects 1 Depth (Default: Reverb Send Level) 0–127 LSB
92	Effects 2 Depth (Default: tremolo Level) 0–127 LSB
93	Effects 3 Depth (Default: Chorus Send Level) 0–127 LSB
94	Effects 4 Depth (Default: Celeste [Detune] Depth) 0–127 LSB
95	Effects 5 Depth (Default: Phaser Depth) 0–127 LSB
Parameter Value Controllers	
96	Data Increment (Data Entry +1)
97	Data Decrement (Data Entry -1)
98	Non-Registered Parameter Number (NRPN)—LSB 0–127 LSB
99	Non-Registered Parameter Number (NRPN)—MSB 0–127 MSB
100	Registered Parameter Number (RPN)–LSB* 0–127 LSB
101	Registered Parameter Number (RPN)—MSB* 0–127 MSB
102–119	Undefined
Reserved for Channel Mode Messages	
120	All Sound Off 0
121	Reset All Controllers
122	Local Control On/Off 0 off, 127 on
123	All Notes Off
124	Omni Mode Off (+ all notes off)
125	Omni Mode On (+ all notes off)
126	Poly Mode On/Off (+ all notes off)
127	Poly Mode On (+ mono off +all notes off)

MIDI channel assignment. The three system message types are system-common messages, system real-time messages, and system-exclusive messages.

System-Common Messages. System-common messages are used to transmit MIDI time code, song position pointer, song select, tune request, and end-of-exclusive data messages throughout the MIDI system or 16 channels of a specified MIDI port.

MTC Quarter-Frame Messages. MIDI time code (MTC) provides a cost-effective and easily implemented way to translate SMPTE (a standardized synchronization time code) into an equivalent code that conforms to the MIDI 1.0 spec. It allows time-based codes and commands to be distributed throughout the MIDI chain in a cheap, stable, and easy-to-implement way. MTC quarter-frame messages are transmitted and recognized by MIDI devices that can understand and execute MTC commands.

A grouping of eight quarter frames is used to denote a complete time code address (in hours, minutes, seconds, and frames), allowing the SMPTE address to be updated every two frames. Each quarter-frame message contains 2 bytes. The first is a quarter-frame common header, while the second byte contains a 4-bit nibble that represents the message number (0–7). A final nibble is used to encode the time field (in hours, minutes, seconds, or frames).

Song Position Pointer Messages. As with MIDI time code, song position pointer (SPP) lets you synchronize a sequencer, tape recorder, or drum machine to an external source from any measure position within a song. The SPP message is used to reference a location point in a MIDI sequence (in measures) to a matching location within an external device. This message provides a timing reference that increments once for every six MIDI clock messages (with respect to the beginning of a composition).

Unlike MTC (which provides the system with a universal address location point), SPP's timing reference can change with tempo variations, often requiring that a special tempo map be calculated in order to maintain synchronization. Because of this fact, SPP is used far less often than MIDI time code.

Song Select Messages. Song select messages are used to request a specific song from a drum machine or sequencer (as identified by its song ID number). Once selected, the song will thereafter respond to MIDI start, stop, and continue messages.

Tune Request Messages. The tune request message is used to request that a MIDI instrument initiate its internal tuning routine (if so equipped).

End-of-Exclusive Messages. The transmission of an end-of-exclusive (EOX) message is used to indicate the end of a system-exclusive message. In-depth coverage of system-exclusive messages will be discussed later in this chapter.

System Real-Time Messages. Single-byte system real-time messages provide the all-important timing element required to synchronize all of the MIDI devices in a connected system. To avoid timing delays, the MIDI specification allows system real-time messages to be inserted at any point in the data stream, even between other MIDI messages.

Timing-Clock Messages. The MIDI timing-clock message is transmitted within the MIDI data stream at various resolution rates. It is used to synchronize the internal timing clocks of each MIDI device within the system and is transmitted in both the start and stop modes at the currently defined tempo rate.

In the early days of MIDI, these rates (which are measured in pulses per quarter note, ppq) ranged from 24 to 128 ppq. However, continued advances in technology have brought these rates up to 240, 480, or even 960 ppq.

Start Messages. Upon receipt of a timing-clock message, the MIDI start command instructs all connected MIDI devices to begin playing from their internal sequences initial start point. Should a program be in midsequence, the start command will reposition the sequence back to its beginning, at which point it will begin to play.

Stop Messages. Upon receipt of a MIDI stop command, all devices within the system will stop playing at their current position point.

Continue Messages. After receiving a MIDI stop command, a MIDI continue message will instruct all connected devices to resume playing their internal sequences from the precise point at which they were stopped.

Active-Sensing Messages. When in the stop mode, an optional active-sensing message can be transmitted throughout the MIDI data stream every 300 milliseconds. This instructs devices that can recognize this message that they're still connected to an active MIDI data stream.

System-Reset Messages. A system-reset message is manually transmitted in order to reset a MIDI device or instrument back to its initial power-up default settings (commonly mode 1, local control on, and all notes off).

System-Exclusive Messages. The system-exclusive (SysEx) message allows MIDI manufacturers, programmers and designers to communicate customized MIDI messages between MIDI devices. It's the purpose of these messages to give manufacturers, programmers, and designers the freedom to communicate any device-specific data of an unrestricted length, as they see fit. In practice, SysEx data is commonly used to communicate real-time controller information (i.e., a remote controller surface will commonly use SysEx to communicate data to/from a MIDI-capable hard- or software device. SysEx can also be used transmit and receive device-specific program, patch parameter and sample data from one instrument or device to another. For example, SysEx can be used to transmit patch and overall setup data between identical make and (most-often) model of synthesizer. Let's say that you have a Brand X Model Z synthesizer and it turns out that you have a buddy across town who also has a Brand X Model Z. That's cool, except your buddy's synth has a completely different set of sound patches that was loaded into their instrument and you want them! SysEx to the rescue! All you need to do is go over and transfer your buddy's patch data into your synth, or into a MIDI sequencer as a SysEx data dump. In order to make life easier, make sure you take your instruction manual along (just in case you run into a snag), and follow these simple guidelines. I'll caution you that you're taking on these tasks at your own risk. Take your time; be patient and be careful during these procedures:

1. Back up your present patch data! This can be done by transmitting a SysEx dump of your synthesizer's entire patch and setup data to your sequencer's SysEx dump utility, or SysEx track on your sequencer (of course, you should get out both the device's manual and your sequencer's manual and follow their SysEx dump instructions very carefully during the process). This is so important that I'll say it again: Back up your present patch data before attempting a SysEx dump! If you forget and download a new SysEx dump, your previous settings could easily be lost.
2. Save the data, according to your sequencer's manual.
3. Check that the dump was successful by reloading it back into the device in question. Did it reload properly? If so, your current patch data is now saved.
4. Next, connect your buddy's device to your sequencer. Dump this data to your sequencer. Save the new patch data (using a new and easily identifiable file name), according to your sequencer's manual and then safely back this data up.
5. Reconnect the sequencer to your synth and load the new data dump into it. Does your synth have a bunch of new sounds? Now reload your original SysEx dump back into your device. Are the original sounds restored?

The transmission format of a SysEx message, Figure 16.12, as defined by the MIDI standard, includes a SysEx status header, manufacturer's ID number, any number of SysEx data bytes, and an EOX byte. On receiving a SysEx message, the identification number is read by a MIDI device to determine whether or not the following messages are relevant. This is easily accomplished, because a unique 1- or 3-byte ID number is assigned to each registered MIDI manufacturer. If this number doesn't match the receiving

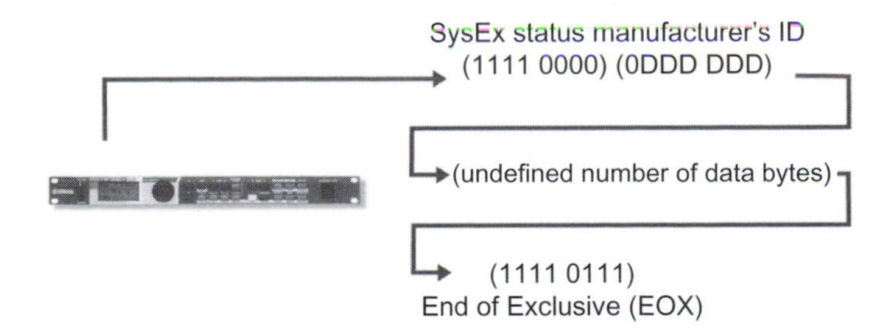

FIGURE 16.12
System-exclusive data (one ID byte format).

MIDI device, the ensuing data bytes will be ignored. Once a valid stream of SysEx data is transmitted, a final EOX message is sent, after which the device will again begin responding to incoming MIDI performance messages.

HARDWARE SYSTEMS WITHIN MIDI PRODUCTION

As a data transmission medium, MIDI is relatively unique in the world of sound production in that it's able to pack 16 discrete channels of performance, controller, and timing information and transmit it in one direction, using data densities that are economically small and easy to manage. In this way, it's possible for MIDI messages to be communicated from a specific source (such as a keyboard or MIDI sequencer) to any number of devices within a connected network over a single MIDI data chain. In addition, MIDI is flexible enough that multiple MIDI data lines can be used to interconnect devices in a wide range of possible system configurations (for example, multiple MIDI lines can be used to transmit data to instruments and devices over 32, 48, 128, or more discrete MIDI channels!)

The MIDI Cable. A MIDI cable, Figure 16.13, consists of a shielded twisted pair of conductor wires that has a male 5-pin DIN plug located at each of its ends. The MIDI specification currently uses only three of the five pins, with pins 4 and 5 being used as conductors for MIDI data, while pin 2 is used to connect the cable's shield to equipment ground. Pins 1 and 3 are currently not in use, although the next section describes an ingenious system for power devices through these pins, using a system that's known as MIDI phantom power. The cables themselves use twisted cable and metal shield groundings to reduce outside interference, such as radio-frequency interference (RFI) or electrostatic interference, both of which can serve to distort or disrupt the transmission of MIDI messages.

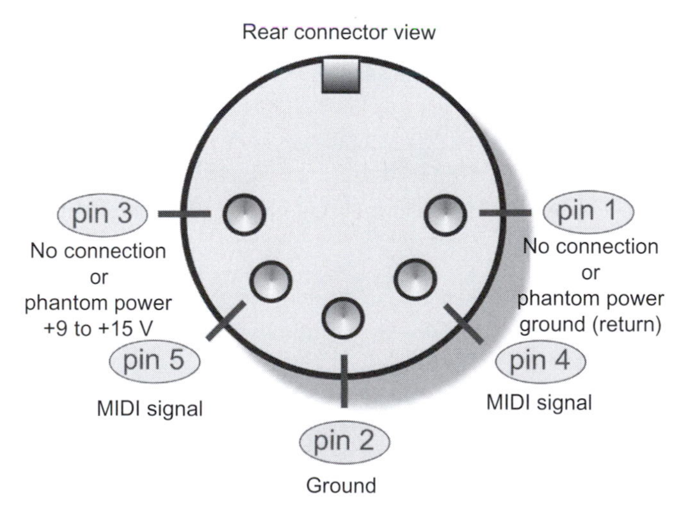

A. Connector wiring diagram

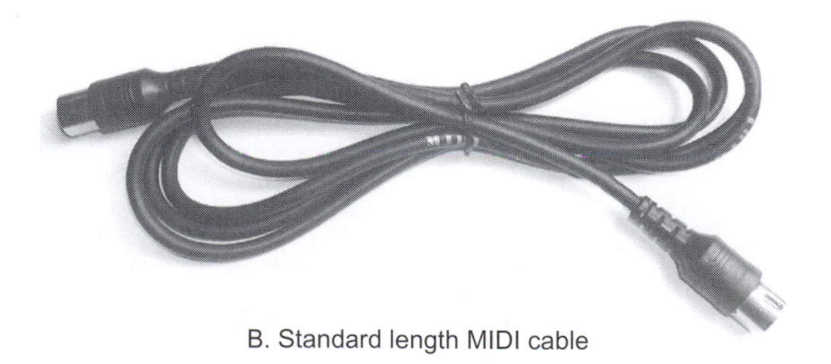

B. Standard length MIDI cable

FIGURE 16.13
The MIDI cable.

MIDI cables come prefabricated in lengths of 2, 6, 10, 20, and 50 feet, and can commonly be obtained from music stores that specialize in MIDI equipment. To reduce signal degradations and external interference that tends to occur over extended cable runs, 50 feet is the maximum length specified by the MIDI specification. (As an insider tip, I found that Radio Shack is also a great source for picking up 3 and 6 feet MIDI cables at a fraction of what you'd sometimes spend at a music store).

MIDI Phantom Power. In December 1989, Craig Anderton wrote an article in *Electronic Musician* about a proposed idea for allowing a source to provide a standardized 12 Vdc power supply to instruments and MIDI devices directly through pins 1 and 3 of a basic MIDI cable. Although pins 1 and 3 are technically reserved for possible changes in future MIDI applications, over the years several forward-thinking manufacturers (and project enthusiasts) have begun to implement MIDI phantom power directly into their studio and on-stage systems.

Wireless MIDI. In recent times, a number of companies have begun to manufacturer wireless MIDI transmitters that can allow a battery-operated MIDI guitar, wind controller, etc. to be footloose and fancy free on-stage and in the studio. Working at distances of up to 500 feet, these battery-powered transmitter/receiver systems introduce very low delay latencies and can be switched over a number of radio channel frequencies.

MIDI Jacks. MIDI is distributed from device to device using three types of MIDI jacks: MIDI In, MIDI Out, and MIDI Thru, Figure 16.14. These three connectors use 5-pin DIN jacks as a way to connect MIDI instruments, devices, and computers into a music and/or production network system. As a side note, it's nice to know that these ports (as strictly defined by MIDI 1.0 Spec.) are optically isolated to eliminate possible ground loops that might occur when connecting numerous devices together.

- **MIDI In**—The MIDI In jack receives messages from an external source and communicates this performance, control, and/or timing data to the device's internal microprocessor, allowing an instrument to be played and/or a device to be controlled. More than one MIDI In jack can be designed into a system to provide for MIDI merging functions or for devices that can support more than 16 channels (such as a MIDI Interface). Other devices (such as a controller) might not have a MIDI In jack at all.
- **MIDI Out**—The MIDI Out jack is used to transmit MIDI performance, control messages or SysEx from one device to another MIDI instrument or device. More than one MIDI Out jack can be designed into a system, giving it the advantage of controlling and distributing data over multiple MIDI paths using more than just 16 channels (i.e., 16 channels $\times$ N MIDI port paths).
- **MIDI Thru**—The MIDI Thru jack retransmits an exact copy of the data that's being received at the MIDI In jack. This process is important, because it allows data to pass directly through an instrument or device to the next device in the MIDI chain. Keep in mind that this jack is used to relay an exact copy of the MIDI In data stream and isn't merged with data being transmitted from the MIDI Out jack.
- **MIDI Echo**—Certain MIDI devices may not include a MIDI Thru jack, at all. Certain of these devices, however, may give the option of switching the MIDI Out between being an actual MIDI Out jack and a MIDI Echo jack, Figure 16.15. As with the MIDI Thru jack, a MIDI echo option can be used to retransmit an exact copy of any information that's received at the MIDI In port and route this data to the MIDI Out/Echo jack. Unlike a dedicated MIDI Out jack, the MIDI

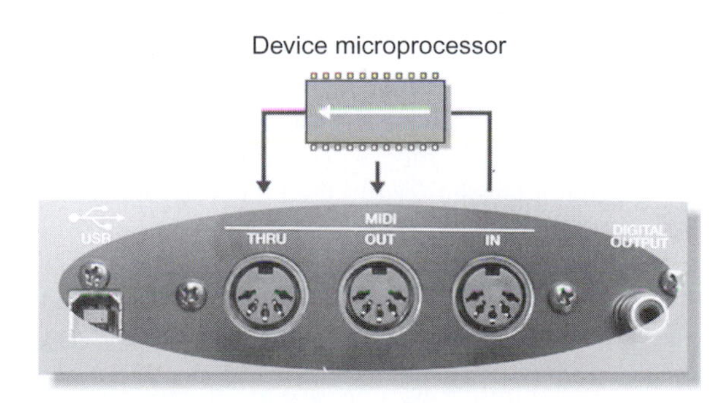

FIGURE 16.14
MIDI in, out, and thru ports, showing the device's signal path routing.

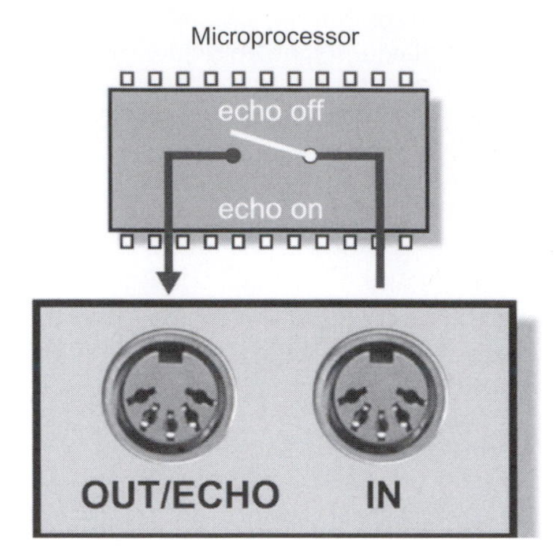

FIGURE 16.15
MIDI echo configuration.

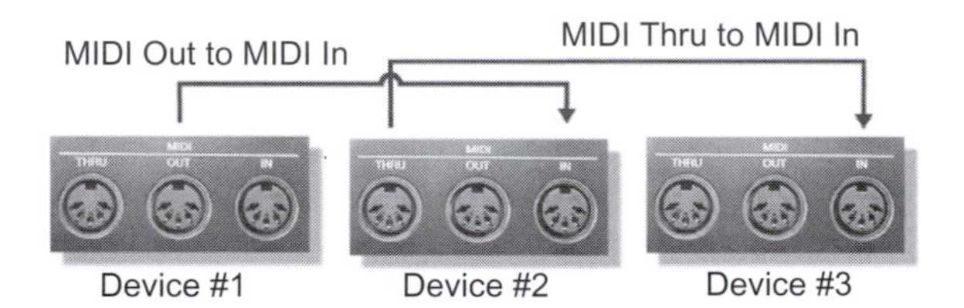

FIGURE 16.16
The two valid means of connecting one MIDI device to another.

Echo function can often be selected to merge incoming data with performance data that's being generated by the device itself. In this way, more than one controller can be placed in a MIDI system at one time. It should be noted that although performance and timing data can be echoed to a MIDI Out/Echo jack, not all devices can echo SysEx data.

Typical Configurations. Although electronic studio production equipment and setups are rarely alike (or even similar), there are a number of general rules that make it easy for MIDI devices to be connected into a functional network. These common configurations allow MIDI data to be distributed in the most efficient and understandable manner possible.

As a primary rule, there are only two valid ways to connect one MIDI device to another within a MIDI chain, Figure 16.16:

1. Connecting the MIDI Out jack of a source device (controller or sequencer/computer) to the MIDI In of a second device in the chain.
2. Connecting the MIDI Thru jack of the second device to the MIDI In jack of the third device in the chain and following this same Thru-to-In convention until the end of the chain is reached.

The Daisy Chain. One of the simplest and most common ways to distribute data throughout a MIDI system is the daisy chain. This method relays MIDI data from a source device (controller or sequencer/computer) to the MIDI In jack of the next device in the chain (which receives and acts upon this data). In turn, this device relays an exact copy of this incoming data out to its MIDI Thru jack, which is then relayed to the next device in the chain. This device can then relay an exact copy of this incoming data out to its MIDI Thru jack, which is then relayed to the next device in the chain… etc. In this way, up to 16 channels of MIDI data can be chained from one device to the next within a connected data network—and it's precisely this concept of transmitting multiple channels through a single MIDI line that makes this concept work! Let's try to understand this concept better by looking at a few examples.

Figure 16.17A shows a simple (and common) example of a MIDI daisy chain, whereby data flows from a controller (MIDI Out jack of the source device) to a synth module (MIDI In jack of the second device in the chain), where an exact copy of this data is relayed from its MIDI Thru jack to another synth

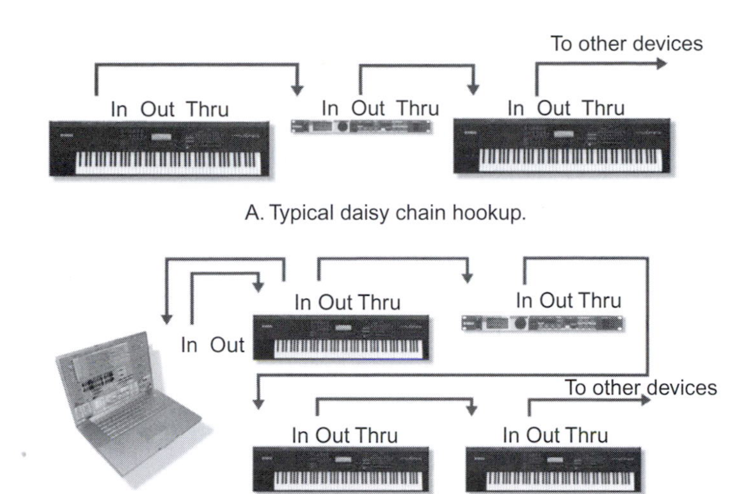

FIGURE 16.17
Example of a connected MIDI system using a daisy chain.

(MIDI In jack of the third device in the chain). It shouldn't be hard to understand that if our controller is transmitting on MIDI channel 2, the second synth in the chain (which is set to channel 2) will ignore the messages and not play while the 3rd synth (which is set to channel 3) will be playing its heart out. The moral of this story is that although there's only one connected data line, a wide range of instruments and channel voices can be played in a surprisingly large number of combinations, all by using individual channel assignments along a daisy chain.

Another example, Figure 16.17b, shows how a computer can easily be designated as the master source within a daisy chain, so that a sequencing program could be used to control the entire playback and channel routing functions of a daisy-chained system. In this situation, the MIDI data flows from a master controller/synth to the MIDI In jack of a computer's MIDI interface—where the data can be played into, processed, and rechannelized through a MIDI sequencer. The MIDI Out of the interface is then routed back to the MIDI In jack of the master controller/synth (which receives and acts on this data). In turn, the controller relays an exact copy of this incoming data out to its MIDI Thru jack, which is then relayed to the next device in the chain. This device can then relay an exact copy of this incoming data out to its MIDI Thru jack, which is then relayed to the next device in the chain, etc. When we stop to think about this second example, the controller is used to perform into the MIDI sequencer, which then is used to communicate this edited and processed performance data out to the various instruments throughout the connected MIDI chain.

The Multiport Network. Another common approach to routing MIDI throughout a production system involves distributing MIDI data through the multiple 2, 4 and 8 In/Out ports that are available on a newer multiport MIDI interfaces or through the use of multiple MIDI interfaces (typically these are USB devices).

In larger, more complex, MIDI systems, a multiport MIDI network, Figure 16.17, offers several advantages over a single daisy chain path. One of the most important is its ability to address devices within a complex setup that requires more than 16 MIDI channels. For example, a 2 × 2 MIDI interface that offers up two-independent In/Out paths is capable of addressing up to 32 channels simultaneously (i.e., port A 1–16 and port B 1–16), whereas an 8 × 8 port interface is capable of addressing up to 128 individual MIDI channels.

The MIDI interface

Although computers and electronic instruments both communicate using the digital language of 1s and 0s, computers simply can't understand the language of MIDI without the use of a device that translates the serial messages into a data structure that computers can comprehend. Such a device is known as the *MIDI interface.*

FIGURE 16.18
M-Audio MIDISPORT 4 × 4 MIDI interface. (Courtesy of M-Audio, a division of Avid Technology, Inc., www.m-audio.com.)

A wide range of MIDI interfaces currently exist that can be used with most computer systems and OS platforms. For the casual and professional musician, interfacing MIDI into a production system can be done in a number of ways. Probably the most common way to access MIDI In, Out, and Thru jacks is on a modern-day USB or FireWire audio interface or instrument/DAW controller surface. It's become a common matter for portable devices to offer 16 channels of I/O (on one port), while multi-channel interfaces often include multiple MIDI I/O ports that can give you access to 32 or more channels.

Another additional option is to choose a USB MIDI interface that can range from devices that include a single I/O port (16 channels) to a multiport system that can easily handle up to 128 channels over eight I/O ports. The multiport MIDI interface, Figure 16.18, is often the device of choice for most professional electronic musicians who require added routing and synchronization capabilities. These rack-mountable USB devices can be used to provide eight independent MIDI Ins and Outs to easily distribute MIDI and time code data through separate lines over a connected network.

Hardware and software electronic instruments

Since its inception in the early 80s, MIDI-based electronic musical instruments have helped to shape the face and sounds of our modern music culture. These devices (along with digital audio and advances in recording equipment technology) have altered music production, through the creation of one of the most cost-effective and powerful tools in the development of music history—the personal project studio.

The following is a sample listing of the many hardware MIDI instrument types that are currently available on the market.

The Synth. A synthesizer, Figure 16.19, is an electronic instrument that uses multiple sound generators to create complex waveforms that can be combined (using various waveform synthesis techniques) into countless sonic variations. These synthesized sounds have become a basic staple of modern music and vary from sounding cheesy, to those that closely mimic traditional instruments all the way, to those that generate rich, otherworldly sounds that literally defy classification.

Synthesizers (also known as synths) generate sounds and percussion sets using a number of different technologies or program algorithms. The earliest synths were analog in nature and generated sounds using additive or subtractive FM (frequency modulation) synthesis. This process generally involves the use of at least two signal generators (commonly referred to as *operators*) to create and modify a voice. Often, this is done through the analog or digital generation of a signal that modulates or changes the tonal and amplitude characteristics of a base carrier signal. More sophisticated FM synths can use up to

FIGURE 16.19
Bass Station analog bass synth. (Courtesy of Novation Digital Music Systems, Ltd.; www.novationmusic.com.)

FIGURE 16.20
Yamaha MOTIF-RACK ES synth. (Courtesy of Yamaha Corporation of America, www.yamaha.com.)

four or six operators per voice and also often use filters and variable amplifier types to alter the signal's characteristics into a sonic voice that either roughly imitates acoustic instruments or creates sounds that are totally unique.

Another technique that's used to create sounds is wavetable synthesis. This technique works by storing small segments of digitally sampled sound into a read-only memory chip. Various sample-based synthesis techniques use sample looping, mathematical interpolation, pitch shifting, and digital filtering to create extended and richly textured sounds that use a very small amount of sample memory.

Synthesizers are also commonly designed into rack- or half-rack-mountable modules, Figure 16.20, that contain all of the features of a standard synthesizer, except that they don't incorporate a keyboard controller. This space-saving feature means that more synths can be placed into your system and can be controlled from a master keyboard controller or sequencer, without cluttering up the studio with redundant keyboards.

Software Synthesis and Sample Re-synthesis. Since wavetable synthesizers derive their sounds from prerecorded samples that are stored in a digital memory medium, it logically follows that these sounds can also be stored on hard disk (or any other medium) and loaded into the RAM memory of a personal computer. This process of downloading wavetable samples into a computer and then manipulating these samples is used to create what is known as a virtual or software synthesizer, Figure 16.21.

In recent years, software synths have grown from being novel and obscure programs that were primarily used by the academic community to their present state of being widely accepted in the production community as a cost-effective musical instrument. These software modules can be used in conjunction with a digital audio workstation to offer up a wide range of complex sounds that can mimic traditional instruments, as well as create sonic textures that are both new and interesting.

Sample re-synthesis software systems are able to take software synthesis to a new level, by allowing the user to build, save, and recall sonic patches that can be built from traditional synthesis building blocks (such as oscillators, voltage-controlled amplifiers, voltage-controlled filters, and mixers). In addition to sound generation, digital audio samples can be imported and re-synthesized in a way that can create sounds of almost any texture or type that you can possibly imagine. All of these software blocks can be combined in a graphic environment that allows these instruments, textures, and soundscapes to be easily saved to disk for later recall.

Using various internal software data communications protocols, it's possible to communicate MIDI, audio, timing sync and control data between an instrument (or effect plug-in) and a host DAW program/

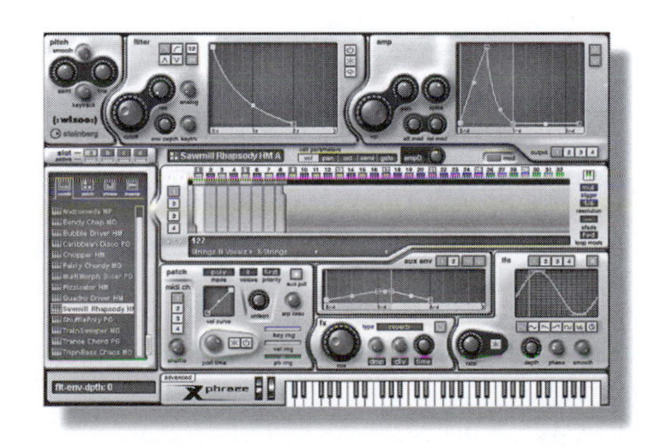

FIGURE 16.21
Steinberg xphrase VSTi software synth. (Courtesy of Steinberg Media Technologies GmbH, a division of Yamaha Corporation, www.steinberg.net.)

FIGURE 16.22
Akai MPC-1000 Music Production Center. (Courtesy of Akai Professional, www.akaipro.com.)

CPU processor. These plug-in protocols make it possible for much or all of the audio and timing data to be routed through the host audio application, allowing the instrument or application to either integrate into the DAW or application or to work in tandem so as to route the audio and performance/control data through the host application with relative ease. A few of these protocols include:

- Steinberg's VST (Virtual Studio Technology)
- MOTU's MAS (MOTU Audio System)
- Propellerheads ReWire.

Samplers. A sampler, Figure 16.22, is a device that can convert audio into a digital form and/or manipulate prerecorded sampled data, using the system's own random access memory (RAM). Once loaded into RAM, the sampled audio can be edited, transposed, processed, and played in a polyphonic musical fashion.

Basically, a sampler can be thought of as a wavetable synth that lets you record, load, and edit samples into RAM memory. Once loaded, these sounds (whose length and complexity are often limited only by memory size and your imagination) can be looped, modulated, filtered, and amplified (according to user or factory setup parameters), in a way that allows the waveshapes and envelopes to be modified. Signal processing capabilities, such as basic editing, looping, gain changing, reverse, sample-rate conversion, pitch change, and digital mixing capabilities can also be altered and/or varied.

A hardware sampler's design will often include a keyboard or set of trigger pads that let you polyphonically play samples as musical chords, sustain pads, triggered percussion sounds, or sound effect events. These samples can be played according to the standard Western musical scale (or any other scale, for that matter) by altering the playback sample rate over the controller's note range. For example, pressing a low-pitched key on the keyboard will cause the sample to be played back at a lower sample rate, while pressing a high-pitched one will cause the sample to be played back at rates that would put Mickey Mouse to shame. By choosing the proper sample rate ratios, sounds can be polyphonically played (whereby multiple notes are sounded at once) at pitches that correspond to standard musical chords and intervals.

A sampler (or synth) with a specific number of voices (i.e., 64 voices) simply means that up to 64 notes can be simultaneously played on a keyboard at any one time. Each sample in a multiple-voice system can be assigned across a performance keyboard, using a process known as *splitting* or *mapping*. In this way, a sound can be assigned to play across the performance surface of a controller over a range of notes, known as a zone, Figure 16.23. In addition to grouping samples into various zones, velocity can enter into the equation by allowing multiple samples to be layered across the same keys of a controller, according to how soft or hard they are played. For example, a single key might be layered so that pressing the key lightly would reproduce a softly recorded sample, while pressing it harder would produce a louder sample with a sharp percussive attack. In this way, mapping can be used to create a more realistic instrument or wild set of soundscapes that change not only with the played keys, but with velocity ranges as well.

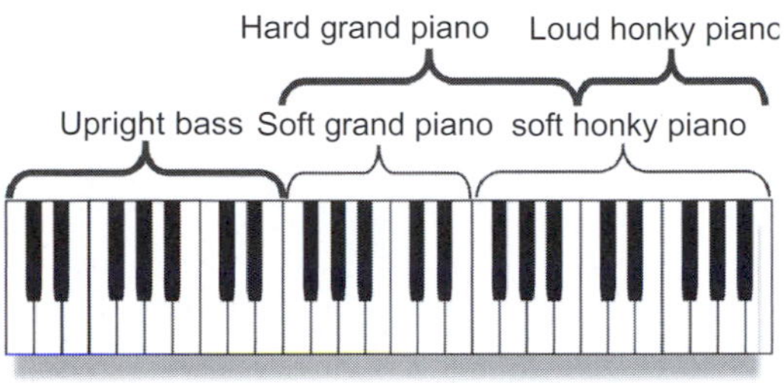

Hard grand piano Loud honky pianc
Upright bass Soft grand piano soft honky piano

FIGURE 16.23
Samples can be mapped to various zones on a keyboard.

FIGURE 16.24
Steinberg's HALion VST software sampler. (Courtesy of Steinberg Media Technologies GmbH, a division of Yamaha Corporation, www.steinberg.net.)

In addition to hardware sampling systems, a growing number of virtual or software samplers exist that use a computer's existing memory, processing, and signal routing capabilities in order to polyphonically reproduce samples in real time.

Offering much of the same functionality as their hardware counterparts, these software-based systems, Figure 16.24, are capable of editing, mapping, and splitting sounds across a MIDI keyboard, using on-screen graphic controls and DAW integration that has improved to the point of equaling or surpassing their hardware counterparts in cost-effectiveness, power, and ease of use.

As with a software synth, software samplers derive their sounds from recorded and/or imported audio data that is stored as digital audio data within a personal computer. Using the DSP capabilities of today's computers (as well as the recording, sequencing, processing, mixing, and signal routing capabilities of most digital audio workstations), most software samplers are able to store and access samples within the internal memory of a laptop or desktop computer. Using a graphic interface, these sampling systems often allow the user to:

- Import previously recorded soundfiles (often in WAV, AIF, and other common formats)
- Edit and loop sounds into a usable form
- Vary envelope parameters (i.e., dynamics over time)
- Vary processing parameters
- Save the edited sample performance setup as a file for later recall.

Software sampler systems are also often able to communicate MIDI, audio, timing sync and control data between a hard- or software instrument and a host DAW program/CPU processor, allowing for a wide range of control and setup recall.

The Drum Machine. The drum machine is most commonly a sample-based digital audio device that can't record audio into its internal memory (although this has changed in recent years, allowing it to import, record, and manipulate sampled audio much like a sampler). Traditionally, these hardware or software systems use ROM-based, prerecorded samples to reproduce high-quality drum sounds.

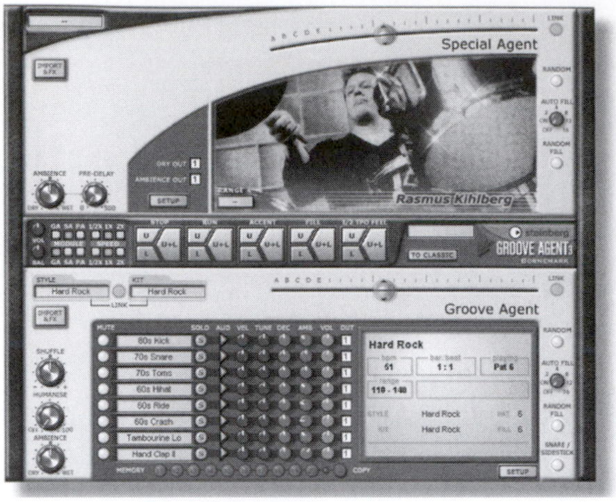

FIGURE 16.25
Groove Agent 3 VST Virtual Drummer. (Courtesy of Steinberg Media Technologies GmbH, a division of Yamaha Corporation, www.steinberg.net).

These factory-loaded sounds often include a wide assortment of drum sets, percussion sets, rare and wacky percussion hits, and effected drum sounds (i.e., reverberated, gated, etc.). Who knows, you might even encounter "Hit me!" screams from the venerable King of Soul—James Brown.

Most hardware drum machines allow prerecorded samples to be assigned to a series of playable keypads that are often located on the machine's top face. This provides a straightforward controller surface that usually includes velocity and aftertouch dynamics. Drum voices can be assigned to each pad and edited using such control parameters as tuning, level, output assignment, and panning position. Multiple outputs are often provided, enabling individual or groups of voices to be routed to a specific output on a mixer or console.

Although a number of hardware drum machine designs include a built-in sequencer, it's more likely that these workhorses will be triggered from a MIDI sequencer. This lets us take full advantage of the real-time performance and editing capabilities that a sequencer has to offer. For example, sequenced patterns can easily be created in step time (where notes are entered and assembled into a rhythmic pattern one note at a time) and can then link together into a song that's composed of several rhythmic patterns. Alternately, performing into a sequencer on-the-fly can help create a live feel or you can combine step- and real-time tracks to create a human-sounding composite rhythm track. In the final analysis, the style and approach to composition is entirely up to you.

In addition to their hardware counterparts, an increasing number of software drum and groove instrument plug-ins have come onto the market that allow for drum patterns to be added to a production in a wide range of pattern and playing styles, Figure 16.25.

Performance and parameter controllers

MIDI performance controllers are used to translate the voicings and expressiveness of a musical performance into MIDI data, while a parameter controller surface is used to alter the control variables of a workstation, device or instrument.

It should be noted that a MIDI controller is expressly designed to control other devices (be they for sound, light or mechanical control) within a connected system. It contains no internal tone generators or sound-producing elements. Instead, it offers a wide range of controls for handling control, trigger and device switching events. In short, controllers have become an integral part of music production, and are available in many incarnations to control and emulate many types of musical instruments.

Keyboard Controller. The MIDI keyboard controller, Figure 16.26, is a keyboard device that's expressly designed to control hard/software synths, samplers, modules and other devices within a connected MIDI system. It contains no internal tone generators or sound-producing elements. Instead, its design includes a performance keyboard and controls for handling MIDI performance, control, and device switching events.

Percussion Controllers. MIDI percussion controllers are used to translate the voicings and expressiveness of a percussion performance into MIDI data. These devices are great for capturing the feel of a live performance, while giving you the flexibility of recording and automating a performance within a DAW/sequencer environment. These controllers vary over a wide range from being a simple and cost-effective setup (i.e., using the pads on a drum machine, keys on a keyboard surface, or pads on an intro-level drum controller) to a full-blown drum kit that mimics its acoustic cousin, Figures 16.27 and 16.28.

FIGURE 16.26
Novation ReMOTE 25SL MIDI Controller/Keyboard. (Courtesy of Novation Digital Music Systems, Ltd, www.novationmusic.com.)

FIGURE 16.27
Trigger Finger 16-Pad MIDI Drum Control Surface. (Courtesy of M-Audio, a division of Avid Technology, Inc., www.m-audio.com.)

FIGURE 16.28
DM5 Electronic Drum Kit. (Courtesy of Alesis, www.alesis.com.)

Wind Controllers. MIDI wind controllers are expressly designed to bring the breath and key articulation of a woodwind or brass instrument to a MIDI performance. These controller types are used because many of the dynamic- and pitch-related expressions (such as breath and controlled pitch glide) simply can't be communicated from a standard music keyboard. In these situations, wind controllers can often help create a dynamic feel that's more in keeping with their acoustic counterparts by using an interface that provides special touch-sensitive keys, glide- and pitch-slider controls, and real-time breath sensors for controlling dynamics.

MIDI Guitars. Guitar players often work at stretching the vocabulary of their instruments beyond the traditional norm. They love doing nontraditional gymnastics using such tools of the trade as distortion, phasing, echo, feedback, etc. Due to advances in guitar pickup and microprocessor technology, it's also possible for the notes and minute inflections of guitar strings to be accurately translated into MIDI data. With this innovation, many of the capabilities that MIDI has to offer are available to the electric (and electronic) guitarist. For example, a guitar's natural sound can be layered with a synth pad that's been transposed down, giving it a rich thick sound that just might shake your boots. Alternately, recording a sequenced guitar track into a session would give a producer the option of changing and shaping the sound later in mixdown! On-stage program changes are also a big plus for the MIDI guitar, allowing the player to radically switch between guitar voices from the guitar or sequencer or by stomping on a MIDI foot controller.

SEQUENCERS

Apart from electronic musical instruments, one of the most important tools that can be found in the modern-day project studio is the MIDI sequencer. Basically, a sequencer is a digital device that's used to record, edit, reproduce, and distribute MIDI messages in a sequential fashion. Most sequencers function using a traditional track-based interface, separating different instruments, voices, beats, etc. in a way that makes it easier for us humans to view MIDI data as though they were linear tracks on a DAW or tape machine.

These virtual tracks contain MIDI-related performance and control events that are made up of such channel and system messages as note on/off, velocity, modulation, aftertouch, and program/continuous-controller messages. Once a performance has been recorded into a sequencer's memory, these events can be graphically (or audibly) edited into a musical performance, played back and saved to a digital storage media for recall at any time.

Integrated Sequencers. Some of the newer and more expensive keyboard synth and sampler designs include a built-in sequencer. These portable keyboard workstations have the advantage of letting you take both the instrument and sequencer on the road without having to drag a computer along.

Integrated sequencers are designed into an instrument for the sole purpose of sequencing MIDI data, and include integrated controls for performing sequence-specific functions. Ease of use and portability are often the advantages of a hardware sequencer, most of which are designed to emulate the basic functions of a tape transport (record, play, start/stop, fast forward, and rewind).

These devices generally offer a moderate amount of editing features, including note editing, velocity and other controller messages, program change, cut and paste and track merging capabilities, tempo changes, etc. Programming, track, and edit information is commonly viewed on a liquid crystal display (LCD) that's often limited in size and resolution and generally limits information to a single parameter or track at a time.

These sequencers often don't offer a wide range of editing tools beyond standard transport functions, punch-in/out commands and other basic edit tools. However, they're often more than adequate for capturing and reproducing a performance and can be integrated with other instruments that are connected in a MIDI chain.

Software Sequencers. By far, the most common sequencer type is the software MIDI sequencer. These programs or integrated components of a digital audio workstation take advantage of the versatility that a computer can offer in the way of speed, flexibility, digital signal processing, memory management, and signal routing.

Computer-based sequencers offer numerous functional advantages over their hardware counterparts. Among these are increased graphic capabilities (which often offers extensive control over track- and transport-related functions), standard computer cut and paste techniques, an on-screen graphic environment (allowing easy manipulation of program and edit-related data), routing of MIDI to multiple ports in a connected system, and the graphic assignment of instrument voices via program change messages (not to mention the ability to save and recall files using standard computer memory media). Now, let's take a look at how these devices function.

A basic introduction to sequencers

When dealing with any type of sequencer, one of the most important concepts to grasp is the fact that these devices don't store sound directly—instead, they encode MIDI messages that instruct an instrument to play a particular note, over a certain channel, at a specific velocity and with any optional controller values. In other words, a sequencer stores music-related data commands that follow in a sequential order, which then tells instruments and/or devices how their voices are to be played and/or controlled. This simple (but important) fact means that the amount of encoded data is far less memory intensive than its hard disk audio or video recording counterparts and that the data overhead that's required by MIDI is very small. In short, a computer-based sequencer can simultaneously operate in a digital audio, digital video, processing environment without placing an additional significant load on a computer's CPU.

As you might expect, many sequencer types are currently on the market, with each offering its own set of advantages and disadvantages. It's also true that each sequencer has its own basic operating feel, and thus choosing the best tool and toy for the job or studio is totally up to you.

Recording. From a functional standpoint, a sequencer is used as a digital workspace for creating personal compositions in environments that range from the bedroom to more elaborate project studios. Whether they're hardware or software based, most sequencers use a working interface that's designed to emulate the traditional multitrack recording environment. A tapelike transport lets you move from one location to the next using standard Play, Stop, FF, REW and Rec command buttons. Beyond using traditional record-enable button(s) to arm selected recording track(s), all you need to do is select the MIDI input (source) and outputs (destination) ports, instrument/voice MIDI channel, instrument patch and other setup information, press the record button, and start playing.

Once you've finished laying down a track, you can jump back to any point in the sequence and listen to your original track while continuing to lay down additional MIDI tracks until the song begins to form.

Almost all sequencers are capable of punching in and out of record while playing a sequence. This common function lets you drop in and out of record on a track (or tracks) in real-time. Although punch-in/out points can often be manually performed on-the-fly, most sequencers can perform a punch automatically, once the in/out measure numbers have been graphically or numerically entered. The sequence can then be rolled back a few measures and the artist can play along, while the sequencer automatically performs the necessary switching functions (usually with multiple take and full undo capabilities).

In addition to recording a performance in a track-based environment, most sequencers let you enter note values into sequence one note at a time. This feature (known as step time) lets you give the sequencer a basic tempo and note length (i.e., quarter note, sixteenth note, etc.) and then manually enter the notes from a keyboard or other controller. This data entry style is often (but not always) used with fast, hi-tech and dance styles, where a real-time performance just isn't possible or accurate enough for the song.

Whether you're recording a track in real-time or in step-time, it's almost always best to select the proper song tempo before recording a sequence. I bring this up because most sequencers are able to output a click track that can be used as an accurate audible guide for keeping in time with the song's selected tempo. It's also critical that the tempo be accurate when trying to sync groove loops and rhythms to a sequence via plug-ins or external instruments.

Editing. One of the more important features that a sequencer (or sequenced MIDI track within a DAW) has to offer is its ability to edit tracks or blocks within a track. Of course, these editing functions and capabilities often vary between hardware and software sequencers.

The main track window of a sequencer or MIDI track on a DAW is used to display such track information as the existence of track data, track names, MIDI port assignments for each track, program change assignments, volume controller values, etc.

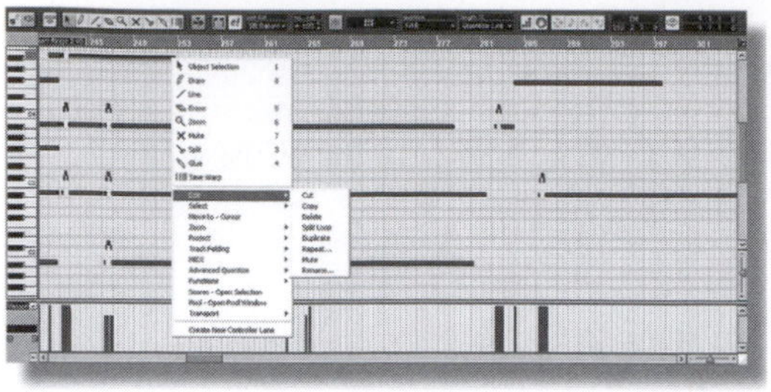

FIGURE 16.29
The presence of MIDI message data will often appear as a series of highlighted areas within a sequence track or a window. (Courtesy of Steinberg Media Technologies GmbH, a division of Yamaha Corporation, www.steinberg.net.)

Depending on the sequencer, the existence of MIDI data on a particular track at a particular measure point (or over a range of measures) is often indicated by the visual display of MIDI data in a piano-roll fashion (showing the general vertical and length placements of the notes as they progress though the musical passage) as shown in Figure 16.29.

By navigating around the various data display and parameter boxes, it's possible to use cut and paste and/or direct edit techniques to vary note, length and controller parameters for almost every facet of a section or musical composition. For example, let's say that we really screwed up a few notes when laying down an otherwise killer bass riff. With MIDI, fixing the problem is totally a no-brainer. Simply highlight each fudged note and drag it to it's proper note location. We can even change the beginning and end points in the process. In addition, tons of other parameters can be changed including velocity, modulation and pitch bend, note and song transposition, quantization, and humanizing (factors that eliminate or introduce human timing errors that are generally present in a live performance), as well as full control over program and continuous controller messages. The list goes on.

Playback. Once a composition is complete, all of the MIDI tracks in a project can be transmitted through the various MIDI ports and channels to plug-ins, instruments, or devices for playback. Since the data exists as encoded real-time control commands, you can listen to the sequence and make changes at any time. For example, you could change instrument settings (by changing or editing patch voices), alter volume and other mix changes, or experiment with such controllers as pitch bend, modulation or aftertouch, and even change the tempo and key signature. In short, this medium is infinitely flexible how a performance and/or set of parameters can be created, saved, folded, spindled, and mutilated until you've arrived at the sound and feel that you want.

Another of the greatest beauties of MIDI production is its ability to be altered at any later point in time. For example, let's say that 5 years ago you laid down a killer synth riff in a song that made it onto the charts. A couple of weeks ago a producer came to you in hopes of collaborating on a remix. Of course, technology marches on and your studio has improved over time. First off, even though a lot of the setup parameters have been saved with the original sequence, let's assume that you were smart enough to keep really good setup notes. One big change, however, is that you have a new software synth that has a patch that sounds better than the original patch. Since the remix is to be used in an upcoming film track, MIDI can be used to tweak things up a bit by splitting the riff into two parts: one that contains the lower notes and another the highs. By sending the lows to one patch on the synth and the highs to another, not only have you improved the overall sound, you've filled it out by expanding the soundfield into surround. Without MIDI, you'd have to arrange for a new session and hope that it all goes well; with MIDI, the performance is exactly the same and improvements are made in a no-brainer environment. This is what MIDI's all about—performance, repeatability, easy editing, and cost-effective power!

I now have to take time out to give you a few pointers that will make your life easier when dealing with MIDI production.

1. Remember to set the session to the proper tempo at the beginning of the session. Although tempo can be changed at a later time, attention to tempo details can help you to avoid later pitfalls.
2. Always name your track before you go into record (this goes for both audio and MIDI tracks). Properly naming your tracks (i.e., with its instrument, patch name) is the first step toward good documentation.

3. You can *never* overdocument a session. Keeping good instrument, patch, settings, musician, studio, and other notes might not only come in handy—it can save your butt if you need to revisit the tracks in the future.
4. Never delete a final take MIDI track from a DAW session. Even though you've transferred the instrument to an audio track, it is *always* wise to archive the original MIDI track with session. Trust me, both you and the producer will be glad you did, should any changes need to be made to the track in the future.

Other software sequencing applications

In addition to DAW and sequencing packages that are designed to handle most of the day-to-day production needs of the musician, other types of software tools and applications exist that can help to carry out specialized tasks. A few of these packages include drum pattern editors, algorithmic composition programs, patch editors and music printing programs.

Drum-Pattern Editor/Sequencers. At any one time, there are a handful of companies that have software or hardware devices that are specifically designed to create and edit drum patterns. In addition, most of the higher-end DAW audio production systems also include a drum pattern editor that relies on user input and quantization to construct and chain together any number of user-created percussion grooves. More often than not, these editors use a grid pattern that displays drum-related MIDI notes or subpatterns along the vertical axis, while time is represented in metric divisions along the horizontal axis, Figure 16.30. By clicking on each grid point with a mouse or other input system, individual drum or effect sounds can be built into rhythmic patterns.

Once created, these and other patterns can be linked together to create a partial or complete rhythm section within a song. These editors commonly offer such features as the ability to change MIDI note values (thereby changing drum voices), note length, quantization and humanization, as well as adjustments to note and pattern velocities. Once completed, the sequenced drum track (or chained patterns) can be imported into a sequence, saved, and/or exported.

Groove Tools. Getting into the groove of a piece of music often refers to a feeling that's derived from the underlying foundation of the piece: rhythm. With the introduction and maturation of MIDI and digital audio, new and wondrous tools have made their way into the mainstream of music production that can help us to use these technologies to forge, fold, mutilate and create compositions that make direct use of rhythm and other building blocks of music through the use of looping technology.

Of course, the cyclic nature of loops can be—repeat repeat—repetitive in nature, but new toys and techniques in looping have injected the notion of flexibility, real-time control, real-time processing, and mixing to new heights that can be used by an artist as a wondrously expressive tool.

Loop-based audio editors are groove-driven music programs, Figures 16.31 and 16.32, that are designed to let you drag and drop prerecorded or user-created loops and audio tracks into a graphic

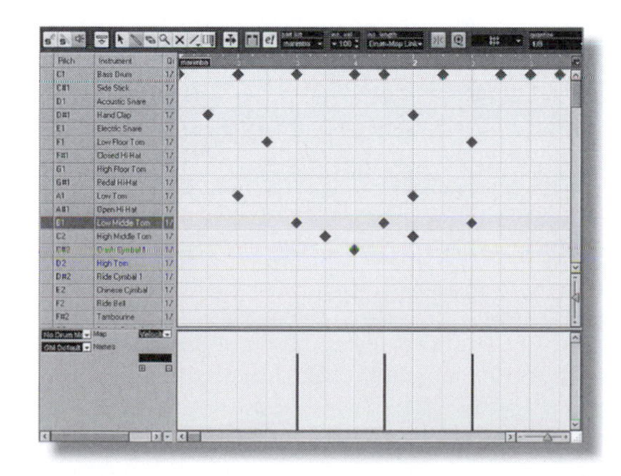

FIGURE 16.30
Steinberg Cubase/Nuendo drum edit window. (Courtesy of Steinberg Media Technologies GmbH, a division of Yamaha Corporation, www.steinberg.net.)

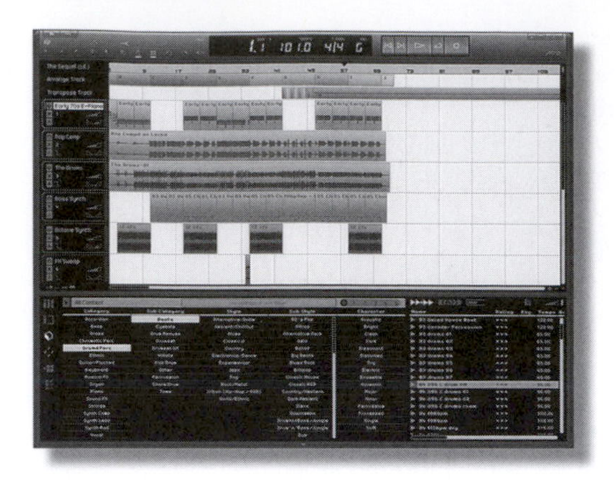

FIGURE 16.31
Steinberg's Sequel music software. (Courtesy of Steinberg Media Technologies GmbH, a division of Yamaha Corporation, www.steinberg.net.)

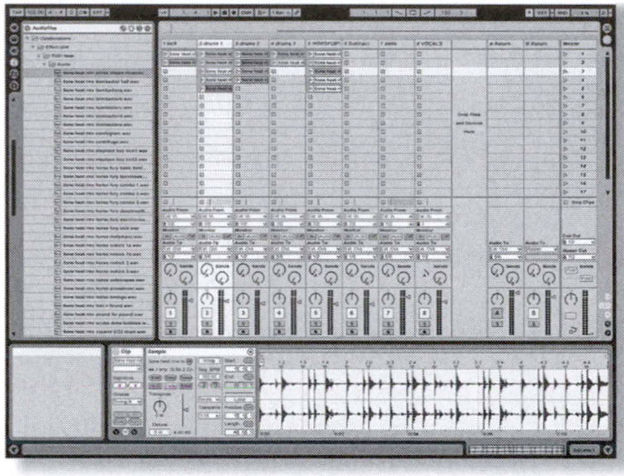

A. Arrangement view.

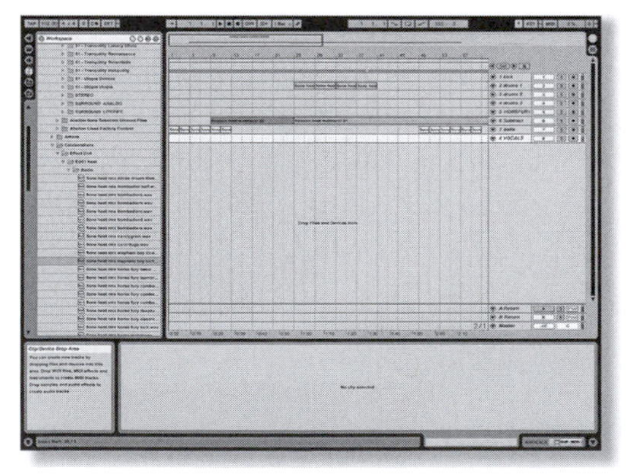

B. Session view.

FIGURE 16.32
Ableton live performance audio workstation. (Courtesy of Ableton, www.ableton.com.)

multitrack production interface. At their basic level, these programs differ conceptually from their traditional DAW counterpart, in that the pitch- and time-shift architecture is so variable and dynamic that even after the basic rhythmic, percussive and melodic grooves have been created, their tempo, track patterns, pitch, session key, etc. can be quickly and easily changed at any time. With the help of custom royalty-free loops (available from the manufacturer and/or third-party companies), users can quickly and easily experiment with setting up grooves, backing tracks, and creating a sonic ambience by simply dragging the loops into the program's main soundfile view where they can be arranged, edited, processed, saved, and exported.

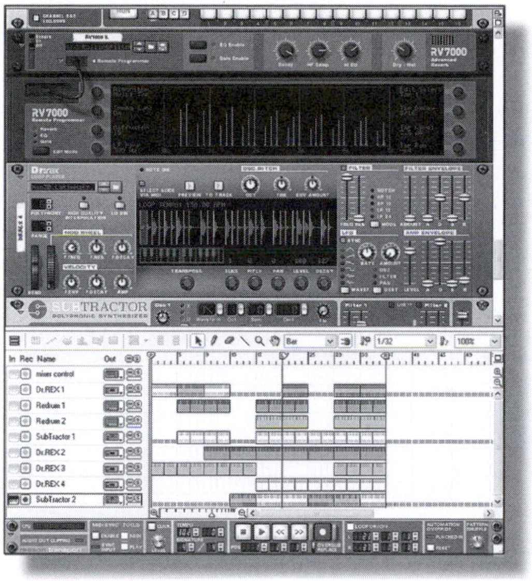

FIGURE 16.33
Reason music production environment. (Courtesy of Propellerheads software, www.propellerheads.se.)

One of the most interesting aspects of a loop-based editor is its ability to match the tempo of a specially programmed loop soundfile to the tempo of the current session. Amazingly enough, this process isn't that difficult to perform, as the program extracts the length, native tempo, and pitch information from the imported file's header and (using various digital time and/or pitch change techniques) adjusts the loop to fit the native time/pitch parameters of the current session. This means that loops of various tempos and musical keys can be automatically adjusted in length and pitch so as to fit in time with previously existing loops. These shifts in time to match a loop to the session's native tempo can actually be performed in a number of ways. For example, using basic DSP techniques to time-stretch and pitch-shift a recorded loop will often work well over a given plus-or-minus percentage range (which is often dependent on the quality of the program algorithms). Beyond this range, the loop will often begin to distort and become jittery. At such extremes, other playback algorithms and beat slice detection techniques can be used to make the loop sound more natural. For example, drums or percussion can be stretched in time by adding additional silence between the various hit points within the loop at precisely calculated intervals. In this way, the pitch will remain the same while the length is altered. Of course, such a loop would sound choppy and broken up when played on its own; however, when buried within a mix, it might work just fine. It's all up to you and the current musical context.

The software world doesn't actually hold the total patent on looping tools and toys; there are a number of groove keyboards and module boxes that are on the market. These systems, which range widely in sounds, functionality, and price, can offer up a wide range of unique sounds that can be quite useful laying a foundation under your production. In the past, getting a hardware grove tool to sync into a session could be time-consuming, frustrating, and problematic, taking time and tons of manual reading. However, with the advent of powerful time and pitch shift processing within most DAWs, the sounds from these hardware devices can be pulled into a session without too much trouble. For example, a single groove loop (or multiple loops) could be recorded into a DAW (at a bpm that's near to the session's tempo), edited, and then imported into the session, at which time the loop could be easily stretched into time sync, allowing it to be looped to your heart's content. Just remember, necessity is the mother of invention. Patience and creativity are probably your most important tools in the looping process.

If there's a software package that has gripped the hearts and minds of electronic musicians in the 21st century, it would have to be Reason from the folks at Propellerheads, Figure 16.33. Reason defies specific classification in that it's an overall music production environment that has many facets. For example, it includes a MIDI sequencer, as well as a wide range of software instrument modules which can be played, mixed, and combined in a comprehensive environment that can be controlled from any external keyboard and/or MIDI controller. Reason also includes a large number of signal processors that can be applied to any instrument or instrument group under full and easily controlled automation.

In essence, Reason is a combination of modeled representations of vintage analog synthesis gear, mixed with the latest digital synthesis and sampling technology. Combine these with a modular approach to signal and effects processing; add a generous amount of internal and remote mix and controller management (via an external MIDI controller); top this off with a quirky but powerful sequencer; and you have a software package that's powerful enough for top-flight production and convenient enough that you can build tracks from your laptop from your seat in a crowded plane. I know that it sounds like I read this from a sales brochure, but these are the basic facts of this program. When asked to explain Reason to others, I'm often at a loss as the basic structure is so open-ended and flexible that the program can be approached in as many ways as there are people who produce on it. That's not to say that Reason doesn't have a signature sound—it often does. However, it's a tool that can be either used on its own or in combination with other production instruments and tools.

Algorithmic Composition Programs. Algorithmic composition programs are interactive sequencers that directly interface with MIDI controllers or imported files to generate a performance in real-time, according to user-programmed computer parameters. In short, once you give it a few basic musical guidelines, it can act as a compositional robot to generate performances or musical parts on its own in order to help you gain new ideas for a song, create an automatic accompaniment, make improvisational exercises, create special performances, or just plain have fun.

This type of sequencer can be programmed to control the performance according to musical key, generated notes, basic order, chords, tempo, velocity, note density, rhythms, accents, etc. Alternatively, an existing standard MIDI file can be imported and further manipulated in real-time, according to new parameters that can be varied from a computer keyboard, mouse, or controller. Often such interactive sequencers will accept input from multiple players, allowing it to be performed as a collective jam. Once a composition has been satisfactorily generated, a standard MIDI file can be created and imported into any sequencer.

Patch Editors. The vast majority of MIDI instruments and devices store their internal patch data within RAM memory. Synths, samplers, or other devices contain information on how to configure oscillators, amplifiers, filters, tuning, and other presets in order to create a particular sound timbre or effect. In addition to controlling sound patch parameters, a unit's internal memory can also store such setup information as effects processor settings, keyboard splits, MIDI channel routing, controller assignments, etc.

Although these settings can be manually accessed from the device's panel controls, another (and sometimes more straightforward) way to gain real-time control over the parameters of an instrument or devices is through the use of a patch editor, Figure 16.34. A patch editor is a software package that's used to provide on-screen controls and graphic windows for emulating and varying an instrument's parameter controls in real-time.

Direct communication between a patch editor and the device's microprocessor commonly occurs through the use of MIDI SysEx messages. Almost all popular voice and setup editing packages include provisions for receiving and transmitting bulk patch data in this way. This makes it possible to save and organize large numbers of patch-data files, vary setting in real-time, and print out patch parameter settings.

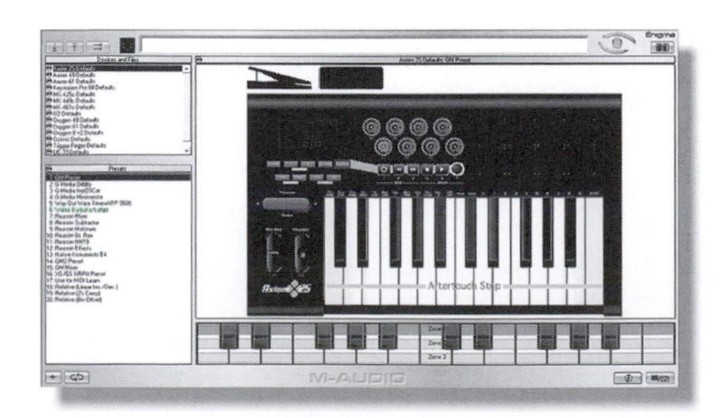

FIGURE 16.34
M-Audio Enigma Software Librarian and Editor. (Courtesy of M-Audio, a division of Avid Technology, Inc., www.m-audio.com.)

In addition to software editing packages, there are also hardware solutions for gaining quick and easy access to device parameters via SysEx. In recent years, MIDI data controllers, Figure 16.35, have sprung onto the market that can control a wide range of instruments and devices using data faders and soft buttons to vary patch, system, and performance parameters, in real-time. In many situations, these controllers can also be used to directly control the volume and mix parameters of a DAW.

Music-Printing Programs. In recent years, the field of transcribing musical scores onto paper has been strongly affected by computer, DAW, and MIDI technology. This process has been enhanced through the use of newer generations of software that make it possible for music notation data to be entered into a computer either manually (by placing the notes onto the screen via keyboard and/or by mouse movements) or via direct MIDI input. Once entered, these notes can be edited in an on-screen environment using a music printing program (or notation app within a DAW) that lets you change and configure a musical score or lead sheet using standard cut-and-paste edit techniques. In addition, most printing programs can play the various instruments in a MIDI system directly from the score. A final and important program feature is their ability to print out hard copies of a score or lead sheets in a wide number of print formats and styles.

These programs or DAW program apps, Figure 16.36, allow musical data to be entered into a computerized score in a number of manual and automated ways (often with varying degrees of complexity and ease). Although scores can be manually entered, most music-transcription programs will generally accept direct MIDI input, allowing a part to be played directly into a sequence. This can be done in real-time (by playing a MIDI instrument or finished sequence into the program), in step-time (entering

FIGURE 16.35
Mackie C4 plug-in and virtual instrument controller. (Courtesy of Loud Technologies, Inc., www.mackie.com.)

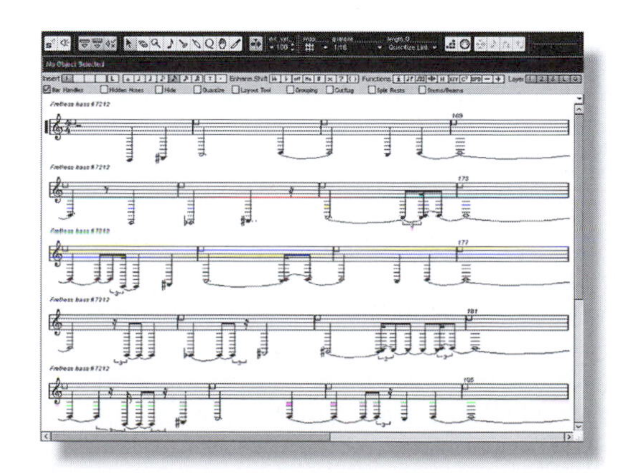

FIGURE 16.36
Score application within Steinberg's Nuendo DAW software. (Courtesy of Steinberg Media Technologies GmbH, a division of Yamaha Corporation, www.steinberg.net.)

the notes of a score one note at a time from a MIDI controller), or from an existing standard or program-specific MIDI file.

Another way to enter music into a score is through the use of an optical recognition program. These programs let you place sheet music or a printed score onto a standard flatbed scanner, scan the music into a program and then save the notes and general layout as a NIFF (notation interchange file format) file.

One of the biggest drawbacks to automatically entering a score via MIDI (either as a real-time performance or from a MIDI file) is the fact that music notation is a very interpretive art. "To err is human," and it's commonly this human feel that gives music its full range of expression. It is very difficult, however, for a program to properly interpret these minute yet important imperfections and place the notes into the score exactly as you want them. (For example, it might interpret a held quarter-note as either a dotted quarter-note or one that's tied to a thirty-second note.) Even though these computer algorithms are getting better at interpreting musical data and quantization can be used to tell a computer to round a note value to a specified length, a score will still often need to be manually edited to correct for misinterpretations.

MULTIMEDIA AND THE WEB

It's no secret that modern-day computers have gotten faster, sleeker, and sexier in their overall design. In addition to its ability to act as a multifunctional production workhorse, one of the crowning achievements of the modern computer is the degree of media and networking integration that has worked its way into our collective consciousness and become known as multimedia.

The combination of working and/or playing with multimedia has found its way into modern computer culture through the use of various hardware and software systems that work in a multitasking environment and combine to bring you a unified experience that seamlessly involves such media types as:

- Text
- Graphics
- Video
- Audio and music
- Computer animation
- MIDI

The obvious reason for integrating and creating these media types is the human desire to create content with the intention of sharing and communicating one's experiences with others. This has been done for centuries in the form of books and more recently by movies and television. In the here and now, the Web has been added to the communications list, in that it has created a vehicle that allows individuals (and corporate entities alike) to communicate a multimedia experience to millions and then allows each individual to manipulate that experience, learn from it, and even respond in an interactive fashion. The Web has indeed unlocked the potential for experiencing multimedia events and information in a way that makes each of us a participant, not just a passive spectator.

One of the unique advantages of MIDI, as it applies to multimedia, is the rich diversity of musical instruments and program styles that can be played back in real-time while requiring almost no overhead processing from the computer's CPU. This makes MIDI a perfect candidate for playing back soundtracks from multimedia games or over the Internet. It's interesting to note that MIDI has taken a back seat to digital audio as a serious music playback format for multimedia. Most likely, this is due to several factors, including:

1. A basic misunderstanding of the medium.
2. The fact that producing MIDI content requires a basic knowledge of music.
3. The frequent difficulty of synchronizing digital audio to MIDI in a multimedia environment.
4. The fact that soundcards often include poorly designed FM synthesizers (although most operating systems now include a higher-quality software synth).

Fortunately, an increasing number of software companies have taken up the banner of embedding MIDI within their media projects and have helped push MIDI a bit more into the Web and gaming mainstream. As a result, it's becoming more common for your PC to begin playing back a MIDI score on its own or perhaps in conjunction with a more data-intensive program or game.

Standard MIDI files

The accepted format for transmitting files or real-time MIDI information in multimedia (or between sequencers from different manufacturers) is the standard MIDI file. This file type (which is stored with a .mid or .smf extension) is used to distribute MIDI data, song, track, time signature, and tempo information to the general masses. Standard MIDI files can support both single and multichannel sequence data and can be loaded into, edited, and then directly saved from almost any sequencer package. When exporting a standard MIDI file, keep in mind that they come in two basic flavors: type 0 and type 1.

- Type 0 is used whenever all of the tracks in a sequence need to be compressed into a single MIDI track. All of the original channel messages still reside within that track; however, the data will have no definitive track assignments. This type might be the best choice when creating a MIDI sequence for the Internet (where the sequencer or MIDI player application might not know or care about dealing with multiple tracks).
- Type 1, on the other hand, will retain its original track information structure and can be imported into another sequencer type with its basic track information and assignments left intact.

General MIDI

One of the most interesting aspects of MIDI production is the absolute setup and patch uniqueness of each professional and even semipro project studio. In fact, no two studios will be alike (unless they've been specifically designed to be the same or there's some unlikely coincidence). Each artist will be unique in having his or her own favorite equipment, supporting hardware, favorite way of routing channels and tracks, and assigning patches. The fact that each system setup is unique and personal has placed MIDI at odds with the need for systems compatibility in the world of multimedia. For example, after importing a standard MIDI file over the Net and loading it into a sequencer, you might hear a song that's being played with a totally irrelevant set of sound patches (it might sound interesting, but it won't sound anything like it was originally intended). If the MIDI file is loaded into a new computer, the sequence will again sound completely different, with patches that are so irrelevant that the guitar track might sound like a bunch of machine-gun shots from the planet Gloob.

In order to eliminate (or at best reduce) the basic differences that exist between systems, a patch and settings standard known as General MIDI (GM) was created. In short, GM assigns a specific instrument patch to each of the 128 available program change numbers. Since all electronic instruments that conform to the GM format must use these patch assignments, placing GM program change commands at the header of each track will automatically configure the sequence to play with its originally intended sound. As such, no matter what sequencer is used to play the file back, as long as the receiving instrument conforms to the GM spec the sequence will be heard using its intended instrumentation. Tables 16.3 and 16.4 detail the program numbers and patch names that conform to the GM format (Table 16.3 for percussion and Table 16.4 for nonpercussion instruments). These patches include sounds that imitate synthesizers, ethnic instruments, and/or sound effects that have been derived from early Roland synth patch maps. Although the GM spec states that a synth must respond to all 16 MIDI channels, the first nine channels are reserved for instruments, while GM restricts the percussion track to MIDI channel 10.

MIDI-BASED SYNCHRONIZATION

Just as synchronization is routinely used in audio and video production, the wide acceptance of MIDI and digital audio within the various media has created the need for synchronization in project studio and midsized production environments. Devices such as MIDI sequencers, digital audio editors, effects devices, and digital mixing consoles make extensive use of synchronization and time code. However, advances in design have fashioned this technology into one that's much more cost-effective and easy-to-use—all through the use of MIDI. The following sections outline the various forms of synchronization that are often encountered in a MIDI-based production environment.

Simply stated, most current forms of synchronization use the MIDI protocol itself for the transmission of sync messages. These messages are transmitted along with other MIDI data over standard MIDI cables, with no need for additional or special connections.

Table 16.3 GM percussion instrument patch map (Channel 10).

35. Acoustic Bass Drum	59. Ride Cymbal 2
36. Bass Drum 1	60. Hi Bongo
37. Side Stick	61. Low Bongo
38. Acoustic Snare	62. Mute Hi Conga
39. Hand Clap	63. Open Hi Conga
40. Electric Snare	64. Low Conga
41. Low Floor Tom	65. High Timbale
42. Closed Hi-Hat	66. Low Timbale
43. High Floor Tom	67. High Agogo
44. Pedal Hi-Hat	68. Low Agogo
45. Low Tom	69. Cabasa
46. Open Hi-Hat	70. Maracas
47. Low Mid Tom	71. Short Whistle
48. Hi Mid Tom	72. Long Whistle
49. Crash Cymbal 1	73. Short Guiro
50. High Tom	74. Long Guiro
51. Ride Cymbal 1	75. Claves
52. Chinese Cymbal	76. Hi Wood Block
53. Ride Bell	77. Low Wood Block
54. Tambourine	78. Mute Cuica
55. Splash Cymbal	79. Open Cuica
56. Cowbell	80. Mute Triangle
57. Crash Cymbal 2	81. Open Triangle
58. Vibraslap	

Note: In contrast to Table 16.3, the numbers in Table 16.4 represent the percussion keynote numbers on a MIDI keyboard, not program change numbers.

MIDI real-time messages

Although no time code-based reference is implemented, it's important to know that MIDI has a built-in (and often transparent) protocol for synchronizing all of the tempo and timing elements of each MIDI device in a system to a master clock. This protocol operates by transmitting real-time messages to the various instruments and devices throughout the system. Although these relationships are usually automatically defined within a system setup, one MIDI device must be designated as the master device in order to provide the timing information to which all other slaved devices are locked. MIDI real-time messages consist of four basic types that are each 1 byte in length:

- **Timing clock**—A clock timing that's transmitted to all devices in the MIDI system at a rate of 24 pulses per quarter note (ppq). This method is used to improve the system's timing resolution and simplify timing when working in nonstandard meters (e.g., 3/8, 5/16, 5/32).
- **Start**—Upon receipt of a timing clock message, the start command instructs all connected devices to begin playing from the beginning of their internal sequences. Should a program be in midsequence, the start command repositions the sequence back to its beginning, at which point it begins to play.
- **Stop**—Upon the transmission of a MIDI stop command, all devices in the system stop at their current positions and wait for a message to follow.

Table 16.4 GM Non-percussion Instrument Patch Map with Program Change Numbers.

1. Acoustic Grand Piano	39. Synth Bass 1
2. Bright Acoustic Piano	40. Synth Bass 2
3. Electric Grand Piano	41. Violin
4. Honky-tonk Piano	42. Viola
5. Electric Piano 1	43. Cello
6. Electric Piano 2	44. Contrabass
7. Harpsichord	45. Tremolo Strings
8. Clavi	46. Pizzicato Strings
9. Celesta	47. Orchestral Harp
10. Glockenspiel	48. Timpani
10. Music Box	49. String Ensemble 1
12. Vibraphone	50. String Ensemble 2
13. Marimba	51. SynthStrings 1
14. Xylophone	52. SynthStrings 2
15. Tubular Bells	53. Choir Aahs
16. Dulcimer	54. Voice Oohs
17. Drawbar Organ	55. Synth Voice
18. Percussive Organ	56. Orchestra Hit
19. Rock Organ	57. Trumpet
20. Church Organ	58. Trombone
21. Reed Organ	59. Tuba
22. Accordion	60. Muted Trumpet
23. Harmonica	61. French Horn
24. Tango Accordion	62. Brass Section
25. Acoustic Guitar (nylon)	63. SynthBrass 1
26. Acoustic Guitar (steel)	64. SynthBrass 2
27. Electric Guitar (jazz)	65. Soprano Sax
28. Electric Guitar (clean)	66. Altto Sax
29. Electric Guitar (muted)	67. Tenor Sax
30. Overdriven Guitar	68. Baritone Sax
31. Distortion Guitar	69. Oboe
32. Guitar harmonics	70. English Horn
33. Acoustic Bass	71. Bassoon
34. Electric Bass (finger)	72. Clarinet
35. Electric Bass (pick)	73. Piccolo
36. Fretless Bass	74. Flute
37. Slap Bass 1	75. Recorder
38. Slap Bass 2	76. Pan Flute

(*Continued*)

Table 16.4 Continued.

77. Blown Bottle	103. FX 7 (echoes)
78. Shakuhachi	104. FX 8 (sci-fi)
79. Whistle	105. Sitar
80. Ocarina	106. Banjo
81. Lead 1 (square)	107. Shamisen
82. Lead 2 (sawtooth)	108. Koto
83. Lead 3 (calliope)	109. Kalimba
84. Lead 4 (chiff)	110. Bag pipe
85. Lead 5 (charang)	111. Fiddle
86. Lead 6 (voice)	112. Shanai
87. Lead 7 (fifths)	113. Tinkle Bell
88. Lead 8 (bass p lead)	114. Agogo
89. Pad 1 (new age)	115. Steel Drums
90. Pad 2 (warm)	116. Woodblock
91. Pad 3 (polysynth)	117. Taiko Drum
92. Pad 4 (choir)	118. Melodic Tom
93. Pad 5 (bowed)	119. Synth Drum
94. Pad 6 (metallic)	120. Reverse Cymbal
95. Pad 7 (halo)	121. Guitar Fret Noise
96. Pad 8 (sweep)	122. Breath Noise
97. FX 1 (rain)	123. Seashore
98. FX 2 (soundtrack)	124. Bird Tweet
99. FX 3 (crystal)	125. Telephone Ring
100. FX 4 (atmosphere)	126. Helicopter
101. FX 5 (brightness)	127. Applause
102. FX 6 (goblins)	128. Gunshot

- **Continue**—Following the receipt of a MIDI stop command, a MIDI continue message instructs all instruments and devices to resume playing from the precise point at which the sequence was stopped. Certain older MIDI devices (most notably drum machines) aren't capable of sending or responding to continue commands. In such a case, the user must either restart the sequence from its beginning or manually position the device to the correct measure.

Song position pointer

In addition to MIDI real-time messages, the Song Position Pointer (SPP) is a MIDI system common message that isn't commonly used in current-day production. Essentially, SPP keeps track of the current position in the song by noting how many measures have passed since the beginning of a sequence. Each pointer is expressed as multiples of six timing-clock messages and is equal to the value of a 16th note.

The song position pointer can synchronize a compatible sequencer or drum machine to an external source from any position within a song containing 1024 or fewer measures. Thus, when using SPP, it is possible for a sequencer to chase and lock to a multitrack tape from any measure point in a song.

Using such a MIDI/tape setup, a specialized sync tone is transmitted that encodes the sequencer's SPP messages and timing data directly onto tape as a modulated signal. Unlike SMPTE time code, the encoding method wasn't standardized between manufacturers. This lack of standardization prevents SPP data written by one device from being decoded by another device that uses an incompatible proprietary sync format.

Unlike SMPTE, where tempos can be easily varied by inserting a tempo change at a specific SMPTE time, once the SPP control track is committed to tape, the tape and sequence are locked into this predetermined tempo or tempo change map. SPP messages are usually transmitted only while the MIDI system is in the stop mode, in advance of other timing and MIDI continue messages. This is due to the relatively short time period that's needed to locate the slaved device to the correct measure position.

MIDI time code

MIDI time code (MTC) was developed to allow electronic musicians, project studios, video facilities, and virtually all other production environments to cost-effectively and easily translate time code into time-stamped messages that can be transmitted via MIDI. Created by Chris Meyer and Evan Brooks, MTC enables SMPTE-based time code to be distributed throughout the MIDI chain to devices or instruments that are capable of synchronizing to and executing MTC commands. MTC is an extension of MIDI 1.0, which makes use of existing message types that were either previously undefined or were being used for other non-conflicting purposes. Since most modern recording devices include MIDI in their design, there's often no need for external hardware when making direct connections. Simply chain the MIDI cables from the master to the appropriate slaves within the system (via physical cables, USB, or virtual internal routing). Although MTC uses a reasonably small percentage of MIDI's available bandwidth (about 7.68% at 30 fr/s), it's customary (but not necessary) to separate these lines from those that are communicating performance data when using MIDI cables. As with conventional SMPTE, only one master can exist within an MTC system, while any number of slaves can be assigned to follow, locate, and chase to the master's speed and position. Because MTC is easy to use and is often included free in many system and program designs, this technology has grown to become the most common and most straightforward way to lock together such devices as DAWs, modular digital multitracks, and MIDI sequencers, as well as analog and videotape machines (by using a MIDI interface that includes a SMPTE-to-MTC converter).

The MTC format can be divided into two parts:

- Time code.
- MIDI cueing.

The time code capabilities of MTC are relatively straightforward and allow devices to be synchronously locked or triggered to SMPTE time code. MIDI cueing is a format that informs a MIDI device of an upcoming event that's to be performed at a specific time (such as load, play, stop, punch in/out, reset). This protocol envisions the use of intelligent MIDI devices that can prepare for a specific event in advance and then execute the command on cue.

MTC is made up of three message types: quarter-frame messages, full messages, and MIDI cueing messages.

- **Quarter**—frame messages—These are transmitted only while the system is running in real or variable speed time, in either forward or reverse direction. True to its name, four quarter-frame messages are generated for each time code frame. Since eight quarter-frame messages are required to encode a full SMPTE address (in hours, minutes, seconds, and frames—00:00:00:00), the complete SMPTE address time is updated once every two frames. In other words, at 30 fps, 120 quarter-frame messages would be transmitted per second, while the full time code address would be updated 15 times in the same period. Each quarter frame message contains 2 bytes. The first byte is F1, the quarter-frame common header, while the second byte contains a nibble (four hits) that represents the message number (0 through 7) and a nibble for encoding the time field digit.
- **Full messages**—Quarter-frame messages are not sent in the fast-forward, rewind, or locate modes, as this would unnecessarily clog a MIDI data line. When the system is in any of these shuttle modes, a full message is used to encode a complete time code address within a single message. After a fast shuttle mode is entered, the system generates a full message and then places

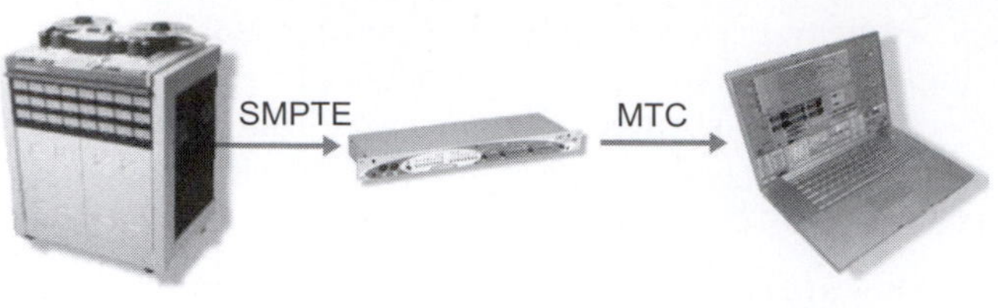

FIGURE 16.37
SMPTE time code can be easily converted to MTC (and vice versa) for distribution throughout a production system.

itself in a pause mode until the time-encoded slaves have located to the correct position. Once playback has resumed, MTC will again begin sending quarter-frame messages.

- **MIDI cueing messages**—MIDI cueing messages are designed to address individual devices or programs within a system. These 13-bit messages can be used to compile a cue or edit decision list, which in turn instructs one or more devices to play, punch in, load, stop, and so on at a specific time. Each instruction within a cueing message contains a unique number, time, name, type, and space for additional information. At the present time, only a small percentage of the possible 128 cueing event types has been defined.

SMPTE/MTC Conversion. Although MTC is commonly implemented within a software or hardware system itself (that's the functional and economic beauty of it), whenever a hardware device doesn't talk MTC (but only a flavor of the SMPTE protocol), a SMPTE-to-MIDI converter must be used, Figure 16.37. These conversion systems are available as stand-alone devices or as an integrated part of a multiport MIDI interface/patch bay/synchronizer system. Certain analog and digital multitrack systems include a built-in MTC port within their design, meaning that the machine can be synchronized to a DAW/sequencing system without the need for any additional hardware, beyond a MIDI interface.

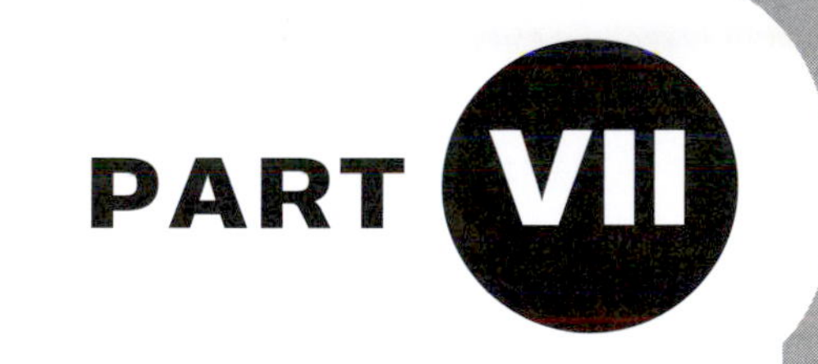

Acoustics and Sound Reinforcement

PREFACE

The seventh section of this book deals with Acoustics and Sound Reinforcement, and the first two chapters come from *Sound System Engineering* by Don and Carolyn Davis. In this first chapter Don and Carolyn cover the complex subject of large-room acoustics, dealing with reverberation, flutter-echoes, and the reverberant and direct sound fields.

They begin by examining the issue of reverberation. In theoretical acoustics, much use is made of the concept of the free field, a space in which there are no reflections to complicate matters. The only way to set up a free field in practice is to put your sound source on a pole outdoors, well above ground level, or to put it in an anechoic chamber. In real-life indoor situations the reflections from the room boundaries, and from other objects in the room, have a profound effect on the acoustical situation. When the sound absorption in the room is relatively low, there are multiple reflections which set up a reverberant (also known as diffuse) sound field. In a perfectly reflective, and so perfectly reverberant, room the time average of the mean square sound pressure is the same in all locations. Practical rooms always have some absorption, and therefore show resonances at low frequencies, and uneven reverberation at higher ones; these are called semi-reverberant sound fields, and represent the practical situations that acoustical practioners have to deal with.

The serious study of large room reverberation dates back to a classic series of experiments carried out by Wallace Sabine in 1885. Harvard University was encountering severe acoustic problems with a lecture room in the new Fogg Art Museum, and Sabine, a young physics professor was called in to investigate. He had several rooms to compare, including the Sanders Theater, which most importantly was a source of seat cushions that could be moved about to vary absorption. Sabine determined that the body of an average adult reduced reverberation time by about as much as six seat cushions. After two years of research he had established the fundamental relationship between the size of a room and the amount of absorption needed for satisfactory results. Sabine formally defined reverberation time, still the most important parameter in current use for assessing the acoustical quality of a room, as the number of seconds it took for the intensity of the sound to fall from the initial level to 60 dB less. This quantity is called the RT60. He also defined the Sabine Equation, which is a simple relationship between RT60 and room volume and absorption area.

Don and Carolyn show in this chapter how the Sabine Equation can be derived mathematically, and how it is used to determine the rate of decay of a reverberant sound field. They go on to describe attempts to improve on the Sabine Equation, such as the rather more complex Norris-Eyring Equation, and the Fitzroy Equation.

From here they go on to describe how absorption is defined and assessed, improved reverberation time calculations, and exactly what reverberation is and what it is not. For example, a flutter-echo is not reverberation.

The rest of the chapter deals with the application of these principles to the design of acoustic spaces, including an examination of the Hopkins-Stryker Equation for assessing acoustic level changes.

CHAPTER 17

Large Room Acoustics

Sound System Engineering by Don and Carolyn Davis

WHAT IS A LARGE ROOM?

Manfred Schroeder has defined a large room frequency (F_L) as the frequency above which a large number of room modes will be excited to vibrate at the source frequency.

$$F_L = K\sqrt{\frac{RT_{60}}{V}} \quad (17.1)$$

where,

F_L is the large room frequency in Hz,

K *is* 2000 in the SI and 11,885 in the U.S.,

RT_{60} is the apparent reverberation time for 60 dB of decay in seconds,

V is the volume of the room in m^3 or ft^3.

If we assume for sound systems:

1. A low frequency limit of 80 Hz for speech systems.
2. A low frequency limit of 30 Hz for music systems.
3. An RT_{60} of 1.6 s approximates the decay time expected for a minimum density sound field, then a large room volume for speech becomes approximately:

$$V = K^2 \frac{RT_{60}}{F_L^{\,2}} = (2000)^2 \frac{1.6}{(80)^2} = 1000\ \text{m}^3$$

or

$$V = K^2 \frac{RT_{60}}{F_L^{\,2}} = (11{,}885)^2 \frac{1.6}{(80)^2} = 35{,}313\ \text{ft}^3$$

and for very wide range music:

$$V = K^2 \frac{RT_{60}}{F_L^{\,2}} = (2000)^2 \left(\frac{1.6}{30^2}\right) = 7111\text{m}^2$$

or

$$V = K^2 \frac{RT_{60}}{F_L^{\,2}} = (11{,}885)^2 \frac{1.6}{(30)^2} = 251{,}116.84\ \text{ft}^3$$

Audio Engineering Explained

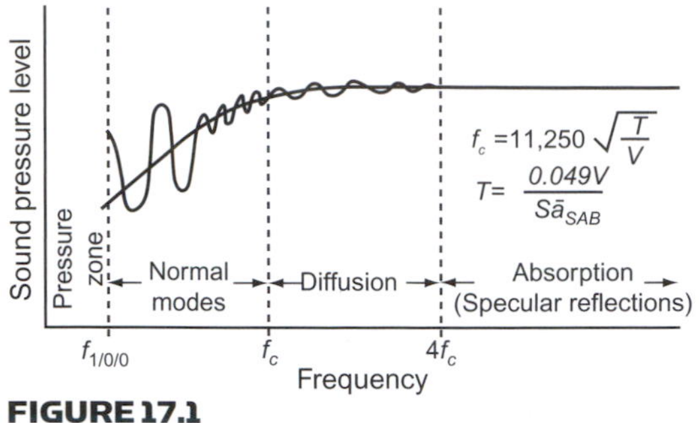

FIGURE 17.1
Controllers of steady-state room acoustic response. In physically small rooms f_c moves to a frequency of 250 Hz–500 Hz. (*Courtesy Bolt, Beranek, and Newman*).

Therefore, in this book a large room will be one in which, for speech, the internal volume is 35,000 ft^3 or greater and, for very wide range music, the internal volume is 250,000 ft^3 or greater.

Since F_L is frequency dependent, we can employ our standard audio analysis technique. First divide the audio spectrum into three decades on a linear frequency scale so that we may treat the first decade as a "small room" acoustic problem while approaching the upper two decades as a "large room" design problem.

Bolt, Beranek and Newman have long utilized an almost identical equation with a K for U.S. system of 11,250 which would, by conversion, be 1,893.14 for the SI, see Figure 17.1. Also, the equation originally used in LEDE control room work yields almost identical results in practical cases.

$$F_L = \frac{3 \times (\text{velocity of sound})}{\textit{Room's smallest dimension}} \tag{17.2}$$

F_L is a transition area and should be viewed as such and not as a rigid fixed frequency. The critical frequency (f_c) is synonymous with F_L and both notations are used in small room literature.

Use of the Sabinian equations in large reverberant spaces

Spaces that qualify as "large rooms" can effectively utilize the myriad of equations based on the original assumptions of Sabine for his reverberation equations. In spaces exceeding these volumes and with an RT_{60} of 1.6 s or greater, we will find mixing homogeneous sound fields of sufficient density to allow accurate engineering estimates of the level of each.

Harvard University found in 1885 that its newly completed Fogg Art Museum had severe acoustical difficulties. President Eliot, head of the university, turned to a young physics professor named Wallace Clement Sabine with the request to "do something" about the problem.

Sabine didn't follow the practices of past generations and hang draperies, place carpets, etc., to "deaden" such a "live" room. Instead he turned from qualitative approaches in finding the solution to a study of the problem on a quantitative basis.

Sabine had at his disposal a number of useful tools to aid in the investigation of the problem. First, there was the troubled lecture room in the Fogg Art Museum. Second, there was nearby Saunders Theater which was considered to have excellent acoustics. Third, the constant-temperature room in the subbasement of the Jefferson Physical Laboratory turned out to be a reverberation chamber. Finally, he had a middle-of-the-road room considered acoustically tolerable, but not much more, in the large lecture room, also in the Jefferson Physical Laboratory building.

With these environments as laboratories, Sabine used the seat cushions from Saunders Theater as his portable absorption, organ pipes and a portable windchest as his sound source, and a stopwatch and his own remarkable hearing as his acoustic test instruments.

After more than two years of intensive research (he often taught classes during the day and did research at night, existing on just a few hours of sleep), Sabine not only had corrected the troubled room by adding the correct amount of acoustical absorption, but as it turned out, he had gathered the raw data for the first important breakthrough in the science of architectural acoustics.

One Saturday evening on the 29th of October, 1898, staring at some of his curves, Sabine called out to his mother (who was living with him at the time), "Mother, it's a hyperbola!" This simple, but inspired, observation took architectural acoustics out of the dark ages of cut-and-try into the sunlight of calculation and measurement.

The insight that came to Sabine, revealing the fundamental relationship between the size of a room and the absorption needed, resulted from his unbelievably precise measurements coupled with his intuitive genius. Thereafter, the reverberation time of a room was calculable prior to construction.

In September, 1975, some 77 years later, W. B. Joyce, in an article entitled, "Sabine's Reverberation Time and Ergodic Auditoriums" in the *Journal of the Acoustical Society of America*, showed the relationship between the second law of thermodynamics and Sabine's equation. This talented Bell Laboratories scientist derived Sabine's equation from a literature search that could have been done at Sabine's time since the necessary thermodynamic concepts were extant by 1895. See box.

SABINE'S REVERBERATION TIME AND ERGODIC AUDITORIUMS

by W. B. Joyce

Published: *J. Acoust. Soc.* Am. Vol. 58, No. 3, pp. 643–655, September 1975

Abstract: It is shown in geometrical acoustics that ergodic specular enclosures do exist and that in such auditoriums, but not in general, 4V/S' is the exact mean directed path length (V is volume and S′ is any part of surface area S). Sabine's expression is then demonstrated to yield the exact reverberation time, provided the enclosure is mixing and provided the inhomogeneous anisotropic surface absorptivity is sufficiently weak. It is further proven that the fundamental form of Sabine's expression cannot be modified so as to become correct for large absorption. In an attempt to reassign credit and reconcile these results with influential findings to the contrary, a short historical account is added. Conditions imposed upon the surface reflectivity—whether the reflectivity be reversible (specular) or other or irreversible (statistical)—by the second law of thermodynamics and by the principle of detailed balance are evaluated. Extensions (e.g., mean length of curved paths in an ergodic auditorium with a thermal gradient) and other applications (electroluminescent diode design) are noted.

In 1929, M. J. O. Strutt considered reverberation by regarding it as a case of free damped vibration of the volume of the air enclosed in a room (this was before computers). The analysis involves the general wave equations, with suitable boundary conditions imposed. Strutt regarded as unsatisfactory the theories which dealt with the paths of separate sound rays (geometric acoustics). The various Eigen Tones or modes of the resonant vibration of the air columns in the room appear in the analysis. This analysis revealed Sabine's law as an asymptotic property toward which the reverberation tends, as the frequency of the (forcing) sound becomes infinitely great compared with the wavelength of the sounds.

Later work at MIT by Philip Morse and Richard Bolt led to the honest but humorous conclusion that "The practical role of wave acoustics is that it can indicate how to design an enclosure for which geometrical acoustics and statistical acoustics are valid, and in which there is no need of wave acoustics."

Figure 17.2 illustrates the typical measurement setup. Pink noise is emitted by the loudspeaker until a steady state level is produced in the enclosure (i.e., the rate of acoustic energy emission is equal to the the rate of acoustical energy absorption). Then the amplifier is disconnected from the loudspeaker. The microphone signal is fed through a bandpass filter (typically either an octave or ⅓ octave), and the decay rate is observed on the display unit (which may be a digital meter, a graphic level recorder, or an oscilloscope screen). When a graphic level recorder is used, Figure 17.3 shows how the trace produced is analyzed.

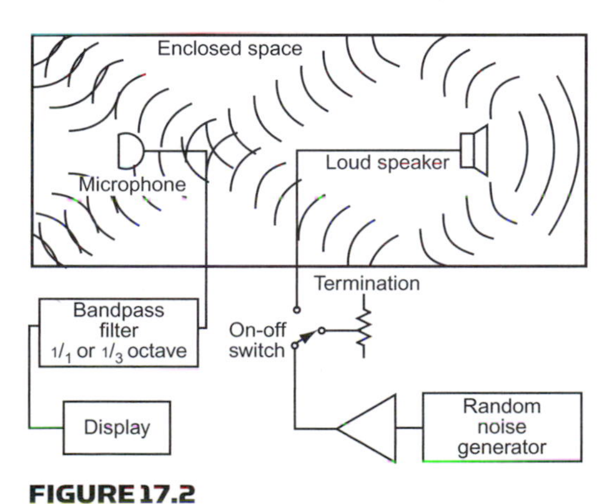

FIGURE 17.2
Measuring the RT_{60} of an enclosure.

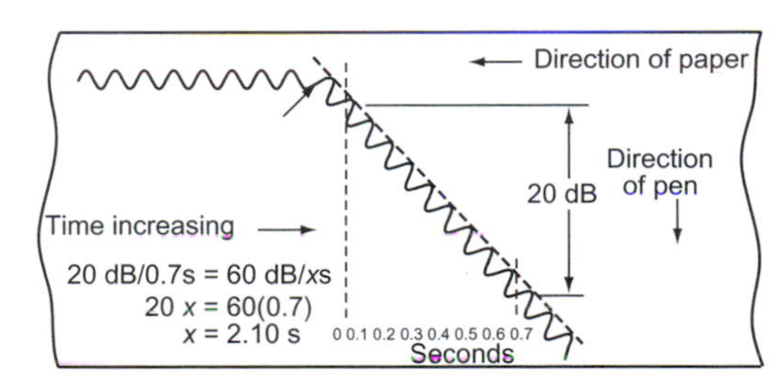

FIGURE 17.3
Chart recorder method of measuring RT_{60}.

The Sabine equation

William Joyce at Bell Telephone Labs has demonstrated that the Sabine Equation is fundamental to the decay of energy in a light emitting diode just as it is in an enclosed acoustic space.

The accuracy of this remarkable equation is dependent upon a space beïng ergodic. There must be a sufficient number of reflections, of a statistical nature, that allow the concept of the mean free path to be meaningful.

In physics, the mean free path is defined as:

$$MFP = \frac{4V}{S} \tag{17.3}$$

where:
V is the volume of the space enclosed in some unit of length, l^3,
S is the internal surface area in l^2. This allows any convenient system to be used, i.e., inches, feet, yards, millimeters, centimeters, meters, etc.
The amount of time it takes a sound field to decay to one-one-millionth of its original energy, -60 dB, after the sound source is turned off is called the reverberation time, RT_{60}.

The principal cause of the energy decay is the amount of acoustic absorption available in sabin, $S\bar{a}$. Materials are rated in absorption units (dimensionless $\bar{a}$) that range from 0.0 (totally reflective) to 1.0 (totally absorptive.) Absorption becomes dimensional when multiplied by the area it is on which makes $S\bar{a}$ in l^2.

We can write an equation that predicts the number (N) of reflections that will occur during a 60 dB decay

$$\begin{aligned} N &= 6\ln(10)\left(\frac{1}{\bar{a}}\right) \\ e^{\ln(10)} &= 10 \\ e^{6\ln(10)} &= 10^6 \end{aligned} \tag{17.4}$$

or

$$\begin{aligned} N &= 2.30\log(10^6)\left(\frac{1}{\bar{a}}\right) \\ \ln(10) &= 2.303 \\ e^{2.303\log(10^6)} &= 10^6 \end{aligned} \tag{17.5}$$

where,
$\bar{a}$ is the average absorption coefficient of all coefficients present in the enclosure:

$$\bar{a} = \frac{s_1a_1 + s_2a_2 + \ldots + s_na_n}{S_T} \tag{17.6}$$

$s_{1,2,3\ldots}$ are the surfaces 1, 2, 3,
$a_{1,2,3\ldots}$ are the coefficients of similarly numbered areas,
S_T is the total boundry surface area.

With these tools at hand it is possible to find the number of reflections per second.

$$RPS = \frac{c}{MFP} \tag{17.7}$$

where,
c is the velocity of sound, light, etc., in l/s.
It then becomes possible to complete the reverberation time by

$$RT_{60} = \frac{N}{RPS}. \tag{17.8}$$

Defined dimensionally

$$\frac{N\,(\text{Dimensionless})}{\cfrac{l/s}{\cfrac{4l^3}{l^2}}} = \text{s} \tag{17.9}$$

For those who work consistently in some preferred dimensional system and prefer to eliminate c from the equation

$$4(2.303)\log(10^6) = 55.26$$

$$RT_{60} = \frac{55.26}{S\bar{a}(c)}$$

$$\frac{55.26}{1130} = 0.049$$

$$\frac{55.26}{344.42} = 0.161$$

results in

$$RT_{60} = \frac{0.049V}{S\bar{a}} \quad \text{U.S. equation}$$

$$RT_{60} = \frac{0.161V}{S\bar{a}} \quad \text{SI equation} \tag{17.10}$$

EXAMPLE:

Let: $V = 500{,}000\ \text{ft}^3$, $S = 42{,}000\ \text{ft}^2$, $\bar{a} = 0.128$,

then

$$MFP = \frac{4(500{,}000)}{42{,}500} = 47\ \text{ft}$$

$$N = 2.303\log(10^6)\left(\frac{1}{0.128}\right) = 108\ \text{reflections}$$

$$RPS = \frac{1130}{47} = 24\ \text{reflections per second}$$

$$RT_{60} = \frac{108}{24} = 4.5\ \text{s}$$

Specific versus statistical

One hundred eight reflections allow a reasonable statistical sample. When small absorptive spaces such as control rooms in recording studios, small classrooms, etc. are computed, the inapplicability of statistical equations becomes apparent because of the low N. Such enclosures do indeed have a finite number of reflections that are best handled by careful Envelope Time Curve (ETC) analysis and specific rather than statistical treatment of the indicated surfaces.

Calculating the rate of decay of reverberant sound energy

The classic Sabine equation is

$$RT_{60} = \frac{0.049\,V}{S\bar{a}} \tag{17.11}$$

Because the reverberation time in seconds is calculated to be the time required for the sound energy to decay to 1/1,000,000th (−60 dB) of its original value prior to switching off the sound source, we can further write decay rate in dB/s = 60 dB $/RT_{60}$ and, because

$$\frac{S\bar{a}}{0.049\,V} = \frac{1}{RT_{60}} \tag{17.12}$$

we can derive the following direct equation for decay rate in dB/s:

$$\frac{60S\bar{a}}{0.049\,V} = \frac{60}{RT_{60}} = \frac{1224.5S\bar{a}}{V} \tag{17.13}$$

For example, in a church that has an RT_{60} of

$$RT_{60} = \frac{0.049(500{,}000)}{9800} = 2.5\text{ s}$$

and a decay rate of

$$\text{dB/s} = \frac{1224.5(9800)}{500{,}000} = 24\text{ dB/s}$$

To check, we can take

$$\frac{60}{2.5} = 24\text{ dB/s}$$

Improved reverberation time calculations

The works of Joyce and Gilbert have increased our appreciation for the fundamental integrity of the original Sabine equation when used in spaces where it can properly be applied. Our work with the TEF analysis process has confirmed that the simple equations that work so well in "live" rooms should simply not be applied in any form in small dead rooms.

One of the confusions that can arise is that absorption is useful for the control of specular reflections in rooms where no real reverberant sound field exists and application of the statistical formulas is nonsensical. Indeed, application of absorption in these cases is immediately audible, often dramatically so, whereas massive absorption in large "live" rooms that meet the classic criteria do indeed provide a lowering of the statistical reverberant field level but is a much more subtle audible affect. In the "gray areas," we find influences from both approaches and care must be taken to use a sufficiently conservative design approach that allows for "worst case" possibilities.

It has been our experience that the majority of listeners who declare a room as "live" or "dead" do so on the basis of initial signal delay gap and the level of the first reflections and not on the level of the reverberant sound field.

Now that we can view instrumentally the density and spectral uniformity, as well as the changes in frequency with time exhibited by sound fields in real spaces, the true cause and effect relationships exerted by the boundary surfaces of a space will become even more accessible than at present. One aspect that is particularly interesting is the length of time required to see the first effect of the presence of absorption in a large hall (see Figures 17.4A and 17.4B for two ETCs of before and after absorption).

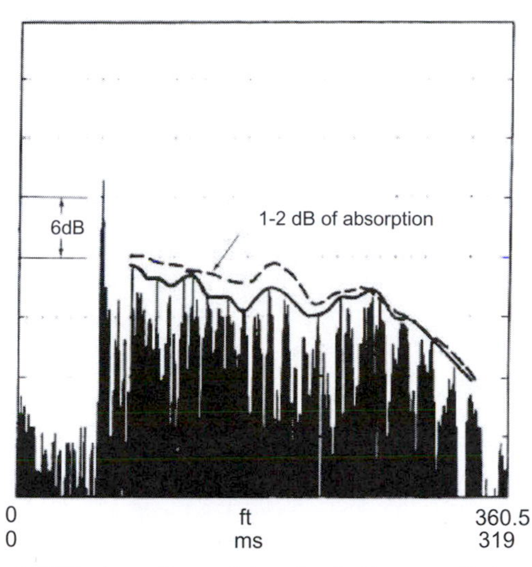

A. With absorption curtain extended in an auditorium (absorption present),1 dB-2dB of absorption is very audible.

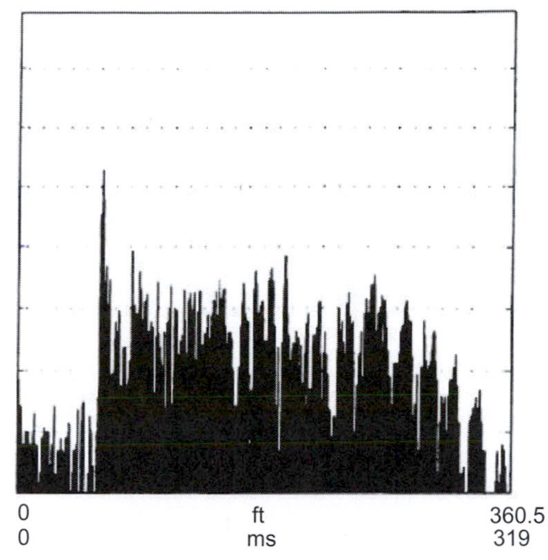

B. Auditorium with curtain retracted (without absorption).

FIGURE 17.4
RT_{60} measurement made with and without absorption.

Importance of Sabine

The work of W. C. Sabine founded the entire field of architectural acoustics and is fundamental to the successful interface of any electroacoustic system to the acoustic environment. A partial list of present day equations directly based on Sabine's work is:

1. Critical distance.
2. Reverberation.
3. Reverberant sound field.
4. Transmission loss.
5. Hopkins-Stryker and its many variations.
6. Articulation loss of consonants.
7. *Q* relative to the adjustment of direct-to-reverberant ratios.

The genius of the man is apparent, important, and yet relatively unheralded. Encyclopedias rarely mention him. Outside of the field of architectural acoustics, students fail to recognize his name. Wallace Clement Sabine deserves our honored respect and acknowledgment.

Limitations of all acoustic equations based on geometry and statistics

It should always be considered that, insofar as the reverberation formulas depend upon statistical averages, they presuppose a complete mixing of sound in the room. In very absorptive rooms, the sound dies away in a few reflections, and the statistical basis of the formulas is weakened. In recent studies done with time energy frequency analysis, typical meeting rooms in hotels have been found in some cases, $RT_{60} = 0.5\,\text{s}$, to develop no reverberant sound field, whereas in others, $RT_{60} = 0.7\,\text{s}$, a field barely appears.

Our experience with time-energy-frequency measurements causes us to state unequivocally that recording studio control rooms are not proper subjects for use of classic statistical equations.

In very large rooms, such as the Astrodome and the Superdome, because the sound cannot cross the room many times during a measured reverberation period of a few seconds, the validity of the formula is affected.

LEVELS DEFINED: SOUND POWER LEVEL (L_W), SOUND INTENSITY LEVEL (L_I), AND SOUND PRESSURE LEVEL (L_P)

Sound power level, L_W

L_W is the total acoustic power level in dB radiated by a sound source.

$$L_W = 10\log\left(\frac{W_a}{10^{-12}\ \text{W}}\right) \tag{17.14}$$

where,

W_a is the acoustic watts,

10^{-12} W is the specified reference.

For an output of 1.0 watt we can write

$$L_W = 10\log(W_a) + 120\text{ dB} = 10\log(1.0\text{ W}) + 120\text{ dB} = 120\text{ dB}$$

This means that a device radiating a total acoustic power of 1.0 watt will have an $L_W = 120$ dB regardless of radius, r, from the source or how confined a directivity factor, Q, happens to be.

Sound intensity level, L_I

If we imagine a sphere with a surface area A of 1.0 m^2, the radius r becomes

$$r = \sqrt{\frac{A}{4\pi}} = 0.282\text{ m}.$$

The 1.0 watt radiating from an omnidirectional point source through a spherical surface area of 1.0 m^2 would have a sound intensity level of

$$L_I = 10\log\left(\frac{\frac{W_a}{m^2}}{\frac{10^{-12}W}{m^2}}\right). \tag{17.15}$$

Which can again be written as

$$L_I = 10\log\left(\frac{W_a}{m^2}\right) + 120 = 10\log\left(\frac{1.0W}{m^2}\right) + 120 = 120\text{ dB}$$

We would also find at the surface of our imaginary sphere a sound intensity of 1 W/m^2.

Sound pressure level, L_P

The root mean square acoustic pressure is given by

$$P_{RMS} = \sqrt{I_a\rho c} = \sqrt{1.0\frac{W}{m^2}(400)} = 20\text{ Pa}$$

where,

I_a is the acoustic intensity in W/m^2,

ρc is the specific acoustic resistance of air. ρ is expressed in kg/m^3 and c is the sound speed in m/s. ρc has a value of 400.

There is a sound pressure reference value of 20 micropascals (μPa) = 0.00002 Pa.

$$L_P\text{ (sound pressure level)} = 20\log\left(\frac{P_{rms}}{0.00002\text{ Pa}}\right)$$

or

$$20\log\left(\frac{P_{rms}}{1\text{ Pa}}\right) + 94\text{ dB}$$

and for 20 Pa

$$L_p = 20\log(20) + 94\text{ dB} = 120\text{ dB}$$

At a radius of 0.282 m, a sphere has a surface area of $1.0\,m^2$ and 1.0 acoustic watt radiating through that surface area produces at that surface an

$$L_W = 120\text{ dB},$$
$$L_I = 120\text{ dB},$$
$$L_P = 120\text{ dB}.$$

This means only that these three parameters are numerically identical.

Next, imagine a hemisphere with a radius of 0.282 m. The source is radiating 1.0 acoustic watt; therefore, the $L_W = 120$ dB. The surface area is now $0.5\,m^2$ and the 1.0 W now produces an intensity of $1.0\,W/0.5\,m^2$ or $2.0\,W/m^2$. This results in $L_I = 123$ dB and an $L_P = 123$ dB. The difference between L_I and L_W is called the directivity index, D_I in dB.

$$10^{\left(\frac{D_I}{10}\right)} = Q$$
$$10^{\left(\frac{3.01}{10}\right)} = 2$$

The directivity factor, Q, describes the increase in power per unit of area that results from confining available power to a smaller area. The comparison is between the Q confined area versus that over the spherical power per unit of area. Today Qs of 50+ are available from devices that cover 1/50 of a spherical surface, thus simultaneously achieving controlled coverage of an audience area and supplying that area with an L_P that required 1/50 the power an omnidirectional device would require for the same L_P, thus yielding a +17 dB advantage (20log 50/1 = 16.99 dB).

Finally, increase the radius of the original sphere by a factor of 2 such that $r = (2)(0.282)$ m.

$$A = 4\pi r^2 = 4\pi\,[2(0.2821)]^2 = 4\text{m}^2$$

The result in a surface area of $4\,m^2$ and 1.0 acoustic watt now produces:

$$L_W = 120\text{ dB},$$
$$L_I = 114\text{ dB},$$
$$L_P = 114\text{ dB}.$$

Or for each doubling of distances we drop 6 dB in level.

$$20\log\,(D_1/D_2) = \text{Change in level of } L_P$$
$$20\log\,(0.282) - 20\log\,(0.564) = -6.02\text{ dB}$$

The rate of change in level is a consequence of the inverse square law that governs radiation from point sources.

A given L_W is independent of both distance and area covered. We can state that L_I and L_P each vary with both distance, r, and directivity, Q as:

$$10\log\,Q \text{ or } 20\log\,r \tag{17.16}$$

$$L_I \text{ or } L_P = L_W + 10\log\,Q \tag{17.17}$$

$$L_I \text{ or } L_P = L_W - 20\log\left(\frac{r}{0.282\text{ m}}\right) \tag{17.18}$$

where,

r is greater than 0.282 m (0.925 ft)

As will be seen further on we can use these identities to predict efficiencies, power, and pressure relationships.

LEVELS IN ENCLOSED SPACES

When a loudspeaker radiates sound into an enclosure, its acoustic performance as determined under open air conditions is modified by the acoustic properties of the space, but the total power L_W radiated is essentially unchanged. When a sound source is turned on in a room, the energy spreads from the source and then strikes the various wall surfaces, S, where it is partially absorbed, a, and partially reflected $1 - a$ to other surfaces which, in turn, absorb and reflect. This process continues until the energy in the room reaches a steady value, i.e., when the rate of energy absorption by the surfaces and in the air becomes equal to the rate of energy emission by the source.

This energy is made up of the total reverberant energy, L_R, assumed uniform in distribution, and the total direct energy L_D. This division is expressed as

$$L_D = L_W + 10\log\left(\frac{Q}{4\pi r^2}\right) + 10.5 \text{ U.S.} \tag{17.19}$$

and

$$L_R = L_W + 10\log\left(\frac{4}{S\bar{a}}\right) + 10.5 \text{ U.S.} \tag{17.20}$$

Interestingly $4V/S$ considered as a radius to a sphere suggests that the volume of such a sphere and the room volume, for all rooms of reasonable proportions, will be nearly equal, thus allowing a simplification of terms. From these relatively simple relationships it becomes evident that L_W and $S\bar{a}$ determine the reverberant levels and that L_W, Q, and r^2 determine the direct sound level at any given position. In fact the design goal in most cases is to insure that any listener receives at the least an $L_D = L_R$ and at the best that $L_D \geq L_R$.

Hopkins-Stryker—U.S. and SI

The interplay of directivity factor, Q, distance from the sound source to the listener, D_2, the total acoustic absorption in the room, $S\bar{a}$, and the expected sound pressure level, L_p at the listener, can all be combined in a single equation called the Hopkins-Stryker.

$$L_P = L_W + 10\log\left(\frac{Q}{4\pi D_2{}^2} + \frac{4}{S\bar{a}}\right) + 10.5 \text{ U.S.} \tag{17.21}$$

$$L_P = L_W + 10\log\left(\frac{Q}{4\pi D_2{}^2} + \frac{4}{S\bar{a}}\right) + 0.2 \text{ SI} \tag{17.22}$$

Figure 17.5a illustrates the levels for the direct sound pressure level, L_D, with the distance, the reverberant sound level, L_R, and the total sound level, L_T, for a $Q = 45$, D_2 from 10 ft to 1000 ft, a total room absorption of 5000 ft^2, and a sound power level, L_W, of 105.6 dB.

Figure 17.5b is a table of log multiplier and ratio exponents for curves that are neither inverse square law, or totally reverberant, but rather fall between −2 dB per doubling of distance to −5 dB per doubling of distance (i.e., for 3 dB per doubling of distance beyond D_c use:

$$9.966 \log D_2$$

or

$$\log D_2{}^{0.996}$$

Finally, Figure 17.5c shows a fully implemented set of modifiers and multipliers for L_D, L_T, −2 dB/doubling of distance, −3 dB/doubling of distance, −4dB/doubling of distance, and −5 dB/doubling of distance.

$L_{EIA} := 71.9$ dB $\quad Q := 45 \quad W := 0.1 \quad D_x := 90 \quad K := 10.5$ (U.S.) (0.2 = S.I.)

$Lw := L_{EIA} - 10 \cdot \log(Q) + 10 \cdot \log(W) + 60.22 \qquad Lw = 105.6$ dB

$Wa := 10^{\left(\frac{Lw}{10}\right)} \cdot 10^{-12}$ *

$a_{aud} := 0.8 \qquad Ma := \dfrac{1}{1 - a_{aud}}$ * $\quad Ma = 5 \qquad Me := 1.0 \qquad Wa = 0.036$

$D := 4..1000$ ft or m $\qquad Sa := 5000$ ft² or m²

$L_{dx} := Lw + 10 \cdot \log\left[\dfrac{Q \cdot Me}{4 \cdot \pi \cdot (D_x)^2}\right] + K$ * $\qquad f(D) := Lw + 10 \cdot \log\left(\dfrac{Q \cdot Me}{4 \cdot \pi \cdot D^2} + \dfrac{4}{Sa \cdot Ma}\right) + K$

$D_c := 0.141 \cdot \sqrt{Q \cdot Sa \cdot Ma \cdot Me} = 149.6$ ft or m $\qquad \%eff := \left[10^{\frac{\left[\left(Lw + 10 \cdot \log\left(\frac{1}{W}\right)\right) - 120\right]}{10}}\right] \cdot 100$ *

$L_{30} := Lw + 10 \cdot \log\left(\dfrac{Q \cdot Me}{4 \cdot \pi \cdot 900}\right) + K$ * $\qquad L_r := Lw + 10 \cdot \log\left(\dfrac{4}{Sa \cdot Ma}\right) + K$

$L_{30} = 92.1$ dB $\qquad L_{dx} = 82.5$ dB $\qquad L_r = 78.1$ dB $\qquad \%eff = 36.2$ %

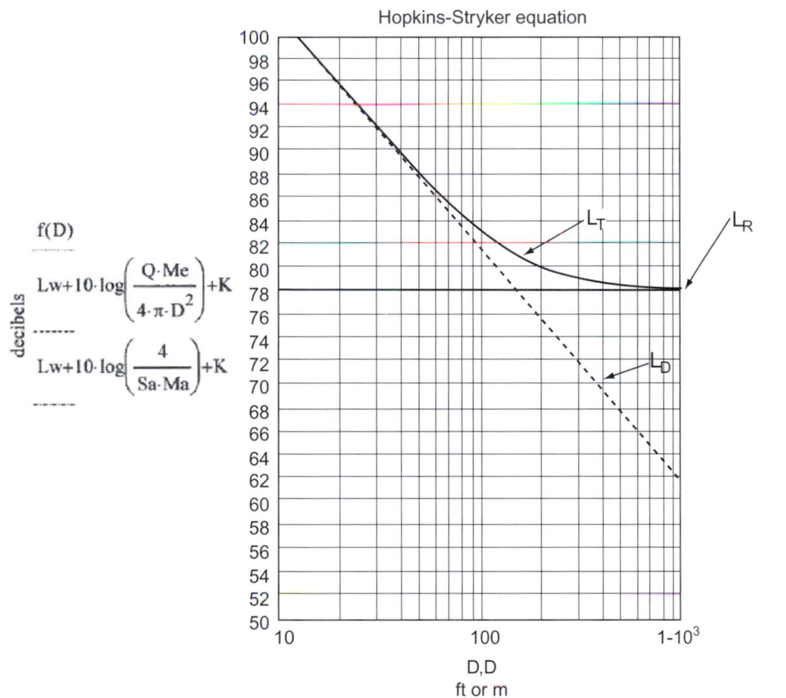

FIGURE 17.5A
MathCAD program of Hopkins-Stryker plot for system design parameters shown above.

Figure 17.5c is the same case as Figure 17.5a with the modifiers added of $N = 1.0$, $Me = 1.5$, and $Ma = 5.0$ where the Hopkins-Stryker equation becomes

$$L_P = L_W + 10\log\left[\frac{Q(Me)}{4\pi D_2{}^2} + \frac{4N}{\bar{S}\bar{a}Ma}\right] + 10.5 \tag{17.23}$$

When knowledge of the dB per doubling of distance has been ascertained either by experience or by measurement, the appropriate multiplier or exponents can be inserted into the calculation. Note that these modifications apply only to distances beyond critical distance (see vertical axis notation on Figure 17.5b).

Logarithim modifiers and exponents for use in Hopkins-Stryker equations in 0.5 dB steps for changes in level per doubling of distance.

Example

For 3.5 dB change for each time distance is doubled

11.62675 log (2) = 3.5 dB or 10 log ($2^{1.16267}$) = 3.5 dB

d := 0.5, 1.0.. 6.0 dB   b := 10

$$f(d) := \frac{d}{\frac{\log(2)}{\log(b)}} \qquad \exp = \frac{f(d)}{b}$$

f(d) =

	1
1	1.66096
2	3.32193
3	4.98289
4	6.64386
5	8.30482
6	9.96578
7	11.62675
8	13.28771
9	14.94868
10	16.60964
11	18.2706
12	19.93157

f(d)/10 =

	1
1	0.1661
2	0.33219
3	0.49829
4	0.66439
5	0.83048
6	0.99658
7	1.16267
8	1.32877
9	1.49487
10	1.66096
11	1.82706
12	1.99316

FIGURE 17.5B .

For those who would like to use solid angle data in place of Q, the Hopkins-Stryker becomes

$$L_T = L_W + 10\log\left[\left(\frac{1}{sr \times D_2{}^2}\right) + \left(\frac{4}{S\bar{a}}\right)\right] + 0.2 \tag{17.24}$$

where,
L_T is the sound pressure level,
L_W is the sound power level,
sr is the solid angle in steradians,
r is the distance in meters,
$S\bar{a}$ is the total absorption in sabins meters squared.
Remember that the solid angle in steradians = $4\pi/Q$.

Other terms derived from Hopkins-Stryker

For those unfamiliar with these terms, the following definitions are useful:

$$D_I = 10\log\left(\frac{4\pi}{sr}\right) \tag{17.25}$$

$$sr = \frac{4\pi}{Q} \tag{17.26}$$

$$Q = \frac{4\pi}{sr} \tag{17.27}$$

$Q := 45 \quad W := 0.1 \quad 10 \cdot \log\left(\frac{W}{0.001}\right) = 20 \;\text{dBm} \quad C := 2.8$

$L_{EIA} := 71.9$ dB

$Lw := L_{EIA} - 10 \cdot \log(Q) + 10 \cdot \log(W) + 60.22 \quad Lw = 105.6$

$Me := 1.0$

$D := 2..1000$ ft or m $\quad K := 10.5$ (U.S.) (0.2 =S.I.)

$N := 1.0 \quad Sa := 5000$ ft² or m² $\quad Ma := 5.0$

$f(D) := Lw + 10 \cdot \log\left(\frac{Q \cdot Me}{4 \cdot \pi \cdot D^2} + \frac{4 \cdot N}{Sa \cdot Ma}\right) + K \qquad D_c := 0.141 \cdot \sqrt{\frac{Q \cdot Sa \cdot Ma \cdot Me}{N}} = 149.6$ ft or m

$Ld_{dc} := Lw + 10 \cdot \log\left(\frac{Q \cdot Me}{4 \cdot \pi \cdot D_c^{\,2}}\right) + K \qquad Ld_{dc} = 78.1 \qquad D_x := 1.0 \cdot D_c..1000$

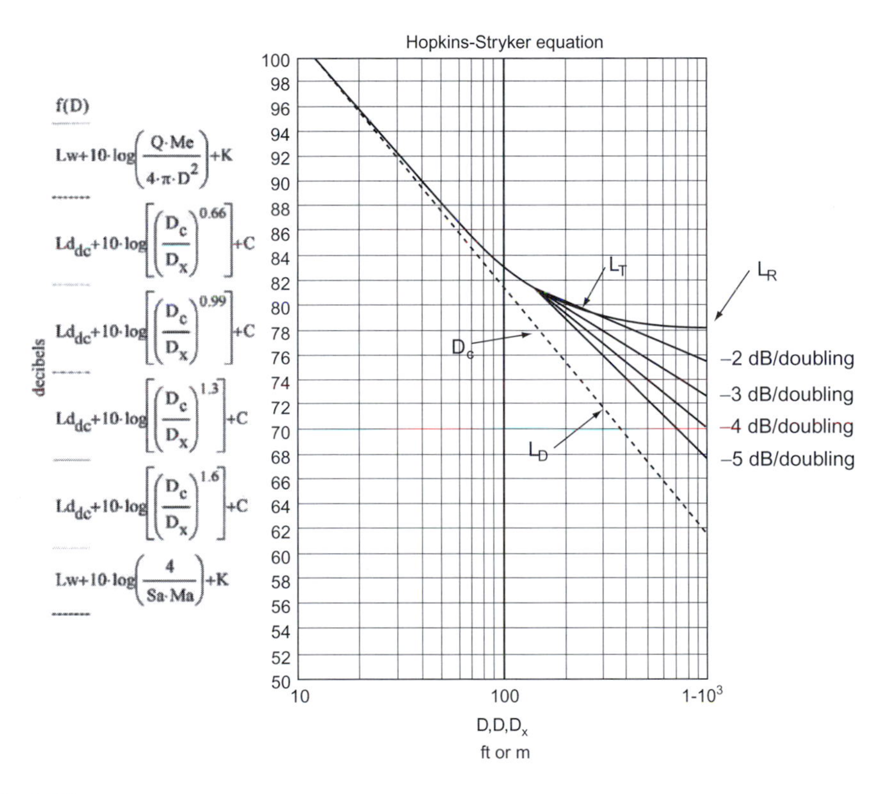

FIGURE 17.5C
MathCAD program depicting Hopkins-Stryker plots for varying values of attenuation with distance doubling for system design parameters in Figure 17.5b.

$$D_c = 0.5\sqrt{\frac{S\bar{a}}{sr}} \tag{17.28}$$

$$\text{Radians} = \textit{Degrees}\left(\frac{\pi}{180}\right) \tag{17.29}$$

$$\text{Degrees} = \textit{Radians}\left(\frac{180}{\pi}\right) \tag{17.30}$$

One advantage of using *sr* is that they can be summed to obtain a total *sr* for two adjoining areas. That this has been widely overlooked is apparent in the book, *Units, Dimensional Analysis and Physical Similarity* by D. S. Massey,

> The magnitude of solid angle may be related to that of plane angle to give the dimensional formula (A^2). However, as the results seem to have no practical use, we shall not discuss further the dimensional formula of solid angle.

It would seem that radar, sonar, and acoustics are not practical applications, in this author's mind. The book, nevertheless, is excellent and worthy of study.

The direct sound level alone can be expressed as

$$L_D = L_W + 10\log\left(\frac{Q}{4\pi r^2}\right) + 10.5\ \text{U.S.} \tag{17.31}$$

$$L_D = L_W + 10\log\left(\frac{1}{(sr)(r^2)}\right) + 0.2\ \text{SI} \tag{17.32}$$

and the reverberant sound level, L_R, from

$$L_R = L_W + 10\log\left(\frac{4}{S\bar{a}}\right) + 10.5\ \text{U.S.} \tag{17.33}$$

$$L_R = L_W + 10\log\left(\frac{4}{S\bar{a}}\right) + 0.2\ \text{SI} \tag{17.34}$$

Finally, critical distance, D_c, is obtained from the fact that when $r = D_c$

$$\begin{aligned} \frac{Q}{4\pi D_c^{\,2}} &= \frac{4}{S\bar{a}} \\ D_c &= \sqrt{\frac{QS\bar{a}}{16\pi}} \quad \text{both in U.S. and SI} \\ D_c &= 0.141\sqrt{QS\bar{a}} \\ D_c &= 0.5\sqrt{\frac{S\bar{a}}{sr}} \end{aligned} \tag{17.35}$$

At D_c the two levels, L_R and L_D are equal; therefore L_T will be +3 dB higher. For a sound system in an enclosed space it is highly desirable to keep as many listeners as possible at or closer to the sound source than D_c.

Bringing the sound source closer to the listener raises the L_D and often, because of being physically closer, inverse square law allows lower power, thus lower L_R. Also raising the directivity factor, Q, of the sound source results in higher L_D for lower power also. Raising the number of sabins in the space lowers L_R.

The accuracy of these equations will get you into the "ball park." They will bring you into the right order of magnitude. Ideally when the space already exists, testing with a sound source of the indicated Q allows much more exact numbers to be computed.

Prediction of sound levels outdoors and in very small non-reverberant spaces will follow inverse square law quite accurately.

Extremely large sporting arenas, domes, etc., have *l*s of such length as to again produce a low number of reflections.

DIFFERENTIATING BETWEEN REVERBERANT LEVEL AND REVERBERATION TIME

Thanks to present day measurement techniques both L_T and L_D are readily accessed.

When non-uniform reverberant sound fields are encountered, measurements made at repeated doubling of distance beyond critical distance, D_c, allows accessing the Hopkins-Stryker modifier for plotting level versus distance.

$$L_P = L_W + 10\log\left(\frac{Q}{4\pi D_x^{\,2}} - \frac{4}{S\bar{a}}\right) + 10\log\left(\left(\left[\left(\frac{D_c}{D_x}\right)^{\exp}\right]\right)\right) + K \qquad (17.36)$$

where,

$$\exp = \left(\frac{\text{dB/Doubling of distance}}{\frac{\log 2}{10}}\right),$$

L_W is the sound power level in dB,

D_x is any distance l beyond D_c,

$K = 10.5$ U.S., 0.2 SI

EVALUATION OF SIGNAL-TO-NOISE RATIO, *SNR*

In my experience you can very seldom turn room noise off, therefore the first measurements are those of room noise levels at various positions with the sound system on but inactive. Subsequent measurements of total level, direct level, and reverberant level can then be corrected for the noise level contribution in those instances where the noise level makes a significant contribution.

Having obtained L_T, L_D, and L_R at any given point, it becomes possible to separate the contribution of the noise level, L_N, by using the measurement of L_T with the noise "on" and then with the noise "off." $L_T - L_N$ is the signal-to-noise ratio expressed in dB. Noise "on" refers to lighting, HVAC, and other man-made noise sources.

In listening to an auditorium where measurements are to be made, always attempt an aural estimate of each of these levels at differing locations at both less than the expected D_c and well beyond D_c. Good sound system design practice tries to minimize the reverberant level while recognizing that changing the reverberation time is a room treatment problem. When a 25 dB *SNR* at 2 kHz is unattainable, recommend the job to a competitor.

ANALYZING REFLECTIONS AND THEIR PATHS

In sound system analysis, sets of linearly spaced notches and peaks in the amplitude response are called "comb filters." These are the result of a reflection or reflections converging with the direct sound from the desired sound source. Even a low resolution ⅓ octave real-time analyzer can see the lower frequency notch. The reflective path distance, *rp*, can be ascertained from:

$$rp = \frac{0.5c}{F_n} \qquad (17.37)$$

where,

c is the velocity of sound. The velocity unit, l/s, can be in ft, inches, mm, M, etc. The reflection distance is in the units chosen,

F_n is the frequency of the first notch.

Because comb filters are spaced linearly, NR_n or NF_p (first peak frequency), this allows the computation of all higher frequency comb filters. For example, a reflection 0.5 mm such as the spacing of a Pressure Zone Microphone, PZM, capsule above its plate results in a first notch at

$$F_n = \frac{0.5c}{diff} = \frac{0.5 \times (344 \times 10^3)}{0.5\,\text{mm}} = 344{,}420\ \text{Hz}$$

which could be disregarded in normal use. If, on the other hand, a low pass filter at 20 kHz, for antialiasing purposes, could be begun acoustically near the bandpass by

$$\text{Spacing} = \frac{0.5c}{diff} = \frac{0.5 \times (13{,}560)}{20{,}000\ \text{Hz}} = 0.34\ \text{inches}$$

Figures 17.6a and 17.6b illustrate the usefulness of comb filters as well how to analyze harmful ones.

Sound system near regeneration

A sound system operating too near positive acoustic feedback amplifies the room's natural reverberation time by many times (as many as four to five times). Be sure that your subjective judgment of the space has not been influenced by the presence of a malfunctioning sound system.

REFLECTIONS AND COMB FILTERS

$d := 75$ L direct sound path $\quad$ $c := 1130$ L /sec

$h := 30$ L source height $\quad$ $a := 0.5$ dimensionless

Comb filters space linearly therefore nf_n and nf_p calculate the higher frequency notches and peaks.

$d_1 := \sqrt{d^2 + (2 \cdot h)^2}$ $\quad$ $d_1 = 96$ L reflected source path

$f_n := \frac{0.5 c}{d_1 - d}$ $\quad$ $f_n = 26.8$ Hz lowest notch frequency

$f_p := \frac{c}{d_1 - d}$ $\quad$ $f_p = 53.7$ Hz first peak frequency

$diff := \frac{0.5 c}{f_n}$ $\quad$ $diff = 21$ L path difference

$d := d_1 - \frac{0.5 c}{f_n}$ $\quad$ $d = 75$ L (used when source distance is unknown but reflected path and f_n are available.)

$d_1 := d + \frac{0.5 c}{f_n}$ $\quad$ $d_1 = 96$ L (used when source distance is known and f_n are available.)

First peak occurs at N λ, first notch at ((2N-1) / 2) λ (Where λ is wavelength)

$10 \cdot \log\left[\frac{d}{d_1}(1 - a)\right] = -4.1$ dB of D/R ratio

Finding distance to nearest reflective surface from lowest notch frequency.

$f_{n1} := 2000$ Hz lowest notch frequency . $\quad$ $c_1 := 13560$ L /sec

$rp := \frac{0.5 \cdot c_1}{f_{n1}}$ $\quad$ $rp = 3.39$ L Reflective path distance (rp) Dimension determined by units chosen for c (ie: ft,in,mm,cm,m,etc.)

A. MathCAD program showing reflections and comb filter equations.

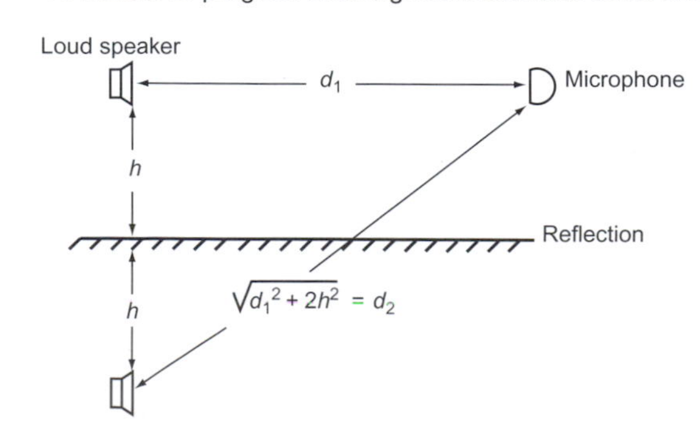

B. How to determine path lengths.

FIGURE 17.6
Reflections and comb filters.

See Figure 17.7 from the work of Bill Snow of Bell Labs. Note that the decay rate changes from 92 dB /s to 22 dB /s as the sound system is brought near sing point (feedback), a 4.2 magnification factor.

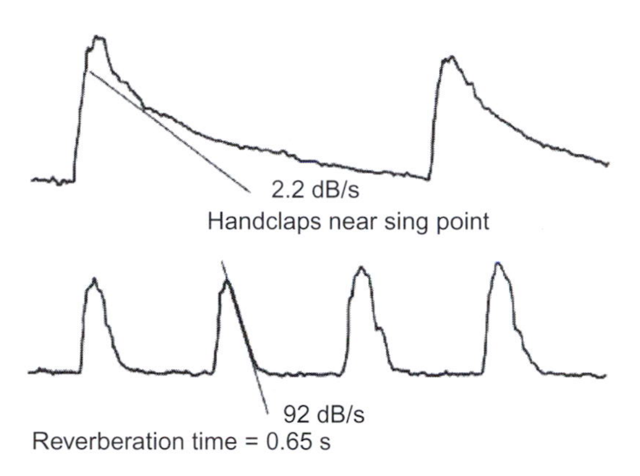

FIGURE 17.7
Sound system near feedback. (*Courtesy Bill Snow*).

How harmful is high RT_{60}?

How harmful is a high RT_{60}? In a truly diffuse acoustic field it is surprisingly not harmful to "live" speech until reverberation times around 3 s to 4 s at 2 kHz are reached. The trouble in many "reverberant" spaces is that it is focused energy returns over long path lengths that are the culprit and not the length of time they take to decay. This can usually be demonstrated by using a large sheet of Sonex (4 ft × 5 ft) and circling the listener while a talker speaks from the podium. When the Sonex passes between the focused energy and the listener, speech intelligibility will return. Remove the Sonex and the speech will again be interfered with. By noting the position of the Sonex relative to the listener and the room surfaces, you can usually detect the offending area. Cupping your ears and moving around the room surfaces will also tell much about reflections in the room.

Troy Savings Bank Concert Hall has a 3+ second reverberation time (empty), yet normal conversation level can be heard clearly from the stage at the back of the upper balcony. Troy Concert Hall has no absorption in a conventional sense (all wood) but is very diffuse.

Effect of reverberation on intelligibility

The effect of reverberations on intelligibility is far less than single late high level reflections, inadequate *SNR*, or comb filters generated by sources within one foot or less of the primary source.

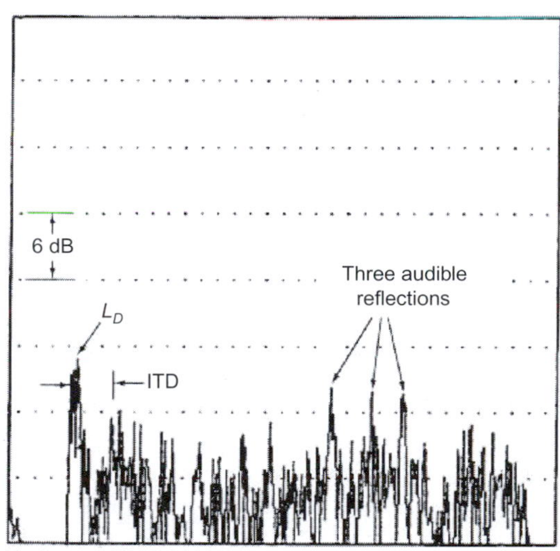

FIGURE 17.8
Three audible late reflections that seriously affected articulation.

Figure 17.8 shows three highly audible late reflections (over 100 ms). We were told that this space was so reverberant that it was difficult to use the reinforcement system. Yet, when we put Sonex around the person in this seat and isolated him from the reflections, one could understand clearly the unaided voice from the stage. Without the Sonex, speech was unintelligible.

Variations in the measurement of reverberation

TEF analysis has revealed some unexpected details in the measurement of reverberation time. Anyone working with high *Q* transducers (*Q*s of 50+) has subjectively experienced the apparent change in RT_{60}. This has always been explained as due to a lower reverberant sound field and consequently what decay of energy was present did not last as long before being masked by the ambient noise floor. Various orientations of the sound source reveal that the way a high *Q* source should be oriented to cover an audience does not excite as many normal room modes as does a lower *Q* device. Thus it would appear from the evidence available that:

1. The classic method is flawed by being *Q* dependent.
2. The higher *Q* sources literally do not excite all the room modes.
3. We possibly need a new, as yet undetermined, method of analyzing the envelope time behavior of large rooms.

Figure 17.9 shows ETCs of the same room excited from the same source location by three different sources. Source number one has a $Q = 1$. Source number two has a $Q = 5$. Source number three has a $Q = 50$.

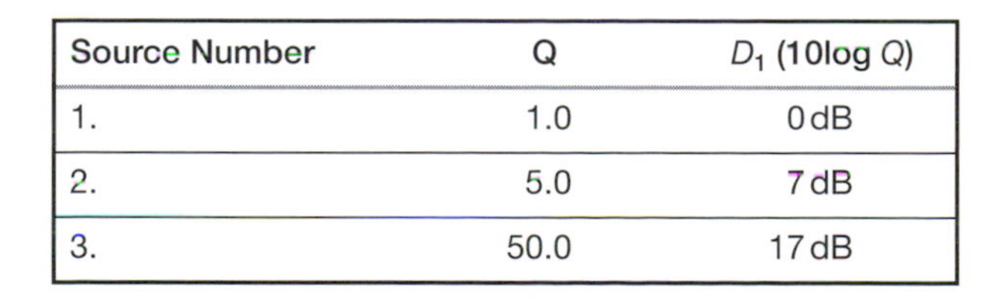

Source Number	Q	D_1 (10log *Q*)
1.	1.0	0 dB
2.	5.0	7 dB
3.	50.0	17 dB

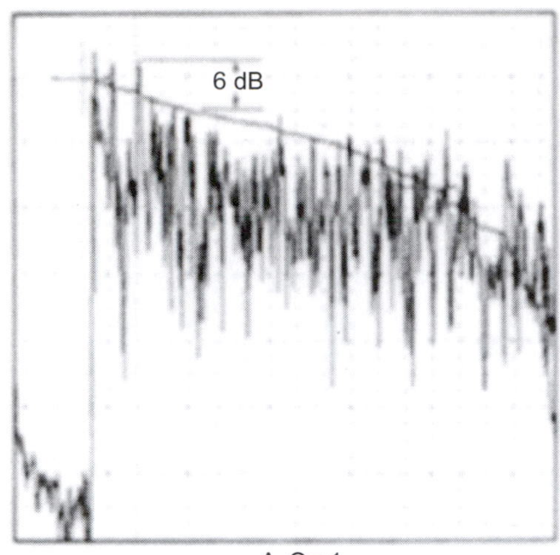

A. Q= 1.

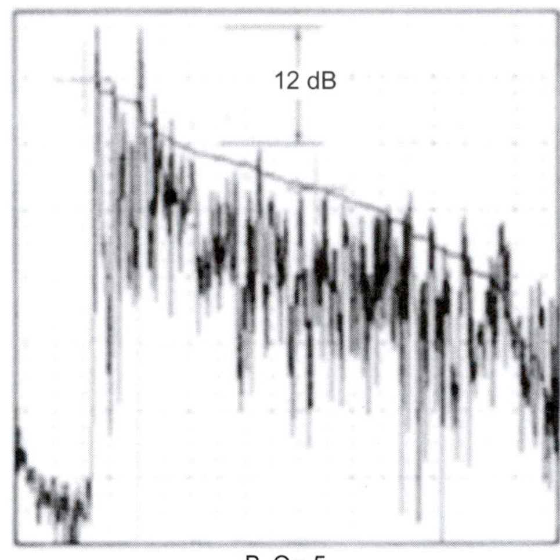

B. Q= 5
(Note the drop in reverberant leve lover Q= 1).

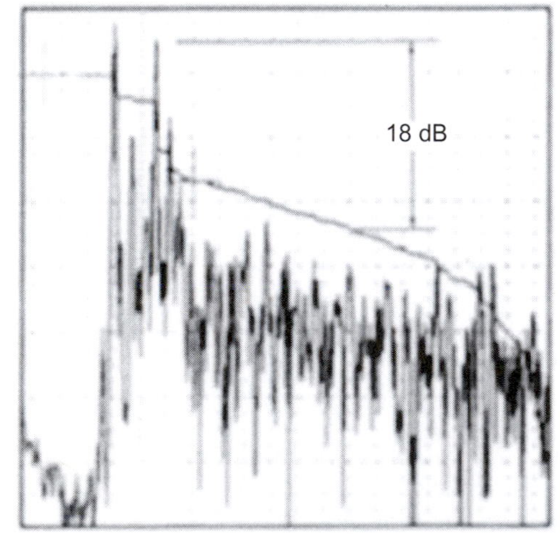

C. Q= 50 with a substantial drop in level of the reverberant field.

FIGURE 17.9
ETCs made in a room excited from the same source location by three different sources.

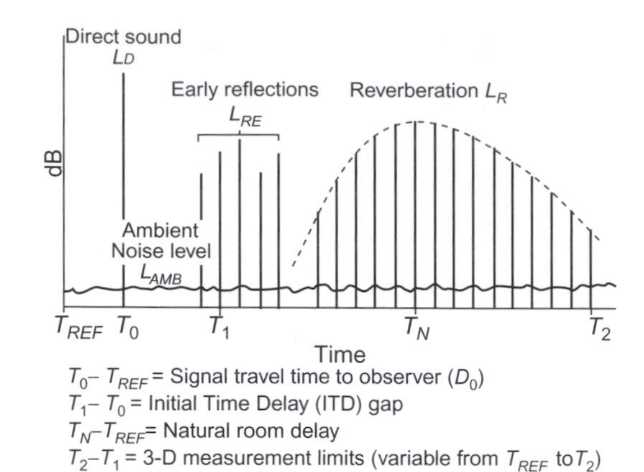

FIGURE 17.10

It cannot be overemphasized that analysis reveals that a large number of smaller volume, acoustically absorptive rooms do not develop a reverberant sound field that rises above the ambient noise floor normally present in a space. In such spaces statistical analysis is an exercise in futility. Study of the fine structure of the early reflection, L_{RE}, is of benefit and leads directly to improved performance.

To obtain accurate information about the presence or lack of a reverberant sound field, both sufficient distance from the source must be established and sufficient time must be provided for it to develop, see Figure 17.10.

CRITICAL DISTANCE

One of the most important concepts regarding the statistical acoustic space is D_c. First, we assume that D_c is within 3 dB of the maximum acoustic separation between the microphones in a given room. Again, if we have a microphone in a steady reverberant field, we could wander all over the reverberant area without encountering a sudden change in level that can cause feedback. In fact, we will make use of D_c in determining the following limits in our design of sound systems:

1. The loudspeaker and the microphone should be at least as far apart as D_c. $D_1 = D_c < 45$ ft, where D_1 is the distance between the loudspeaker and the microphone.
2. In rooms with a reverberation time exceeding 1.6 s you will not be able to have any listener beyond 3.16 D_c. As the time raises more, this multiplier will become even lower. (Also see Figure 17.5c.)

Q versus $S\bar{a}$ for controlling D_c

In examining the equation for D_c, it is apparent that both Q and $S\bar{a}$ have the same relative weight. This means that in a space that requires a doubling of $S\bar{a}$ to be acceptable, we could just as well leave $S\bar{a}$ alone and double Q. In typical church systems, for example, the doubling of $S\bar{a}$ can easily cost $100,000 and change the entire visual appearance of the structure as well as making the music director very unhappy. Doubling Q usually costs under $10,000. While the array may be huge, it does occupy only one spot and not whole walls and ceilings. This is a relatively new concept and not widely practiced, though certain acoustic consultants have effectively used the general idea for years. We can now enumerate a few of the factors proceeding from the existence of D_c in a space:

1. D_c determines maximum acoustic separation hence maximum acoustic gain.
2. D_c determines the ratio of direct-to-reverberant sound.

3. D_c determines the required directivity of the loudspeaker in an already existing room.
4. D_c can determine the required room characteristics in a space being planned if a chosen loudspeaker is desired.

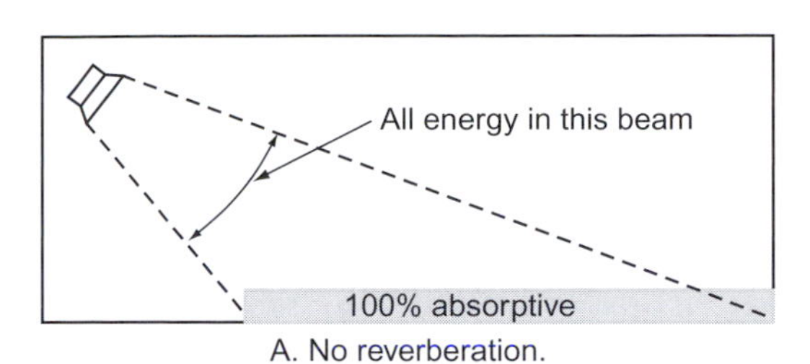

A. No reverberation.

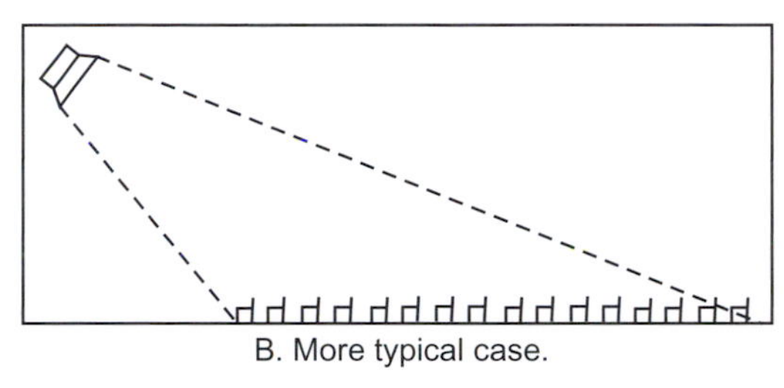
B. More typical case.

FIGURE 17.11
D_c multiplier, *Ma*, illustrated.

D_c multipliers and dividers

Just as N operates as a D_c divider, there are factors that can operate as D_c multipliers. First, let's take an extreme case in which the $C_\angle$ contains all the useful energy. It is aimed at an audience area that is 100% absorptive as shown in Figure 17.11a. For example, if the loudspeaker has an axial Q of 5, the room has an $\bar{a}$ of 0.01, and the audience area has an $\bar{a}$ of 1, the apparent Q would be

$$Q_{App} = Q_{Axial}\left(\frac{1-\bar{a}\text{ of total room}}{1-\bar{a}\text{ of audience area}}\right) = 5\left(\frac{1-0.01}{1-1}\right) = \infty \tag{17.38}$$

Even if the room were highly reverberant, this loudspeaker would cause no reverberation.

A more typical case is shown in Figure 17.11b. In this case the loudspeaker still has an axial Q of 5, the room has an $\bar{a}$ of 0.16, and the audience area has an $\bar{a}$ of 0.32. The apparent Q would be

$$Q_{App} = Q_{Axial}\left(\frac{1-\bar{a}\text{ of total room}}{1-\bar{a}\text{ of audience area}}\right) = 5\left(\frac{1-0.16}{1-0.32}\right) = 6.18$$

Theory of *Ma*

The D_c modifier (*Ma*) results from the removal of additional energy from the signal emitted upon its first encounter with a selected absorbent boundary surface than would have been expected had the same energy first encountered a surface possessing the average absorption coefficient of the space as a whole.

In the limiting case, if the area the sound energy first encountered were 100% absorptive and if none of the energy encountered any other surface, there would be no reverberation. Thus, such an *Ma* makes the source act as if it were in a free field. It would appear that *Ma* is only of interest for some very high Q devices.

Consider what is actually happening at a listener's ears. As *Ma* increases the ratio of direct-to-reverberant sound heard increases if the listener is situated on the absorbing surface that the energy first encounters. The main purpose of increasing Q or *Ma* is to increase this ratio at the listeners ears.

Another parameter available to us that allows us to accomplish the same results is to move the loudspeaker closer to the listener. If we move a loudspeaker of any given Q in a room of any given *Ma* to half its former distance from the listener's ears, we raise the direct-to-reverberant ratio by 6 dB. This could also be accomplished by leaving the loudspeaker at its original position and raising its Q by a factor of four.

The relation of the loudspeaker's angular discrimination relative to a microphone's angular discrimination is referred to as an electroacoustic modifier of D_c (*Me*).

Additionally, a tilted rear wall may raise the apparent Q at the rear seats in an auditorium even more. When you measure Q in a room with a calibrated loudspeaker, the difference between what you calculate and the calibration includes all the multipliers and dividers. Therefore, such a calibrated test source allows you to do several things:

1. Measure the room absorption, $S\bar{a}$, by the reverberation time method and calculation. Then measure the apparent Q. This will include all multipliers and dividers. By changing such variables as position of materials and position of sources, you can investigate the effect of such phenomena.
2. Measure the room absorption, $S\bar{a}$, by the critical-distance method using the Q of your calibrated sources in your calculations. By using this room absorption, which includes all the multipliers and dividers, you can accurately measure the axial Q of unknown loudspeakers.

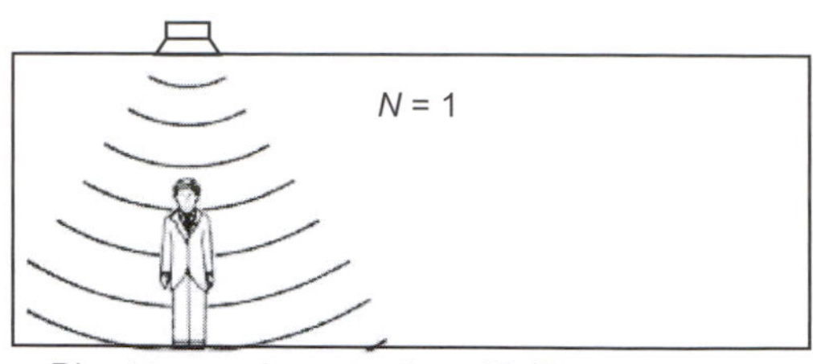

Direct to reverberant ratio = 17 dB

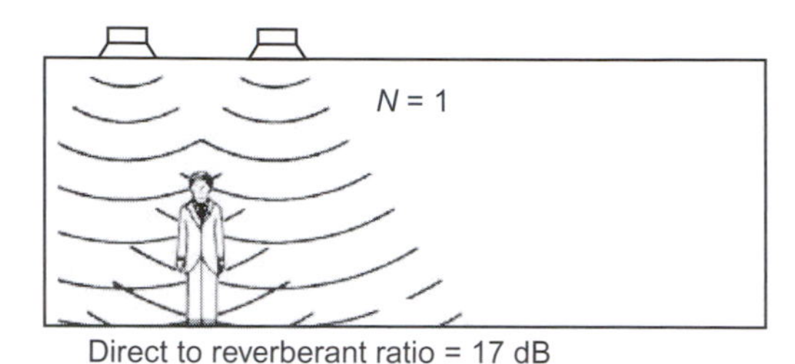

Direct to reverberant ratio = 17 dB

N = 2

Direct to reverberant ratio = 14 dB

FIGURE 17.12
Visualization of *N*.

The effect of the N factor

N is the ratio of the acoustic power produced by all loudspeakers to the acoustic power produced by the loudspeaker or loudspeaker groups providing the listener with direct sounds. In its simplest form as shown in Figure 17.12, two loudspeakers are furnishing direct sound to the listener and there are four loudspeakers producing equal acoustic power. Therefore

$$N = \frac{\text{Total number of loudspeakers}}{\text{Number of loudspeakers producing direct sound to the listener}} = 2.$$

To demonstrate the "*N* " effect, sound fields were compared using a loudspeaker on stage as a substitute for a talker into the sound system microphone, Figures 17.13 and 17.14.

1. Loudspeaker on stage only (a talker from the stage).
2. Loudspeaker on stage plus a center cluster ($Q = 11$).
3. Loudspeaker on stage plus two low-*Q* stereo loudspeakers ($Q = 2$).
4. All loudspeakers on at the same time.

The effects of the complex *N* factor are clearly evident. Some interesting effects from discrete early reflections that affect the ratio of direct (defined here as the first 50 ms) sound level versus reverberant sound level are also apparent. The effect of *N* is a 10log function. If all four sources developed equal acoustic power then the expected deterioration would be 10log $5 \approx 7$ dB.

This rule follows within a few decibels at all sampled locations, see Table 17.1.

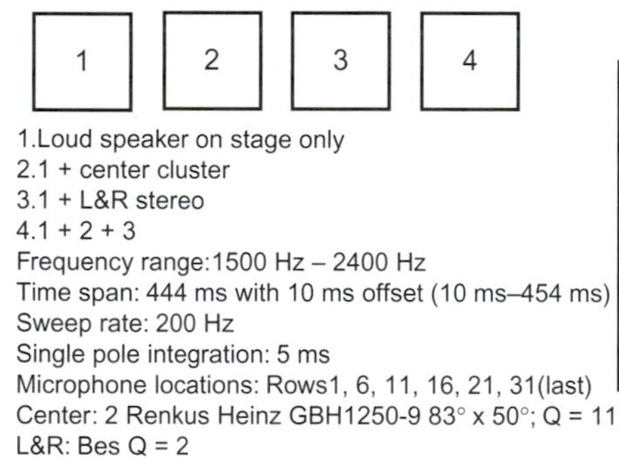

Measurement parameters.

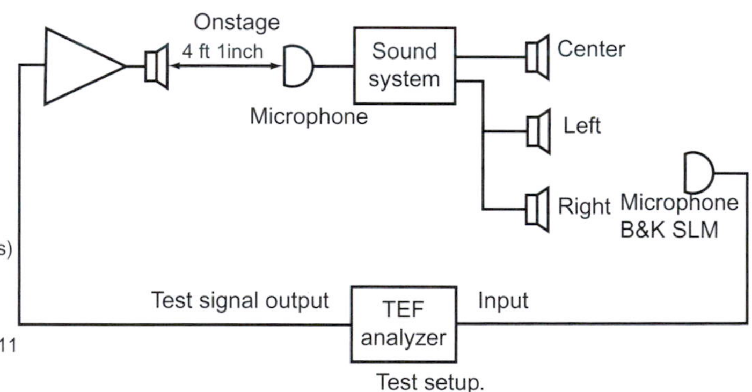

Test setup.

FIGURE 17.13
Measurement parameters for sound system study in Fig. 17.14.

Table 17.1 Summary of Direct to Reverberant Ratios

Row	Stage	Center	Stereo	All
1	14	6	4	4
6	14	7	3	3
11	13	5	5	4
16	12	4	5	3
21	10	9	0	4
26	9	6	4	3
31	8	4	1	1

Direct = 0–50 ms
Reverberant = 50 ms–444 ms

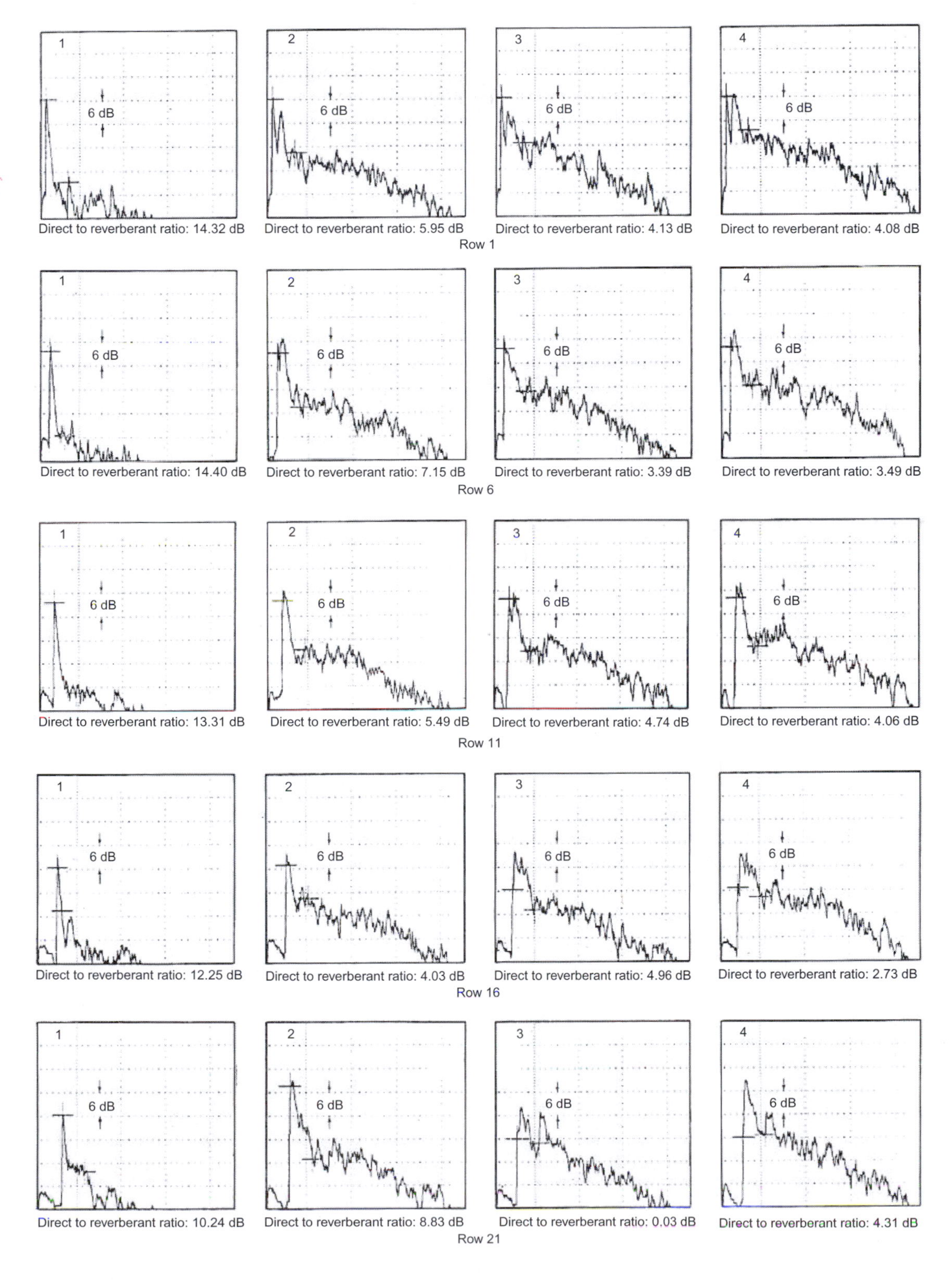

FIGURE 17.14
Complete study of a sound system where the direct-to-reverberant ratio is degraded as each loudspeaker is added. *(Prepared by Rollins Brook of BBN).*

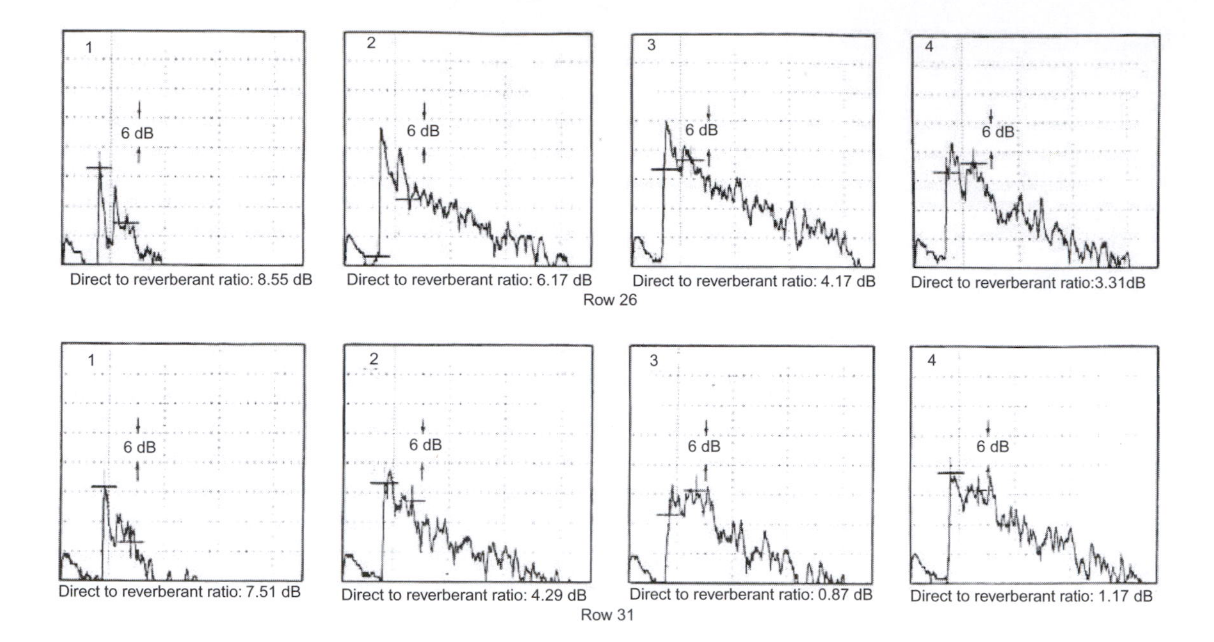

FIGURE 17-14
(continued).

Factors to watch for in rooms

The following factors can be serious trouble for the sound system if they are not properly controlled:

1. Curved surfaces, especially concave curved surfaces.
2. Absolutely parallel walls. Such walls cause flutter-echo. A splay of 1 inch per foot will avoid this problem.
3. Absorption on the ceiling. Unless the ceiling is very high (over 60 feet), the placement of absorption on it means the sound system has lost some useful reflecting surfaces. Absorption belongs on rear walls (large spaces only), rear ceilings, in the seats, etc.
4. Potential ambient noise sources—air handlers, unenclosed machinery, etc.
5. Extra wide or round audience seating.

CONCLUSION

The meaningful use of acoustical absorption is not limited to its statistical application. Indeed sound systems are more often installed in spaces where the statistical equations are invalid than in spaces where they are valid.

In semireverberant spaces and in very "dead" spaces the use of absorption to control discrete specular reflections is quite valid in spite of the application of the material having no real meaning in a statistical sense. Even in spaces where the statistical equations are valid, intelligibility can be, and often is, degraded by a specular reflection which must, of course, be isolated and corrected directly, not statistically. Therefore, it is well to bear in mind that many "large rooms" have some small room properties at certain frequencies, especially with regard to specular reflections.

BIBLIOGRAPHY

Davis, D., 1980. Contemporary electro-acoustic investigations. Syn-Aud-Con Tech Topics 7 (11).

Davis, D., 1980. Uses, abuses, and misuses of the critical distance concept. Syn-Aud-Con Tech Topics 7 (12).

Davis, D., Davis, C., 1985. What reverberation is and what it is not. Syn-Aud-Con Tech Topics 12 (13).

Joyce, W.B., June 13, 1979. Power series for the reverberation time. Paper presented at 97th Convention of the Acoustical Society of America. Cambridge, MA.

Joyce, W.B., 1975. Sabine's reverberation time and ergodic auditorium. J. Acousti. Soc. Am. 58, 643–655.

Klepper, D.L., Aug. 1970. Sound systems in reverberant rooms for worship. J. Audio Eng. Soc. 18.

Klepper, D.L., 1979. Improved reverberation time calculations. Syn-Aud-Con Tech Topics 6 (14).

Norris, R.F., 1932. Appendix II: a discussion of the true coefficient of sound absorption—a derivation of the reverberation formula. In: Architectural Acoustics. Wiley, New York, pp. 603–665. . .

Rayleigh, B., 1945. The Theory of Sound, second ed. Dover, New York.

Sabine, W.C., 1922. Collected Papers on Acoustics. Harvard Univ. Press, Cambridge, MA.

Schroeder, M.R., 1969. Computers in acoustics: symbiosis of an old science and a new tool. J. Acoust. Soc. Am. 45 (5).

Shankland, R.S., Oct. 1973. Acoustics of greek theaters. Phys. Today pp. 30–35.

Snow, W.B., April 1955. Frequency characteristics of a sound reinforcing system. J. Acoust. Soc. Am..

PREFACE

The second chapter in this section on Acoustics and Sound Reinforcement is also by Don and Carolyn Davis, and deals with the problem of equalizing the sound system and its associated acoustic environment to obtain the desired frequency response, and suitable freedom from feedback (or "howlround") and transient ringing. This topic is of fundamental importance, because if a system has an irregular and peaky gain response, the tallest peak will put the limit on how much gain can be used before feedback. If the response is smoothed then the overall gain can be increased at all frequencies and the system is not only being used with much greater efficiency, but its sound will be far more acceptable.

Don and Carolyn first examine the ways in which the system response can be measured. The response of a sound system and its environment is usually assessed with a real-time analyzer, which displays the audio spectrum as a series of levels from a set of filters that divide up the audio band, for example into thirds of an octave. Not all acoustic problems can be solved simply by adjusting the frequency response with filters; for example the response irregularities caused by physical offset of loudspeaker units can only be compensated for by a time delay.

They then move on to look at the filters used to perform the equalization. These are usually bandpass or band-reject (notch) filters, but transversal filters, which work by combining the outputs from a tapped delay line, have also been used. While the first transversal filter appears to have been described in 1940 by Kallmann, constructing them in the analog domain is awkward and expensive. With the advance of digital audio, the transversal filter, now more often called an FIR or Finite-Impulse-Response filter, has become much, much easier to realize.

When a bank of filters are used for equalization, they are very often divided into third-octave bands, as for a real-time analyzer. These are more commonly called graphic equalizers, as cut or boost is determined by a row of vertical slide controls. Since their filter bands are fixed in frequency, graphic equalizers are often supplemented by a parametric equalizer, similar to those used in mixing consoles, which have infinitely variable parameters, i.e., cut/boost, center frequency and bandwidth. This allows a notch or boost curve to be placed exactly where it is required in the audio band.

Don and Carolyn then consider what it takes to make a good filter system, and the importance of using minimum-phase filters, which introduce as little phase-shift as possible for a given amplitude correction response. They next deal with filter transient responses, and how they relate to the critical bands shown by human hearing. These bands are about one-third-octave wide, which explain why that bandwidth is so popular both in analyzers and filters.

Finally, they go through the extra equalization problems that can occur when dealing with a sound-reinforcement system with microphones. Microphones have their own frequency-response irregularities, and also exhibit proximity effects, such as the well-known rise in bass response as the talker gets closer to the microphone. There can also be stability problems because reflections from the talker's body create a comb-filter effect. Don and Carolyn make the point that it is essential to check that microphone polarity is correct before wading in with equalization.

Sound System Equalization

Sound System Engineering by Don and Carolyn Davis

The original one-third octave band rejection filter set utilizing summing circuitry was first used by one of the authors in 1967 and the patent 3,624,298 was filed in March 1969 and issued in November 1971. Since its inception the basic problems in its correct use have been two fold: first, the ability to design a sound system capable of benefiting from the use of an equalizer, and second, the attempt to equalize the unequalizable. These problems are with us more than 40 years later.

SYSTEM CRITERIA

Equalization can't solve loudspeaker coverage problems. Equalization can't signal align loudspeakers. Equalizers can't raise acoustic gain unless the system has adequate power available to support the gain increase. Equalizers are of no use in controlling reverberation, discrete echoes, etc.

Careful practice can minimize aggravating these problems via regeneration through the sound system. An equalizer can adjust the direct sound pressure level of the loudspeaker's minimum phase output frequencies. This is accomplished by providing the conjugate amplitude and phase response to any minimum phase aberrations in the loudspeaker's direct sound level, L_D.

Proper equalization adjusts both amplitude and phase to a more uniform response. A delay in microseconds is introduced by the insertion of the filters. The measurement of both amplitude and relative phase is essential in the process of equalization. Group delay for a phono record can be years. The delay through some adaptive digital filters can be appreciable, +30 ms.

The triumvirate of proper equalization, signal synchronization, and seamless coverage is a very powerful tool used to create extraordinary sound quality.

If the system is designed capable of benefiting from equalization, constructed and installed so that coverage is of the proper density, the electrical power is adequate and matched to sufficiently efficient transducers able to absorb it, and the entire system is free from hum, noise, oscillations, and RF interference, you are ready to equalize this system in its acoustic environment to ensure the specified tonal response and acoustic gain at each listener's ears. To do this requires insertion of the necessary filters into the sound system and the taking of meaningful acoustic measurements.

EARLY RESEARCH ON EQUALIZATION

Insofar as the authors can discover, the earliest researcher to correctly perform meaningful sound system equalization was Dr. Wayne Rudmose, who at the time of the work to be described was at the Southern Methodist University, Dallas, Texas. Dr. Rudmose published a truly remarkable paper in the journal, *Noise Control* (a supplementary journal of the *Acoustical Society of America)* in July 1958.

Audio Engineering Explained

We feel the two most authoritative papers ever published on this subject that have retained their fundamental integrity many years later are Dr. Rudmose's and that of one of his students and later an employee of Tracor, Inc. of Austin, Texas, William K. Conner.

Conner's paper "Theoretical and Practical Considerations in the Equalization of Sound Systems" first given at the 1965 AES Convention, appeared in the April 1967, Vol. 15, No. 2 issue of the *Journal of the Audio Engineering Society.* Amid all the nonsense written on this subject, these two papers stand as bedrock for the serious investigator.

In the fall of 1967, one of the authors gave the first paper on a ⅓ octave contiguous equalizer. Wayne Rudmose was the chairman of the session and when the author referred to ⅓ octave as "broad band" Rudmose raised his eyebrow and said *"broad band??"* In 1969, a thorough discussion of acoustic feedback that possessed absolute relevance to "real life" equalization appeared in the *Australian Proceedings of the IREE.* "A Feedback-Mode Analyzer/Suppressor Unit for Auditorium Sound System Stabilization" by J. E. Benson and D. F. Craig, illustrated clearly the step function behavior of the onset and decay of regeneration in sound systems. The aforementioned papers relate the most directly to modern practice and the application of current devices to the adjustment of sound system response. These four sources constitute the genesis of modern sound system equalization.

Rudmose went on to not only introduce ⅓ octave band analyzers, but to describe correctly the cause of "howlback," room resonance effects, cancellations, and the "ringing" encountered in sound systems. In this same paper Rudmose described the devastating effects of transducer misalignments—large holes in the amplitude response and the basic importance of obtaining uniformity of distribution prior to trying to equalize a sound system.

William K. Conner's paper, originally given in 1965, clearly delineated the role of Q and distance on the L_D and $S\bar{a}$ and L_w on L_R. The first thoroughly correct statements regarding useful ratios of L_D/L_R and how to achieve them are presented here. Conner's paper identified the effect of humidity on the sound system's response. Also he pointed out that directional microphones have little to no effect on the power loop gain of the system.

Feedback defined

> Feedback: *"The return to the input of a part of the output of a machine, system, or process."*

Feedback can be positive or negative in sound systems. Positive feedback, carefully employed, can raise gain. Negative feedback in amplifiers can lower distortion. Hearing yourself speak over a sound system is one form of feedback. It can be beneficial over a monitor loudspeaker or detrimental when the delay causes the talker to stutter. This is a psychoacoustic effect. Some talkers instinctively lower their level upon hearing the monitor; others raise their level, usually a matter of previous exposure.

Oscillatory feedback occurs when the signal from the loudspeaker returns in phase to the open microphone at a level equal to the normal input level resulting in a single frequency "howling" tone. Shock excitation of a sound system on the threshold of sustained feedback may excite many simultaneous tones. The decay of these tones following excitation may be observed on a fast real time analyzer. These tones can then be compensated for one at a time by sequentially introducing equalization at the appropriate frequencies beginning with the tone exhibiting the slowest decay.

The mathematical description given here is a carefully specified single frequency example to illustrate the complexity of the circuitous path. The advent over the past generation of more controlled frequency and polar responses has led to less need of equalizers for the control of feedback. Engineering trade-offs, such as controlling directional response at the expense of smooth frequency response, can lead to the legitimate use of an equalizer.

Benson and Craig's detailed explanation of the fundamental mechanism behind acoustic feedback was first published in the March 1969 issue of the *Proceedings of the IRE of Australia.* The paper was entitled "A Feedback-Mode Analyzer/Suppressor Unit for Auditorium Sound System Stabilization." Their analysis made use of the results from an earlier paper by H. S. Antman entitled "Extension to the Theory of Howlback in Reverberant Rooms" that was published in *J. Acoust. Soc. Am.*, Vol. 39, No. 2, February 1966, p. 399 (Letters). A careful study of both of these papers is highly recommended. An abbreviated discussion of the more salient points of these papers related to the transient nature of the acoustic feedback process is presented in the next section.

THE TRANSIENT NATURE OF ACOUSTIC FEEDBACK

Acoustic feedback occurs whenever an open system microphone is exposed to the sound field of the system's loudspeaker array. This means that except for very unusual microphone locations, acoustic feedback is always occurring. What is important, apparently, is not whether acoustic feedback is occurring but rather of the type and degree of the feedback. It is well known that negative or degenerative feedback when judiciously applied in an amplifier can have a desirable stabilizing influence on the amplifier's performance at the expense of reducing the amplifier's overall gain. On the other hand, positive or regenerative feedback applied to an amplifier destabilizes the amplifier's performance, increases its gain, and, if present to a sufficient degree, leads to sustained oscillation. Acoustic feedback in a sound system behaves in a very similar way. The major distinctions between the amplifier case and the sound system case have to do with the feedback path. In the amplifier case the feedback path is usually well defined and the transit time through the feedback path is usually negligibly small. In the sound system case the feedback paths are numerous with appreciable transit times that are dependent upon physical path length.

Consider the simplified situation depicted in Figure 18.1.

Growth

In Figure 18.1, μ represents the feed forward transfer function of the system. It is made up of the product of the transfer functions of the microphone, amplifier, and loudspeaker. In the absence of any feedback, if p_1 is the acoustic pressure at some reference distance from the microphone, then the acoustic pressure at a similar reference distance from the loudspeaker would be $p_2 = \mu p_1$. In the general case μ would be a complex function of frequency. Similarly, in Figure 18.1, β represents the feedback transfer function. It, too, is a complex function of the frequency in the general case. As mentioned earlier there are normally many parallel feedback paths, each with their own particular value of β. In our simplified example we will consider only a single feedback path along which there exists a transit time of amount $\Delta t = d/c$ where d is the path length and c is the sound speed. When p_1 is first applied to the microphone, the loudspeaker almost instantaneously produces an acoustic pressure $p_2 = \mu p_1$. After the elapse of an interval of time Δt, a feedback signal of amount $\beta p_2 = \mu\beta p_1$ arrives at the microphone.

Assuming that the original acoustic pressure is still present, the signal at the microphone now becomes $p_1 + \mu\beta p_1 = p_1(1 + \mu\beta)$ and the output now instantaneously jumps to $p_2 = \mu p_1(1 + \mu\beta)$. After the elapse of a second time interval Δt, the input becomes $p_1 + \mu\beta p_1 + (\mu\beta)^2 p_1 = p_1[1 + \mu\beta + (\mu\beta)^2]$ and the output instantaneously becomes $\mu p_1[1 + \mu\beta + (\mu\beta)^2]$. At this point it is safe to generalize the result for an arbitrary number of delay intervals N. After N delay intervals or a total time $N\Delta t$, the output pressure will be given by $\mu p_1[1 + \mu\beta + (\mu\beta)^2 + (\mu\beta)^3 + \ldots + (\mu\beta)^N]$. The analysis so far has placed no restrictions on either μ, β, or on the product $\mu\beta$.

They can each be either real or complex. The remainder of the discussion is greatly eased by requiring only that $|\mu\beta| < 1$, that is, the absolute magnitude of the product of μ with β be less than 1. When this is true the series describing the output converges to the value $\mu p_1[(1 - (\mu\beta)^{N+1})/(1 - \mu\beta)]$. Remember that this output pressure was originally the result of an input pressure signal of p_1. Now if N becomes very large, the term $(\mu\beta)^{N+1}$ becomes vanishingly small and the output divided by the original signal input, system transfer function or gain, becomes $\mu/(1 - \mu\beta)$. This last expression is exactly of the same mathematical form as that for a feedback amplifier. We can simplify the analysis from here on and still obtain meaningful results by letting p_1 be the acoustic pressure associated with a single frequency tone such as 1000 Hz and require that the feedback path is of such length that the return signal is in phase with p_1. This would constitute a case of pure positive feedback. With such a restriction,

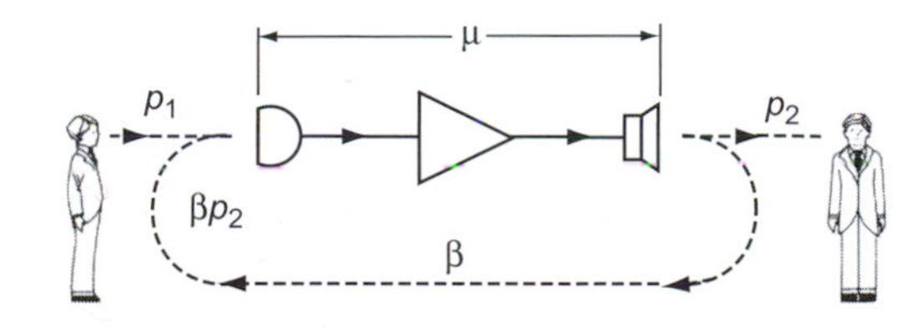

FIGURE 18.1
Schematic diagram of an elementary sound system with an acoustic feedback path.

both μ and β can be taken as real quantities. Let $\mu = 20$ and $\beta = 0.03$. The steady state gain with feedback becomes $20/[1 - (20 \times 0.03)] = 50$. The gain in the absence of feedback is of course μ or in this case 20. The gain ratio is the gain with feedback divided by the gain without or $50/20 = 2.5$. The gain increase brought about by this positive or regenerative acoustic feedback expressed in decibels is $20 \log 2.5 = 8$ dB. The effect of this regenerative feedback has been two-fold. Not only has the steady state gain been increased by a really significant amount but also the output arrives at its ultimate value through a series of steps.

Decay

The decay is initiated by the removal of the input p_1. When p_1 is removed, the output drops suddenly by an amount μp_1 while the input continues to be supplied by the feedback component βp_2 that had left the loudspeaker before p_1 had been removed. This continues for a time Δt during which the output remains at the value $\mu\beta p_0$ where now p_0 is the value of the output at the time p_1 was removed. At the end of the interval Δt, the input suddenly drops to $\mu\beta^2 p_0$ thus producing a new output $(\mu\beta)^2 p_0$. This new output persists again for a time Δt. In this fashion the output falls in a series of steps such that after the elapse of N intervals it has become $(\mu\beta)^N p_0$ and the gain ratio has become $(\mu\beta)^N p_0/\mu p_0$.

Figure 18.2 displays the growth and decay of regenerative acoustic feedback for $\mu = 20$ and $\beta = 0.03$. These values produce a steady state gain ratio of 2.5.

In Figure 18.2 the exciting signal is removed at $N = 20$. Note the slow growth and prolonged ringing brought about by the large amount of positive feedback. The system will become completely unstable if $\mu\beta$ becomes 1. In the complete absence of feedback, the gain ratio would step up to 1 at 0 and immediately step down when the exciting signal is removed. Figure 18.3 shows the situation that exists when the positive feedback is less. In this instance, $\mu = 20$ and $\beta = 0.01$. These values produce a steady state gain ratio of 1.25.

In comparing Figures 18.2 and 18.3 one can conclude that a larger amount of positive feedback increases the gain ratio and brings about a slower stepwise approach to steady state conditions followed by a longer decay after the removal of the exciting signal.

Negative acoustic feedback also has some undesirable acoustic consequences. This is brought about by the relatively long transit time, Δt, around the feedback loop. In a practical case Δt can easily be several milliseconds. Figure 18.4 illustrates the case where a large amount of negative acoustic feedback is present. In this instance the feedback signal is always of the opposite polarity to that of the exciting signal.

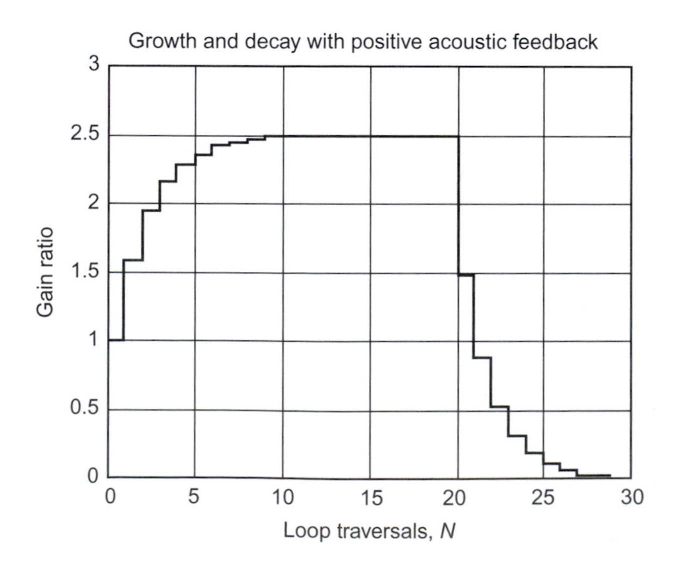

FIGURE 18.2
Gain ratio versus, N, the number of traversals around the feedback loop.

The steady state gain ratio for the conditions of Figure 18.4 is given by $1/(1 - \mu\beta) = 0.625$. When the initiating signal is turned on at $N = 0$, the gain ratio first overshoots its steady state value of 0.625 and then approaches the steady state value by means of a set of oscillatory steps. At $N = 20$, the exciting signal is removed and the gain ratio now approaches zero through a sequence of constantly diminishing steps. If no feedback had been present, the gain ratio would have stepped immediately up to one at $N = 0$ and would have stepped down immediately to 0 at $N = 20$ when the exciting signal was removed. This situation should be compared with a case of a smaller amount of negative acoustic feedback as illustrated in Figure 18.5.

In Figure 18.5, the steady state value of the gain ratio is 0.8333. It should be evident that even negative acoustic feedback distorts the time behavior of the original acoustic signal. Negative acoustic feedback, however, can never produce sustained system oscillation as is true in the positive acoustic feedback case. In the numerical examples we have taken both μ and β to be real numbers just for the sake of simplicity. The general equations are equally valid when they are complex quantities. In such an instance, positive feedback occurs when $|1 - \mu\beta| < 1$ and negative acoustic feedback when $|1 - \mu\beta| > 1$. The case where $|1 - \mu\beta| = 0$ is to be avoided at all costs.

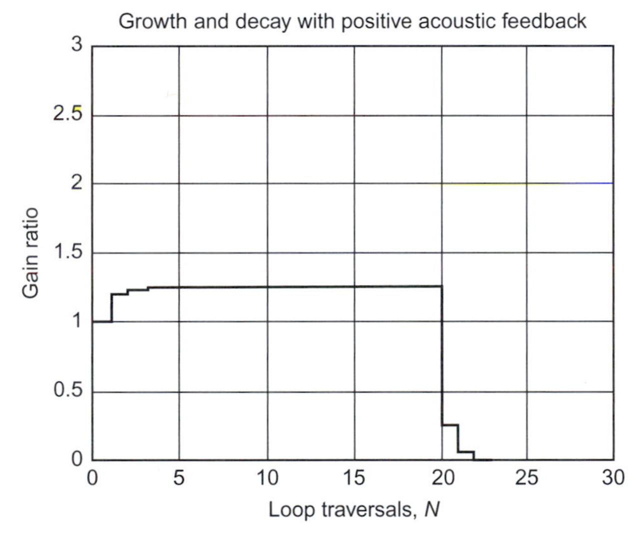

FIGURE 18.3
Growth and decay with a reduced amount of positive acoustic feedback.

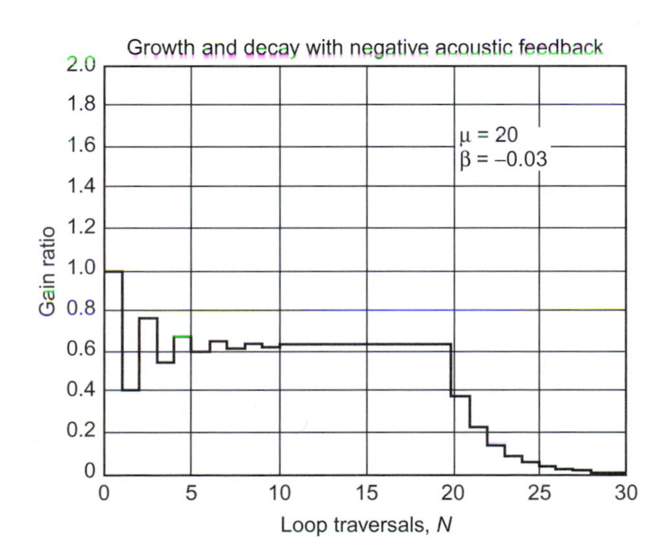

FIGURE 18.4
Behavior with a large amount of negative acoustic feedback.

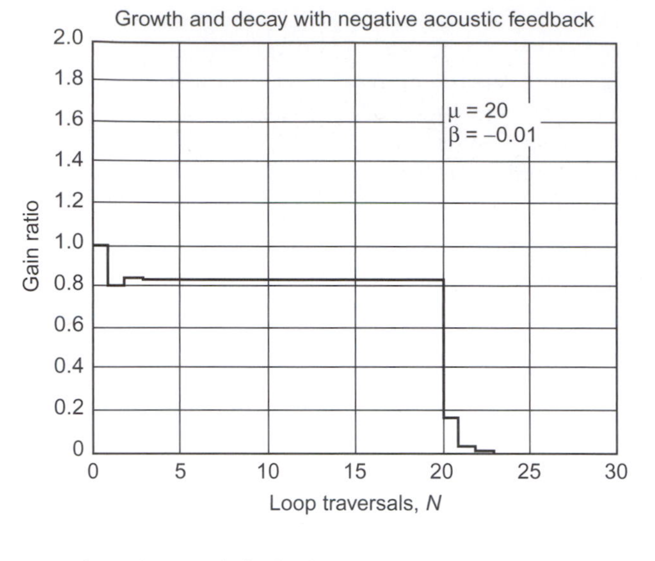

FIGURE 18.5
The behavior with a small amount of negative acoustic feedback.

Early practitioners

Equalization was employed by many early experimenters including Kellogg and Rice in the early 1920s, Volkmann of RCA in the 1930s, and the Boners in the 1960s.

Dr. C. P. Boner provided a valuable example of repeated application of the principles of equalization by adjusting sound systems that were considered unsatisfactory before equalization into sound systems that satisfied their owners and operators. Often Dr. Boner's work had as much to do with correcting wiring faults, coverage patterns, impedance mismatches, etc., as with equalizing, but in common with all who did such work at that early period, all improvements were attributed to the magic "equalizer."

Until real-time analyzers were available for under $5000 (1970), early equalization work was predominately done by repeatedly raising the gain of the sound system until feedback occurred and adjusting the appropriate filter to reduce the system amplitude at that frequency. One or 2 dB at the very most are necessary to bring the system back to stability. By repeating this for 20–30 feedbacks, it is possible to raise the acoustic gain of a sound-reinforcement system to within a few decibels of unity gain at all frequencies. While this method is an excellent way to increase gain with a nominal amount of test equipment, it leaves the overall tonal balance to the ear of the practitioner. Those with perfect pitch often exhibit a nonuniversal taste, and those with a taste that agrees with the majority of listeners are not often gifted with perfect pitch.

INTRODUCTION OF REAL-TIME ANALYZERS

Equalization of sound reinforcement systems with the end purpose of increased acoustic gain and enhanced acoustic quality became universal with the introduction of the ⅓ octave equalizer by one of the authors in 1967.

Figure 18.6 shows one of the authors training a class of sound contractors in January of 1968 in the use of a step-by-step ⅓ octave analyzer for making frequency response measurements for equalization work. The analyzer being used was the GenRad 1564A. In May of 1968 Hewlett Packard delivered to one of the authors the first 8054A ⅓ octave real time analyzer for $10,000 (that's 1960 dollars). The 8054A quickly led to a special stripped down version called the H23-8054A, which came without the nixie tube read out and the frequency selective push buttons, Figure 18.7. The success of each of the early instruments with the sound contractors who received them led to collaboration with HP which reduced the size and price. Approximately 500 instruments went to the audio industry for use in early equalization work. Other analyzers followed in the 1970s but these were the only ones for several years.

The ability to see, in real time, the result of the equalizer adjustment, in addition to hearing the result, changed the nature of the activity dramatically. The change from 45 minutes per house curve to 1 s had to be lived through to be truly appreciated.

FIGURE 18.6
An early equalization training class.

(a) H23-8054A

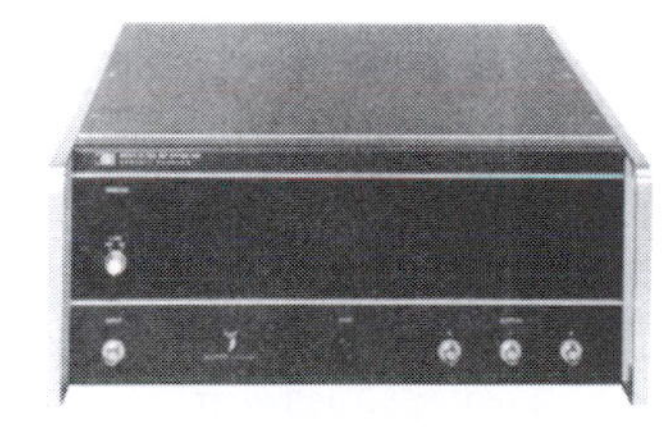

(b) 8056

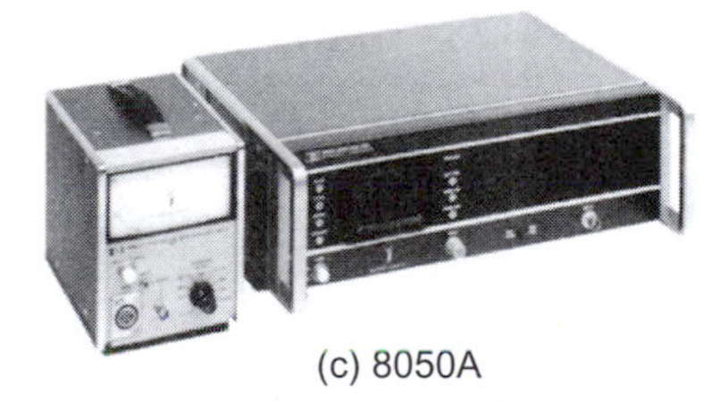

(c) 8050A

FIGURE 18.7
Early RTA equipment.

Today the 1/3 octave analyzer is still one of our most usable audio tools. Heyser-based analysis is more detailed, but there are many cases where the resolution and accuracy of a 1/3, 1/6, and 1/12 octave real time analyzers are ideally suited to the job, such as checking coverage.

The authors feel fortunate indeed to have been the first users in audio to employ and apply 1/3 octave real-time analysis to the study of sound system performance.

Forty years later

The authors have each spent over 40 years making precision acoustic measurements on sound systems and nearly that long in the practice of equalizing them. The opinions we put forth here are not intended to be final judgments on the matter but rather it is hoped that our suggestions learned from this experience will save you repeating some of the same errors we committed.

Today, the knowledge gained is available and useful in contemplating the next steps. New analyzers are daily revealing the possibilities of manipulating the phase characteristics of a system more directly than in the past, and current digital technology promises full control over the time domain, as well as the frequency domain. As in the past, we expect the future to be shaped by those with the analysis capability coupled to everyday "real-life" exposure to sound system design, installation, and operation.

Simple filter networks and their interaction with real-life electroacoustic transducers are a complex technical study. To quote Richard C. Heyser,

> As many technically trained people are prone to do, I naively presumed when I first began analyzing loudspeakers... that I could bring contemporary communication theory to bear and simply overwhelm the poor loudspeaker with technology... I soon found the error in my thinking. The evaluation of the acoustics of loudspeakers and the room containing them proved to represent a microcosm of all the difficult problems in wave propagation. A wavelength range of over 1,000 to 1 is bad enough but the physical extent of the important dimensions in a single experiment range from one thirtieth wavelength to greater than thirty wavelengths for many practical loudspeaker systems.

The wonder is that the simple tools we employ work as well as they do. It's very fascinating to find that techniques we had to work out intuitively can now be measured accurately and justified as engineering facts.

Equalizers are now omnipresent on the sound system scene. Unfortunately, many are designed as "program equalizers" rather than specifically for the adjustment of systems to maximum acoustic gain. The electronic circuit designer designs to his termination resistor free of all self-doubts or mathematical uneasiness. Then they wonder why some of us are less than enamored of their latest electronic non-solution to our very real acoustic problems.

RTA applications

The real-time analyzer has been used by audio engineers for almost 40 years and has proved invaluable in the following areas:

1. House curves made with measuring microphone.
2. House curves made with sound system microphone.
3. Examining distribution of all frequencies at differing locations at the same time.
4. Examining house curve at the performer's location.
5. Response curves of the filter settings.
6. Detecting feedback frequencies.
7. Frequency response of microphones to be used in the sound system.
8. Examining crosstalk between lines.
9. Setting levels throughout sound system areas both electrically and acoustically.
10. Detecting resonating surface areas by observing the effect of manual damping of the vibrating surface.

Pink noise (equal energy per octave) is used rather than white noise (equal energy per hertz) because the bandpass filters used in the typical RTA are constant percentage bandwidth rather than constant bandwidth. This means that a white-noise signal put into a constant-percentage bandwidth analyzer would have a +3 dB/octave rise with increasing frequency. The filters grow wider with increasing frequency, thereby summing more energy at the same level. Pink noise on a per hertz basis decreases 3 dB/octave (10 dB/decade); therefore, pink noise matches constant-percentage bandwidth response, allowing a flat response across the screen of an RTA.

Figure 18.8 shows an example of how unavailable acoustic gain, due to feedback caused by highest amplitude present, can be made available by equalizing all frequencies, making them equal in amplitude response.

BAND-REJECTION, BANDPASS, AND BAND-BOOST FILTERS

All types have been used to equalize sound systems. The authors' preference is for band-rejection filters. Boosting a narrow band of frequencies is not a natural acoustic phenomenon. Any two frequencies can come together in a room to a maximum (in the practical case) of +3 dB but may combine to complete cancellation. The only thing "narrow band" going on in an acoustic environment is rejection, never

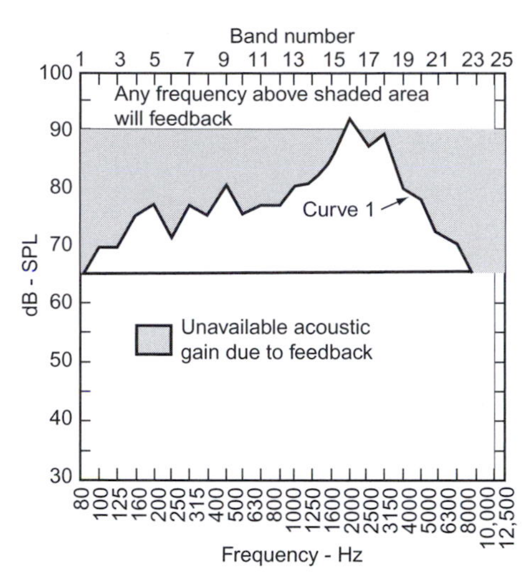

(a). Typical situation that can occur in an auditorium. Curve 1 shows the sound pressure output of the loudspeaker if a signal equal in level at all frequencies is connected to the input of the sound system. The irregularity of the output is partly due to the inability of the loudspeaker to respond perfectly uniformly to a uniform input signal, and partly—the major part—due to the effect of the room itself on the acoustic output of the loudspeaker. Feedback will occur first at 2000 Hz, since any attempts to raise the gain will find this peak to be the first frequency to push above the limit line. To follow this example, assume the acoustic gain is 10 dB.

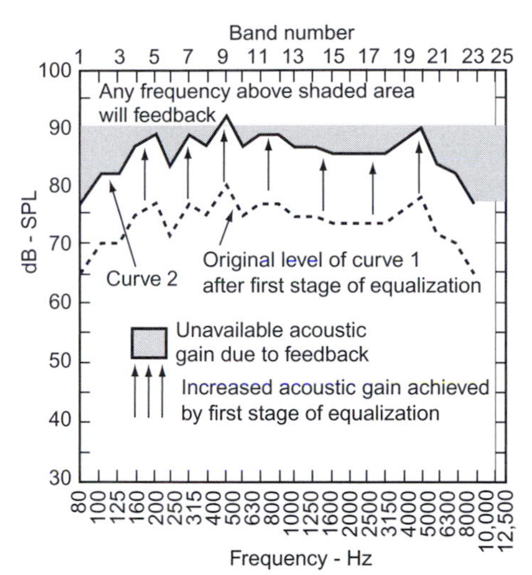

(b). Curve 1, after the peaked area between 2000 Hz and 3150 Hz, has been equalized to the majority of the other frequencies. The arrows indicate how all these frequencies may now be raised simultaneously in gain before the new peak of 400 Hz pushes above the hatch lines and causes feedback. The number of decibels at each frequency between curve 1 and curve 2 respresents the increased acoustic gain made possible by the first stage of sound system equalization.

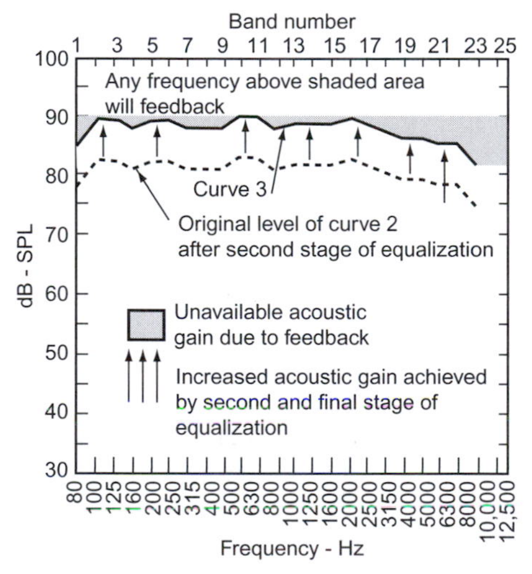

(c). Additional smoothing of the curve can allow greater acoustic gain at the majority of frequencies. However, further smoothing, even if perfectly done, would yield either 1 dB or 2 dB at the very most throughout the frequency region of critical importance for speech. By comparing curve 2 with curve 3 we can see that through the vital frequency response area for speech, for example, the acoustic gain at all frequencies is increased from 300 to 3000 by, typically, 10 dB or more. Originally, only 2000 Hz could be brought to 90 dB L_p before feedback occurred; now all frequencies can be brought to 90 dB L_p before feedback.

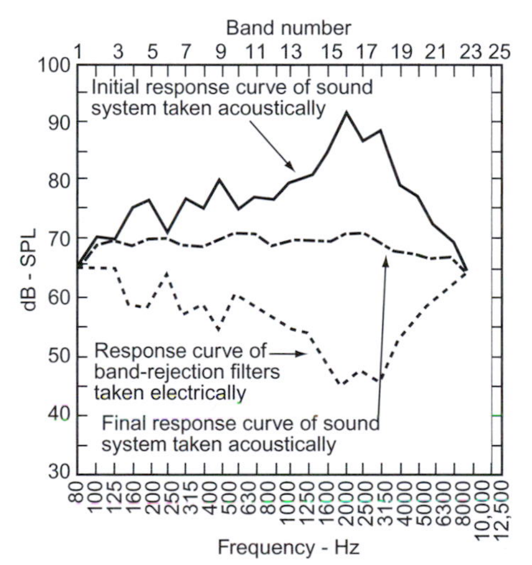

(d). The electrical response curve of the critical band rejection filters (bottom curve) join together to form the inverse of the loudspeaker room response curve (top curve). Because this inverse filter response curve is included in the total sound response, the smoothed overall acoustic response shown by the middle curve results.

FIGURE 18.8
How equalization raises acoustic gain.

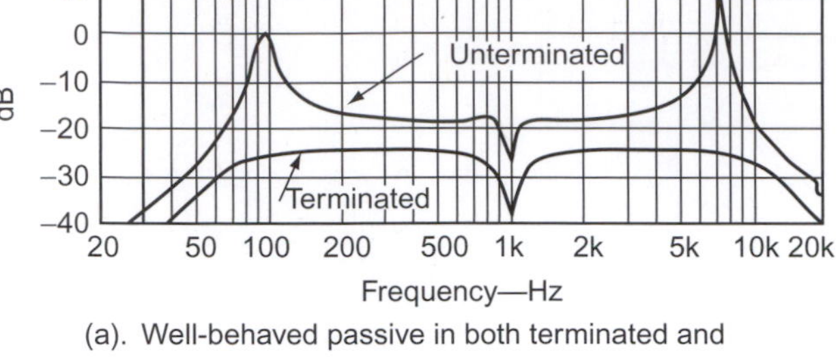

(a). Well-behaved passive in both terminated and unterminated states.

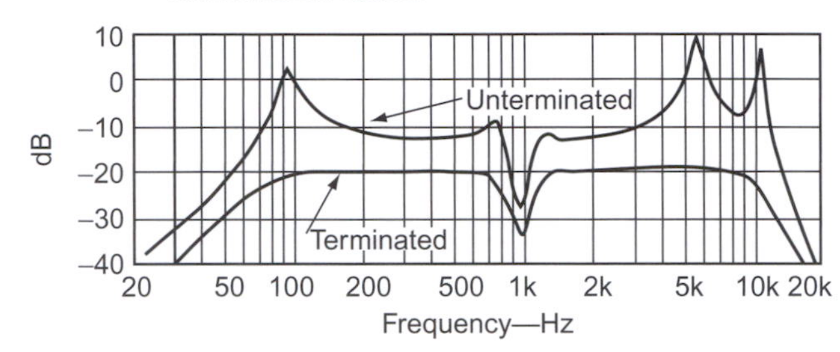

(b). Well-behaved passive in both terminated and unterminated states.

FIGURE 18.9
Frequency response curves for a passive equalizer (1000 Hz filter set at −14 dB with high pass frequency at 80 Hz and low pass frequency at 10 kHz.

summing, though acoustic focusing may be mistaken for summing. Those who think there are narrow acoustic "peaks" in the environment are the same ones who advocate equalizing low frequency room modes, i.e., usually by trying to boost the null frequency. Very narrow peaked responses can and do lurk in the transducers used in sound systems. Such devices should be replaced, not "equalized." Again, we have often witnessed skilled electronic circuit designers trying to correct a deep notch in the frequency response with a very sophisticated electronic filter when the cause of the notch is one driver out of alignment with another.

Interesting to end users, but a matter of disgust to creative circuit designers, the simpler, almost sloppy, filters seem to work the best.

Criteria for band-rejection filters

So far as the authors are concerned, if the following criteria are met, it should be a satisfactory filter:

1. Minimum phase response.
2. Combining, sometimes called summing.
3. Minimum excess delay.
4. Not narrower than 1/6 octave bandwidth with 1/3 octave preferred.
5. Band rejection-type with maximum depth of 14 dB, preferably in 1 dB steps.

This is not to say that there are not interesting differences in designers' choice of circuits. Figure 18.9 shows two exceptional fine passive equalizers wherein the only observable difference is in the unterminated state. Though largely not manufactured today, these passive units never wear out. It would be a pity to discard them when newer electronic equipment is installed, but one needs to be aware of the need of the "buildout" and "termination" requirements.

Filter parameters

Figures 18.10a and 18.10b show the parameters of a very narrow-band filter for both amplitude and phase. The meaning of the bandwidth of a filter is shown in Figure 18.11.

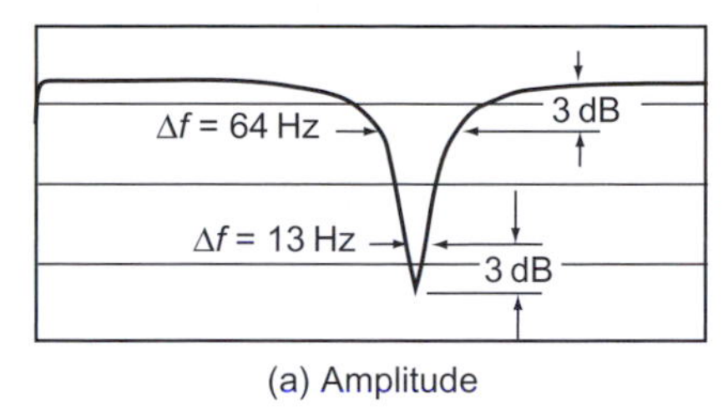

(a) Amplitude

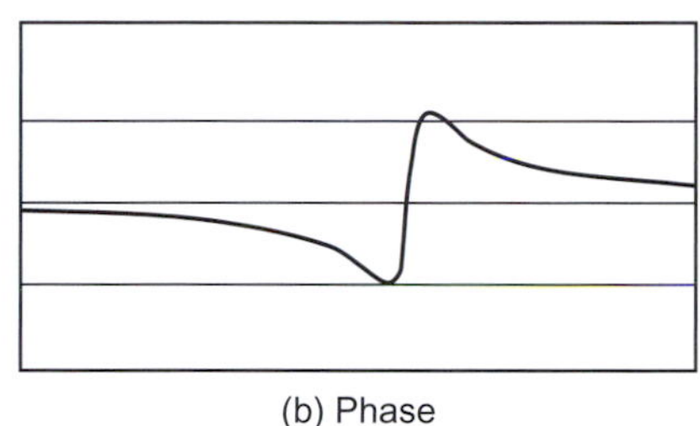

(b) Phase

FIGURE 18.10
A very narrow-band filter.

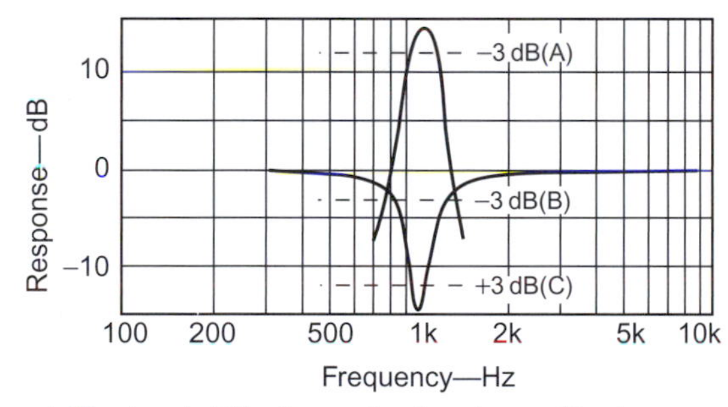

1. The bandwidth of an active band pass filter is measured 3 dB below its center-frequency level (A).
2. The bandwidth of a band-rejection filter is measured 3 dB below the normal level before the filter is inserted (B).
3. On occasion individuals have chosen to define the bandwidth of a band-rejection filter as "up 3 dB from the center notch."

FIGURE 18.11
The bandwidth of an active bandpass filter.

One of the earliest methods of sound-system-room equalization employed individual broadband networks to shape a rough inverse of the house curve, Figure 18.12a. This was followed by the insertion—one at a time—of very narrow notch filters at the predominant feedback frequencies, Figure 18.12b. When the two sets of filters were electrically combined, the response was like Figure 18.12c. Figure 18.12d is the replacement tuning finally put in the job after the advent of 1/3 octave-spaced bridged-T filter sets. Note how the same overall gain restoration is provided by either type of filter, but that the 1/3 octave-spaced filter set removed feedback at many frequencies by changing the slope rate instead of depressing the amplitude. It is important to remember that the sound system deviations from uniformity in a system worth equalization are quite correctable by the slope-rate changes available in the 1/3 octave-spaced filter sets.

Characteristics of successful filters

Let us compare an ISO 1/3 octave bandpass filter with a 1/3 octave spaced band-rejection filter, Figure 18.13. By further comparing the two basic bandpass filter shapes, both 1/3 octave and 1/10 octave, we can see that the inverse of the 1/10 octave response most closely approximates the response of the band-rejection filter, Figure 18.14. Looking at a single filter section and recording each of its 1 dB steps, we get a series of curves as shown in Figure 18.15.

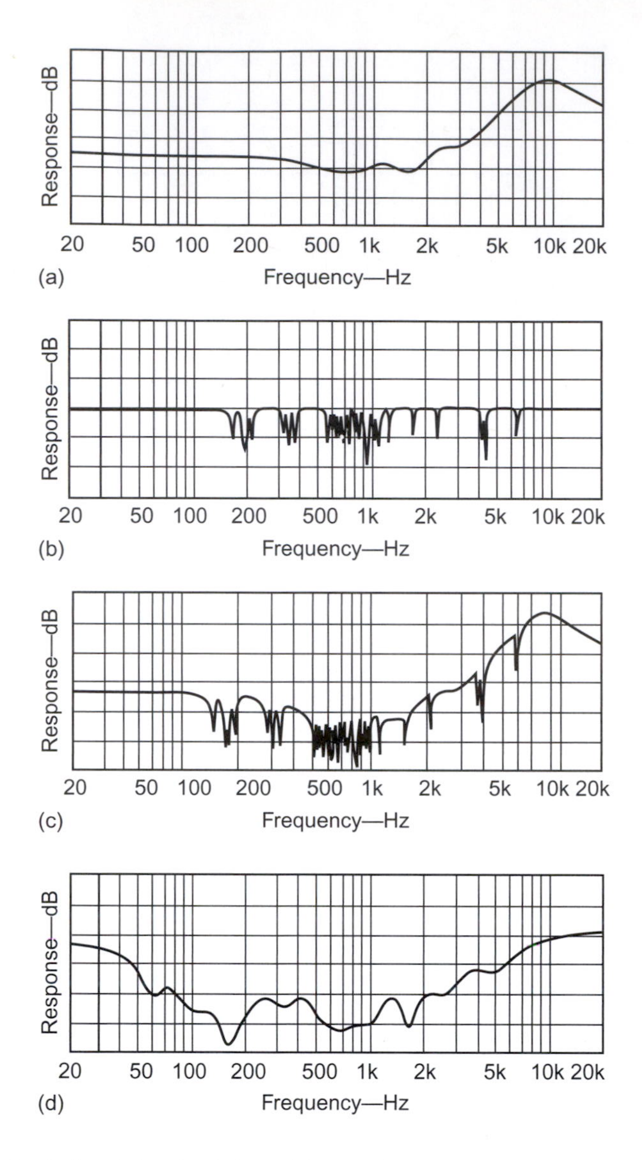

FIGURE 18.12
Very narrow-band equalization (A, B, and C) and typical combining-type filter equalization of the same system (D).

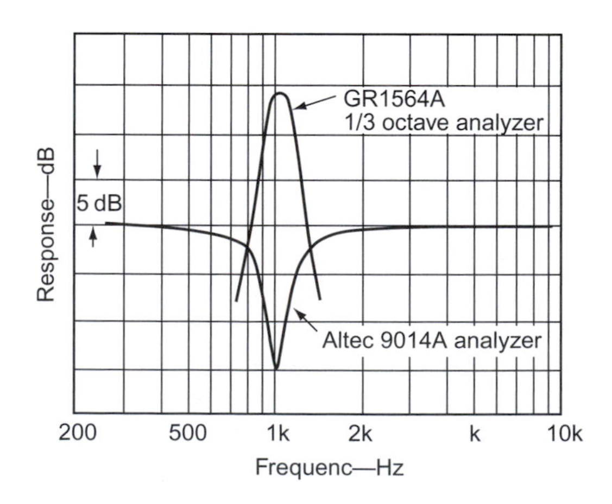

FIGURE 18.13
Broad-band combining-type band-rejection filter section compared with ⅓ octave active band-pass filter section.

If we were to record each filter section in a set of 24 such filters, by setting each filter for a maximum rejection and then restoring it to zero before recording the next filter section set at its maximum rejection, we would obtain the set of curves shown in Figure 18.16. If we were to turn all 24 filter sections to maximum rejection at the same time, we would discover their most important property—they combine,

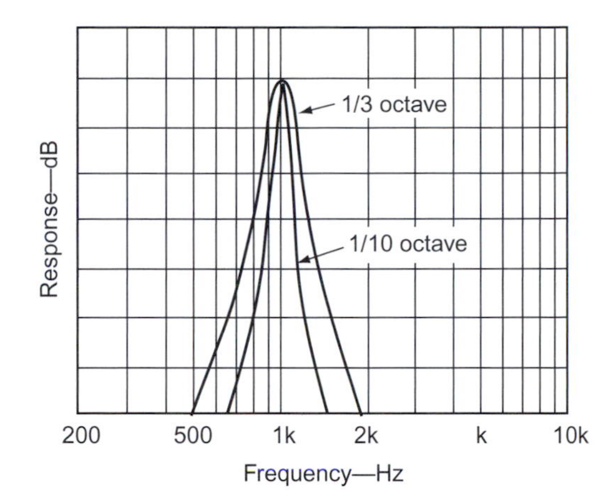

FIGURE 18.14
Comparison of a ⅓ octave bandpass filter with a ⅒ octave bandpass filter.

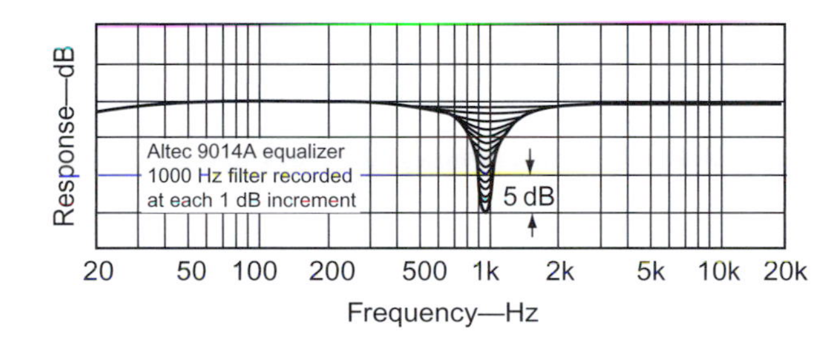

FIGURE 18.15
Response of a band-rejection filter at each of its 14 steps.

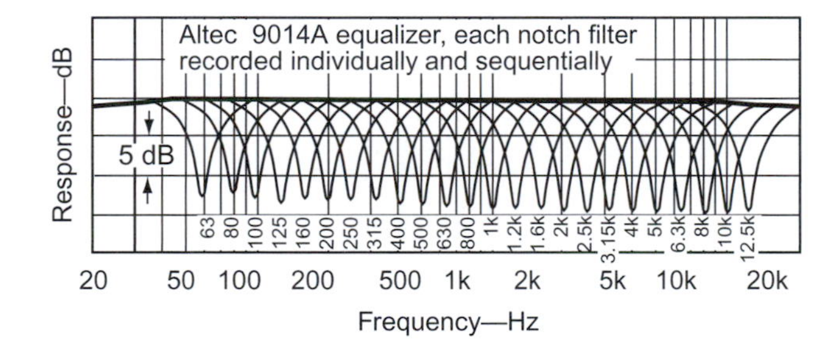

FIGURE 18.16
A series of band-rejection sections recorded sequentially one at a time.

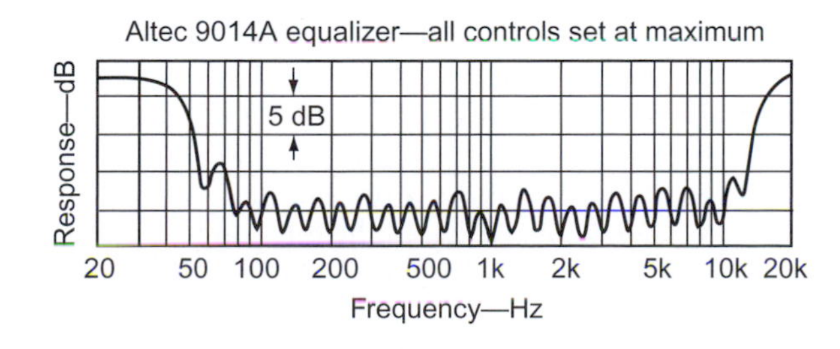

FIGURE 18.17
Series of band-rejection sections all simultaneously turned to maximum attenuation.

Figure 18.17. Note that when they combine they are essentially additive (their combined depth exceeds 20 dB at the bottom of the ripple).

Figure 18.18 shows in detail how they combine. The narrower curve is 1000 Hz set at −6 dB. The wider curve is 800 Hz at −2 dB, 1000 Hz at −2 dB, and 1250 Hz at −2 dB. Here they have combined to

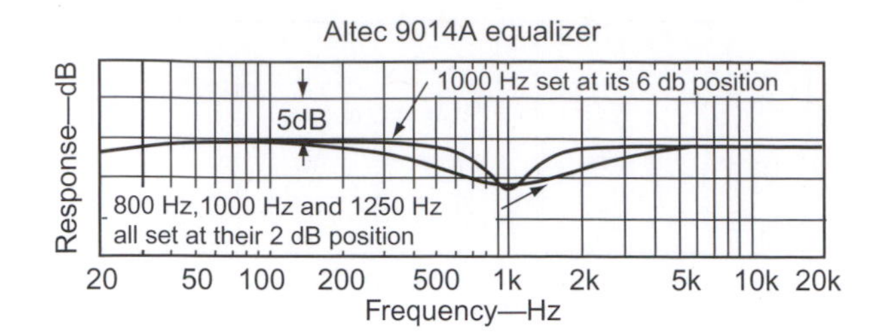

FIGURE 18.18
Response of a 1000 Hz section with 6 dB attenuation compared with the response of a combination of 800 Hz, 1000 Hz, and 1250 Hz sections with 2 dB attenuation each.

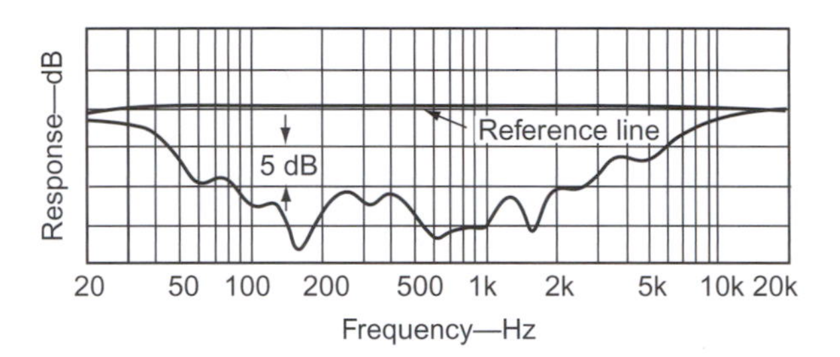

FIGURE 18.19
Example of band-filter section in combination to form inverse of raw house curve.

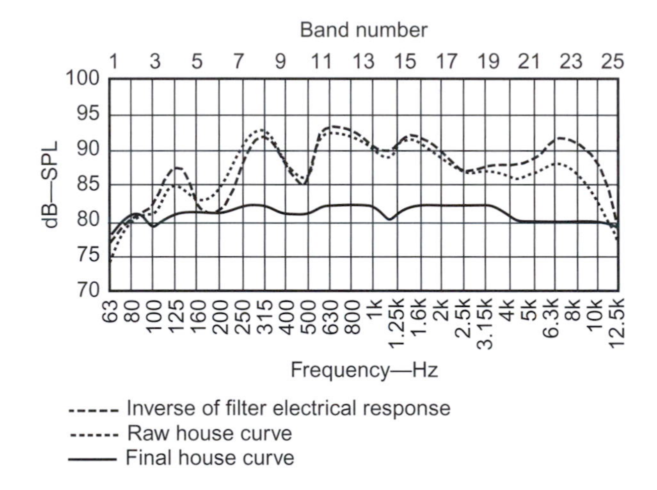

FIGURE 18.20
Example of correlation between raw house curve and inverse of filter electrical response curve.

become 6 dB deep, and the center of the curve is at the middle of the three. It is not difficult to imagine the complexity of combining from 14 to 24 of these sections all at different levels to appreciate that some form of real-time observation is required to comprehend thoroughly what is going on. Through such combining, the smoothest conjugate phase response is achieved. Figure 18.19 shows the electrical response of a set of filters on an actual job. Fourteen sections were employed at the frequencies and levels indicated. Note that at 160 Hz the filter is at −10 dB but the curve is at −18.5 dB due to combination effects. The real test is to compare the inverse of the filter response before equalization. This is done on a ⅓ octave basis in Figure 18.20. Filters producing the type of response we have just looked at can be either passive or active.

Figure 18.21 compares the electrical amplitude response of a set of very narrow-band filters and a set of critical-bandwidth combining filters after adjustment on the same job.

Filter transfer characteristics

The transfer characteristics of band-rejection, minimum-phase filters are shown in Figure 18.22. In the previous example of the combining power of these filters, in terms of amplitude, we see an example of their combining power in terms of phase, Figure 18.23. The steeper the amplitude slope rate, the steeper the rate of phase change.

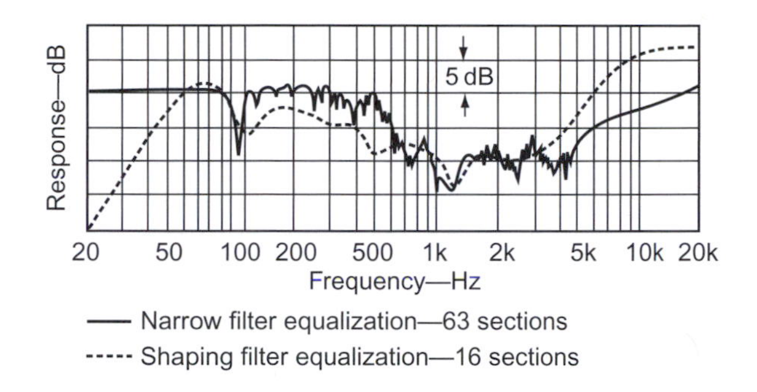

FIGURE 18.21
Comparison of response of a set of very narrow-band filters and a set of critical-bandwidth combining filters.

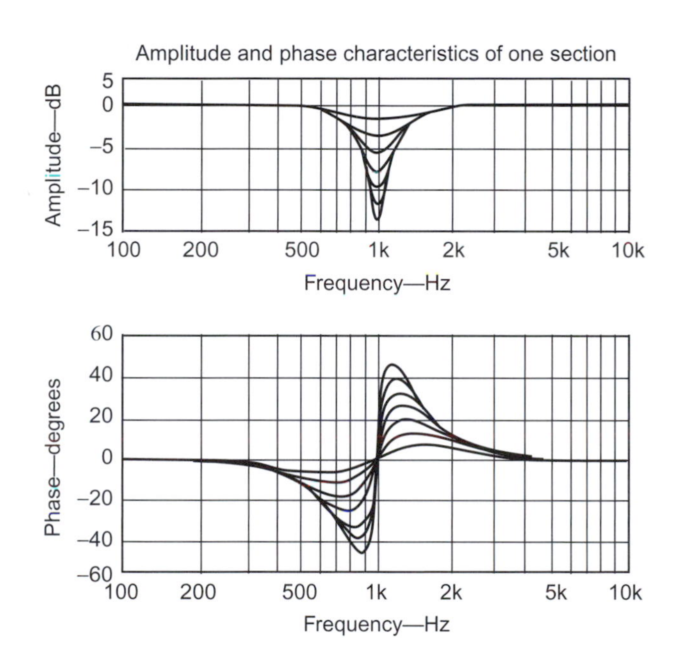

FIGURE 18.22
Bridged-T configuration for tandem filter sections.

Figure 18.24 shows all the filters at −1 dB amplitude and the resultant phase characteristic. Figure 18.25 shows the same information for −4 dB settings.

Taking a practical example, Figure 18.26a shows the equalized and unequalized response of the left channel of a monitor system. Figure 18.26b shows the same information for the right channel. Figure 18.27a shows the electrical amplitude response of the corrective filters, and Figure 18.27b shows the phase response of each channel.

Minimum-phase filters

A minimum-phase filter introduces the minimum possible phase shift but still retains the corrective amplitude change. It is also obvious that the relative phase between channels is virtually identical. It is this careful band-by-band resolution of phase that so many listeners have dubbed as the "sharp focus" that equalization seems to produce in sound systems already relatively smooth in an amplitude sense.

Richard C. Heyser has pointed out that:

> Highly important to a loudspeaker's ability to produce accurate sound, when it has been properly equalized, is that of minimum phase change. A minimum-phase-change loudspeaker is one in which, when all amplitude response variations are removed by conventional resistance, capacitance, and inductance networks, it has the minimum possible phase shift over the frequency spectrum. Properly designed equalizers for balancing the amplitude response will also automatically balance the phase response for a minimum phase loudspeaker.

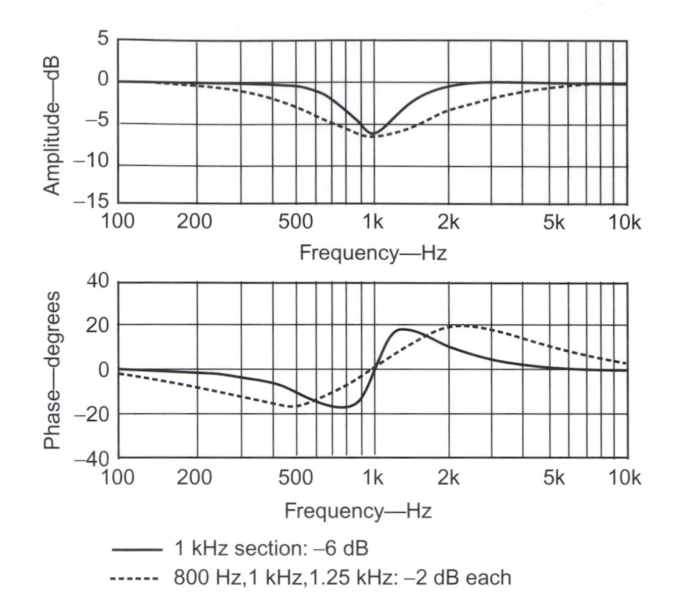

FIGURE 18.23
Active configuration for tandem filter sections.

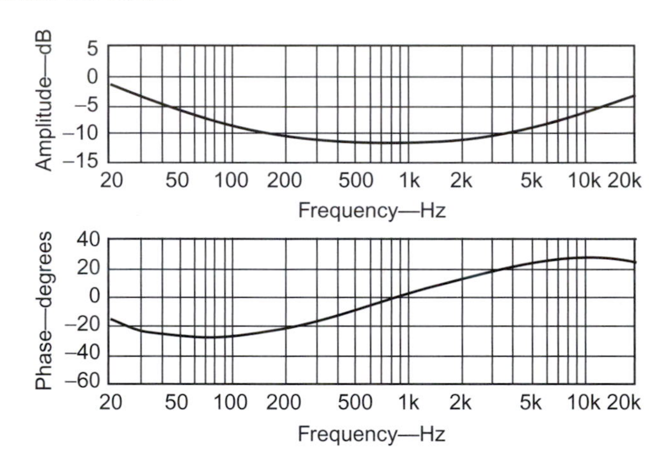

FIGURE 18.24
Combined phase and amplitude response, all sections set for −1 dB.

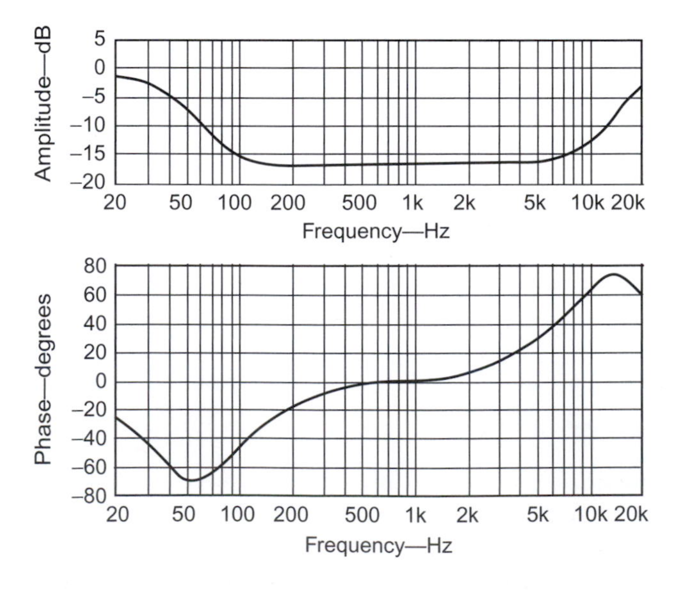

FIGURE 18.25
Combined phase and amplitude response, all sections set for −4 dB.

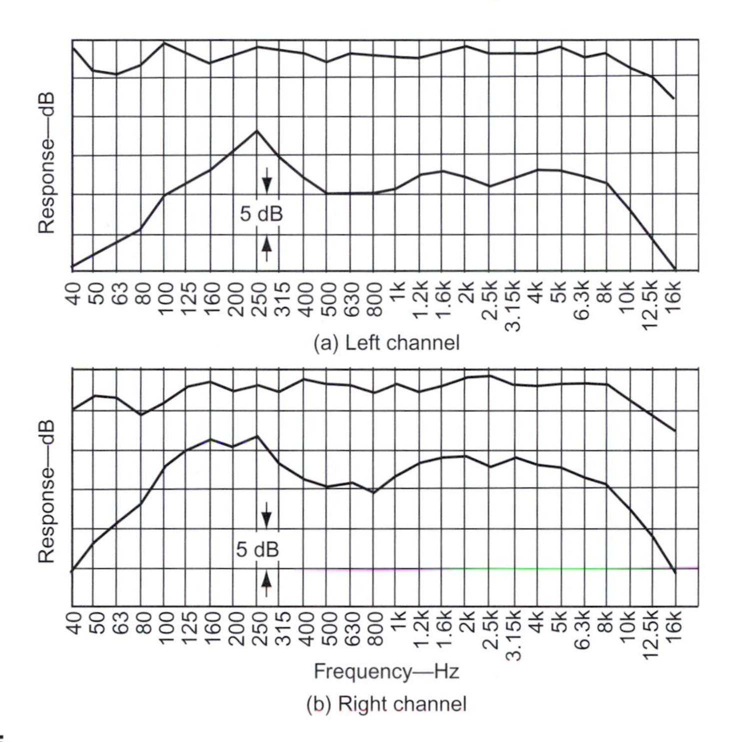

FIGURE 18.26
Equalized and unequalized response of a monitor system.

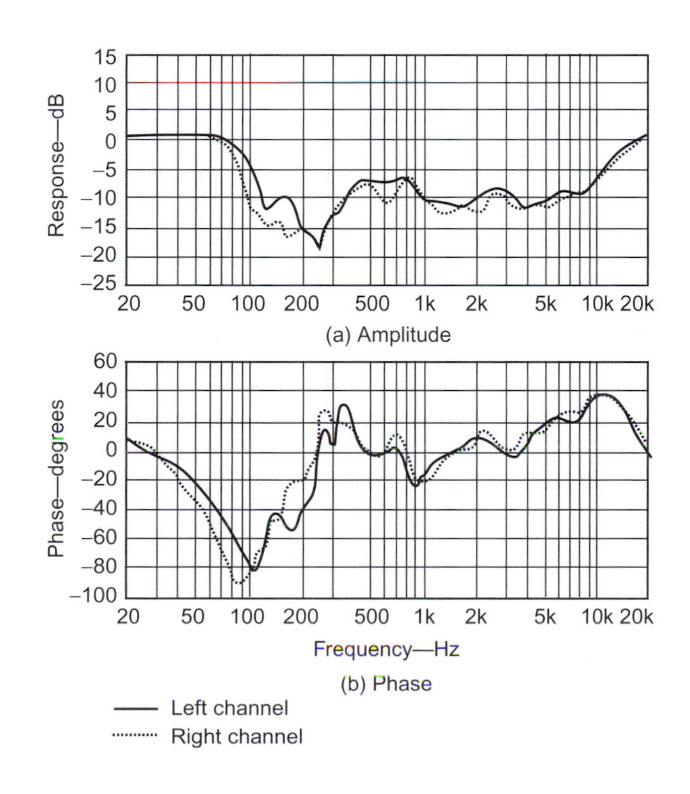

FIGURE 18.27
Response of equalizers.

Also:

> A nonminimum-phase loudspeaker will usually exhibit frequency-response difficulties which can be associated with signal delay effects which, in turn cannot be corrected with conventional passive or active equalization.

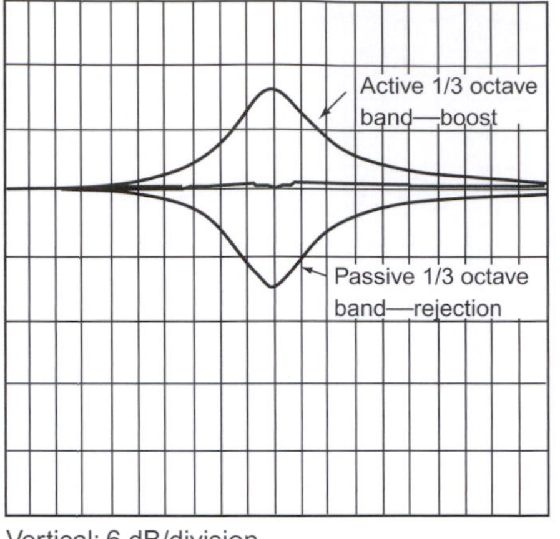

Vertical: 6 dB/division
Horizontal: Auto 0.00—2000.24 Hz
Resolution: 1.0010 E + 01 Hz
(a) EFC

BBF
Zero degrees
BRF

Vertical: 45 degrees/division
Horizontal: Auto 0.00—2000.24 Hz
Resolution: 1.0010 E + 01 Hz
(b) Phasefrequencycurve,PFC

FIGURE 18.28
Notch filter in opposition to boost filter.

Figure 18.28a shows an active ⅓ octave band-boost filter (BBF) raised 9 dB, a passive ⅓ octave band-rejection filter (BRF) lowered 9 dB, and the resultant smooth amplitude response. Figure 18.28b shows the phase response of the BBF and the BRF as well as the resultant smooth phase response.

The distinction between a bandpass filter (BPF) and a band-boost filter (BBF) is that the "skirts" of a BPF continue on to minus infinity ($-\infty$). The skirts on a BBF return to zero reference after passing through the peak. Such filters are found in active parametric filter sets and extreme caution is advised in their use in live system stability adjustment. One must exercise caution when it comes to program effects' adjustment.

TEF ANALYSIS IN EQUALIZATION

With TEF analysis in use for almost all serious equalization work today, the engineer can view L_D, L_R, and L_{RE} (early reflection levels), separately or together as desired. The level, direction, and time of travel for each reflective interference can be observed. An aberration in L_T can be segregated and if it is caused by L_D equalized, if caused by L_{RE} blocked, and if due to L_R, it can be treated in the statistical manner. Formerly well hidden transducer aberrations, particularly in phase and in time behavior, are now strikingly evident and consequently easily prevented.

As a consequence of this enhanced ability to actually "see" what's going in the total system, both rapidly and accurately, the design procedures are modified to incorporate this new knowledge and we are finding that less and less equalization is required in the newer systems. Where and when equalization is needed, it is invaluable. Misapplied, it can create harsh sounding high frequency distortion, instability in the system, and, worst of all, a belief that equalization has solved a problem that in actual fact is still unaddressed.

Figure 18.29a is the Envelope Time Curve, ETC, of a packaged system with a separation between the low frequency unit and the high frequency of 0.17 ft, 0.15 ms. Figure 18.29b is the magnitude and phase of the same loudspeaker with the phase response made as smooth as possible. Figure 18.29c is the Nyquist display of the same loudspeaker with the cursor set at a high frequency that has encircled the origin of the display indicating a *non-minimum phase* system. Failure of a cursor in a Nyquist plot to rotate clockwise as frequency increases indicates a mis-selected signal arrival time at the measurement microphone. Harry Nyquist of the Bell Telephone Labs in the old days was truly a genius.

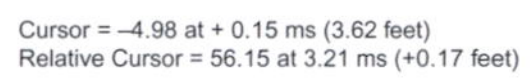

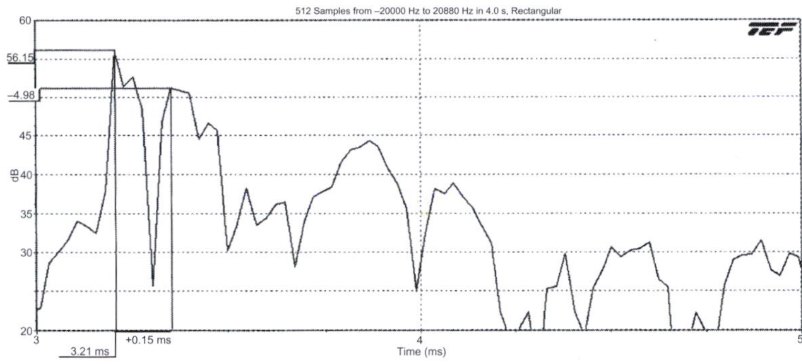

A. Envelope time curve of a packaged two-way loud speaker system arrival times and relative distances plus levels.

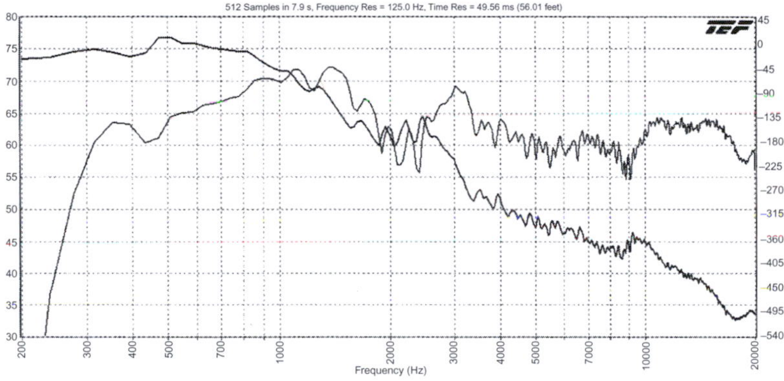

B. Magnitude and phase of a packaged two-way loud speaker system.

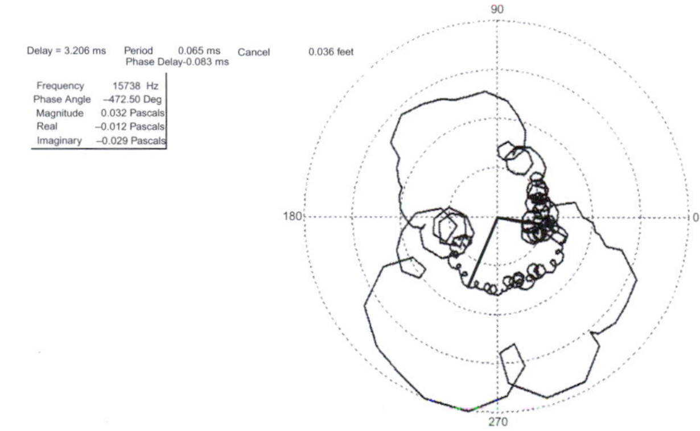

C. Magnitude is length of the cursoratany frequency. The angle of the cursor is the phase. Encirclement of the origin reveals nonminimum phase behavior.

FIGURE 18.29
Envelope Time Curve, magnitude, and phase with the phase response made as smooth as possible and the Nyquist display of a loudspeaker.

HOW TO APPROACH EQUALIZATION

Gently! Slowly! These are key words for key attitudes when utilizing equalizers. After every adjustment, listen carefully to the remaining sounds. The goal is to improve sound quality in sound reproduction systems as well as increase acoustic gain in the case of reinforcement systems. In any type of system, stop tuning and examine the system with care when the equalizer causes a detrimental change in the sound quality.

In using equalizers, we can borrow from the medical fraternity and say, "First, do no harm."

Anything within a system that requires steep slope rates (over 18 dB/octave) or excessive amplitude change (in excess of 3 dB) in order to control a given frequency increment (on the order of 1/3 of an octave) should be subject to serious consideration as to its replacement. In today's marketplace there is a sufficient number of very well-behaved electronics, microphones, loudspeakers, and interconnection networks available to allow avoiding the use of inferior products. Increasing understanding of the Heyser Transform and its ability to allow us to understand the transformations in time at given frequencies coupled to the newer analyzers to measure such parameters should quickly lead to both better components and more skillful application and adjustment of them.

When to use an equalizer

Equalization, in common with most sound system components, can be misapplied. Therefore, a discussion of when and where it is appropriate to use an equalizer and when and where it should not be employed is useful.

Let's first imagine a very special case wherein we have a loudspeaker that already has the acoustic response we desire and its coverage pattern exactly covers our audience area so that each listener is receiving the identical level and $\%AL_{CONS}$. Let's further assume that no sound reflects off this audience into the reverberant space. In such an ideal case, we would require no equalization unless for some reason we desired program equalization or the deliberate distortion of program material for some departure from this ideal case. This clearly identifies the fact that only some departure from this ideal case might require correction. A deviation in L_D might then require an equalizer. Other deviations would require different remedies.

Sources of feedback that should not be equalized

Equalizers are most misused by end users in the following areas:

1. Used to correct instabilities caused by comb filters designed into the system rather than removing the cause of the comb filter.
2. A microphone on a desk stand near a hard surface.
3. Excessive insertion loss per filter.
4. Use of the filter to control a problem caused by mechanical feedback.
5. Use of the filter to control feedback caused by crosstalk between circuits.
6. Use of the equalizer to adjust the steady-state response of devices whose transient response is an undamped resonance.

There are many very critical areas where the distinction between acoustic gain equalizers and program equalizers becomes quite hazy. Our increasing appreciation of right brain-left brain influences in the perception of the received signal by the listener causes us to proceed with caution in interpreting the technical data inundating users of analyzers. Perhaps we should separate electroacoustic transducer magnitude and phase adjustment from other forms of signal processing that are quickly coming on the scene (e.g., signal synchronization).

If we were to reserve the word "equalizer" for those devices that made system gain equal at all frequencies and used the words "signal processor" or "adjuster" for those devices intended to shape specifically the transfer function of the system for any purpose other than optimum acoustic gain, we might forestall at least a little of the confusion.

Feedback is a single frequency

Acoustic feedback is indeed single frequency. The fallacy that is usually being defended by that statement is the implication that since feedback is single frequency, so should the compensating filter be single frequency.

Nyquist, Waterfall, and Antman have clearly shown the amplitude, phase, and signal delay path causes of feedback. A single frequency feedback is often less than 0.1 dB greater amplitude at that frequency than the surrounding frequencies.

The only extremely narrow-band effects the authors have ever observed in real-life sound systems is band rejection such as phase cancellation or diaphragmatic absorption. Recall that the acoustic environment is passive, not active.

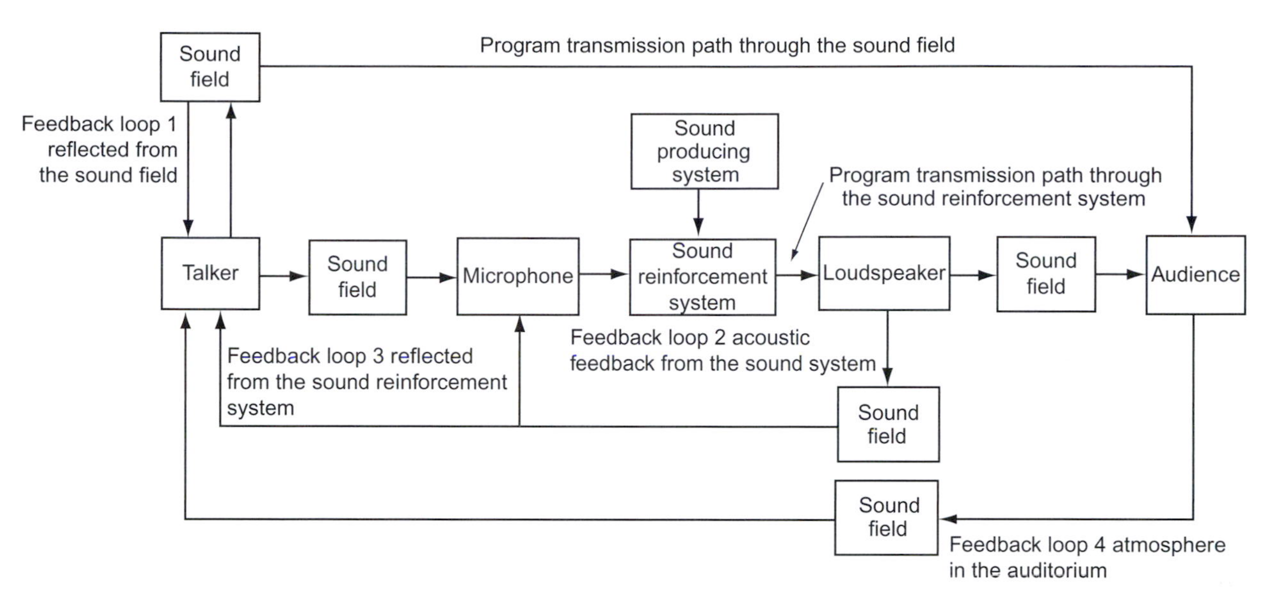

FIGURE 18.30
Example of the multiplicity of sound fields that can affect how you choose to equalize.

Which sound field is the microphone in?

Figure 18.30 hints at the multiplicity of sound fields that might be encountered in a single acoustic environment in connection with the operation of a sound reinforcement system. Being alert to each of these acoustic fields can solve many problems that seem mysterious when considered in the context of a single field. Often, you will equalize in more than one of the fields at the same time. For example, you will equalize the main system for the audience, the foldback system for the entertainer, a delayed under-balcony system for a distributed system, and separate equalization for a signal-delayed portion of the main system. Some of these are in the direct field and some in the reverberant field. No one, as yet, has all the answers to applying equalization to the fantastic variety of sound systems being designed today.

Figure 18.31a illustrates the time dependency of each of the sound fields. Figure 18.31b depicts the frequency dependency of the characteristics of a sound field. Finally, Figure 18.31c shows the level dependency of sound fields relative to their distance from the sound source.

In almost all sound system measurements, we avoid the "near field" of an acoustic source. Most measuring microphones used with equalization measurements are placed in the far reverberant sound field. Equalization of monitor loudspeakers in recording control rooms almost always are in the far free field.

When using modern analysis capable of separating L_D from L_R, the operator is able to choose between direct sound levels, L_D, early reflected sound levels, L_{RE}, reverberant sound levels, L_R, and total sound levels, L_T (L_T is comprised of L_D, L_{RE}, L_R, and L_N, where L_N is the ambient noise sound field).

One of the most frequent errors made in using ⅓ octave analyzers, or FFTs which measure L_T, is the failure to note the level of L_N separately followed by observation of its effect on L_T.

In complex multi-loudspeaker arrays the best practice is to turn on one loudspeaker at a time for equalization. If two loudspeakers share a common coverage area they are first looked at individually and then adjusted combined.

WHAT CAN AN EQUALIZER EQUALIZE?

The question, "What can an equalizer equalize?" needs to be asked. Some claim to equalize the room. Is this possible? We think not. When an electronic or passive equalizer is installed in between a mixer and a power amplifier we need to know that all we can equalize is the electrical signal being sent to the loudspeaker.

What can an audience do to affect L_D from a sound source? The answer, of course, is absolutely nothing. Therefore, it is clear that the audience can only alter L_{RE}, L_R, and L_N. Now, ask yourself the

FIGURE 18.31
Acoustic level versus distance and time.

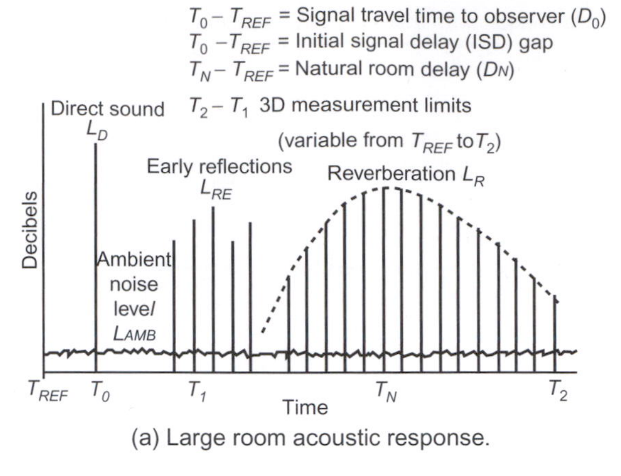

(a) Large room acoustic response.

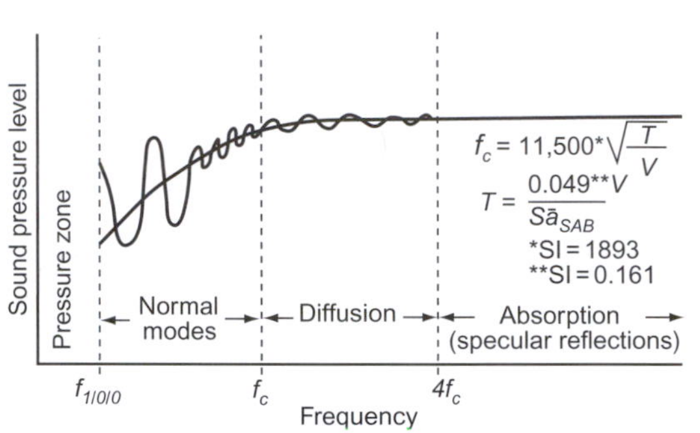

(b) Controllers of steady-state room acoustic response.

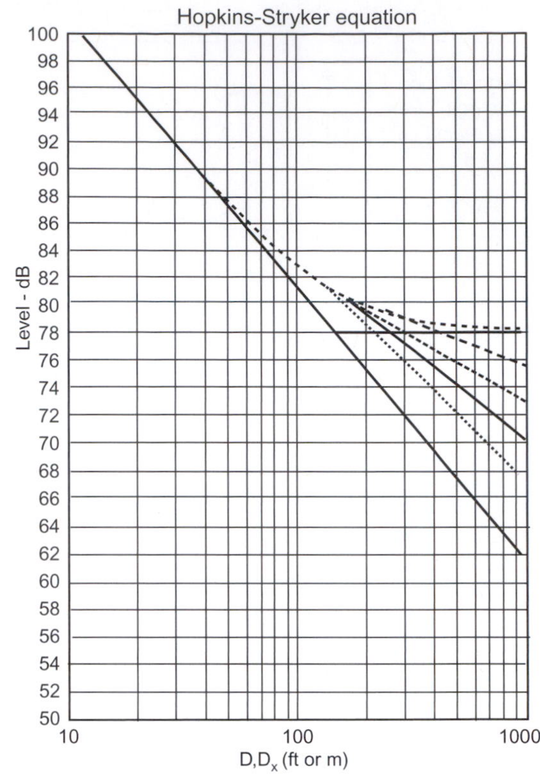

(c) Acoustic level versus distance.

question, "how can an equalizer adjust L_{RE}, L_R, and L_N?" The answer is that it cannot. I would hesitate to mention such obvious facts except for the remarkable number of times in the popular press that claim the contrary.

If one has only an RTA, one needs to equalize the loudspeaker's L_D. Often, due to the location of the loudspeaker; the microphone is in the reverberant sound field. Knowledge of the free field response of the loudspeaker allows for intelligent "guesses" as to what aberrations in the reverberant sound field should not be responded to with the equalizer, i.e., focused reflections.

If the facility is already built and functions are taking place, be sure to attend a "performance" before installing the equalization and follow up with a visit at a performance after equalization. It is important to test the equalization with speech. Often only music will be played which gives little idea of speech clarity and often a separate equalization is required for speech and music.

In one case, whenever the equalization was made uniform to the desired house curve, there was not enough acoustic gain in the audience area. When the acoustic gain was raised by the feedback method, the shape of the house curve in the audience area was unacceptable. Analysis with an RTA revealed that the loudspeaker array mounted in the proscenium area was not properly shock mounted and was causing the arch structure to reradiate a signal downward to the microphones, causing premature feedback. When this large array was properly shock mounted and properly isolated in the proscenium arch area, then the house curve could be shaped as desired, and the acoustic gain could be brought to its potential at the same time.

Another cause of the same effect is when the rear of the proscenium is open to the stage house and the rear of the array has substantial radiation of its own into the microphone. The house curve in the audience may be a good response, yet the microphone on stage has severe feedback problems because it is receiving a strong radiation from the loudspeaker, usually in the bass region.

Mother nature's way

There are in audio and acoustics both natural and unnatural distortions. An example of a natural distortion is harmonic distortion because we hear harmonics in nature. An example of an unnatural distortion is print through on tape recordings where we hear the echo first followed by the desired sound. This never occurs in nature, thus our brain is extremely sensitive to its occurrence. So it is with "boost filters."

In nature any two signals can combine and go to various depths depending upon the relative phase angle between them. Under ideal conditions, they can only add to 3 dB greater levels and psychologically we notice the presence of something far more than its absence. Subjectively, over more than 40 years of system equalization in the field, qualified investigators report slope rates in excess of 18 dB/octave as audible. Therefore, filter sets (equalizers) should be designed to avoid steep slope rates, be combining, and not introduce unnatural distortions (boosting).

A REAL-TIME REGENERATIVE-RESPONSE METHOD OF EQUALIZING A SOUND SYSTEM

One of the earliest demonstrations of the effect of operation near regeneration on the measured frequency response of a sound reinforcement system was by William B. Snow before the February 1954 meeting of the Audio Engineering Society in Los Angeles. Snow recorded on a high-speed graphic level recorder the dramatic amplitude changes that occurred in the overall amplitude-versus-frequency response as the reinforcement system was brought nearer and nearer regeneration.

This same method is still used with a manually operated oscillator to identify those frequencies whose amplitude responded unduly to the approach of the regeneration point of the sound system. Shock excitation was employed to observe the increased decay periods of frequencies that otherwise would not feed back upon being increased in gain but were unduly affected by the approach of the system regeneration point. Snow's paper had also demonstrated this point, proving that such frequencies, when shock excited near regeneration, could take as much as 4 to 6 times as long to decay as the same frequencies required when they were excited well below the regeneration point (−12 dB).

Over 40 years of active participation in equalization of sound systems has shown us that making regenerative response curves is one of the most useful techniques applicable to equalizer adjustment. More often than not, the frequencies that ring as they are swept do not come up into steady-state-feedback yet they interfere with speech intelligibility and the overall acoustic gain of the system. The regenerative response curve technique allows careful analysis of both the electrical and the acoustic responses while allowing both the regenerative and degenerative frequencies to be identified.

Where to put the microphone for regenerative response?

Where to place the microphone for viewing the equalization with an analyzer is one question. The second question is where do I place the microphone that is to cause the regeneration of the signal in the sound system?

One good practice is to place a measuring microphone out in the main coverage pattern looking at the array. Then place the regenerative microphone (the microphone that is to be used to cause the system to go into acoustic feedback during the test) in the location where it is to be employed for normal system usage.

Degree of correction necessary

Interconnect the instruments and the sound system as illustrated in Figure 18.32. The instrument labeled "analyzer receive" can be a 1/3 octave of 1/1 octave real-time analyzer, a TEF analyzer, a high-quality wave analyzer, or a suitable FFT. The instrument labeled "analyzer send" can be a logarithmic sweep oscillator (for use with the constant percentage bandwidth analyzers), a linear sweep oscillator (for use with the TEF or FFT), or a random noise generator. In fact, it can be just about any controlled source that is capable of exciting the sound system's entire bandpass.

Clearly evident in this process is the fact that below regeneration the amplitudes of the frequencies involved are of the order of 1 or 2 dB higher than other nearby frequencies. As these frequencies approach regeneration, their amplitudes can "swell" to +20 dB or more. Naturally the sound system must supply the power for this regeneration, as the room cannot—it is passive. The room can provide nulls and peaks, but not amplification. If, before regeneration is approached, one of these frequencies has its amplitude brought to uniformity with the remainder, then upon approaching regeneration again, this frequency will not "swell" in amplitude.

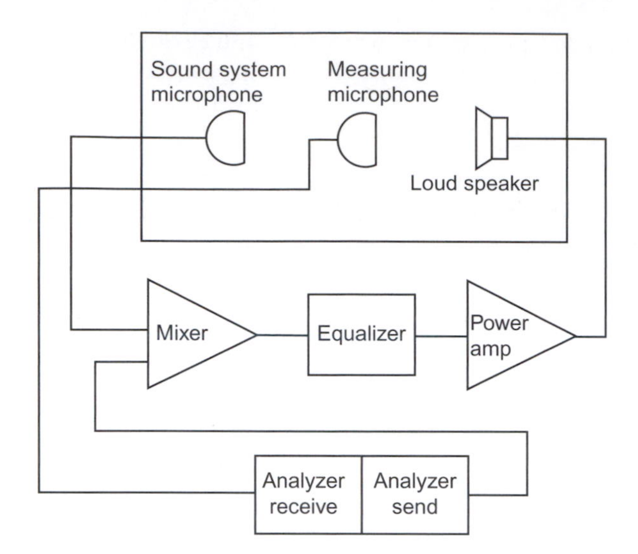

FIGURE 18.32
Regenerative response tuning.

By using sweeps from an oscillator, it is then possible to watch what happens to the sound system's electrical and acoustic responses as well as to listen to the transient response. (Measure the transient response effects if a TEF analyzer is available.)

The ability to make the needed controlling corrections at a magnitude associated with the nonregenerative state of the system was first observed by the authors in 1966 and was presented as a paper.*

Using sweep oscillators

Rapid sweep oscillators allow very effective regenerative response tuning. By rapid rate we mean a full sweep from 20 Hz to 20,000 Hz at about one-half the time it takes mid-range sound to decay 60 dB in the environment where the tuning takes place. Typically, sweep rates of 10 kHz/s are useful. If searching for a low frequency anomaly, use a logarithmic sweep. If searching for a high frequency anomaly, use a linear sweep.

This rapid sweep will cause all the bands on a 1/3 octave equalizer to jump up on the screen and then drop at the integration rate of the analyzer (usually 0.1 s at the "fast" rate). A "ringing" band will drop slower than the rest of the bands being shock excited by the sweep. The sound created by the sweep passing through the system and shock exciting the nonlinear frequencies has a "gong-like" tone not unlike that heard in older department stores to summon or alert personnel. As the equalizer's amplitude is adjusted slowly while listening to the "gonging," it turns into a gong-in-a-pillow sound, indicating sufficient attenuation is present at that frequency.

Pink noise for regenerative excitation is more useful at lower frequencies. Bringing the system to within a few dB of regeneration using pink noise allows the room and sound system to "display" which frequencies are unduly sensitive to approaching regeneration including those due to phase as well as amplitude.

EQUALIZING FOR PLAYBACK

The real-time regenerative response method can be startlingly effective in the equalization of sound reproduction systems or sound synthesizing systems. In this case, instead of using a performer's microphone to achieve regeneration, the calibrated measuring microphone is simultaneously used for both room-response measurement and regeneration. Multichannel systems tuned using this technique are characterized by superior spatial geometry as well as improved tonal response. The improved reproduction of geometry is believed to be due to the better acoustic phase response between channels at the listener's position, and, while only a small area so benefits, it is usually only the mixer's general area that has to be covered in the typical studio monitoring room situation. Some unusually extended-range systems may now feed back at frequencies well above audibility (in one case above 30,000 Hz), and

*D. Davis. "Adjustable 1/3 Octave Band Notch Equalizer for Minimizing Detrimental Interaction Between a Sound System and Its Acoustic Environment," presented at SMPTE Meeting in Chicago (Sept. 1967) and AES Meeting in New York (Oct. 1967).

care must be taken to use a low-pass filter in conjunction with either the sound system or the measuring system. Of course, attention must be paid to avoid tuning in the null of a standing-wave pattern. A short walk with the microphone of the analyzer, especially in a small control room, is fascinating, instructive, and necessary.

AN IMPROPER USE OF REAL-TIME ANALYSIS IN MONITORING MUSIC AND SPEECH

Constant percentage bandwidth filters have absolute widths that increase in direct proportion to the center frequency of the filter. When performing spectrum analysis with instruments based on such filters it is necessary to employ a random noise source whose spectrum has constant energy per octave, i.e., pink noise as opposed to a noise source that has constant energy per unit bandwidth, i.e., white noise.

A system possessing a uniform or flat response on a per unit bandwidth basis that is excited with pink noise will produce a flat display on a constant percentage bandwidth analyzer. Such a system excited with white noise would produce a response that rises at 3 dB/octave on a constant percentage bandwidth analyzer. Therefore, when constant percentage bandwidth analyzers are employed to study the spectra of program material where it is desired to determine the response displayed on a per unit bandwidth basis, it is necessary to precede such an analyzer by a filter that has a response that falls at the rate of 3 dB/octave. Any evaluation of program material without such a device is invalid.

It is the authors' belief that this uncorrected error is why so many professional mixing engineers still use meters and indicators in place of the much more useful real-time analyzer. Trained ears didn't agree with the uncorrected visual display. The noise control people made their criteria constant percentage bandwidth based, thereby judging relative results. The recording engineers, home hi-fi enthusiasts, and other researchers did not realize the need and therefore failed to compensate for it.

DIAPHRAGMATIC ABSORBERS

Care should be observed in the handling of dips in the response of a loudspeaker and a room caused by diaphragmatic action of some boundary surface. This is identifiable when, after all the bands around the dips are brought down, they still fall the same number of dB below the surrounding bands. Do not chase it on down, because that will only increase the insertion loss of the total equalization with but negligible improvement in the response. The correct method is to drive the loudspeaker room combination with a tunable bandpass filter and observe the effect on the real-time analyzer to find the frequency where the absorption of the signal is greatest; then use your fingertips and feel all the surfaces of the space, including walls, doors, windows, etc. You will feel the offending surface vibrating in sympathy with the test signal, in one case a walled-up window area.

At a famous recording studio during a demonstration of equalizing monitor loudspeakers, a diaphragmatic absorption was traced to a loose "sound-lock" door. Upon holding the door tightly shut, an 80 Hz notch in the house curve disappeared.

Room absorption at specular frequencies

It has been common practice for the past 50 years to adjust the high frequency response of sound systems to compensate for high frequency absorption in the room. Analysis suggests that the loss of high frequencies being compensated for does not occur as a result of the absorption present, but as a result of L_w lowering rapidly at about the same frequency Q increases, with the resultant illusion that the response is uniform, but duller. The most common cause of radical high frequency loss in sound systems is either device misalignment or a high level, very early reflection (i.e., within 2 ms or less).

House curves

A famous acoustician once was heard to say that the "house curve" (i.e., the response as viewed on a 1/3 octave real-time analyzer) should be down 10 dB at 10 kHz referenced to 1 kHz. What his listeners forgot to consider was where he was standing. It was 70 feet to 80 feet out in front of a horn-type loudspeaker system. At that distance, when you take into account air absorption, microphone diffraction characteristics, and high-frequency distortion components, 10 dB down seems quite sensible. When you are in a control room 10 feet from a loudspeaker the 2 dB to 3 dB typical of microphone diffraction at 10 kHz is more logical and air absorption is not a factor. When necessary to err, then err on the side of a little extra rolloff

DON'T EQUALIZE FOR HEARING LOSS

Many times there is a tendency to attempt to adjust the amplitude response of a sound system to make it the inverse of the hearing-loss curve. This is not a good idea for several reasons:

1. Young people with normal hearing will be annoyed.
2. Older people have made mental compensation for the gradual onset of the loss and would also be annoyed. They usually desire the overall level higher.
3. Available high-frequency drivers would have their distortion increased noticeably with such a boost.

PROXIMITY MODES

The microphone proximity effect, traditionally referred to in the technical literature, is the effect of increased bass response in the microphone as the talker gets closer to the unit. This remains true of most unidirectional microphones today and is often effectively used by trained performers to enhance their otherwise weak bass tones. Since the advent of sound system equalization, however, we have become aware of still another effect of the proximity of large bodies (performers) on a typical cardioid microphone. That is the increased tendency to feedback at some key midrange frequency where the system is otherwise stable until the microphone is approached. You can use your hands cupped around the microphone to bring the system into feedback and can adjust the level of feedback by "playing" the microphone. In adjusting the appropriate filter, care should be taken not to carry the adjustment too far. The idea is to correct the tendency of the microphone to cause instability when it is approached by the performer and not to remove all tendency toward feedback even when the microphone is completely encircled by a closed hand. TEF analysis has shown that this instability is caused by "comb filters" produced by reflected sound from the performer.

One classic example was Dan Seals who is a left-handed guitar player and was having trouble with acoustic feedback whenever he turned to his left at the microphone. He allowed us to make a measurement with the same setup as when he was performing. Figure 18.33 shows our measurement. The guitar reflection and the hat brim reflection combined acoustically at the microphone to cause a genuine excess gain problem. When he turned to the left, the body of his guitar reflected the left monitor towards the microphone and his hat brim reflected the right monitor to the same place. When Dan Seals saw the measurement, he said, *"We have met the enemy and they is us!"*

CHECKING MICROPHONE POLARITY

Surprisingly, one minor checkout prior to equalization time that often is overlooked is the poling of the microphones in a multimicrophone system. The old way was to arbitrarily assume that the first

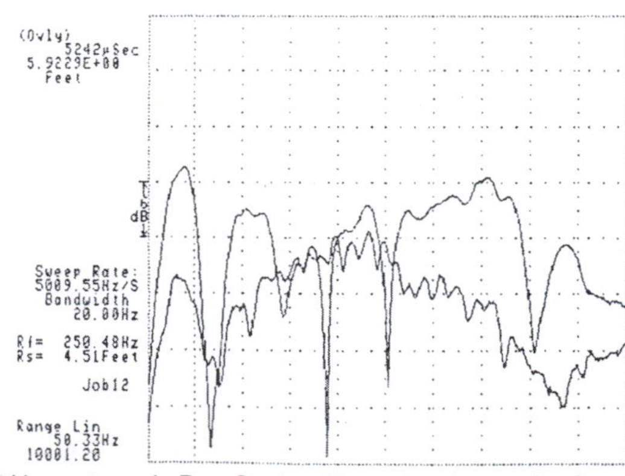

FIGURE 18.33
Effects of surfaces on feedback.

microphone you picked up was correctly poled. Holding it in one hand and the second microphone in the other hand, and bringing them closer and closer together while talking into them (such as *"one-one-one"*). They were in polarity if the apparent bass response increased as they were brought closer together in front of your mouth. They were out of polarity if the bass weakened as they were brought together. In one memorable case the "first" microphone was reversed and this simple process reversed all the others. The arrival on the scene of a polarity checker revealed the error. In any case, be sure to check this important factor before equalizing. A polarity reverser is invaluable in this work. Today we know to check for absolute polarity as it has been repeatedly demonstrated that it is audible on speech. TEF analyzer phase measurements instantly indicate the correct polarity (as well as, in the TEF case, the difference between polarity and phase). Be careful before you rewire microphones; the patch cords could be miswired.

LOUDSPEAKER POLARITY

In examining the "raw" response of a loudspeaker array, pick the poling that gives the most usable response through the crossover region. True phasing can enter in here, as well as polarity, and great care should be exercised in the relative positioning of horns to each other, especially the spacing and positioning of the high-frequency elements in relation to the low-frequency elements. Remember, out in the audience area there will be phase relationships between direct and reflected sound as well as those between two direct sound sources. The real-time analyzer is invaluable for examining the potential variations and their effects on the audience area. Today, through TEF analysis we have identified signal misalignments of from fractions of an inch to about one foot as particularly hazardous to speech quality.

SUMMARY

The advent of practical sound system equalization in situ in the late 1960s coincided with the development of portable 1/3 octave constant percentage bandwidth real-time analyzers which led to a revolution in the design, installation, and operation of sound reinforcement systems.

The availability of equalizers and analyzers quickly led to the training of large numbers of alert sound contractors, consultants, and operators in sound system measurements. Proper design led to much more powerful loudspeaker arrays constructed by those knowledgeable about directivity factor, comb filter interference, and signal delay and synchronization. Manufacturers responded with vastly improved data.

Today we have unimagined design aids, loudspeaker data, and a cadre of knowledgeable users. Today equalization is a small component in the cornucopia of tools available, but it does have the satisfaction of having been the catalyst to dramatic improvement in the design and installation of outstanding sound systems.

BIBLIOGRAPHY

Benson, J.E., Craig, D.F., Mar. 1969. A feedback-mode analyser/suppresser unit for auditorium sound-system stabilisation. Proc. IREE Australia.

Connor, W.K., Apr. 1967. Theoretical and practical considerations in the equalization of sound systems. Audio Eng. Soc. 15.

Davis, D., 1969. Facts and fallacies on detailed sound system equalization. Audio.

Davis, D. May 1973. A real time regenerative response method of equalizing a sound system. AES Paper.

Patronis, G., Elementary system theory Spring 1994. Syn-Aud-Con Tech Topic 21 (3).

Rudmose, W., July 1958. Equalization of sound systems. Noise Contr. 24.

Schroeder, M.R., Sept. 1964. Improvement of acoustic-feedback stability by frequency shifting. J. Acoust. Soc. Am. 36.

Snow, B., Apr. 1955. Frequency characteristics of a sound-reinforcing system. J. Audio Eng. Soc. 3.

Stanley, G., 1978. Minimum phase: defined and illustrated. Syn-Aud-Con Tech Topic 5 (10).

PREFACE

The final chapter in this section on Acoustics and Sound Reinforcement comes from *Sound Reproduction* by Floyd Toole. It deals with the propagation of sound in spaces, and how that propagation interacts with human perception.

The most important phenomenon in this area is the precedence effect, otherwise known as the Haas effect, after Helmut Haas who first described it in his Ph.D. thesis. It was first published in English in 1949. The essentials of the effect are that when two identical sounds, that is, identical sounds of the same perceived intensity, originate from two sources at different distances from the listener, the sound from the closest location is heard first, and this creates the impression that the sound comes from the first location alone. The Precedence Effect occurs only when arrival times of the sounds differ by more than 1ms and less than 30–40ms. When the arrival time difference is less, the sound is perceived as coming from somewhere between the two actual sources, as in stereo reproduction through a pair of loudspeakers. As the arrival time differences increase beyond 40ms, the sounds will begin to be heard as distinct; in other words with an echo.

After a very clear and informative description of the experiments that Haas performed, Floyd goes on to emphasize that the precedence effect is *not* a masking effect, where the first sound completely suppresses the later one so it might as well not have happened. The process is better described as fusion or integration, whereby the later sound *is* perceived, because it adds to the aggregate volume that is apparently concentrated in the first sound, but does not register itself as a distinct acoustic event.

He goes on to show how a single reflection from a wall relates to the precedence effect, describing several important experiments that have been done on this subject, and then covers the situation when there are multiple reflections from different directions.

The final section of the chapter deals with the measurement of sound reflections, and their presentation as three-dimensional "waterfall" curves that plot energy and frequency against elapsed time.

The precedence effect is of fundamental importance in sound reinforcement because it can ensure that the perceived location of the original sound remains as desired. In many cases, and particularily when dealing with large areas of listeners at outdoor concerts, some of the loudspeaker clusters must be placed at some distance from the stage, because if all the sound originated from one point the sound pressure levels near to it would be intolerable. The signal to the remote loudspeakers can then be electronically delayed for a period equal to or slightly greater than the time taken for the original sound to travel to the position of the remote loudspeaker. This means that the sound is still perceived as coming from the point of origin—the stage—rather than from a loudspeaker that is nowhere near it. The operation of perceptible thresholds in the precedence effect is such that the level of the delayed signal can be up to 10dB above that of the original signal at the listener's location without disrupting this extremely useful illusion.

CHAPTER 19

Reflections, Images, and the Precedence Effect

Sound Reproduction by Floyd Toole

AUDIBLE EFFECTS OF A SINGLE REFLECTION

Investigations of these effects go back many decades, and observations of our ability to localize a source of sound in an acoustically hostile—that is, reflective—environment were first recorded more than a century ago.

In audio in the past, the terms *Haas effect* and *law of the first wavefront* were used to identify this effect, but current scientific work has settled on the other original term, *precedence effect*. Whatever it is called, it describes the well-known phenomenon wherein the first arrived sound, normally the direct sound from a source, dominates our impression of where sound is coming from. Within a time interval often called the "fusion zone," we are not aware of reflected sounds that arrive from other directions as separate spatial events. All of the sound appears to come from the direction of the first arrival. Sounds that arrive later than the fusion interval may be perceived as spatially separated auditory images, coexisting with the direct sound, but the direct sound is still perceptually dominant. At very long delays, the secondary images are perceived as echoes, separated in time as well as direction. The literature is not consistent in language, with the word *echo* often being used to describe a delayed sound that is not perceived as being separate in either direction or time.

Haas was not the first person to observe the primacy of the first arrived sound so far as localization in rooms is concerned (Gardner, 1968, 1969, 1973 describes a rich history), but work done for his Ph.D. thesis in 1949, translated from German to English in Haas (1972), has become one of the standard references in the audio field. Sadly, his conclusions are often misconstrued. Let us review his core experiment (Figure 19.1).

Figure 19.2 shows the essence of the experiment. On the hemi-anechoic space provided by the flat roof of a laboratory building, a listener was positioned facing loudspeakers that had been placed 45° apart. The Haas (1972) translation describes the setup as "at an angle of 45° to the left and right side of the observer" (p. 150). This could be construed in two ways. Gardner (1968), however, in a translation of a different Haas document, reports "loudspeakers... at an angle of 45°, half to the right and half to the left of him...." When Lochner and Burger (1958) repeated the Haas experiment, they used loud speakers that were placed 90° apart. So there is ambiguity about the angular separation.

A recording of running speech was sent to both loudspeakers, and a delay could be introduced into the signal fed to one of them. In all situations except for Figure 19.2d, both signals were radiated with the same sound level.

Figure 19.2a shows summing localization. When there is no delay, the perceived result was a phantom (stereo) image floating midway between the loudspeakers. When delay was introduced, the center image moved toward the loudspeaker that radiated the earlier sound, reaching that location at delays of about 0.6–1.0 ms. This is called summing localization, and it is the basis for the phantom images

Audio Engineering Explained

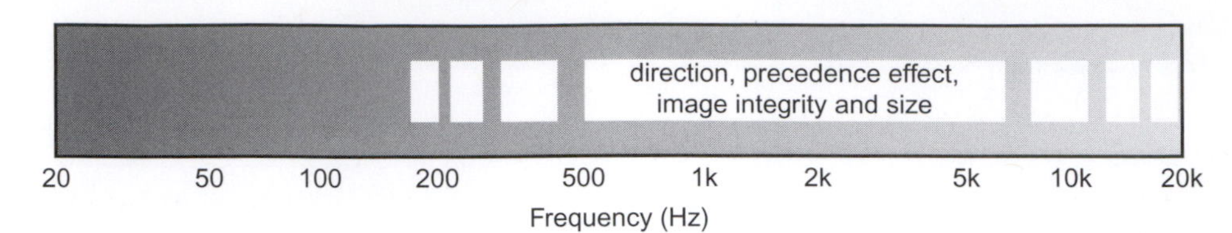

FIGURE 19.1
The approximate frequency range over which reflections appear to influence perceptions of the direction of a sound source and the apparent size of that source. In some circumstances, reflections may be audible as separate “images” of the sound source.

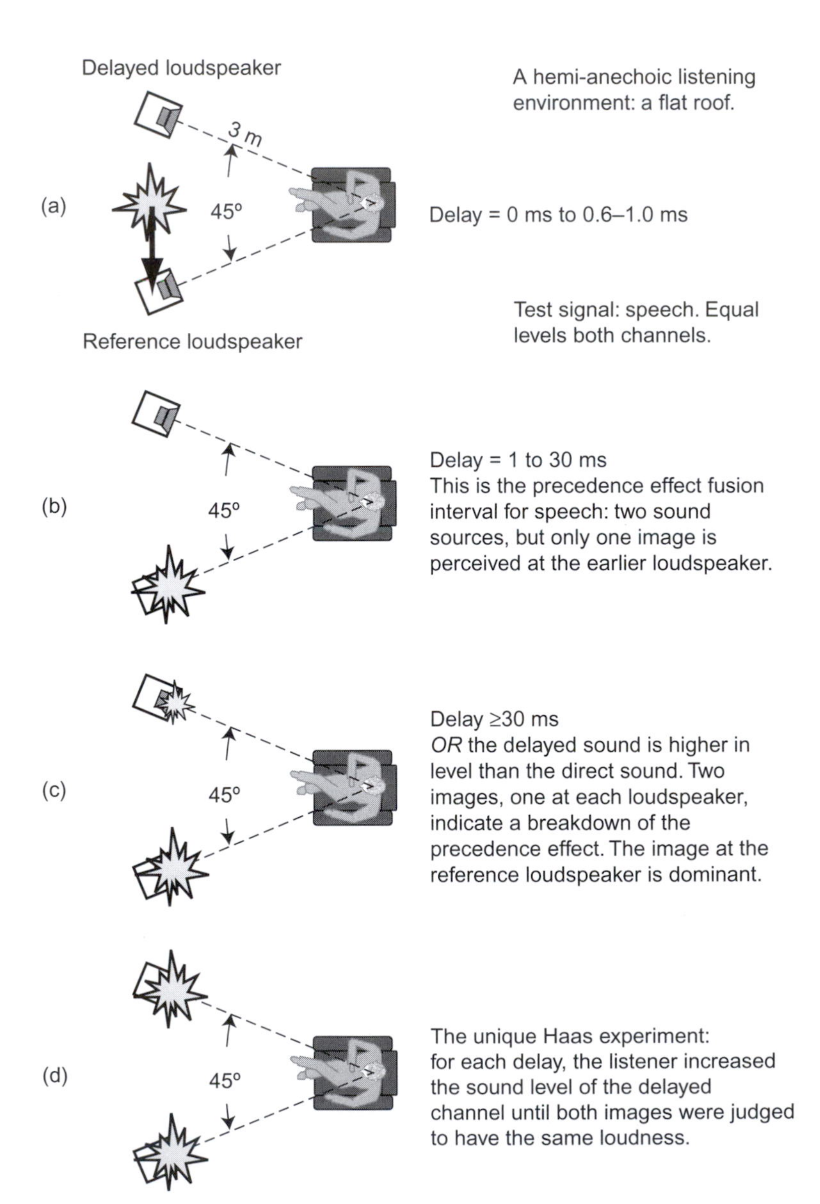

FIGURE 19.2
A progression of localization effects observed in the experimental setup used by Haas, including stereo (summing) localization, the precedence effect, and the equal-loudness experiment. Because the experiments were done on a flat roof, to minimize the effect of the roof reflection, Haas placed the loudspeakers directly on the roof, aimed upward toward the listener’s ears. He found, though, that there was no significant difference if the loudspeakers were elevated to ear level, and that is the configuration used for the experiments.

that can be positioned between the left and right loudspeakers in stereo recordings, assuming a listener is in the "sweet spot" (Blauert, 1996).

Figure 19.2b shows the precedence effect. For delays in excess of 1 ms, it is found that the single image remains at the reference loudspeaker up to about 30 ms. This is the precedence effect—that is, when there are two (or more) sound sources and only one sound image is perceived. It needs to be noted that the 30 ms interval is only for speech and only for equal level direct and reflected sounds.

Figure 19.2c shows multiple images—the breakdown of precedence. With delays greater than 30 ms but certainly by 40 ms, the listener becomes aware of a second sound image at the location of the delayed loudspeaker. The precedence effect has broken down because there are two images, but the second image is a subordinate one; the dominant (louder) localization cue still comes from the loudspeaker that radiated the earlier sound.

Figure 19.2d shows the Haas equal-loudness experiment. In the first three illustrations, the first and delayed sounds had identical amplitudes. Obviously, this is artificial because if the delayed sound were a reflection, it would be attenuated by having traveled a greater distance. But Haas moved even farther from passive acoustical realities and deliberately amplified the delayed sound, as would happen in a public address situation. His interest was to determine how much higher in sound level the delayed sound could be before it became the dominant localization cue—in other words, subjectively louder. To do this, he asked his listeners to adjust the sound level of the delayed loudspeaker until both of the perceived images appeared to be equally loud. This is the balance point, beyond which the delayed loudspeaker would be perceived as being dominant. The objective was to prevent an audience from seeing a person speaking in one direction and being distracted by a louder voice coming from a different direction. As shown in Figure 19.3, over a wide range of delays, the later loudspeaker can be as much as 10 dB higher in level before it is perceived to be equally loud and therefore a major distraction to the audience. Naturally, this would depend on where the audience member is seated relative to the symmetrical axis of the two sound sources.

Haas described this as an "echo suppression effect." Some people have taken this to mean that the delayed sound is masked, but it isn't. Within the precedence effect fusion interval, there is no masking—all of the reflected (delayed) sounds are audible, making their contributions to timbre and loudness, but the early reflections simply are not heard as spatially separate events. They are perceived as coming from the direction of the first sound; this, and only this, is the essence of the "fusion." The widely held belief that there is a "Haas fusion zone," approximately the first 20 ms after the direct sound, within which everything gets innocently combined, is simply untrue.

Haas observed audible effects that had nothing to do with localization. First, the addition of a second sound source increased loudness. There were some changes to sound quality "liveliness" and "body" (Haas, 1972, p. 150) and a "pleasant broadening of the primary sound source" (p. 159). Increased loudness was a benefit to speech reinforcement, and the other effects would be of concern only if they affected intelligibility.

Benade (1985) contributed a thoughtful summary under the title "Generalized Precedence Effect," in which he stated the following:

1. The human auditory system combines the information contained in a set of reduplicated sound sequences and hears them as though they were a single entity, provided (a) that these sequences are reasonably similar in their spectral and temporal patterns and (b) that most of them arrive within a time interval of about 40 ms following the arrival of the first member of the set.

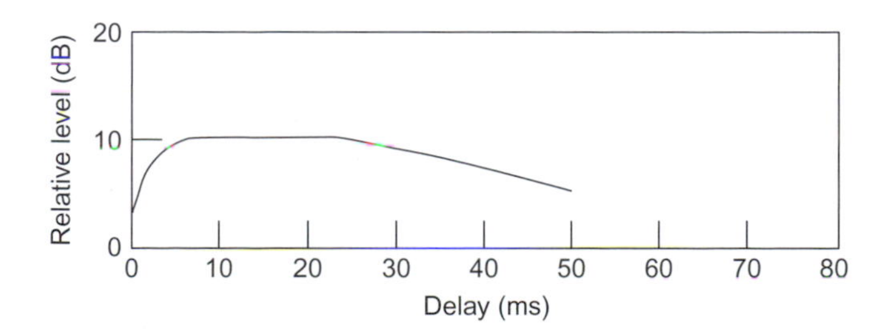

FIGURE 19.3
The sound level of the delayed sound, relative to that of the first arrival, at which listeners judged the two sound images to be equal in loudness.

2. The singly perceived composite entity represents the accumulated information about the acoustical features shared by the set of signals (tone color, articulation, etc.). It is heard as though all the later arrivals were piled upon the first one without any delay—that is, the perceived time of arrival of the entire set is the physical instant at which the earliest member arrived.
3. The loudness of the perceived sound is augmented above that of the first arrival by the accumulated contributions from the later arrivals. This is true even in the case when one or more of the later signals is stronger than the first one to arrive—that is, a strong later pulse does not start a new sequence of its own.
4. The apparent position of the source of the composite sound coincides with the position of the source of the first-arriving member of the set, regardless of the physical directions from which the later arrivals may be coming.
5. If there are any arrivals of sounds from the original acceptably similar set which come in after a delay of 100–200 ms, they will not be accepted for processing with their fellows. On the contrary, they will be taken as a source of confusion and will damage the clarity and certainty of the previously established percept. These "middle-delay" signals that dog the footsteps of their betters may or may not be heard as separate events.
6. If for some reason a reasonably strong member of the original set should come in with a delay of something more than 250–300 ms, it will be distinctly heard as a separate echo. This late reflection will be so heard even if it is superposed on a welter of other (for example, reverberant) sounds.

It is important to notice that these very strongly worded categorical statements all emphasize that there is an *accumulation of information* from the various members of the sequence. It is quite incorrect to assume that the precedence effect is some sort of masking phenomenon which, by blocking out the later arrivals of the signal, prevents the auditory system from being confused. Quite to the contrary, those arrivals that come in within a reasonable time after the first one actively contribute to our knowledge of the source. Furthermore, members of the set that are delayed somewhat too long actually disrupt and confuse our perceptions even when they may not be consciously recognized. If the arrivals are later yet, they are heard as separate events (echoes) and are treated as a nuisance. In neither case are the late arrivals masked out.

Effects of a single reflection

This is the "begin at the beginning" experiment, in which the number of variables is minimized. The listening environment is anechoic, the signal is speech, and only a single lateral reflection is examined. It is not data that can be applied to real-world circumstances listening to music or movies, but it is scientific data that establishes a baseline for further research (Figure 19.4).

In Figure 19.5, the lowest curve describes the sound level at which listeners reported hearing any change attributable to the presence of the reflection. This is the "absolute threshold"; nothing is perceived for reflections at lower levels. Most listeners described what they heard as a sense of spaciousness (Olive and Toole, 1989). Although the experiment was conducted in an anechoic chamber, a single detectable reflection was sufficient to create the impression of a (rudimentary) three-dimensional space. Throughout, listeners reported all of the sound as originating at the location of the loudspeaker that reproduced the first sound, meaning that the precedence effect was working. As the sound level of the delayed sound was increased, the impression of spaciousness increased.

The next higher curve is the level at which listeners reported hearing a change in size or position of the main sound image, which the precedence effect causes to be localized at the position of the loudspeaker that reproduced the earlier sound. This was called the "image shift" threshold. In general, these changes were subtle and noticeable in these controlled A versus B comparisons, but it is doubtful that they would be detected in the context of a multiple-image music or movie soundstage. As the sound level of the delayed sound was further increased, the impression of spaciousness also increased.

With the two curves that portray the third perceptual category, a major transition is reached, because it is at this sound level that listeners report hearing a second sound source or image, simultaneously coexisting with the original one (we have not reached the long delays at which there is a sense of a temporally as well as a spatially separate echo). Data from Lochner and Burger (1958) and Meyer and Schodder (1952). This means that the precedence effect directional "fusion" has broken down.

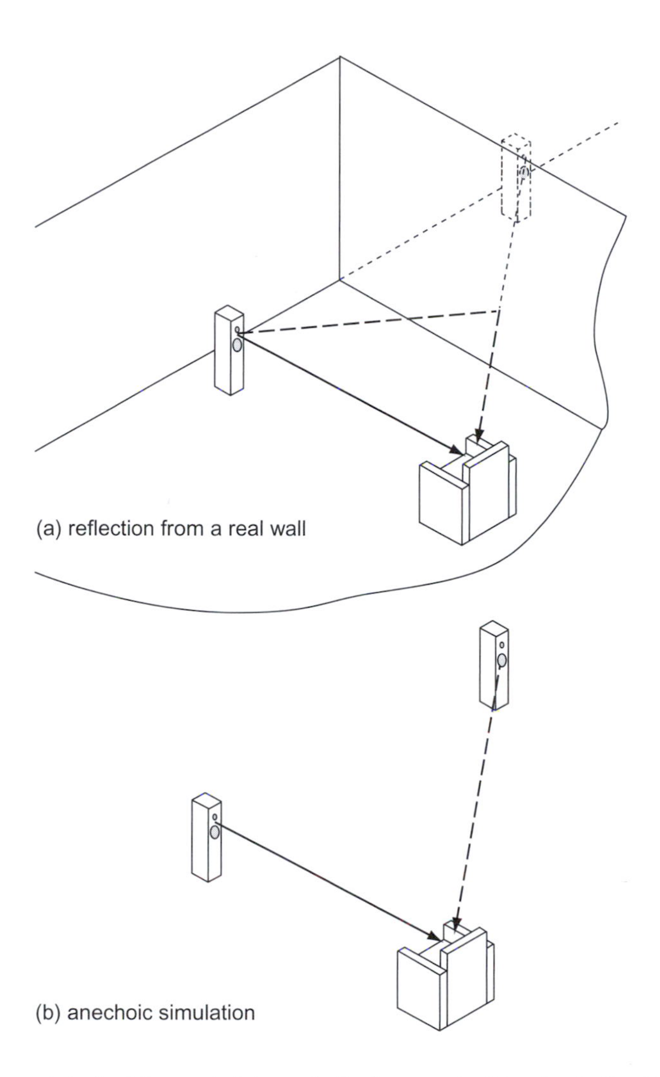

FIGURE 19.4
An explanation of how an anechoic simulation can imitate a reflection from a real—flat and perfectly reflective—wall. The anechoic setup uses a real loudspeaker to simulate the "mirror image" loudspeaker in the room situation. This is the experimental method that has been used in numerous experiments conducted over the decades. Electrical adjustments of delay, amplitude, and frequency response of the signal sent to the "reflection" loudspeaker allow the simulation of different geometries and reflective surface types.

Although the original source remains the perceptually louder, spatially dominant source, there is a problem because two spatial events are perceived when there should be only one.

The top curve is from the well-known work by Haas (1972) in which he asked his listeners to adjust the relative levels of the spatially separate images associated with the direct and reflected sounds until they appeared to be equally loud. This tells us that in a public address situation, it is possible to raise the level of delayed sound from a laterally positioned loudspeaker by as much as 10 dB above the direct sound before it is perceived as being as loud as the direct sound. It is important information in the context of professional audio, but it is irrelevant in the context of small-room acoustics.

All of the data points are thresholds—the sound levels at which listeners detected a change in their perceptions. As we will see later, some of the perceived changes are beneficial and, up to a point, listeners find that levels well above threshold provide greater pleasure. For example, the perception described at threshold as "image shift or spreading" may seem like a negative attribute, but when it is translated into what is heard in rooms, it becomes "image broadening" or apparent source width (ASW), which are widely-liked qualities. Even "second-image" thresholds can be exceeded with certain kinds of sounds, expanding the size of an orchestra beyond its visible extent in a concert hall or extending the stereo soundstage beyond the spread of the loudspeakers. In reproduced sound, the picture is more confused because some techniques in the recording process can achieve similar perceptions. Because

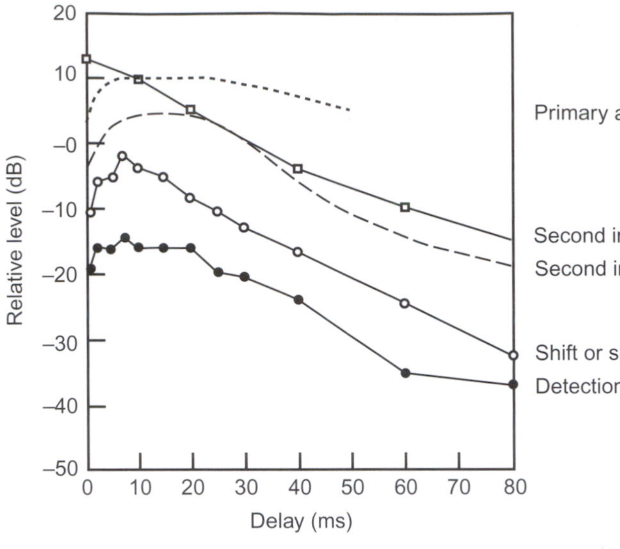

FIGURE 19.5
An illustration of the several audible effects that occur when a single lateral reflection is added to a direct sound, in an anechoic simulation similar to that shown in Figure 19.4b. All of these curves were determined using speech as a signal. In the experiments, at each of several delays, the sound level of the reflected sound was adjusted to identify those levels at which each of the described perceptions became apparent. The bottom two curves are from Olive and Toole (1989), in which the direct sound was at 0° and the lateral reflection arrived from 65°. Meyer and Schodder (1952) had their reflection arrive from 90°, and listeners reported when the echo was not perceived at all. Lochner and Burger (1958) employed a direct sound arriving from −45° and a delayed sound from +45°, and their listeners reported when the second source was just audible. Adapted from Toole (1990), with additional information from Cremer and Müller, 1982, Figure 1.25.

all of these factors are influenced by how the recordings are made as well as how they are reproduced, these comments are observations, not judgments of relative merit. Some evidence suggests that even these small effects might be diminished by experience during listening within a given room (Shinn-Cunningham, 2003), another in the growing list of perceptual phenomena we can adapt to.

Another view of the precedence effect

If we extract from Figure 19.5 those things that are relevant to sound reproduction from a single loudspeaker, we end up with Figure 19.6. The Haas data have been removed because amplified delayed sounds do not exist in passive acoustics. The "second-image" data (Lochner and Burger, 1958; Meyer and Schodder, 1952) have been combined into a single average curve for simplicity. There is some justification for doing this, as one curve expresses a "just audible" criterion and the other a "just not audible" criterion.

The area under the "second-image" curve has been shaded. This is the real-world precedence effect fusion zone for speech, within which any delayed sound will not be perceived as a spatially separate localizable event. This perspective is very different from most discussions of the precedence effect fusion interval. Normally, only a single number is stated, and that number normally relates to direct and delayed sounds at the same sound levels. This is a correct description of the results of a laboratory experiment but is simply wrong as guidance about what may happen in the real world.

The fusion interval for speech is often quoted as being around 30 ms. This is true for anechoic listening to a single reflection that has the same sound level as the direct sound, as can be seen in Figure 19.6b. This is how the classic psychoacoustic experiments were conducted, but these circumstances are far from the acoustical realities in normal rooms. For reflections at realistically lower levels, the fusion interval is much longer. So far, in small rooms, the precedence effect is undoubtedly the dominant factor in the localization of speech.

Reflections from different directions

Figure 19.7 shows more data from Olive and Toole (1989), in which it is seen that the thresholds for the side wall and the ceiling reflections are almost identical. This is counterintuitive because one

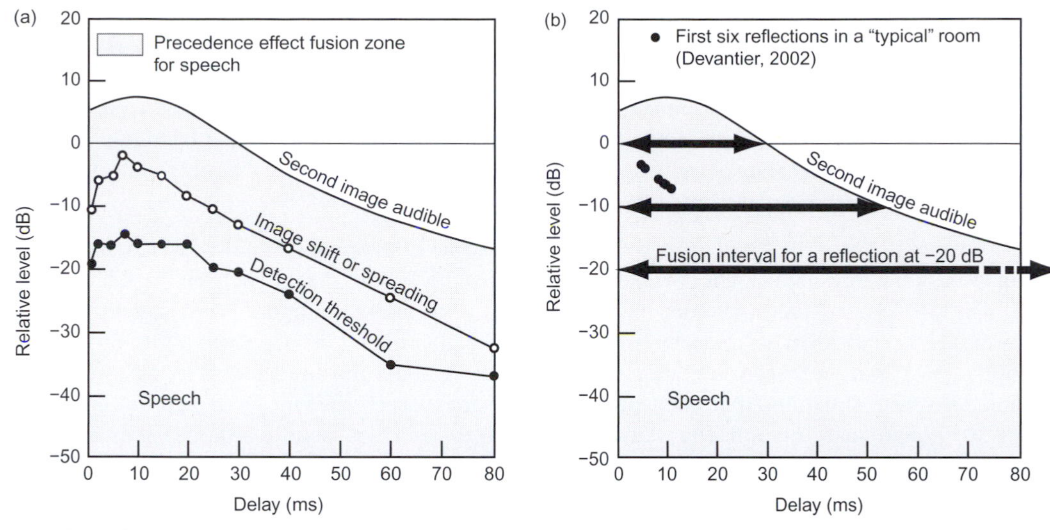

FIGURE 19.6

(a) A simplification of Figure 19.5 in which only data that are relevant to sound in small rooms are preserved and a shaded area representing the precedence effect fusion zone for speech is identified. This is the range of amplitudes and delays within which a reflected sound will not be identified as a separately localizable event. From Toole, 2006. (b) The precedence effect fusion intervals for delayed sounds at three sound levels. The classic experiments much quoted from psychoacoustic literature generally used equal-level direct and delayed sounds. This is the highest large arrow at 0 dB, showing an interval of about 30 ms. In rooms, delayed sounds are attenuated by propagation loss, typically −6 dB/double distance, and sound absorption at the reflecting surfaces. As the delayed sound is reduced in level, the fusion zone increases rapidly. The set of black dots shows the delays and amplitudes for the first six reflections in a typical listening room (Devantier, 2002), indicating that in such rooms the precedence effect is solidly in control of the localization of speech sounds.

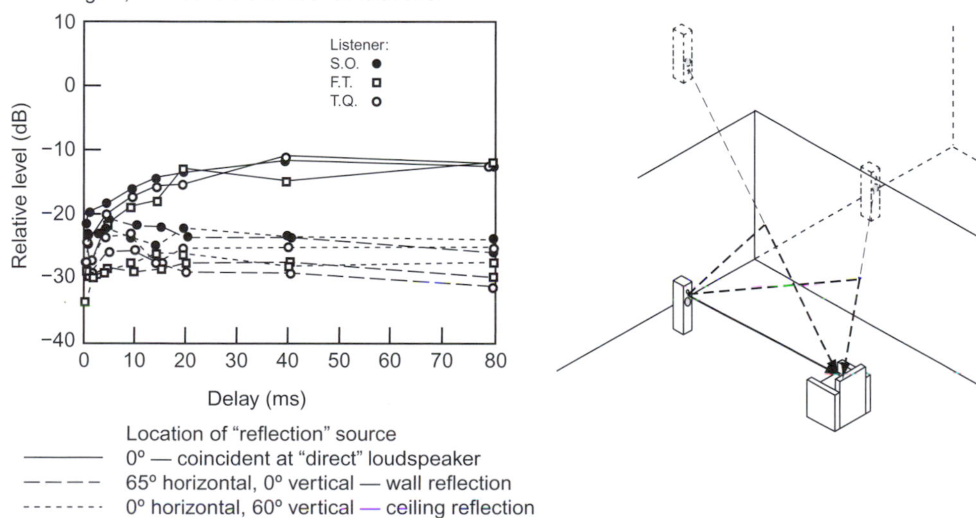

FIGURE 19.7

The detection thresholds for delayed sounds simulating a wall reflection, a ceiling reflection, and one arriving from the same direction as the direct sound. The test signal was pink noise. Adapted from Olive and Toole, 1989.

would expect a lateral reflection to be much more strongly identified by the binaural discrimination mechanism because of the large signal differences at the two ears. For sounds that differ only in elevation, we have only the spectral cues provided by the external ears and the torso (HRTFs). Although the threshold levels might be surprising, intuition is rewarded in that the dominant audible effect of the lateral reflection was spaciousness (the result of interaural differences) and that of the vertical reflection was timbre change (the result of spectral differences). The broadband pink noise used in these tests would be very good at revealing colorations, especially those associated with HRTF differences at high frequencies. On the other hand, continuous noise lacks the strong temporal patterns of some other sounds, like speech.

This makes the findings of Rakerd et al. (2000) especially interesting. These authors examined what happened with sources arranged in a horizontal plane and vertically on the front-back (median sagittal) plane. Using speech as a test sound, they found no significant differences in masked thresholds and echo thresholds sources in the horizontal and vertical planes. In explanation, they agreed with other referenced researchers that there may be an "echo suppression mechanism mediated by higher auditory centers where binaural and spectral cues to location are combined." This is another example of humans being very well adapted to listening in reflective environments.

Another surprise in Figure 19.7 is that delayed sounds that come from the same loudspeaker are more difficult to hear; the threshold here is consistently higher than for sounds that arrive from the side or above, slightly for short delays, and much higher (10 + dB) at long delays. Burgtorf (1961) agrees, finding thresholds for coincident delayed sounds to be 5–10 dB higher than those separated by 40–80°. Seraphim (1961) used a delayed source that was positioned just above the direct-sound source (~5° elevation difference) and found that, with speech, the threshold was elevated by about 5 dB compared to one at a 30° horizontal separation. The relative insensitivity to coincident sounds appears to be real, and the explanation seems to be that it is the result of spectral similarities between the direct sound and the delayed sound. These sounds take on progressively greater timbral differences as they are elevated (or, one assumes, lowered) relative to the direct sound. For those readers who have been wondering about the phenomenon of "comb filtering," it is worthy of note that this evidence tells us that the situation of maximum comb filtering, when the direct and delayed sounds emerge from the same loudspeaker, is the one for which we are least sensitive. (Encouraging news!)

All this said, it still seems remarkable that a vertically displaced reflection, with no apparent binaural (between the ears) differences, can be detected as well as a reflection that arrives from the side, generating large binaural differences. Not only are the auditory effects at threshold different—timbre versus spaciousness—the perceptual mechanisms required for their detection are also different.

A REFLECTION IN THE PRESENCE OF OTHER REFLECTIONS

Working with a single reflection allows for intensely analytical investigations, but, inevitably, the tests must include others to be realistic. A long-standing belief in the area of control room design is that early reflections from monitor loudspeakers must be attenuated to allow those in the recordings to be audible. Consequently, embodied in several standards, and published designs, are schemes to attenuate or eliminate the first reflections from a loudspeaker using deflecting reflectors, absorbers, or scattering surfaces (diffusers).

Olive and Toole (1989) appear to have been the first to test the validity of this idea. Figure 19.8 shows the results of experiments that examined the audibility of a single lateral reflection simulated in an anechoic chamber with 3 ft (1 m) wedges. For the second experiment, the same physical arrangement was replicated in a typical small room in which the first wall, floor, and ceiling reflections had been attenuated using 2-in. (5 cm) panels of rigid fiberglass board. A third experiment was conducted in the same room with most of the absorption removed (midfrequency reverberation time = 0.4 s). The idea was to show the effects, on the perception of a single reflection, of increasing levels of natural reflected sound within a real room.

THE IEC ROOM

The listening room used in these experiments was the prototype room underlying IEC 268–13–1985. It was constructed at the National Research Council, in Ottawa, within an existing laboratory space (which explains the dimensions). There was little real science to guide the choice of dimensions and acoustical treatment, so the resulting room became one of the variables in ongoing experiments. Of course, at that time stereo was the standard reproduction format. The room was 6.7 m × 4.1 m × 2.8 m (22 ft × 13.5 ft × 9.2 ft) with a midfrequency reverberation time of 0.34 s. More description and measurements are shown in the appendix of Toole (1982). The original concept of the standard was to specify a room that could be duplicated so test results from different laboratories could be compared. In subsequent editions of the standard, the requirements were relaxed so more rooms could qualify, which is a different and significantly less useful objective but much more popular among users who want to claim IEC compliance.

The large changes in the level of reflected sound had only a modest (1–5 dB) effect on the absolute threshold or the image shift threshold of an additional lateral reflection occurring within about 30 ms

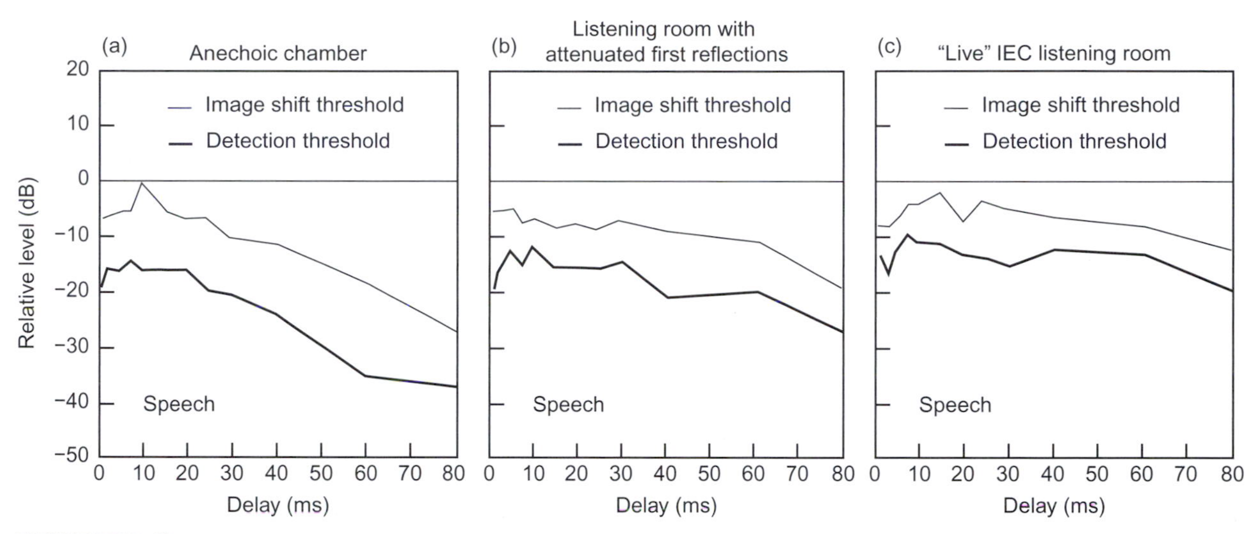

FIGURE 19.8
Detection and image shift thresholds as a function of delay for a single reflection auditioned in three very different acoustical circumstances: (a) Anechoic; (b) A normal room in which the first-order reflections were attenuated with 2-in. (50 mm) fiberglass board; (c) The same room in a relatively reverberant configuration (midfrequency reverberation time = 0.4 s). From Olive and Toole, 1989.

of the direct sound. At longer delays, the threshold shifts were up to about 20 dB, a clear response to elevated late-reflected sounds in the increasingly live rooms. This is a good point to remember, as we will see it again: the threshold curves become more horizontal when the sound—in this case, speech—becomes prolonged by reflected energy (repetitions).

Figure 19.9 shows a direct comparison of the thresholds with the ETC (energy-time curve) measured in each of the three test environments. Here the huge variations in level of the reflections can be clearly seen, in contrast with the relatively small changes in the detection thresholds within the first 30 ms or so.

Real versus simulated rooms

In a large anechoic-chamber simulation of a room of similar size, Bech (1998) investigated the audibility of single reflections in the presence of 16 other reflections, plus a simulated "reverberant" sound field beginning at 22 ms. One of his results is directly comparable with these data. The figure caption in Bech's paper describes the response criterion as "a change in spatial aspects," which seems to match the image shift/image spreading criterion used by Olive and Toole. Figure 19.10 shows the image shift thresholds in the "live" configuration of the IEC room for two subjects (the FT data are from Figure 19.9; the SO data were previously unpublished) and thresholds determined in the simulated room, an average of the three listeners from Bech (1998). The similarity of the results is remarkable considering the very different physical circumstances of the tests. It suggests that listeners were responding to the same audible effect and that the real and simulated rooms had similar acoustical properties.

Bech separately examined the influence of several individual reflections on timbral and spatial aspects of perception. In all of the results, it was evident that signal was a major factor: Broadband pink noise was more revealing than male speech. In terms of timbre changes, only the noise signal was able to show any audible effects and then only for the floor reflection; speech revealed no audible effects on timbre.

Looking at the absorption coefficients used in modeling the floor reflection (Bech, 1996, Table II) reveals that the simulated floor was significantly more reflective than would be the case if it had been covered by a conventional clipped pile carpet on a felt underlay. Further investigations revealed that the detection was based mainly on sounds in the 500 Hz–2 kHz range, meaning that ordinary room furnishings are likely to be highly effective at reducing first reflections below threshold, even for the more demanding signal: broadband pink noise.

ETCs at the listening location compared with the detection thresholds for a single reflection in that same space. Signal: speech.

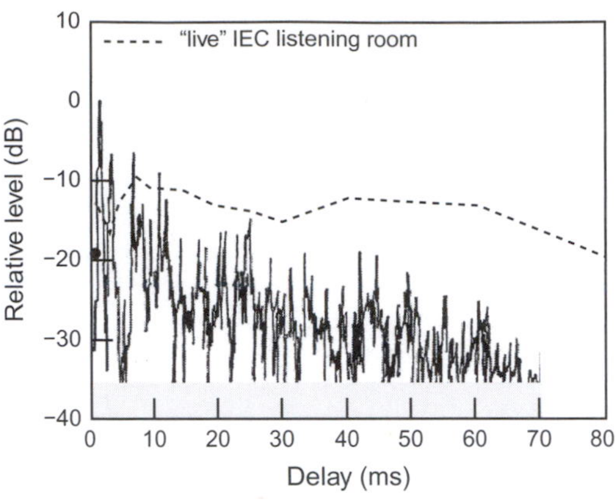

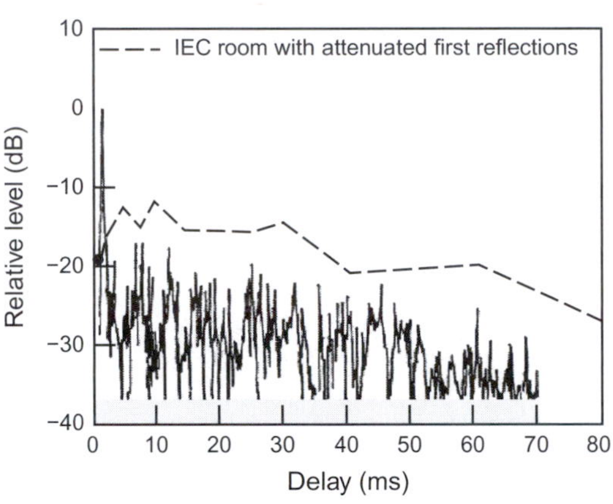

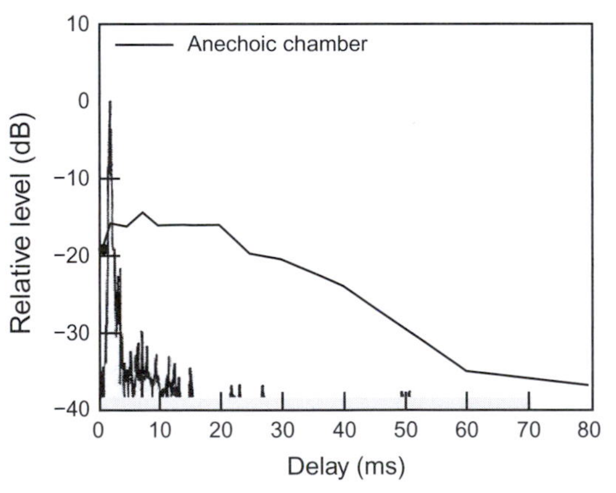

FIGURE 19.9
A comparison of the absolute thresholds shown in Figure 19.8, with measured energy-time curves (ETCs) for the three spaces within which the tests were done. All data from Olive and Toole, 1989.

In terms of spatial aspects, Bech (1998) concluded that those sounds above ~2 kHz contributed to audibility and that "only the first-order floor reflection will contribute to the spatial aspects." The effect was not large, and, as before, speech was less revealing than broadband noise. Again, this is a case where a good carpet and underlay would appear to be sufficient to eliminate any audible effect.

In conclusion, it seems that the basic audible effects of early reflections in recordings are well preserved in the reflective sound fields of ordinary rooms. There is no requirement to absorb first reflections to allow recorded reflections to be heard.

The "Family" of thresholds

Figure 19.11 shows a complete set of thresholds, like those shown in Figure 19.5, determined in an anechoic chamber but here determined in the "live" IEC listening room. The curves are slightly irregular because the data were based on a small number of repetitions. As expected, the curves all have a more horizontal appearance than for speech auditioned in an anechoic environment. It is significant that all the curves have the same basic shape from detection at the bottom to the Haas-inspired equal-loudness curve at the top.

A COMPARISON OF REAL AND PHANTOM IMAGES

A phantom image is a perceptual illusion resulting from summing localization when the same sound is radiated by two loudspeakers. It is natural to think that these directional illusions may be more fragile than those created by a single loudspeaker at the same location. The evidence shown here applies to the simple case of a single lateral reflection, simulated in a normally reflective room with a loudspeaker positioned along a side wall, as shown in Figure 19.12. When detection threshold and image-shift threshold determinations were done first with real and then with phantom center images, in the presence of an asymmetrical single lateral reflection, the differences were insignificantly small. It appears that concerns about the fragility of a phantom center image are misplaced.

Examining the phantom image in transition from front to surround loudspeakers (±30° to ±110°), Corey and Woszczyk (2002) concluded that adding simulated reflections of each of the individual loudspeakers did not significantly change image position or blur, but it did slightly reduce the confidence that listeners expressed in the judgment.

EXPERIMENTAL RESULTS WITH MUSIC AND OTHER SOUNDS

A good introduction to investigations that used music is Figure 19.13, the widely reproduced illustration from Barron (1971), in which he combines several subjective effects for a single lateral reflection simulated in an anechoic chamber using a "direct sound" loudspeaker at 0° (forward) and a "reflected sound" loudspeaker at 40° to the left, both at 3 m distance. For different electronically introduced delays, listeners adjusted the sound level of the "reflection," reporting what they heard while listening to an excerpt from an anechoic-chamber recording of Mozart's *Jupiter* symphony. They heard several identifiable effects, as shown in the figure and described in the caption. There is more to this matter, but this important paper provides a good summary of research up to that point and some new data contributed by Barron.

There is a lot of information in this diagram, but most of it is familiar from the discussions of perceptions in experiments using speech. In the Barron paper, much emphasis is placed on spatial impression because of the direct parallel with concert hall experiences. These days, discussions of spatial impression would be separated into listener envelopment (LEV) and apparent source width (ASW). The discussions here appear to relate primarily to ASW, but the quote in the caption includes the remark "the impression of being in a three-dimensional space," indicating that it is not a hard division. In any event, Barron considers spatial impression to be a desirable effect, as opposed to "tone coloration."

On the topic of "tone coloration," it was suggested that a contributing factor may be comb filtering, the interference between the direct and reflected sound, but Barron further noted that this is mostly a "monaural effect" because "the effect becomes less noticeable as the direct sound and reflection sources are separated laterally." The "tone coloration ... will frequently be masked in a complex reflection sequence," meaning that in rooms with multiple reflecting surfaces, tone coloration is not a concern. More recent evidence supports this opinion.

We will discuss the matter of timbre changes later, and we will see that tone coloration can be either positive or negative, depending on how one asks the question in an experiment. Again, we will go back to the quote in the caption that with the addition of a reflection, "the music [begins] to gain body and fullness," which can readily be interpreted to be tonal coloration but of a possibly desirable form.

Threshold curve shapes for different sounds

It is useful to go back now and compare the shapes of the threshold contours determined by Barron for music with those shown earlier for speech, both in anechoic listening conditions. Figure 19.14 shows such a comparison, and it is seen that curves obtained using the anechoically recorded Mozart excerpt are much flatter than those for speech.

These data suggest two things. First, it appears that the slower paced, longer notes in the music cause the threshold curves to be flatter than they are for the more compact syllables in speech. This "prolongation" appears to be similar in perceptual effect to that occurring due to reflections in the listening environment (Figure 19.8). Second, it appears that the slope of the absolute threshold curve is similar to that of the second-image curve, something that was foreshadowed in Figure 19.11.

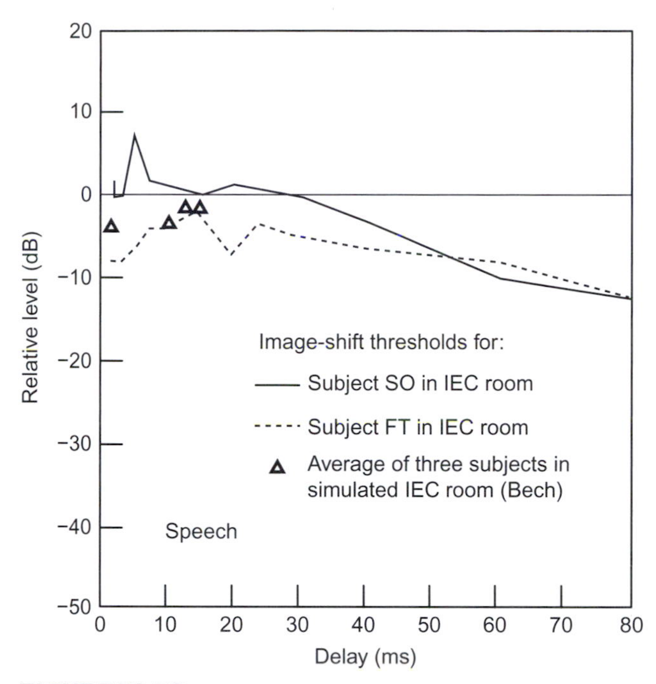

FIGURE 19.10
Image-shift thresholds as a function of delay for two listeners in the "live" IEC room (FT data from Figure 19.8) and averaged results for three listeners in a simulation of an IEC room using multiple loudspeakers set up in a large anechoic chamber (Bech, 1998).

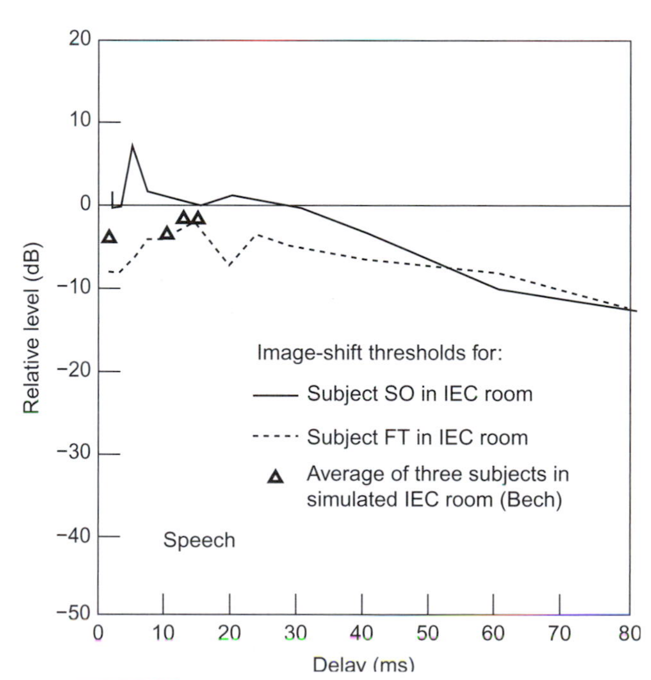

FIGURE 19.11
The full set of thresholds, as shown in Figure 19.5, but here obtained while listening in a normally reflective room rather than an anechoic chamber. One listener (SO). Unpublished data acquired during the experiments of Olive and Toole, 1989.

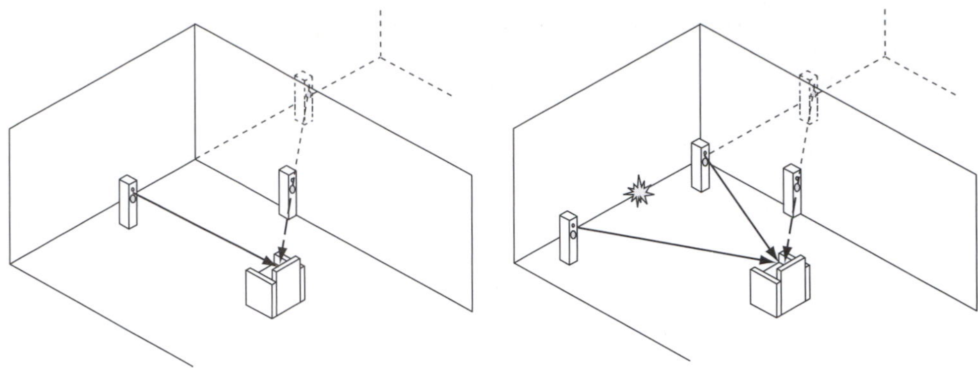

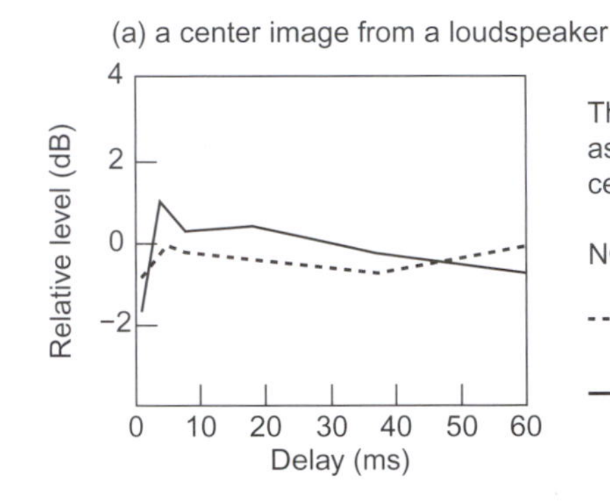

FIGURE 19.12
An examination of how a real and a phantom center image respond to a single lateral reflection simulated by a loudspeaker located at the right side wall. The room was the "live" version of the IEC listening room used in other experiments. Note that the vertical scale has been greatly expanded to emphasize the lack of any consequential effect. The signal was speech.

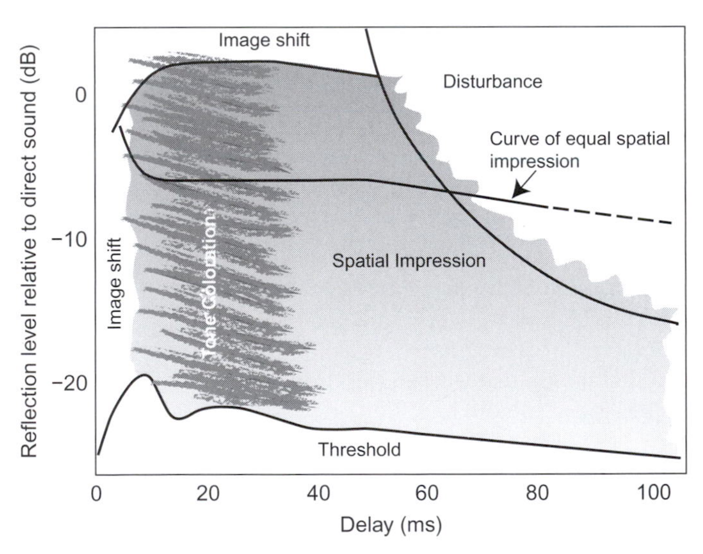

FIGURE 19.13
Subjective effects of a single reflection arriving from 40° to the side, adjusted for different delays and sound levels. An important unseen effect is an increase in loudness, which occurs when the reflected sound is within what is colloquially called the "integration interval": about 30 ms for speech and 50 ms or more for music, all depending on the temporal structure of the sound. In this figure, the lowest curve is the hearing threshold. Above this, at short delays, listeners reported various forms of image shift in the direction of the reflection. At all delays larger than about 10 ms, listeners reported "spatial impression" wherein "the source appeared to broaden, the music beginning to gain body and fullness. One had the impression of being in a three-dimensional space" (Barron, 1971, p. 483). Spatial impression increased with increasing reflection level, a fact illustrated in the figure by the increased shading density. The "curve of equal spatial impression" shows that at short delays, levels must be higher to produce the same perceived effect. At high levels and long delays, disturbing echoes were heard (upper right quadrant). At intermediate delays and at all levels, some degree of tone coloration was heard (darkened brushstrokes). The areas identified as exhibiting "image shift" refer to impressions that the principal image has been shifted toward the reflection image. At short delays, this would begin with summing localization—the stereo-image phenomenon in which the image moves to the leading loudspeaker. At longer delays, the image would likely be perceived to be larger and less spatially clear. Finally, at longer delays and higher sound levels, a second image at the location of the reflection would be expected to add to the spatial illusion. From this presentation it is not clear where these divisions occur. From Barron, 1971, Figure 5, redrawn.

Figure 19.15 shows detection thresholds for sounds chosen to exemplify different degrees of "continuity," starting with continuous pink noise and moving through Mozart, speech, castanets with reverberation, and "anechoic" clicks (brief electronically generated pulses sent to the loudspeakers). The result is that increasing "continuity" produces the kind of progressive flattening seen in Figures 19.8 and 19.9. The perceptual effect is similar if the "continuity" or "prolongation" is due to variations in the structure of the signal itself or due to reflective repetitions added in the listening environment. In any event, pink noise generated an almost horizontal flat line, Mozart was only slightly different over the 80 ms delay range examined, speech produced a moderate tilt, castanets (clicks) with some recorded reverberation were even more tilted, and isolated clicks generated a very compact, steeply tilted threshold curve.

Assuming that the patterns seen in previous data for speech and Mozart apply to other sounds as well, Figure 19.16 shows a compilation and extension of data portraying detection thresholds and second-image thresholds for Mozart, speech, and clicks. To achieve this, the second-image curve for clicks had to be "created" by elevating the click threshold curve by an amount similar to the separation of the speech and music curves. Absolute proof of this must await more experiments, but it is interesting to go out on a (strong) limb and speculate.

Looking at the 0 dB relative level line—where the direct and reflected sounds are identical in level—it can be seen that the precedence-effect interval for clicks appears to be just under 10 ms. According to Litovsky et al. (1999), this is consistent with other determinations (<10 ms), and the approximately 30 ms for speech is also in the right range (<50 ms). They offer no fusion interval data for Mozart, but it is reasonable to speculate based on the Barron data that it might be substantially longer than 50 ms. The short fusion interval for clicks suggests that sounds like close-miked percussion instruments might, in an acoustically dead room, elicit second images.

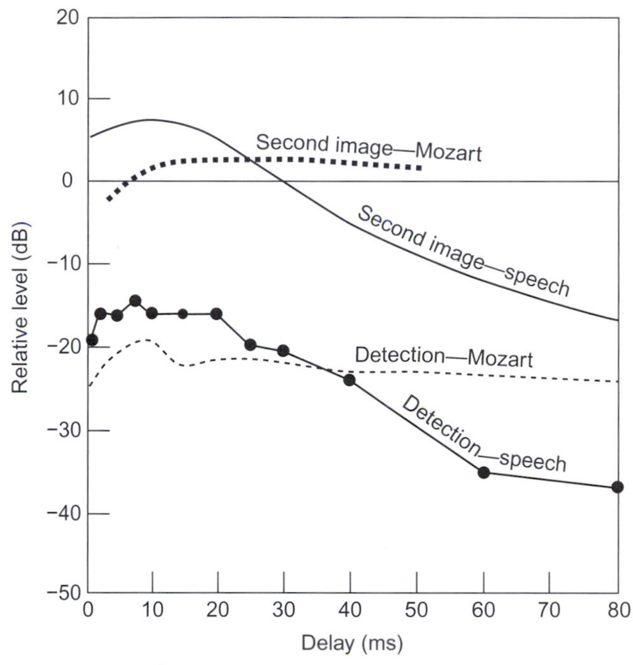

FIGURE 19.14
Data from Figure 19.6a showing thresholds obtained using speech and data from Figure 19.13 showing thresholds obtained using Mozart. The upper curve for music was described as that at which the "apparent source moved from direct sound loudspeaker toward reflection loudspeaker." This could be interpreted as being equivalent to the Olive and Toole "image shift" threshold, but the pattern of the data in the comparison suggests that it is more likely equivalent to the "second image" criterion.

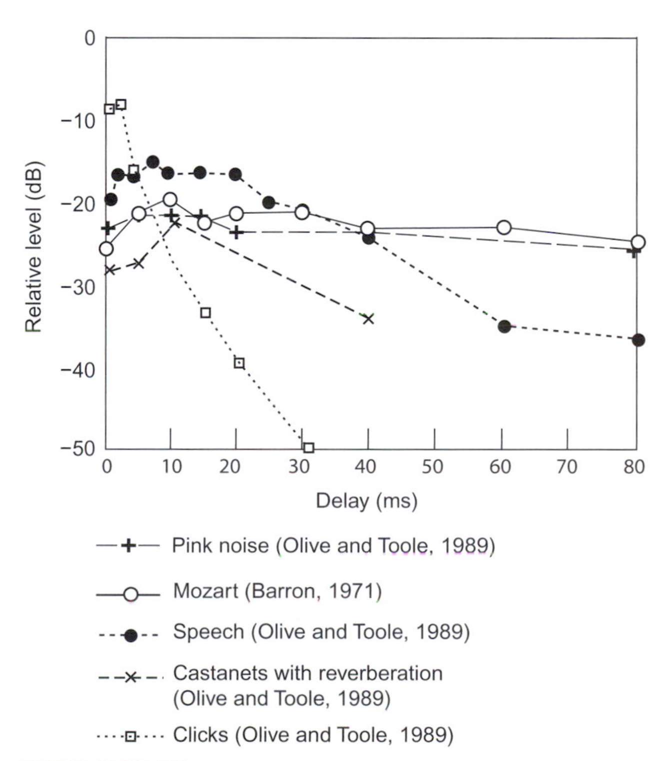

FIGURE 19.15
Detection thresholds for a single lateral reflection, determined in an anechoic chamber for several sounds exhibiting different degrees of "continuity" or temporal extension.

SINGLE VERSUS MULTIPLE REFLECTIONS

So far, we have looked at some audible effects of single reflections when they appear in anechoic isolation and when they appear in the presence of room reflections. Now we will look at some evidence of how a sequence of reflections is perceived.

Cremer and Müller (1982) provide a limited but interesting perspective. Figure 19.17 shows a microphone picking up the direct sound from a loudspeaker and either a single large or three smaller reflections in rapid sequence. The middle layer of images displays sound pressure, showing the direct sound followed by the reflections. The bottom layer of images portrays what Cremer and Müller call an "ear-imitative" function, which is a simple attempt to show that the ear has a short memory that fades with time—a relaxation time. The point of this illustration is that events occurring within short intervals of each other can accumulate "effect," whatever that may be. The sequence of three smaller reflections can be seen to cause the "ear-imitative" function to progressively grow, although not to the same level as that for the single reflection.

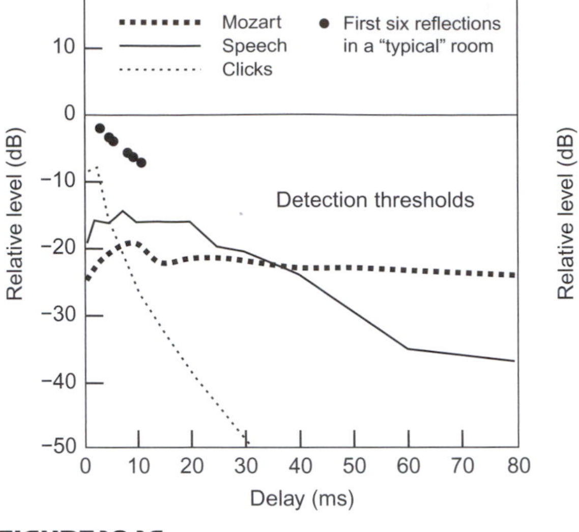

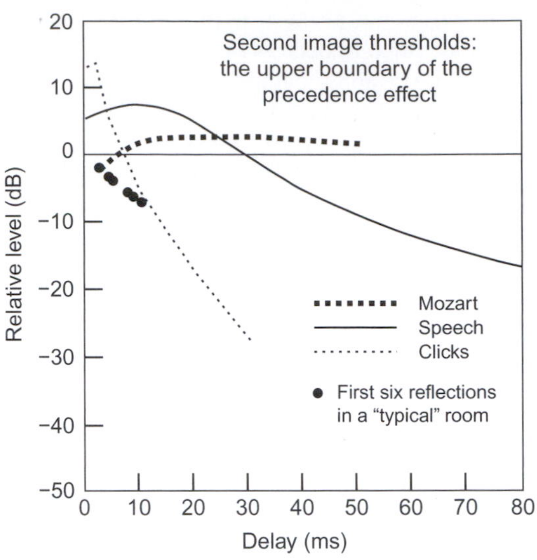

FIGURE 19.16
Using data from Figures 19.16 and 19.17, this is a comparative estimate of the detection thresholds and the second image thresholds (i.e., the boundary of the precedence effect) for clicks, speech, and Mozart. The "typical room reflections" suggest that in the absence of any other reflections, the clicks are approaching the point of being detected as a second image. However, normal room reflections would be expected to prevent this from happening because the threshold curve would be flattened (see Figures 19.8 and 19.9).

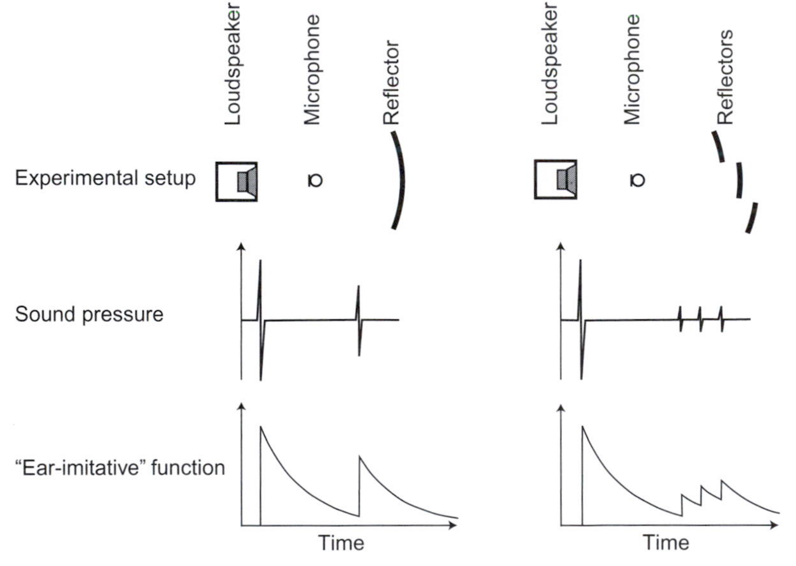

FIGURE 19.17
A comparison of a single large reflection with a sequence of three lower-level reflections. From Cremer and Müller, 1982, Figure 1.16.

However, when the authors conducted subjective tests in an anechoic chamber, they found that the sequence of three low-level reflections and the large single reflection were "almost equally loud." The message here is that if we believed the impulse response measurements, we might have concluded that by breaking up the large reflecting surface, we had reduced the audible effects. This is one of the persistent problems of psychoacoustics. Human perception is usually nonlinear, and technical measurements are remarkably linear.

Angus (1997, 1999) compared large, single lateral reflections from a side wall with diffuse—multiple small—reflections from the same surface covered with scattering elements. There were no subjective tests, but mathematical simulations showed some counterintuitive results—namely that although the amplitudes of individual reflections were attenuated (as seen in an ETC), the variations in frequency responses measured at the listening position were not necessarily reduced. If the Cremer and Müller perceptual-summation effect is incorporated, the multiple smaller reflections seen in the ETC may end up being perceived as louder than anticipated. It is suggested, however, that a diffuse reflecting surface may make listening position less critical.

So there are both subjective and objective perspectives indicating that breaking up reflective surfaces may not yield results that align with our intuitions. It is another of those topics worthy of more investigation.

MEASURING REFLECTIONS

It seems obvious to look at reflections in the time domain, in a "reflectogram" or impulse response, a simple oscilloscope-like display of events as a function of time or, the currently popular alternative, the

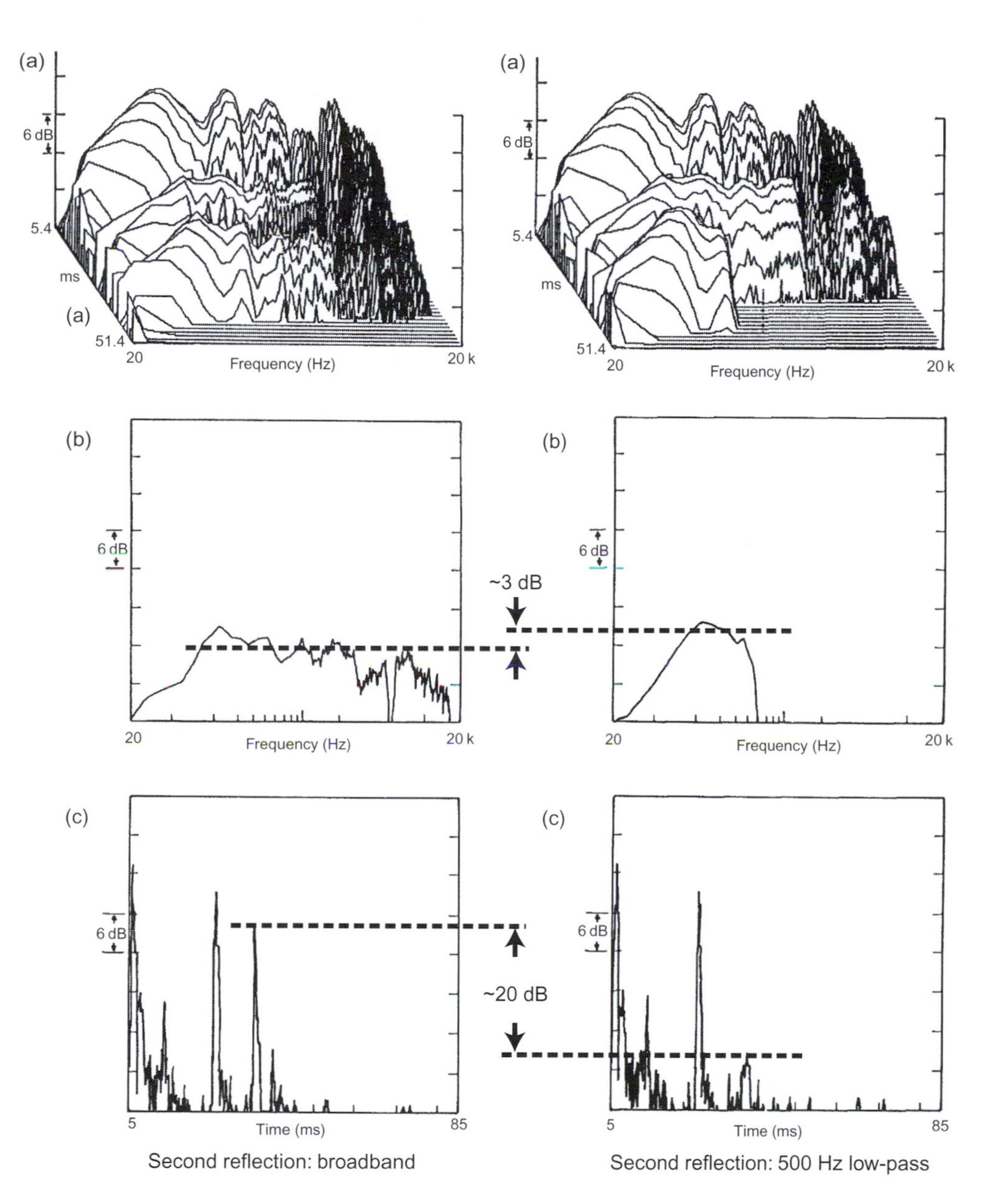

FIGURE 19.18
The left column of data shows results when the second of a series of reflections was adjusted to the threshold of detection when it was broadband; the right column shows comparable data when the reflection was low-pass filtered at 500 Hz. (a) Shows the waterfall diagram, (b) the spectrum of the second reflection taken from the waterfall, and (c) the ETC measured with a Techron 12 in its default condition (Hamming windowing). The signal was speech. The horizontal dotted lines are "eyeball" estimates of reflection levels. From Olive and Toole, 1989, Figures 18 and 19.

ETC (energy-time curve). In such displays, the strength of the reflection would be represented by the height of the spike. However, the height of a spike is affected by the frequency content of the reflection, and time-domain displays are "blind" to spectrum. The measurement has no information about the frequency content of the sound it represents. Only if the spectra of the sounds represented by two spikes are identical can they legitimately be compared.

Let us take an example. In a very common room acoustic situation, suppose a time-domain measurement reveals a reflection that it is believed needs attenuation. Following a common procedure, a large panel of fiberglass is placed at the reflection point. It is respectably thick—2 in. (50 mm)—so it attenuates sounds above about 500 Hz. A new measurement is made, and—behold!—the spike has gone down. Success, right? Maybe not.

In a controlled situation, Olive and Toole (1989) performed a test intended to show how different measurements portrayed reflections that, subjectively, were adjusted to be at the threshold of detection. So from the listener's perspective, the two reflections that are about to be discussed are the same: just at the point of audibility or inaudibility.

The results are shown in Figure 19.18. At the top, the (a) graphs are waterfall diagrams displaying events in three dimensions. At the rear is the direct sound, the next event in time is an intermediate reflection, and at the front is the second reflection, the one that we are interested in. It can be seen that the second reflection is broadband in the left-hand diagram and that frequencies above 500 Hz have been eliminated in the right-hand version. When that particular "layer" of the waterfall is isolated, as in the (b) displays, the differences in frequency content are obvious. The amplitudes are rather similar, although the low-pass filtered version is a little higher, which seems to make sense considering that slightly over 5 octaves of the audible spectrum have been removed from the signal. Recall that these signals have been adjusted to produce the same subjective effect—a threshold detection—and it would be logical for a reduced-bandwidth signal to be higher in level.

In contrast, the (c) displays, showing the ETC measurements, were telling us that there might be a difference of about 20 dB in the opposite direction; the narrow-band sound is shown to be lower in level. Obviously, this particular form of the measurement is not a good correlate with the audible effect in this test.

The message is that we need to know the spectrum level of reflections to be able to gauge their relative audible effects. This can be done using time-domain representations, like ETC or impulse responses, but it must be done using a method that equates the spectra in all of the spikes in the display, such as bandpass filtering. Examining the "slices" of a waterfall would also be to the point, as would performing FFTs on individual reflections isolated by time windowing of an impulse response. Such processes need to be done with care because of the trade-off between time and frequency resolution. It is quite possible to generate meaningless data.

All of this is especially relevant in room acoustics because acoustical materials, absorbers, and diffusers routinely modify the spectra of reflected sounds. Whenever the direct and reflected sounds have different spectra, simple broadband ETCs or impulse responses are not trustworthy indicators of audible effects.

REFERENCES

Angus, J.A.S., 1997. Controlling early reflections using diffusion. Audio Eng. Soc. 102nd Convention Preprint 4405.

Angus, J.A.S., 1999. The effects of specular versus diffuse reflections on the frequency response at the listener. Audio Eng. Soc. 106th Convention Preprint 4938.

Barron, M., 1971. The subjective effects of first reflections in concert halls—the need for lateral reflections. J. Sound Vib. 15, 475–494.

Bech, S., 1996. Timbral aspects of reproduced sound in small rooms, II. J. Acoust. Soc. Am. 99, 3539–3549.

Bech, S., 1998. Spatial aspects of reproduced sound in small rooms. J. Acoust. Soc. Am. 103, 434–445.

Benade, A.H., 1985. From instrument to ear in a room: direct or via recording. J. Audio Eng. Soc. 33, 218–233.

Blauert, J., 1996. Spatial Hearing: The Psychophysics of Human Sound Localization. MIT Press, Cambridge, Mass.

Burgtorf, W., 1961. Untersuchungen zur wahrnehmbarkeit verzögerter schallsignale. Acustica 11, 97–111.

Corey, J., Woszczyk, W., 2002. Localization of lateral phantom images in a 5-channel system with and without simulated early reflections. Audio Eng. Soc. 113th Convention Preprint 5673.

Cremer, L., Müller, H.A., 1982. Principles and Applications of Room Acoustics (T.J. Schultz, Trans.) vols. 1,2. Applied Science Publishers, London.

Devantier, A., 2002. Characterizing the amplitude response of loudspeaker systems. Audio Eng. Soc. 113th Convention Preprint 5638.

Gardner, M., 1968. Historical background of the Haas and/or precedence effect. J. Acoust. Soc. Am. 43, 1243–1248.

Gardner, M., 1969. Image fusion, broadening, and displacement in sound localization. J. Acoust. Soc. Am. 46, 339–349.

Gardner, M., 1973. Some single- and multiple-source localization effects. J. Audio Eng. Soc. 21, 430–437.

Griffin, J.R. (2003). "Design Guidelines for Practical Near Field Line Arrays,"http://www.audiodiycentral.com/resource/pdf/nflawp.pdf.

Litovsky, R.Y., Colburn, H.S., Yost, W.A., Guzman, S.J., 1999. The precedence effect. J. Acoust. Soc. Am. 106, 1633–1654.

Lochner, J.P.A., Burger, J.F., 1958. The subjective masking of short time-delayed echoes by their primary sounds and their contribution to the intelligibility of speech. Acustica 8, 1–10.

Meyer, E., Schodder, G.R., 1952. On the influence of reflected sound on directional localization and loudness of speech. Nachr. Akad. Wiss. Gottingen, Math. Phys. Klasse IIa 6, 31–42.

Olive, S.E., Toole, F.E., 1989a. The detection of reflections in typical rooms. J. Audio Eng. Soc. 37, 539–553.

Olive, S., Toole, F.E., 1989b. The evaluation of microphones—part 1: measurements. Audio Eng. Soc. 87th Convention Preprint 2837.

Rakerd, B., Hartmann, W.M., Hsu, J., 2000. Echo suppression in the horizontal and median sagittal planes. J. Acoust. Soc. Am. 107, 1061–1064.

Seraphim, H.-P., 1961. Über die wahrnehmbarkeit mehrerer rückwürfe von sprachschall. Acustica 11, 80–91.

Shinn-Cunningham, B.G. (2003). Acoustics and perception of sound in everyday environments. Proc. 3rd Int. Workshop on Spatial Media, Aisu-Wakamatsu, Japan. http://cns.bu.edu/~shinn/pages/RecentPapers.html.

Toole, F.E., 1982. Listening tests—turning opinion into fact. J. Audio Eng. Soc. 30, 431–445.

Toole, F.E., 2006. Loudspeakers and rooms for sound reproduction—a scientific review. J. Audio Eng. Soc. 54, 451–476.

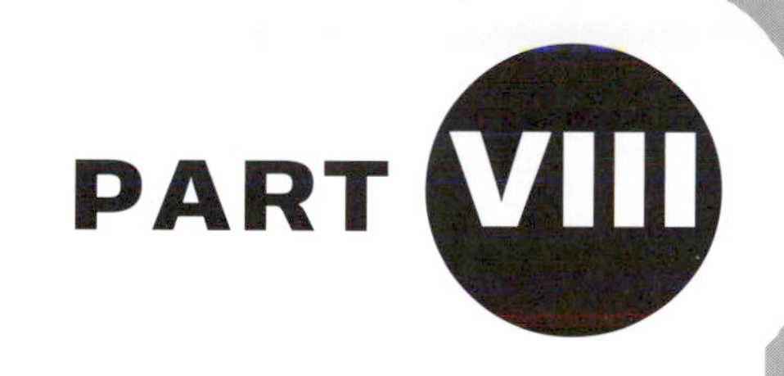

Recording Studios

PREFACE

The final section of this book "Recording Studios" consists of two chapters taken from Philip Newell's *Recording Studio Design*. The correction of a poorly-thought-out sound isolation scheme is very likely to be difficult, expensive, and highly disruptive to the business of a recording studio, and it is essential to consider all the issues and lay the plans very carefully before starting construction.

Philip begins by looking at the basics of air and vibrational behavior. The first question that arises is the fundamental one of why sound waves in air are so difficult to stop. The answer is that air is heavier than most people think; a cubic meter at sea level weighs about 1.2 kilograms. Philip puts forward the remarkable fact that when a jumbo jet allows its internal cabin pressure to drop to the equivalent of 8000 feet, it not only avoids a lot of extra stress on the fuselage, but also reduces the aircraft's weight by about half a ton, with a resulting improvement in fuel economy. It is the weight of air that allows vibrations in it to move a brick wall sufficiently for the sound to be heard on the other side.

We move on to examine what happens when a sound wave strikes a surface. There are three possibilities: the sound energy may be reflected, absorbed, or transmitted, and usually a combination of all three occurs. A rigid wall, such as a really thick mass of concrete, will reflect almost all the sound that hits it, and absorb virtually nothing. In contrast, a wall of open-cell foam or mineral wool suitably constructed to present a gradual impedance change would absorb almost all the incident sound energy, converting it to heat and giving minimal reflection. It is the sound that is transmitted that causes the trouble when isolation is inadequate.

Philip points out that reflection is the most important isolation technique, because walls that were purely absorbent would have to be of stupendous size to be effective at low audio frequencies; at 20 Hz you would need a thickness of several meters.

He goes on to consider the basic concepts of sound isolation, which are mass, rigidity, damping, and distance. The greater the mass of a wall, the less the air vibrations can move it, because of the greater impedance mismatch, and so the less it will re-radiate on the other side. Mass alone however is not that useful, as doubling the thickness of a wall can only increase isolation by 6 dB, and the weight of wall required for a given figure rapidly becomes impractical.

This leads to the idea of floating structures for isolation. There is an inner and an outer wall, with damping in between them to prevent sound propagating in solid building materials. Philip shows how this method of construction can achieve an isolation with just 35 cm of thickness that would otherwise need a solid wall two meters thick. An important point is that one of the walls needs to rest on a resilient support to prevent sound coupling to it through the floor.

Philip next considers how much sound isolation is required for various applications, a selection of methods for floating floors, the problems of ceiling isolation, and the issues of reciprocity and impact noise control. Finally, he considers the simple use of distance to ease isolation problems, but it is of course not always practical to site your studio in the middle of a deserted valley.

Sound Isolation

Recording Studio Design by Philip Newell

The weight of air. Interaction of the air vibrations with room boundaries. Reflexion, transmission and absorption. Reflexion and absorption as means of isolation. The mass law, damping and decoupling. Frequency dependence of isolation needs. Level dependence of isolation needs *vis-à-vis* the non-uniform ear response. Floor, wall and ceiling isolation. Weight considerations. Material densities. The journey through a complex isolation system. Considerations regarding impact noises. Matters influencing studio location choice. The behavior of mass/spring floor systems and the characteristics of fibrous and cellular base layers.

VIBRATIONAL BEHAVIOR

Sound waves in air can be remarkably difficult to stop. Air is a fluid of considerable substance. It can support 500-tonne aeroplanes and blow down buildings. In fact, it is much heavier than most people think. On earth, at sea level, air weighs about 1.2 kg per cubic meter, which is a very good reason for pumping any unnecessary excess of it out of aeroplanes (de-pressurizing). This process is partially carried out to reduce the pressure differential stresses on the aircraft fuselage when flying at altitude in air of lower pressure (and density), but the aircraft also use less fuel by not having to carry excessive quantities of air over their entire journeys. A jumbo jet can actually reduce its load by around half a tonne by de-pressurizing to an equivalent pressure altitude of around 8000 feet (2500 m). Air is also a rather springy substance, which makes it useful in air-pistols, shock absorbers and car tires.

When any material is immersed in a vibrating fluid, it will itself be set into vibration to some degree. The characteristics of the resulting vibration will depend on the properties of the material, especially as, unlike air, the speed of sound *is* frequency dependent in many materials. The "laser gun" sound sometimes heard in railway tracks when a train is approaching is due to phase dispersion within the steel tracks. The high frequencies travel faster than the low frequencies when the wheel flanges excite the rail modes, and so arrive first, with the lower frequencies following later.

An acoustic wave when traveling through air will continue to expand until it reaches a discontinuity, which may be a solid boundary or a porous material. It will then encounter a change in acoustic impedance that will cause some of the energy to be reflected, some to be absorbed, and some to be transmitted beyond the discontinuity. The degree of reflexion will be determined by the lack of acoustic permeability of the boundary and the degree of internal losses that will tend to convert the acoustic energy into heat. A rigid, heavy, impermeable structure, such as could be made from a one-meter thickness of reinforced concrete with a well-sealed surface, would reflect back perhaps 99.9% of the incident energy. Conversely, a wall of one meter wedges of open-cell foam or mineral wool would present a very gradual impedance change which would allow the acoustic wave to enter with minimum reflexion, perhaps *absorbing* 99.9% of the incident energy. Mechanisms such as internal friction, tortuosity, and adiabatic losses convert acoustic energy into heat. The internal friction results in losses as the materials, on a molecular or particulate level, rub together. The energy required for these motions is taken from the acoustic energy.

Tortuosity relates to the degree of obstruction placed in the way of the air particles as they try to move under the influence of the acoustic wave. Air, in some circumstances, can act like a rather sticky, viscous fluid, and the viscous losses are increased as the degree of tortuosity increases. The fibers of a medium density mineral wool, for example, present a very tortuous path for the air vibration to negotiate. Remember, though, that there is no net *airflow* associated with these vibrations. The air particle motion is very localized, and is dependent upon the amplitude and frequency of the vibration. Adiabatic losses occur when the heat of compression and the cold of rarefaction, which both normally contribute to the sound propagation, are conducted away from the air by the close proximity of a porous medium in which the air is dispersed.

Because the above losses are proportional to the speed with which a particle of vibrating air tries to oscillate within the material, their absorption coefficients tend to be greater as the frequency rises due to the particle movements being more rapid. In addition, as the acoustic wave propagation must stop and reverse when it reaches a solid boundary, the particle motion will tend to zero when the boundary is approached. The absorbent effect of fibrous materials is therefore greater when the materials are placed some distance away from the reflective boundary. The particle velocity is greatest, and hence fibrous absorption is greatest, at the quarter and three-quarter wavelength distances from the boundary, on the *velocity anti-nodes*, which correspond with the pressure *nodes* of an acoustical wave.

Wave motion can be thought of like a swinging pendulum. When the pendulum is at its maximum height, the movement is zero, and when the movement is at its maximum velocity, the pendulum is swinging through its point of minimum height. The height is therefore at its maximum when the speed is at its minimum, and vice versa. Similarly, in an acoustic wave, the velocity component is at its maximum when the pressure component is at its minimum, and vice versa, and the energy in the propagation is continuously passing from one component to the other. When absorbent materials are used in isolation systems, they tend to be used more as mechanical springs and for acoustic damping. One reason for this is that fibrous/porous absorbers in general are of little effect at low frequencies except in great depth and away from a reflective boundary. When we are considering sound isolation for recording studios, we tend to be considering treatment principally *at* low frequencies and *near* to boundaries, because preventing the acoustic energy from bass guitars and bass drums entering the boundaries is what we are largely seeking to achieve. Once acoustic waves are allowed to enter a structure, things may become highly unpredictable. The disturbance will travel through the materials of the structure at their characteristic speeds of sound, sometimes with very little loss, then re-radiate into the air from the vibrating surfaces with unfathomable phase relationships that can produce "hot spots" of sound energy in some very unexpected places. The speed of sound in solids is usually much greater than in air. For example, in concrete it is around 3500 m/s, in steel over 5000 m/s, and in water about 1500 m/s. In woods, the speed of sound can be dependent on the direction of propagation. Taking the case of beech, sound propagates along the grain at around 4500 m/s, yet at only just over 1100 m/s across the grain. These speeds are all approximate because the materials are somewhat variable in nature—fresh water and seawater, for instance—and there can be different speeds at different frequencies.

Relevance to isolation

If isolation is the goal, then it is the degree of transmission *through* the boundaries that is important. From this point of view absorption and reflexions are often lumped together, because isolation will be achieved either by reflecting the energy back from a boundary or by absorption within it. Reflexion is by far the most important control technique, because absorbent walls would need to be of enormous thickness to be effective at low frequencies—several metres at 20 Hz, for example. An ane-choic chamber with one-meter wedges would absorb around 99.9% of the sound power down to 70 or 80 Hz or so, but such absorption is minimal in terms of isolation. Ninety-nine point nine per cent of the power being absorbed leaves only one part in 1000 of the original energy, but still only represents a 30 dB reduction (10 dB for each 10 times reduction). For measurement purposes, this renders reflexions insignificant in most cases, but 30 dB of isolation around a drum kit and bass guitar could still leave over 80 dB SPL on the other side of a 99.9% absorbent wall. To isolate by 60 dB we cannot allow more than one part in *one million* of the sound power to escape. Ten-meter thick walls of 70 kg/m^3 mineral wool may suffice at 20 Hz, but it is hardly a practical solution, especially when one realizes that the ceiling would need the same treatment.

BASIC ISOLATION CONCEPTS

There are essentially four aspects to sound isolation: mass, rigidity, damping and distance. Taking the last one first, if we can get far enough away from noise sources and our neighbors, then we will have solved our isolation problems. At least that seems to be rather obvious, but obvious things in acoustics are rather rare. The characteristics of mass, rigidity and damping are a little more complex.

All other things being equal, it takes more energy to move a large mass than a small one. Consequently, a large mass subjected to an acoustic wave will tend to reflect back more energy than a small mass, because it has more inertia. It has more acoustic impedance because it has a greater tendency to impede the path of the wave. However, if the mass is not rigid, and has a tendency to vibrate at its natural frequencies, energy may be absorbed from the acoustic wave that can set up resonances in the structure. Once the whole mass is resonating, its surfaces will be in movement and will act as diaphragms, re-radiating acoustic energy. If this mass were a wall, then the outer surface would selectively re-radiate the sound which was striking the inner surfaces. Isolation would therefore be dependent on the degree of the freedom of the mass to resonate.

If the mass were perfectly rigid, then resonances could not occur, because vibration implies movement, and infinite rigidity precludes this. Theoretically, of course, if a sealed room were made from an infinitely rigid, lightweight material, then because it could not vibrate it would be sound proof unless the whole thing could be set in movement *en masse.* In the latter case, the inertia of the air inside the room would resist the motion of the shell and set up pressure waves from the boundaries. Unfortunately, lightweight infinitely rigid materials do not exist, so the only way we can normally achieve high degrees of sound isolation over short distances is by the use of highly rigid, massive structures.

Damping and the mass law

A great influence on the ability of any structure to provide sound isolation is that of damping. Damping is the degree to which a propagating wave within a material or structure is internally absorbed, normally by the conversion of the vibrational energy into heat. The damping of a material or structure can also be achieved to some degree by the addition of a damping material to its surface—Plasticine on a bell, for example. An acoustically very "lossy" (highly damped), massive structure can in many cases, for the same degree of isolation, be less rigid, because the passage of the vibrational waves through the structure is severely attenuated before the waves can re-radiate from another surface.

Limpness, to some degree, achieves the same end as rigidity—the inability of the structure or material to vibrate sympathetically—but unsupported limp materials are incapable of forming a structure—they have inertia, but no stiffness. This leads us to partitions that are essentially controlled by the *mass law,* which roughly states that when the inertia of a panel, rather than its stiffness, is the dominant principle for sound transmission loss, that loss increases by 6 dB for each doubling of the mass per unit area, and by 6 dB for each doubling of frequency. At least, that is, for plane waves at a given angle, this is why the mass law is only an approximation to normal circumstances.

Floating structures

In practice, the means by which isolation is usually achieved is by mechanically decoupling the inner structure of a room from the main structure of the building. As has been discussed, isolation by pure absorption is very unwieldy—rooms made from 10 m thick mineral wool walls and ceilings are not an option. As no lightweight super-rigid structures are readily available, then neither is rigidity alone a practicable solution. To some degree, mass *is* used, but if it were to be used alone, then it too can become unrealistic in its application. For example, let us presume that a concrete block wall of 20 cm thickness and 40 dB isolation were to be augmented, by mass alone, to achieve 60 dB of isolation at low frequencies. The mass law would add *at the most* around 6 dB of isolation for each doubling of the mass per unit area, though the increasing rigidity of the more massive structure could tend to raise the isolation to above the 6 dB mark. Under ideal circumstances, which may not be realized in practice, doubling the thickness to 40 cm would yield 46 dB. Doubling this to 80cm would result in 52 dB, and doubling this yet again to 1 m 60 cm would still only provide 58 dB of isolation. We would therefore require walls of around 2 m thickness to achieve 60 dB of low frequency isolation if we were to rely on increasing the mass alone, and even this may be compromised by internal resonances within the structure.

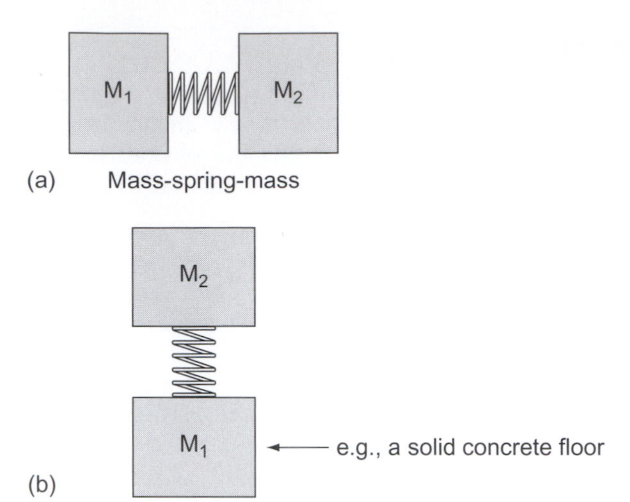

FIGURE 20.1

(a) A resonant system. (b) If we now "earth" M_1 we will have a good representation of a floated floor, where M_1 becomes of almost infinite impedance, such as would be the case for a heavy base slab on solid ground. If mass M_2 were to be set in motion, say by pressing down on the mass and releasing it, the system would oscillate at its resonant frequency until all the energy was dissipated. The length of time for which the system oscillates depends on the resistive losses—the damping. If the mass M_2 were caused to vibrate at a frequency above the resonant frequency of the system, the spring would effectively decouple the vibrations from the earthed mass. M_1.

The practical answer to the isolation problem lies in decoupling the inner and outer structures. This can be achieved by many means, such as by steel springs, rubber or neoprene blocks, fibrous mats, polyurethane foams and other means. There have even been cases of rooms floated on shredded car tires, and even tennis balls, but the problem with tennis balls is that the air will gradually leak out over time. Re-inflatable air bags use the same principle, and this is another practical solution sometimes used.

The floating relies on the mass/spring/mass resonant system as shown in Figure 20.1. If the first mass is set in motion, the force exerted on the spring will be resisted by the inertia of the second mass, and above the resonant frequency of the system the spring will be heated by the vibrational energy. This converts the acoustic energy into thermal energy. Such isolation systems work down to about 1.4 times the resonant frequency of the system, below which the decoupling ceases to become effective. The resonant frequency is a function of the mass that is sprung and stiffness of the springs. Increasing the mass and decreasing the spring stiffness both tend to reduce the resonant frequency.

The effect of this decoupling of the masses is to render the two systems (floated and structural) acoustically independent. Hence if we had a structure of 20 cm sand-filled concrete blocks, with 40 dB of low frequency isolation, we would only need an internal floated structure with 20 dB of isolation in order to achieve 60 dB of total isolation. This may be achieved by internal floated walls of 10 cm thickness, such as of sand-filled, concrete blocks, spring isolated from the floor, with a 5 cm mineral-wool lined air space between the two walls. Such a system, as depicted in Figure 20.2, would achieve in a total thickness of 35 cm what could only be achieved by two meters thickness of blocks that were mechanically connected.

The way that this works is that the internal wall attenuates the sound from inside the room by 20 dB, which is the resultant level on the outside of the inner wall. The air between the walls acts as a spring and it is not capable of efficiently pushing or pulling the much greater mass of the inert outer wall. If the walls were in contact, the masses would be reasonably comparable, so the vibration in the inner layers would have relatively little trouble progressing through the structure. In the isolated wall system, it is only the sound pressure which has already been attenuated by 20 dB which impinges on the outer wall, which then attenuates the sound by another 40 dB before reaching the outside of the building, thus achieving the 60 dB of isolation.

Floating system choices

The choice of what to use to float the inner structure depends on the mass to be floated and the lowest frequency to which isolation is needed. Two identical studio rooms, one for the recording of bass

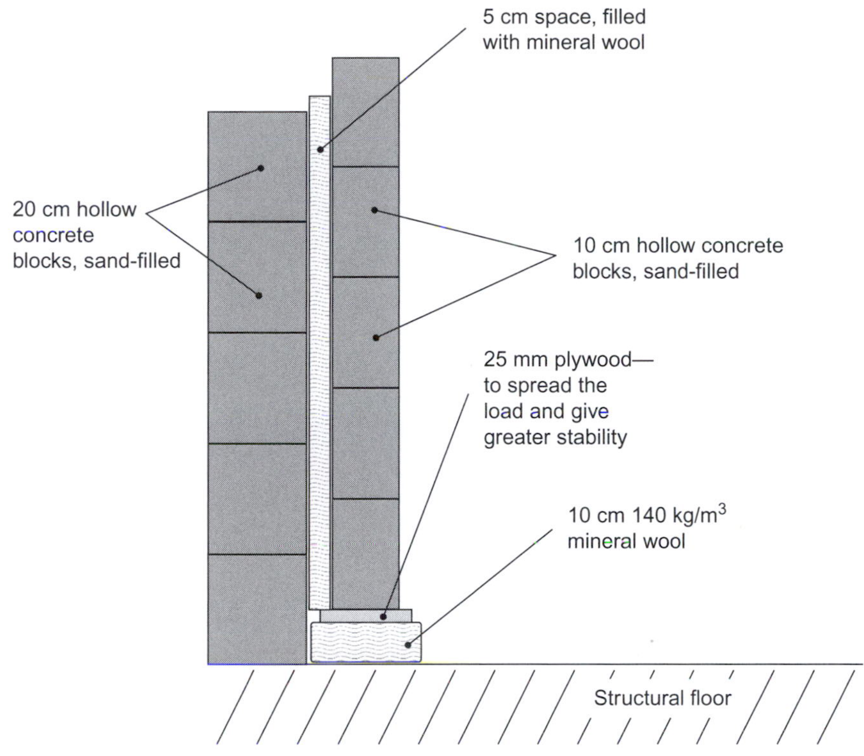

FIGURE 20.2
Floating a wall—an earthed mass-spring-mass system in practice.

guitar and one for the recording of speech, both requiring the same degrees of isolation within the frequency range of their intended usages, would need rather different suspension systems if they were to be built in economically efficient ways. The former room would be much more expensive to build, and it would be a total waste of money to make a room such as this if so much low frequency isolation was unnecessary. For simplicity, let us make a comparison of two such rooms in which we are only interested in the sound leakage in the outward direction to a neighbouring bedroom.

Taking the vocal recording room first, most of us will have experienced how the sound of speech can travel through walls. Noisy neighbors talking till late in the night whilst we are trying to get some sleep, either at home or in a holiday hotel, for example. This problem is so easy to solve that it is a disgrace that so many buildings have been constructed with so little thought about the problem. However, if a huge number of people want to go away on holiday as cheaply as possible, it is not too surprising that the hotels that they stay in will have been *built* as cheaply as possible. You often get what you pay for, but not much more.

People talking relatively loudly are likely to produce sound pressure levels in the order of 80 dB SPL (or the same dBA as there is little low frequency content) at the boundary of a room, a couple of meters from the source of the sound. (The speech frequencies, whose lower frequencies begin around 100 Hz, are little affected by the low frequency roll-off of the dBA weighting.) The deepest tones of some male voices extend down to 80 Hz or so, but they tend to be weak in level, and even with a little attenuation through a wall soon reduce to levels that become imperceptible. At low frequencies at low levels the equal loudness curves close together. They do so to such a degree that only 5 dB or less reduction in SPL will produce a halving of subjective loudness as opposed to the 10 dB a mid frequency would require to produce the same effect. Twenty decibels of isolation at 70 Hz and 50 dB of isolation at mid frequencies would render speech all but inaudible above a quiet background noise in an adjacent room. Figure 20.3 shows how this could typically be achieved. A person recording drama in an otherwise residential building would be unlikely to make the neighbors aware of their activities if such a treatment of the recording room were to be undertaken.

A musician wishing to practice playing bass guitar at home, in the same building as our drama studio, would not be so lucky. In this case, the fundamental frequencies of the new five-string basses can go

10 cm thickness of 80 kg/m3 reconstituted open-cell polyurethane foam

Structural dividing wall

18 mm plasterboard

Foam glued to wall, and plasterboard glued to foam, with contact (impact) adhesive

Structural floor

(a)

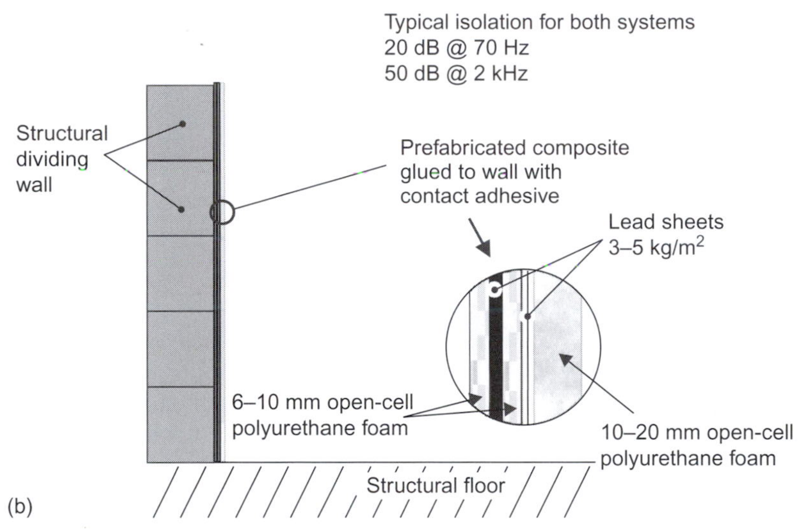

FIGURE 20.3
Isolation treatments for direct application to walls. (a) With reflective surface to room, (b) With absorbent surface to room. Typical isolation for both systems: 20 dB at 70 Hz; 50 dB at 2 kHz.

down as far as 35 Hz. Enthusiastic playing could produce 100 dB SPL or more if it was for recording, where the sound of amplifier compression at high level can become an integral part of the sound. What is more, the relatively sustained notes of a bass guitar can exist for a sufficient period to excite strong structural resonances in an acoustically non-specifically designed construction, such as an apartment building.

Below about 55 dB SPL, 35 Hz is inaudible. If the bass guitarist were playing at a level of 80 dB at that frequency, then whether the room had 30 dB of isolation or 40 dB of isolation would make no difference, because in either case the resultant leakage at 35 Hz would be inaudible. Thirty dBA is usually considered to be an acceptable noise level for sleeping, and at 35 Hz, 30 dBA would actually be about 70 dB SPL so in practice, even 10 dB of isolation at 35 Hz would suffice if the intention was not to annoy the neighbors. At 500 Hz, if the guitar were also producing 80 dB, then 50 dB of isolation would be the minimum needed in order not to annoy the neighbors with more than 30dB of "noise".

If the musician increased the volume by 30 dB, so that both the 35 Hz and 500 Hz components of the bass guitar were producing 110 dB, 30 dB extra isolation would be needed at both 35 Hz and 500 Hz to reduce the 110 dB down to a tolerable 30phon (the 30 dBA curve equivalent) for the neighbors. However, because of the ear's tendency to increase its sensitivity to low frequency level changes as the frequency descends, the provision of only 15 dB of extra isolation at both frequencies would not have an equal subjective effect. At 500 Hz, the 15 dB excess leakage would somewhat more than double the subjective "noise" level for the neighbor (10 dB doubles loudness at mid frequencies) which during the daytime and early evening may be tolerated. Not so at 35 Hz, though. At 35 Hz, 70 dB SPL lies about the 30phon level. Increasing the SPL to 85 dB (the 15 dB extra leakage) now places 35 Hz on the 60phon level. In other words, it will have subjectively doubled in loudness from the 30 to 40phon curves, doubled again between the 40 and 50phon curves, and doubled yet again between the 50 and 60phon levels. The result would be an eight times increase in apparent loudness, which the neighbors simply may not tolerate.

What this means is that if the drama group begin to scream and shout and produced 30 dB more sound during the day, then they may well get away with an increase of isolation of only 15 dB in the mid and low frequencies, as they produce little low frequencies. However, the bass guitarist would *not* remain a friendly neighbor if the volume were to be increased by 30 dB with only 15 dB extra isolation being provided. What is worse is that low frequency isolation in structures is much less effective than high frequency isolation because of the aforementioned mass law. Therefore, not only will the bass player need more isolation, but also the extra low frequency isolation needed will be much heavier than the equivalent mid frequency isolation. In a domestic building, the basic structure may simply not support the necessary weight of isolation materials, and so being a good neighbor whilst playing a bass guitar at 110 dB SPL may simply not be possible unless the other occupants of the building are deaf.

This variability in hearing sensitivity with level causes so much confusion for non-specialists in the understanding of sound isolation requirements. Simple figures are not possible, as everything must relate to frequency and level. In effect, only curves of isolation requirements can be specified, dependent on the frequency range to be isolated and the levels to which they either will produce sound, or need to be isolated from it.

Figure 20.4 shows an actual construction that was necessary to prevent bass guitars at around 110 dB from annoying the neighbours in a bedroom above. Looking at this and Figure 20.2 will show clearly the degree to which the isolation of low frequencies at high levels can get wildly out of hand in inappropriately chosen buildings. The room shown in Figure 20.4 weighs about 40 tonnes, and has a floor area of about 27 m^2, so had it been on anything other than the ground floor-and on *solid* ground, with no basement -then it would not have been feasible to construct it in any normal domestic building. The only exception would have been if acoustic engineers could have got to the architects before the building was constructed, in order to suggest the steelwork type of reinforcement. If the premises were to be used for commercial recording, as in the case in Figure 20.4, then such isolation work might just be realistic, but for purely rehearsal, it would normally (in fact in almost all cases) be economically out of the question. What often fails to be appreciated is that the sound waves do not know *why* they are being produced, so the isolation requirements do not change according to the use or the economic viability of their constraint. It is remarkable how many people believe that if they "only" want to make the sound for personal fun, the isolation will somehow cost less than if they were making the sound for commercially serious recording.

PRACTICAL FLOORS

Figure 20.5 shows a selection of typical floor systems. The range of choice is necessary because of the range of isolation needs and the structural conditions encountered. Figure 20.1 shows the principle of the mass/spring/mass system, and Figure 20.6 shows the need for the ideal deflexion range of the spring material. In this case, it is a rubber block, but the principle applies equally to whatever type of spring is used.

It can therefore be seen that the springs cannot be chosen without an accurate knowledge of the weight that they will be supporting. In turn, the weight cannot be calculated until the structural requirements and the lowest frequency of required isolation are known. The choice of isolating springs may also depend on the floor which supports it, because obviously the springs which are spaced out over the surface of the floor present much higher point-loads on the supporting floor than do the "blanket" coverings of fibrous materials or polyurethane foam panels [Compare Figures 20.5(a) and (b)].

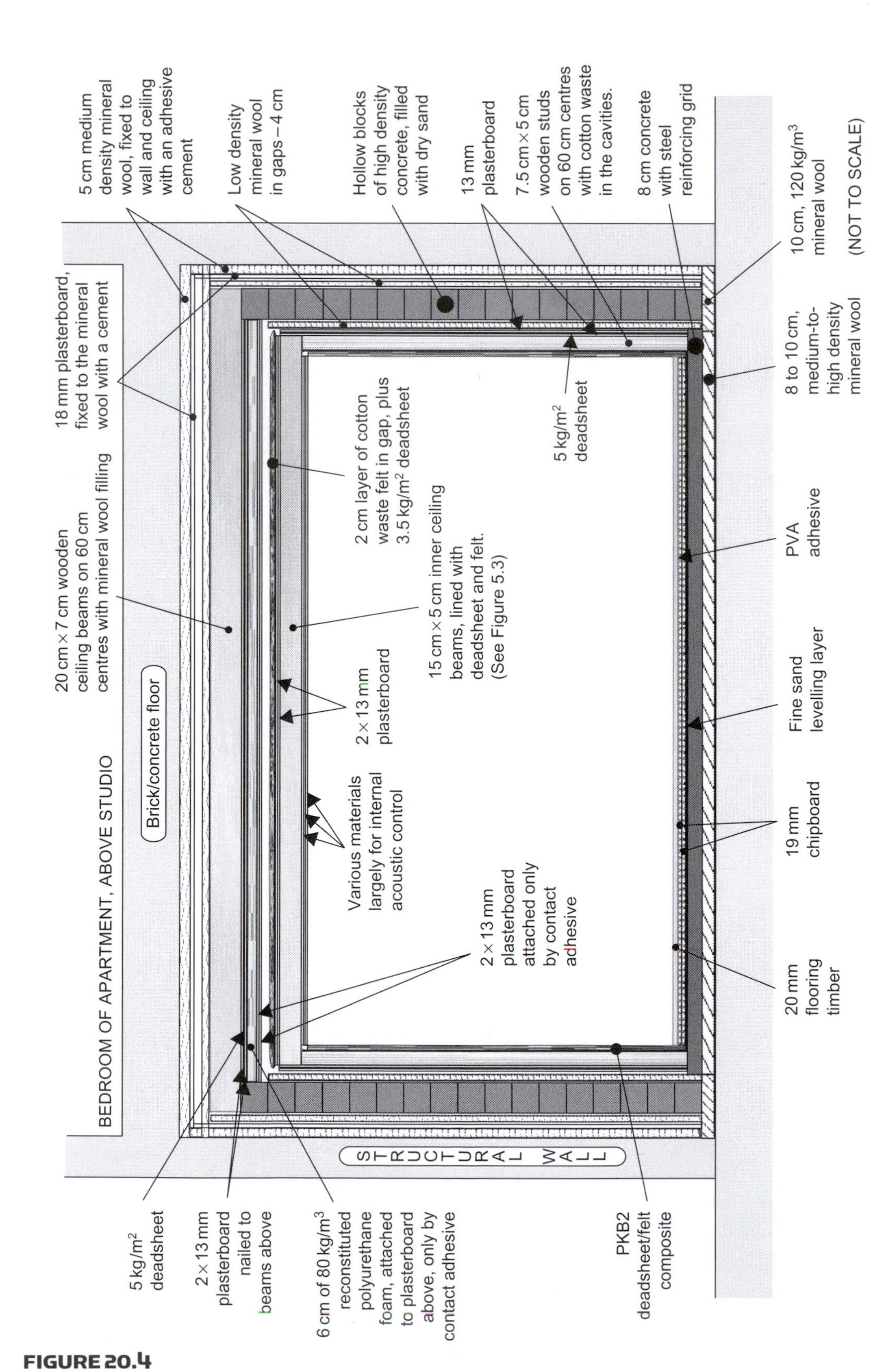

FIGURE 20.4
Triple isolation shell in a weak domestic building.

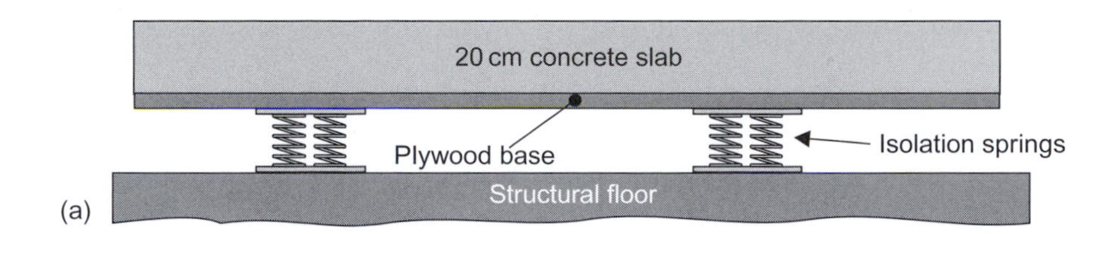

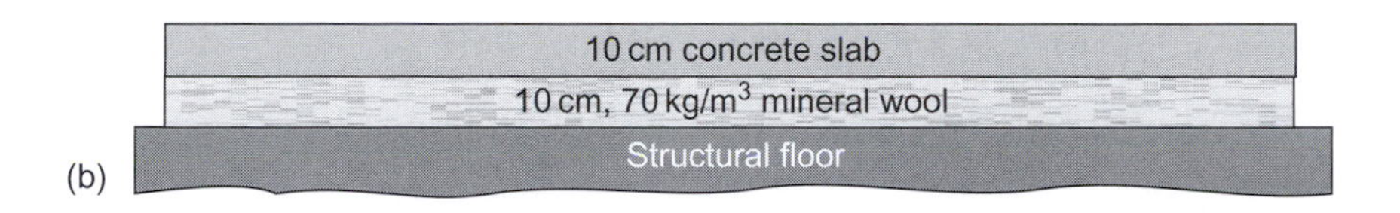

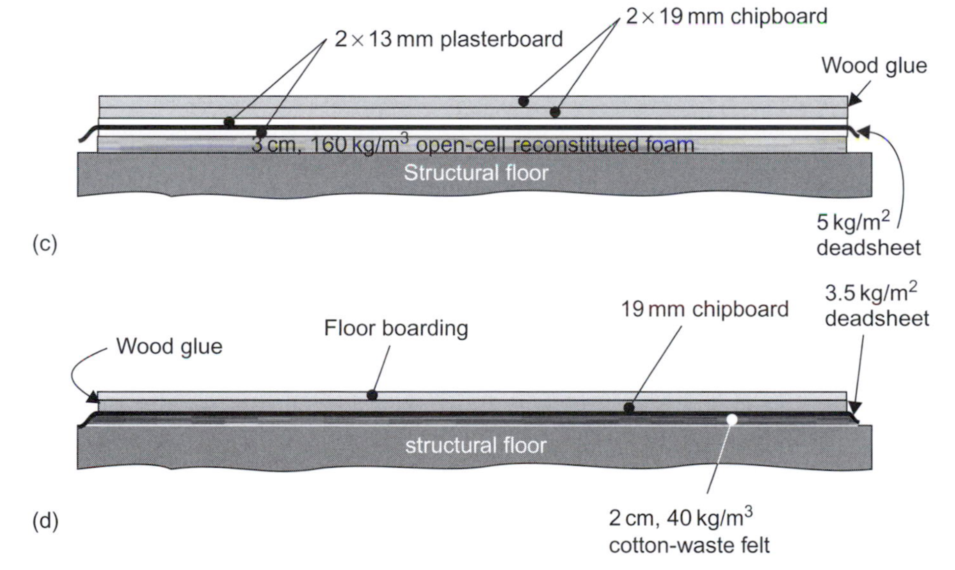

FIGURE 20.5
A selection of floor-floating options.

Figures 20.7 and 20.8 show the practical problems which may be encountered if the floor floating spring material is inappropriately chosen, even though the isolation achieved may be entirely adequate. These problems are not restricted to lightweight foams, however, as there have been cases of rooms built on 40 cm concrete slabs, and floated on very low frequency steel springs, which have tilted alarmingly when a heavy mixing console has been placed in an off-center position when the rooms were equipped ready for use.

Vibration conduction via the ground itself can also be considerable, but it is so dependent on the nature of the local terrain that it is difficult to predict or describe. Therefore, even a very massive rock and concrete floor can be set into vibration by a bass guitar amplifier in contact with it, so, in all but the rarest of cases, a floated floor is mandatory for a professional recording studio. This is even more necessary if there is a potential for vibrations to *enter* the studio, such as may be the case with main roads or railways in the vicinity, or neighboring factories with heavy vibrating machinery.

As previously discussed, the low frequency isolation in terms of both the cut-off frequency and the degree of isolation will be dependent upon the type of recording to be undertaken. The isolation system shown in Figure 20.5(a) is typical of the large studios for full frequency range recording. The 20 cm concrete slab weighs about half a tonne per square meter, and the springs are chosen to support this weight in their mid-compression range, when the whole system will probably have a resonance of 5 to 8 Hz. The bases of the spring mounting frames (or boxes) will be likely to have an area of about

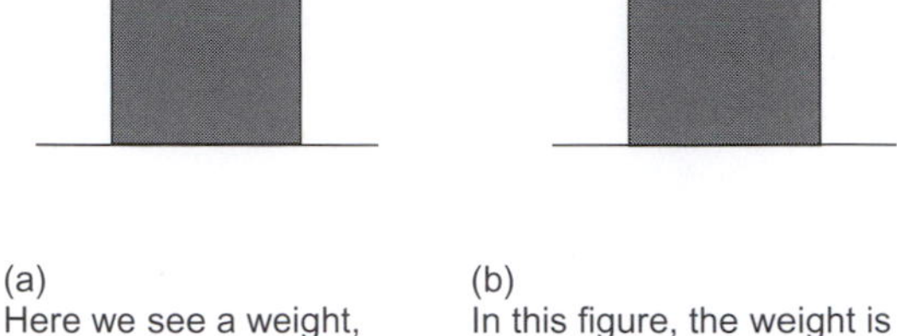

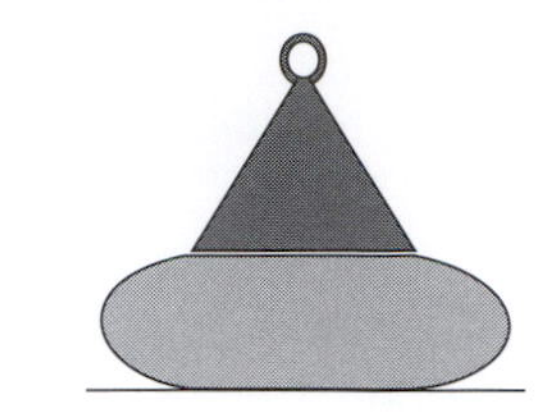

(a)
Here we see a weight, suspended above a block of material intended for isolation.

(b)
In this figure, the weight is resting on the block, but the block shows no sign of compression. Its stiffness must therefore be very great, effectively making a rigid coupling between the weight and the ground. Vibrations in the weight (if, for example, it was a vibrating machine) will be transmitted to the ground.

(c)
In the above example the isolation material is too soft, and the weight has compressed it down to a very thin layer. Its density and stiffness will thus also increase, and the effect will be an almost rigid coupling with the same effect as in (b).

(d)
Above, the weight can be seen to have compressed the isolation material to about half its original thickness. The system is at rest due to the equilibrium being found between the gravitational down-force on the weight and the elasticity of the material.
Vibrations in the weight will be resisted by the mass of the floor and the elasticity of the isolation material. Much of the vibrational energy will be turned into heat by the internal losses in the isolation material, and hence will not be transmitted into the ground.

FIGURE 20.6
Float materials need to be in the center of their compression range for the most effective isolation.

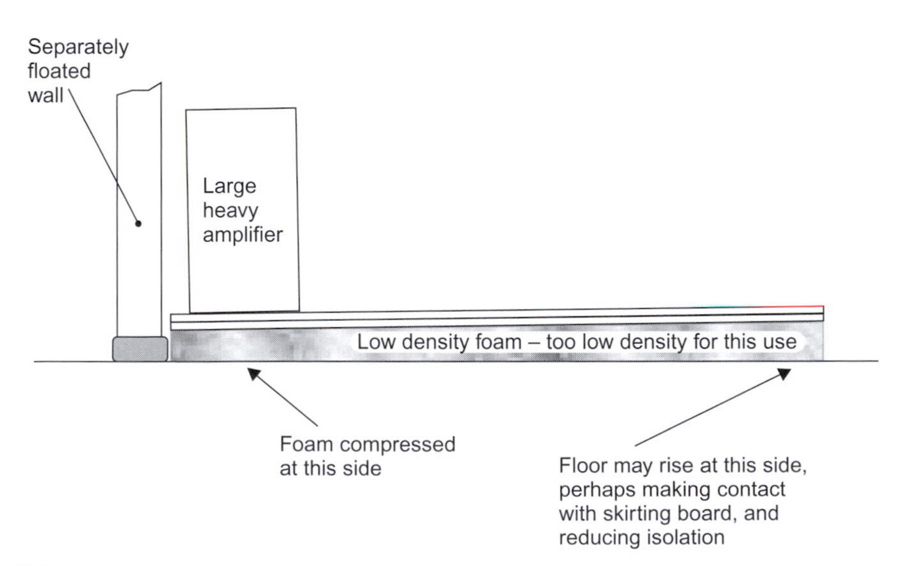

FIGURE 20.7
Effect of uneven floor loading on low-density float material.

400 cm^2, or 0.04 m^2. If there were one spring for every 1 m^2, then the load on each 0.04 m^2 base would be 500 kg, which equates to a pressure of 12.5 tonnes per square meter immediately below the bases. Things like this must be taken into account when designing floor-floating systems. If such spot loads cannot be supported, then something must be done to spread the load. Likewise the floated slab itself must usually have some load spreading, to avoid the *tops* of the springs punching through it. However, this is usually automatically a part of the construction, because the slab will need a continuous base over its entire area, above the springs, before it is cast. (See the plywood base in Figure 20.5(a).)

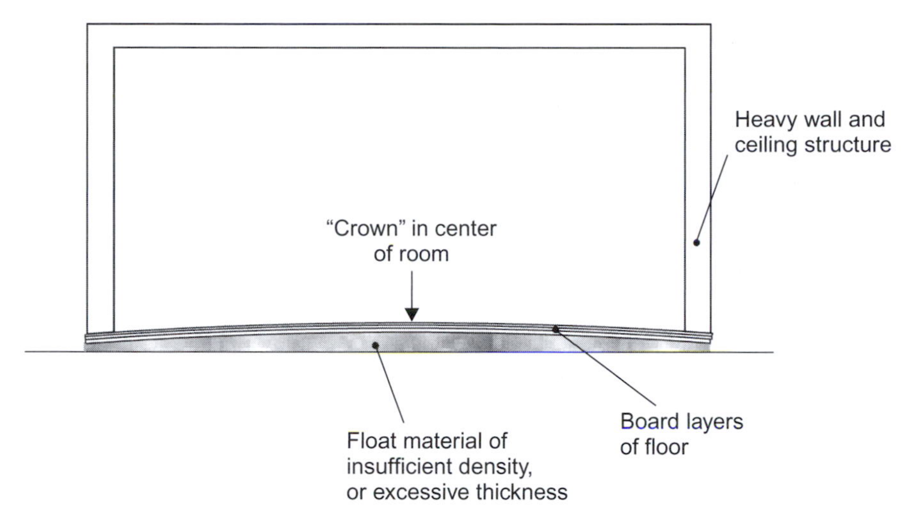

FIGURE 20.8
Crown effect.

Whether the floated slab supports walls around its perimeter is largely dependent upon the weight of the perimeter walls. For a heavy concrete block isolation-wall structure, such as shown in Figure 20.4, it would be typical to float the heavy walls separately, and only to build the lighter weight acoustic control shell walls directly on the floated slab. Too much weight around the perimeter could lead to the crown effect shown in Figure 20.8 unless extra spring reinforcement were to be placed close to the edges of the slab. It is definitely preferable, however, with heavy structures, to float the walls separately on their own spring base, which makes the calculations and the construction much easier. It is also more flexible should modifications need to be made later. Furthermore, the heavy isolation walls may also have to carry the weight of a roof structure, so the downloads below the isolation wall can easily reach 5 or 10 tonnes per square meter, which is excessive in terms of being supported on the edges of a floated slab.

Figure 20.5(b) shows a 10 cm reinforced concrete slab floated on a mineral wool base "spring." Using this technique, the load is evenly distributed over the entire supporting area. It is relatively simple to cast tubes in this type of floor to carry cables, even though the concrete is relatively thin, as shown in Figure 20.9. There can be advantages and disadvantages to this type of floating because, depending on circumstances, the continuous blanket of mineral wool can transmit more energy into the base slab by virtue of the greater area of contact, or, it may improve matters due to the extra damping of the base slab. Certainly, if the floated floor is not on solid ground, such as if it were above a basement, the more even weight distribution of the continuous spring material may be advantageous structurally, even if not acoustically.

Figure 20.5(c) shows a composite floor made from multiple layers of heavy and limp materials, floated on a reconstituted polyurethane foam base. (The polyurethane foam is the sponge [open cell] type, not the aerosol type of closed cell expanding polyurethane foam, which has entirely different properties.) This system is useful where wet-work (cement) is impracticable. Such a floor sandwich would weigh about 50 kg per square meter, which is about one-fifth of the weight of a concrete slab of 10 cm thickness. For stability, and to avoid the effect shown in Figure 20.7, the foam spring layer would typically be 3 cm of 120 kg/m^3 sponge. This type of floated floor exhibits less low frequency isolation than the concrete floors, but can achieve what is often an adequate degree of isolation in circumstances where the floor loading is restricted. Many domestic buildings may only have total floor loading capabilities in the region of 120 kg/m^2, which would be incapable of supporting even a 10 cm concrete floated slab. The density of concrete is typically around 2500 kg/m^3 so a one-meter square of 10 cm concrete slab would weigh around 250 kg. The typical light industrial loading for a building would be in the order of 400 kg/m^2 (or 100 pounds per square foot in imperial measure). Table 20.1 shows the densities of some typical acoustic materials, plus a few others to provide some points of reference.

Figure 20.5(d) shows a very light floating system. It consists of a wooden floor surface glued on top of a 19 mm layer of chipboard, floated on top of a composite of a 3.5 kg/m^2 deadsheet and a 2 cm layer of

(a)

(b)

FIGURE 20.9

Cable tubes. (a) If the 10 cm of mineral wool is composed of two layers of 5 cm, the upper layer can have channels cut into it to allow the tubes to be lowered on to the lower layer, thus positioning the tube half in the mineral wool and half in the concrete. (b) The tubes and mineral wool are covered in plastic sheeting and steel reinforcing mesh. The tube mouths can be seen protruding. They are blocked with mineral wool plugs prior to the pouring of the 10 to 15 cm concrete slab. (Tio Pete Studios, Bilbao, Spain, 1998).

cotton-waste felt. Deadsheets are essentially heavy, limp membranes used as damping layers. This type of floor is principally used to reduce the effect of impact noises from entering a structure. They are particularly useful above a weak ceiling where impact noises from footsteps could enter a structure and be rather difficult to treat from below, especially when the room below has only limited height, and hence not much room for treatment. Figure 20.10 shows a range of isolators used for structural floating. It should now be apparent that the range of possibilities for floor floating is enormous, which is just as well because the range of different circumstances requiring floated floors is equally enormous.

Floors on weak sub-floors

The situation of floor floating is complicated enormously when the structural floor is itself significantly resonant. Such can typically be the case when a ground floor studio is sited over an underground car park, or when studios are stacked one above the other in the same building. If the site under consideration as a potential studio building also has low ceilings, then the sound isolation problems can become impossible to solve without major reconstruction of the building.

Table 20.1 Density of materials

Cork	0.25
Pine	0.45
Typical hardwoods	0.60–0.75
Plywood	0.60–0.70
Plasterboard	0.75
Chipboard	0.81
Water	1.0
Dry sand	1.5
Brick (solid)	1.8
Glass	2.4
Concrete	1.8–2.7
Aluminium	2.7
Granite	2.7
Slate	2.9
Steel	7.7
Iron	7.8
Lead	11.3
Gold	19.3
Osmium	22.4
(the densest element known)	

The figures are in grammes per cubic centimeter, but they also represent tonnes per cubic meter.

When structural floors are weak, it is very difficult to prevent low frequency noises (such as caused by a bass drum on a floated floor) from passing through the spring layer and exciting the resonances in the floor below. If the floor is resonant, then it only requires a relatively small amount of energy to excite the natural frequencies of resonance. In many cases, the only way to overcome such difficulties would be to support another floor, on steel beams, above the main floor, but this can only be done in cases where the walls are strong enough to support the extra weight and when there is sufficient headroom to accept the loss of height. In the following section, these problems will arise again when we look at the problem of low headroom and weak ceilings, and it will also be discussed later in the chapter.

Essentially, a fundamental requirement for high isolation, unless large amounts of money or space are to be consumed, is that the entire outer shell of the chosen building should be massive, rigid, and well damped. Steel, and cast concrete, often fail to provide the last of the three requirements; they can have characteristic resonances that weaken the isolation. However, steel-reinforced concrete floors can be excellent when they are sufficiently strong, as the steel rods encased in the concrete can be effective in providing damping, but the situation can be improved further by the addition of damping agents to the wet concrete mix if the floors are to be cast from new. "Concredamp," for example, is one proprietary brand of such a compound.

CEILING ISOLATION

Reference, once again, to Figure 20.4 will show a range of techniques used for ceiling isolation. The need for this seemingly excessive collection of techniques was precipitated by the somewhat absurd decision by four Andalusians to site a recording studio, for 24 hours per day use by rock bands, directly under the bedroom of an unsympathetic neighbor in a building of unsuitably weak structure.

(Taken from the publicity material of Eurovib Acoustic Products Ltd. [central group], Fabreeka UK and Mason UK Ltd.) [www.eurovib.co.uk / www.fabreeka.com / Mason P.O. Box 190, GU9 8XN, UK]

FIGURE 20.10
Range of isolation mounts.

Although, for clarity, the drawing is not exactly to scale, neither is it too far from scale. It does give, therefore, a reasonably realistic view of the amount of space lost to isolation and acoustic control.

Such a construction was only possible because the original space had (barely) adequate height-about 3.5 m-allowing about 1 m for isolation and acoustic control before the inner fabric ceilings at 2.5 m. There is currently a tendency towards smaller and smaller studios as the perceived size requirement comes down by using hard-disc-based recording and mixing systems. However, the size of human beings and acoustic wavelengths is *not* coming down, and as virtual studios do not replace conventional ones, there is a limit to how small one can go. The other force that is bearing down on studio size is an economic one. In many cases, as the real price of recording equipment of good sound quality comes down, an enormous number of people now seem to expect that the cost of recording acoustics should somehow fall accordingly. It is unlikely to do so, though; simply because economics have no bearing on the speed of sound or any other physical law, and it is the physical laws of acoustics that determine whether a room is well isolated or not—and also whether it is going to have good internal acoustics or not. There is definitely is a tendency for people to underestimate the need for adequate ceiling height in order to make a good studio.

Given a ceiling height of only 3 m, it is almost impossible to install adequate ceiling isolation if the floor above is either noise sensitive or liable to radiate noises that can be prejudicial to the operation of the studio. The only exception is if the floor is truly massive, but this may only be of use in constraining sound within the studio. In the reverse direction, the vibrations in a floor that has considerable mass can be extremely difficult to stop from entering the space below unless an isolated ceiling can be employed which is suspended from floated walls, and not from the ceiling above. In Figure 20.4, an isolated plasterboard inner ceiling, above the acoustic control shell, can be seen suspended from a 6 cm layer of polyurethane sponge (reconstituted foam) which is in turn glued to a plasterboard/dead-sheet/plasterboard sandwich fixed below wooden beams. These are supported on the floated concrete walls. In addition, to augment the transmission loss and to prevent resonances between the upper plasterboard layer and the plasterboard suspended from the upper ceiling, a mineral wool infill was used between the beams. There is therefore no physical contact between the structural ceiling and the intermediate isolation ceiling except via the air, and ultimately the ground, but this contact is decoupled by way of the suspension materials between the floated concrete isolation walls and the main structure. Furthermore, on the underside of the structural ceiling (the floor of the bedroom), a layer of 5 cm medium density mineral wool had been attached by a cement. Below that, also by means of a cement, had been suspended a layer of 18 mm plasterboard. A further isolation layer of two 13 mm plasterboards had been laid on top of the inner isolation structure.

One reason why so many different materials and suspension systems were used was to avoid any coincident resonances that can result from using too much of too few materials. Any resonances inherent in any structure or material will tend to reduce the isolation, perhaps seriously, at the resonant frequency. By using a variety of isolation techniques, the resonances will tend to occur at different frequencies, and hence any weak spots will be covered by the other material combinations in the different layers. A brief "ride" on a sound wave attempting to propagate through the isolation ceiling may be usefully informative, as we can discuss the isolation mechanisms involved as we come across them.

A trip through the ceiling

Before reading this sub-section, it may be worthwhile photocopying Figure 20.4 and keeping it visible during the discussion. The description of the passage of an acoustic wave through the ceiling system should be informative because it will encounter the many different types of obstacles that have been built into this ceiling in order to try to achieve the desired isolation. An understanding of something of the feel for these isolation concepts is essential, because experience and knowledge of the practical application of these principles are perhaps the only real way of deciding on the most effective combination.

Even great academic acousticians acknowledge that "the theoretical analysis of the sound transmission through double leaf partitions is far less well developed than that of single leaf partitions, and that consequently greater reliance must be placed on empirical information." (Fahy, 2001) Clearly, the ceiling construction that we are describing here is even more complex than a double leaf partition, so if any readers get any bright ideas about using their computers to solve these complex interactions—forget it! There are simply no programs to deal with this sort of thing, and even experts need to use a lot of discretion when using what limited computerized help that they have. In the words of Fahy "The reason is not hard to find; the complexity of construction and the correspondingly large number of parameters, some of which are difficult to evaluate, militate against the refinement of theoretical treatments." (Fahy, 2001) Bearing this in mind, please forgive the following wordy treatment of the subject.

An acoustic wave originating in a recording space shown in Figure 20.4 and traveling in the direction of the ceiling would first pass through the lower layers of the inner ceiling, whose purpose is primarily for controlling the character of the sound within the room itself. The wave would next encounter the two layers of plasterboard fixed on top of the support beams of the inner acoustic shell. Plasterboard consists of a granular textured plaster core, covered with a layer of paper on each side to provide a degree of structural integrity. It is more resistant, less dense and less brittle than pure plaster sheet. High frequencies striking this surface would be almost entirely reflected back, because the plasterboard is reasonably heavy, having a density of about 750 kg/m^3. The weight of a double layer, around 18 kg/m^2 would be sufficient for the mass law, alone, to ensure that the highest frequencies were reflected back. As the frequencies lower, they will be progressively more able to enter the material and set it into vibration. To some degree, these vibrations will set the whole layer in motion, and the opposite surface will then re-radiate the vibrations into the air above. However, to a considerable degree, the particulate/granular nature of the core material will cause frictional losses as the vibrations pass through the material, turning acoustic energy into heat energy. The amount of energy available to re-radiate from the upper surface of the plasterboard will therefore have been reduced, and a degree of isolation will have been achieved. Figure 20.11 shows a typical loss versus frequency plot for this type of ceiling. However, the plot can only be taken as approximate, because the isolation provided by such structures is dependent upon their absolute surface area, their rigidity, and numerous other factors.

Not all the isolation is due to absorption, of course, because some is due to the energy being reflected back down towards the floor, but there will be progressively more absorption, and transmission via re-radiation, as the frequency lowers and the reflectivity reduces. The overall transmission loss will include the *reflexion* and *absorption*, and the overall coefficient of absorption figure will include *absorption* and *transmission*. Essentially the isolation is what is not transmitted, and the absorption coefficient is what is not reflected. In fact, absorption is traditionally measured in units called sabins, after Wallace Clement Sabine, the pioneer of reverberation analysis, who himself first used "equivalent open window

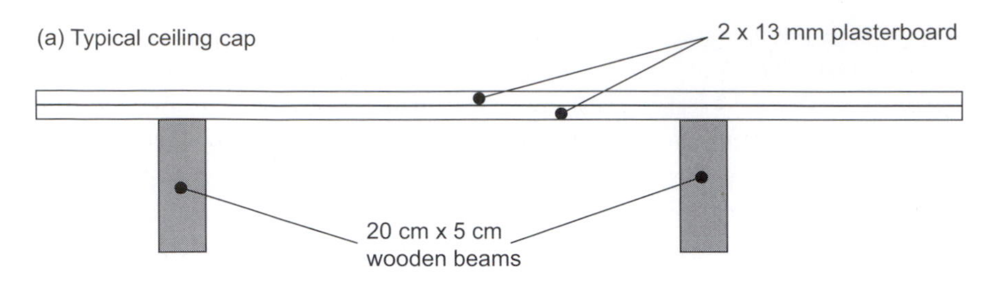

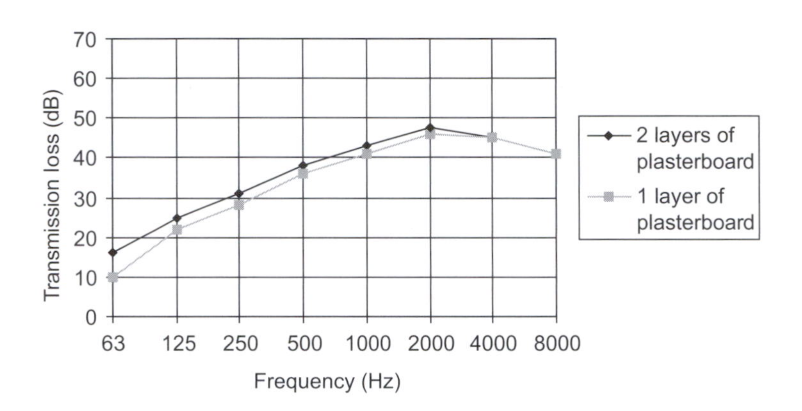

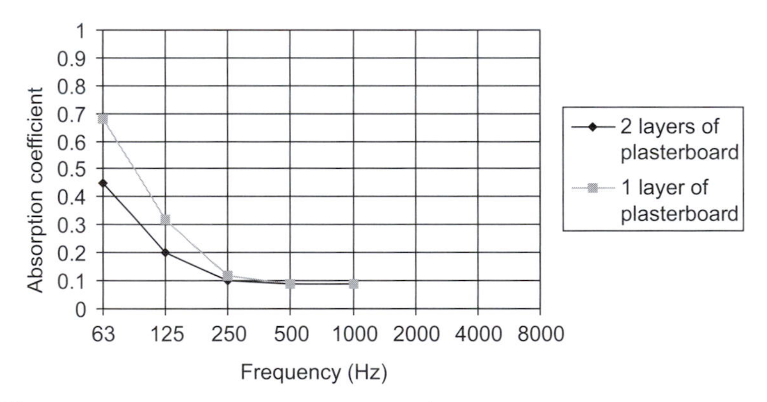

FIGURE 20.11

Acoustic performance of a typical ceiling cap. (a) Typical ceiling cap, (b) transmission loss versus frequency, (c) absorption versus frequency. Note how the addition of an extra layer of plasterboard ***increases*** the transmission loss (isolation), as shown in (b), but ***decreases*** the absorption, as shown in (c). The addition of a second layer adds more mass to the system and tends to make it more rigid. The effect is that the two layers become more reflective than the single layer. The additional layer therefore improves the isolation by means of reflecting more energy back into the room, but this will of course make the room more reflective or reverberant. If the decay time of the room with only one plasterboard installed was to be maintained after the addition of the second layer, more absorbent materials would need to be introduced into the room. Isolation and absorption are often confused by non-specialists, but this situation clearly demonstrates the difference.

area" as a unit of absorption. From this concept, it can be clearly understood how absorption can include transmission, because an open window converts almost no acoustic energy into heat. It transmits almost all of it to the other side.

One hundred per cent transmission = 100% absorption = zero reflexion. Zero transmission, achieved either by 100% absorption, 100% reflexion, or some combination of the two, still amounts to 100% isolation. The loosely used terminology can be a little confusing, but absorption does get lumped into isolation and transmission figures, dependent upon which one is of greatest importance in the context of the problems. Just to help to clarify things, a room of one-meter thick, smooth, solid concrete walls and ceiling would have excellent isolation, but very little absorption. It would tend to have along reverberation time because the sound energy would be trapped inside of it. Almost nothing would escape. Conversely, a room of tissue paper in the middle of a desert would have very high absorption (and consequently almost no reverberation time), but almost no isolation, because all the energy would be either lost in the sand, or would be free to escape to the outside air.

Anyhow, let us now return to our discussion of Figure 20.4. The vibrational energy that is transmitted through the double plasterboard acoustic shell cap would then proceed across the air gap to the lower surface of the isolation shell ceiling. The acoustic wave would first pass through a layer of 2 cm cotton waste felt, bonded to a 3.5 kg/m^2 layer of an acoustic deadsheet. Here we encounter an additional, loosely coupled mass layer with the absorbent felt layer uppermost, in the air gap. The function of these layers is to damp resonances in both the plasterboard and the air. The high viscous losses in the deadsheet tend to suppress any high Q resonances in the plasterboard, and the fibrous absorption characteristics of the felt will tend to damp down any cavity resonances, caused by the hard parallel surfaces of the upper surface of the control shell and the lower surface of the isolation shell.

The lower surface of the isolation shell is a classic mass/spring/mass combination, formed by a double layer of 13 mm plasterboard (the lower mass) connected, by contact adhesive only, to a 6 cm layer of reconstituted open cell polyurethane foam of 80 kg/m^3 (the spring). This is in turn connected, by contact adhesive only, to the upper mass. In this case, the upper mass consists of a sandwich of two layers of plasterboard with a 5 kg/m^2 deadsheet in-between, nailed to a relatively heavy and rigid series of wooden support beams. The lower plasterboard layers are not in contact with the concrete block walls-there is a gap of a few millimetres all around. This mass layer is therefore free to vibrate, with its vibrations alternately applying compression and expansion forces on the polyurethane foam spring. Such a system might seem a little insecure, but in fact the foam can withstand traction forces of over 3 tonnes per square metre. With a good adhesive, there is little chance of the plasterboard causing problems, as the double layer weighs less than 20 kg per square meter—less than 1% of the traction limit.

Once again, the lower plasterboard layers will tend to reflect back the incident wave to a degree according to the frequency, and will absorb some of the energy in the internal frictional losses. However, this time its upper surface is not so free to radiate, because it is glued to a spongy foam. The foam will have the action of damping the vibration in the plasterboard, again rather like sticking Plasticine on a bell. The vibrations that do enter the foam layer will be strongly resisted by the mass and rigidity of the upper mass layer. The tendency, therefore, will be for the foam to compress and expand, rather than to move as a whole, and the internal frictional losses in the cellular construction of the foam will be very effective in converting the acoustical energy into heat. This action will be effective down to the resonant frequency of the spring thus loaded.

The upper mass layer acts as a low frequency barrier, and its effect is augmented by the rigidity imparted by the wooden beam structure. In this instance, the two layers of plasterboard sandwich a layer of a 5 kg/m^2 dead-sheet, the combination constituting what is known as a constrained layer system. The principle of the damping effect and additional acoustic losses due to the constrained layer are shown in Figure 20.12. Such a system will be reflective down to lower frequencies than the plasterboard alone, but it will also produce more absorption down to lower frequencies. It will certainly transmit less.

Whatever acoustic energy is left to propagate above the constrained layer will then cross the air gap between the wooden beams and proceed on to the layer of 18 mm plasterboard, attached to a layer of medium density mineral wool by a cement. The mineral wool is itself then attached to the structural ceiling by the same cement-type adhesive. The air gap again contains mineral wool to suppress any cavity resonances, but it also acts as a velocity component absorber.

The plasterboard/mineral wool/structural ceiling combination again is a mass/ spring/mass system. As mentioned previously, the reason for using different materials and thickness in this upper system was to avoid any weak spots caused by resonances that coincided with the resonances in the lower mass/spring/mass system. Once again, the upper mass of the structural floor acts to resist the bodily movement of the mineral wool spring as it is set in motion by the vibrational energy impinging on the lower mass of 18 mm plasterboard. By this time, the acoustic energy from the source in the recording room will have been reduced to such a degree that what remains to re-radiate into the bedroom is insufficient to create any disturbance to the sleep of the neighbors. By now, also, it will only consist of low frequencies-the highs will be virtually non-existent. If this all seems a little extreme, it must be borne in

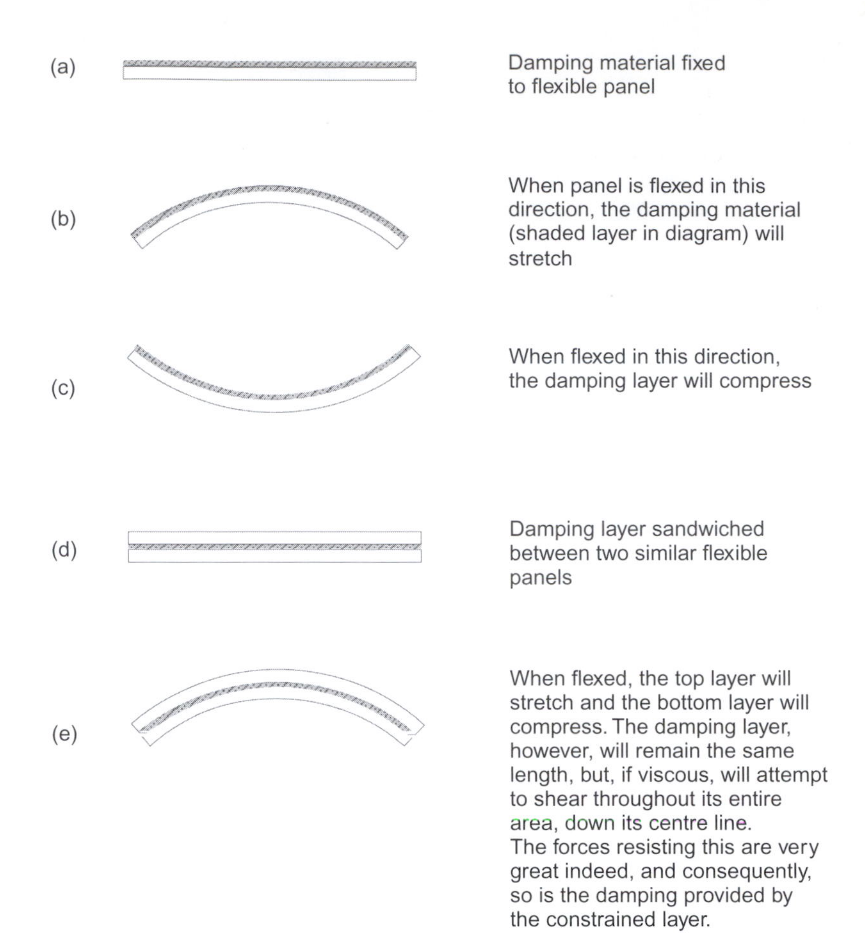

FIGURE 20.12
Constrained-layer damping principle.

mind that in order to achieve even 60 dB of isolation, only one one-millionth of the sound power can be allowed to penetrate the barriers.

SUMMING THE RESULTS

So, from the individual isolation figures of the individual isolation systems we should be able to calculate the total isolation, but there are some practical circumstances which make calculations difficult unless many other factors are known.

If we have a noise source in the recording room, then it is of little help if we only know its output SPL in free-field (anechoic) conditions. If we know that a guitar amplifier can produce 110-dB SPL at one meter distance in a relatively dead room, that does not mean that it cannot produce more in other circumstances. If we were to take it into a very live room, the level of the direct signal would be augmented by all the reflexions, which could superimpose their energy on the direct sound and easily produce 6 or 8 dB more SPL. Therefore, we need to know the sound pressure levels that will be produced by instruments *in the actual rooms* which we need to isolate, and from this, it will be obvious that for any given sound source, a live room will need more isolation than a dead room. Quite simply if things sound louder in an acoustically live room, then the sound isolation requirements to an adjacent room will be correspondingly greater.

Conversely, if the receiving room is reverberant it will amplify, by means of reflected energy, the noises entering from an adjacent room. In a small bedroom, a double bed provides significant absorption, as do furniture and curtains. In the case that we have been discussing, the bedroom was well furnished, and the noise level measurements were taken at a point 50 cm above the bed. Had the room been stripped of all furnishings it would have become much more reverberant, so the 30 dB A measured after the isolation may then have risen to 35 dBA or so, which could be deemed to be unacceptable

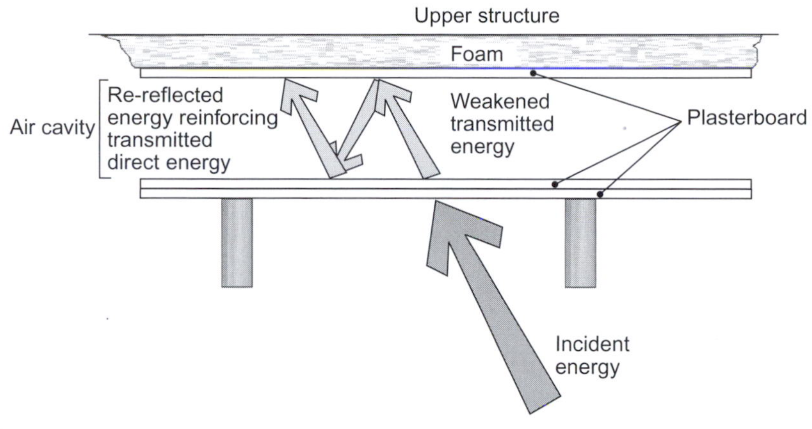

FIGURE 20.13
Re-reflexion between layers. Resonant build-up in an air cavity can reduce isolation in a similar way to which the build-up of reverberant energy in a space can make a source louder than it would be in free-field conditions.

for the neighbors to sleep easily. Similarly, if the acoustic control treatment were to be removed from the recording room, rendering it much more reverberant, the maximum SPL produced in the room by typical instruments would also increase. So many variables exist in acoustic control work, which is why it is often so difficult for acoustics engineers to give simple answers to what may appear to be simple questions.

Internal reflexions

There is another reason why the simple summation of the calculated isolation provided by each individual section of the complete structure cannot be relied upon to give the isolation of the whole system. It is that internal reflexions can be set up between the layers, which can add re-reflected energy to the forward-going propagation. This is shown diagrammatically in Figure 20.13. The reduction of this effect requires fibrous layers in the various air spaces. Especially in existing buildings, where the exact nature of the structure is not known, a degree of "cut and try" can be involved, even for very experienced acoustic engineers.

WALL ISOLATION

In general, walls are easier to isolate than floors or ceilings. Figure 20.14 shows a relatively simple solution, the addition of a heavy internal wall with a cavity between the isolation wall and the structural wall. There is usually little problem constructing walls of significantly greater mass per square meter than can be used in ceilings, for example. Floors, of course, will be in some sort of physical contact with the structure, via either springs or mats, but heavy ceilings need to be supported with massive beams or suspension systems, which can often be quite difficult or impracticable. A wall, however, as long as it has a solid floor below it, can usually be both very heavy *and* simple to construct. The only problem with the wall system in Figure 20.14(a) is the flanking transmission via the contacts at the floor and ceiling. This is simply avoided by the insertion of spring seals as shown in Figure 20.14(b).

Wall isolation is sometimes also made easier by the fact that the existing structural walls are often more massive than the floors (other than those built directly on the ground) and ceilings. Very effective wall isolation can be achieved by the use of floated concrete block walls filled with sand. These are especially effective if the structural walls are also massive and well damped. In new buildings, it can be advantageous to ask the architects to use sand-filled concrete blocks in the main structure. The combination of structural and floated wall systems so made, with a 5 or 10 cm air space between, can easily produce more than 70 dB A of isolation, which is extremely difficult to achieve in floor or ceiling isolation systems unless these can be used simultaneously, both above and below the structural ceilings and floors. In shared buildings, therefore, where access to the walls, floors and ceilings of neighboring premises is not possible, wall isolation is by far the easiest to achieve in one-side-only treatments.

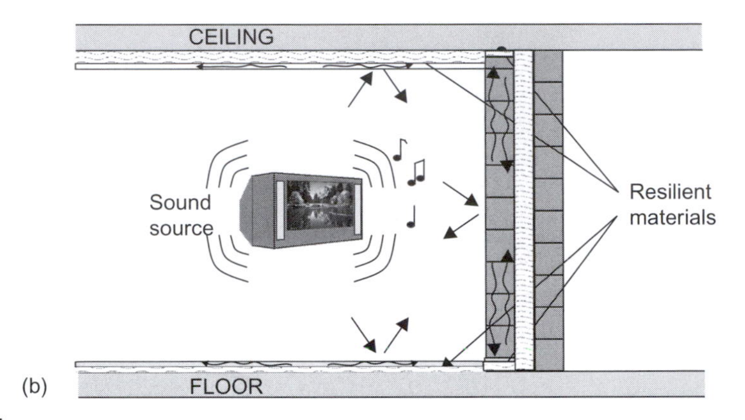

FIGURE 20.14
The benefits of floating surfaces. (a) The addition of an inner wall to aid isolation between the rooms can be largely ineffective if flanking transmission via the floor and ceiling pass round it. What is more, resonances within the inner wall structure can themselves pass additional energy into the floor and ceiling, perhaps making matters worse at some frequencies. (b) Protecting the floor and ceiling from direct sound, plus floating the inner wall away from the walls, floor and ceiling, can greatly reduce the amount of sound passed into the structure, and hence to the neighbor.

LIGHTER WEIGHT ISOLATION SYSTEMS

In some circumstances, either only a limited degree of isolation is necessary, or moderate isolation is needed in buildings with weak floors. In such cases, systems such as those shown in Figure 20.15 can be used. The technique simply involves lining the entire room with open-cell polyurethane foam (or mineral wool) of suitable density and thickness, and lining the interior with plasterboard. With low density spring materials (foam or mineral wool) on the walls and ceiling, the system resonances can be kept quite low. On the floor, however, low density materials would lead to instability, such as shown in Figure 20.7, so to keep the system resonances sufficiently low with the higher density spring, a correspondingly greater weight needs to be added. This can be achieved by laying concrete paving slabs on top of the floor covering, or even casting a reinforced concrete slab if it is feasible.

The results with this type of isolation system can be surprisingly good, though their resistance to the ingress of impact noise from outside the structure can be poor. The other benefit is that the weight is distributed about the room surfaces, and the whole load is not imposed on the floor. Even the walls carry some of the load, which can be very advantageous in weak buildings.

RECIPROCITY AND IMPACT NOISES

Notwithstanding certain restrictions, if we were to build two adjacent rooms with 60 dB of isolation between them, it would not make any difference which room contained the source of the sound and which was the receiving room. The isolation would remain the same, even though the order of materials in the isolation system may be asymmetrical, as shown in Figure 20.16. However, this only remains true for airborne sound, where the solid materials present an acoustic impedance much greater than that of the air, because their densities are enormously different to that of air.

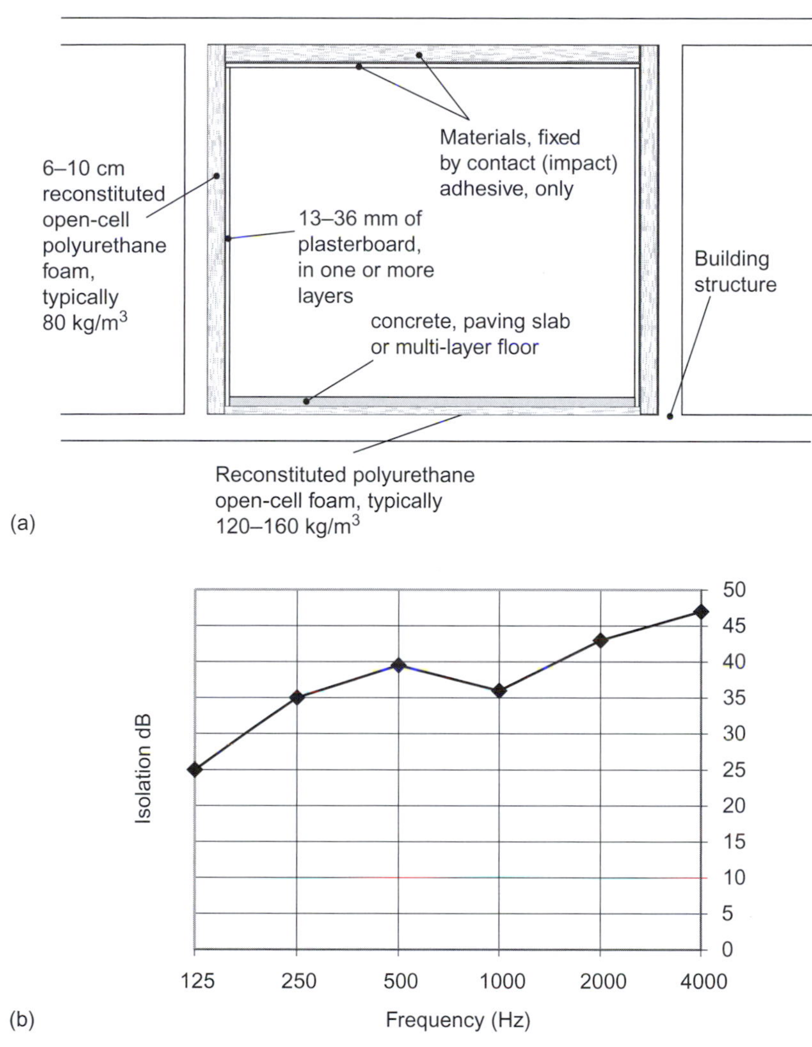

FIGURE 20.15

Distributing the isolation load. (a) Typical construction. Isolation of this type can be very effective. The walls and ceiling bear much of the load, which can make this technique useful in premises with weak floors. Alternatively, a similar system can be used with mineral wool and a cement-based adhesive. (See Figure 20.4.) Contact adhesive gives a more instant bond, without the need for support during setting, but the noxious fumes can be a problem during construction. (b) The plot shows the typical extra isolation achieved by the sort of system shown in (a), above that provided by the basic building structure.

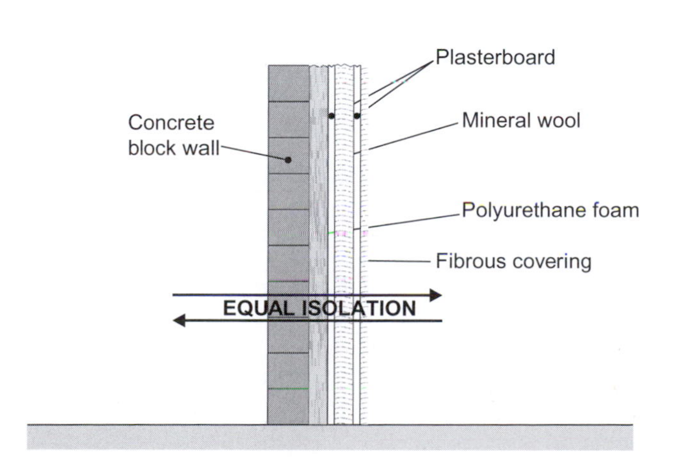

FIGURE 20.16

Non-directionality of airborne sound isolation. Despite the asymmetric distribution of the isolation materials, the isolation for airborne sounds will be equal in either direction.

Figure 20.17 shows the vertical separation of two rooms, the lower one having an open-cell foam/plasterboard ceiling isolation, which is typical in many rooms with limited ceiling height because it provides a means of, in many cases, adequately protecting the rooms above from airborne noise from below. Once again, 50 dBA of isolation could be achieved, irrespective of the direction, if the sounds reaching the horizontal surfaces were airborne.

The problem arises when people, and especially when wearing high, slim heeled shoes, walk on the upper floor. In this case, the weight on a small area when the heel strikes the floor presents a totally different case than if somebody were to strike the lower plasterboard ceiling with a similar shoe heel. The difference between this situation, where the isolation is *not* reciprocal, and the case of airborne noise isolation, where reciprocity exists, lies in the nature of the impedance differences.

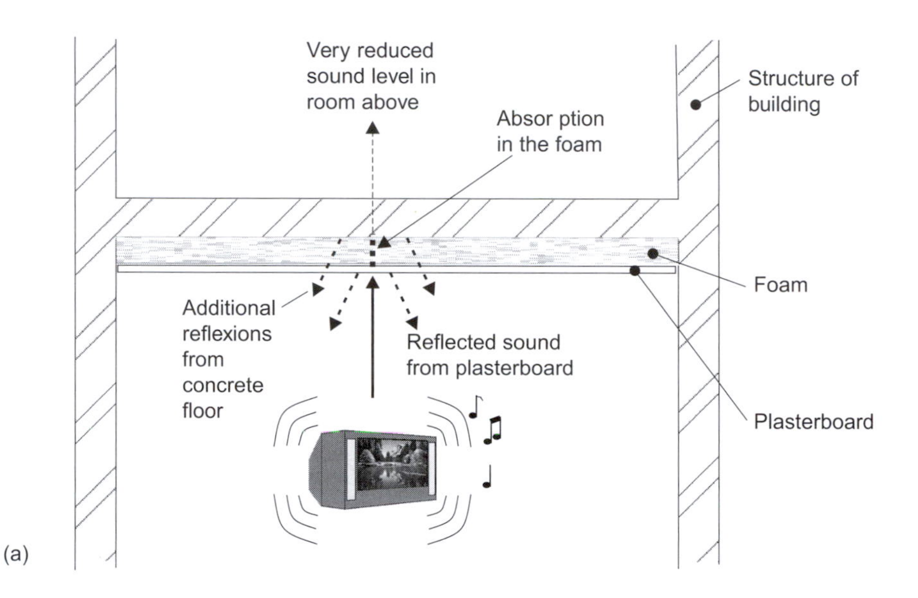

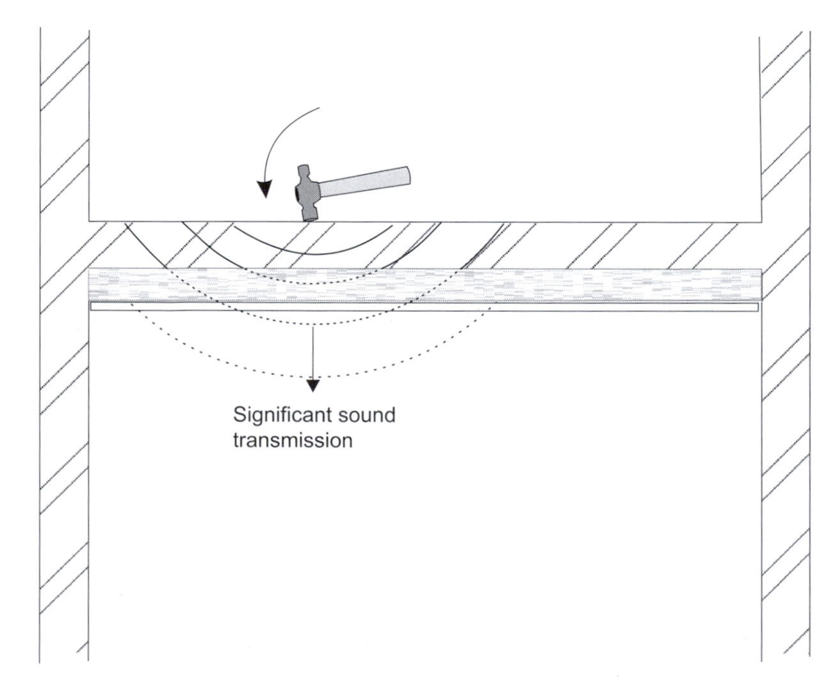

FIGURE 20.17
Airborne and impact noise isolation. (a) For airborne sound, (b) for impact noises. Direct impacts on the structure can set the whole thing in motion, including anything that may be attached to it. What is more, a very wide range of frequencies are produced by the impact, which are capable of exciting many resonances.

When a shoe heel strikes the hard surface of the upper floor, the heel is brought to rest almost instantly. The transmission of the impact energy into the floor is great because the termination is abrupt and resistive (solid). If the heel were to strike the plasterboard surface, below, the surface presents a less abrupt termination to the impact. The plasterboard on the foam will tend to give, acting like a shock absorber, and the springiness will rebound and return some of the energy to the heel. There will be a sensation of a bounce-back: the termination will be more reactive. Such solid object-to-solid surface contacts are *not* typical of the gas/solid/gas contacts of normal airborne sound isolation, and they behave differently.

Without making a complete break in the solid materials it is very difficult to prevent the structural impact noises from penetrating most practical ceiling isolation systems. This is where the problem arises when structural ceilings are too low, and there is too little room to use a separate set of ceiling joists on the floated walls to float a completely separate ceiling structure. One solution to the problem is to use a simple floated floor system, such as shown in Figure 20.5(d), on the floor above, but this may not be possible when the upper floors are occupied by different owners. In any case it may not be possible to float the floors because of the need to change doors; and their frames if the raised floor reduces headroom below legal limits.

In general, adequate isolation may be difficult if not impossible to achieve when chosen premises do not have adequate ceiling height.

THE DISTANCE OPTION

The degree to which sound isolation is required in any proposed recording studio building is dependent upon several factors: the construction of the building (whether light or heavy); the proximity of noise sensitive neighbors; the proximity of external noise sources; also the nature of the recording to be undertaken. All affect the noise control calculations. At one extreme, let us imagine a synthesizer-based musical group, who wanted to make recordings in a studio on the ground floor of a massively built farm building in a valley. If this building were sited miles away from any neighbors, and not subject to any farm machinery or heavy transport noise, then the sum total of isolation needed would probably be to fit double-glazing with 40 cm between the panes of glass, and change the doors to a type with good acoustic isolation.

At the other extreme, let us consider the isolation needs for a recording studio for orchestral and choral use, sited in an office building in a location above an underground railway and surrounded by streets with heavy traffic. In this case, it would be unacceptably expensive for noise-induced delays to affect the recording of a 100-piece orchestra due to underground train rumble during quiet passages. It would also be unacceptable for the fortissimo passages to disturb people who may be concentrating on their work in adjacent offices. The isolation work may require the construction of an entirely floated inner isolation shell, of considerable weight, which may need the reinforcement of the floor in order to support it. In turn, this new hermetically sealed box would need to be penetrated by HVAC (heating, ventilation and air-conditioning) systems, which themselves would require considerable isolation. The costs could be very high indeed.

In the first of the cases discussed above (the studio for the synthesizer band), probably the only acoustic recordings would be of close-mic'd vocals and the occasional guest musician. Neither the sound egress nor ingress would be particularly problematical, because there would be nobody to annoy by any external leakage of the sound, and except for the rare coincidences of extreme weather conditions during the recording of an even rarer acoustic guitar, the signal to noise ratio of recordings would generally be excellent. The only problem in this type of situation is that it would be restricted in its use. To some degree, though, all recording studios *are* restricted in their use; none are universal in their applications.

In the second case, after due isolation work and internal acoustic treatment, the studio would be capable of recording just about any type of music at any time of day. The main restriction would be financial. After spending so much money on the studio building and preparation, the hourly rate would need to be prohibitively high for the general overdubbing of vocals, for example. That high hourly rate may be inconsequential though, compared with the cost of an orchestra's expenses when traveling to a studio in a distant out of town location.

It is true, therefore, that an isolated location, by virtue of its distance from noise related problems *is* an option for reducing isolation costs, because the physical, geographical isolation *is* acoustic isolation.

However, when overall convenience of access is important, this geographical isolation may be completely impractical, so the only solution may lie in the choice of an inner-city site and a considerable amount of acoustic engineering.

DISCUSSION AND ANALYSIS

Clearly, isolation is not a simple subject to grasp, and no simple computer programs can solve the problems. Isolation is a subject for specialists, and where critical situations exist, the cost of calling in a specialist before construction will surely be less than the cost of trying to fix the problems after construction. In fact, the problems may lie deep within the construction, in which case total rebuilding may be required. Intuitively, leaving out a layer of fibrous material between two of many layers in an isolation system may seem insignificant, but if the cavity resonates without the lining, the air at resonance can be a remarkably strong coupling medium.

The addition, or not, of a small component in a complex structure may thus make a disproportionate difference to the results, and even one decibel of improved isolation may be important. Experience has shown that it is better to design for inaudibility for the neighbors, rather than to legal limits, because peace with the neighbors can be crucial to their tolerance of the occasional nuisance. Nevertheless, ultimately the law will decide if the worse comes to the worst, and that is where that one decibel can mean make or break.

In fact, if the room shown in Figure 20.4 had failed to achieve its low frequency isolation target by 1 dB, then where would one begin to look for the weakness? What would one do to plug the 1 dB leak? If the room had been completed, with all its decorations, and was considered by all concerned to sound good, then how could an extra decibel of isolation be found without perhaps destroying all the finishes, changing the sound within the room, and severely straining the finances of the owners? These are not easy questions to answer; not even for experts, so to fail by 1 dB could be disastrous.

Efficient acoustic engineering implies that the maximum effect is achieved with the minimum use of resources, but isolation in difficult situations does not always lend itself to precise analysis and the presentation of finely tuned solutions. Despite the fact that isolation can be expensive, there needs to be a safety margin in the calculations and assessments of situations, because the science is not exact and the cost of failure can be so grave. It is always better to err on the safe side.

Hopefully, this chapter will have given a good idea of the problems to be aware of and the likely solutions available, but isolation is nevertheless something that should not be undertaken without adequate knowledge of the subject, because much of it is very counter-intuitive, and the potential for disappointing results and wasted money is very great.

Fibrous and cellular springs-thicknesses and densities

To highlight some of the above points, it may be informative to discuss the seemingly simple subject of the resilient layers below floated floors. However, despite the fact that it may seemingly be simple, there are many factors which need to be taken into account in order to choose the most appropriate density and thickness for any given application.

As shown in Figure 20.6, if the elasticity of the floor floating springs is either too great or too little, good isolation will not be achieved. Below the resonant frequency of the mass/spring system, such as that formed by a floated concrete slab and a mineral wool layer beneath it, most of the isolation is lost. At and around the resonant frequency, the isolation can actually be worse than with no treatment at all because of the amplifying effect of the resonance. In general, isolation is considered to begin to become effective above about 1.4 ($\sqrt{2}$) times the resonant frequency; thus if isolation down to 20 Hz is required, the floor resonance should be below 14 Hz or so (14 Hz $\times$ 1.4 = 19.6 Hz). Nevertheless, *effective* isolation does not tend to occur until the resonant frequency is around half of the lowest frequency to be isolated. In practice, for full frequency range isolation, 12 Hz is a good target to aim for.

Using a weaker spring will lower the resonant frequency, as will increasing the weight placed upon it, but, as shown in Figures 20.7 and 20.8, if the spring material is *too* weak then floor stability and flatness may be lost. If, for example, 70 kg/m^3 mineral wool were to be used as the base, then doubling the density to 140 kg/m^3 would *increase* the resonant frequency (by virtue of providing a stiffer spring) by $\sqrt{2}$ (1.414). Halving the density to 35 kg/m^3 would reduce the resonant frequency by the same amount. If the higher density was necessary in order to give increased stability, then the only way to reduce the resonance frequency to its original value would be to double the weight resting upon it.

If the resonance frequency were required to be halved, whilst using the same mineral wool spring (or, in fact, any given spring), then the weight would need to be quadrupled. That is, if the frequency of resonance of a 10 cm concrete slab floor resting on a given mineral wool base spring were to be halved, then 40 cm of the same concrete would be necessary. Adjusting the springs is therefore often a more practical (and height saving) option than adjusting the masses. Of course, one could always increase the density of the concrete if the height increase from 10 cm to 40 cm was unacceptable, but even a density change from 2 tonnes per cubic meter to 3 tonnes per cubic meter would still need a thickness of about 27 cm.

Note, here, that the *weight* (mass × gravity) is what is important. This gives rise to the force acting down upon the spring. In the case of the mineral wool, the density affects the elasticity of the spring, whereas the concrete density affects the weight (and hence force) *applied* to the spring for any given thickness of slab.

However, as we shall see below, we cannot reduce the mineral wool density too far or it may become so compressed that it looses its elasticity and ceases to behave like the required spring. Furthermore, if the mineral wool is compressed to a thickness commensurate with any irregularities in the surface of the floor on which it is laid (or the concrete that may be poured on top of it), then at some places rigid contacts may be made as some points penetrate the material. This problem can be ameliorated by placing a layer of plywood, for example, on top of the mineral wool, and smoothing the floor below it, but such treatments tend to be wasteful of time and materials compared to simply using a thicker layer of mineral wool and/or a higher density.

The general situation with masses and springs

Figure 20.18 shows a graph of the relationship between the mass and the resonance frequency when the mass is loaded on a moulded panel material made from glass fiber, sold under the trade name of Acustilastic. The aforementioned 12 Hz resonance can be seen to be achieved with a load of about 230 kg/m^2, which would signify a 10 cm slab of 2.30 density concrete (2.3 tonnes per cubic meter). Another type of graph which is commonly encountered is shown in Figure 20.19. In this case, it is for a rubber matting known as Acustisol used typically under brick or concrete block walls, where the loads can be very great. The graph shows the degree of compression under load-up to 300 *tonnes* per square meter. Although this does not directly show the resulting resonant frequency, it can be calculated from applied force (the weight) and the degree to which the material is compressed under static load as a proportion of its original thickness. In both cases (Figures 20.18 and 20.19), it can be seen that the relationships are described by curves, and not by straight lines, because as the materials are compressed

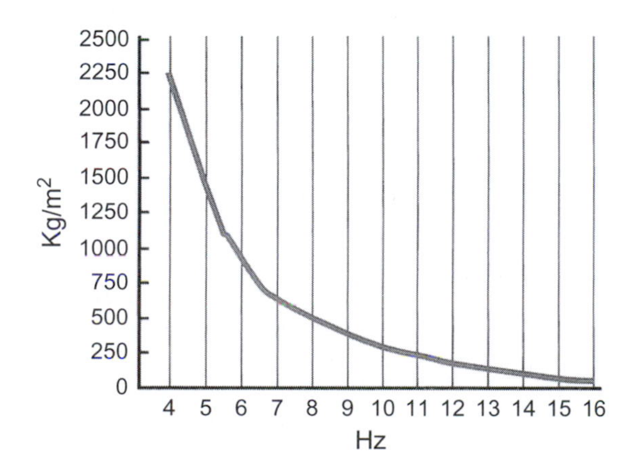

FIGURE 20.18
Load versus resonance frequency graph for the material Acustilastic. It can be seen that the weight of the load must be quadrupled in order to halve the resonance frequency; hence to *quarter* the resonance frequency the load must be multiplied by *16*.

their densities increase. Consequently, a 70 kg/m^3 mineral wool base compressed to 50% of its original thickness would then have a density of 140 kg/m^3, and so would resist more a given extra force (weight placed upon it) than would the uncompressed mineral wool. In other words, if a 50 kg weight over a given area compressed the original 70 kg/m^3 mineral wool by 4 mm, the addition of an extra 50 kg/m^2 would only compress the material by around a further 2 mm.

In the case of Figure 20.18, the graph stops at 2250 kg/m^2. This material (Acustilastic) has an allowed maximum deformation of 30% (which corresponds to a resonant frequency of 4 Hz with the 2250 kg/m^2 load) because above around 2500 kg/m^2 the material goes beyond its elastic limits and begins to permanently deform due to the breaking of individual fibers. The recommended usable load for this material is 2000 kg/m^2. However, if significantly heavy loads are expected to be placed on such a floor, the usable load would need to include the extra load, and cannot be consumed by the floated slab alone. The resonant frequency would be a function of the *total* load. In the case shown in Figure 20.8, where the walls and ceilings are also loaded on the slab, their extra weight will also contribute to the overall load and so they must also be taken into account when calculating the final resonant frequency of the floor when the whole room is finished and loaded with equipment.

On the subject of the thickness of a "spring" material, consider the examples shown in Figures 20.20, 21 and 22. In Figure 20.20, a single spring is loaded by a given weight and can be seen to compress by 20% of its original length. In Figure 20.21, two similar springs are placed in parallel and loaded by the same weight, where the deflexion can be seen to be less for each spring. In Figure 20.22, however, the two springs are placed one on top of the other—in series—and as the force acting upon them is equal (i.e., they are each carrying the same load—it is *not* divided between them) they each compress to the same degree as the single spring shown in Figure 20.20. The proportional compression of each spring

(a)

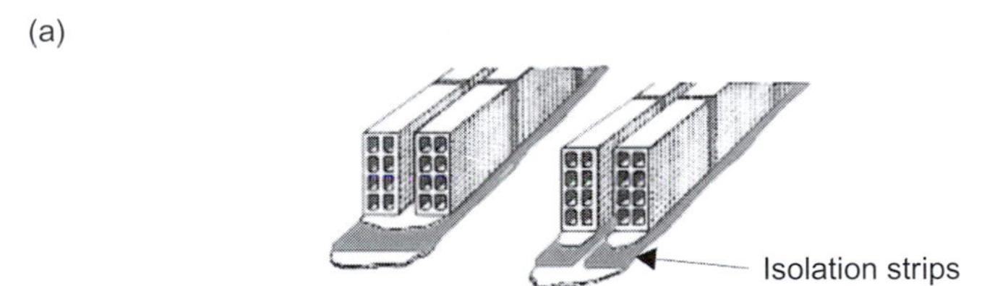

(b)

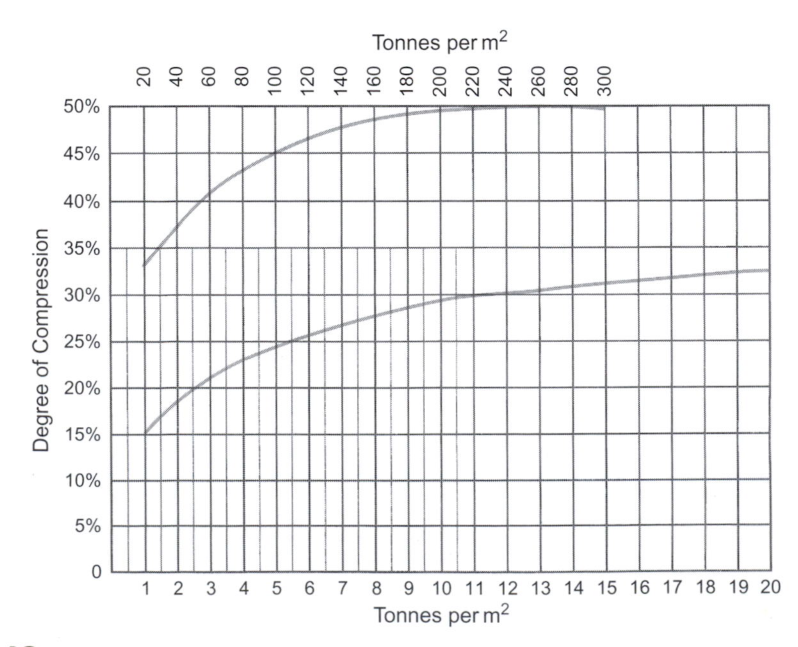

FIGURE 20.19

(a) Typical application of the material for floating heavy walls. (b) Plots showing the degree of compression of the material with different loads.

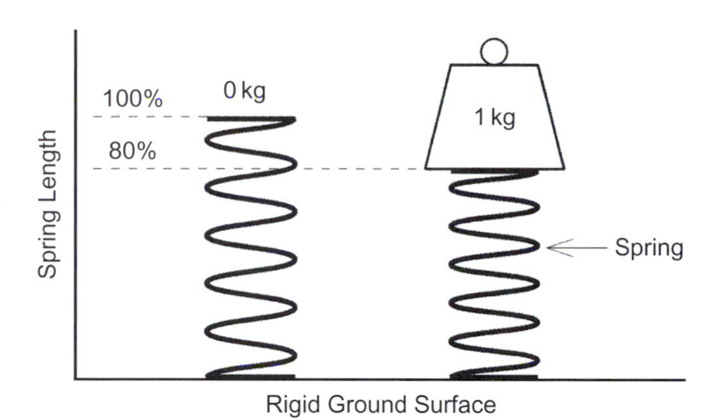

FIGURE 20.20
In the example shown, a weight of 1 kg has compressed the spring to 80% of its unloaded length.

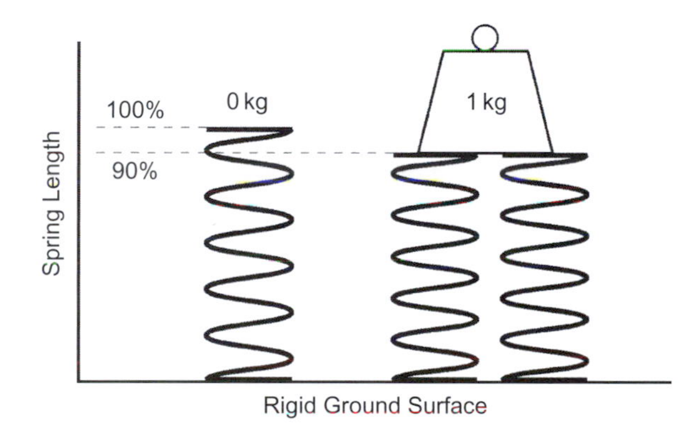

FIGURE 20.21
In the case of the two parallel springs, each identical to the single spring shown in Figure 20.20, the compression for each spring is half of the compression of the single spring: the force on each spring is half of the force on the spring shown in the previous figure. The resonance frequency will rise by $\sqrt{2}$ (1.414) compared to the resonance frequency shown in Figure 20.20 because the stiffness has been increased.

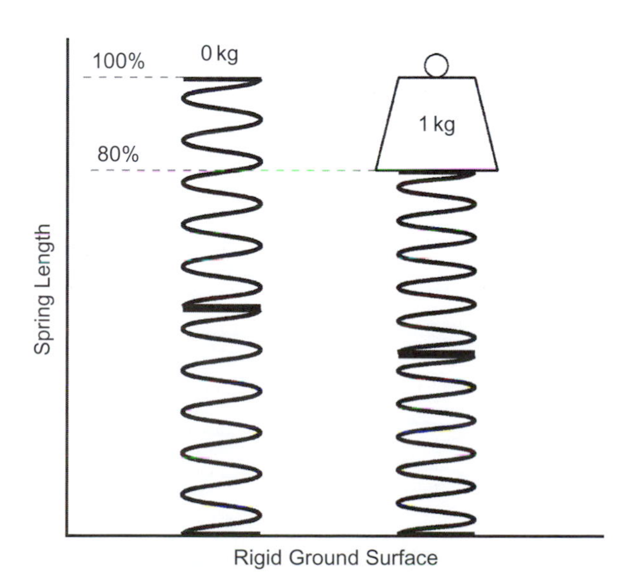

FIGURE 20.22
In the above example, two springs, each identical to the spring shown in Figure 20.20, are placed in series. The force acting on each spring is the same, and is also the same as the force acting on the spring in Figure 20.20, so the compression of each spring will also be the same. However, the overall effect is that of a softer spring, as the total compression *distance* is double that of a single spring, and the resonant frequency will be that of Figure 20.20 divided by $\sqrt{2}$ (1.414).

would therefore be the same as for the single spring. Nevertheless, twice the deflexion would be experienced, so effectively the spring would be softer. The stiffness would be halved, so the resonant frequency of the mass/spring system would be the resonance of the same weight on a single spring *divided* by $\sqrt{2}$ i.e., 1.414. (The same work cannot compress two springs as many times per second as it can compress only one spring.)

Consequently, *doubling* the thickness of a mineral wool base below a concrete slab would reduce the resonant frequency by a factor of about 1.4. *Halving* the thickness would increase the resonant frequency by about 1.4. Doubling the density or halving the density of the mineral wool would also increase or reduce the resonant frequency by a factor of about 1.4, respectively. However, as previously mentioned, if the initial thickness of the material is too little, and/or the density is too low, the material may be compressed to the extent shown in Figure 20.6(c), in which case the spring effect may be lost, and with it much of the isolation if it were the base for a floated floor.

Figure 20.23 shows a somewhat more complicated case. Here, two springs of different stiffnesses are placed in series, one on top of the other. Despite being in series physically, their stiffnesses behave like two electrical resistors in parallel—the resultant stiffness is always less than the least stiff spring, just as the combined resistance of two parallel resistors is always less than the lower of the two resistances. For resistors in parallel, the combined resistance is calculated by dividing the product by the sum; in other words:

$$\frac{R_1 \times R_2}{R_1 \times R_2} \tag{20.1}$$

For two parallel resistors of 10 ohms and 20 ohms:

$$\frac{10 \times 20}{10 + 20} = \frac{200}{30} = 6.666\,\text{ohms}$$

(In fact, the springs in series behave like two *capacitors* in series, but the concept of resistors is perhaps more widely understood by people not so conversant with electrical theory.)

The resonance frequency of the system shown in Figure 20.23 with respect to that shown in Figure 20.20 can be calculated from the change in stiffness, but will always be *below* the resonance frequency

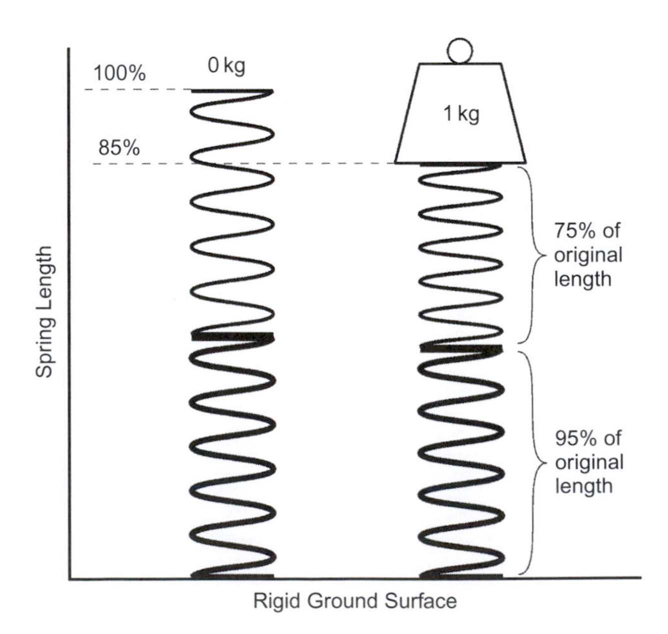

FIGURE 20.23
In this example, two dissimilar springs have been placed in ***series.*** The result is still a single resonance frequency, because the springs combine to yield a total stiffness which can be calculated from the product divided by the sum, as for ***parallel*** resistors or inductors, or capacitors in series.

of either of the single springs loaded by the same weight. However, if we separate the springs by another mass, as may occur if a floated floor is laid on top of an already resonant floor—which has its own mass and spring components—the situation begins to become rather complicated. The concept is shown diagrammatically in Figure 20.24. If the two resonant systems share the same resonant frequency, individually, then the interactive coupling will move the resonances apart. Prediction of *exactly* what will happen in practical structures is very complicated, as many of the relevant parameters will not be known.

On the other hand, if the resonances of the two systems are well separated, then one resonance may dominate, or the pair may behave chaotically, especially under transient excitation. Parallels can be drawn with the behavior of double pendulums and their "strange attractors" described in chaos theory. In the automobile industry, such complicated multiple spring/mass systems *are* sometimes used, where the huge research and development costs of modeling and measuring can be amortized over a long production run of identical vehicles. Conversely, most acoustic calculations for architectural sound isolation purposes relate to one specific job, so it is usually better to stick to the use of more easily predictable systems, and to avoid double mass/spring systems because the research costs usually cannot be supported.

In multiple mass/spring systems, factors such as the relative masses or stiffnesses will also have to be taken into consideration, as will any damping or loading effects which one system may impose on the other. What is more, if the masses are reasonably similar to each other they may interact in a way which is very different to the case if they were greatly different. The importance of the various rules which govern the behavior of such systems can therefore change with the relative proportions of each system.

As a practical example of this type of double system, one can imagine a case where a floated isolation floor needs to be laid over a structural floor at ground level, but with a subterranean car park below. It is not uncommon to find such floors with natural resonances in the region of 20 or 30 Hz. In general, if a load-bearing, structural floor exhibits a relatively high resonance frequency, then a floated floor of soft springs and high mass (low stiffness, high inertia) may be needed in order to achieve much isolation below the resonant frequency of the structural floor. If that floor cannot support such a load, then achieving useful low frequency isolation may be out of the question. These are the sorts of situations where the experience of acoustics engineers is invaluable on a case by case basis. There tend to be too many variables in this type of equation to expect a simple computer analysis to predict the results.

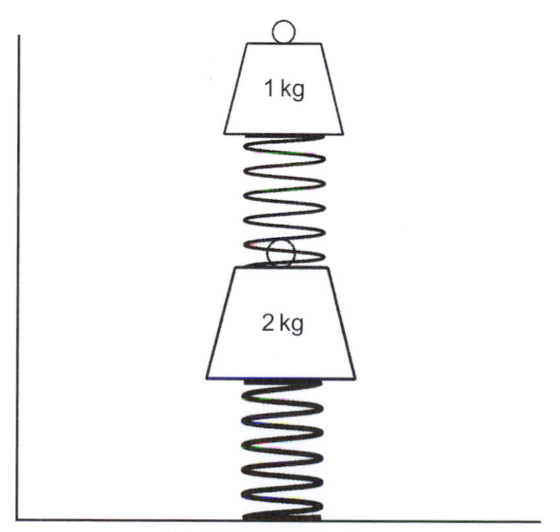

FIGURE 20.24
This can be a difficult problem to calculate, because, in practice, combined structures are rarely pure springs or pure masses. The springs in this example are separated by a mass, so *two* resonant systems exist. There are two degrees of freedom, and interactions of accelerations. Had the two springs and masses in the above case been equal, they would still not resonate at the same frequency because the resonance would be displaced by the interactions.

In practice therefore, combinations of slab thickness and density both affect the *weight* (mass), and changes of the thickness and density of the mineral wool will affect the *stiffness* of the spring. All of these changes will have effects on the resonance frequencies of the systems. When considerations such as floor height or strength must be taken into account the parameters will be chosen for the most practical result. For example, a mineral wool under-layer which is doubled in thickness and halved in density will lower the resonance frequency by half; achieving a reduction of $\sqrt{2}$ from the doubling of the thickness and a further $\sqrt{2}$ from the doubling of the density. The resonance could be returned to its original frequency by quartering the weight of the concrete slab. Such a change may be necessary if a lighter weight or less high floor was required, but the floor would also, obviously, be much weaker with only one quarter of the thickness of concrete. On the other hand, rigidity could perhaps be increased by choosing a thicker layer of a lower density concrete and using more steel reinforcement mesh. Such are the decisions of an acoustics engineer.

Measured characteristics of various suspension materials

In order to show the effects of the practical application of a selection of commonly used materials for floating floors, a series of measurements was undertaken specifically for this book. Materials were cut into samples of 10 cm squares (i.e., 100 cm^2). On each sample, a 1 kg weight would equate to approximately 100 kg/m^3, allowing for small differences due to the different ratios of surface area to edge. The material samples were placed in a device which allowed different weights to be loaded on to them, and to be distributed evenly over their surfaces by means of a light but rigid plate of the same size as the samples. In each case, the height of each material sample was measured unloaded, then measured again at increments of 1 kg between 1 kg and 10 kg, the latter corresponding to approximately 1 tonne per square meter.

Figure 20.25 shows the static loading effect on two samples of a reconstituted polyurethane, open-cell foam, known as Arkobel, and having a density of 60 kg/m^3. In three of the measurements the sample is the same, but it has been rotated in each case through 90 degrees to measure each axis separately. There is one axis of recommended use, preferred because of the direction in which the foam is cut after compression during the manufacturing process. The preferred axis can be seen to exhibit a curve which is a little less steep overall. However, what is interesting is that when the sample was unloaded, from 10 kg back to 0 kg, the curve did not follow the compression curve. The material was exhibiting hysteresis under static conditions, but this should not introduce any unwanted effects under dynamic conditions because the vibrational movements would be much too small to exhibit non-linearities.

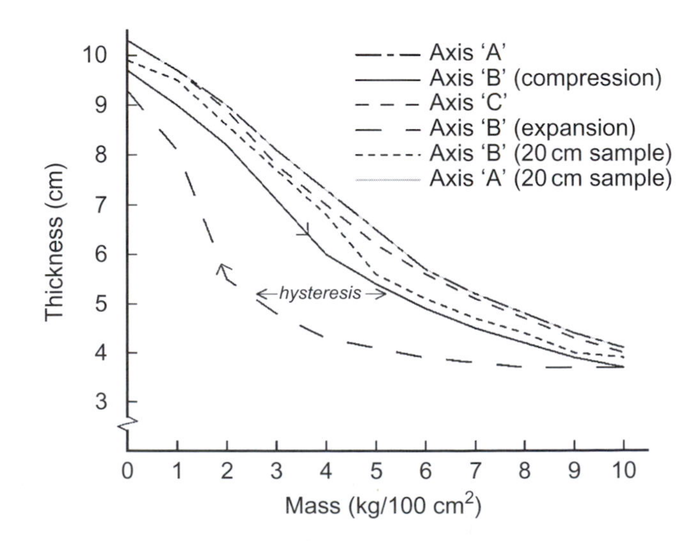

FIGURE 20.25

Compression versus load for 10 cm and 20 cm Arkobel (a reconstituted polyurethane foam) of 60 kg/m^3 density. The thicknesses of the 20 cm samples have been divided by two on the plots in order to make easier comparisons with the 10 cm samples. In fact, the 20 cm sample compressed in 'axis A' coincides so well with the 10 cm sample in the same axis that the two lines are almost indistinguishable. Axis B is the normal axis of use. The hysteresis loop shows the different behavior in compression and expansion.

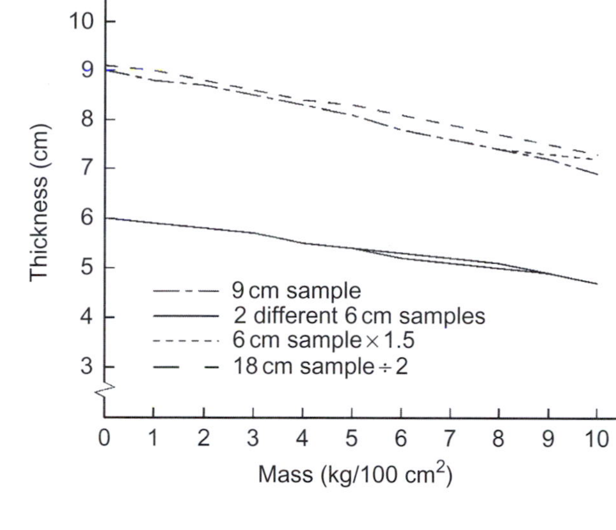

FIGURE 20.26
Compression versus load for different thicknesses of Arkobel (reconstituted polyurethane foam) of 120 kg/m³ density. The 6 cm measurement multiplied by 1.5 is overlaid on the measurement for the 9 cm sample. The two plots correspond very closely up to 8 kg, after which the 6 cm sample begins to compress less, indicating that it is beginning to "bottom-out." The 18 cm sample (÷2) and the 9 cm sample show no such effect.

A 20 cm sample was then tested in the "A" and "B" axes, with the figure showing the results divided by two, which overlays the 10 cm and 20 cm curves for easy comparison. In the "A" axis, except for the section between 0 kg and 1 kg, the plots overlap exactly, showing the practical results of the effect shown in Figure 20.22 relating to springs in series. The "B" axis of the material, in both the 10 cm and 20 cm samples, exhibits a slightly less regular curve, but the agreement at low and high loadings is relatively consistent.

Figure 20.26 shows the plots for a type of Arkobel with double the density of the first sample—120 kg/m³. Samples of 6, 9 and 18 cm were tested. Normally, this higher density is used in smaller thicknesses, such as 3 cm, but for these tests the deflexions would be too small for reliable measurement, so greater thicknesses were chosen. The 18 cm sample measurements have been divided by two for easier comparison of the plots. It can be seen that the proportional compression is much less than for the 60 kg/m³ sample. Nine point seven centimeters of 60 kg Arkobel compresses to about 3.7 cm under a 10 kg load, whereas 9.1 cm of 120 kg Arkobel compresses to only 6.9 cm under a similar load—a deflexion of 2.2 cm as opposed to 6 cm.

An interesting observation from Figure 20.26 is the comparison of the multiplication of the 6 cm sample measurement by 1.5, overlaid on the 9 cm sample plot. At loadings above 8 kg, the 6 cm sample curve begins to flatten out, showing that the thinner sample is beginning to bottom; that is, compress beyond its elastic limits. This effect is also observable with the thinner samples in Figures 20.27 and 20.28. The bottoming occurs when the material begins to be compressed into a solid mass, and the mechanisms which give rise to the springiness begin to cease to operate. (See Figure 20.6(c)).

The Blancot (a cotton-waste felt material) shown in Figure 20.27 exhibits a different shape of curve to the foam materials in the two previous figures. In this case, the 9 layer measurement divided by 3 and overlaid on the 3 layer measurement shows that the 3 layers begin to bottom at loads above 4 kg (400 g/m³), and between 8 and 10 kg there is a horizontal straight line, showing that the material has ceased to act as a spring. The density of this Blancot is the same as that of the first Arkobel sample, 60 kg/m³, but a comparison of Figures 20.25 and 20.27 shows that the way in which the two materials compress is very different, as shown by the different curve shapes, particularly in the lower half of the weight scale.

The mineral wool plots shown in Figure 20.28 are different from both the Blancot and the Arkobel. Of the lower density samples, the plots show a high initial rate of compression which rapidly begins to flatten out, and indeed bottoms out in the case of the thinner sample. The 10 cm sample of the higher density (60–70 kg) is very typical of what is often used in studio isolation floors. Finally, Figure 20.29

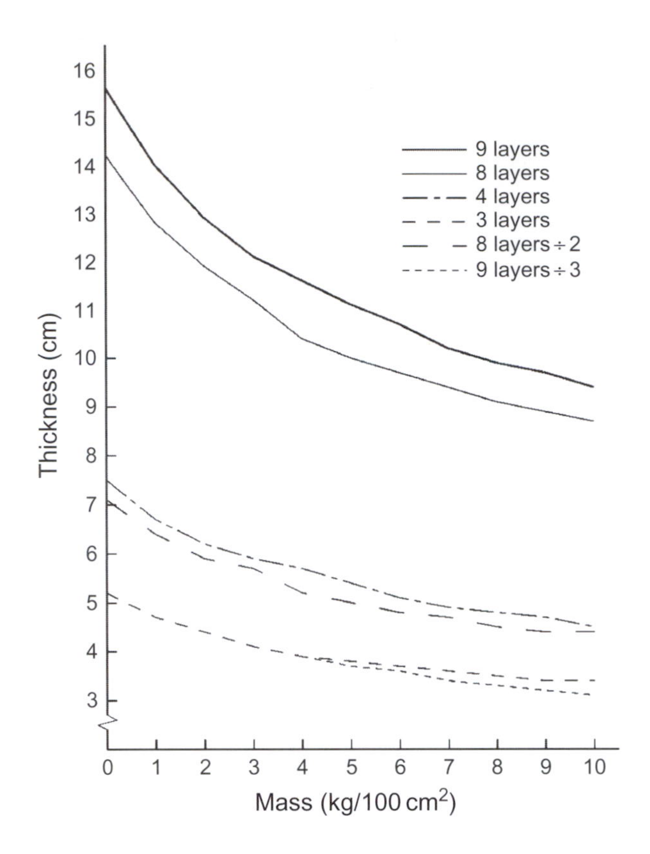

FIGURE 20.27
Compression versus load behavior of various thicknesses of Blancot, a cotton-waste felt material of around 60 kg/m^3 density. The 3 layer plot and the 9 layer plot ÷3 are identical up to 4 kg loading, after which the 3 layer plot shows evidence of bottoming out as the thinner sample begins to compress proportionately much less than the thicker, 9 layer sample.

shows an interesting comparison of the different slopes and shapes of the different materials as measured for the previous figures. Clearly, there are some very different mechanisms at work which are giving rise to their elasticity.

Calculation of resonance frequency

From the measured deflexions of the materials, the resonance frequency for each mass/spring combination can be calculated. If the spring is perfect, i.e., 2 × load = 2 × deflexion, the resonance frequency is given by:

$$f_{\text{res}} = \frac{1}{2\pi}\sqrt{\frac{g}{x}} \approx \frac{1}{2\sqrt{x}} \quad (20.2)$$

where x is the deflexion in meters and g is the acceleration due to gravity (9.81 ms^{-2}). In general, as shown above, practical materials do not behave as perfect springs and the stiffness of the spring changes with changing load. Under these conditions, the resonance frequency for a particular load may be estimated from the *slope* of the load/deflexion curve, thus:

$$f_{\text{res}} = \frac{1}{2\pi}\sqrt{\frac{g}{m\left(\frac{dx}{dm}\right)}} \approx \frac{1}{2\sqrt{m\left(\frac{dx}{dm}\right)}} \quad (20.3)$$

where dx/dm is the slope (gradient) of the load/deflexion curve at the applied mass m (note that for a perfect spring, $dx/dm = x/m$ and the two expressions become identical).

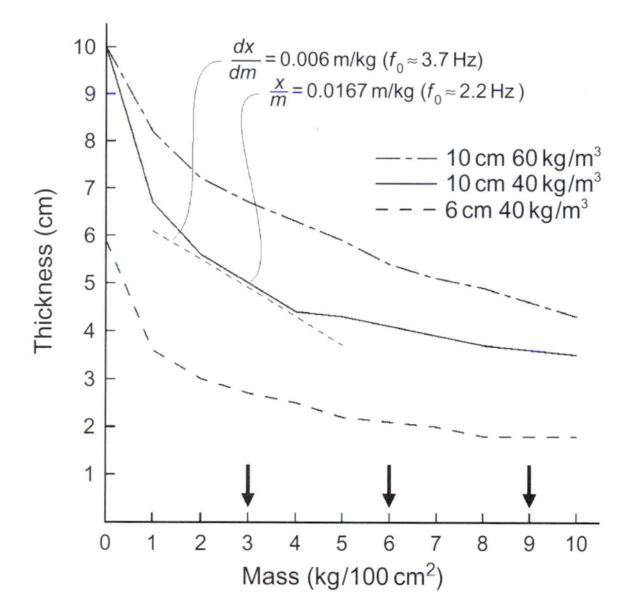

Material	Load (kg/100 cm^2)	Resonance (Hz)
10 cm 60 kg/m^3	3	4.4
	6	3.4
	9	3.0
10 cm 40 kg/m^3	3	3.7
	6	4.5
	9	5.3
6 cm 40 kg/m^3	3	5.4
	6	6.0
	9	15.0

The fact that the resonance frequency *rises* as weight is added to the 40 kg/m^3 mineral wool samples means that even at 3 kg/100 cm^2 (or 300 kg/m^2) the material is already stiffening to the point where it is losing its spring effect.

FIGURE 20.28
Compression versus load for different thicknesses of 40 kg/m^3 and 60 kg/m^3 mineral wool. Again, bottoming-out is evident above 8 kg/m^2 loading for the thinner sample of the lower density material. The resonance calculations relate to the descriptions given on page 598. Of the two samples which had a thickness of 10 cm when unloaded, the lower density material, as expected, compresses much more as weight is added, up to about 4 kg/100 cm^2, but thereafter compresses less than the higher density sample. This suggests the beginning of a bottoming-out process for the lower density sample under higher loads. In the resonance/load table, note how the resonance frequency begins to rise abruptly once the bottoming-out begins, showing that the spring is becoming very stiff.

By way of example, marked on Figure 20.28 are two estimates of the resonance frequency of a material (which is clearly not a perfect spring) loaded with a mass of 2 kg/100 cm^2, one based on the static deflexion and the other on the slope of the curve. The static deflexion estimate is shown to be less than half of the value of the more reliable estimate based on the slope of the curve.

SUMMARY

- When anything is immersed in a sound field, it will, to some degree or other, be set into vibration. Any vibrating structure can, depending upon the degree of isolation that it can achieve, receive the vibrations on one of its surfaces and re-radiate them from another.
- When a sound wave strikes any surface, the three possibilities are that the energy in the wave can be reflected, transmitted, or absorbed.
- Absorption can be achieved in several ways, such as by tortuosity, particulate friction, and internal viscous losses.
- Porous absorbers work best in situations of maximum particle velocity, such as at pressure nodes and at high frequencies.

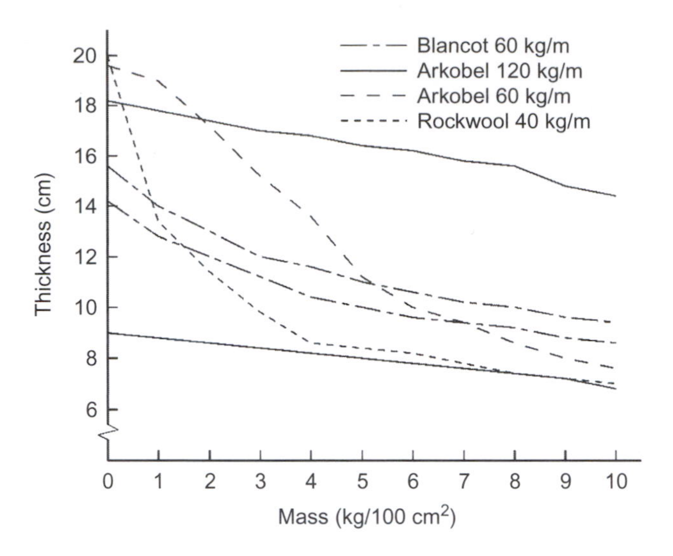

FIGURE 20.29
Comparison of the shapes of the compression versus load curves for the different fibrous and cellular materials. The cellular materials can be seen to behave more uniformly, which is evident from the straighter nature of the curves. Clearly, the mechanisms giving rise to the spring-like behavior are different in each material.

- The differences in the speed of sound through the different materials in the structure of a building can give rise to great difficulties in predicting the structural vibration and re-radiation.
- Isolation is best achieved by reflexion, because isolation by absorption would generally be too bulky to be practical.
- The three basic means of sound isolation are mass, rigidity and distance; but acoustic damping can play a prominent part.
- The use of springs to isolate internal and external structures is usually the best way to achieve high isolation; by *decoupling*.
- A good hint to remember is that springs do not like to change their length, and masses do not like to change their velocity. When something tries to vibrate a mass/spring system, both the mass *and* the spring try to resist the vibration, for their own individual reasons. Working together, they can make very efficient isolation systems.
- The degree of isolation needed at any given frequency will be dependent upon circumstances and the type of recordings to be made. Low frequencies are generally much more difficult to isolate than high frequencies, though the general reduced sensitivity of the human ear to low frequencies at low levels helps to mitigate the problems incurred.
- For floated isolation systems to be effective, their frequency of resonance must be *below* the lowest frequency to be isolated.
- In weak buildings, adhesive mounted isolation systems can be attached to all parts of the internal surfaces.
- Complex isolation structures are difficult to analyse, and a degree of empirical experience is often needed to achieve reliable predictions of the outcome of a design.
- Impact noises can be much more difficult to deal with than airborne noises.
- If possible, careful choice of location can also be instrumental in reducing isolation requirements.
- Mass/spring systems are often used for making isolation floors by means of concrete slabs over mineral wool bases.
- Doubling or halving the mass, whilst maintaining the same spring, will respectively lower or raise the resonant frequency of the system by $\sqrt{2}$. Quadrupling or quartering the mass will respectively halve or double the resonance frequency.

- Doubling or halving the stiffness of the spring, such as by means of doubling or halving the density of mineral wool (the initial thickness remaining the same), will respectively raise or lower the resonance frequency of the system by $\sqrt{2}$. Quadrupling or quartering the stiffness will double or halve the resonance frequency.
- Doubling or halving the thickness of a mineral wool spring will lower or raise the resonance frequency by $\sqrt{2}$. Quadrupling or quartering the thickness will respectively halve or double the resonance frequency.
- Mass/spring floated floors begin to become usefully effective in their isolation about one octave above their resonant frequency.
- Resonant frequencies can be calculated from the applied force (mass $\times$ gravity, or weight) and the proportional compression of the spring.
- Two identical springs in series (such as doubling the thickness of a mineral wool base) will *reduce* the resonance frequency by $\sqrt{2}$.
- Two identical springs in parallel (such as achieved by doubling the *density* of a mineral wool base) will *increase* the resonance frequency by $\sqrt{2}$.
- Springs of different stiffness, placed in series, will behave like electrical resistors in parallel, and the combined stiffness will always be below the stiffness of the weaker spring.
- Systems with two degrees of freedom, such as mass/spring/mass/spring, can behave in complex ways. If the two masses and springs are identical, the resonances will *not* be: they will move apart. If the masses and springs are not identical, then their behavior can depend on their relative proportions. Prediction of the overall behavior can be very difficult to achieve: one resonance may dominate, or the two resonant systems may even behave somewhat chaotically under shock excitation.

REFERENCE

Fahy, F. 2001. Foundations of Engineering Acoustics, Academic Press, London and San Diego, p. 331.

BIBLIOGRAPHY

Gréhant, B., 1994. Acoustics in Buildings, London, UK: Thomas Telford Publishing, 1996, Originally published as Acoustique en Bâtiment, Lavoisier Tec & Doc, Paris, France.

Parkin, P.H., Humphreys, H.R., Cowell, J.R., 1979. Acoustics, Noise and Buildings, 4th edn, London, UK: Faber and Faber.

PREFACE

The studio environment is vitally important if you want to get the best artistic results, the best reputation, and hopefully lots of profitable repeat bookings. The second chapter by Philip Newell examines the environmental issues of lighting, ventilation, decoration, and the provision of headphone foldback and clean mains supplies.

Philip begins by considering the issue of daylight in recording studios. A few years ago, the idea of a recording studio having windows to the outside world, with all the extra sound isolation problems that that implied, would have been considered quite bizarre. However, the conclusion is inescapable that customers like windows, and the extra constructional expense involved is well justified.

He moves on to consider the general issue of studio lighting. Fluorescent tubes are not used because they give a harsh and unpleasant light, and worse still, are prone to emitting electrical noise, and even acoustic noise from their ballast chokes. Incandescent bulbs are preferred (though how the forthcoming UK ban on light bulbs will affect this situation remains to be seen) and are typically dimmed by noiseless rotary transformers rather than electrically noisy triac dimmers. It is better to have too much light rather than too little as you can always turn it down.

Good ventilation is also essential to customer comfort, and this obviously needs very careful design as the air-handling system must not generate appreciable noise in action. One factor in this is duct size; Philip gives an example where increasing the duct diameter from 20 cm to 30 cm halves the air flow rate; since the noise generated by air turbulence falls at something like a sixth power law, the noise reduction is considerable.

Philip next examines the need for an effective headphone foldback system, so that musicians can hear what has been done and what they are doing. He makes the telling point that for a musician playing in headphones, the foldback system constitutes the whole of his acoustical environment, and it is worth expending a good deal of effort to set up an effective system that musicians are comfortable with.

The next topic is that of studio and control-room decoration; this may seem frivolous compared with the complexities of acoustics, but it is an integral part of the overall impression that the customer takes away with them. Philip issues an urgent warning against allowing interior decorators to compromise the studio acoustics in order to obtain whatever look is fashionable that month.

Another necessity for a well-functioning studio is the provision of a good clean mains supply. A primary requirement is that all the audio equipment should be run from the same phase, for otherwise the hazardous voltages possible in fault conditions are increased, and there is likely to be much more trouble with ground loops. This runs counter to the normal electrical practice of trying to put roughly equal loads on each phase, and is just one example of the way that the special needs of the recording studio do not line up with standard electrical installation procedures. For this reason there can be difficulties in dealing with contractors unless they specialize in studio work.

CHAPTER 21

The Studio Environment

Recording Studio Design by Philip Newell

Lighting. Ventilation and air-conditioning. Foldback systems. Studio decoration. AC mains supplies. Earthing.

SOME HUMAN NEEDS

Much has been written about the studio rooms being, first and foremost, rooms for musicians to play in. Obviously, all the individual tastes of every musician cannot be accommodated by any one studio design, so the "ultimate" environment cannot be produced. Nonetheless there are a few general points which are worthy of consideration. Light colors, for example, make spaces feel larger than they would do if finished in dark colors. In general, they tend to create less oppressive atmospheres in which to spend long periods of time. Daylight is also now being recognized as growing in popularity.

Daylight

Traditionally, it has been the out of town studios which have been more inclined towards allowing daylight into the recording areas. Initially it was probably easier to do this in the relative tranquillity of rural life, as the sound isolation problems of inner city locations were not so great. The knowledge of the changes from day to night, and more unpleasantly after very late sessions from night to day, along with an awareness of the changing seasons, are all instrumental in helping a person's well-being. This fact is perhaps more medically recognized now than it was only a few years ago. Once it had been established that the sound isolation in studios with outside windows was adequate, daylight soon became widely accepted as a desirable asset for studios.

Artificial light

The use of imaginative lighting to create moods can be highly beneficial to the general ambience of a studio. Fluorescent lighting, apart from the "hardness" of its light, is generally taboo in recording studios because of the problems of mechanically and electrically radiated noises. The mechanical noise problem can sometimes be overcome by the remote mounting of the ballast chokes (inductors), which prevent the tubes from drawing excess current once they have "struck" (lit), but the problem of the radiated electrical noise can be a curse upon electric guitar players. In some large studios, fluorescent lights can be used without apparent noise problems in very high ceilings (6 m), well above the instruments, but in general they are likely to be more trouble than they are worth.

Lighting is a personal thing, of that there is no doubt, and the design of studio lighting systems is quite an art form in itself. Many people do not like the small, halogen lights, as they find their light too stark.

Audio Engineering Explained

Argon filled, tungsten filament bulbs are still a very practical solution, though interior designers seem to eschew them.

Lighting is controllable not only by the switches of individual groups of lights (or even single lights in some cases) but also by "Variacs." The Variacs are continuously variable transformers, which despite their bulk and expense (roughly €80 each for the 500 VA (watts) variety, and 15 cm in diameter by 15 cm deep) are the ideal choice of controller in many instances. They cannot produce any of the problems of electrical interference noise which almost all electronic systems are likely to create from time to time. Variacs simply reduce the voltage to the bulbs, without the power wastage and heat generation of rheostats, which dim by resistive loss.

Electronic dimming systems using semiconductor switching can be a great noise inducing nuisance, both via the mains power (AC) and by direct radiation. It is simply not possible to switch an AC voltage without causing voltage spikes. So-called "zero voltage switching" may still not be zero current switching, as the loads which they control will be unlikely to be purely resistive. A theoretically perfect sine wave can only exist from the dawn of time to eternity, and cannot be switched on or off without transient spikes. Even if you find this hard to believe, believe it! It *is* possible by more complex electronic means to produce a "clean", variable AC, but the methods of doing so tend to be even larger and even more costly than Variacs, whose pure simplicity is a blessing in itself.

From the early 1970s to the mid-80s many studios installed colored lighting as well as the white light for music reading and general maintenance. Suddenly, they became *passé* for a decade or so; but recently, they have again begun to occasionally emerge. Such are the cycles of fashion. Anyhow, whatever lighting system is used, the general rule should be "better too much than too little." One can always reduce the level of lighting by circuit switching or by voltage reduction; but if there is insufficient total lighting for music reading or maintenance, work can become very tiresome.

Ease and comfort

General comfort is also an important issue, as again, comfortable musicians are inclined to play better than uncomfortable ones, and few things seem to kill the sense of comfort and relaxation more than an untidy mess of cables and other "technical" equipment. Musicians should never be made to feel in any way subservient to the technology of the recording process. The studio is there to record what *they* do; they are not there to make sounds for the studio to record. Sufficient sensitivity about this subject is all too frequently lacking in studio staff, especially in many of those having more of a technical background than a musical one.

Easy access to the studio, for the musicians, is another point worth considering well when choosing the location for a studio. Studios have been sited on the fourth floor of buildings with no lifts, but their existences have usually been rather short lived. Especially for musicians with busy schedules, easy access and loading, together with easy local parking, can be the difference between a session being booked in a certain studio or not. It is also useful to have some facility for the storage of flight cases and instrument cases, *outside* of the recording room. Apart from the problems of unwanted rattles and vibrations which the cases may exhibit in response to the music, when a studio looks like a cross between a warehouse and a junk room it is hardly conducive to creating an appropriately "artistic" environment in which to play.

The points made in this section are not trivial, though they are not given their due attention in all too many designs. Experienced studio owners, operators, engineers, users and designers, appreciate these points well, but in the current state of the industry, a large proportion of studios are built for first-time owners. As so many of them fail to realize the true importance of these things, they are frequently the first things to be trimmed from the budget, especially when some new expensive electronic processor appears on the market midway through the studio construction. Most long-established studios do provide most of these things though, which is probably one of the reasons why they have stayed in business long enough to *become* long-established.

VENTILATION AND AIR-CONDITIONING

It is certainly not only in studios that people complain about the unnatural, and at times uncomfortable, sensation of air-conditioned rooms. Unfortunately though, with sound isolation usually being good thermal isolation, and given that people, lights and electrical equipment produce considerable amounts of heat, air-conditioning of some sort is more or less a mandatory requirement in all studios. In smaller rooms, such as vocal rooms, where few people are likely to be playing at one time, sometimes a simple ventilation system is all that is required. In fact, it can be advantageous in many instances to provide a separate, well-filtered, ventilation-only system in addition to any air-conditioning. It is remarkable how many times these systems have refreshed the musicians without having them complain of dry throats.

Ventilation

In order to get the best out of a ventilation system there are a few points worth bearing in mind which are of great importance. One of the greatest rules is never to only extract air from the room. If an extraction-only system is used, the room will be in a state of under-pressure; a partial vacuum. Whenever a door is opened (or for that matter, via any other available route, such as via air-conditioning *outlets*), air will be drawn into the room. This air will be dirty, because it cannot be filtered. Dust, dirt, and general pollution will enter the room along with the air, perhaps affecting the throats of singers and leaving dust everywhere. The air-conditioning *inlets* will almost certainly be filtered, but in almost all cases the outlets will not be filtered, so air which an extraction system pulls in through an air-conditioning *outlet* may be dirty. It may even pick up more dirt from the outlet duct itself, which may be full of dust and tobacco deposits.

In rooms with only a low volume of air flow, such as vocal rooms, it is often possible to simply use an input-only fan, with the air being allowed to find its own way out via an outlet duct of suitable dimensions. When high volumes of air-flow are necessary, extraction fans *are* usually fitted, but there are some basic rules which should be followed. For the reasons mentioned in the previous paragraph, the extraction fans should never be operating without the inlet fans, as they would constitute an extract-only system and would suffer all the drawbacks. What is more, the flow rate of the extraction system should never exceed that of the inlet system, as this would still produce an under-pressure in the room. It is prudent to restrict the flow rate of the extraction system to between 60 and 80% of the inlet flow, dependent largely upon the degree to which air can find its way out of the rooms by routes other than the intended outlet system. In purpose-designed studios, the extraction flow rate can usually be almost as high as the inlet flow rate, because the rooms are normally more or less hermetically sealed at all points other than via the air-change system(s). However, multi-functional rooms, which may serve as studios from time to time, are often somewhat more leaky through their doors, windows, and roofs, in particular. In these cases a proportionally lower extract flow rate would be desirable, as over-pressures will be more difficult to achieve.

The normal way to ventilate is to draw air in from outside through a filtration system. This would usually consist of either a single filter or a series of filters, with removable filter elements which can easily be cleaned and/or replaced. Air will then pass to an in-line fan and on to a silencer, or series of silencers, before entering the room. This way, the room is kept in over-pressure. Therefore, any time a door is opened, the clean, filtered air in the room will escape through the door, keeping the dirty, outside air from entering through the doorway. Such systems keep the rooms cleaner, and ensure that the air which enters the rooms is always clean. Outlets will normally pass first through silencers (both to prevent street noise from entering and studio noise from leaving) and then on out to the outside air, perhaps via a back-draught damper. The damper is a one-way flap arrangement which only allows air to flow out. If the ventilation system is switched off, and the wind is in an unfavorable direction, then should air attempt to enter via the outlet, the flaps close off the ducts. This also prevents dirt from entering via the unfiltered outlets. Fire dampers are sometimes also installed, which close the air ducts completely should the room temperature rise above a pre-determined limit. By this means, if the temperature rise is due to a fire, the oxygen supply to the fire will be cut off, thus slowing the spread of the fire even if not extinguishing it completely. Another point to remember is to turn off the ventilation system when the studio is unoccupied, so that if a fire should begin it will not be supported by a constant supply of oxygen. A typical ventilation system outside of a studio with three rooms is shown in Figure 21.1.

The ventilation ducts will often be of the "acoustic" type, having a thin, perforated aluminium foil liner, then a wrapping of 5 cm of fibrous absorbent material, and finally an outer metalized foil covering. In this type of ducting the air is allowed to pass through a relatively smooth inner tube, which presents only minimal friction, and hence loss of air flow. This smooth inner lining must be as acoustically transparent as possible to allow the sound to be absorbed by the fibrous layers around it. If the sound cannot easily penetrate the inner lining of the duct, it will travel very effectively along the tube.

FIGURE 21.1
The ventilation system of a studio with three rooms. On the shelves are, from left to right, the filter boxes, the axial inline fans, and the 90 cm long silencers. All the flexible ducting is of the acoustic type, which itself acts as an in-line silencer, except for the extensions on the outlet ducts to the right, which are simple, flexible aluminium tubes. The air inlet is where the cable tubes can be seen, at the bottom left of the picture. Normally, to comply with building regulations, the inlets and outlets should be at least 5 m apart, to prevent re-circulation of exhausted air. Audio Record Studios, Morón de la Frontera, Spain (2001).

FIGURE 21.2
"Acoustic" ventilation tubing passing through the absorbent rear "trap" of a control room, during construction, before the mounting of the principal absorption materials. Sonobox, Madrid, Spain (2001).

Sound can travel along kilometers of smooth bore tubing with remarkably little loss, as no expansion of the wave can take place. (The double distance rule does not apply.) This is the principle by which the speaking tubes of the old ocean liners were able to provide excellent communications, often over long distances and in noisy surroundings; bridge to engine-room, for example. Figure 21.2 shows an installation of acoustic ducts running through the back of an absorber system.

Duct *size* is also very important. For any given quantity of air to pass, the air flow down a duct of large diameter will be much slower than down a narrower duct. A 20 cm diameter duct has an approximate cross-sectional area of 315 cm^2, and a 30 cm diameter duct an area of 709 cm^2, which is more than double. Therefore, for any given rate of air flow, the speed of the air down the 30 cm duct need only be just under half of that down the 20 cm duct. As the noise caused by turbulent air flow follows something like a sixth-power law, then for any given flow-rate, ventilation system noise will rise rapidly with

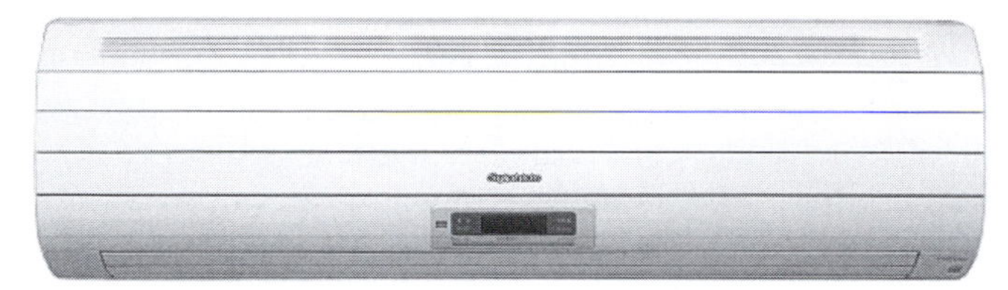

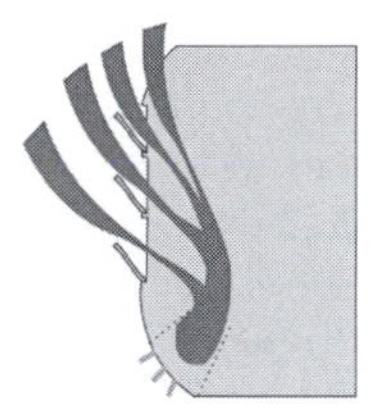

FIGURE 21.3
Towards quieter, split air-conditioning. If split air-conditioning systems *must* be used, then it is wise to choose units with aerodynamically streamlined inlet and outlet vents, to avoid the higher levels of turbulence noise associated with the more common grills.

falling duct diameters. For adequate air flow, even in the smaller rooms, 20 cm seems to be about the minimum usable duct diameter. Appropriate fans with a flow rate of about 700 m^3 per hour on full speed will usually suffice for this diameter of tube. Speed control should again be provided by variable or switchable transformers, and not by electronic means, to avoid interference.

Air-conditioning systems and general mechanical noises

Traditionally, professional recording studios have used conventional, ducted air-conditioning systems, and noise floors of NC20 or less have been achievable, even with quite high rates of air flow. Such systems are still the only way to properly air-condition a studio, but as has always been the case, they are relatively expensive.

Since the early 1980s, the real cost of studio equipment has been falling. Partly as a consequence of this, and somewhat unrealistically, the charges per hour for the studios have plummeted. At a time when one-and-a-half million euros is spent on the recording equipment for a studio, to have to pay €150,000 for an air-conditioning system may not seem too disproportionate. On the other hand, when manufacturing technology has made it possible to buy for €150,000 an entire set of equipment that will not perform far short of the equipment costing €1,500,000, people seem to baulk at similar charges for good air-conditioning. Competition between studios has forced real prices ever downwards, and a state of affairs has now developed where the cost of ducted air-conditioning systems, for most of the mid-priced studios, has become insupportable. It must be said, though, that this is a situation driven by commercial realities, and that the need for good conventional air-conditioning systems is as much a professional requirement today as ever it was.

It is a strange situation, exactly analogous to that which exists with sound isolation costs. People seem to expect sound isolation to be cheaper if they "only" want to use a room to practice drums, as opposed to the more "serious" purpose of recording. Similarly, people seem to expect to have to pay less for the air-conditioning systems if they are paying less for the recording equipment. There is simply no logic in any of this.

Largely for economic reasons therefore, there has been a great increase in the number of "split" type air-conditioning systems coming into studio use. Although these are by no means ideal for this purpose, they are many times cheaper than the ducted systems, but as the heat exchangers and their fans are in the studio, with only the compressors remaining outside, there is an attendant noise problem. In control rooms the units can usually be left running in "quiet" mode, as the noise which they produce is often less than the disc drives and machine fans which may also be in the room (but which really should not be there), but in the studio rooms they usually must be turned off during quiet recordings. Unfortunately this intermittent use can lead to temperature fluctuations which may not be too good for the consistency of the tuning of the instruments. Nevertheless, despite their problems, there are now many split systems in studio use. The use of multiple, small, quiet heat exchangers is normally preferable, where practicable, to the use of larger single units. Some small units now produce less than 30 dBA of background noise at a distance of one meter when used on low speed. Such a unit is shown in Figure 21.3.

Studio customers have become used to inexpensive recording, and all too few of them now want to pay for facilities which can provide all the right conditions for optimal recordings. Unfortunately, as this has become widespread, it has also apparently become *acceptable* in these "market forces" days. Air-conditioning systems have in very many cases been tailored to a "reasonable" proportion of the recording equipment budget, which has led to unsatisfactory air-conditioning, but this reality exists.

There are limits, however, below which the lack of suitable environmental control will pose serious problems, not only for the musicians but also for instruments such as pianos and drums. Draughts of air are generally disliked, most musicians would agree about that. However, there may be some considerable differences in their preferences of optimal temperature for their comfort. If a studio is block-booked for a week or so, then the chosen temperature can be selected at the start of the session and maintained, but pianos should not be tuned until the temperature has had time to stabilize. With shorter bookings, the temperature should be kept at a suitable compromise, as frequent changes in air temperature will cause great problems with the tuning of any permanently situated pianos, and many other instruments for that matter.

Humidity is another factor which needs consideration. If maintained too low it can dry out the throats of singers, and may cause piano sound-boards to crack. If it is too high it can be uncomfortable for the musicians and corrosive to instruments. Sixty or seventy per cent humidity is a good level for most purposes, and in the better studios regular attention is paid to the maintenance of the appropriate levels. Unfortunately, in an enormous number of smaller or less professionally operated studios, little or no attention whatsoever is paid to humidity control. Once again, the relentless driving down of studio prices has rendered it impossible for many "professional" studios to provide the sort of environmental controls that are musts for *truly* professional studio operations.

There is now a huge "consumer" recording market, which, although providing standards below what the "professional" market was accustomed to, has blurred with the professional market to such an extent that the lower sets of standards have begun to influence the professional world. Some of this has no doubt been due to the enormous influence of electronically based music, where studio acoustics and air-conditioning noise have not been problems. But it is difficult to understand how some people can hear any detail at all in control rooms whose noise floors are made ridiculously high by the presence of hard-disk drives and numerous equipment fans. Fortunately though, there seems to be a swing back to the use of acoustic and electrified (as opposed to electronic) instruments, which is just as well if much of the passed down experience that exists in the recording industry is not to be forgotten through the loss of a whole generation of recording staff to the computer world. The provision of acoustically isolated machine rooms, with good temperature control, is now an important feature of many studios, in order to keep the control room background noise level below acceptable limits. In these cases the "split" air-conditioning systems are ideal, and the background noise levels which they may add are of no practical consequence. In fact, given their efficiency they are probably the *preferred* option.

When very low noise levels are required in studios, a very viable option is to employ batteries of water coils in the ventilation ductwork. Figure 21.4 shows such a system above a film dubbing theater, and Figure 21.5 shows the noise inside with all systems running. The rise towards 1 kHz is due to—somewhat perversely—the Dolby processors; when the requirement for low noise in the room was made by Dolby, themselves. The low frequency noise from the HVAC (*h*eating, *v*entilation and *a*ir-*c*onditioning) system can be seen to be around NC10 (approximately 10 dBA). Normal, "split" air-conditioning systems, and other systems using ammonia, or similar halogen-free refrigerants tend to operate in the cooling mode with very low temperatures, often below 0 °C in their heat exchangers. These low temperatures are efficient coolers, but they tend to condense too much of the humidity out of the air. This is what often gives rise to the well-known discomfort which many people feel when working long hours in air-conditioned rooms, and it can also badly affect the voices of singers and actors. When water is used to cool the heat exchangers, and high flow rates can be allowed, temperatures as high as 14 °C can be employed in order to cool rooms down to 20 degrees, even in relatively hot weather. High volumes of relatively cool water dry the air much less than low volumes of refrigerants operating below zero. Where possible, when water systems are used, it is best to try to site the heat exchangers in places where leaks will not cause disasters (i.e., not directly over the studio rooms) and where there is no danger of them freezing up in winter conditions, which will probably split the pipes and spring leaks once they thaw out. Turning them off during the winter, weekends or holidays, is not an option if outside temperatures dip below 0 °C, unless the systems are first drained of their water. Nevertheless, using water as a temperature controlling medium can be beneficial. In winter, the systems can be reversed to provide heat for the rooms, and normally the sensation is of a very pleasant natural heat. In general, these systems are very effective, but they tend to be more expensive than some of the other options, especially as all the water pipe work usually needs to be made of stainless steel. All the ductwork shown in Figure 21.4 is Sonodec, which is both flexible and also acts as a silencer. The inner tube must be stretched as much as possible to open the "accordion" folds and avoid the introduction of turbulence noise into the system at higher air flow rates.

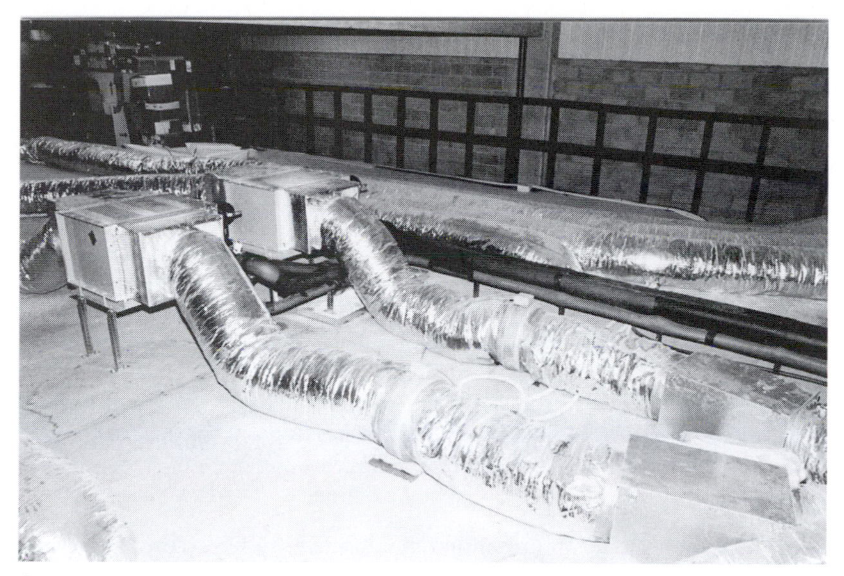

FIGURE 21.4
In the system shown, the air enters through the rectangular filter boxes, on the extreme right of the photograph, before passing through the fans. The "water battery" heat exchangers are in the large rectangular boxes, which also act as expansion chambers and quite effective silencers. The air then loops round through the "acoustic" ducting before entering the room. An external heat pump controls the water temperature between about 28°C in winter and 12°C in summer (dependent upon the outside air temperature) in order to maintain a constant temperature of around 20°C in the studio. Stepped transformers control the fan speed, thus regulating the air flow through the room.

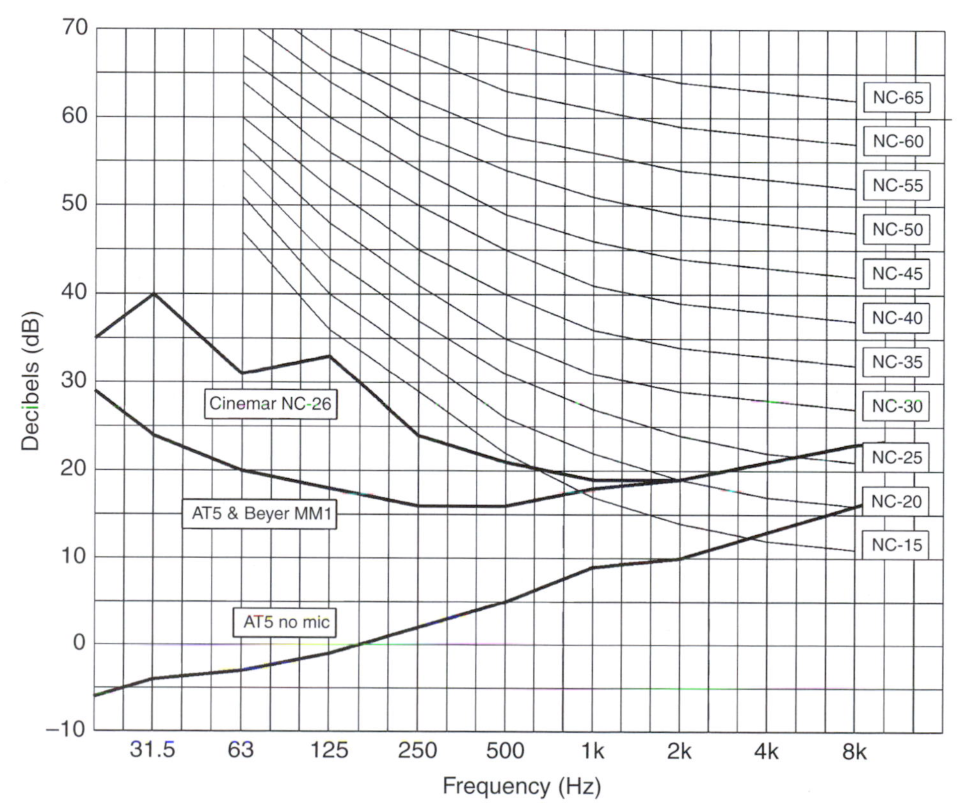

FIGURE 21.5
The above plot shows the noise level inside the dubbing theater whose temperature control system is shown in Figure 21.4. The upper trace shows the noise level with all systems running. The middle trace shows the noise floor of the measuring microphone. Somewhat ironically, the noise in the region around 500 Hz was due to the ventilation fan in the Dolby processor, mounted close to the mixing console. Without the processor the noise floor, below 1 kHz, remains always less than NC I5.

HEADPHONE FOLDBACK

Foldback is an extremely important part of the studio environment. It can be the entire acoustic reality for the musicians during their performances, because in by far the majority of cases musicians will record whilst wearing headphones. In these instances, the acoustics of the studios can only be heard via the microphones, mixing console and headphones, so the musicians can find themselves in a totally alien world if due consideration is not given by the recording engineers to the creation of the right foldback ambience.

If musicians need to hear what they need to hear in the studio, then they need to hear it in their headphones as well. Many musicians play off their own tone, so if they cannot hear themselves with their usual sound they may "force their tone," or perhaps hold back, and neither is satisfactory for optimal recording. If musicians need to hear richness in the direct sound, then they ideally would need to hear it in their headphones; if they need the reinforcement of lateral reflexions, then they need to hear those in the headphones also. At times, it can be considered beneficial to put up ambience microphones which will be used in the foldback only. If this helps to improve the sense of "being there" for the musicians then it is surely worthwhile, though it is rarely done.

There are two big problems with the optimization of foldback. First, few recording engineers have spent enough time as recording *musicians* to be sufficiently aware of the complexities and importance of the requirements of different musicians in terms of foldback. They cannot be blamed for this, as they cannot spend their lifetimes doing two things at once. The reverse of course applies. So many musicians are used to poor foldback that they fail to realize what *could* be achieved. As with the case of the recording engineers, the musicians have also not spent a working lifetime doing the other end of the foldback job. There are times when musicians are given their own foldback mixer for use in the studio, but they rarely have access to the reverberations that are available from the control room, and which can greatly aid their perceptions of space.

Second, however, if too much of the foldback burden is loaded on to the musicians, then they can become distracted from their primary job: playing. In fact, one significant restriction on foldback balance is time. The foldback mix cannot be set up until the musicians are playing, but once they *are* playing, too much time cannot be spent setting up the foldback before they "go off the boil," and lose their motivation to play. Furthermore, too much fiddling with the foldback, with levels going up and down and things switching in and out, is absolutely infuriating for the musicians. Remember, effectively it is their whole audible environment that is being disturbed. For them, it is like an artist trying to paint a picture with the lights going on and off randomly.

Where possible, foldback should be very carefully considered. A stereo foldback system is much easier to hear clearly than a mono one. In stereo, even things which are perhaps a little too low for the ideal balance for a mono mix can be perceived much more easily due to the spacial separation which stereo provides. Whenever possible, it seems beneficial to provide the facility of systems where the recording engineer can monitor exactly what the musicians are hearing, which includes listening at the same level. Obviously, on systems where all, or at least many, of the musicians are in a position to make their own balances, this is hardly practicable, but in cases where many headphones are driven from a common power amplifier, it is. In these cases, it can be beneficial for the engineer to have a line in the control room which is connected to the same power amplifier output as the studio headphones, and, where possible, to monitor on the same type of headphones that are being used by the musicians. There can in this way be little doubt that the engineer is monitoring the selfsame space that the musicians are immersed in, and there is therefore much less chance of misunderstandings.

In many cases where the control room foldback monitoring system is only via headphones plugged into the mixing console, or where it is heard on loudspeakers, the recording engineers can never truly know what the musicians are hearing. This has often led to time wasting, or, if the problems have not been pursued, to the musicians having to try to play with the problems still unresolved. In either case, the musical performance will probably suffer. The musicians also seem to feel an added sense of being understood and appreciated when they feel that their environment is being shared by its controller, and this helps to allay any insecurities which they all too frequently feel.

It is imperative that the design of any studio extends to ensuring that sufficiently flexible foldback systems are available, because the "virtual" spaces in which the musicians may have to perform can be just as important to the recording process as the very real spaces in which they physically play. A full understanding of this is a fundamental requirement of being a good studio designer or a good recording engineer, and good foldback systems are an equally fundamental part of a good studio. In effect there are therefore three aspects to the acoustics of a recording space. First, the acoustic as heard in the room; second, the acoustic as collected by the microphones; and third, the combination of the two as perceived by the musicians if they must use headphones. All should be considered very carefully in both the design and the use of the rooms.

Choice of open or closed headphones can also modify this environmental balance. Closed (sealed) headphones tend to add to the feeling of isolation, and take the musicians one step further away from their "real" acoustic space. Nonetheless, there are times when closed headphones are very necessary. Drummers, for example, may need closed headphones in order to avoid having to use the painfully high foldback levels which may be necessary so that the other instruments can be heard over the acoustic sound of the drums. In this case, closed headphones are used to keep unwanted sounds *out*. Conversely, vocalists, or the players of quiet, acoustic instruments and, in particular, those instruments which require the positioning of a microphone close to the head of the musician, may also need to use closed headphones. This is especially the case during overdubbing, when the "tizz tizz" reminiscent of sitting next to somebody using a personal stereo system may be picked up by the microphone. It may subsequently be difficult to remove these sounds (especially timing "clicks") from an exposed track. In this instance the closed headphones are used to keep the unwanted sound *in*.

It is a pity that foldback is so often subject to so much compromise, but in away such is its nature. As so much modern music has developed out of domestic recording technology it is not surprising that aspects of the limitations of the prior technology should be carried along. The simple headphone systems often used in small studios get used in bigger studios, for which they are often inappropriate, when studio companies expand without the necessary experience for a larger operation. It is quite amazing how many people in relatively large studios are totally ignorant of many excellent practices that were well-established 30 years ago. So many people are now self-taught, and they can lack so much knowledge of many simple things which make recording practices so much more effective. But, it is imperative to remember that whatever wonders may be created in the acoustic design of the studio, often for the benefit of the musicians, as soon as they put on a pair of headphones those musicians can be in a different world, and it is best not to leave them feeling lost in it. This is one very important difference between designing live performance spaces and designing recording spaces. In the latter case, it is not only how the musicians hear the space which predominates in the design considerations, but how *microphones* react to the space. This point should never be forgotten.

Loudspeaker foldback

Many studios *and* musicians find the facility of being able to provide foldback via a loudspeaker to be a useful asset. The concept of the "tracking loudspeaker" goes back almost to the dawn of electrical recording. Many vocalists found that it was easier for them to perform without headphones, so that they could clearly hear their own, natural voices. A loudspeaker would be positioned facing the vocalist, and the backing track would be played back through the loudspeaker at the minimum level needed to allow the singer to perform well, but not so loud that the overspill into the microphone would cause problems. The directional characteristics of a cardioid microphone were employed in order to reject as much of the loudspeaker output as possible. This was further aided by placing non-reflective screens behind the vocalists, to prevent the playback signal from bouncing off a wall behind and subsequently entering the microphone.

These loudspeakers also became useful as a means of playing a recording back to the musicians, without them having to leave their positions and going into the control room; which would in any case perhaps be too small to house them all. This was never a particularly good means of assessing the *sound* of the recording, but was a useful tool for discussing performance quality or mistakes with the producer, conductor, or musical director. They were also useful in orchestral recordings, when the conductor needed to highlight some point in the performance to the musicians, as taking a whole symphony orchestra into the control room would clearly be out of the question. Once again, ease and comfort go a very long way towards getting the best performances out of musicians, so they should be given great attention when designing, *or* operating, a studio. The recording process begins with the musicians, and they are the foundation on which the rest of the process is built. If these foundations are weak, then all the subsequent proceedings may never achieve their potential strength or quality.

COLORS, AND GENERAL DECORATION

These are the areas over which a studio designer often has least control. They are very subjective; very personal. Each studio owner or manager seems to want to make their own input to the design, and this is an area where they feel free to do so. Nevertheless, some guidance can be given by the designers, based on previous experiences, which may help the studio chiefs to avoid making any great errors.

Dark colors tend to make spaces feel smaller. In large studios this may help to create a more intimate atmosphere, but in small studios the effect can be claustrophobic. Very often, of necessity, the ceilings of control rooms and small overdubbing booths are not very high, and light colors can go a long way towards lifting any sense of oppression. Light colors also reflect more light back into the room, so they tend to diffuse the light, making it less hard on the eyes. Also, the extra reflected light means that less overall power of lighting is needed, therefore less heat is generated and less air-conditioning is necessary.

In a studio in Watford, England, the owner began seriously running out of money towards the end of the construction. Wisely, and it was a decision he never regretted, he continued to invest more in the acoustics than the decorations, which he felt could be easily improved at a later date. He went to a local outdoor market, looking for inexpensive fabric, and returned with several short rolls of leftover material which he had bought cheaply. The comments of the building crew when he showed them the fabric that they were to work with would not be printable in any decent publication. He had bought white fabric with yellow stripes, blue flowers on a white background, red dots on a white background, and various other designs. Nevertheless, once the different fabrics had been juxtaposed carefully, the result was a small studio with a spacious feel and a remarkably happy and pleasant atmosphere to work in. Even the most severe of the original critics admitted that the result was, surprisingly, very agreeable.

The other extreme of this situation is when a studio owner employs the services of an interior designer who is allowed to severely compromise the acoustics. This happens in a great number of cases, especially when marketing and financial people, in their customary absolute ignorance of recording needs, believe that if a studio looks absolutely great, but sounds only mediocre, it is a better business option than sounding fantastic but perhaps lacking the "designer" touch. It must also be said that some studio designers *will* bend to such compromises if it ensures more work for their business, and hence more profit.

There is, of course, no law against any of this, and some adherents to market forces philosophies may even applaud the existence of this state of affairs, but it really does nothing for the advancement of studio functionality. This is being mentioned here to stress the fact that a studio which looks great in publicity photographs, and has the name of a reputable studio designer attached to it, does not necessarily mean that it is either a good studio, *or* that the named designer had as much control over the acoustics in the way that the publicity is implying. One of the problems of the "low profile" of acoustic work is that good acoustics can be easily sacrificed to other, more visible trivia. Cases abound such as where money has been short for good anti-vibration mounts for an air-conditioning system because the owner decided to spend €60 each on 24 door handles. Or there is an inadequate electrical installation, because the owner spent €100 per square meter on a hardwood floor that would largely be covered by a mixing console and effects racks. These are examples of common occurrences.

There is perhaps a philosophically different outlook between studio designers who make money from building good studios, and studio designers who are in business to make money. The latter will probably be more easily swayed into being subjugated by interior designers. For a proposed studio owner looking for a designer, it is perhaps a case of caveat emptor—let the buyer beware—because he or she alone is responsible if the results are disappointing. It also must be said that some owners *do* want something which looks great, even if the acoustic and operational results are compromised, but how to achieve that would be a totally inappropriate subject for discussion in a book such as this.

The aesthetics of recording studios are a very important aspect of studio design, but good design *can* incorporate aesthetically pleasing ideas. However, the aesthetic demands of some interior designers are frequently acoustically insupportable. An air of respectful cooperation between the specialists is the obvious solution.

AC MAINS SUPPLIES

Appropriate AC mains supplies are a fundamental part of the infrastructure of any good recording studio, but general electrical contractors are often totally unfamiliar with the needs of high quality recording studios, and they are often greatly resistant to suggestions which they consider to be out of the ordinary. Sometimes there is good reason for this, because standardized systems of installation and power distribution have been developed as a safety measure. Standardized systems also mean that any qualified electrician can make modifications or tests on a system which will not upset the rest of the installation, or show up as fault conditions.

However, standardized industrial or domestic power installations may be inadequate for highly sensitive recording equipment. Hospital power supplies which feed delicate equipment upon which lives depend are another example of installations where normal power distribution techniques can be inadequate. The life or death circumstances which exist in medical facilities ensured that hospitals and the like have had exemptions in many countries, for many years, which have allowed special AC supply techniques to be used. Electricians who work on medical installations will be trained in the use of these special techniques. However, wiring systems in recording studios are generally not so well regulated or supervised. The recording industry is not usually perceived as being a particularly responsible or necessary profession, so such exemptions have been slow in arriving.

Indeed, there have been many cases where special AC power installations *have* been incorporated into new studio designs, and where much care and attention has been paid to detail. Nevertheless, within a few months of the opening, perhaps where a studio manager or head of maintenance has been replaced, other electricians have arrived, working to "the book." Under such circumstances, chaotic systems have begun to develop, sometimes with dangerous results. Studios have not done themselves any favors by this sort of lax behavior, so it is little wonder that they have not been granted their technologically necessary exemptions. Even in a technically advanced country like the USA, electrical codes were not modified until the 1990s to allow balanced power installation, and even these were restricted to studios with a competent, qualified electrical supervisor. It was recognized that despite its advanced technology, the USA was also the home of the very market forces which would seek to cut costs by *not* employing a qualified electrical supervisor if at all possible. Technological advancement does not always go hand in hand with responsible business attitudes.

It is therefore quite dangerous and irresponsible in a book such as this to try to explain how to make AC power installations. It is also not practicable, because the rules and regulations in not only each country, but often in each state or region, can be very different. What *can* be done, however, is to discuss a few of the main ideal requirements. Nevertheless it would still be up to any individual wishing to pursue these to find a knowledgeable and flexible electrical engineer or contractor who could then interpret them in such a way that would not break local installation rules. Whilst it is true that many qualified electricians are neither knowledgeable of electrical engineering theory nor are they willing to change their habits, it is usually possible to find specialized electrical engineers who *can* make the appropriate arrangements. Remember also that a non-standard installation can be dangerous if it is worked on during later visits by electricians who, quite justifiably, expect the installation to be standard. Adequate and permanently accessible documentation, that will *not* be lost when staff change, is not something which one can expect to exist in 99% of the recording industry. Technologically advanced countries such as the UK, France and the USA are no exceptions, either.

Phase

The ideal number of phases to which all the audio equipment should be connected in *any* studio, or complex of studios which can be interconnected, is ONE! All audio equipment, together with any other equipment to which it may be connected, including computers, video, film or radio equipment, should all be connected to the same power phase. The only exceptions are where equipment, such as radio transmitters, are fed via galvanically isolated audio connections; i.e., through transformers with adequate insulation. Electricians often do not like to put all the equipment on one phase, because they prefer for good power engineering reasons to balance the load. Nevertheless, the reasons to connect to one phase only are first, safety, and second, noise.

In countries where the nominal mains supply voltage is in the 230 volt region, the voltage between the live connections of any two phases will be in the region of 380–400 volts. Under certain fault conditions, connecting the audio cables between two pieces of equipment on different phases can have instantaneously lethal results. Under normal circumstances this should not happen, but with weak insulation on two devices, even if the insulation holds on monophase operation, the extra potential when two pieces of such equipment are connected together can initiate a fatal breakdown.

What is more, when audio signals (analog or digital) pass repeatedly through equipment with different ground plane potentials, as is almost inevitably the case if interconnected pieces of equipment are connected to different mains supply phases, signal contamination is likely to result. It is true that extremely well-designed equipment can be immune to these effects, but unfortunately even some very *expensive* audio equipment does not come under the heading of "extremely well designed." Capacitive coupling in power supply transformers, for example, can easily lead to fluctuating ground plane voltages. In general, the higher the gain of any equipment, the more sensitive it will be to supply-phase differences.

If three phases are available it is best to dedicate the cleanest, most stable phase to the audio equipment. Lighting and ventilation equipment can use a second phase; and refrigerators, coffee machines, and general office equipment can be connected to a third phase. General-purpose wall outlets outside of the recording area can be connected to the least heavily loaded of the second and third phases, but some local regulations may modify this. Although this may not seem to balance the phases very well in terms of equal current drain on each phase, electricians frequently fail to realize just how variable is the current drawn by a recording studio. Equipment such as power amplifiers often draw current in pulses, dependent on the musical drive signal. Musicians connect and disconnect their instruments and amplifiers, lighting gets turned up and down, air-conditioning currents are temperature dependent, tape recorder motors stop and start; many things change during the course of a session. There is a tendency for non-specialist electricians to look at the maximum power consumption of each piece of equipment, then try to distribute it presuming that the current drain is constant. The reality is nothing like this, and phases sensibly balanced for technical reasons can easily be made to give a distributed average load that is better balanced than many electricians would achieve by distributing things according to their "standard" calculations. The problem is often how to convince them about this when they perceive *themselves* as being the experts on such matters.

Power cabling

Guitar amplifiers and audio power amplifiers have a tendency to draw current in pulses. For this reason oversized cable sections are often specified. General electricians will size cable according to the heat produced by the cable resistance in response to the power being consumed by the equipment. To many electricians, anything much beyond what such power requirements would need for reasons of safety is usually seen as a waste of money. However, pulse-drawing equipment *needs* oversize cables. There are two reasons for this. Excess resistance in the power supply cable may restrict the peak level of the required current pulse, which can limit the high-level transient response of audio power amplifiers. Furthermore, cables lacking sufficient cross-section for the current pulses can create problems in other equipment, and can even crash computers. If a current pulse causes a voltage drop on a power cable, then even if this cable is not shared by any other piece of equipment the voltage drop can cause harmonic distortion on the supply line. This can then bypass the power supply filtering of some sensitive equipment and interfere with the proper operation.

In fact, the problem of harmonic induced interference can also be caused by uninterruptible power supplies (UPSs) which do not have waveform feedback. Not all audio computers are happy working on all UPSs. It would seem a pity for a UPS first to save a recording when the mains power fails, only to subsequently crash the computer due to its bad voltage waveform. Many computer crashes that are blamed on software problems can be traced to bad power installations or inappropriate UPSs.

Balanced power

In some hostile power line environments, one of the best solutions to avoid interference problems coming in via the electricity supply can be to balance the power. This involves the use of power transformers with center-tapped secondary windings, producing, for example, 115 V-0- 115 V in place of a single 230 V supply. Electricians should be consulted if this sort of measure is needed, in order to avoid making any illegal installations, but the technique can reject interference in exactly the same way that balanced audio lines are less sensitive to interference than unbalanced lines. Signal-to-noise ratio improvements of 15–20 dB have been reported as a result of power balancing.

In many countries, it may be illegal to supply balanced power to normal wall-socket power outlets. Special installation work may be required, by expert engineers. Many electricians may simply refuse to do it, as they cannot justifiably put their signature to something with which they are not familiar. Nevertheless, specialist engineers can usually find legal ways of installing such systems if there appears to be no other option for interference suppression. It should also be stressed that with such a specialist job as studio electrics, it is *all* really a job for electricians with some experience in this field if the highest standards of performance are required.

Mains feeds

It is important to try to ensure that the main cables feeding the studio in-coming fuse-board are connected as close as possible to the main distribution board for the building. This ensures that the large-section cables will continue to the street, and onwards to the principal supply. If this is not done, then any common cable between the point of connection of the studio to the mains supply of the building, and the main feed connecting the building to the street, will be a common impedance. Any currents and noises generated elsewhere in any part of the building which shares the same feed cables will superimpose themselves on the supply to the studio, and the supply will not be clean. In addition, if these cables are only rated in terms of the power consumption, they may not have the cross-section sufficient for the non-distortion of surge currents. A low impedance supply is highly desirable.

Once again, independent supply cables to the main feed to the building may be inconvenient for the electricians, and they may resist it on the grounds of being unnecessary because they are still thinking in terms of power consumption and standard cable sizes. Nonetheless, it needs to be impressed upon them that a recording studio is no ordinary installation in terms of its power cable requirements. Again, an experienced specialist electrical engineer may need to be called in, if only to convince the electrical contractors that the necessity is a very real one.

Earthing

A good earth is essential as a safety measure, and it is normally also needed for a good signal to noise ratio, but if very elaborate earthing systems seem to be the only way to reduce noises the implication is that there are problems in the system wiring; either the power wiring or the audio wiring. Normally, in buildings with steel frames or steel reinforced concrete in the foundations, connection of the earthing (grounding) system to the steel provides an excellent earth (ground).

In the first reference at the end of this chapter, the first 113 pages of the book are dedicated to power and grounding systems, so it should be evident that a few paragraphs, here, cannot deal with the subject. However, what *can* be made clear is that similar rules apply to the earthing cables as to the power feed cables. The studio power system, and its technical earth, should be connected to the best earth point via the shortest cable run and with the largest practical cross-section of cable.

Another point worth mentioning, which is often overlooked, is the degree to which the earthing system of a studio may be being polluted from within. Obviously, sharing the long tail of an earth cable with other users of a building may result in a lot of electrical noise on the common (shared) section of the earth cable. The studio earthing system therefore needs to be connected *directly* to the building safety earth, and *not* via any shared lengths of cable. However, even from *within* the studio, the earth can be polluted by bad choices of electrical filters. Many filters of inappropriate design, which frequently still find use in studios, do little other than remove the interference from the live and neutral cables and dump it all on the earth, which in some cases is *more* sensitive to the noise than the live and neutral are. Power balancing can be a big help in these cases, because the noises on the live and neutral *cancel* on their way to earth.

Very well designed systems using well-designed equipment can often work well with the most simple of earthing systems. Unfortunately, all recording equipment is not so cleverly designed, so in buildings with earth noise problems, and on ground which is geologically bad for earthing, expert advice may be needed. Standard electrical contractors may only know of safety earths, and perhaps a little, learned by hearsay, about technical earths. It is not always wise to rely on their advice. In cases of persistent problems, a specialist engineer should be sought.

SUMMARY

- In the majority of cases, the presence of daylight in a studio is a desirable asset.
- Cho0ice of artificial lighting should be made carefully, so as not to risk compromising the electro-acoustic performance of a studio.
- Variable transformers tend to be the best form of lighting control.
- Good ventilation, air-conditioning and humidity control is usually essential. Extraction-only ventilation systems should *not* be used. There has been a great increase in the number of split air-conditioning units being used in studios. These are not ideal, but if they *must* be used, the quietest, aerodynamically profiled ones should be chosen.
- Foldback systems are very much part of the studio environment for the musicians, and the necessity of seeking the most appropriate systems for the musicians' needs should not be neglected.
- Colors and general decoration are a very important part of a studio environment, but interior design should never be allowed to degrade or limit the work of the acoustic designer.
- Studio power wiring needs are often beyond the experience of normal electrical contractors. Specialist advice may be needed.
- When any non-standard installation is completed, documentation should be thorough, and available for consultation by any electrical contractors who may do work in the studio in the future.
- Electrical regulations can change from country to country, and even from state to state or region to region, so local advice should always be sought.
- It is strongly advisable to connect all audio and associated equipment to the same electrical power phase. This is for reasons of both safety and noise.
- The main studio breaker-board/fuse-board should be connected to the incoming supply of the building, without sharing any cable runs with other tenants. Oversized cable sections should also be used, to help to provide the lowest possible source impedance.
- In very problematical installations, from the point of view of electrical supply noise, balancing the power can be a very effective cure, but it should only be done under the supervision of an electrical engineer experienced in such techniques.
- Good earthing systems should be installed, again avoiding common cabling with other tenants. Direct connection to the steelwork in building foundations can make a good technical earth. Despite concrete being perceived to be a good insulator, it still grounds well the steel buried inside concrete foundations.
- Many mains filters can often be a *source* of noise on the earthing system.

REFERENCE

Giddings, P., 1990. Audio Systems-Design and Installation, Boston, USA, Oxford, UK: Focal Press.

INDEX

M

N

O

P

Q

R

T